D1797148

North Western Medical Physics

Christie Hospital, Manchester

Catalogue No.

657

MEDICAL RADIOLOGY

Diagnostic Imaging and Radiation Oncology

Radiation Therapy Physics

Contributors

M.D. Altschuler · P. Bloch · F.J. Bova · A. Brahme
D.J. Buchsbaum · G.T.Y. Chen · P. Fessenden · B.A. Fraass
Z. Fuks · J.M. Galvin · J.W. Hand · M.S. Huq · D. Jette
G. J. Kutcher · D.D. Leavitt · S.A. Leibel · C.C. Ling · D. L. McShan
R. Mohan · R. Nath · C.A. Pelizzari · I.I. Rosen · M.C. Schell
T.E. Schultheiss · S. Shalev · K.A. Weaver · B.W. Wessels
J.F. Williamson · A. Wu

Edited by
Alfred R. Smith

Foreword by
Luther W. Brady and Hans-Peter Heilmann

With 235 Figures, Some in Color

Springer-Verlag
Berlin Heidelberg New York
London Paris Tokyo
Hong Kong Barcelona
Budapest

ALFRED R. SMITH, PhD

Radiation Biophysicist
Department of Radiation Medicine
Massachusetts General Hospital
Cox Bldg., 3rd Floor
Boston, MA 02114
USA

MEDICAL RADIOLOGY · Diagnostic Imaging and Radiation Oncology

Continuation of
Handbuch der medizinischen Radiologie
Encyclopedia of Medical Radiology

ISBN 3-540-55430-0 Springer-Verlag Berlin Heidelberg New York
ISBN 0-387-55430-0 Springer-Verlag New York Berlin Heidelberg

Library of Congress Cataloging-in-Publication Data. Radiation therapy physics/with contributions by M.D. Altschuler ... [et al.]; edited by Alfred R. Smith; foreword by Luther W. Brady and Hans-Peter Heilmann.   p.   cm. – (Medical radiology) Includes bibliographical references and index. ISBN 3-540-55430-0 – ISBN 0-387-55430-0 1. Medical physics. 2. Radiotherapy. 3. Radiation dosimetry. I. Altschuler, M. D. II. Smith, Alfred R. III. Series. [DNLM: 1. Radiotherapy, High-Energy. 2. Radiotherapy Planning, Computer-Assisted. 3. Radiotherapy Dosage. 4. Dose-Response Relationship, Radiation. WN 250 R1287 1994] R895.R27 1994 615.8´42 – dc20 DNLM/DLC for Library of Congress 94-25784

Typesetting: Thomson Press (I) Ltd., New Delhi

SPIN: 10551980   21/3111/SPS – 5 4 3 2 1 – Printed on acid-free paper

Foreword

The discovery of X-rays in 1895 by Roentgen set in motion a long period of discovery and rapid progress in technology. Investigators inadvertently played leapfrog – with one laboratory then another announcing new records for the highest voltage generated for the most intense X-ray beam. Technological advances were first driven by the desire for better "pictures" and then separately by the need for more penetrating radiation to treat deep-seated tumors. Concomitantly, new dosimetric methods were continually introduced to keep pace with the ever-increasing energy of the X-rays produced by newer and newer machines.

Initially, there was little or no distinction between diagnostic and therapeutic X-ray equipment. Eventually, however, the divergent uses resulted in different, very specialized designs, with function dictating form. The radiation therapy tubes had to run continuously for many minutes and they needed to deliver large, up to 25 cm, uniform, flat and symmetric beams at distances ranging from 15 to 100 cm. Additionally, the power supplies and control units had to support these operating conditions. These new therapy designs accommodated several different kinds of X-ray therapy arbitrarily categorized as Grenz-ray, contact, superficial (Roentgen), intermediate (medium), deep (orthovoltage), and supervoltage therapies. The resulting beams had to be collimated and delivered accurately to the intended target. Thus, the delivery technology evolved from bare tubes, to bare tubes with lead sheets on the patients, to mountable cones and, ultimately, to adjustable collimators with light localizers. Additionally, beam modifiers such as filters were designed to maximize therapeutic ratios, that is, deep-target dose to skin-surface dose. Meanwhile, dosimetry evolved from describing X-rays only by voltage and current to using half-value layers to quantify the X-ray spectrum. Beam intensity measurements also evolved from photographic plates darkening per minute to current data derived from ionization chamber measurements. Exposure and dose measurements evolved from "skin erythema" dose to film densitometry to various kinds of ionization measurements, first in air and then in phantom.

As radiation oncology matured, X-ray tubes and transformer-based power supplies were gradually replaced by other, more efficient and effective X-ray production machines at high energies, including electrostatic Van de Graaf generators, betatrons, and, eventually, linear accelerators.

The years from 1950 to 1970 were a time of both consolidation of previously developed technologies and of continued development. While progress and discovery continued, they did not occur at the same pace as during the earlier period. Technological advances continued to be driven by the desire for "better" dose distributions within the patient. Concomitantly, dosimetric methods evolved with improved understanding of the underlying physics and with international standardization of description and methods. While the evolution of technology continued across the spectrum of therapy-related endeavors, perhaps this period is best characterized by the clinical introduction of medical linear accelerators, Linacs, and by that of their direct competitors, the Cobalt-60 teletherapy gamma ray units.

The physics community worked toward producing higher and higher energy accelerators for use in basic research into the nature of matter. The ancillary technologies developed for these basic research atom smashers enabled the manufacturers of medical accelerators to increase the efficiency and reduce the size of the clinical machines.

The book by Smith and his colleagues demonstrates clearly the application of these foundations in contemporary clinical radiation oncology. It covers the broad array of basic data, instrumentation, clinical technologies, and their application in the highly sophisticated management of the patient with cancer. The authors have produced a landmark text serving the basic foundations in terms of physics applications in clinical radiation oncology.

Philadelphia/Hamburg, November 1994   L.W. BRADY
                                                                 H.-P. HEILMANN

Preface

During the last decade there has been a virtual explosion of advances in the field of radiation therapy physics. It is difficult to keep abreast of the literature during periods of rapid change and individuals are apt to learn more and more about increasingly fewer topics. Thus our field, like others, has tended to be characterized by increasing specialization which makes it less likely that optimum integration and generalization of knowledge will take place.

The goal of this volume the Springer-Verlag series in Medical Radiology: Diagnostic Imaging and Radiation Oncology, is to promote the integration of the various areas of radiation therapy physics by reviewing most of the recent advances and discussing how these new technologies have been implemented and utilized in the service of radiation oncology. The authors were asked to present up-to-date reviews of their topics for an audience of professionally mature medical physicists. Therefore this volume is not intended to be an introduction to the field of radiation therapy physics however anyone with sufficient background and serious intent will find a wealth of information herein. Moreover, the comprehensive literature citations will enable interested readers to pursue their learning to broader scopes and deeper levels. The material in this volume covers topics ranging from tumor localization to treatment verification. The chapters cover advances in imaging, dose calculation models, treatment planning, biological modeling, accelerators, devices and techniques, radio-chemistry, dosimetry, and quality control, and describe how these advances have enabled us to improve the delivery of radiation therapy. We have not always hit the mark and, as in every like venture, there is not uniformity in quality and depth of coverage among the chapters, however this was not caused by any lack of the authors' abilities or intent but was, in part, a product of the happenstances of fate: people changed jobs, labored through personal and professional trials, and kept up with their busy lives while working hard on these chapters and each author has given us something of particular value.

It will be quite some time before we are able to fully exploit the multiplicity of technologies which recent research and development efforts have placed at our disposal. Our challenge is to use these new tools wisely and to design even more effective and efficient means to use radiation in cancer therapy. We present this book with the hope that the information we have provided herein will assist readers in their work toward achieving these ends.

Boston, November 1994   ALFRED R. SMITH

Contents

1 The Role of Imaging in Tumor Localization and Portal Design

GEORGE T.Y. CHEN and CHARLES A. PELIZZARI

CONTENTS

1.1 Introduction

The development of three-dimensional (3D) medical imaging modalities has profoundly affected the diagnosis and treatment of cancer. The ability to visualize the location, character, and extent of lesions has facilitated the development of more precisely directed therapies, most clearly radiotherapy (HENKELMAN 1990; LICHTER et al. 1988; ROSENMAN 1991) but also chemotherapy and surgery (ENGEL et al. 1990; KELLY 1986). Patients typically undergo multiple 2D and 3D imaging studies for diagnosis, therapy planning, and follow-up. Extensive quantitative use of these images to define tumor location in relation to normal anatomy has been made in planning radiation therapy for a number of years (GOITEIN et al. 1983; MCSHAN et al. 1979). The goals of a higher tumor control probability and reduced treatment-related morbidity are postulated to be fea-

sible with precision irradiation techniques (SUIT and DuBOIS 1991), which are heavily image based.

Radiation oncologists depend critically on radiographic imaging to help delineate the location of primary and metastatic tumors (LICHTER et al. 1988). Radiographs, computed tomography (CT), magnetic resonance imaging (MRI), isotope scans, and ultrasonography are (alone or combined) routinely obtained to assess the extent of tumor and its geometric relationship with adjacent normal structures. Image-based treatment planning requires the accurate segmentation of tumor and normal anatomic structures from 3D data sets. Rapid and accurate image segmentation of multiple anatomic structures is essential in the radiation treatment planning process.

Once the tumor volume is defined, it is expanded to create a target volume, to which the prescribed dose is delivered. The margins applied to create target volumes are based on estimates of uncertainty in patient positioning from fraction to fraction, physiologic motion, and the need to account for microscopic tumor extension. A further expansion of the volume accounts for finite beam penumbra.

These structures can then be viewed with display techniques which simulate the treatment geometry (known as beam's eye view) and are used in planning conformal radiotherapy, where the high-dose region is tailored to the 3D shape of the tumor, thereby sparing normal tissues (GEHRING et al. 1991; Low et al. 1990, 1992; MCSHAN et al. 1979; ROSENMAN et al. 1989; SHEROUSE and CHANEY 1991; VIJAYAKUMAR et al. 1992).

An adequate dose must be delivered to the target volume in the presence of a number of spatial uncertainties. To what accuracy can one define the 3D coordinates of tumor and target volumes of interest? Sources of spatial error (SUIT and DuBOIS 1991; SVENSSON 1984) in treatment planning include a uncertainty associated with definition of the tumor boundaries in the imaging data, due to fundamental limitations of oncologic imaging, (b) uncertainty in patient alignment during fractionated treatment,

GEORGE T.Y. CHEN, PhD, CHARLES A. PELIZZARI, PhD, Department of Radiation and Cellular Oncology, The University of Chicago Medical Center, 5841 South Maryland Avenue MC 0085, Chicago, IL 60637, USA

Table 1.1 Sources of uncertainty and possible improvements

Uncertainties in targeting tumor	Possible improvements
Abnormal mass: tumor or edema	Improvements in diagnostic-oncologic imaging specificity
Inability to visualize microscopic extent of tumor	Increase sensitivity
Patient positional variation during fractionated radiotherapy	Improve immobilization
Geometric distortions in MRI	Scanner calibration, post processing to remove distortions
Finite slice thickness, partial volume averaging	Thin slices, more slices
Physiologic motion-induced artifacts	Study physiologic motion Determine anisotropic margins Faster data acquisition methods

and (c) uncertainties due to physiologic motion. These and other factors affecting the design of the radiation portal are listed in Table 1.1, along with possible improvements to reduce the uncertainty. Under special conditions, such as stereotactic radiosurgery or particle therapy of the eye, millimeter or submillimeter accuracy may be an attainable goal. However, in general, this level of accuracy is not possible.

This chapter summarizes the current state of oncologic imaging with particular emphasis on the relative advantages and disadvantages of CT and MRI for various sites of interest. This is followed by a detailed examination of some of the technical aspects of target delineation and the magnitudes of margins to be applied in order to adequately cover the target volume in fractionated radiotherapy.

1.2 Oncologic Imaging

Oncologic imaging involves the detection, diagnosis, staging, and follow-up of cancers. Comprehensive reviews of oncologic imaging can be found in recent conference proceedings and cancer textbooks (BRAGG and THOMPSON 1992; HOLLAND et al. 1993). A general discussion is provided here which emphasizes the differences between CT and MRI in treatment planning, and describes how these modalities meet the needs of treatment planning. While oncologic imaging is generally discussed from the perspective of the diagnostic radiologist, there is considerable overlap in the information needed by the radiation oncologist.

Images provide information on the size and location of abnormalities. These abnormalities guide the radiation oncologist in the clinical definition of the tumor and its extensions. Ideally, such images uniquely identify tumor, and early studies with MRI demonstrated that tumors have a characteristically longer relaxation time than corresponding normal tissues (HENKELMAN 1990). It was hoped that these differences could be used to uniquely identify tumor. These expectations were, however, not fully realized, and it is accepted today that MRI cannot uniquely separate tumor from surrounding edema or necrosis. Nor can MRI or CT detect the presence of tumor in normal-sized lymph nodes.

The most commonly used tomographic imaging technique in image-based planning is CT. There are several reasons for this predominance. CT is a more mature imaging modality and has been available for a longer period of time. CT is more accessible since there are a larger number of scanners available, and the cost of an imaging study is less. Scan times for CT can also be significantly shorter than for MRI when large areas are to be scanned. CT scans provide a measure of electron density data useful in inhomogeneity corrections to dose calculations. CT data can be processed to generate digitally reconstructed radiographs which are similar to conventional radiographs, used for portal alignment. Geometrically, CT is more accurate than MRI. Several radiation medicine departments with interests in conformal radiotherapy have acquired CT scanners.

In recent years, MRI has become the diagnostic imaging modality of choice for a number of sites. MRI has many characteristics which make this modality appropriate for tumor delineation. The appropriate application of both CT and MRI in oncologic imaging continues to be an area of active research. The Radiological Diagnostic Oncology Group (HEELAN 1993) is conducting national trials to quantitate the accuracy of these imaging modalities. In certain cases, a combination of MRI and CT may provide target and normal tissue delineation significantly more accurate than either modality alone.

1.2.1 Advantages of Magnetic Resonance Imaging

The primary advantage of MRI over CT in delineation of tumor is its excellent soft tissue visualization. Tumors have characteristically longer MR relaxation times than normal tissues (DAMADIAN

1971; HOLLIS et al. 1973), and the ability to obtain both T1- and T2-weighted images provides additional information useful in diagnosis.

Delineation of the target volume requires knowledge of tumor involvement in adjacent lymph nodes. Noncontrast MRI studies can differentiate between nodes and vessels because of MRI's sensitivity to flow. Unfortunately, nodal involvement is generally deduced from size; generally, neither CT nor MRI can detect presence of tumor in normal-sized lymph nodes.

Both MRI and CT demonstrate high spatial resolution. The voxel sizes of CT and MRI planning scans are comparable and are approximately 1 mm × 1 mm × 3 mm or less for head scans. CT slice thickness on some scanners can be as small as 1.5 mm, while a typical MR image slice thickness is 4 mm. With special surface coils, high spatial resolution can be attained. Specially designed surface coils (SCHNALL et al. 1989, 1991) have been used to acquire highly zoomed images of the prostate and seminal vesicles.

Another important advantage of MRI over CT is its ability to image in an arbitrary plane. Clinicians consider this capability to be useful in more easily visualizing anatomy. Sagittal and coronal plane images of the pelvis delineate anatomy clearly for tumor staging (HRICAK 1992). When standard orthogonal portals are used (e.g., lateral or AP fields), corresponding sagittal and coronal images can facilitate the transfer of tumor location to simulator radiographs.

MRI may also provide information on tumor metabolism. Measurement of multiple nuclear MR spectra (both hydrogen and potassium) from a well-defined region of interest may be used to determine metabolite concentration in tumors (HENDRIX et al. 1990). Integrated anatomic images and spectroscopy potentially could be used to monitor response of malignant neoplasms to therapy (CAO et al. 1993). These advantages have led Goitein to speculate that MRI may supplant CT over the next decade as the imaging modality of choice for radiation therapy treatment planning (GOITEIN 1991).

1.2.2 Disadvantages
of Magnetic Resonance Imaging

There are several disadvantages of MRI relative to CT for treatment planning. While MRI has superior soft tissue delineation, it provides no information on electron density. Lack of electron density information in MRI studies hinders radiation dose calculations and the assessment of dose perturbations by density inhomogeneities.

Geometric distortions in MR images have been a concern. Imaging for treatment planning requires accurate geometry, and early MRI scanners exhibited sufficiently distorted images that postprocessing of these images was necessary (SCHAD et al. 1987b). Distortions can be inherent in the tissues due to a concentration of ferromagnetic compounds, or due to implanted structures (clips, prostheses).

Bone and air cavities are poorly visualized, reducing the clarity with which bony landmarks can be identified for field alignment. Furthermore, calcification is some lesions provides data useful in diagnosis and tumor delineation. MRI's inability to clearly visualize calcium is a disadvantage in comparison to CT.

1.2.3 Imaging Selected Sites

The ACR Categorical Course Syllabus reviewed oncologic imaging, and consisted of more than 20 review articles spanning the body. RDOG has also conducted trials to determine the accuracy of MRI and CT in diagnosis and staging of cancers. Selected sites of interest and the imaging modality of choice are now summarized.

1.2.3.1 Head and Neck Tumors

Several investigators have compared the usefulness of MRI and CT in defining the extent of head and neck tumors and found MRI to be superior. DILLON et al. 1984 reported that MRI defined the extent of nasopharyngeal tumors as well or better than CT in 10 of 12 cases. He observed that MRI provided superior detailed images of tumor margins, especially with squamous cell carcinoma, chordoma, and histiocytic lymphoma. T2-weighted images were optimal for emphasizing differences in intensity between tumor and normal tissue, and adenopathy in 2 of 12 cases was seen with MRI but not detected on CT. In patients with malignancies of the nasopharynx, parapharyngeal space, and infratemporal fossa, tumor infiltration of bone marrow was appreciated best on MR images (TERESI et al. 1987) and correlated well with CT. Interfaces between tumor and muscle and brain were better visualized on MRI, and intracranial and extracranial soft tissue tumors were better seen on MRI. HONIG (1992) studied MRI and CT imaging of 16 patients with head and neck masses. In all cases examined, imaging of tumor masses was judged to be vastly superior on MRI. Honig found that MRI is the method of choice for the demarcation of infiltrative tumor growths, because of the superior differentiation of soft tissue structures such as muscle, fat, and vascular structures. MRI was found to be superior to CT in the delineation of lateral neck masses, orbital tumors, and edge of tongue and floor of mouth tumors. CT was found to be superior for maxillary sinus tumors. The authors stated that MRI is indicated for solid and cystic processes of the tongue, floor of mouth, salivary glands, and orbits. Yet there are situations where MRI does not provide sufficient diagnostic information. When it is essential that subtle bone destruction or new bone formation be recognized, CT is superior (LUFKIN 1993).

MRI can alter the target volume defined by thin slice CT (BROWN et al. 1989). After examination of all of the clinical data, Brown et al. outlined target volumes and normal tissues on the CT study. MRI scans were then analyzed to determine whether additional information altered targets. Additional information from MRI was reported to have complemented the CT data. In one case study, the nasopharyngeal tumor was seen to breach the nasopharynx by MRI, which was not detected on CT. MRI was also helpful in defining radiation-sensitive normal structures such as the optic chiasm, brain stem, and other structures.

Image texture of nodes also provides information of prognostic value in head and neck disease. Based on a retrospective examination of CT images, MUNCK et al. (1991) classified lymph nodes into two groups: nodes with less than 33% of their area hypodense relative to normal adjacent muscle and nodes with >33% hypodense area. Patients with a lesser amount of necrotic center in nodes had a complete response (CR) rate of 68% and median survival of 32 months, compared with 8% CR and median survival of 13 months for those with a greater necrotic nodal area. Correlation with other known prognostic factors was analyzed, and node density was found to be the significant prognostic factor for complete node response. Such information on the size and texture of nodes can in principle be extracted from treatment-planning scans with appropriate software.

1.2.3.2 Brain

In general, MRI is effective in detecting intracranial tumors but cannot differentiate between edema and tumor. Initially it was hoped that tissue characterization in brain tumors could be achieved by determining relative T1 and T2 values. However, there is a wide overlap between the signals of tumor and edema, and neither tumor type nor degree of malignancy can be determined. Inability of MRI to demonstrate calcification is also a disadvantage in characterizing some brain tumors (SCHAD et al. 1992).

MRI is considered to be superior to CT, particularly in defining tumor extent when the bony base of skull produces beam hardening artifacts on CT which obscure tumor. Target volumes defined on MRI may then be transferred to CT for radiation dose calculations using image correlation techniques (discussed later).

The delineation of brain tumor boundaries from MRI is still under active investigation. Investigators (YUH et al. 1992) have recently reported that the extent of tumor involvement as seen on T2-weighted MR images is a function of gadolinium contrast dose. Higher doses of contrast agent improved tumor delineation. Stereotactic sampling verified that areas of contrast uptake at high dose were consistent with tumor in this region.

The volumes of tumor/target for brain lesions derived from CT and MRI are not necessarily consistent. Some investigators have reported (TEN HAKEN et al. 1992) that on the average, MRI volumes are larger than the CT volumes. However, these differences in volume between different modalities were comparable to variations in defining the volumes by various radiation oncologists. We (MYRIANTHOPOULOS et al. 1992) have reported in a limited series that the CT volumes are consistently larger. When both modalities are available and correlated, the combined target volumes are sometimes used. Which imaging modality best describes tumor extent remains as a research question.

1.2.3.3 Thorax

Radiographs are still the most valuable and cost-effective diagnostic imaging examination for detection of thoracic lesions (BATRA 1993). However, CT, through cross-sectional display, is essential in delineating the tumor's location and size as well as lung tissue density for image-based 3D planning. The detection of calcium, visualized by CT, is also of diagnostic importance. CT is reported to have a sensitivity of 95% in detecting mediastinal adenopathy, but its specificity or accuracy is low (50%) since enlarged nodes may be tumor free. MRI cannot differentiate malignancies through T1/T2 analysis, but coronal or sagittal MRI can be helpful in detecting the superior extent of tumors at the lung apex.

Accuracy by CT or MRI in distinguishing early stages of lung tumors (T1-T2) from late stages (T3-T4) showed essentially no difference between CT and MRI, using surgical and pathologic findings as "truth" (HEELAN 1992). The RDOG concludes that CT is the standard evaluation method. MRI does not provide additional useful information. Abnormally enlarged lymph nodes must still be verified histologically. CT is the choice over MRI because: (a) CT can detect localized invasion of the chest wall or mediastinum, (b) CT can detect invasion of the mediastinal organs, including bone, vessels, bronchi, trachea, esophagus, and heart, and (c) CT can detect hilar and mediastinal metastases.

MRI has been shown to be equivalent to CT in detecting enlarged mediastinal lymph nodes and masses. There is evidence to suggest that MRI may be helpful in distinguishing postradiation fibrosis from residual recurrent disease.

1.2.3.4 Abdomen and Pelvis

Imaging has provided accurate delineation of tumors and adjacent structures in the abdomen and pelvis (HALVORSEN and THOMSON 1993). The choice of CT or MRI to evaluate abdomin-opelvic abnormalities depends on the organ system and patient's condition.

A disadvantage of MRI vs CT of the abdomen is that a longer scan time is needed. MRI studies also cover only a limited anatomic area. For example, a thorough MRI study of the liver may require several pulse sequences, and will only provide images of the upper abdomen. To evaluate the entire abdomen and pelvis, two or three additional scans may be required, leading to excessive scan acquisition times. Additionally, patients with cardiac pace-makers are not candidates for MRI imaging.

Delineation of the prostate and seminal vesicles for conformal radiotherapy involves CT. Rectal and bladder contrast are helpful in defining these normal structures. A urethrogram has been used to identify the inferior border of the prostate, which is often difficult on a transverse cut. Image-based beam's eye view techniques have resulted in changes in 40% of the portal apertures, in comparison to conventional simulator methods of prostate target delineation (SANDLER et al. 1991).

TEN HAKEN et al. (1989) describe CT based contouring for prostate conformal therapy. Patients are immobilized using alpha cradle type molds. The patient is then CT scanned with radio-opaque catheters to mark reference points as defined in the simulation. Both GI contrast and IV contrast are used to define bladder and rectum. Scans with 5-mm slice thickness are obtained through the prostate and scans of other normal anatomy are acquired at 1-cm intervals. The gross tumor volume is taken to be the prostate, seminal vesicles, and puboprostatic ligament. The clinical target volume is defined by isotropically enlarging this volume by 5 mm, to account for patient setup uncertainties and motion. This is further enlarged to account for beam penumbra effects, which define the geometry of the blocks and aperture. The target volume is treated to the 95% isodose line. Patient alignment on the treatment couch and the plan is facilitated through matching bony anatomy during the final simulation and treatment verification films.

Staging of prostate cancer has recently been performed with endorectal surface coil MRI (SCHNALL et al. 1989, 1991). In this technique, a highly zoomed field of view (FOV) permits improved pretherapy staging. An FOV in the range of 10–12 cm can be obtained vs the normal 36-cm FOV for standard pelvic imaging. Schnall et al. demonstrated that an endorectal surface coil placed within the rectum could be used to obtain high-resolution images of the prostate, with excellent visualization of the gland structure. Endorectal surface coil imaging of the prostate was found to be more accurate in staging than the conventional digital rectal examination and MRI body coil images. The authors report an increase in accuracy of 16% using the new coil technique. The current technique involves inflating the rectum with an air-filled balloon during the image acquisition process. This distorts the anatomy, resulting in movement of normal organs and prostate. Thus, such zoomed images may not be directly useful in delineating the target volume, but primarily serve to increase staging accuracy.

1.2.3.5 GI Malignancies

Controversies still exist as to the optimal imaging studies for GI malignancies (HALVORSEN and THOMPSON 1991). Imaging modalities to stage and follow patients with neoplasms of the GI tract have included MRI, CT, and ultrasonography. CT is most frequently used for detecting liver and lung metastases. CT is the imaging study of choice for esophageal carcinoma, but is less helpful in staging patients with gastric or colorectal cancer. MRI is limited by lack of an adequate oral intraluminal contrast agent, and degradation of images due to motion. Ultrasonography, especially endoscopic ultrasonography, appears promising for the detection of local invasion by GI tract malignancies. The initial hope that MRI could differentiate scar from recurrent tumor has not been realized.

1.2.4 Future Trends in Oncologic Imaging

The past two decades have seen tremendous technological advances in imaging, through the development of computed axial x-ray tomography and MRI. Experts have speculated that improvements in oncologic imaging will continue at a slower rate (HENKELMAN 1990). Imaging currently provides the accurate anatomic data needed for precision radiation treatment. The next goal of imaging research is to advance beyond geometric definition, and to provide information on tumor physiology, response to therapies, and follow-up. If tumor characterization is realized, it will contribute to many aspects of cancer management. MR spectroscopy is a tool which provides a noninvasive means of determining certain concentrations of metabolites which may provide a more specific signal of tumor activity.

Radionuclide imaging of cancer (WAXMAN 1993) may increase in importance in the future. Radiolabeled monoclonal antibodies in principle can be used in the detection of malignancies since this technique identifies regions with higher concentrations of tumor-associated antigen. If limited-resolution single-photon emission computed tomography (SPECT) or positron emission tomography (PET) images are registered with high-resolution anatomic data (CT and MRI), it may be possible to improve diagnostic specificity (BLACK et al. 1989; KNOPP et al. 1990; KRAMER and NOZ 1991). As more specific antibodies are developed, and spatial resolution of SPECT and PET improved through algorithmic and hardware improvements, radionuclide studies will provide improved quantitative information (LEICHNER et al 1993).

Future software developments may also permit the use of images data sets not specifically taken for radiation therapy treatment-planning purposes. Scans of the head can generally be used for image-based planning if the assumption that it is a rigid body is valid. Under these conditions, the 3D target and normal tissue localization data are appropriately rotated and translated to the beam's eye view. The warping and displacement of organs in the torso under slightly different positioning are much more complex. As algorithms are developed to handle such distortions and displacements of organs, it may be possible to utilize images not taken with immobilization in treatment position, and use them in ways useful for planning and treatment assessment.

1.3 Technical Aspects of Target Delineation

1.3.1 Acquisition of Image Data for Planning

Image-based treatment planning requires the data be acquired in the treatment position. Typically, volumetric data are acquired, with slice thickness and table increments of 0.3–0.5 cm for head and neck tumors, and 0.5–1 cm slice thickness for tumors of the thorax, abdomen, and pelvis (Collaborative Working Group in Photon Radiotherapy 1991b). With coarser sampling, rapidly changing anatomy may not be adequately defined and thus inadequately reconstructed. Thinner slice thickness results in smoother target delineation, but must be balanced with the numbers of slices to be contoured. For planning nonaxial beams, the superior and inferior scan limits are extended to cover the entrance area of any possible extremely oriented beams. Reference materials with known electron densities have been included on immobilization devices or within the scanner couch to provide calibration data for inhomogeneity corrections. Radio-opaque catheters may be used on the patient surface to mark points and axes.

Occasionally, CT or MRI scan data are available only on film hard copy. Furthermore, the retrospective analysis of patients may involve a dosimetric reconstruction without availability of the image data in digital format. Investigators (WEINHOUS 1993) have developed methods to input tomographic images from film hard copy into the treatment-planning system by laser digitization. BOXWALA and ROSENMAN (1993) have shown that such digitization can be accurately performed without excessive distortion of dose volume histograms.

1.3.2 Compensating for Inaccuracies in Patient Setup

Variations in patient setup in fractionated radiotherapy are usually compensated for by increasing the field margin. Two approaches can be implemented to reduce the margin size, thereby decreasing the amount of normal tissue irradiated: (a) the patient can be more carefully immobilized, with the expectation that positioning variations will be reduced and (b) real time imaging devices can be used to determine patient positioning on a daily basis, with appropriate repositioning relative to the beam.

The image data for planning are acquired with the patient in the immobilization to be used during treatment. Effective immobilization can influence the radiation portal design by reducing the margins included in aperture design that account for uncertainty in patient repositioning. Immobilization devices also facilitate the image correlation of different data sets in target volume definition. Techniques to appropriately immobilize the patient have been reported in the literature, selected citations of which are described below.

Immobilization for head and neck lesions is highly developed. A number of immobilization techniques (VERHEY et al. 1982) permit precise daily repositioning for fractionated radiotherapy. Patients immobilized with a bite block and mask have been reported to be reproducibly repositioned to a mean value of one half mm. Other reports indicate that an individually formed mask (LYMAN et al. 1989; THORNTON et al. 1991) can be used to reposition to about 1 mm.

Masks and immobilization with markers also facilitate diagnostic imaging with multiple modalities (THORNTON et al. 1991). Both tomographic image data and radiographic films are used by the radiation oncologist to define appropriate target volumes (FRAASS et al. 1987; GLATSTEIN et al. 1985) and their alignment relative to landmarks during treatment. Markers on the mask facilitate the integration and correlation of the various modalities used in imaging.

With the current interest in conformal therapy of the prostate, several immobilization studies have been reported for this site. Daily setup variations of pelvic target volumes are found to be markedly improved with the use of alpha cradle (SOFFEN et al. 1991). Variations in field placement relative to bony landmarks were reduced to about 1 mm rather than 3 mm without immobilization.

Real time portal imagers have been developed by commercial vendors and research institutions (LEONG and STRACHER 1987; VAN HERK and MEERTENS 1988). Software to register the acquired images with simulation films has also been developed, with the intent of repositioning before each therapeutic fraction. Initial analysis of the impact on

dosimetry has suggested that about 10%-15% of the high dose delivered to normal structures can be reduced through real time imagers (BALTER et al. 1992).

1.3.3 Nomenclature of Volumes of Interest

Cooperative interinstitutional trials evaluating 3D planning and delivery will necessitate uniformity in defining tumor and target volumes. The nomenclature associated with volumes of interest for radiation therapy planning have evolved over the past decade. Collaborative Working Groups in Charged Particle Radiotherapy (1987) and Photon Radiotherapy (1991a) initially developed the nomenclature defined below:

Biological target volume (BTV) was defined as the volume of tissue including the tumor volume (Gross disease) and regions known to have or considered to be at risk for containing microscopic extension of disease, including regional lymph nodes.

Mobile target volume (MTV) described the volume which includes the BTV and exceeds its dimensions to account for factors such as variations in patient positioning, respiration, moveable internal organs.

The need for these definitions arose from the perceived need to separate the geometric characteristics of different radiations (e.g., sharp penumbras of particle beams vs broader penumbras of neutron beams) and different immobilization methods used at various institutions which could decrease the margins needed for tumor coverage.

With the rapidly growing interest in conformal radiation therapy, and the corresponding requirement for multi-institutional cooperative trials, there is a need to define consistent volumes of interest in 3D image data sets. The ICRU (1993) has recently proposed new definitions of target and tumor volumes which include:

Gross tumor volume (GTV): the gross palpable or visible extent of the tumor.

Clinical target volume (CTV): the volume which contains microscopic and subclinical disease. The CTV is generally a large volume which is considered to contain disease. Clinical target volume 2 (CTV2) includes areas of gross nodal disease as seen on CT or MRI with adequate margins. Clinical target volume 3 (CTV3) is the volume containing the GTV and subclinical microscopic malignant disease around the GTV. CTV3 corresponds to the volume of tissue receiving the final boost dose.

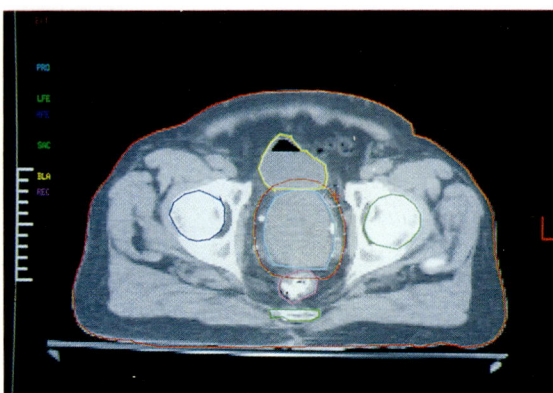

Fig. 1.1. GTV (*green contour*) and CTV (*red contour*) for a prostatic carcinoma. The bladder and rectum have also been manually contoured. Technically, the GTV is not visible, since that portion of the prostate with tumor is not visualized on CT. The GTV is defined to be the entire prostate gland

Planning target volume (PTV): encompasses the clinical target volumes plus additional margins which fold in the effects of organ and patient motion as well as inaccuracies in beam and patient setup. Beam penumbra effects are not included.

The general approach to defining and delineating targets as described is appropriate for many situations, but not all. Modification and adaptation of these definitions will be required in specific situations. Consider, for example, the delineation of volumes of interest for conformal therapy of prostatic carcinoma. The GTV is not strictly visible in this situation. Rather, the tumor volume is interpreted to be the entire prostate gland, as indicated by the green contour of Fig. 1.1. Depending upon the stage of disease, the seminal vesicles may be included in the GTV. CTVl, which nominally includes gross nodal disease in addition to the tumor, is not treated at many centers. CTV3, the final boost, includes the prostate and or seminal vesicles, depending on the stage. The PTV includes margins for movement and setup error. A margin of 1.5 cm around the prostate and seminal vesicles is currently used at several centers. An example of the delineation of the GTV and CT3V (red contour) for a carcinoma of the prostate is shown in Fig. 1.1.

1.3.4 Critical Structures and Landmarks

While this chapter primarily deals with the delineation of target volumes, such volumes are impacted by their geometric relationship to adjacent radiosensitive critical organs. The delineation of critical structures and the requirement to avoid irradiation

of such organs may impact the delineation of the field aperture. The tools for delineation of target volumes must also adequately accommodate delineation of normal organs and landmarks used for field alignment.

In addition to placing a margin on the tumor, a margin may also be placed around critical structures to account for possible setup errors (SHEROUSE 1993). If the clinical goal is to fully exclude a critical structure from irradiation (possibly due to previous irradiation to organ tolerance), the final aperture may be the exclusive combination of the two structures from the beam's eye view.

Delineation of landmarks or bony structures to facilitate field alignment is essential in contouring software. The software must be capable of demarcating points, lines, curves, and other markings needed to correlate objects seen on tomographic sections with plan radiographs used in final portal alignment. An advantage of CT over MRI is that digitally reconstructed radiographs can be generated from CT data sets, which are similar in appearance to conventional simulator radiographs. Such comparisons are of use in field alignment.

1.3.5 Software Tools for Contouring

The software capabilities needed to efficiently delineate volumes of interest for 3D treatment planning have been summarized in the recent literature (Collaborative Working Group in Photon Radiotherapy 1991a,b). Desirable features include:(a) a high degree of software interactivity to permit display adjustment for optimal viewing, (b) ability to display 2D sections, both standard orthogonal planes such as axial, sagittal, and coronal, and arbitrary oblique section, (c) capability of superposition of contours of interest, (d) the ability to delineate volumes of interest on different imaging modalities, (e) capability of registration of CT with MRI and other modalities, (f) dynamic displays of anatomy such as movie loops, which provide a kinetic view of anatomy to convey the sense of three dimensionality, and (g) display of 3D structures through shaded graphics or other display techniques. Once the volumes are defined, the radiation oncologist can understand the spatial relationships of the organs and tumor, and verify the consistency of objects contoured.

The number of tomographic slices for 3D treatment planning can be large, ranging from 20 to 100 cuts. The tedium of manually delineating an organ or

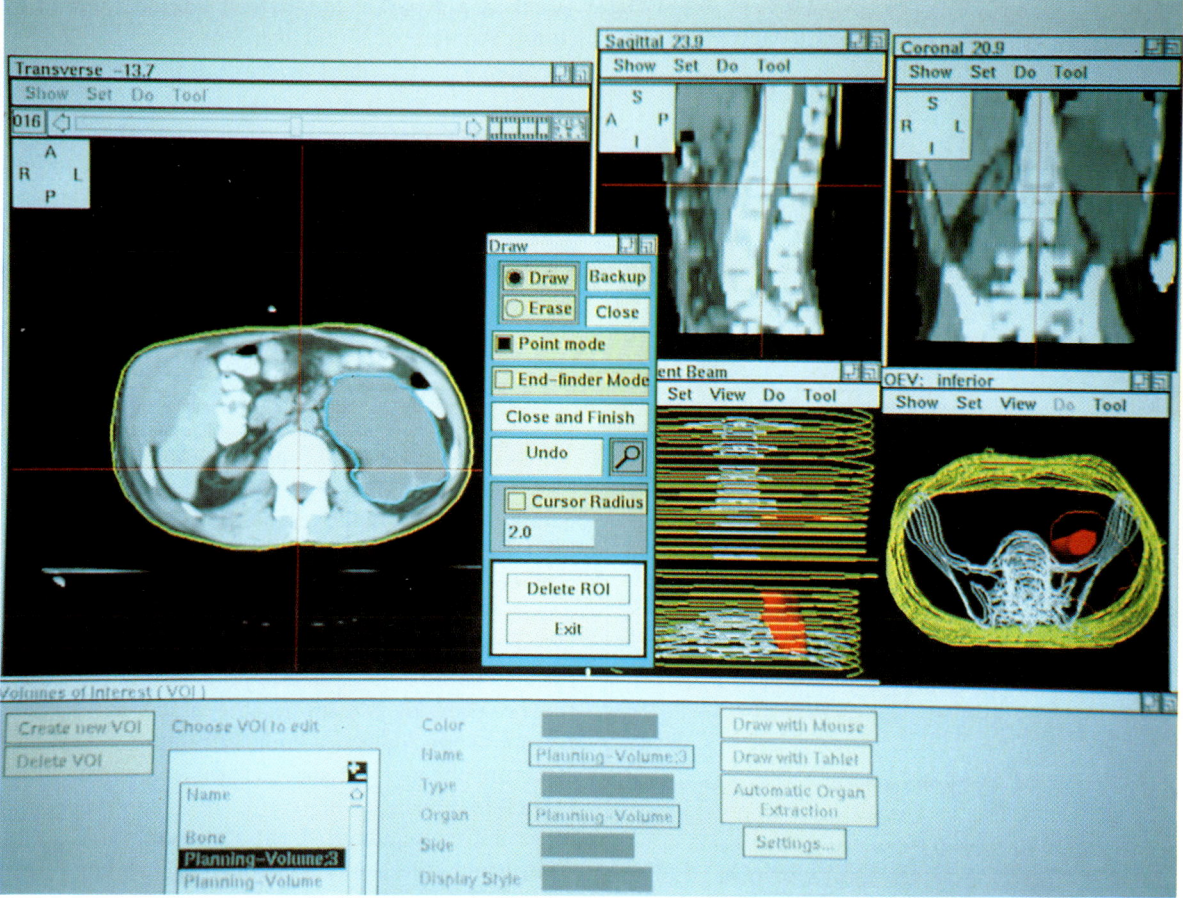

Fig. 1.2. AXIOM screen displays available during the manual contouring process. Views include the axial, sagittal, and coronal sections through the CT volume. A room's eye view and beam's eye view can also be displayed

target volume over many slices can be reduced in some planning systems through the capability of interpolation of volumes of interest. If the structure is smoothly varying, interpolation can be effectively used to decrease the required time. The systems implemented usually provide for the contourer to visually check the adequacy of the interpolated contour, and to edit if needed.

1.3.5.1 Manual Contouring

Much of the effort expended in 3D treatment planning involves manual delineation of volumes of interest. Most soft tissue organs with densities near water, such as kidneys or liver, are currently manually contoured. In this process, the user interactively pages through axial slices on a workstation display and enters points (or draws a line) around the organ boundary with a mouse or digitizing puck. A typical X-window display of a 3D treatment planning sys-

tem is shown in Fig. 1.2. The screen displays the program's volume of interest entry mode, where organs can be outlined interactively through a mouse. The upper smaller windows show coronal and sagittal multiplanar reconstructions while the large window in the screen center displays the axial slice. To its right are shown a beam's eye view and room's eye view of the anatomy. Multiple cuts through the volumetric imaging data assist in the visualization of 3D anatomy. Structures drawn in any of the axial, coronal, and sagittal windows will be displayed in corresponding positions in the other views.

Segmentation of 3D anatomic structures from CT and MRI scans is one of the limiting steps in implementation of 3D conformal therapy. In part, the efficiency of manually contouring is dependent on the type of hardware for contouring (DOWSETT et al. 1992). LI et al. (1991) analyzed the amount of time needed to segment single or multiple structures with manual and semiautomated contouring tools. Each contour in the treatment planning system is

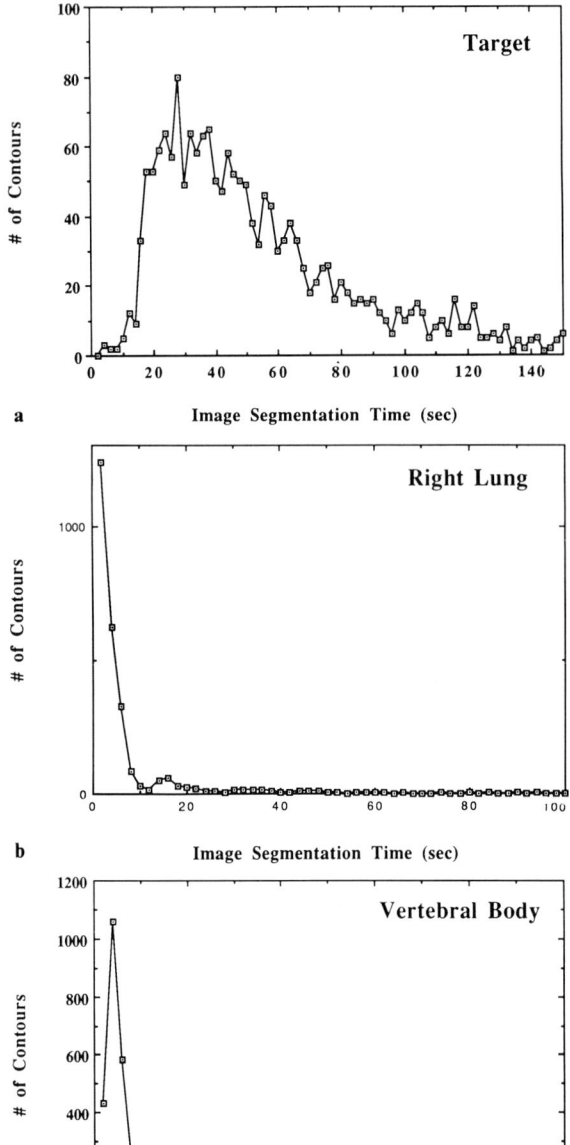

a   Image Segmentation Time (sec)

b   Image Segmentation Time (sec)

c   Image Segmentation Time (sec)

Fig. 1.3a–c. Histograms of times required to manually contour various structures using in-house developed software. Most vertebral body and lung contours are defined automatically or semi-automatically, and the modal time required is a few seconds per contour. Conedown target is manually entered by the radiation oncologist, and exhibits a broad distribution with peak at 30 s/slice. The mean time for defining a target contour per slice is approximately 50 s

saved as a separate file, with a unique time date stamp. Analysis of 30 000 such contours generated in our department yields average times and numbers of contours for common sites of interest.

High-contrast objects such as external contour, lungs, and vertebral body can be contoured at a rate of approximately 2–4s per contour per slice. The histograms of times required to manually contour the boost target CTV3 and initial target CTV1 are shown in Fig. 1.3. The mean time is about 40–60s per contour on a given slice. A similar distribution is seen for the GTV.

1.3.5.2 Semiautomated Segmentation Tools

Semiautomated image segmentation tools require limited human input. This approach is typically applied to normal structures with input from the dosimetrist or physicist. An example of such a process is the initiation of lung contouring through the depositing of a single seed point within the lung; the program then identifies the lung contour by threshold edge detection and continues to map the structure through all slices. These tools generally request human help when confusing or complex contours are encountered. Following semiautomated contouring, the appropriateness of the contours must be checked.

Tools for semiautomatic segmentation of brain images to separate white and gray matter have been developed, the output of which is applied clinically in the production of 3D volume-rendered images of the brain for diagnostic and surgery-planning purposes (LEVIN et al. 1988,1989). This software is routinely used to segment the brain from 3D volume MRI studies. Highly optimized algorithms for brain segmentation have been implemented utilizing a priori information such as the characteristic image intensity gradient at the brain surface. Morphologic operations such as erosion and dilation are provided to quickly correct most suboptimal segmentations.

1.3.5.3 Automated Image Segmentation of High-Contrast Objects

More advanced computer vision concepts are being developed to automatically identify and contour structures. Fully automated image feature detection is under development at the University of North Carolina (CHANEY and PIZER 1993), sponsored by the

NCI software tools contract (KALET et al. 1990). An automated image segmentation algorithm uses aspects of computer vision to automatically recognize and extract the boundaries of soft tissue organs.

Other approaches to extract information potentially useful for automated segmentation include: (a) the development of an automated model-based segmentation method for the extraction of normal organ contours from CT slices, (b) texture analysis, which may be useful in the classification of tumor, necrosis, edema, and normal tissue in both CT and MRI images, and (c) enhancement of detectability of soft-tissue tumors using linear combinations of MR images and multispectral feature space classifications based on multiple MR sequences (HERRMANN and LEVIN 1986; LEVIN et al. 1987).

Images can require editing before they can be used for treatment planning (MOHAN et al. 1988), and thus user-friendly software capabilities to modify contours are needed. In particle therapy, GI contrast may require editing of image pixel values rather than contours. Tissue-equivalent Hounsfield units replace contrast in bladder, rectum, and other parts of the GI tract in order to make appropriate density assessments for charged particle penetration. Similar efforts may be needed for pixel by pixel electron beam treatment planning, and to a lesser extent photon planning.

1.3.6 Size and Stage

Delineation of the gross tumor volume also provides data useful for AJCC staging. Current American Joint Cancer Commission criteria for TNM staging usually include specification of the largest tumor dimensions. With a 3D quantitation of tumor size, such data become significantly more precise in three dimensions. After outlining the tumor volume, a contouring system could easily determine tumor size and categorize the tumor volume and provide supplementary information useful in TNM staging.

1.3.7 Quality Assurance

Peer review of target volumes and facilitation of manual contouring have led some groups to explore the utility of large-screen projection-type display of imaging data during manual contouring. The University of Michigan group (FRAASS 1993) described a wall-mounted digitizer over a projected image and reported that such hardware speeds the process of target and normal tissue entry by factors

of 4–10. Display of the target volumes for peer review could also be important in quality assurance of contour definitions, particularly for multi-institutional cooperative trials.

1.3.8 Image Correlation

Image registration, the synthesis of information from more than one image study, is useful in the treatment planning process because tumors and normal tissues often may be most completely defined by imaging with multiple-modality studies (CHEN 1985; CHEN and PELIZZARI 1989; CHEN et al. 1990; KESSLER et al. 1991; PELIZZARI et al. 1989; SCHAD et al. 1987a). For example, brain scans with both CT and MRI, including axial, sagittal, and/or coronal slices, may be used to define the relationship of a tumor to surrounding bony anatomy. Registration of serial scans also provides information on the movement of organs, which determines the limits of radiation field margins. The estimation of variation of prostate and seminal vesicle position during a protracted course of fractionated radiotherapy can provide insights into the appropriate size of field margins.

Registration methods have also been proposed to align the patient in the treatment room with the beam, in order to match the delivered plan with the computer-modeled plan (PELIZZARI et al. 1991). Furthermore, image registration is useful in the analysis of follow-up scans to detect changes as a result of therapy. Correlating the dose distribution calculated from a pre-therapy scan with changes observed on a follow-up scan facilitates the analysis of possible radiation effects.

1.3.8.1 Approaches to Image Correlation

The approaches to registration of multiple 3D imaging scans of a single subject may be divided into two classes: those which (a) rely on repeated accurate patient immobilization or (b) extract the interscan coordinate transformation from information in the images. The latter group further may be classified as "prospective" or "retrospective".

Prospective methods utilize an external coordinate system attached to the subject, to which the orientation of each scan may be related. The use of a rigid stereotactic frame is an example of such a method, as is the procedure of attaching a number of point landmarks to the subject in identical locations for each scan, and using the visualized positions of these marks in the images to calculate the interscan coordinate transformation.

Retrospective methods recover the interscan coordinate system using information intrinsic to the images, and in principle require no special preparation either in immobilizing the subject or in attaching external or implanted markers. These methods vary according to the type of anatomic information utilized to supply constraints in finding the interscan transformation. Methods using anatomic point landmarks (EVANS et al. 1989), surfaces or volumes (PELIZZARI et al. 1992), planes (KAPOULEAS et al. 1989), and combinations of these elements have been described. PELIZZARI et al. (1989) developed a retrospective method of 3D image registration based on surface matching. This method has been applied to the registration of CT, MRI, PET, and SPECT images of the brain, pelvis, abdomen, lungs, and extremities.

A retrospecive image registration method is useful in target volume delineation, because it utilizes image data available but not specifically required for conformal therapy. Since such a technique imposes no special conditions on image studies, archival data may be processed as well as those newly acquired. Furthermore, to successfully become integrated into everyday clinical practice, it is important that an image registration method not require that cumbersome additional steps, e.g., patient immobilization or placement of fiducial marks, be added to frequently performed procedures such as patient CT or MRI scans.

The original surface fitting algorithm has been modified to incorporate additional constraints based on multiple pairs of surfaces identified in each scan, and to calculate the distance between surfaces more accurately for highly anisotropic objects such as bones (PELIZZARI et al. 1992), lungs, or kidneys. 3D image registration involving CT, MRI, PET, and SPECT images for a number of clinical and research problems has been described elsewhere (HOLMAN et al. 1991; KASSAEE et al. 1991; MAZZIOTTA et al. 1991; PELIZZARI et al. 1991, 1992; ROESKE 1992; TURKINGTON et al. 1991, 1993). Capabilities for synthesis and display of integrated 2D and 3D MRI/PET brain images have been used for planning neurosurgery (GEHRING et al. 1991) and brain tumor cases (HU et al. 1989, 1990; Levin et al. 1988, 1989) Image registration has also been used to produce 3D brain surface models combining MRI and phosphorus-31 MR spectroscopy (CAO et al. 1992) and to register information from biplane radiographs with MRI (GRZESZCZUK et al. 1992).

A clinical example of the utility of image registration can be useful. The MRI image (Fig. 1.4a) is an axial cut through the patient's eyes, with an adenocystitic lesion posterior and lateral to the right globe. The patient then underwent resection of this lesion with gross tumor in the margins. The primary objective of correlating the MRI and postsurgical CT scan in treatment position was to locate on the latter scan the original tumor location. The case was further complicated by the existence of a second lesion, a meningioma near the foramen. The treatment plan-

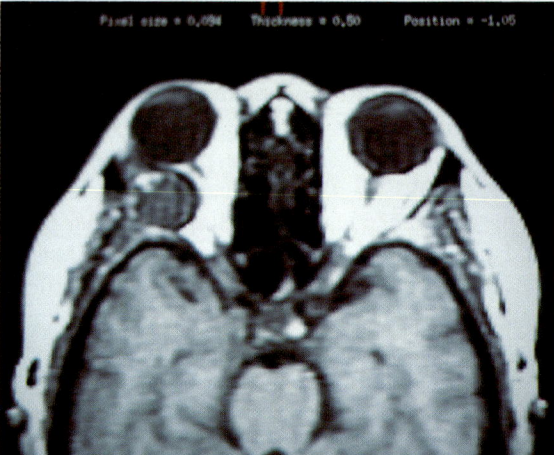

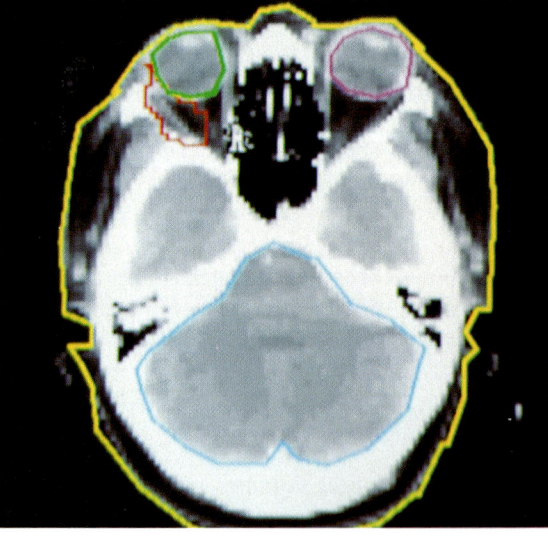

Fig. 1.4a,b. Image correlation is used to define the target volume post surgical resection. **a** Preoperative MRI scan of patient with a tumor in the left eye. **b** CT scan post surgical resection, with contour transferred from MRI study. This process was useful in defining the original target volume. The correlation was performed through surface fitting the inner table

ning problem involved external beam treatment of the residual disease in the eye, while avoiding over-irradiation of the retina and contralateral eye and avoiding the external beam fields used to irradiate the base of skull lesion.

1.3.9 Organ Movement and Deformation

There are limited quantitative data in the literature on organ movement, which provide some guidance in determining margins for organ motion. Whether the magnitude of movement varies significantly between patients is not quantitatively well known. Several of the studies do not specifically include cancer patients; movement of tumor partially fixed to bony structures clearly will be variable. Additional studies are needed.

1.3.9.1 Thorax

Patients with intrathoracic neoplasms have been studied with high-speed CT (Ross et al. 1990). In applying radiation portals designed by conventional simulation to the observed lung tumors visualized by cine CT, major geographic misses were detected in 15% of patients. The greatest tumor displacement was seen in tumors near the heart, aorta, or diaphragm. Motion of hilar lesions showed an average value of 9 mm due to cardiac motion. Not all lung tumors showed motion. Tumors in the chest wall showed no measurable motion. Motion in this study was analyzed only in the transverse plane; no direct measurement of tumor movement in the longitudinal axis was performed.

Physiologic motion can cause streak artifacts in CT scans, particularly in the chest, where cardiac and breathing motions are present (RITCHIE et al. 1992). Artifacts are also caused by peristalsis in the abdomen. These motions, especially when contrast is present, produce streaks and loss of resolution as well as anatomic distortions.

Ultra-high-speed CT has been used to study the movement of vessels and heart wall, which may indicate similar movement in tumors near the heart. Vessels near the cardiac wall moved 6 mm radially from the heart. Vessel motion during respiration was observed to be primarily AP-PA in the upper chest but radially in the lower chest, with a magnitude estimated to be approximately 2 cm in the upper chest and 3–4 cm in the lower chest. These values are for normal patients.

1.3.9.2 Abdomen

The quantitation of organ movement in the abdomen (HARUAZ and BRONSKILL 1979) has been of concern in nuclear medicine for many years. Nuclear medicine scanning of hepatic lesions requires tens of minutes and is subject to blurring by respiration. To recover sharper images, a correction signal proportional to the center of activity of the liver has been calculated. Average respiratory frequency is about 20 per minute, and the amplitude observed is about 14 mm peak to peak. The observed amplitudes range from 28 mm to 10 mm, with 18 mm including most observed values.

More recent MR studies have verified the magnitudes of organ motion in the abdomen (KORIN et al. 1992). Motion along two principal axes was determined simultaneously, e.g., superior/inferior motion vs AP motion. Liver motion in the superior/inferior axis was significantly greater than right/left motion. The peak to peak superior/inferior excursion of the dome of the liver was 1.7 cm for quiet breathing vs 3.9 cm for deep breathing. Movement in the right/left axis was approximately 3 mm, and small motions in the AP axis were also recorded. The predominant translation during respiration of upper abdominal structures is a translation in the superior/inferior axis, with little movement either right/left or AP. This study would suggest significantly anisotropic field margins for irradiation.

Some of these organ movement studies did not involve patients with tumors, nor were the patients imaged positioned and immobilized in treatment position. While the amplitudes quoted are general guidelines, and illustrate the anisotropy of organ motion, additional studies must be performed before specific recommendations for field margins due to organ motion are adopted.

1.3.9.3 Pelvis

In a study of 18 patients with prostate cancer, Mayo Clinic investigators (SCHILD et al. 1993) quantified prostate movement when the bladder and rectum were distended. Distention of the rectum displaced the prostate anteriorly by as much as 1 cm, although the median movement was only 1 mm. Similar displacements were observed with distention of the bladder, displacing the prostate posteriorly.

Forman et al. (1993) are analyzing the positional variation of the prostate, seminal vesicles, bladder, and rectum by CT scanning patients approximately once a week during the course of external beam

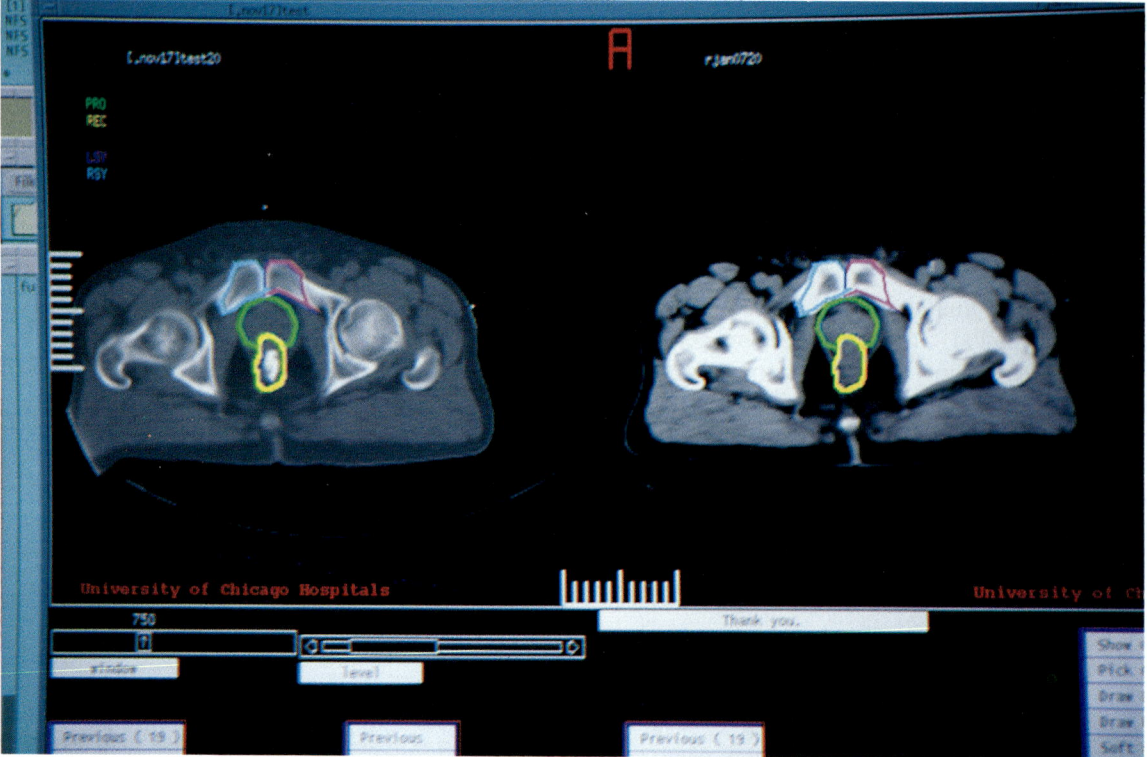

Fig. 1.5. Variations in rectal contours at two different times during a course of fractionated radiotherapy. Bony pelvis was correlated using surface fitting, and images were resliced along comparable planes. The contour of the rectum is drawn on the *left*, and mapped to the later scan on the *right*. Movement and deformation shown is approximately 1 cm in this slice

radiotherapy. In the analysis, the bony pelvis of the serial scans are aligned using surface matching. The organ contours are then mapped into the same coordinate system, and analyzed to determine their range of motions. Preliminary findings indicate movements of 1 cm or greater for all structures. Figure 1.5 shows an initial CT scan before therapy, and one taken several weeks later. The contour of the rectum, drawn on the left, is mapped to the image on the right. In this example, there is a significant displacement and shrinkage of the rectal contour. The dosimetric consequences of such variations in target and normal tissue positions on conformal therapy are under analysis at several centers.

1.4 Assessment of Response

In addition to defining the target for treatment planning, tools for volume of interest delineation can also be used to quantitate response to therapy. The importance of reliable tumor delineation in assessment of response to therapy has been recognized. GROSSMAN and BURCH (1988) observed "the ability to compare results of clinical trials depends upon consistent definitions of tumor response and the use of reliable and reproducible methods of tumor measurement." Despite this recognition, in practice classification schemes used to evaluate patient response remain semiquantitative and ambiguous. The response of tumors to localized or systemic therapy is commonly classified into one of four categories: complete response, partial response, stable disease, or progressive disease. Each of these is associated with an approximate quantitation of tumor volume.

Such commonly applied indices of tumor response are subjective, since the oncologist or radiologist either estimates the volume changes by eye or measures with a ruler from films the length/width dimensions of one or a few prominent lesions identified in a pretreatment scan. An attempt is made to identify the same lesions and measure at similar cross-sections in subsequent studies. Generally speaking, a dramatic response categorized as CR is easily detectable. Similarly, significant progression of the disease under a treatment regimen is readily detectable. In intermediate situations, classification may be uncertain due to difficulty in determining whether a substantial change in tumor volume has occurred. Due to difficulty in the assessment of tumor extent changes, and the resulting variance in classification in intermediate cases, results of the same therapy have sometimes appeared quite different in the literature (DAVIS et al. 1980). Clinically, the decision whether to maintain a chemotherapeutic regimen that has been begun, or to change to another, is often based on the oncologist's and/or radiologist's perception as to whether subtle changes in tumor size have occurred. More accurate quantitation of response will be of considerable use to clinical oncologists in this decision-making process.

Tumor volume has been correlated with local control (GILBERT 1987), and the rate of tumor volume reduction post therapy has been documented with the intent of determining its prognostic significance (To et al. 1990). The analysis of follow-up CT and MRI has also led to the quantification of damage from brain irradiation (CONSTINE et al. 1988), with the hope of leading to a better understanding of the adverse effects of radiation therapy. However, altered appearance post radiation therapy can be confused with tumor, and must be taken into account (CHAN and KRESSEL 1991).

References

Balter J, Pelizzari CA, Chen GTY (1992) Correlation of projection radiographs in radiation therapy using open curve segments and points. Med Phys 19: 329–334

Batra P (1993) Imaging neoplasams of the thorax. In: Holland J, Frei III E, Bast RC et al. (eds) Cancer Medicine. Lea and Febiger, Philadelphia

Black KL, Hawkins RA, Kim RT, Becker DP, Lerner C, Marciano D (1989) Use of thallium-201 SPECT to quantitate malignancy grade of gliomas. J Neurosurg 71: 342–346

Boxwala AA, Rosenman JG (1994) Retrospective reconstruction of three dimensional radiotherapy plans from two dimensional planning data. Int J Rad Onc Bio Phys 28: 1009–1016

Bragg DG, Thompson WM (1992) Categorical course on imaging of cancers: diagnosis, staging and follow-up challenges. American College of Radiology, Reston, VA

Brown AP, Urie MM, Chisin R, Suit HD (1989) Proton therapy for carcinoma of the nasopharynx: a study in comparative treatment planning. Int J Radiat Oncol Biol Phys 16: 1607–1614

Cao YG, So J, Gregory CD, Dawson MJ, Raidy T, Lauterbur PC, Pelizzari CA (1993). Integrated 3-D display of gyral anatomy and MR spectra on the brain surface. Mag Res Imag 11: 1043–1049

Chan TM, Kressel HY (1991) Prostate and seminal vesicles after irradiation: MR appearance. JMRI 1: 503–511

Chaney E, Pizer SM (1993) Defining anatomical structures from medical images. Semin Radiat Oncol 2: 215–225

Chen GTY, Pelizzari CA (1989) Image correlation techniques in radiation therapy treatment planning. Comput Med Imaging Graph 13: 235–240

Chen GTY, Kessler ML, Pitluck S (1985) Structure transfer in three dimensional imaging studies. National Computer Graphics Assn 3: 171–177

Chen GTY, Pelizzari CA, Levin DN (1990) Image correlation in oncology. In: Devita V. Hellman S, Rosenberg S (eds) Important advances in oncology 1990. J.B. Lippincott, Philadelphia, pp 131–142

Constine LS, Konski A, Ekholm S, McDonald S, Rubin P (1988) Adverse effects of brain irradiation correlated with MR and CT imaging. Int J Radiat Oncol Biol Phys 15: 319–330

Collaborative Working Group in Charged Particle Radiotherapy (1987) In: Zink S (ed) Evaluation of treatment planning for particle beam radiotherapy.

Collaborative Working Group in Photon Radiotherapy (1991a) Evaluation of high energy photon external beam treatment planning: project summary. Int J Radiat Oncol Biol Phys 21: 3–8

Collaborative Working Group in Photon Radiotherapy (1991b) Three dimensional display in planning radiation therapy: a clinical perspective. Int J Radiat Oncol Biol Phys 21: 79–89

Damadian R (1971) Tumor detection by MR. Science 171: 1151–1153

Davis HL, Multhauf P, Klotz J (1980) Comparisons of cooperative group evaluation criteria for multiple drug therapy for breast cancer. Cancer Treat Rep 64: 507–517

Dillon WP, Mills CM, Kjos B et al. (1984) Magnetic resonance imaging of the nasopharynx. Radiology 152: 731–738

Dowsett RJ, Galvin JM, Chen E et al. (1992) Contouring structures for 3D treatment planning. Int J radiat Oncol Biol Phys 22: 1083–1088

Engel JP, Lufkin R, Behnke EJ (1990) Multimodality imaging of brain structures for stereotactic surgery. Radiology 175: 433–441

Evans AC, Marrett S, Collins L, Peters TM (1989) Anatomical-functional correlation. SPIE Medical Imaging III – Image Processing. 1092: 264–274

Forman JD, Mesina CF, He T, Devi SB Ben-Josef E, Pelizzari C, Vijayakumar S, Chen GT (1994) Evaluation of changes in the location and shape of the prostate and rectum during a seven week course of conformal radiotherapy. Int J Rad Onc Bio Phys 27: 222 (abstract)

Fraass BA (1993) Clinical application of 3D treatment planning. In: Purdy JA (ed) Advances in radiation oncology physics. AAPM Monograph 19, AIP, New York, pp 967–997

Fraass BA, McShan DL, Diaz RF et al. (1987) Integration of magnetic resonance imaging into radiation therapy treatment planning. Int J Radiat Oncol Biol Phys 13: 1897–1908

Gehring M, Mackie TR, Kubsad S, Paliwal BR, Mehta M, Kinsella T (1991) A three-dimensional volume visualization package applied to stereotactic radiosurgery treatment planning. Int J Radiat Oncol Biol Phys 21: 491–500

Gilbert RW, Birt D, Shulman H, Freeman J, Jenkin D, MacKenzie R, Smith C (1987) Correlation of tumor volume with local control in laryngeal carcinoma treated by radiotherapy. Ann Otol Rhinol Largngl 96: 514–518

Glatstein E, Lichter AS, Fraass BA, Kelley BA, Van de Geign J (1985) The imaging revolution and radiation oncology: use of CT, ultrasound, and NMR for localization treatment planning, and treatment delivery. Int J Radiat Oncol Biol Phys 11: 299–314

Goitein M (1991) CT simulation: an overview. CT simulation for radiotherapy. Medical Physics Publishing, Madison, Wiscon, pp 161–171

Goitein M, Abrams M, Rowell D, Pollari H, Wiles J (1983) Multidimensional treatment planning: II. Beam's eye view, back projection, and projection through CT sections. Int J Radiat Oncol Biol Phys 9: 789–797

Grossman SA, Burch PA (1988) Quantitation of tumor response to anti-neoplastic therapy. Seminars in Oncology. 15: 441–454

Grzeszczuk R, Alperin N, Levin DN, Cao Y, Tan KK, Pelizzari CA, Mojtahedi S (1992) Multimodality intracerebral angiography: registration and integration of conventional angiograms and MRA data. Magn Reson Imaging (Abstract) IEEE Engineering in Med and Biol 7: 2783

Halvorsen RA, Thompson WM (1991) Primary neoplasms of the hollow organs of the gastrointestinal tract. Cancer 67: 1181–1188

Halvorsen RAJ, Thompson WM (1993) In: Holland J, Frei III E, Bast RC et al. (eds.) Imaging neoplasms of the abdomen and pelvis. Cancer medicine. Lea and Febiger, Philadelphia

Haruaz G, Bronskill MJ (1979) Comparison of the liver's respiratory motion in the supine and upright positions. J Nucl Med 20: 733–735

Heelan RT (1992) Primary lung cancer RDOG trials: MRI vs CT. Categorical course on imaging of cancers: diagnosis, staging and followup challenges. Amer. College of Radiology, Reston, VA, pp 13–19

Hendrix RA, Lenkinski RE, Vogele K, Bloch P, McKenna WG (1990) 31-P localized magnetic resonance spectroscopy of head and neck tumors – preliminary findings. Otolaryngology 103: 775–783

Henkelman RM (1990) New imaging technologies: prospects for target definition. Int J Radiat Oncol Biol Phys 22: 251–257

Hermann A, Levin DN (1987) Segmentation and 3D display of brain and muskuloskeletal tissues in magnetic resonance imaging. Radiology 165: 345 (abstract)

Hollis DP, Economou JS, Parks LC et al. (1973) NMR studies of several experimental and human malignant tumors. Cancer Res 33: 2156–2160

Holland J, Frei III E, Bast RC et al. (eds) (1993) Cancer Medicine. Lea & Febiger, Philadelphia

Holman BL, Zimmerman RE, Carvalho PA et al. (1991) Computer-assisted superimposition of magnetic resonance and high resolution Tc-99m HMPAO and Tl-201 SPECT images of the brain. J Nucl Med 32: 1478–484

Honig JF (1992) Evaluation of magnetic resonance tomography with regard to the assessment of solid and cystic processes in the head and neck region. Electromedica 60: 55–56

Hricak H (1992) Imaging of gynecologic malignancies; staging, post treatment follow up and evaluation of radiation injury. Categorical course on imaging of cancers: Diagnosis, staging, and followup challenges. Amer. College of Radiology, Reston, VA, pp. 63–72

Hu X, Tan KK, Levin DD et al. (1989) In: Upson C (ed) Volumetric rendering of multimodality, multivariable medical imaging data. Proc Volume Visualisation Workshop (Chapel Hill, NC, May 18–19). Department of Computer Science, University of North Carolina, Chapel Hill, NC, pp 379–379

Hu X, Tan KK, Levin DN, Pelizzari CA, Chen GTY (1990) A volume rendering technique for integrated three dimensional display of MR and PET data. 3D imaging in medicine. Springer, Berlin Heidelberg New York

ICRU (1993) New ICRU dose specification report for external radiotherapy report #50.

Kalet I, Chaney E, Purdy J, Zink S (1990) Radiotherapy treatment planning tools: first year report NCI, Bethesda, technical Report 90–1

Kapouleas I, Alavi A, Alves WM, Gur RE, Weiss DW (1989) Registration of three-dimensional MR and PET images of the human brain without markers. Radiology 181: 731–739

Kassaee A, Balter J, Pelizzari CA, Sutton HG, Chen GTY (1991) MRI in treatment planning of pelvic tumors. Med Phys 18: 615 (abstract)

Kelly PJ (1986) Computer assisted stereotaxis: new approaches for the management of intracranial and intra-axial tumors". Neurology 36: 427–439

Kessler ML, Pitluck S, Petti P, Castro J (1991) Integration of multimodality imaging data for radiotherapy treatment planning. Int J Radiat Oncol Biol Phys 21: 1653–1667

Korin H, Ehman R, Riederer SJ, Felmlee JP, Grimm RC (1992) Respiration kinematics of the upper abdominal organs: a quantitative study. Magn Reson Med 23: 172–178

Kramer E, Noz ME (1991) CT-Spect fusion for analysis of radiolabeled antibodies: application in GI and lung carcinoma. Int J Rad Appl Instrum [B] 18: 27–42

Leichner PK, Koral KF, Jaszczak RJ, Green AJ, Chen GTY, Roeske JC (1993) Imaging techniques and treatment planning in radioimmunotherapy. Med Phys 20: 569–578

Leong JC, Stracher MA (1987) Visualization of internal motion within a treatment portal during a radiation therapy treatment. Radiother Oncol 9: 153–156

Levin DN, Herrmann A, Spraggins T, Collins PA, Dixon LB, Simon MA, Stillman AE (1987). Musculoskeletal tumors: improved depiction with linear combinations of MR images. Radiology 163: 545–549

Levin DN, Pelizzari CA, Chen GTY, Chen CT, Cooper MD (1988) Retrospective geometric correlation of MR, CT and PET images Radiology 169: 817–823

Levin DN, Hu X, Tan KK et al. (1989) The brain: integrated three-dimensional display of MR and PET images. Radiology 172: 783–789

Levin DN, Hu X, Tan KK et al. (1990) Integrated 3D display if MR, CT, and pet images of the Brain. 3D imaging in medicine. Springer, Berlin Heidelberg New York

Li C, Spelbring DR, Chen GTY (1991) An analysis of image segmentation times for beam's eye view planning. Med Dosim 8: 119–124

Lichter A, Fraass B, McShan DL (1988) Recent advances in radiotherapy treatment planning. Oncology 2 (5): 43–57

Low NN, Vijayakumar S, Rosenberg I, Rubin S, Virudachalam R, Spelbring DR, Chen GTY (1990) Beam's eye view based prostate treatment planning: Is it useful? Int J Radiat Oncol Biol Phys 19: 759–768

Low NN, Vijayakumar S, Myrianthopoulos LC, Sutton H, Krishnasamy S, Rubin S, Chen GTY (1992) Practical applications of beam's eye view based treatment planning to head and neck sites. Int J Radiat Oncol Biol Phys 22: 1975–1082

Lufkin R (1993) Imaging neoplasms of the head and neck and CNS In: Holland J, Frei III E, Bast RC et al., Cancer Medicine. Lea and Febiger, Philadelphia, pp. 455–459

Lyman JT, Phillips MH, Frankel KA, Fabrikant JI (1989) Stereotactic frame for neuroradiology and charged particle Bragg peak radiosurgery of intracranial disorders. Int J Radiat Oncol Biol Phys 16: 1615–1621

Mazziotta JC, Pelizzari CA, Chen GTY, Bookstein FL, Valentino D (1991) region of interest issues: the relationship between structure and function in the brain. J Cereb Blood Flow Metab 11: A51–A56

McShan DL, Silverman A, Lanza DN, Reinstein LE, Glicksman AS (1979) A computerized three-dimensional treatment planning system utilizing interactive color graphics. Br J Radiol 52: 478–481

Mohan R, Barest G, Brewster L, Chiu CS, Kutcher GJ, Laughlin JS, Fuks Z (1988) A comprehensive three-dimensional radiation treatment planning system. Int J Radiat Oncol Biol Phys 15: 481–495

Munck JN, Cvitkovic E, Piekarski JD, Benhamou E, Recondo G (1991) Computed tomographic density of metastatic lymph nodes as a treatment-related prognostic factor in advanced head and neck cancers. J. Natl Canc Inst 83: 569–575

Myrianthopoulos LC, Vijayakumar S, Spelbring DR, Krishnasamy S, Blum S, Chen GTY (1992) Quantitation of treatment volumes from CT and MRI in high-grade gliomas: implications for radiotherapy. Magn Reson Imaging 10: 375–383

Pelizzari CA, Chen GTY, Spelbring DR, Weichselbaum RR, Chen CT (1989) Accurate three-dimensional registration of PET, CT and MR images of the brain. J Comput Assist Tomogr 13: 20–27

Pelizzari CA, Tan KK, Levin DN, Chen GTY, Balter J (1991) In: Colchester ACF, Hawkes DJ (eds) Interactive patient image registration. Information processing in medical imaging. Springer, Berlin Heidelberg New York, pp 132–141

Pelizzari CA, Chen GTY, Du JZ (1992) In: Morucci JP, Plonsey R, Coatoueux JL, Laxminarijan S (eds) Registration of multiple MRI scans by matching bony surfaces. Proc IEEE Eng Med Biol, IEEE, Piscataway, NJ, pp 1972–1973

Ritchie CJ, Godwin JD, Crawford CR, Stanford W, Anno H, Kim Y (1992) Minimum scan speeds for suppression of motion artifacts in CT. Radiology 185: 37–42

Roeske JC (1992) Dosimetry of radiolabeled monoclonal antibodies. PhD thesis, The University of Chicago.

Rosenman J (1991) 3D imaging in radiotherapy treatment planning. In: Udupa J, Herman GT (eds) 3D imaging in medicine. CRC press, Boca Raton, pp 313–330

Rosenman J, Sherouse GW, Fuchs H et al. (1989) Three-dimensional display techniques in radiation therapy treatment planning. Int J Radiat Oncol Biol Phys 16: 263–269

Ross CS, Hussey DH, Pennington EC, Stanford W, Doornbos JF (1990) Analysis of movement of intrathoracic neoplasms using ultrafast computerized tomography. Int J Radiat Oncol Biol Phys 18: 671–677

Sandler H, McShan D, Lichter AS (1991) Potential improvement in the results of irradiation for prostate carcinoma using improved dose distribution. Int J Radiat Oncol Biol Phys 22: 361–367

Schad L, Boesecke R, Schlegel W et al. (1987a) Three dimensional image correlation of CT, MR, and PET studies in radiotherapy treatment planning of brain tumors. J Comput Assist Tomogr 11: 948–954

Schad L, Lott S, Schmidtt F et al. (1987b) Correction of spatial distortion in MR imaging: a prerequisite for accurate stereotaxy. J Comput Assist Tomogr 11: 499–505

Schad LR, Gademann G, Knopp M, Zabel H, Schlegel W, Lorenz WJ (1992) Radiotherapy treatment planning of basal meningiomas: improved tumor localization by correlation of CT and MR imaging data. Radiother Oncol 25: 56–62

Schild SE, Casale HE, Bellefontaine LP (1993) Movements of the prostate due to rectal and bladder distension: implications for radiotherapy. Med Dosim 18: 13–15

Schnall MD, Lenkinski RE, Pollack HM, Imai Y, Kressel HY (1989) Prostate: MR imaging with an endorectal surface coil. Radiology 172: 570–574

Schnall MD, Imai Y, Tomaszewski J, Pollack HM, Lenkinski RE, Kressel HY (1991) Prostate cancer: local staging with endorectal surface coil imaging. Radiology 178: 797–802

Sherouse GW (1993) Images and treatment simulation In: Purdy JA (ed) Advances in radiation oncology physics. AAPM Monograph 19, AIP, New York, pp 925–947

Sherouse GW, Chaney EL (1991) The portable virtual simulator. Int J Radiat Oncol Biol Phys 21: 475–482

Soffen EM, Hanks GE, Hwang CC, Chu JCH (1991) Conformal static field therapy for low volume low grade prostate cancer with rigid immobilization. Int J Radiat Oncol Biol Phys 20: 141–146

Suit H, duBois W (1991) The importance of optimal treatment planning in radiation therapy. Int J Radiat Oncol Biol Phys 21: 1471–1478

Svensson G (1984) Quality assurance in radiation therapy: physics efforts. Int J Radiat Oncol Biol Phys 10: 23–29

Ten Haken RK, Perez-Tamayo C, Tesser RJ, McShan DL, Fraass BA, Lichter AS (1989) Boost treatment of the prostate using shaped fixed fields. Int J Radiat Oncol Biol Phys 16: 193–200

Ten Haken RK, Thornton AF, Sandler HM et al. (1992) A quantitative assessment of the addition of MRI to CT based 3D treatment planning of brain tumors. Radiother Oncol 25: 121–133

Teresi L, Lufkin R, Wortham D et al. (1987) MR imaging of the intratemporal facial nerve by using surface coils. Am J Roentgenol 148: 589–594

Thornton AF, Ten Haken RK, Gerhardsson A, Correll M (1991) 3 Dimensional motion analysis of an improved head immobilization system for simulation, CT, MRI and PET imaging. Radiother Oncol 20: 224–228

To YC, Lufkin RB, Rand Robinson JDR, Hanafee W (1990) Volume growth rate of acoustic neuromas on MRI post-stereotactic radiosurgery. Comput Med Imag Graphics 14: 53–59

Turkington TG, Jaszczak RJ, Greer KL, Coleman RE, Pelizzari CA (1991) Correlation of SPECT images of a 3D brain phantom using a surface fitting techinque. IEEE Nuclear Science Symposium and Medical Imaging Conference, pp 2154–2157

Turkington TG, Jaszczak RJ, Pelizzari CA, Harris CC, MacFall JR, Hoffman JM (1993) Accuracy of registration of PET, SPECT, and MR images of a brain phantom. J Nucl Med 34: 1587–1594

van Herk M, Meertens H (1988) A matrix ionisation chamber imaging device for on-line patient setup verification during radiotherapy. Radiother Oncol 11: 369–378

Verhey L, Goitein M, McNulty P, Munzenrider JE, Suit H (1982) Precise positioning of patients for radiation therapy. Int J Radiat Oncol Biol Phys 8: 289–294

Vijayakumar S, Chen GTY, Low NN, Myrianthopoulos LC, Chiru P, Rosenberg I (1992) BEV based radiation therapy: description of methods. Radiographics 12(5): 961–968

Waxman AD (1993) Radionuclide imaging in cancer medicine. In: Holland J, Frei III E, Bast RC et al. (eds) Cancer medicine. Lea and Febiger, Philadelphia, pp 483–487

Weinhous MS (1993) Radiation oncology imaging and image processing. In: Purdy JA (ed) Advances in radiation oncology physics. AAPM Monograph 19, New York

Yuh WT, Mayr NA, Atlas SW, et al. (1992) Glioma delineation with contrast enhanced MR: effect of gadolinium dose (abstract) Radiology 185: 297

2 The CT-simulator and the Simulator-CT: Advantages, Disadvantages, and Future Developments

JAMES M. GALVIN

CONTENTS

2.1 Introduction

The idea of using a CT unit in place of a conventional gantry-mounted radiographic/fluoroscopic x-ray simulator is not new. A converted CT unit, called a *CT-simulator*, capable of marking treatment field outlines on a patient, has been available in the United States since 1981 (originally manufactured by Pfizer Medical Corp. and now available through Medical High Technology International, Inc., Clearwater, Fl.). (GALVIN et al. 1982). A short time later similar equipment was introduced in Japan (ENDO et al. 1982). There is currently renewed interest in this technique. Two additional manufacturers in the United States now offer something called a CT-simulator [CT-simulators marketed by Siemens Medical Systems, Inc., Island, N.J. and the unit manufactured by Picker Medical Systems (marketed by Varian, Palo Alto, Calif.)] and two others are showing work-in-progress for their CT-simulator projects (GE Medical Systems, Milwaukee, Wisc. and Philips Medical Systems North America, Shelton, Conn.).

James M. Galvin, PhD, New York University Medical Centre, Department of Radiation Oncology, 566 First Avenue, New York, NY 10016, USA

The marking systems on these devices depend on CT scans of the patient to define the skin surface in three-dimensional (3D) space. This information is used to compute the intersection of projected fields with this surface, and a computer-controlled laser beam points to field reference positions on the skin (ENDO et al. 1982; GOITEIN and MENTO 1982; NISHIDAI et al. 1990; RAGAN et al. 1993a)

A related process, called *virtual simulation* (SHERHOUSE et al. 1990b; SHEROUSE and CHANEY 1991), differs somewhat from the earlier implementation of CT-simulation. Virtual simulation does not use a computer-controlled patient marking system. Instead, sophisticated immobilization equipment is employed so that the patient can be returned to the same position for treatment, and marks placed on the immobilization device are used as reference. The virtual simulator also incorporates digitally reconstructed radiographs (DRRs) (SHEROUSE et al. 1990a; GOITEIN et al. 1983) for field verification. The early implementations of DRRs were slow, and considerably faster techniques have only recently been introduced (CULLIP et al. 1993; GALVIN et al. 1993b).

This report will include a brief discussion of an additional device called the *simulator-CT* (HARRISON and FARMER 1978). This equipment uses the image intensifier output of a conventional simulator to produce a transverse CT image. The advantage of this approach is that all the capabilities of the conventional simulator are maintained.

The following section describes in detail the different features and functions of the conventional simulation. This description is aimed at providing information that can be used to evaluate new simulation approaches by comparison with traditional techniques.

2.2 The Conventional Simulator

The conventional simulator (FARMER et al. 1963) has been used routinely for radiation therapy since the late 1960s. This equipment handles a wide range of tasks important for planning a patient's treatment. Some features of this equipment are standard, while other capabilities are used less frequently. Initially, the simulator was the primary tool for tumor localization. More recently, since the introduction of the CT scanner, the simulator has shared this function with diagnostic modalities capable of producing sagittal, coronal, or cross-sectional views of the anatomy.

Another important use of the conventional simulator is the generation of a diagnostic quality radiograph for comparison to the poorer quality portal image produced with the treatment unit. This "verification" radiograph matches the perspective of the portal image and is helpful for assessing correct positioning of treatment fields. The quality of the portal radiograph obtained with high-energy photons from a treatment unit is not always adequate to assure correct field placement. For this reason side-by-side comparison with a higher contrast conventional simulator radiograph is an important step in evaluating the portal image. This comparison allows the viewer to extract vague structures by seeing where they *should* appear in the portal image. The recent availability of digital portal images introduces the possibility of contrast enhancement, but it does not appear at this time that the processed images are good enough to bypass the comparison with images of normal diagnostic quality (ROSENMAN et al. 1993.)

A radiograph taken with a conventional simulator freezes anatomic motion to one point in time. The image intensifier (II) tube adds the ability to determine the movement of structures and, in some situations, is used to define mobile volumes for treatment planning. For example, movement of the chest wall due to respiration is evident on the TV display of the II tube when tangential breast fields are simulated. It is important to assess this movement to determine that adequate margin is maintained for all parts of the respiratory cycle. The simulator is equipped with a divergent *visible light* field that mimics the projection of the *x-ray* field onto the patient's skin. This light field, together with wall- and ceiling-mounted lasers, is used to mark reference points and field outlines on the patient. When the simulator is fitted with a block support mechanism which matches the target-to-tray distance on the treatment unit, it can be used to determine clearance as the gantry rotates. Additionally, when a blocking tray is attached, field shaping can be checked by obtaining a radiograph with the blocks in place. The idea of *exact simulation* (used on conventional simulators manufactured by

Table 2.1 Functions of a conventional simulator

1. Localize target and critical normal structures
2. Generate verification radiograph
3. Determine mobile volumes
4. Mark field outlines and reference points on patient
5. Demonstrate feasibility of using planned fields
6. Check block position

Varian, Palo Alto, Calif.) modifies the patient support system so that accessories used during treatment can be employed for the simulation procedure. This includes the addition of radio-opaque sections on the simulator table top so that the position of support bars that provide strength for the treatment unit table top appear on radiograph and image intensifier views. Table 2.1 summarizes the functions of the conventional simulator. With these capabilities in mind, the features of the CT-simulator can be compared.

The CT-simulator can handle most but not all of these same tasks. In some cases the CT-simulator is the device of choice; other functions are not easily accomplished with this device. The following sections discuss the use of the CT-simulator for each of the categories listed in Table 2.1

2.3 The CT-simulator

2.3.1 The CT-simulator as a Localization Device

As mentioned above, soon after its introduction, the conventional simulator became an essential device for target volume localization. It is not always possible to see the tumor or, if seen, its true extent on the planar radiograph, but its location can often be inferred by viewing the position of nearby bony landmarks or low-density lung tissue and air cavities. The localization capabilities of the conventional simulator are still important, but the superior ability of the CT unit to define the patient's 3D anatomy and to position the tumor relative to normal structures has made this diagnostic modality a superior tool for target localization. The planar radiograph produced by the conventional simulator superimposes overlying anatomy, while cross-sectional images produced by CT scanning do not have this problem. This argument applies also to sagittal or coronal sections made from the CT dataset, and to transverse, sagittal, or coronal magnetic resonance (MR) images. In general, thin-section images are easier to interpret than standard radiographs.

The CT-simulator potentially replaces the conventional simulator with its standard radiographic output by relying instead on digital information obtained by the CT scan process. The superior localization capabilities of the CT unit for both tumor and normal tissues makes the idea of CT-simulation attractive, but significant questions about the wisdom of this change remain. This manuscript attempts to identify the pros and cons of CT-simulation relative to conventional simulation, and tries to provide information helpful for evaluating the new technology. Current practice combines conventional simulator radiographs with MRI or CT for tumor localization. However, it should be pointed out that a relatively small number of patients are scanned in the treatment position, and integration of the CT information (or MR images) into the planning process still relies heavily on the conventional simulator as a localization device. That is, CT or MRI information obtained in a nontreatment position is transferred to simulator radiographs taken in the treatment position. As access to MRI improves and the number of CT units dedicated for radiation therapy treatment planning increases (due to either special arrangements with diagnostic radiology departments or purchase of the equipment by the radiation therapy department), it is reasonable to question the need for a conventional simulator.

2.3.2 The Verification Image

Since the CT-simulator is built around a CT unit, it can directly produce either transverse images or a *scout* view (also known as a scanogram or pilot view) of the patient. The scout is a digital image that differs from the conventional simulator radiograph in that it is nondivergent in the patient's inferior/superior direction and is considerably more divergent than the standard treatment portal film in the transverse direction (x-ray source to patient center of about 40 cm compared with the standard treatment geometry of about 100 cm to the patient's midline). The scout view is helpful for some aspects of the CT-simulation process, but the different divergence prevents its use for portal film verification.

Using the CT dataset, it is possible to create an image having the same geometric perspective as the treatment unit portal image by summing the voxels along divergent rays extending from the radiation source to the imaging plane (GOITEIN et al. 1983). This reformatting of the CT data produces an image commonly referred to as a digitally reconstructed radiograph (DRR). The DRR algorithms interpolate among surrounding voxels as each ray passes

through the CT dataset. The more sophisticated methods use trilinear interpolation (SHEROUSE et al. 1990a) at each node point through the object. Trilinear interpolation averages the CT values for the eight voxels surrounding each node. If the CT-simulator is to replace the conventional simulator, the DRR must provide the information needed to verify positioning of the treatment portal.

It has been shown that the quality of the DRR is poor compared to a simulator radiograph (GALVIN et al. 1987, 1993b). The difference is due to the decreased spatial resolution of the DRR compared to the excellent resolution of a radiograph taken with an x-ray tube with a small focal spot. The resolution of the DRR is limited by the voxel size of the CT elements from which it is made. Within a transverse section the size of the pixel element is determined by the number and size of the individual detectors and the field of view (FOV) used for scanning. The dimension of a side of the square pixel for a transverse section is typically in the range of 0.5 mm (small FOV used to scan the head) to 1.5 mm (large FOV used when scanning the trunk of the body of larger patients). The scan thickness sets the length of the voxel, and this length determines the amount of anatomy covered when contiguous sections (recommended for radiation therapy planning) are obtained. Typically, 3-mm-thick sections are used for the head and 5-mm-thick slices for the body. Simulator radiographs taken with a small focal spot can resolve about 2.0 line pairs (lp)/mm. The best possible resolution for a DRR will be in the cross-table direction for head scans (transverse section pixel dimension of 0.5 mm). This resolution will correspond to about one pixel per line in a bar pattern test object, or approximately 1.0 lp/mm. The resolution in the direction of couch movement will be even poorer.

Decreasing slice thickness will reduce the voxel size and improve resolution, but making this dimension too small will increase the total number of scans and the time to complete the scan series when the total amount of anatomy covered is kept constant. It is also possible to hold the number of scans constant when the scan thickness is reduced, but this will reduce the amount of anatomy covered. An additional problem occurs in that decreasing the scan thickness reduces the number of photons contributing to the image and increases noise. For scans of the head, a total of 50 contiguous slices 3 mm thick will cover 15 cm of anatomy. For the trunk of the body, the same number of scans 5 mm thick will cover only 25 cm. These total lengths are small compared to the anatomy shown on typical simulator films and some surrounding structures will not be seen on the DRR.

It is not always possible to routinely obtain large numbers of CT sections (say 100 or more slices). This is because tube cooling on most CT units adds about a 10-s delay between scans so that 100 sections obtained with a 2-s scan time will take about 20 min from the beginning to the end of the scan portion of the procedure. This scan time is long when similar lengths of time are used for positioning the patient prior to the start of scanning. In some cases the patient is not able to remain immobile for the required time and movement is evident on the DRR. If the number of scans is increased even further to decrease the voxel length, the tube cooling problem can be more severe. Some CT units are not equipped with a high heat capacity x-ray tube and the 10-s delay time continues to climb beyond about 60 sections. In this case a total of 100 scans can take from 40 to 50 min depending on the exposure technique used. Because of the importance of the resolution of the DRR, high heat capacity x-ray tubes (about 1.5 million heat units) are mandatory for CT scanning for radiation therapy treatment planning. Tubes with this heat capacity will reduce the total scan time, but the increased cost of the CT unit can be significant.

The 3- and 5-mm slice thicknesses given above are the recommended values for head and body scans, respectively, when the CT procedure is performed for radiation therapy planning. Scans of the head and neck extending down to the supraclavicular region are usually obained with 5-mm-thick sections. Although varying the scan thickness for different regions in the same study is common for diagnostic studies, most reformatting algorithms currently available for radiation therapy planning cannot handle this situation. One exception is the DRR available on the Picker ACQSIM unit. An additional problem exists in that none of the DRR reformatting algorithms allow a change in the FOV as scanning moves from the neck to the shoulders. Restricting studies to 5-mm scan thickness and a large FOV in this situation further reduces the quality of the reformatted images relative to scans of the head/neck region not including the shoulders.

The limited resolution of the DRR can make some important tasks difficult. For example, if the CT slice thickness for the dataset used as input for the DRR is greater than 5 mm, identifying the vertebral body anatomy is not always possible. However, other advantages of the DRR relative to simulator radiographs tend to compensate for this poorer resolution. Since the DRR is a digital image, it can be manipulated in various ways. The viewer's ability to discern different structures within the vertebrae (e.g., pedicles, transverse processes, rib facets, spinous processes) can be enhanced by using only those voxels corresponding to bone in the summation to obtain the DRR. Viewing the DRR with a number of different windows is helpful, but an extremely fast reformatting algorithm is needed if sequential images are compared. This is a problem because the generation of DRRs can take on the order of minutes to produce each image (SHEROUSE et al. 1990a). Using a three-field plan to treat an esophageal lesion as an example, it is not unusual to use nine different DRRs for comparison: three soft tissue images, three bone images, and three air window images for viewing the bronchus. Reformatting CT data in 5–10 s is required if this comparison is to be accomplished in a reasonable amount of time (GALVIN et al. 1993b). Additionally, the highest possible resolution should be obtainable within this time range. This means that a trilinear interpolation should be used and the DRR should be presented as at least a 512 × 512 image. The 5- to 10-s reformatting time should apply to input datasets of 50 512 × 512 CT images.

The question of breath control is also important. Scanning patients during quiet respiration has the advantage of being easier for seriously medically compromised patients. The large number of scans required for the DRR makes suspension of breathing for each scan difficult. The disadvantage is that the patient's anatomy moves during the time needed to accumulate data for a single scan so that image quality is degraded relative to scanning with breath control. Additionally, for sequential scans, moving structures are captured in different phases of movement so that scalloping occurs at the edge of some organs and structures as seen on the DRR. Significant scalloping is evident at the diaphragm when a patient breathes during scanning. Also, the superior pole of each kidney exhibits significant movement with breathing. This same movement occurs during treatment, but the problem is that the scalloping tends to degrade the DRR so that structures such as the vertebral bodies that do not move are more difficult to discern. One method of avoiding this problem is to view the DRR with a bone window so that overlying anatomy is not seen. Scalloping at the heart/lung boundary is evident with or without breath control. Degradation of the quality of the DRR relative to the conventional simulator radiograph is particularly important when non-traditional fields such as beams with center axis out of the transverse plane are used. Portal images of these fields are hard to interpret and require the highest quality image for comparison. However, the DRR has an advantage over the conventional simulator radiograph in that some views can be constructed that cannot be filmed because of collision of the image intensifier with the base of the patient support couch. Obviously, this is only valuable for those situations where a less bulky cassette holder on the treatment unit allows a port film to be taken.

Extremely fast CT units are now available. These devices are referred to as helical, spiral, or continuous rotation units. The 1.0-s scan times currently achievable can eliminate many of the problems discussed above. This dramatic increase in speed allows for larger numbers of scans to improve resolution of the DRR while also reducing the time the patient must remain immobile on the CT table. A second advantage is that 15 or so cross-sections can be obtained with a single breath hold so that breathing motion is reduced for the DRR. The price of these units remains high and, at least for the near future,

it is not likely that they will be purchased as dedicated equipment for radiation therapy treatment planning.

The DRR must be presented with the treatment field superimposed. The image should include the rectangular field outline and the irregular shape of the blocked opening. A center cross together with tick marks for determining distance should be shown on the DRR. Outlines of critical structures and target volumes (a schematic beam's eye view) should be included in the presentation. Switching these outlines on and off is extremely helpful. The outlines are obtained by contouring structures on the individual CT slices and are placed with the correct geometric perspective on the DRR. They can be used when a structure obvious on the port film is not easily seen on the DRR. An example is the air shadow of the trachea. This structure is not always seen on the DRR, but outlines can be superimposed on this image to show its position. Figure 2.1 shows an example of a DRR. The figure includes the field outline and *cross hairs* through the field center.

2.3.3 Field Outlines and Field Reference Marks

An important step in the CT-simulation process is the placing of marks on the patient to facilitate transfer to subsequent procedures such as the initial setup on the treatment unit. Various techniques have been devised for using modified CT hardware to mark field outlines and reference marks on a patient's skin surface and, as pointed out above, one piece of commercial equipment for this purpose has been sold in the United States for more than 10 years. Two additional units, also designed specifically for radiation therapy use, have been introduced in the past few years. The different marking techniques employed by the various manufacturers are presented below and discussed in the following sections. Since most institutions do not employ special CT hardware for incorporating CT scanning into the treatment planning process, a description of the standard marking technique is included for comparison to the new approaches.

Method 1. General reference marks (Fig 2.2). (a) Use "scout" or "pilot" views on the CT unit to place radio-opaque markers on patient's skin surface prior to cross-sectional CT scanning. The scout views are used to find the cross-section at the *approximate* target volume center in cephalad/caudad direction. The markers are positioned (usually one on the skin away from the table top and one on each lateral surface) to identify the center of the body for this cross-section. (b) Remove

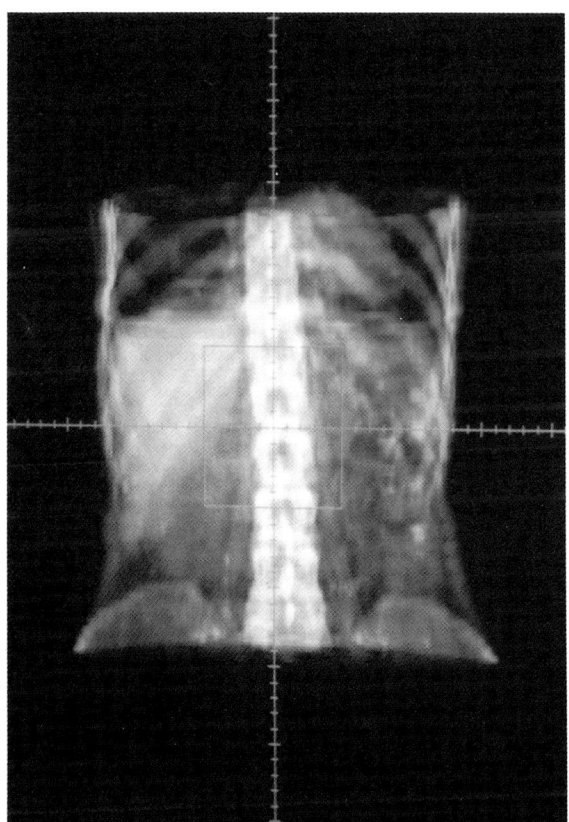

Fig. 2.1. DRR for the abdomen. Image was constructed from 61 contiguous CT slices of 5 mm thickness and trilinear interpolation was used. The Picker AcQSim System was used to obtain this DRR

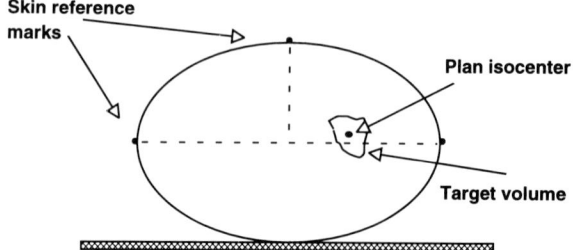

Fig. 2.2. Method 1. Schematic representation of patient showing radio-opaque markers centered on patient prior to CT scanning. Note that the plan isocenter, as determined by the treatment planning process, is at a different location

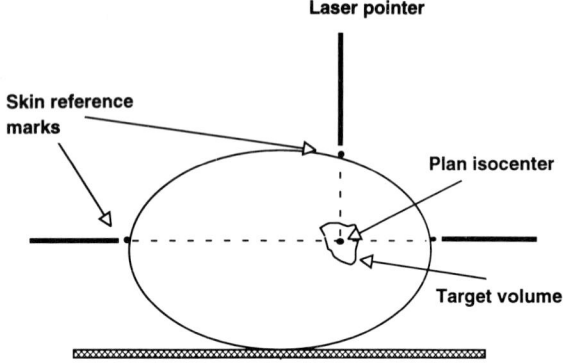

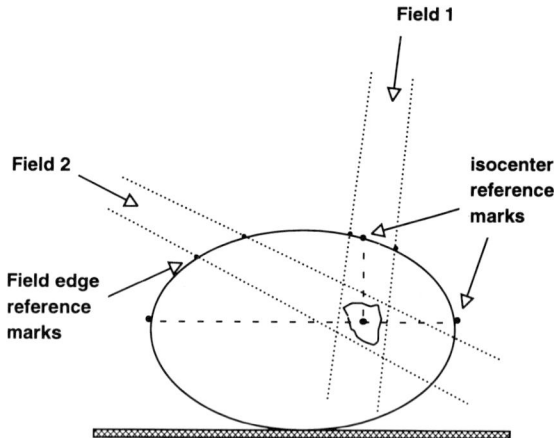

Fig. 2.4. Method 3 and 4. Reference marks for both the plan isocenter and field outline. A completed treatment plan is needed to determine the location of each mark

Fig. 2.3. Method 2. Skin reference marks indicating the *approximate* plan isocenter. These marks are determined by quickly contouring the external skin surface and the target volume

patient from CT unit. (c) Complete treatment planning. (d) Using conventional simulator and initial reference marks, place final isocenter reference marks and field marks on patient. Obtain verification radiographs.

Advantages. Does not require a dedicated or modified CT unit. Treatment planning carried out without pressure of patient waiting on CT table.

Disadvantages. Shift from general reference marks to final reference marks can introduce inaccuracies. Errors can result from slight differences in patient position for CT scanning compared to simulation session used for verification filming and skin marking.

Example. Technique used by most institutions doing 3D treatment planning. Similar to virtual simulation technique.

Method 2. Place general reference marks on patient skin using special marking system on CT unit (Fig 2.3). (a) CT scan patient and *quickly* contour target volume and external skin surface. (b) Use special laser system supplied with CT scanner to locate target volume center as it projects to skin surface. (c) Remove patient from CT unit. (d) Use CT information for *detailed* contouring of target volume and all critical normal structures. Complete treatment planning to obtain final isocenter position. (e) Use coordinate shifts from approximate isocenter position to final isocenter position for daily treatment.

Advantages. General reference marks should be very close to final reference marks and inaccuracies are reduced.

Disadvantages. Special hardware must be added to CT scanner. There are some situations, such as when the patient's skin is loose and easily shifts position, where marks near the center of the patient's body may prove to be more helpful than reference marks identifying the target center. DRR may not be adequate for verification in some cases.

Example The AcuSim system produced by Picker International. Inc. and marketed by Varian Associates.

Method 3. Place final reference marks and field outlines on patient skin using special marking system on CT unit and single session (Fig 2.4). (a) CT scan patient and *carefully* contour target volume, external skin surface, and all critical normal tissues. (b) Without moving patient, complete treatment planning using CT information. (c) Using special CT hardware, place final isocenter reference marks and field marks on patient.

Advantages. Total consistency of data provided patient does not move during procedure.

Disadvantages. Patient must lie still for long period. Treatment planning, including detailed contouring of structures, must be completed in a relatively short period.

Example. The CT-simulator manufactured by Medical High Technology or the unit produced by Siemens.

Method 4. Two-session marking of reference marks and field outlines with special CT marking system (Fig 2.4). (a) CT scan patient with simple repositioning marks. (b) Remove patient from CT unit. (c) Complete treatment planning using CT information. (d) Return patient to CT unit and reestablish treatment position. (e) Place final isocenter reference marks and field marks on patient using hardware on the CT unit.

Advantages. Relative to single-session approach above, patient does not have to remain immobile for contouring and treatment planning.

Disadvantages. Possible inconsistency of data when patient position for scanning is slightly different than position for marking. Particularly confusing when verification images (DRRs obtained from first session) do not correspond exactly with skin marks (placed during second session).

Example. The CT-simulator manufactured by Medical High Technology or unit produced by Siemens.

2.3.3.1 The Use of Fiducials as a Reference for CT-scanning (Method 1 and Fig. 2.2)

The first technique described above is the one typically used for 3D treatment planning. Since modification of the basic CT unit hardware is not necessary, this method is conveniently adapted for either a dedicated CT unit placed within the radiation oncology department or equipment located in a department of diagnostic radiology. With some slight modifications, this method describes the process of virtual simulation.

The exact position of the isocenter for the treatment fields is not known prior to completion of both CT scanning and treament planning, but an *approximate* location can be determined from anterior and lateral CT scout views together with any previously obtained diagnostic information. Although the scout images do not match the divergence of the treatment unit portal radiograph, they can be used to identify a CT section approximately centered on the target volume in the cephalad/caudad direction. For the transverse section near the target center, skin reference markers (radio-opaque catheters or lead bb's) are placed to identify the patient's center anterior-to-posterior and left-to-right. The advantage of placing the anterior reference point (posterior when the patient is prone) on a midsagittal plane is that the vertebral bodies are often a good anatomic reference for both cephalad/caudad and left/right positioning. In those cases where the vertebral bodies can be visualized on subsequent imaging studies (e.g., during conventional simulation) a double-check on the positioning of the reference point is possible. The standard laser lights available on most CT units can be used to adjust the lateral marks so that they define a true horizontal line, or a ruler can be used to adjust the position of the marks to equal height from the table top. An *initial* internal point is defined by these marks as the intersect of the line joining the lateral points and a vertical line through the anterior point. A *final* position (the isocenter for the selected fields) is determined by the treatment planning process and is located as a series of moves from the initial internal point. The shift from the initial to the final position can involve a combination of changes in depth, inferior/superior position, and lateral position.

2.3.3.2 The CT-Simulator as a Devise for Placing Skin Reference Marks (Method 2 and Fig. 2.3)

The procedure described above can be used without placing radio-opaque markers on the skin if a special laser system capable of pointing to positions on the patient's surface is available as an attachment on the CT unit. This is the approach used for one of the commercially available CT-simulator units (Picker/ Varian AcQSim, Highland Heights, Ohio). In this case the process described in the previous section is reversed in that an internal reference point is found first and external reference points on the patient's skin identified second. An approximate target volume center is identified using this procedure. This center can be identified in a relatively short time by *quickly* contouring the target without reference to nearby critical structures. This center is projected to the patient's surface (anterior point for supine positioning and posterior for prone, and two lateral points) and marked using the laser system. The patient is removed from the CT table and the CT scans are used to produce more *detailed* contours. This contouring process includes careful outlining of the target and all other relevant structures. Field directions and shapes are selected and dose distributions generated. This procedure establishes the final isocenter for the treatment and results in a set of shifts from the skin marks to the desired point within the patient. These shifts should be small and can be made each day the patient returns for treatment. This method of placing reference marks can sometimes cause inaccuracies. This is the case when overlying skin is not fixed in position and can easily be moved. When this happens it is preferable to locate skin marks away from the region of the isocenter for the treatment fields. This placement of skin marks is popular for treating the breast with tangential fields and for older patients with loose skin. In these cases midline marks are typically used.

Method 2 depends on the quality of the DRR for verification of the placement of the treatment fields. The usefulness of the DRR image has been discussed in a previous section. The DRR available on the ACQSIM is extremely fast and allows for rapid changes of the display window so that the field position relative to bone or air shadows can easily be compared to a soft tissue display. However, there are some cases where the DRR will not provide acceptable information for field verification. This is the case when the patient moves or fidgets during scanning or when a smaller number of cross-sections are

obtained in order to speed the procedure for uncooperative patients. In these situations the patient
must be placed on a conventional simulator to obtain
verification radiographs. This step can be accomplished rather quickly when the skin reference marks
are available for patient positioning.

2.3.3.3 Marking Field Outlines
with the CT-simulator
(Methods 3 and 4 and Fig. 2.4)

Methods 3 and 4 are extensions of the idea of marking simple skin reference points to locate an isocenter
for selected treatment fields (method 2). These methods use the exact target volume shape to draw complete field outlines on the patient (ENDO et al. 1982;
GOITEIN and MENTO 1982; NISHIDAI et al. 1990;
RAGAN et al. 1993a). Two commercial companies
now offer equipment with this capability (Medical
High Technology International, Inc., Clearwater,
Fl. and Siemens Medical Systems, Inc., Island, N.J.).
This approach requires a final treatment plan prior
to field marking. Careful contouring of the target
volume and all relevant critical normal structures
must be completed, and treatment fields must be
selected prior to marking. Methods 3 and 4 are similar to method 2 in that they do not require the use of
a conventional simulator.

Method 3 completes the scanning, contouring,
and field selection in a single session while the patient
is immobile on the CT table. Method 4 allows the
patient to move after scanning. There are two possible implementations of method 4. The two
approaches differ in the way the patient's position
is reestablished for the second session on the CT
table. One approach (method 4a) rescans the patient
to determine positioning. The second approach
(method 4b) uses reference marks to bring the patient
back to the same position used for the initial scan session. The projection of the standard laser alignment
lights on the CT unit are marked on the patient during the first session and used for repositioning at the
time of the second session. The patient is moved on
the CT table to match these marks prior to the start
of the second session.

The two commercial companies with a laser
marking technique to define the entire field outline
use different hardware. One (Medical High Technology, Inc.) uses the laser to point to a series of discrete skin positions to describe the field and the other
(Siemens uses a fast scanning laser to show a continuous trace of the outline on the patient.

2.3.3.4 Comparison of Different Marking Methods

Method 1 does not require special CT hardware and
can be adapted for use with any available CT unit.
Method 3 has the advantage of completing scanning,
contouring, field selection, and marking in a single
session so that inaccuracies due to moving from one
procedure to the next are reduced. However, it is
clear that 3D treatment planning takes a considerable amount of time and the patient must remain
immobile for this entire period when method 3 is
used. It has been shown that the contouring portion
of the procedure alone requires nearly 1 h (DOWSETT
et al. 1992) and the total time for generating complex
plans can be many hours. For this reason method 4 is
more useful than method 3 for many typical treatment situations.

Dividing the procedure into two sessions when
method 4 is used is similar to the technique of obtaining orthogonal films and contours in one session on
a conventional simulator, and having a second session for filming fields obtained through the planning
process. However, major differences exist and must
be discussed. When a patient is rescanned at the start
of the second session to reestablish positioning,
increased time is an issue. A full set of CT scans can
add 10–20 min depending on the scan technique and
the CT equipment available. This increased time
makes it hard to justify placing the patient on the CT
unit for marking instead of using a conventional simulator. The possibility of gathering fewer scans for
the second session has not been investigated. This
would reduce the time for scanning, but could also
reduce accuracy of the definition of the skin surface.

Method 4b avoids the problem of increasing the
time on the CT table for the second session by eliminating the additional scanning and using reference
marks to reposition the patient. In this case a problem can occur when the final marks on the patient's
skin do not correspond exactly with the verification
DRRs used to check positioning of treatment unit
portal images. As pointed out above, two sessions
are often used for conventional simulation, but there
is a major difference in that it is common to obtain a
full set of radiographs at the time final reference
marks are placed on the patient's skin. Separating
the processes of generating verification images and
marking can introduce an additional inaccuracy.
This is because the patient's position may not be
exactly the same at the two different points in time.
Difficulties in explaining differences between
verification images and portal images are then
magnified. Many studies are now available [see, for

example, ROSENTHAL et al. (1992)] to show that repositioning errors are significant in the best of circumstances. The narrow couch and restricted work area within the CT aperture can only serve to make this problem more severe.

If a significant number of cases must be handled with two CT sessions, it is hard to argue that a modified CT system that places *final* reference and field marks on the patient is useful. The justification for this type of system depends on a complete replacement of the conventional simulator with a CT- based system. Without this change, there is little reason to use the CT-simulator instead of the conventional simulator for the second session. Additionally, the conventional simulator produces a verification radiograph of superior quality to a DRR made from the CT data. With the ability to combine the CT data in the form of a schematic BEV with these radiographs either electronically or using transparent overlays (GALVIN et al. 1986), the best of both worlds is obtained. The argument against replacing the conventional simulator with a CT-simulator is that this new technology is not convenient for simple field. Simple parallel opposed fields can be established in a short time using a conventional simulator, but a full set of cross-sectional scans is needed to define the patient in 3D space for the CT-based approach. For more complex cases, many years of experience have shown that the marking process with the conventional simulator can be performed very quickly once the isocenter position and field directions are known. The inaccuracies that result whenever serial procedures are performed will apply for any combination of CT and conventional simulation, but it is hard to match the speed and efficiency of conventional simulation for the second session.

The question of which marks should be placed on the patient's skin has not been clearly addressed in the literature. In fact, some investigators have emphasized the importance of marking the immobilization device instead of the patient (SHEROUSE et al. 1990b). Using tangential field treatment of the breast or chest wall as one example, it is common to use a reference mark that is positioned at the patient's center on the anterior skin surface (GALVIN et al. 1993) This point is usually 10 or more cm from the isocenter for the tangential fields. In this case none of the techniques (methods 2, 3, or 4) for marking the target volume center is particularly helpful. A simple radio-opaque marker at the patient's midline (easily determined by viewing the patient's skin surface) and centered in the cephalad/caudad direction (determined using scout views) is sufficient for scanning with a

standard CT unit using method 1. Also, a standard method for representing treatment fields does not exist. Some institutions mark the center and all four corners of the open rectangular field. Another approach abbreviates this potentially large number of points. In some cases a setup field not used for treatment is marked. With the widespread availability of rigidly mounted custom fabricated shielding blocks, it is not obvious that it is necessary to paint the irregularly shaped field on the patient. Additional marking might include reference points used to reproduce the patient's position on the treatment table. For example, lateral skin marks might be placed when treatment fields from this direction are not part of the patient's treatment plan.

The appeal of method 2 is the relatively short time the patient must remain motionless on the CT couch. Initial contouring of the target volume is aimed at locating an internal reference point that is within a few centimeters of the final plan isocenter. Following the description of method 2 given above, the final isocenter is localized and marked on the patient's skin. It is also possible to use the approximate isocenter for patient's daily setup each day. In this case the established shifts from the approximate isocenter are made each day. This technique avoids the second simulation session entirely, but the individual fields would not be marked on the patient as part of a simulation process. They could be marked using the treatment machine at the time of the first treatment. Obviously, any of the CT-simulators can be used in this way. That is, the step of marking field outlines on the patient using special hardware on the CT scanner in not used.

The additional advantage of method 2 is the simplicity of the laser marking system. Standard lasers similar to those used for a conventional simulator are mounted on the two side walls. The CT table is moved up/down and left/right under computer control to bring these lasers into alignment (see Fig. 2.3). An additional laser is mounted on the wall at the foot of the CT couch. This laser is mounted on a worm drive and is also moved by computer to achieve alignment.

2.3.4 Definition of Mobile Volumes

The CT-simulator does not give clear information on the movement of structures due to respiration or cardiac motion. This represents an important limitation relative to the conventional simulator with an image intensifier attachment. However, it may be possible to extract limited information from the initial CT dataset reformatted as sagittal or coronal slices. When transverse sections are obtained *without* breath control, structures near the heart and lungs are stopped in different positions for each slice. Reformatting these images results in scalloping at the edge of moving structures, and this scalloping can be used to estimate movement during treatment. This raises the question of the best technique for CT scanning patients for radiation therapy treatment planning. Breath holding for each scan when more than 40 cross-sections are obtained is difficult for some patients. However, diagnostic scanning is seldom performed without suspending respiration. This is because the quality of the cross-sectional images is significantly improved when breathing is stopped during scanning. As mentioned above, the availability of fast helical scanners will make it possible to obtain a large number of scans in a single breath hold. It will be interesting to see the role this new technology will play in radiation therapy treatment planning.

2.3.5 Other Issues

Additional limitations of the CT-simulation process may be solved in future implementations. For example, CT scanning defines the patient in 3D space and collision avoidance using computer simulation of treatment should be possible. This same technique can be used to determine when bars in the patient support couch will intercept the beam and perturb the dose distribution. It is also possible to use computer simulation to check different positions on the treatment couch to predict the best portion of the table (center spin or side supports) to use for treatment. These potential positives must be balanced against the negative of a smaller opening for the CT unit compared to the circle swept out by the conventional simulator. This difference means that some patient positions cannot be duplicated. For example, treatments with one or both arms up are problematic for CT scanning.

The design of the CT table does not lend itself to direct attachment of immobilization equipment. It is possible to use support and immobilization devices with skid-proof undersurfaces placed directly on the CT couch. This equipment must be constructed without the use of metal parts so that artifacts are not produced during scanning.

The use of the conventional simulator to verify block position is not particularly popular. This is because final approval of the block position must be made on the treatment unit and filming blocks on the simulator represents an extra step. It is possible to use CT information to produce templates for block fabrication and these templates can be generated with a reduced magnification on clear sheets for mounting blocks on the support trays. The use of templates for block mounting reduces block placement error and further strengthens the argument that it is not necessary to film blocks as part of the simulation process.

2.4 The Simulator-CT

2.4.1 Description

Simulator units modified to produce cross-sectional images have been available in the united States for more than 10 years. An early unit (KIJEWSKI et al. 1984) manufactured by Oldelft (Fairfax, Va.) used an optical filtering technique and back-projection to produce transverse images from the output of the imaging tube (a modified version of a standard image intensifier). This device was not stable and it was difficult to obtain images of acceptable quality on a routine basis. Digital systems have been introduced more recently by Oldelft and Varian. The Varian device is described as an example of this type of equipment. The x-ray beam for a standard simulator is collimated to a narrow slit extending across the patient. The beam is collimated to one side and the image intensifier (II) is offset to include the rotational isocenter and one side of the patient. The output of the II is viewed with an array of 500 light-sensitive diodes. An additional diode array is mounted on the face of the II tube to extend the length of the field and the amount of anatomy included in the image. Since data from only half the object are gathered for any gantry position, a full 360° rotation is needed to fully sample the patient. The gantry rotates at 1.0 rpm as the transmission data are gathered.

The rotation speed is limited by government regulations for an open gantry system. The time to accumulate the transmission data for this equipment is much longer than the 2–3 s used for standard CT units, and image quality is compromised in the thorax and abdomen, where respiration degrades the image.

2.4.2 Effective Aperture, Heat Loading and Number of Scans

A major advantage of this approach is that the *effective* aperture for the simulator-CT unit is larger than a standard CT unit. However, this is also a disadvantage in that the x-ray tube is further from the detector system and heat loading is a problem. Since the simulator is an open system it is hard to compare the effective aperture for the simulator-CT with the opening for a standard CT unit. This is because the patient can be safely positioned to pass within a few centimeters of the side of the doughnut for a CT unit while a safety margin must be used for the rapidly moving simulator gantry and II tube. The II tube is bulkier than the x-ray head and limits the safe zone when the simulator-CT is used. The II tube is positioned 40 cm below the mechanical isocenter when the collision guard is included in the measurement. This gives an effective aperture of 80 cm or about 10 cm greater than a conventional CT unit. It is possible to position the II tube about 5 cm nearer the floor without collision. It is anticipated that this modification will be made by the manufacturer in the future so that the effective aperture is increased to 90 cm.

The x-ray target for the Varian unit in its present configuration is on the order of 150 cm from the detector system. A conventional CT unit has the detector system positioned at about 90–100 cm from the x-ray source so that many more photons are available to form the image. This means that less heat is produced in the target of the conventional CT for the same slice thickness and noise level. The final result is that the number of scans obtainable with the simulator-CT is limited by heat production, and slice thicknesses of less than 0.8 cm for the body and 0.5 cm for the head are not availabe. Heating of the x-ray tube increases the time from the start of one scan to the beginning of the next to somewhat more than the 1.0 rpm speed for the gantry. The first six scans can be obtained in about 8 min. Thereafter, each scan takes 2.25 min. These times limit the total number of scans to about 15. This number falls short of what is needed for true 3D treatment planning. It should be pointed out that the number of scans needed for the simulator-CT device is less than the number needed for the CT-simulator. The number needed when the CT-simulator is used is determined by the requirements for the production of the DRR. When the simulator-CT is used it is not necessary to construct a DRR because a gray-scale beam's eye view is available as a digitized version of the image intensifier. Thus, the number of cross-sections is determined by the number needed to accurately define critical structures for dose-volume histogram analysis and not the number needed to produce a DRR. This reduces the requirement to about 20 scans. New x-ray tube technology may allow for an increase in the number of

scans, but the 1.0 rpm limit forms a second barrier that is not easily overcome. After a lengthy simulation session, it is not reasonable to have the patient lie on the table for more than about an additional 20 min for CT scanning.

2.4.3 Slice Thickness and Image Noise

A related issue is the slice thickness. As the number of scans goes up, slice thickness can be decreased to improve the resolution of the DRR. Since the simulator-CT does not require the production of DRRs, the 0.8 and 0.5 cm limits in slice thickness for the body and head, respectively, may not prove to be a significant problem. This is certainly true for body scans, where a slice thickness of 1.0 cm is typically used for body scanning for diagnostic CT. However, sections of 0.3 cm are usually employed for scanning the head and using 0.5 cm will result in volume averaging and a degradation of image quality. Thus, the ability to obtain thinner sections when using the simulator-CT device is desirable. This reduction in slice thickness should be accomplished without compromising image quality due to increased noise. That is, the number of photons reaching the detector system should be increased proportionately.

2.5 Discussion

The above sections discuss the various capabilities of the CT-simulator. In each section a comparison with the conventional simulator is made. Some of the tasks previously performed with the conventional simulator are handled better by the CT-simulator. For example, tumor localization is usually more accurate when CT scans are available. This is well known and few patients are treated without the aid of CT scans (or MR images) for planning. Other tasks are not as easily performed using the CT-simulator. The best example is the determination of mobile volumes. When a structure or volume is visible on the image intensifier display, the conventional simulator is ideally suited for showing the movement of structures during treatment. Although methods described in this manuscript for using the CT scanner to monitor movement with respiration and heartbeat may gain popularity in the future, they are not now routinely employed.

At present the quality of the verification image produced by reformatting the CT data is not always acceptable as a replacement for the conventional simulator radiograph. The quality of these images varies with the part of the body viewed, the scan thickness, and the FOV. In general, DRRs for the head and neck are superior to images for the thorax and abdomen. It is not clear whether future developments will change this situation. As pointed out above, faster continuous rotation helical scanners are now available. These CT devices can increase the number of CT sections so that thinner slices are available to improve the quality of the reformatted image. However, this technology is extremely expensive at this time and it is not reasonable to consider having this equipment dedicated for radiation therapy treatment planning. Additional tricks can be used to make the reformatted images more useful. Some systems can produce DRRs in a short enough time (about 10 s) to allow the comparison of different types of presentations (an air window, bone window, or soft tissue window, etc.) as a technique of improving the usefulness of these images (GALVIN et al. 1993b). Progress is also being made in the area of improving the quality of the portal image. Digitization of portal films allows the possibility of contrast enhancement to improve image quality (ROSENMAN et al. 1993). Developments in the area of portal imaging could decrease reliance on the verification image.

The above discussion does not address an important question raised when the idea of CT-simulation was first introduced. Is it acceptable to have a CT-simulator as the only mode of simulation in a radiation oncology department? After using one of the early CT-simulator devices on more than 750 patients between January 1982 and the end of 1986, this investigator believes that a CT-simulator is not appropriate technology for many patients treated in an active radiation oncology deparment. Using a CT scanner to plan a simple AP/PA field arrangement is too time consuming, and the CT-simulator is not suited for seriously medically compromised or uncooperative patients. Filming fields with a conventional simulator can be accomplished extremely rapidly, and the patient's position can be reestablished at various times as the procedure progresses. This is not the case for the CT-simulation process because a new CT dataset must be obtained whenever the patient moves. It is hard to define the patient population that can best benefit from this new technology. Certainly simple palliative fields are not suited for CT-simulation. Taken as a percentage of simulations performed in a typical radiation therapy department, between 15% and 30% of simulations could be handled with a CT-simulator. This number is arrived at

by eliminating not only the palliative group, but also one simulation for all patients receiving cone-down fields. These patients would have CT scans either at the beginning of treatment or at the time the fields are reduced in size. This approach to CT scanning limits the market for the CT-simulator to institutions with patient loads large enough to require more than one simulation device.

It is not clear that departments of average size or even major institutions can fully utilize a dedicated CT unit. It is interesting to relate the experience using a dedicated CT unit at one major radiation therapy department. A state-of-the-art GE 9800 CT unit was installed in this department for beam's eye view treatment planning at the beginning of 1987. Two planning systems [an in-house software and hardware package, and the G.E. Target system (GE Medical Systems, Milwaukee, Wisc.)] were available for CT-based planning, and there was a general interest in 3D treatment planning among the physicians in the department as a result of participation in the NCI-funded Photon Contract (GOITEIN et al. 1991). The average treatment load was on the order of 90 patients per day. This CT scanner was not equipped with a special laser marking system and the technique used for marking closely followed method 1, described above. For a period of 4 years extending to the end of 1990, an average of 275 patients were CT scanned in the treatment position each year. This is just over 1.0 patients scanned each working day. Since eight patients on the average were simulated at this institution each day, 12.5% of simulated patients were scheduled for CT scanning. This number needs to be viewed with some caution. This is because different departments can have widely varying philosophies about the simulation process. For example, one department might send all patients back to simulation for each field cone down, while other institutions use the same isocenter and simply add blocking and close collimators to reduce the field size without resimulation. The general trend in this department was to resimulate patients at the time of cone down. Thus, in general, the simulation load at this institution was high. This helps explain the 1 out of 8 ratio for patients receiving 3D treatment planning. The 15%–30% number used above is more reasonable for an institution that does not simulate all cone-down fields, and adjusts for the increased interest in 3D treatment planning over the past few years. It should also be added that many of the institutions with dedicated CT scanners for radiation therapy treatment planning have unusual referral patterns. A good example is the institutions cooperating in the

NCI study of prostate treatment using conformal 3D treatment techniques. These institutions can have a much higher percentage of patients being CT scanned specifically for planning radiation treatments. Statistics for these departments should not be used to project the usage of CT equipment in smaller institutions. It is tempting to try to increase the number of patients receiving CT scans for radiation therapy treatment planning, but the labor-intensive nature of the 3D planning process makes this impractical. Due to the problems of staffing a CT unit for only one to two procedures per day, scanning at the institution described was carried out 2 days each week and the CT unit was not used for the other 3 days.

As an additional justification, it may be possible to perform other procedures on the CT unit that are not strictly part of the 3D planning process. For example, when clips are inserted at the time of lumpectomy for breast cancer, the CT unit can be used to determine depth for local boost with electrons. This procedure alone, however, will not greatly increase the utilization of the CT unit. It is also tempting to use a scanner for the routine diagnostic CT needs of the radiation oncology department, but this is not without problems. In addition to the obvious political considerations, there is the problem that diagnostic CT scans are performed differently than scans acquired for 3D treatment planning. The use of IV contrast agents is more common when information is gathered for diagnosis, and tilting the gantry on the CT unit is standard practice for head studies. Most diagnostic CT scans in the abdomen and thorax are obtained with suspended respiration, while treatment planning CT scanning usually is not performed using breath control. These details must be worked out with the diagnostic radiologist reading the CT images, and the radiation oncology department must employ individuals capable of obtaining *diagnostic* quality CT scans when the unit is to be used for this purpose.

An additional consideration when purchasing a CT unit dedicated for radiation therapy applications is the rapid evolution of this technology. There is the possibility that a dedicated CT unit will become obsolete and underutilization will prevent a timely upgrading to more modern equipment. The example of this is the current interest in helical scanners mentioned previously. The speed of these units may be advantageous for some radiation therapy applications, but the cost of this technology is high and it is not likely that dedicated scanners of this type will be used for radation therapy planning in the near future. The obvious solution to many of the problems identified here is to use the equipment available in the institution's diagnostic radiology department. This means scheduling blocks of time long enough to obtain the special information needed for radiation therapy treatment planning, and it is essential that radiation therapy personnel participate in the CT scanning process to guarantee that the patient's position is correct.

2.6 Conclusions

The conventional simulator has been used for radiation therapy treatment planning for about 25 years. In recent years the CT scanner has become an important tool for treatment planning, and it is not surprising that attempts have been made to integrate this technology more fully into the simulation process. However, at least for CT-simulation equipment that attempts to replace the conventional simulator, it is not clear at this time that the advantages of this new approach to simulation outweigh disadvantages. Equipping radiation therapy departments with CT scanners instead of conventional simulators in not advised at this time because the CT-simulation approach is too labor intensive for simple treatments. Additional research is needed to determine whether having a CT-simulator as a second simulation device is cost-effective. The few clinical studies carried out thus far (NAGATA et al. 1990; RAGAN et al. 1993b) have not been designed to demonstrate the advantage of using a CT-simulator. Research aimed at answering this important question is urgently needed. The simulator-CT could represent a useful modification of the conventional simulator, but additional research is also needed to demonstrate the efficacy of this device. For the immediate future, until certain modifications are incorporated in the design of the simulator-CT, the usefulness of this equipment is limited by image quality and the number of cross-sections conveniently obtained.

References

Cullip TJ, Symons JR, Rosenman JG, Chaney EL (1993) Digitally reconstructed fluoroscopy and other interactive volume visualizations in 3D treatment planning. Int J Radiat Oncol Biol Phys 27: 145–151

Dowsett R, Galvin JM, Cheng E, et al. (1992) Contouring structures for 3-dimensional treatment planning. Int J Radiat Oncol Biol Phys 22: 1083–1088

Endo M, Kutsutani-Nakamura Y (1982) Patient beam positioning system using CT Images. Phys Med Biol 27: 301–305

Farmer FT, Fowler JF, Haggith JW (1963) Megavoltage treatment planning and the use of xeroradiography Br J Radiol 36: 426–435

Galvin J, Heidtman B, Cheng E, Bloch P, Goodman R (1982) The use of a CT scanner specially designed to perform the functions of a radiation therapy treatment unit simulator. Med Phys 9: 615

Galvin JM, Turrisi AT, Cheng E (1986) Treatment simulation using a CT unit. In: Kereiakes JG, Elson HR, Born CG (eds) Radiation oncology physics. American Association of Physicists in Medicine Monograph Series. American Institute of Physics, New York, p 462

Galvin JM, Waxler G, Sontag M., Smith R (1987) Use of reformatted CT images for radiation therapy treatment planning. Med Phys 14: 459

Galvin JM, Powlis W, Fowble B, Goodman RL (1993a) A new technique for positioning tangential fields. Int J Radiat Oncol Biol Phys 26: 877–881

Galvin JM, Sims C, Dominiak, GS, Cooper JS (1993b) The use of digitally reconstructed radiographs for 3D treatment planning and CT-simulation. Int J Radiat Oncol Biol Phys 27 (Suppl 1): 141

Goitein M, Mento D (1982) An optical scanner as an aid in simulating treatment with CT data. J Comp Tomogr 6:1201–1204

Goitein M, Abrams M, Rowell D, Pollari H, Wiles J (1983) Multidimensional treatment planning. II. Beam's eye-view back projection and projection through CT sections. Int J Radiation Oncol Biol Phys 9: 789–797

Goitein M, Laughlin J, Purdy J et al. (1991) Evaluation of high energy photon external beam treatment planning: project summary. Int J Radiat Oncol Biol Phys 21: 3–8

Harrison RM, Farmer FT (1978) The determination of anatomical cross-sections using a radiotherapy simulator. Br J Radiol 51: 488–453

Kijewski MF, Judy PF, Svensson GK (1984) Image quality of an analog radiation therapy simulator-based tomographic scanner. Med Phys 11: 502–507

Nagata Y, Nishidai T, Abe M et al. (1990) CT simulator: a new 3-D planning and simulating system for radiotherapy. Part 2. Clinical application. Int J Radiat Oncol Biol Phys 18: 505–513

Nishidai T, Nagata Y, Takahashi M et al. (1990) CT simulator: a new 3-D planning and simulating system for radiotherapy. Part 1. Description of system. Int J Radiat Oncol Biol Phys 18: 499–504

Ragan D, He T, Liu X (1993a) Correction for distortion in a beam outline transfer device in radiotherapy CT-based simulation. Med Phys 20: 179–185

Ragan D, He T, Mesina CF, Ratanatharathorn V (1993b) CT-based simulation with laser patient marking. Med Phys 20: 379–380

Rosenman J, Roe C, Cromartie R, Muller K, Pizer S (1993) Portal film enhancement: technique and clinical utility. Int J Radiat Oncol Biol Phys 25: 333–338

Rosenthal SA, Galvin JM, Goldwein JW, Smith AR, Blitzer PH (1992) Improved methods for determination of variability in patient positioning for radiation therapy using simulation and serial portal film measurements. Int J Radiat Oncol Biol Phys 23: 621–625

Sherouse GW, Chaney EL (1991) The portable virtual simulator. Int J Radiat Oncol Biol Phys 21: 475–482

Sherouse GW, Novins K, Chaney EL (1990a) Computation of digitally reconstructed radiographs for use in radiation treatment design. Int J Radiat Oncol Biol Phys 18: 651–658

Sherouse GW, Bourland DJ, Reynolds K, McMurry HL, Mitchell T, Chaney EL (1990b) Virtual simulation in the clinical setting: some practical considerations. Int J Radiat Oncol Biol Phys 19: 1059–1065

3 Three-Dimensional Photon Beam Calculations

Peter Bloch and Martin D. Altschuler

CONTENTS

3.1 Introduction:
The Importance of Accurate Dosimetry

In radiation treatment of cancer it is necessary to: (a) accurately define the extent of disease, (b) customize radiation treatment delivery of required dose to diseased tissues while minimizing dose to normal surrounding tissue, and (c) verify that the delivered dose is the amount planned. Attention to all these issues is essential to design a treatment regimen that controls the disease without causing serious complications to normal critical structures. In most clinical cases direct measurement of the dose delivered to the tumor and normal tissue is not possible. Thus the radiation oncologist depends on the calculated dose distribution to evaluate the appropriateness of a particular treatment plan. Clinical studies have documented that small changes of ±5% in the dose delivered can result in significant differences in complication-free local control of disease (ICRU Report 24, 1976).

To achieve 5% accuracy in dose delivery requires reproducible precision in both machine calibration and patient setup, and 2%–3% accuracy in the calcu-

lated dose at any point in an irradiated volume (Mijnheer et al. 1987; Purdy 1992). To achieve such precision in the dosimetry of treatment planning requires evaluation of the dose throughout the 3D treatment volume.

Dose calculation algorithms for external photon beams have been developed for determining the dose distribution throughout the 3D irradiated volume. The physical basis and assumptions employed in these algorithms will be reviewed. To determine dose at any point in a photon-irradiated volume, the primary (or first collision) dose and the dose due to scattered radiation are usually evaluated separately (Clarkson 1941; Cunningham 1972).

3.2 Primary Dose Calculation

Important theoretical and experimental work has focused on the transport of photons in tissues (Johns and Cunningham 1983; ICRU Report 33, 1980). The kinetic energy per unit mass released to matter by the first or primary photon interaction at a depth d in the media, $\mathrm{Kerma_p}(d)$, can be calculated from first principles:

$$\mathrm{Kerma_p}(d) = \int^{E_{max}} \Phi_p(E,d)\big(\mu_k(E)/\rho\big)dE, \quad (3.1)$$

where $\Phi_p(E,d)dE$ is the energy fluence spectrum (energy/area) at depth d, E_{max} is the maximum energy of the bremsstrahlung radiation, and $(\mu_k(E)/\rho)$ is the mass energy-transfer coefficient (area/mass) for the irradiated media at energy E.

The energy fluence at d is given by

$$\Phi_p(E,d)dE$$
$$= \Phi(E,0)\exp\left[-\big(\mu_a(E)/\rho\big)\int^d \rho(x)dx\right]dE, \quad (3.2)$$

where $\Phi(E,0)\,dE$ is the photon energy fluence at energy E in air, $(\mu_a(E)/\rho)$ is the mass attenuation coefficient at energy E, and $\rho(X)\,dX$ is the areal mass

Peter Bloch, PhD, Martin D. Altschuler, PhD, Hospital of The University of Pennsylvania, Department of Radiation Oncology, 3400 Spruce Street, 2 Donner, Philadelphia, PA 19104, USA

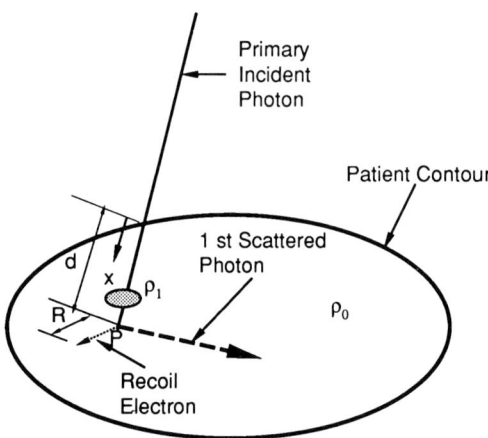

Fig. 3.1. Primary photon interacting at point P in the body producing a scattered photon and recoil electron having range R. The variation of electron densities ρ_0, ρ_1 are employed to determine the areal mass or radiographic depth traversed by the primary photon

or radiographic depth (mass/area) traversed by the ray in the media. The integral is along the track traversed by the X-ray taking into account the variation of electron density $\rho(x)$ along this path (Fig. 3.1).

The energy spectrum of the incident photons, $\Phi(E,0)dE$, requires detailed information on both the bremsstrahlung spectrum emanating from the target as well as the differential absorption in the flattening filter placed in the beam to achieve the desired distribution of energy fluence over a large irradiated area. Experimental and Monte Carlo simulation have been employed to determine the energy spectrum for clinical photon beams (LEVY et al. 1974; MOHAN et al.1985; ALTSCHULER et al. 1992).

In regions of charged particle equilibrium (CPE) the absorbed dose, D_p, in tissue due to primary photon interaction is the same as the collisional kerma if bremsstrahlung production by the secondary charged particles is neglected. Thus

$$D_p(d) \approx \text{Kerma}_p(d) \approx \Delta E_d / \Delta m$$
$$= \int^{E_{max}} \Phi(E,0) \exp\left[-\left(\mu_a(E)/\rho\right)\int^d \rho(x)dx\right]$$
$$\cdot \left(\mu_k(E)/\rho\right)dE, \qquad (3.3)$$

where ΔE_d is the energy deposited in mass Δm at depth d.

Implicit in most 3D dose calculation algorithms presently employed in commercial radiotherapy treatment planning algorithms is the assumption that the recoil electrons produced by photoelectric, Compton, and/or pair processes are absorbed at the point of collision. This approximation can be applied to evaluate dose in regions of CPE but it breaks down

in transition zones between tissue types and between air and tissue when CPE does not exist. The physical width of the transition zone at these interfaces depends on the range of the recoil electron, which can be several centimeters for high-energy photon beams.

The increasing clinical use of high-energy x-ray beams from linear accelerators with consequently wider transition zones of non-CPE makes critical the use of improved 3D algorithms that can calculate dose to tissues without assuming CPE. In particular, doses to tissues adjacent to air cavities in treating lesions of the upper respiratory tract and/or lung tissue (when breast lesions are being treated with lateral photon fields) require special care when high-energy photons are used. Experimental results reported by several investigators have shown dose perturbations of 10%–15% at tissue interfaces due to the finite range of the recoil electrons (DUTREIX et al. 1965; DUTREIX and BERNARD 1966; EPP et al. 1977; WERNER et al. 1987). More recent developments in 3D photon dose calculation algorithms to be discussed later take into account the transport of these secondary electrons.

Recoil electrons arising from secondary Compton scattered photons, however, are of much lower energy. Thus in calculating the scatter dose component, the approximation that all the energy of the recoil electron is deposited at the point of interaction should not induce appreciable dosimetry error at tissue interfaces (Woo et al. 1990).

3.3 Scatter Dose Calculation

The absorbed energy associated with Compton scattered photons can be separated into two components: first scatter photons and mulitiple scattered photons. The dose due to the first Compton scattered photon in a media can also be calculated from first principles.

The first scattered photon dose component at Q (Fig.3.2) requires integration of the scatter contribution over the entire irradiated volume. The primary energy fluence at each volume element ΔV and the scattered energy fluence from ΔV to point Q need to be determined.

The primary photon energy fluence $\Phi_p(E,d)$ at ΔV (depth d) is evaluated by integrating Eq. 3.2 over the incident energy fluence spectrum.

The first Compton scattered photon energy fluence, $\Phi_s(E)$, at energy E originating from ΔV and reaching point Q is

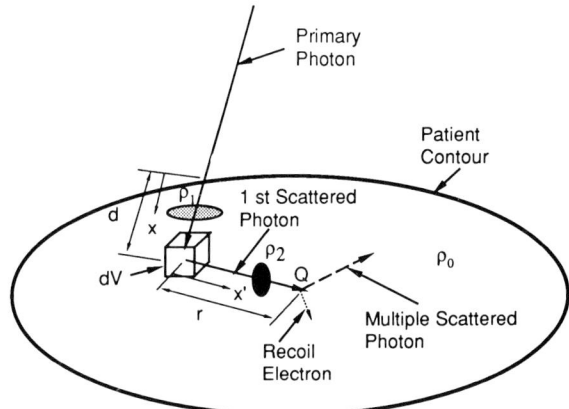

Fig. 3.2. First Compton scattered photon dose at Q arising from primary photon interactions in volume element dV. The variation of electron densities ρ_0, ρ_1, and ρ_2 traversed by the primary and scattered photons are used to determine the radiographic distances d and r respectively

$$\Phi_s(E,d)dE$$
$$= \Phi_p(E,d)\left(r_0^2/2\right)\int^\Omega\left(1+\cos^2\theta\right)F_{KN}$$
$$\cdot \exp\left[-\left(\mu_a(E_s)/\rho\right)\int^r \rho(x')dx'\right]dEd\Omega, \qquad (3.4)$$

where r_0 is the classical electron radius e^2/mc^2, and F_{KN} is the Klein-Nishina probability that a photon of energy E is scattered at angle θ into solid angle $d\Omega = 2\pi\sin\theta d\theta$ as given by the formula

$$F_{KN} = \left\{\frac{1}{1+\alpha(1+\cos\theta)}\right\}^2$$
$$\cdot \left\{1+\frac{\alpha^2(1-\cos\theta)^2}{[1+\alpha(1-\cos\theta)(1+\cos^2\theta)]}\right\},$$
$$\qquad (3.5)$$

where $\alpha = E_0/m_0c^2$ and E_s is the Compton scattered photon energy given by the Compton shift equation,

$$E_s = E_0/\left(1+\alpha(1-\cos\theta)\right), \qquad (3.6)$$

with $\mu_a(E_s)/\rho$ the mass attenuation coefficient for the Compton scattered photon at energy E_s.

The energy transferred to the recoil electrons by the first Compton scattered energy fluence, Kerma$_{fs}(d)$, is

$$\text{Kerma}_{fs}(d) = \int^{\text{Vol}}\int_{E_{min}}^{E_{max}} \Phi_s(E,d)$$
$$\cdot \left(\mu_k(E_s)/\rho\right)dEdV, \qquad (3.7)$$

where $[E_{min}, E_{max}]$ is the range of Compton scattered photon energy obtained from Eq. 3.6 for scatter in the range of 0–180°, $(\mu_k(E)/\rho)$ is the mass energy transfer coefficient for the Compton scattered photons with energy E_s. The total first scatter dose component at P is evaluated from the entire irradiated volume.

The primary and first Compton scattered dose at a depth of 15 cm in water irradiated with a 10-cm-diameter cobalt-60 beam was shown to account for 90% of the total dose, leaving 10% attributable to multiple scattered photons (WONG et al. 1981; WONG and HENKELMAN 1983). For higher energy beams the multiple scattered dose component is relatively less. For a 15-MV clinical beam (Varian 2100C) a Monte Carlo-derived photon spectrum at isocenter was used to calculate the first scattered dose component as a function of field size and depth in a water phantom and compared with measured scatter phantom ratios (Fig. 3.3). The difference between the calculated and measured scattered dose is the proportion of dose attributed to multiple Compton scattered photons, which from the figure is approximately 4% for a 10-cm-radius field at a depth of 15 cm.

WONG and HENKELMAN (1983) showed in addition that a fraction of the multiple scattered dose component (where the second Compton photon is scattered less than 45°) could be calculated as a first scatter-like dose. In general, however, the multiple scatter dose component cannot be evaluated readily from first principles. Rather it requires tracing the path of each scattered photon through the media using Monte Carlo simulations.

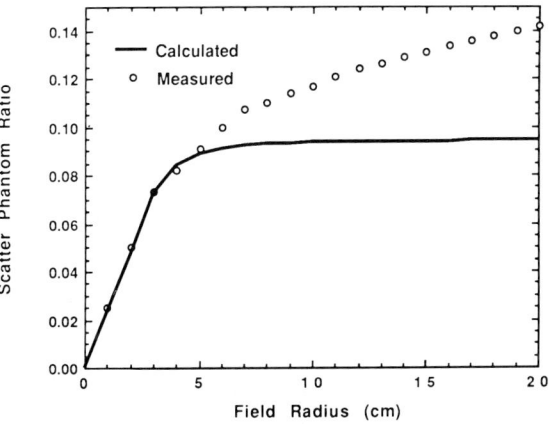

Fig. 3.3. Comparison of calculated first scatter dose and measured scatter phantom ratio at a 15 cm depth in water for a 15-MV photon beam. The difference between the measured and calculated scatter dose for large field radii is attributed to multiple scattered photons

Traditionally a more empirical heuristic approach has been employed for the radiotherapy photon dose calculation. The depth-dose characteristics of a clinical beam are measured in a homogeneous water phantom as a function of field size. From the measured data the zero area tissue phantom ratio, TPR(d,0), and the scatter phantom ratios, SPR (d,r), are derived as a function of depth and circular field size of radius r. No attempt, however, is made to partition the scattered dose arising from single or multiple scattering events. The primary and scatter dose component at a point in a homogeneous irradiated volume is evaluated using these derived parameters:

$$D_q(d,r_0) = D_{prim} + D_{scat}$$
$$= D_{ref}[f(r_0)\text{TPR}(d,0)$$
$$+ \int^{2\pi} d\theta \int^{R_0} f(r_0 + r)$$
$$\cdot (\text{dSPR}(d,r)/dr)dr], \qquad (3.8)$$

where D_{ref} is the measured dose at a reference depth and $f(r_0)$ is the attenuation of the beam including changes in beam profile at a distance r_0 from the central ray. The differential scatter phantom ratio, dSPR(d,r)/dr, derived from the measured TPR values, is used to determine the scatter dose from an annular sector of width $d\theta$ at a radial distance r from r_0, where $f(r_0 + r)$ is the incident beam profile value at the annular sector from which the scatter photons originate, evaluated at a distance from the central ray corresponding to the vector sum $(r_0 + r)$. The double integration provides the scatter dose component at the point (d, r_0) from the total irradiated volume.

Algorithms for calculating the dose distribution for clinical 2D and/or 3D photon treatment planning typically employ Eq. 3.8 to evaluate the primary and scatter dose components. These algorithms require measured beam parameters in water including depth dose or tissue phantom ratios as a function of field size and measured beam profiles (for both open or nonblocked fields and, when applicable, for beam modifiers such as wedges or tissue compensators). The algorithm can be employed for dose calculation in clinical situations taking into account the anatomic shape of the patient and irregularly shaped radiation fields. It has also been used to account for tissue heterogeneity by scaling both depth and radial distances by the regional electron densities traversed by the photon beam obtained from computer axial tomography scan data (O'CONNOR 1956; SONTAG and CUNNINGHAM 1977).

These algorithms require that the energy of the recoil electrons deposit all their energy at the point of photon-electron collision. Neglecting the transport of the recoil electrons limits the calculation of dose to regions of tissue where CPE exists. The dose deposition to tissues in transition zones such as air–tissue junctions cannot be evaluated. WOO et al. (1990) modified the calculated primary dose component by depositing the energy of the recoil electron over a finite region rather than at the point of origin. Recent developments in photon dosimetry employing photon pencil beam or point energy deposition kernels use first principles to evaluate the dose in tissue when CPE dose not exist. In addition, these dosimetry developments permit the evaluation from first principles of the total scatter dose contribution, including first and multiple scattered photons.

3.4 Recent Advances in Photon Dosimetry

The evaluation of the photon beam dose distribution in commercial treatment planning algorithms assumes that the recoil electrons produced by photon interactions in tissue deposit all their energy at the point of interaction. This assumption is justified when CPE exists; however, it cannot be employed for calculating the dose in the buildup region or in tissues at the interface between air and lung. Before discussing recent developments in photon dosimetry it is worth understanding the present limitations associated with evaluating a dose distribution from first principles by Monte Carlo simulations.

Monte Carlo employs a stochastic technique to simulate both (a) the photon transport, including the mulitiple scattered photons, and (b) the transport and energy deposition of the recoil electrons produced by the photon interactions in tissues (NELSON et al. 1985; HALBLEIB et al. 1992). To obtain statistically significant results from Monte Carlo simulations requires following a large number of primary photons. In principle, a Monte Carlo calculation for each photon beam employed in treating a patient can be performed by following the energy deposition of several billion photon histories. Accurate Monte Carlo codes are available to compute the random path of a photon through a known heterogeneous medium, and to account for the primary and secondary collisions and the ensuing particle tracks and kinetic energy depositions. Repeated for a billion or so such photons, the Monte Carlo calculation would provide the spatial dose deposition distribution within a patient to the required clinical accuracy. Such computations are exceedingly time consuming even on today's fastest computers, and only the advent of massively parallel computers orders of magnitude faster than those presently available would make practical the routine clinical use of Monte Carlo methods. However, even with expected developments in software and hardware, Monte Carlo simulations would probably be too slow for the demanding dose-distribution calculations necessary to solve the inverse problem, i.e., determine the optimal beam placements (including beam orientations, shapes, weights and modifiers, i.e., wedges, compensators) to deliver the desired dose to the target volume while limiting the dose to surrounding normal critical structures.

The transport of photons and recoil electrons in a media can also be obtained from first principles by solving the Boltzmann transport equation for discrete spatial and energy ordinates employing a deterministic method based upon finite differences. This avoids the statistical limitations imposed by the stochastic Monte Carlo technique and thus has the potential of substantially decreasing the CPU time required to calculate a full 3D dose distribution. The photons are treated as a source term while the spatial diffusion and the energy deposition of the recoil electrons are derived from the Boltzmann transport equation for multiple energy groups. MORREL (1987) showed that a modification of the Boltzmann-Fokker-Planck transport equation BFP could be employed to solve the multidimensional, multienergy group transport equation:

$$\frac{\mu \partial \Psi}{\partial z} = \frac{\alpha}{2} \frac{\partial \left[1 - \mu^2\right]}{\partial \mu} \frac{\partial \Psi}{\partial \mu} + \frac{\partial (S\Psi)}{\partial E} + Q \qquad (3.9)$$

where z is the space co-ordinate, μ the direction cosine of the transport particle with respect to the z-axis, Ψ the angular flux, α the momentum transfer, S the stopping power of the media, and Q the source term. The angular diffusion operator $\partial / \partial \mu$ causes a particle to continuously diffuse in space rather than undergo discrete scatter, while the continuous slowing down operator causes the particle to lose energy continuously rather than by discrete scattering events. The solution of the transport equation for a large number of photon and electron energy intervals is required for calculating a 3D photon dose distribution in clinical cases involving heterogeneous tissues. Preliminary work at the Radiation Transport Group X-6 at the Los Alamos National Laboratory (MORREL 1987) showed that the deterministic solution of the transport equation produced similar results to Monte Carlo simulations with a reduction in CPU time of approximately two orders of magnitude.

A clinically useful computer program which follows the transport and energy deposition of recoil electrons directly by a deterministic or stochastic solution of the Boltzmann transport equation and evaluates the 3D photon dose distribution in irradiated tissues is not available presently. Developments in photon dosimetry have recently occurred that permit evaluation of the dose in transition zones between tissue types where CPE is not present. These developments involve either (a) convolution of the recoil electron energy deposition point spread function with the collisional kerma at each point in tissue, or (b) fast superposition of the energy deposition kernels of incident photon pencil beams.

3.5 Photon Dosimetry by Convolution Methods

A convolution involves filtering an input distribution with a kernel function. The kernel is a (blurring) function whose integral over the domain of the distribution is finite (and usually normalized) and whose functional form is invariant to shifts in space (or time). If a problem can be formulated in terms of a convolution, considerable simplification may occur in calculation because the Fourier transform of the output distribution is then the product of the Fourier transforms of the input distribution and the kernel. One-dimensional applications of convolution are well known in signal processing (temporal variation) for such applications as filtering, smoothing, and correlation; two-dimensional applications involve the blurring or deblurring of an image by the point spread function (kernel) of a detector. If the kernel is very localized (essentially a delta function), then convolution of the input distribution by the kernel reproduces the input distribution; otherwise, the output distribution is blurred relative to the input distribution.

3.6 Application of Convolution to Radiation Treatment Planning

Radiation therapy treatment planning requires accurate information about how dose will be distributed within a 3D heterogeneous patient volume bounded by an irregular surface when several beams are used, each modified by a window of irregular shape and an incident energy fluence that may vary as a function of position. Further, for high-energy photon beams the range of the recoil electron must also be taken into account, since the region of electron non-equilibrium becomes extended over several centimeters and may involve not only the patient surface (dose-buildup) regions but the radiosensitive lung surfaces as well.

The energy deposition of a recoil electron energy transfer from a primary collision site in a uniform medium can be calculated from first principles using Monte Carlo simulation (MACKIE et al. 1988). If these dose distributions are relatively invariant in geometric form (if not scale) to spatial shifts within the patient volume, then they can be used as a convolution kernel in the calcualtion of complex 3D dose distributions for individual patients. The output dose distribution would then correspond to the convolution of an input distribution of say energy fluence and a Monte Carlo-derived kernel for energy deposition. In practice, the effects of density heterogeneity, surface contour irregularities, radiation depth, and beam hardening (or spectral variations) introduce spatial dependence so that (depending on the kernel) modifications from simple convolution may be needed, which in turn remove some of the calculational advantages. Alternatively, the advantages of convolution may be retained with the caveat that the validity of the dose distribution should be questioned for certain critical volumes. In either case, convolution provides a conceptual simplicity for complex 3D radiotherapy dose calculations through the paradigm of an input energy distribution convolved with a kernel for the energy deposited per unit of input. In the following, we discuss several kernels and convolution approaches that have been used or proposed for use in the clinical calculation of 3D dose distributions in patient volumes.

MACKIE et al.(1985) used Monte Carlo to derive a kernel called the "primary dose spread array" corresponding to the spatial distribution of deposited charged particle energy that originates from the site of a primary photon interaction. The kernel is normalized so that for each volume element (voxel) in the kernel the energy deposited is divided by the total energy of charged particles set in motion at the interaction site and lost to electron collisions (i.e., the dose deposited in a voxel due to electron collisions per unit kinetic energy released per unit mass at the site of the primary photon interaction voxel). The primary dose spread array is elegant proof of the need to consider recoil electrons in higher energy teletherapy beams. In a homogeneous medium, this primary dose spread array is convolved spatially with the kinetic energy released at each site of primary photon–electron interaction. In a heterogeneous medi-

um, the dose spread arrays must be modified. A density correction factor is needed for each line between a primary photon interaction voxel and a dose deposition voxel. That factor involves the average density on that line. Because such a density correction is not invariant to a shift in space, the heterogeneous case cannot be considered as a convolution, but rather as a superposition of distributions.

In addition to the dose kernel due to electron kinetic energy strewn from primary interaction sites, the authors also calculated kernels for scattered photons and for charged particles produced by interactions of those scattered photons. A kernel for the first Compton scattered photon dose and a kernel for dose deposited due to multiple photon scatter (and residual first scatter) are calculated separately. The total dose is the dose obtained by summing the doses obtained from the individual convolutions.

Convolution which takes advantage of the fast Fourier methods was advanced by BOYER et al. (1989). Here the kernels are of the type described above by Mackie. Five energies are used to model the 18-MV beam, and a kernel for each energy is calculated and stored together with its spectral weight. The kernels include all the interactions in the radiation shower that result from a photon interacting at a point in water. For each of the five energies in the beam model, a kernel is convolved with a primary fluence distribution. The primary fluence is obtained by tracing the ray through the phantom surface to the point in depth, and the use of a different mass attenuation factor for each of the beam energy components. Correction terms are used for different collimator sizes. The calculations done were for a water phantom.

DESOBRY et al. (1991) derived a convolution kernel as the sum of pencil beams irradiating the center of a cylindrical phantom uniformly from all angles. The ("rotational") kernel is defined as the dose distribution due to a set of pencil beams converging on a point and interacting with the medium along the rays to that point. In a heterogeneous medium, radiation distances to the point are calculated and the surface fluence then weighted to provide the kernel with equal fluence from all directions. To make use of the fast Fourier transform convolution, the kernel must be invariant to spatial shifts. That is, for any point in the phantom the dose distribution generated for the rotational kernel must not differ significantly from that generated for any other point. Tests made in the beam central axis plane showed that, relative to the maximum kernel dose, the offset kernels at a large distance (30 cm) from the axis of rotation agreed with the central kernel within 2% for 6-MV beams and within 15% for 18-MV; for both energies, the maximum difference between calculation and measurement in any histogram interval was less than 2% relative to the histogram maximum. These findings indicate that when 360° rotation therapy is used or else a large number of fields (say 16), the dose distribution in the plane of the gantry rotation can be accurately described using a convolution of the rotational kernel.

BOYER et al. (1991) applied their rotation kernels to implement in 3D the method of BRAHME (1988) for the inverse problem of radiotherapy. The desired dose distribution is assumed to be known, and constrained iterative deconvolution is used to asymptotically approach the desired dose distribution. The rotationally symmetric kernel is not completely general, so accuracy in the low-dose sectors at the margins of fixed beams is sacrificed for a rapid calculation. In addition, the buildup region of the beam near the phantom surface is not accounted for with the rotational kernel.

Recently BORTFELD et al. (1993) have achieved very rapid calculation times using pencil beam convolution, which involves 2D convolutions of the primary fluence with (a series of) pencil beam dose spread kernels for various depths.

For homogeneous media the convolution method is a mathematically elegant and computationally efficient method for evaluating the 3D dose distribution. When heterogeneity is introduced, most of the simplicity and speed is lost. Methods to account for heterogeneous media involve the collapsed cone approach of AHNESJO (1989) and the rapid superposition method of BLOCH (1988) and ALTSCHULER et al. (1992).

The adaptation of convolution methods to heterogeneous 3D volumes by scaling of radiation distance can be computationally intensive. With heterogeneity, the computational advantage of convolution (which derives from the spatial invariance of a kernel and the option of using fast Fourier methods) no longer applies. To improve the efficiency of heterogeneous 3D dose calculation, AHNESJO (1989) introduced the "collapsed cone" algorithm, which maintains the conceptual simplicity of convolution algorithms. Dose distributions are obtained as polyenergetic energy deposition kernels (derived from the beam spectrum) convolved with the total energy released per unit mass. All energy released into coaxial cones of equal solid angle from volume elements on a cones axis is rectilinearly transported, attenuated, and deposited in elements on that axis. Kernel scaling for the heterogeneous medium is performed during the calculation. The method uses an analytic representation for the dose point spread function, which allows several integrals to be partially calculated with the results in the form of recursion relations on a rectilinear grid. The radiant energy for a particular direction can then be calculated with one evaluation of the exponential factors per direction and point of dose calculation. By arranging the collapsed cone lines into a lattice covering the irradiated volume, each voxel is passed only once for each direction. Direct comparison of calculations with the collapsed cone convolution method with the Monte Carlo calculations indicate that the collapsed cone method in heterogeneous media is very accurate for regions of charged particle equilibrium.

3.7 Photon Dosimetry by Superposition Techniques

To obtain accurate 3D dose distributions for both heterogeneous media and regions of electron nonequilibrium, with the goal of routine clinical use, BLOCH (1988) and ALTSCHULER et al. (1992) adapted a fast superposition method. A (nonconvolution) kernel is defined as the distribution of energy density absorbed in a uniform medium when a normally incident pencil delivers one unit of energy to the surface. The goal is to use such kernels to calculate and merge the 3D internal dose contributions of all the pencils of all the beams impinging on the irregular surface of the patient.

The method puts all of the microscopic physics into the kernel. A photon kernel for a clinical beam is obtained directly from Monte Carlo calculations. The beam spectrum at isocenter is obtained from Monte Carlo simulations by determining the bremsstrahlung spectrum produced at the target and attenuated by the flattening filter in the beam. A kernel includes the physics of electron nonequilibrium. (For comparison with older methods, kernels can also be created which deposit the energy at the photon collision site.) For rapid calculation, kernels are replaced by cumulative distribution tables which tell how much of the pencil energy must be distributed along and normal to the pencil axis at each radiation depth, and for the radial distribution at each depth.

The beam is specified by setting the usual parameters for gantry angle, source target distance, collimator orientation, and field size. The modulation of the beam (due to blocks, wedges, multileaf collimators, etc.) is represented by the different weights of the pencils. In practice, a plane of calculation is provided with a grid which expands in depth in proportion to the beam divergence from the source. Grid pixels at different depths with the same planar indices correspond to a pencil. The calculation involves sweeping the plane of calculation along the beam direction, and strewing energy from pixel to pixel in the calculation plane in accord with cumulative distribution ratios derived from the kernel. Thus all the input beam energy is tracked and accounted for as the calculation plane proceeds along the beam. For a heterogeneous medium, radiation depth is calculated and updated separately for each pencil, as is radiation distance lateral from each pencil, in accord with the electron densities derived from CT. To obtain a beam edge, a convolution is performed in the calculation plane for a point spread function derived from measurements of the edge spread function.

3.8 Clinical Example of a Full 3D Photon Dose Calculation

The implication of taking into account the spatial energy deposition of the recoil electron in treatment planning is demonstrated by the following clinical example. A patient with a right-sided breast tumor was planned for treatment with two opposite opposed 15 MV photon beams. Three-dimensional dose calculations were performed using Monte Carlo-derived pencil beam (energy deposition) kernels evaluated for the 15-MV photon spectrum. Two cases were calculated (a) assuming all the recoil electron energy was deposited at the point of origin, and (b) taking into account the energy deposition of the recoil electrons. Figure 3.4a shows the isodose distribution in the central plane with no lung density correction but with the recoil electron energy deposition taken into account. This distribution is similar to the isodose distribution calculated by evaluating the primary and scatter components from measured tissue phantom ratios (Eq. 3.8) except for small differences at the edge of the beam. If, further, the electron transport is not taken into account (Fig. 3.4b), the extension of the 10% isodose line into the lung tissue is reduced. As expected, a photon pencil beam kernel that does not account for electron transport cannot predict the dose in the buildup region between air and soft tissue. Lung density corrections were employed by scaling both depth and the radial distance using the electron densities along the path traversed by the beam. Figure 3.4c shows the calculated dose distribution with both lung density correction and transport of recoil electrons. The increase in lung

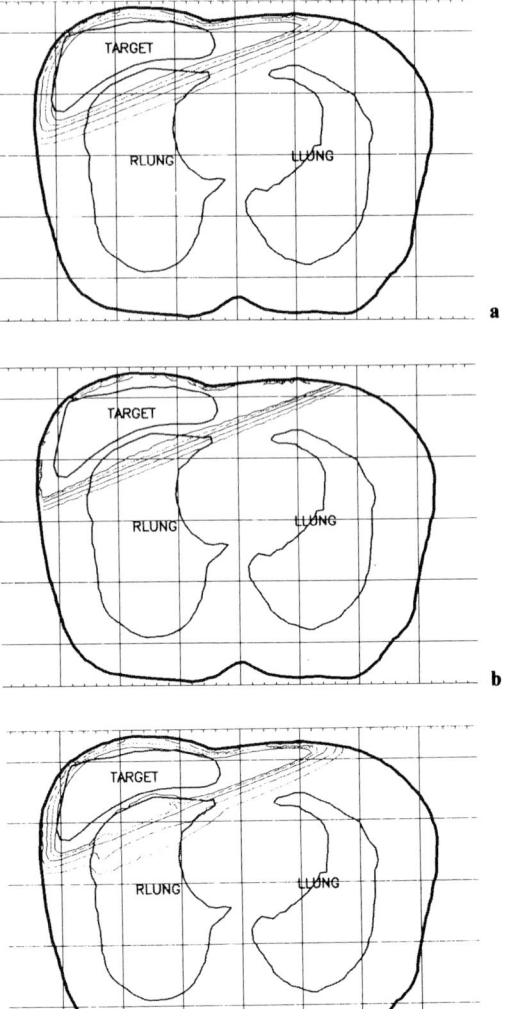

Fig. 3.4a–c. Isodose distributions for a primary breast tumor treated with opposite opposed 15-MV beams. Shown are the 100%, 90%, 70%, 30%, and 10% isodose lines in the central plane. **a** No lung density correction but energy deposition of recoil electrons taken into account. **b** No lung density correction, and electron energy deposited at point of origin. **c** Lung density correction with energy deposition of recoil electrons taken into account. (Wide fields were used here to treat mediastinal nodes.)

dose associated with the finite path of the recoil electron can be appreciated by comparing Figs.3.4a–c. The dose–volume histogram for the target volume (Fig.3.5) indicates only a small difference in the target dose distribution when the electron transport is taken into account. However, a large increase in the lung volume which receives 20%–30% of the target dose occurs when the transport of the recoil electrons is included in the calculation. It may be important clinically to know the dose distribution in the

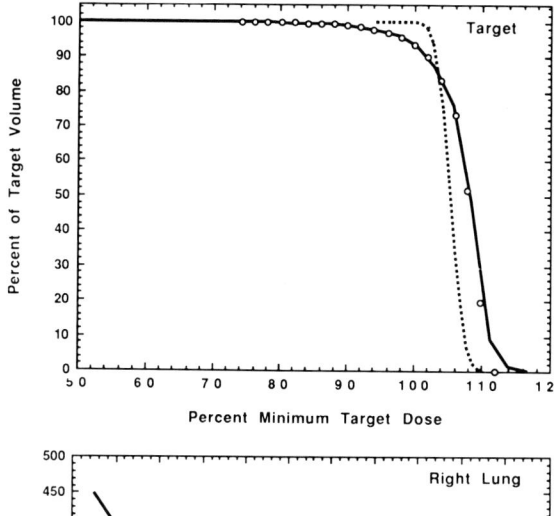

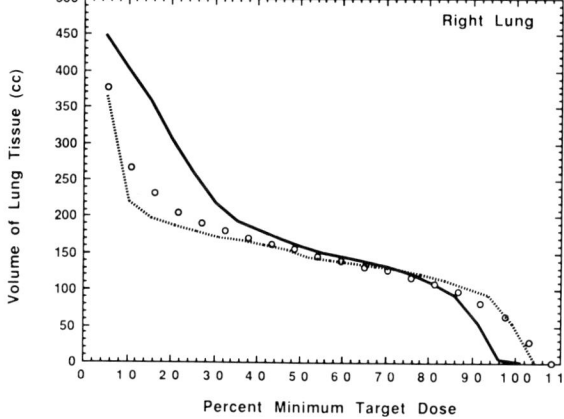

Fig. 3.5. Cumulative dose-volume histograms for the target and right lung volumes calculated for: (a) lung density correction with recoil electron energy deposition (*solid line*), (b) no lung density correction and recoil electron energy deposition (*open circles*), and (c) no lung density correction with electron energy deposited at point of origin (*dotted line*)

lung accurately since radiation pneumonitis is one of the complications associated with irradiation of breast lesions.

3.9 Conclusion

Photon dosimetry for treatment planning was until recently based solely on clinical photon beam measurements in water. The data extracted from measured central axis depth-dose or tissue-phantom ratios for square or circular fields and beam profiles and penumbra were employed in algorithms to calculate the photon dose distribution in irradiated tissues. This approach can be used to calculate accurately the dose to homogeneous tissues in charged particle equilibrium. However, for high-energy photon beams, transition zones between tissue types, e.g., lung/muscle or bone/muscle, may not be regions of CPE. Removing the constraint of CPE required a more basic approach to photon dosimetry. Recent developments in photon dosimetry have addressed this problem by accounting for both the transport of photons and the energy deposition of the recoil electrons in the irradiated tissues. The energy deposition of the recoil electron transferred from the collision site in a homogeneous media can be calculated from first principles using Monte Carlo simulation.

Modern methodologies for fully three-dimensional treatment planning with photon beams involve relatively independent conceptual and computational modules for (a) microscopic physics, (b) geometry of the patient (CT electron density distribution, and contours of the body and critical organs), (c) construction of the beam setup, (d) calculation of 3D dose for homogeneous or inhomogeneous density distributions and for electron nonequilibrium with transfer of electron kinetic energy away from the photon collision site, (e) weighting and autoweighting of beams for optimization of the global dose distribution relative to the prescribed dose and dose constraints, and (f) analysis and display of the dose distribution.

The modularization of the treatment planning problem is more than a conceptual simplification. All methodologies employ approximations and assumptions in the physical formulations, the mathematical formulae, the computational method, the discretization between real and integer values, and the resolution scales of different segments of the problem–CT data grid, calculational grid, display grid, etc. If all these approximations are not carefully separated and identified, it is not possible to verify, evaluate, improve, maintain, or even trust the output of a method. The approximations may even amplify error if undersampled distributions are used for calculations of dose at the beam edge, or for a heterogeneous calculation, or for the optimization of global dose, or for beam compensator construction. Thus efforts at analyzing and understanding all aspects of the treatment planning problem are essential for better dose delivery and medical interpretation.

Acknowledgment. The authors would like to express their appreciation for the valuable technical assistance of Seetha Ayyalasomayajula and to Peg Flynn for preparation of the manuscript.

References

Ahnesjo A (1989) Collapsed cone convolution of radiation energy for photon dose calculation. Med Phys 16: 577–591

Altschuler MD, Bloch P, Buhle El Jr, Ayyalasomayajula S (1992) 3 D dose calculations for electron and photon beams. Phys Med Biol 37: 391–441

Bloch P (1988) A unified electron/photon dosimetry approach. Phys Med Biol 33: 373–379

Bortfeld T, Schlegel W, Rhein B(1993) Decomposition of pencil beam kernels for fast dose calculations in three-dimensional treatment planning. Med Phys 20: 311–318

Boyer Al, Zhu Y, Wang L, Francois P (1989) Fast Fourier transform convolution calculations of x-ray isodose distributions in homogeneous media.Med Phys 16: 248–253

Boyer AL, Desobry GE, Wells NH (1991) Potential and limitations of invariant kernel conformal therapy. Med Phys 18: 703–712

Brahme A (1988) Optimization of stationary and moving beam radiation therapy techniques. Radiother Oncol 12: 129–140

Clarkson JR(1941) A note on depth dose in fields of irregular shape. Br J Radiol 14: 265

Cunningham JR(1972) Scatter-ratio. Phys Med Biol 17: 43–51

Desobry GE, Wells NH, Boyer AL (1991) Rotational kernels for conformal therapy. Med Phys 18: 481–487

Dutreix J, Bernard M(1966) Dosimetry at interfaces for high energy X and gamma rays. Br J Radio 10: 177–190

Dutreix J, Dutreix A, Tubiana M (1965) Electronic equilibrium and transition stages. Phys Med Biol 10: 177–190

Epp ER, Boyer AL, Dopple KP(1977) Underdosing of lesions resulting from lack of electronic equilibrium in upper respiratory air cavities irradiated by 10 MV X-ray beams. Int J Radiat Oncol Biol Phys 2:613–619

Halbleib JA, Kansek RP, Mehlhorn TA, Valdez CD, Selter SM, Berger MJ (1992) ITS Version 3.0 the Integrated TIGER Server of Coupled Electron/Photon Monte Carlo Transport Codes", SAND 91–1634

ICRU, Report 33 (1980) International Commission on Radiation Units and Measurements Radiation Quantities and Unit ICRU Report 33, Washington, DC

ICRU, Report 24 (1976) Determination of the absorbed dose in a patient irradiated by beams of X or gamma rays in radiation therapy procedures. International Commission on Radiation Units and Measurements, Washington, DC

Johns HE, Cunningham JE(1983) The physics of radiology, 4th edn. Charles C. Thomas, Springfield, ILL., Chap 7

Levy LB, Waggener RG, McDavid WD, Payne WH (1974) Experimental and calculated Bremsstrahlung spectra from a 25 MeV linear accelerator and a 19 MeV Betatron. Med Phys 1: 62–67

Mackie TR, Scrimger JW, Battita JJ (1985) A convolution method of calculating dose for 15 MV X rays. Med Phys 12: 188–196

Mackie TR, Bielajew AF, Rogers DWO, Battista JJ (1988) Generation of photon energy deposition kernels using the EGS4 Monte Carlo code. Phys Med Biol 33: 1–20

Mijnheer BJ, Battermann JJ, Wambersie A (1987) What degree of accuracy is required and can be achieved in photon and neutron therapy? Radiother Oncol 8: 237–252

Mohan R, Chui C, Lidolsky L (1985) Energy and angular distribution of photons from medical accelerators. Med Phys 12: 726–730

Morrel JE (1987) Boltzmann-Fokker-Planck calculations using standard discrete ordinate codes. Los Alamos National Labs LA-UR Report #83–229

Nelson WR, Hirayama H, Roger DWO (1985) The EGS code system Stanford linear accelerator center. Internal Report SLAC 265

O'Connor JE (1956) The variation of scattered x-rays with density in an irradiated body. Phys Med Biol 1: 352–369

Purdy JA (1992) Photon dose calculations for three-dimensional radiation treatment planning. Semin Radiat Oncol 2: 235–245

Sontag MR, Cunningham JR(1977) Corrections to absorbed dose calculations for tissue inhomogeneities. Med Phys 4: 431–436

Werner BL, Das IJ, Khan FM, Meigooni AS(1987) Dose perturbations at interfaces in photon beams. Med Phys 14: 585–594

Wong JW, Henkelman RM (1983) A new approach to CT-pixel-based photon dose calculations in heterogeneous media. Med Phys 10: 199–208

Wong JW, Henkelman RM, Andrews JW (1981) Effect of small inhomogeneities on dose in a Co-60 beam. Med Phys 8: 783–791

Woo MK, Cunningham JR, Jezioranski JJ (1990) Extending the concept of primary and scatter separation to the condition of electronic disequilibrium. Med Phys 17: 588–595

4 Three-Dimensional Photon Beam Treatment Planning

BENEDICK A. FRAASS and DANIEL L. McSHAN

CONTENTS

4.1 Introduction

4.1.1 What Is Treatment Planning?

Treatment planning, in the most general sense, consists of all procedures which are used by the radiation oncologist (and other staff) to help determine the plan with which the patient will be treated. In practice, however, the term "treatment planning" has two different usages:

1. Treatment planning is often considered to be the act of entering the patient shape and beam locations into a computer system (treatment planning system), and then generating a calculated dose distribution which predicts what one expects the actual dose distribution to be if the patient is treated with the chosen plan.

2. A more general definition is that treatment planning refers to the quantitative parts of the process (at least) by which the patient's plan is determined. Thus, in this second definition, the use of computerized tomography (CT) and/or magnetic resonance imaging (MRI) to determine the patient anatomy, the methods used to input the physician's expectations and requirements, interactive or automated optimization of the desired beam arrangement to try to improve the dose distribution, the methods used to verify the consistency of the input data, and the ability to reproducibly and accurately treat the patient in the same way each day are part of the generalized process of treatment planning.

BENEDICK A. FRAASS, PhD, DANIEL L. McSHAN, PhD, University of Michigan Medical Center, Radiation Oncology, Rm. B2C490, 1500 E. Medical Center Drive, Ann Arbor, MI 48109-0010, USA

In recent years, a new class of computerized radiation therapy treatment planning systems have been developed and are being clinically implemented. These systems typically attempt to address the entire treatment planning process, rather than the limited view of "treatment planning" which is stated in the first usage listed above. In this chapter, the latter definition of treatment planning is the one which is discussed.

These developments have made clear the differences between two philosophically different approaches which are used clinically: (a) "dose display" and (b) "treatment planning." In many clinical situations, the physician determines the "treatment plan" (the technique which will be used to treat the patient) by using the simulator. Then, to document the plan, or to determine exactly what the dose distribution will look like, the plan is input into the computerized treatment planning system, and the dose distribution is generated and displayed. This process is most often used in older-style two-dimensional (2-D) planning, and can be called "dose display." In contrast to this procedure, the term "treatment planning" can be used. When treatment planning is performed, the computerized planning system is used to try a number of different beam arrangements while the planner optimizes the plan to accomplish the aims set out by the physician. In this chapter, the term "treatment planning" is used in the latter context throughout.

4.1.2 Differences Between 2-D and 3-D Treatment Planning Systems

Two-dimensional treatment planning has been developed and used since the late 1950s (TSIEN 1955; VAN DE GEIJN 1965; CUNNINGHAM 1972). As generally implemented, the treatment planning system used for 2-D planning has a number of limitations, some of which are listed in Table 4.1. Display of a single contour of the patient is used as a backdrop for graphics which show the radiation beam outlines, the dose distribution, or other information. Some systems may replace the manually entered contours with one or more CT images; however, few other imaging modalities or orientations are available. Treatment beams are usually displayed on one slice at a time and are often movable only in the plane of the axial slice. Field shaping blocks are typically not completely modeled (and often are not modeled at all). Generation of treatment plans which use nonaxial and/or noncoplanar beams is difficult or

Table 4.1. Differences between 2-D and 3-D treatment planning

Subject	2-D	3-D
Definition of anatomy		
Orientation of image slices	Axial only	Arbitrary
Number of slices	Small (1–10)	Arbitrary (>100)
Use 3-D structures	No	Yes
Use of MRI, PET, etc.	No	Yes
Integrated use of Sim/Tx treatment portal films	No	Yes
Beam design and display		
Use of BEV for beam shaping	No	Yes
Beam orientation	In axial planes	Arbitrary
Beam display	On axial slice	3-D divergent object
Dose calculations		
3-D shape and density of patient	No	Yes
3-D divergence/beam geometry	No	Yes
3-D beam flatness/symmetry	No	Yes
3-D scatter effects of blocks, etc.	No	Yes
3-D inhomogeneity corrections	No	Yes
Dose display and plan evaluation		
Dose display	One slice at a time	3-D isodose surfaces
Plan analysis tools (DVH, etc).	No	Yes
Plan comparison tools	No	Yes

impossible. 2-D dose calculations often neglect out-of-plane divergence, patient and beam shape changes, scatter, and other variables, and are rarely evaluated on more than a few 2-D slices. Evaluation of 2-D treatment plans rarely incorporates any quantitative evaluation tools, such as dose-volume or dose-area histograms, because the 2-D plans are typically evaluated solely by visual inspection of the isodose line displays.

"Three-dimensional" (3-D) is a much-used term which conveys many different meanings with respect to treatment planning. 3-D can refer to the description of the anatomy, degrees of freedom in beam or source location, ability to calculate a volumetric grid of dose points (even if using a 1-D or 2-D calculation algorithm), or the ability to display some part of the treatment planning information using solid surface graphics techniques. "Fully three-dimensional treatment planning" should mean that the planner is freed from all the 2-D limitations listed in Table 4.1 and has full freedom to explore any possible treatment technique in an accurate and practical way. GOITEIN et al. (1983b) have used the term "multi-dimensional" to include the use of multiple imaging modalities.

In order to fulfill the above-described goals, the "fully three-dimensional treatment planning system" should contain the following basic capabilities:

1. Use of a complete 3-D description of the anatomy.
2. Diagnostic imaging (CT, MRI, etc.) should be integrated into the planning process in a quantitative way.
3. The radiation beam and/or sources must be defined fully in a 3-D coordinate system. It must be possible to move the beam in all directions that the treatment machine can be moved.
4. Dose calculations must be performed in a 3-D matrix of points throughout the volume of interest in the patient.
5. External beam dose calculations must accurately consider the following effects: (a) the 3-D shape of the patient's external surface, (b) the 3-D electron densities provided by CT and their effect on the primary beam, (c) the 3-D location and shape of the radiation field/source, (d) 3-D beam divergence, (e) 3-D beam flatness/symmetry, (f) 3-D scatter and effects of beam modifiers such as wedges, blocks, and compensators, and (g) 3-D scatter effects of inhomogeneities.
6. Dose display and evaluation tools must work in a three-dimensional way so that the 3-D results of the dose calculations can be appreciated by the treatment planner. This must include 3-D displays, dose-volume analysis, and other plan comparison tools.
7. The system must be fast and interactive so that it is reasonable to use the three-dimensionality in a routine clinical setting.
8. For 3-D treatment plans obtained from a 3-D planning system to be used clinically, one must be able to verify the accuracy of the 3-D plans as they are implemented. Numerous tools and techniques useful for plan verification and confirmation must be included in the planning system to assist in practical implementation of the plans.

It is only in the last few years (GOITEIN and ABRAMS 1983; GOITEIN et al. 1983; FRAASS and MCSHAN 1987; MOHAN et al. 1988; SHEROUSE et al. 1987) that treatment planning systems have begun to satisfy most of these criteria.

4.1.3 Standard Procedures
Used for Treatment Planning

The set of procedures used as part of the treatment planning process are summarized below. Although the actual implementation of each of these procedures may differ significantly from institution to institution, nearly all treatment planning includes at least some reference to each of the processes listed.

1. *Immobilization and localization.* The first step in treatment planning is the establishment of the treatment position for the patient. Immobilization is a critical step for the application of 3-D treatment planning, especially with conformal techniques. The procedure also involves defining an origin and a coordinate system for use in future planning procedures.

2. *Imaging.* In most state-of-the-art treatment planning, the next step is to perform a number of imaging procedures. Most common is the use of CT scanning (as described below in Sect. 4.2). However, MRI or other procedures are also used. It is critical that the original positioning, immobilization, and coordinate system which were determined at the localization procedure be used for these procedures.

3. *Anatomic definition.* After the patient is imaged, the next major step is the definition of the anatomic model to be used inside the computer as the basis of the treatment planning. This is described in detail in Sect. 4.2. An important part of this process will be the geometric registration of the various imaging modalities, if more than one is used. In principle, this always happens at least once, since the CT scans must be registered with the simulator data obtained at the initial localization. If the localization/immobilization procedure occurs at the same time as the CT scanning, as proposed in the virtual simulation paradigm (SHEROUSE et al. 1987), then this image registration step is not required until one registers the treatment plan information with the simulator information which is obtained at the plan verification step (see below).

4. *Treatment plan development and optimization.* In this part of the process, the planner (a) determines the beam technique, (b) calculates the dose distribution, (c) evaluates the doses, (d) iterates through various different plan techniques or otherwise attempts to optimize the plan, and (e) performs a final evaluation of the plan including review of the dose distribution, dose-volume histograms, and/or normal tissue complication and tumor control probabilities. Hard-copy output of the plan is generated, and the physician decides to implement the plan.

5. *Treatment device preparation.* Field shaping blocks, compensators, bolus, or other devices used in the treatment plan must be fabricated and verified.

6. *Plan verification simulation.* After the computerized treatment plan has been developed and decided upon by the physician, one typically performs a plan verification simulation. In its most complete form, this procedure should include (a) positioning the patient on the simulator in the treatment position, (b) verifying the setup of the original isocenter using beam's eye view (BEV) or digital reconstructed radiograph (DRR) representations of the patient anatomy as they should be seen by the simulator using films or fluoroscopy, (c) verifying the location of the new plan isocenter in the same fashion, and (d) checking the location and orientation of each of the planned treatment fields using the BEV and/or DRR method again.

7. *Treatment verification.* On the first day of treatment, and at least once a week after treatment commences, a treatment verification procedure is followed. This process should be similar to the plan verification simulation, except that only the current plan isocenter and treatment fields must be verified. In principle, the use of modern on-line imagers and sophisticated software should replace the current use of megavoltage portal films and analog (by eye) verification inspection of portal and simulator images.

4.2 Anatomic Representation of the Patient

The representation of anatomy and the display of that information is a major component of treatment planning. The anatomic representation of the patient is used throughout the process of treatment planning:

1. To illustrate the clinical situation at the beginning of the planning session
2. To help the planner and physician appreciate the relationships between target and normal tissue
3. To help suggest treatment approaches
4. For the display of beams or radioactive sources
5. For the display of dose

One of the critical features that makes 3-D treatment planning three-dimensional is the 3-D definition and display of patient anatomy.

4.2.1 Types of Anatomic Representations

There are three major kinds of representations used routinely in treatment planning to describe the patient anatomy. In modern treatment planning, much of the basic knowledge about the anatomy of the patient comes from CT scans and other kinds of imaging. CT images, in particular, are very useful for treatment planning, as described in Sect. 4.2.3 below. However, the direct use of images alone cannot satisfy all of the needs of treatment planning. A very important part of the initial phase of computerized treatment planning involves the definition by the physician and planner of various anatomic areas that are of special importance in designing or evaluating the treatment plan. These areas include the macroscopic tumor(s), the target(s) to which the physician would like to deliver a high dose, various dose-limiting critical structures, and other important anatomic landmarks. In standard 2-D treatment planning, these objects are identified by outlining (contouring) the structures (see Sect. 4.2.4). In 3-D planning, full 3-D objects are defined and displayed, as discussed in Sect. 4.2.5. A third class of representations which are used, at least in 3-D systems, are volumetrically oriented. The use of a 3-D matrix of CT-based density values is an important part of dose calculations if one is to be able to take into account the heterogeneity of the patient. Another kind of volumetric representation used in 3-D systems is a volumetric region of interest file, in which each voxel is coded to define the structures of which the voxel is a part. This kind of file is very useful for dose distribution analysis tools, particularly for dose-volume histograms.

4.2.2 Patient Localization and Immobilization

The first step in treatment planning is the establishment of the treatment position for the patient. Since all future procedures are based on positioning the patient in the same fashion, this position must be one that allows good reproducibility, comfort for the patient, a position which is useful for all the imaging and treatment procedures which will follow, and in which the patient is well immobilized. As conformal and 3-D treatment planning are utilized, the immobilization of the patient becomes critical. Numerous types of immobilization devices, including casts (SEWCHAND et al. 1986), mask systems (THORNTON et al. 1991a; VERHEY et al. 1982), foam cradles (SHEROUSE et al. 1990; JAKOBSEN et al. 1987), bite blocks (VAN DE GEIJN et al. 1983), and stereotactic-like frames (SCHWADE et al. 1990), have been created for this task. After positioning and immobilization are completed, the patient is "localized" using orthogonal radiographs or CT scanograms to establish a coordinate system which can be used throughout the planning process. This coordinate system is then marked on the patient (typically using marks to define anterior and lateral marks that correspond with wall-mounted lasers which are used to define the isocenter of the machine).

4.2.3 Use of Imaging Information

The use of various kinds of imaging studies to help define the anatomic model of the patient is an essential part of modern treatment planning. The use of CT for radiotherapy planning began nearly as soon as CT became widely available (McCULLOUGH 1987), and many early studies showed the impact that CT had on the planning process (GOITEIN et al. 1979). An excellent summary of the early use of CT for treatment planning is contained in the book by LING et al. (1983). While CT has been used as the basis for 2-D type treatment planning, it is only relatively recently that the CT data have been used in a fully three-dimensional way. The development of 3-D systems has also allowed the use of other imaging studies. MRI has proven to be very valuable in certain sites. Other studies, like positron emission tomography (PET), single-photon emission computerized tomography (SPECT), and ultrasonography, have also been used (GLATSTEIN et al. 1985).

4.2.3.1 Computerized Tomography

The basic information used to generate the volumetric description of the patient which is used for 3-D planning almost always comes from the use of CT. CT is invaluable for treatment planning since it provides (a) geometrically accurate imaging of the internal anatomy of the patient, (b) electron density values for every pixel in each image, which are used for the incorporation of the heterogeneity of the patient into the dose calculation results, (c) accurate knowledge of the 3-D location of each slice of imaging information, and (d) basically operator-independent imaging (as opposed to MRI, ultrasonography, etc). A full CT scan series dataset is obtained for the patient, scanning through the entire volume of interest in the patient. This is performed with the patient in an immobilization device, in the treatment position, and with skin marks or other alignment points made visible on the CT images by the use of radiopaque tubing (LICHTER et al. 1983). The CT scan protocols which should be used are tumor site dependent and also depend on the ability of the planning system to use the information. An example of the kinds of protocols which are defined is shown in Table 4.2.

For 2-D planning, only a few CT slices are typically used. However, for 3-D planning it is critical to obtain the CT data throughout the volume of interest and to include the whole volume of any organs which are to be evaluated using dose-volume histogram analysis. Another factor which increases the extent of the CT scan is the use of image registration techniques, either to allow the use of MRI or other imaging studies (see next section) or for detailed patient alignment checks (see Sect. 4.6). The CT protocol must scan an area significantly larger than the area directly involved with the tumor in order to give the registration techniques enough information for them to work accurately. A final reason for an extended range of CT scanning is that it is possible that the beam arrangement will include nonaxial beams. In this case, the anatomy of the

Table 4.2. Clinical CT scan protocols

Site	Area	Length (cm)	Spacing (cm)	No. of slices
Brain	Top of head–below chin	20	0.3–0.5	40–55
Orbit	Top of head–top of orbit	7	0.3–0.5	
	Top of orbit–bottom of orbit	4	0.3	55
	Bottom of orbit–below chin	7	0.5	
Lung	Neck–below diaphragm	40	0.5–1.0	35–70
Liver	Carina–midpelvis	50	0.5–1.0	40–60
Prostate	L4–Top of acetabulum	25	1.0	
	Top of acetabulum–1 cm below ischial			30–50
	tuberosities		0.5	
Cervix + para aortics	Like prostate	25	1.0–0.5	25–50
ENT	Top of head–top of orbit	7	1.0	
	Top of orbit–to superclav	23	0.5	40–60

patient must be known accurately well above and/or below the area containing the target volume.

4.2.3.2 Use of MRI for Treatment Planning

One of the disadvantages of CT imaging, at least for treatment planning, is the generally poor soft tissue contrast, and in particular the difficulty in imaging the tumor. In contrast, the imaging of the tumor and normal anatomy which can be provided by MRI for many sites makes its use extremely attractive for radiation therapy treatment planning (RTTP) (FRAASS et al. 1987a). Although several early reports on the use of MRI for RTTP noted substantial potential problems with its use (HENKELMAN et al. 1984; SONTAG et al. 1984; COFFEY et al. 1984), MRI has proved to be a very useful addition to CT in a number of situations (FRAASS et al. 1987a; THORNTON et al. 1992). MRI scanning for RTTP is much more involved than CT scanning, since there are so many different variables available (scans in axial, sagittal, and coronal planes, T2- and T1-weighted scan protocols, scans with and without gadolinium contrast, etc). In order to make effective use of the time the patient spends in the MRI scanner, and to make sure that the images acquired can be (a) registered geometrically with the CT data and (b) useful for tumor, target, and/or normal tissue identification, a series of MRI scan protocols have been designed specifically for use for radiotherapy planning in various sites in the body (Table 4.3) (YANKE et al. 1991). Although the quantitative use of MRI in treatment planning can be difficult, various workers have addressed the technical issues associated with the use of MRI for treatment planning (FRAASS et al. 1987a; KESSLER 1987; McSHAN and FRAASS 1987; PELIZZARI and CHEN 1987; SCHAD et al. 1987).

Analysis of a recently published study on the use of MRI for treatment planning for brain tumors is illustrative of some of the technical problems associated with the use of MRI for planning. As part of a study comparing CT-based 3-D treatment

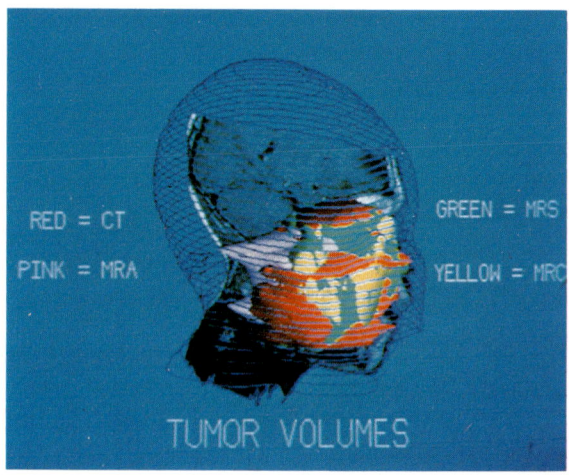

Fig. 4.1. Different target volumes derived from axial (*CT and MRA*), sagittal (*MRS*), and coronal (*MRC*) scans

planning to 3-D planning based on both CT and MRI, TEN HAKEN et al. (1992) analyzed 15 patients planned with MRI and CT in detail. The brain tumor treatment planning protocol used for this study (as well as for routine clinical planning) makes use of T2-weighted MR images, T1-weighted scans with gadolinium contrast, and contrast CT images. The study compares target volumes (for the original treatment plan and for the high-dose boost plan) which are generated with CT data alone, versus the volumes defined when the various sets of MR data are also considered. The situation is fairly complicated, since target volumes drawn on axial, sagittal, and coronal MR images each differ (Fig. 4.1), even when the different datasets are based on the same MR pulse sequences. The study shows that neither MRI or CT information alone is adequate for the definition of the target volume, and that great care should be taken in attempting to define the 3-D target volume in a consistent fashion, based on all the available imaging data.

4.2.3.3 Use of Multiple Imaging Modalities

A number of treatment planning capabilities and/or procedures are needed to implement the use of multiple imaging modalities, regardless of the kinds of modalities being considered. We have experience with input and use of the following kinds of data: CT, MRI, electromagnetic digitizer (used for digitization of mechanically obtained patient contours), video-digitized and laser-digitized radiographs or other projective x-ray information, PET and SPECT

Table 4.3. RTTP MRI scan protocols for brain

Sequence	Plane	Separation (mm)	Size
T1-locator	Sagittal	10	Around midline
T2	Axial	5	Cranial contents
T1	Coronal	10	Cranial contents
T1-gadolinium	Coronal	5	CT target + margin
T1-gadolinium	Axial	5	CT target + margin

images, digitized autoradiographs, and the keyboard. Problems associated with attempting to directly use these various kinds of images include (a) each manufacturer has their own method for formatting and storing image data, and (b) each image source reflects its own set of geometry, resolution, image dimensionality, calibration, and dynamic range parameters. One way to make the use of these different imaging systems rather straightforward is to store and use all images in a standard image format. This format must include a geometric description of the orientation and scale of the image with respect to the standard coordinate system of the imaging device, image format, pixel size, number of significant bits per pixel, and other details. Other relevant issues are addressed below: the bookkeeping required to maintain these image sets uniquely (the "dataset" concept) is summarized in Sect. 4.2.5.1, and dataset ("image") registration is discussed in Sect. 4.2.8.

4.2.4 2-D Analog: Contours

In most 2-D treatment planning, contours defined on individual axial slices (or planes) are the only anatomic objects which are used. For non-CT planning, one or more manual contours of the external surface of the patient will be obtained (using solder wire, plaster casts, a pin contouring device, or other manual contouring methods), and these contours are digitized into the computer using a sonic

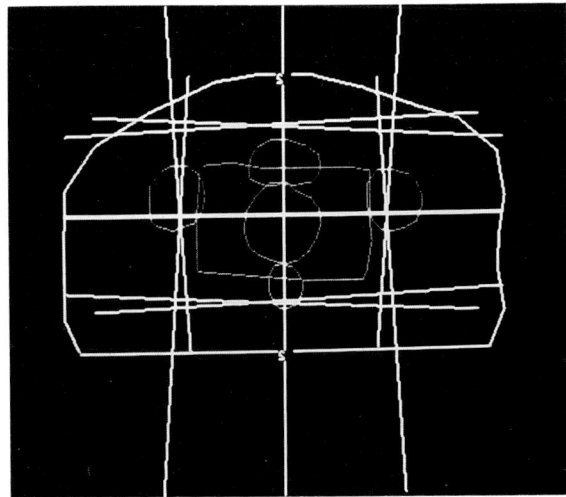

Fig. 4.2. Manual contour display for prostate case. Contours for target, bladder, rectum, and femurs shown in addition to a planned 4 field box beam arrangement

or electromagnetic digitizer. Several more contours (representing other anatomic structures) may also be digitized for each of these slices. These contours are typically obtained by protection from orthogonal radiographs. The target volume (or contour, in this case) will sometimes be drawn, based on film or other information. These contours are then used as the basis for the development of the plan, for dose calculations, and for analysis of the plan. The shape of the contours, the external surface of the patient, and/or the dose distribution on any particular slice will typically be unrelated to the information on the next slice. Figure 4.2 illustrates this kind of anatomic description, as used for treatment of the prostate.

4.2.5 3-D Structures, Surfaces, and Contours

In contrast to the simple 2-D approach, a 3-D approach to anatomic representation of the patient involves a great deal more information than just a few more contours and slices. Simple contours are replaced by fully three-dimensional objects. It must be possible to derive and/or define the 3-D shape and other features of each object in a very efficient and useful way, since the creation of these abstractions of the original data is one of the more time-consuming parts of the treatment planning process. As an example of one way to handle the complexity associated with this issue, the anatomic objects and terminology used in the treatment planning system developed at the University of Michigan will be used so that concrete examples of the important issues may be discussed. Note that there are clearly other ways to organize an approach to the 3-D description of the patient, but that the problems or issues highlighted by this organization must be dealt with by any treatment planning system.

A number of constraints exist which directly impact how the anatomic data should be handled inside a computerized planning system:

1. The data obtained from different imaging studies (CT, MRI, etc.), as well as other sources of anatomic information, should all be coalesced into a combined, self-consistent set of data upon which treatment planning can be based.
2. All manually entered or derived (from images) anatomic information should be defined with specific 3-D coordinate information, so it can be used for 3-D treatment design, dose calculations, and plan evaluation.
3. The planning system will have to make quantitative use of image data for region analysis and

segmentation, and for inhomogeneity corrections for dose calculations.

4. Sets of data which are obtained in a self-consistent way (for example, a CT scan set) must be maintained in a way that takes advantage of the expected consistency of the data contained in the dataset.

4.2.5.1 Datasets

The bookkeeping associated with the use of multiple sources of geometric information can be based on the "dataset" concept. A dataset is defined to encompass all the data obtained from one self-consistent procedure, for example, one CT scan set. In this example, all CT data from a single scan session are tied to a single 3-D coordinate system, the "dataset reference system." This coordinate system is then related to the "treatment reference system," the coordinate system in which treatments occur, by a transformation, as illustrated in Fig. 4.3. The 3-D transformation which is used may need to describe rotation, translation, scaling, and perspective transforms. A single 4 × 4 matrix is used to describe all four factors (HALL 1987). Using this kind of dataset representation, all of the geometric data related to the dataset, including images, cuts, and structures, move as one unit when the dataset coordinate system is moved relative to the treatment reference system by changing the dataset to treatment reference system transform. This feature can be used, for example, to realign a CT dataset whose x-y (axial plane) origin is nominally in the center of the field of view to the actual treatment setup origin defined by an initial localization study. Since each different

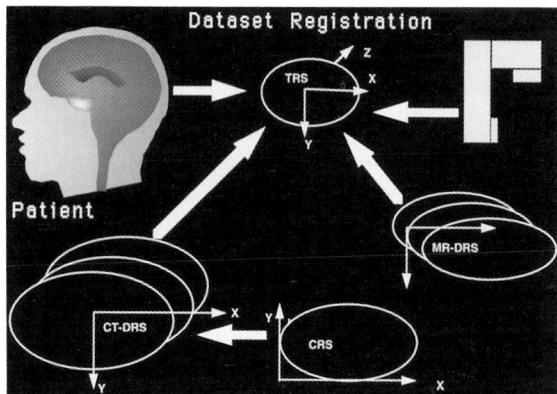

Fig. 4.3. Different serial slices defined within a cut reference system (*CRS*) are geometrically related to their respective image datasets (*CT-DRS and MR-DRS*) which are in turn registered to a common treatment reference system (*TRS*)

dataset is related to the common treatment reference system by its own transform, multiple datasets can be aligned three-dimensionally to one another through modifications of these transforms.

4.2.5.2 Cuts

Within a dataset, a number of "cuts" or slices can be defined. Each cut represents a 2-D plane (cross-section) of arbitrary orientation. In general, any image-based cross-sectional slice can be defined as a cut within a dataset. The typical dataset is a CT dataset with a number of cuts, consisting of parallel axial CT slices, and several sagittal, coronal, or oblique CT reconstructions. The University of Michigan treatment planning system, "U-MPlan" (FRAASS and McSHAN 1987; FRAASS et al. 1987b; McSHAN and FRAASS 1987; McSHAN et al. 1990), uses a separate coordinate system (the cut reference system) to describe the coordinate system associated with a set of parallel cuts. Then, the transform between the set of parallel cuts and the CT dataset (for example) can be handled by another transformation matrix. With this scheme, sagittal, coronal, and oblique reformatted CT images are referenced in the same way that the original axial slices are referenced. Several different types of cuts can be defined, including axial, sagittal, coronal, and oblique image planes, point source (radiograph-type images), and infinite source (CT scout view images, at least in one direction, can be considered to be infinite source projection images). This image type classification is used to determine whether a point defined in the plane reflects a true point or a ray projected through a volume.

A large amount of information must be tied to each cut. For each cut defined by a 2-D image, the image filename and location must be stored, as must the geometric relationship of the cut to its dataset and the overall geometric limits of any associated contour data (used for scaling and centering plots). Note that if the CT data are obtained with a tilt in the CT gantry, as is often used for head scans, then the CT slices are not aligned and parallel, so separate transforms must be used for each cut. The gray-scale window and level and the display zoom and roam status are useful to store so that the image is recalled with the same kind of display parameters each time.

4.2.5.3 Structures

In any 3-D system, there must be a 3-D "structure" defined which corresponds to each anatomic object which is to be used in planning. Examples of structures include the external surface, left lung, spinal cord, and target volume. A structure is a three-dimensionally defined object with a set of attributes describing the dataset to which the structure is attached, characteristics which should be used for graphic display of the structure, surface formation characteristics, parameters for auto-structure recognition (discussed later), and bulk density assignments. Structures are generally classified into several different types:

1. The external structure (patient surface) is critical to the dose calculation results. It is often used to determine the limit of dose calculation windows as well as the limit of densities which are considered in the system.
2. Internal structures include objects like the target volume, spinal cord, and other structures which are used for visualization purposes only. These structures do not affect dose calculations.
3. Inhomogeneity structures can be used to define a region (for example, a lung) to which a bulk density may be assigned. Density values based on CT data do not rely on this concept, but the use of structures whose density is defined as a bulk value is often useful.
4. Bolus is another type of structure which can be used to affect the patient and density description. However, boluses are typically defined during the planning process, and may be removed, or even used for just one beam in a plan.
5. For bookkeeping and display reasons, isodose surfaces are often treated as 3-D anatomic structures.

As described immediately below, the shape of the structure is typically defined using a series of contours which are then formed into a 3-D surface. The structure-defining data must include parameters which define for the system how the surface is formed from the contours, and in particular how it is completed at the top and bottom of the contours (see Sect. 4.2.5.5).

4.2.5.4 Contours

A structure can be defined geometrically by a collection of cross-sectional contours (closed or unclosed) or as a "tiled" surface. Most structures are initially defined by drawing contours of the structure in multiple axial cuts (since planning is typically based on CT information). Cross-sectional contours

are identified by cut number and structure, so that they can be meshed into 3-D surface descriptions.

Contours are defined in one of three ways, generally. For regions which are identified by a large gradient in the density values in the image (for example, the lung–tissue interface on a CT scan), relatively simple boundary tracking routines can be used to define the contour. These autotracking routines are currently effective on the external surface of the patient on lungs, and on some bones. Contours on other structures are typically drawn by the treatment planner or physician, using a mouse or joystick to draw directly over the displayed image. Specialized hardware/software systems (a large screen digitizer system (McSHAN et al. 1993), for example) have been developed to speed up the contour-drawing process, which is one of the most time-intensive parts of treatment planning. Numerous researchers are working on automatic structure recognition (IEEE Computer Society Press 1990) and 2- and 3-D image segmentation (LIFSHITZ and PIZER 1990) schemes to more automatically define the contours (and in fact 3-D surfaces) associated with anatomic objects. Finally, contours can be extracted from the 3-D surface descriptions. Once contours are drawn on a series of axial cuts and made into a surface, then that surface is "sliced" onto sagittal, coronal, or oblique cuts so that the 3-D extent of the surface is displayed on those images.

4.2.5.5 Surfaces

Once a number of serial contours are defined for a given structure (illustrated in Fig. 4.4A), a tiled 3-D "surface description" is generated by evenly distributing mesh points around each of the contours, aligning the points, and then connecting the mesh points (Fig. 4.4B). The surface descriptions are often maintained in a database, since the creation of the surface description is calculationally intensive. Once the surface description exists, any number of computer graphics techniques can be applied to change the color, lighting, transparency, texture, and perspective that are used in the display of the object(s). The 3-D surface description is also critical in many of the calculations associated with dose-calculation models. Note that if the dataset concept is employed, the entire surface description moves with any change of the dataset to which it belongs, since it is part of a "structure" which is defined for a particular dataset. This allows the definition of different surfaces (a CT-based brain surface and the analogous

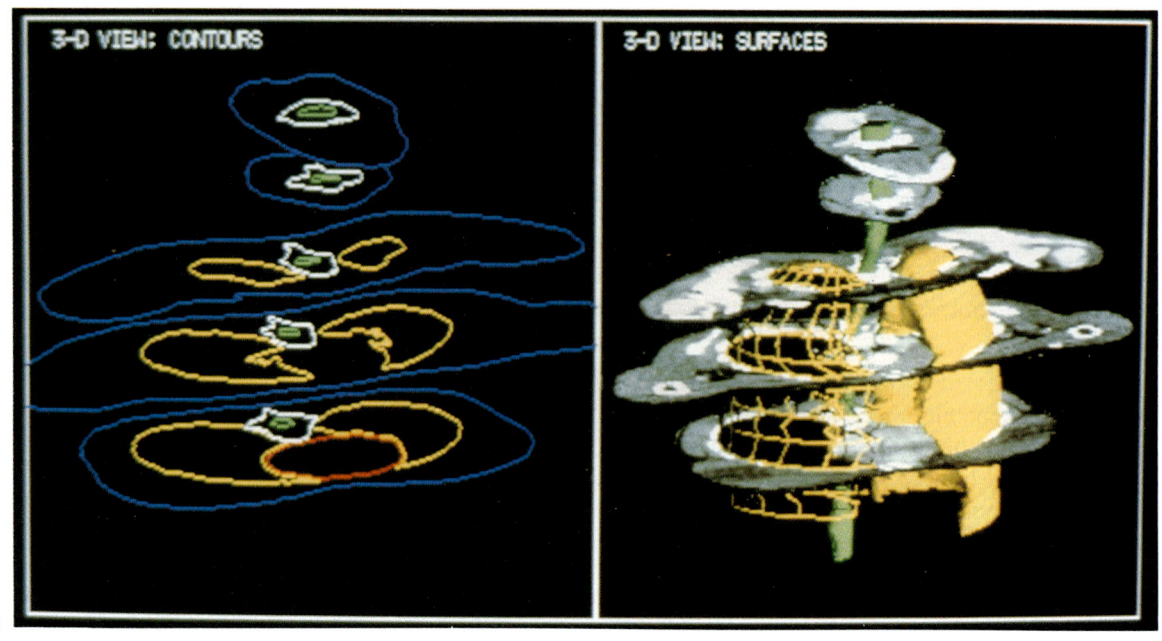

Fig. 4.4. A Axonometric view of a stack of cross-sectional contours derived from CT of the upper body. **B** Contours are used to generate surfaces which can be rendered as a mesh (ex: right lung) or as solid, shaded surfaces (left lung)

MR-based brain surface) which can be used for registration of the various datasets (see Sect. 4.2.8).

4.2.5.6 Labels, Reference Points, and Calculation Points

Labels, reference points, and calculation points are useful geometric entities which are defined in three dimensions. Labels provide a way to annotate particular points or other objects. One way to define labels is to maintain the definition of two points which are specified in 3-D: the position of the text string label (and one end of a pointer line), and the position of the other end of the pointer line. Reference points can be used to help register different datasets if they are defined in 3-D, and tied to a particular dataset. If one defines the same reference point in several datasets (for example, CT and MRI), then reference points with like names, but tied to different datasets, can be used to determine the transform between one dataset and the other. Calculation points are defined similarly to reference points, but are defined in the treatment reference system only, so that they can be used for dose calculations.

4.2.5.7 Volumetric Regions of Interest

For volumetric assessments of dose delivery to target and other defined anatomic structures, it is useful to define a volumetric region of interest (VROI) matrix. For each voxel contained in the volume to be studied, the system determines which structures the voxel is contained in. This matrix can then be used to easily perform dose-volume histograms (see Sect. 4.5.2) for each structure individually. A separate VROI matrix is often used because the conversion from surfaces to discrete volumetric elements (voxels) is somewhat time consuming, and this calculation need be done only once for each patient when a separate VROI matrix is used.

4.2.6 Target Volumes

One of the most critical parts of the input required for treatment planning is the delineation of the target for the radiation. Many groups and organizations have spent a significant amount of effort trying to standardize the semantics associated with the definition of the "tumor" and the "target." Rather than repeat some of these very detailed and thoughtful discussions of this subject [see for example ICRU 29 (ICRU 1987), the more recent revision by the ICRU (ICRU 1992), and the recent work by the NCI Cooperative Photon Treatment Planning Working Group (URIE et al. 1991)], the concepts from those reports which appear to be the most used in the course of routine 3-D planning will be discussed.

One of the most used concepts is typically called the *tumor volume*. In practice, the tumor volume is often outlined directly using the parts of the images which appear to be macroscopic tumor or other abnormal and suspicious-appearing tissue. The definition of this area is usually performed by the radiation oncologist, often with the assistance of a radiologist who can help determine the extent of the macroscopic tumor. For more information see Chap. 1 by Chen and Pelizzari.

The tumor volume gives an indication of the macroscopic region which contains obvious tumor, but the area which should be treated with a high dose of radiation is often significantly different than the tumor volume. This area, which the oncologist wants to treat to high dose, is typically called the *target volume*. The treatment planner attempts to generate a treatment plan which will deliver a high (and typically uniform) dose to the target volume.

However, the target volume is a complex entity. Uncertainty in the biologic margins that should be used has given rise to the definition of a *biologic target volume* (URIE et al. 1991), which incorporates the biologic uncertainties but not the more technically related parts of the uncertainty. However, the experience of many workers is that the physicians actually already incorporate several different kinds of uncertainties when they draw the initial target volume. It is quite clear, however, that explicit introduction of specific margins for patient motion and other technically related uncertainties is important. In our clinic, for example, the drawn target volume is usually in principle the biologic target volume, and a 3-D expansion of that surface is used to determine the final target volume (Fig. 4.5), where the margin for expansion is chosen by the physician to take into account motion and other setup uncertainties. In some sites, the additional margins which are needed are often patient specific (TEN HAKEN et al. 1991a; LAWRENCE et al. 1992), and are individually determined. Although various institutions use terms like tumor and target volume differently, consideration of biologic, setup, and motion uncertainties is critical if treatment planning is to be performed appropriately. See, for example, the recent work by Kutcher et al. on effects of uncertainties on conformal treatment plans (HUNT et al. 1989).

4.2.7 Normal Tissues

In treatment planning (particularly 3-D planning), normal structures are defined for two different reasons. The first just involves alignment and consistency checks of the planning process. By contouring

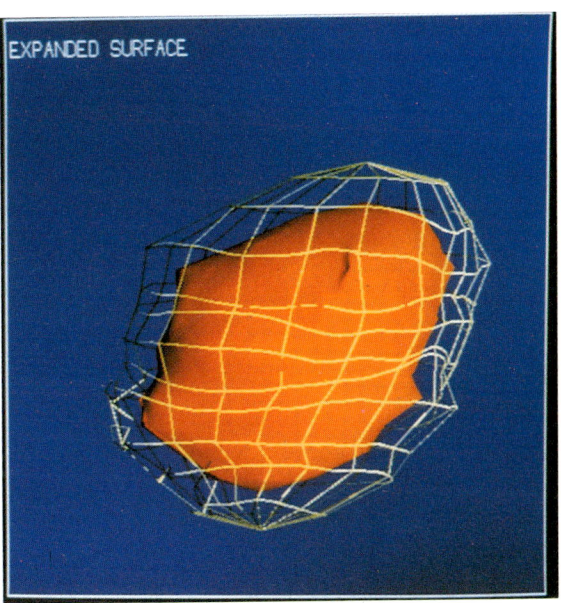

Fig. 4.5. A target volume derived from cross-sectional contours is shown as solid shaded figure. Expansion along normals for each vertex are used to generate an expanded volume described by wire-frame surface

various bony structures, or other easily identified (on both CT and radiographic images) landmarks, the accuracy with which a treatment plan is actually implemented on the patient can be checked. This process is described in more detail in Sect. 4.6. Normal structures are also defined so that the planner can determine how the dose distributions and beams relate to critical normal structures. These critical structures are structures which limit the dose that can be delivered to the target, since they are receiving some dose and they have a relatively low tolerance to that radiation. These descriptions of normal tissues are used to assist in geometric beam placement (particularly with the use of beam's eye view displays), and for the analysis of the dose to critical organs using dose-volume histograms.

4.2.8 Dataset Registration

Quantitative use of multimodality images, for example CT and MRI, requires the registration of the geometry of the different datasets. Given the description of the various anatomic constructs described above, the "transfer" of the MRI-defined target volume information to a CT dataset is straightforward:

1. Establish a separate dataset for each kind of data (for example, CT and MRI)
2. Define the relevant structures, images, cuts,

contours, and reference points which are associated with each dataset

3. Register one dataset (typically the MR dataset) with the base dataset (CT, typically)

Once the geometric transform between the two datasets is known, then any information which is contained in one dataset can be transferred or viewed with respect to images or other information from the second dataset. This registration problem is most often associated with the use of MRI (along with CT). However, other situations also require registration:

1. Use of a manual contour with CT data
2. Use of simulator films with a CT plan (the simulator dataset must be registered with the CT dataset)
3. Analysis of portal images, since each day's portal images (at least) are in reality a different dataset which must be registered with the CT plan
4. Alignment of the CT with the localization simulation data, or alignment of the CT dataset with the position of the patient in the plan verification procedure performed on the simulator
5. Use of presurgery, or diagnostic images of any type (CT, MRI, etc.)

Dataset registration, often called image registration, has been performed in numerous ways. As summarized by KESSLER et al. (1991), points, contours, surfaces, and volumes can all be used to determine when the two sets of data have been registered. In this discussion, however, we will concentrate on just a few of the most common methods of dataset registration.

One of the most powerful methods, within certain constraints, is the use of point markers on the various datasets. If markers (which can be imaged by all imaging modalities of interest) can be placed on the same points of the patient's surface during the acquisition of each imaging dataset, then it is relatively straightforward to determine the transform which aligns all the markers. By defining the location of reference points to be the same as the markers as imaged in each dataset, then the computer can calculate a least squares fit to the tie lines connecting the reference points in the two datasets, thereby calculating the most likely transform between the two datasets (FRAASS et al. 1987a). This system has worked well for MRI and CT, for SPECT and CT (KORAL et al. 1990), and for PET, MRI, and CT (THORNTON et al. 1991a) and is really the basis for the alignment of CT scans with the basic coordinate

system which is established in the simulator using lateral and anteroposterior laser marks on the patient, since in the CT the laser marks are typically marked off with radiopaque markers. This method can also be used if there are easily identifiable internal anatomic points which can be identified in each dataset. However, the major problem with this method is that points are very difficult to unambiguously identify in three dimensions using just 2-D images, since there is always some volume-averaging uncertainty or artifact which is apparent.

A very sophisticated technique which has been successfully applied for registration of CT and MR images of the head has been named the "head and the hat" technique. This technique, which was developed by KESSLER et al. (1991) and CHEN and PELIZZARI (1989), matches surface models generated from the two datasets. Typically using the inner table of the skull as imaged on CT and MRI, the model calculates a least squares fit which minimizes the separation between the two surfaces. Once the surfaces are defined, this method can run automatically. However, it has a number of disadvantages. This method has so far only been used in the head. It is also sensitive to the regions in which the surface (of one of the two datasets) is not defined since the image scan set was not done throughout the entire region.

The final registration technique discussed here has been the registration technique which has been used on the great majority of the more than 125 clinical MRI treatment planning cases in our clinic in the last several years. This interactive technique borrows the use of surfaces, as used in the head and the hat technique, and adds to it the interactivity and intuitive feeling which is associated with a more interactive method which "drags" one dataset around in space with respect to the other. One begins by generating a display which includes axial, sagittal, and coronal cuts through the region of interest, as well as a 3-D view (if desired) (Fig. 4.6A). The contour shown on each panel of the display is the contour cut from the MR surface along the plane of the cut. Final agreement is reached by dragging (rotation is also possible) the MR dataset in each plane. Finally, to perform a last check on the reasonability of the registration, the 3-D surfaces of both the CT and MRI (in this case) structures are displayed in solid surface graphics views (Fig. 4.6B). These solid images show whether large areas of one or the other surface are entirely outside the other (bad registration); if the two surfaces are intermingled, it shows the correctness of the registration. Note that with all the semi-interactive methods of

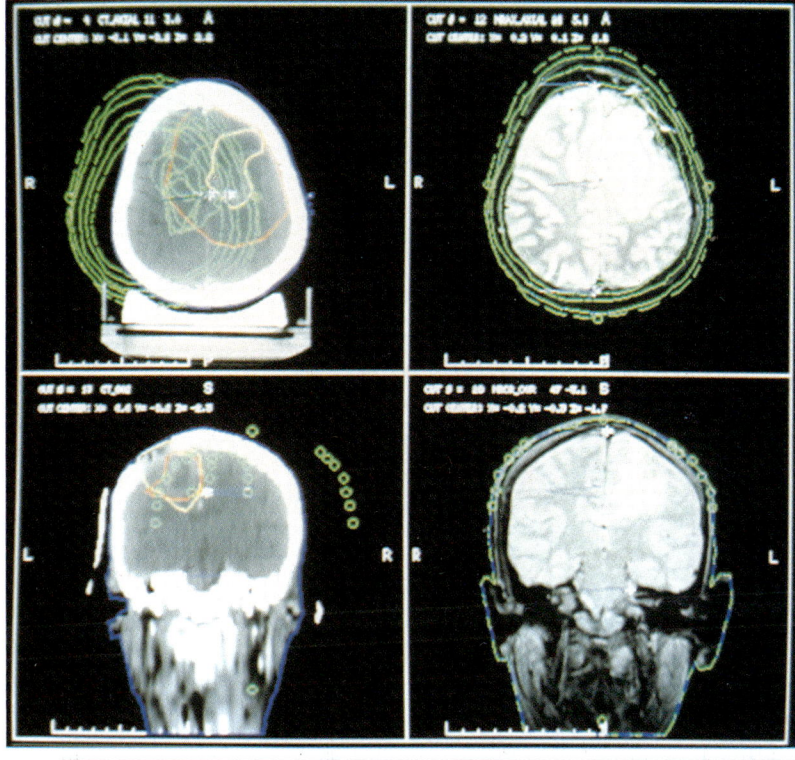

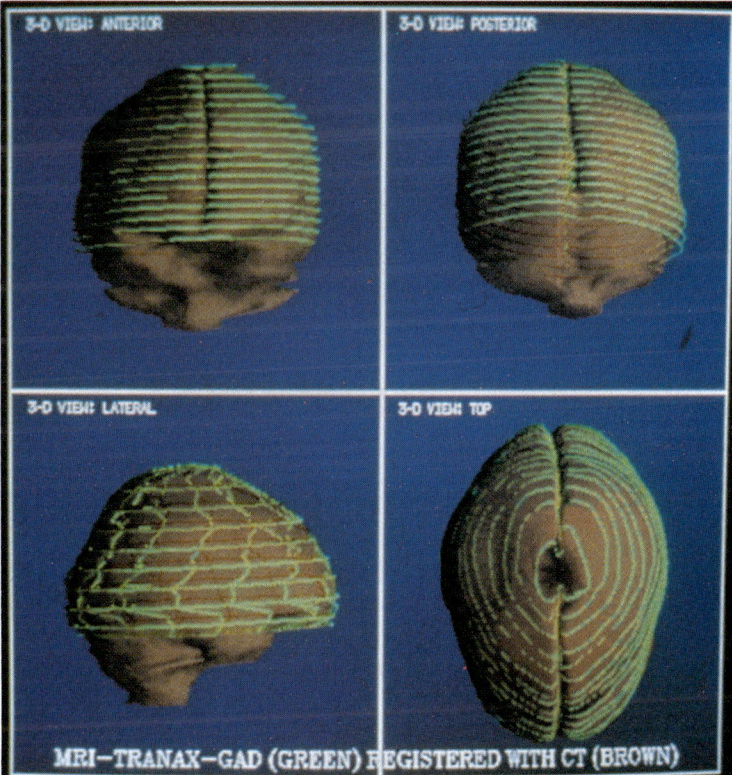

Fig. 4.6. A 4-panel view of CT and MR images, with MR contours overlaid. Contours can be dragged and rotated to match CT image. **B** 4-orthogonal views of CT derivered surface (*brown*) displayed with MR derived surface (*green wire-frame*) after registration

registration, there is no hard number which can be used as a measure of the precision of the registration. If there were such a number, one could minimize the value of that number automatically, thereby making the registration method noninteractive.

4.2.9 Use of Density Information

Another part of the three-dimensional description of the patient anatomy which is needed for treatment planning is a 3-D matrix which contains the CT-derived electron densities of every voxel contained within the area of interest. This information is obtained from CT values, converted to Hounsfield units using CT-scanner-specific conversion routines, and then converted to electron density values using published or empirically determined conversion tables (MUSTAFA and JACKSON 1983). However, many different schemes for handling this data can be considered, since the amount of data required is large. CT scans are typically obtained with 512×512 pixels per cut, which is on the order of 1 mm resolution (body scans). If one wanted to create a density matrix with a resolution of 1 mm, one would have to include about $512 \times 512 \times 512$ voxels, which is more than 100 megabytes of memory per matrix: too large a number to be used routinely for every patient. Since the original CT slices typically have a thickness of 0.5–1cm, and density-related scatter effects vary somewhat slowly with distance, a reasonable compromise may be to use a density file based on 128×128 density values for each CT slice obtained (for photon calculations, at least).

4.2.10 Anatomic Display

The display of the three-dimensional anatomic model which is used to describe the patient is a critical part of the treatment planning process, since it is through this mechanism that the three dimensionality of the patient is transmitted to the physician and planner so that they can create a plan which optimally treats the disease. With 2-D treatment planning, the display problem was essentially straightforward: one axial cut was displayed, if contours were drawn on that cut, they were displayed, and if CT was used, the CT gray-scale image was displayed under all the other graphics. Now, with a multitude of different kinds of anatomic and other information to display, the challenge is to create displays which present the important information in ways that help the planners assimilate the information. A number of general techniques which have proven useful are described below.

4.2.10.1 2-D Display of Cuts and Images

The most commonly used type of display is the planar display of a single slice of CT or other image information. Gray-scale images are displayed overlaid with computer graphics which show contours, beams, doses, and other relevant information. In order to display the important parts of the image, it is necessary to use variations of the gray-scale window and level to modify the displayed gray-scale image. CT images contain about 12 bits of gray-level information, while typical display systems display at most 8 bits, and sometimes as little as 5 bits (when an 8-bit workstation display is used to display gray-scale images overlaid with several bits of graphics) of gray-level differentiation (KESSLER et al. 1992). Standard labels for each 2-D cut typically should include the cut name and number and the location of the cut. In a fully 3-D system, this can be difficult, since there are generated oblique images and cuts which need to be referenced. One solution is to use the coordinates of the center point in the image, in the treatment reference system (TRS). In this way, there is at least one unambiguously defined point in the reference coordinate system for each cut. The old-fashioned use of "z" to describe the third dimension location of a slice is not useful when one has multiple datasets with multiple origins which do not coincide, and Z-axes which are not necessarily parallel.

4.2.10.2 Multiple Windows

With a single window display of a single planar image, it is difficult to get a sense of the three dimensionality of the anatomy. Although movie loop displays have been used in this context (STERLING et al. 1971), another widely used technique is a multiple window display that allows the planner to view a number of different kinds of 2-D images simultaneously. Multiple-panel displays, as illustrated in Fig. 4.6, allow the planner to view the three dimensionality of the problem while continuing to use the planar gray-scale images which are still the basis for much of the treatment planning process. The use of axial, sagittal, coronal, and oblique cuts and images is a very useful and intuitive way to interact with anatomy, beams, and other objects used in treatment planning. This multiple window system also allows the use of 3-D views, beam's eye view displays, and other more three-dimensional displays to be used in conjuction with the gray-scale images. If different kinds of images (CT and MRI, for example) are to be used in different windows on the same display, then the display system must be able to display different gray-scale window/level settings for each of the different windows on the display separately.

4.2.10.3 3-D Solid Surface Graphics

Three-dimensional computer graphics are used to display the 3-D surfaces with a three-dimensional perspective. A number of different surface renderings can be used to display the selected structures, including contours only, wire frames, solid or transparent surfaces, and variable lighting and/or textures, as illustrated in Figs. 4.4 and 4.5. Front, back, or both faces of the surface can be displayed. Cuts may be used as clipping planes in order to slice into the 3-D display. A very important feature of these kinds of displays is the ease with which all of the various display parameters may be changed, and the display regenerated. A second capability which greatly enhances the usefulness of the display is the ability to add motion. Although precalculated motion sequences can be very valuable, the most useful technique is the ability to have the planner or physician control the perspective of the display. With the new powerful workstations which are becoming commercially available, it is now possible to generate these 3-D views and rotate the objects, under user control, more or less in real-time.

4.2.10.4 Axonometric Displays and 3-D Views

As with axial cuts, orthogonal and oblique cuts can be examined individually in order to assess dose distributions. However, because of the need to coalesce the individual results into a mental three-dimensional picture of the overall distribution, other techniques may be preferable. A very useful technique which aids in the assimilation of the available gray-scale image information is to produce an axonometric display of multiple 2-D images (cuts), as displayed in Fig. 4.7. Contour and solid surface displays can be added to these images in order to try to display all of the data involved in the situation.

4.3 Beam Arrangements and Field Shaping

4.3.1 Interactive Planning

Once the anatomy and target are defined, the next stage in planning is typically to determine the technique to be used to treat the patient. In 2-D planning,

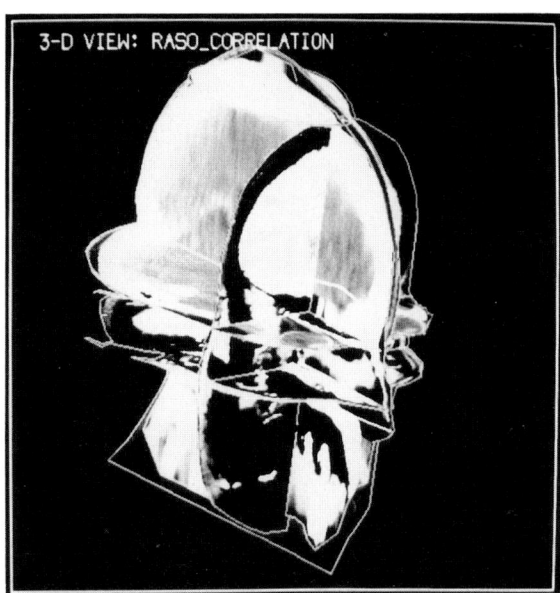

Fig. 4.7. Axonometric view, multiple datasets and images, with graphics overlaid

this is relatively easy to accomplish, since one must pick a 2-D slice to look at while one determines the number of beams, their field sizes and gantry angles, and their isocenters. Wedges and other modifiers can also be applied to improve the dose distribution. Simple interactive graphics can be used to overlay the beam axis and edges over the CT image (see Fig. 4.2).

In 3-D treatment planning, a number of limitations to the above situation are removed. The planner should have full 3-D positioning capability for any beam, rather than being limited to the axial plane only. The anatomy is defined in 3-D, and the planner must be able to pick and understand the direction, location, and relationship of the beam to the anatomy, in all dimensions. Once planners can see clearly what needs to be changed in order to improve the plan, they must also have the tools to accomplish the changes that are so clearly identified by the 3-D displays. The correct beam projections are needed on all kinds of displays which are used, including axial, sagittal, coronal, and oblique cuts (from the different imaging modalities), 3-D perspective views, and beam's eye view displays. In order to accurately project all beam and block edges, the beam display algorithm must be based on a divergent solid surface description of the beam and block edges which is then intersected with any 3-D object, plane, or surface which is contained in the display. The effect which appears to be most surprising to many people is the actual shape of curved block edges when projected on cuts some distance from the cut which

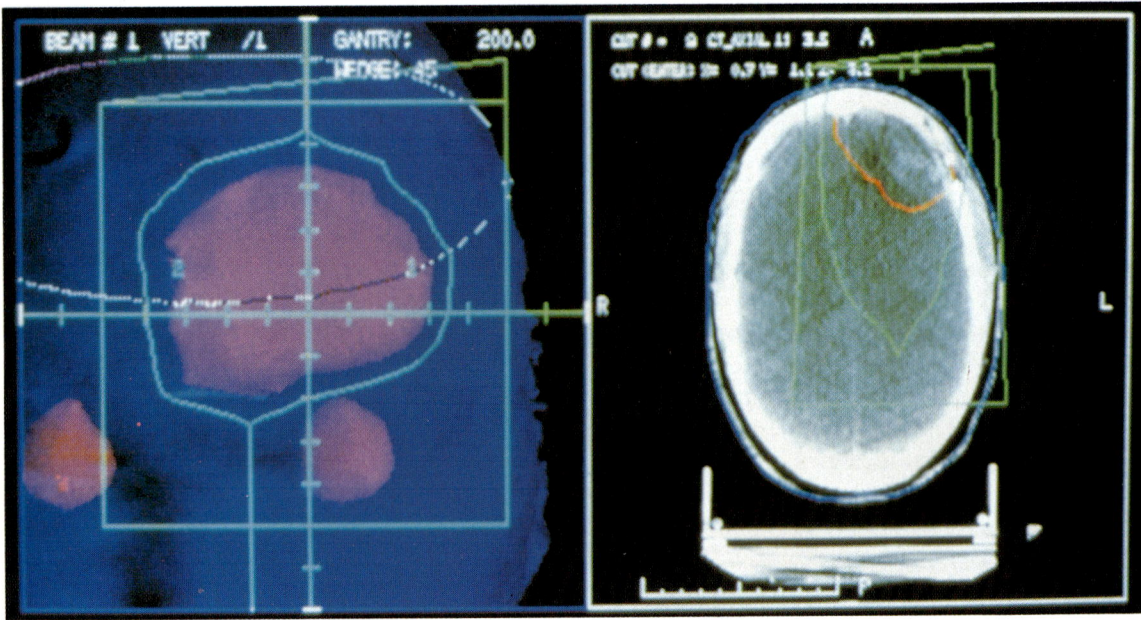

Fig. 4.8. *Left panel* shows beam's eye view of shaped field. The *right panel* shows the intersection of the block aperture on an off-axis CT slice

contains the beam central axis. As shown in Fig. 4.8, this can lead to unexpected block shapes when divergence and projection are correctly calculated and displayed.

Multiple window displays are very useful when defining the beam technique, as a full appreciation of the 3-D situation is easily obtained while at the same time retaining the familiarity and usefulness of 2-D gray-scale images. For this display to be useful, however, one must be able to move the beam(s) in any or all windows, and all windows must be constantly updated. This allows CT and other images to be used for beam placement determinations in three dimensions by moving beams using displays of axial, sagittal, and coronal cuts and the BEV display for the active beam at the same time.

The planner must be able to move beams in 2-D, in 3-D, and/or on oblique planes, using a selection of different movement controls. However, one should not need to know which particular combination of gantry, table, and collimator angle rotations will cause a beam to move to the correct angle to make a wedged pair on an oblique plane. Figure 4.9 illustrates the generation of a simple wedged pair beam arrangement on a sagittal-oblique cut, in which the planner can move the beams using the mouse just like for axial planning, but without the requirement to figure out which actual motions are needed, since the system calculates the appropriate gantry and table motions which are required.

4.3.2 Beam's Eye View

The beam's eye view (BEV) display is one of the most powerful tools which is available to treatment planners with 3-D treatment planning capabilities. This display shows the anatomy of the patient as viewed from the point of view of the source of the radiation. Divergence of the radiation from the source is taken into account, so that the relationship of the collimator and/or block shape with respect to the anatomy is displayed correctly for the divergent beam. As illustrated in Fig. 4.10, this view is accomplished by calculating a projection through the patient data along the divergent rays of radiation which start at the source of the radiation. Typically the anatomic information is displayed as contours or solid surface graphics in a BEV display (Fig. 4.11). If a projection is made through the CT data, then the image which is obtained is called a digital reconstructed radiograph (DRR) (GOITEIN et al. 1983) as illustrated in Fig. 4.12. Note that the geometry of the BEV and DRR is identical to the geometry of the standard radiographs which have been used for many years for field shaping. The only difference is that the BEV and DRR displays are calculated from the computerized patient data which are used for treatment planning, while the radiograph is a real projection image obtained with film and the patient.

Although the BEV concept was introduced many years ago (McSHAN et al. 1979), it is only relatively

Fig. 4.9. A Non-Axial wedge beam pair relative to patient anatomy (derived from axial CTs) **B** Wedge pair plan on an oblique CT slice, illustrating planning in oblique planes

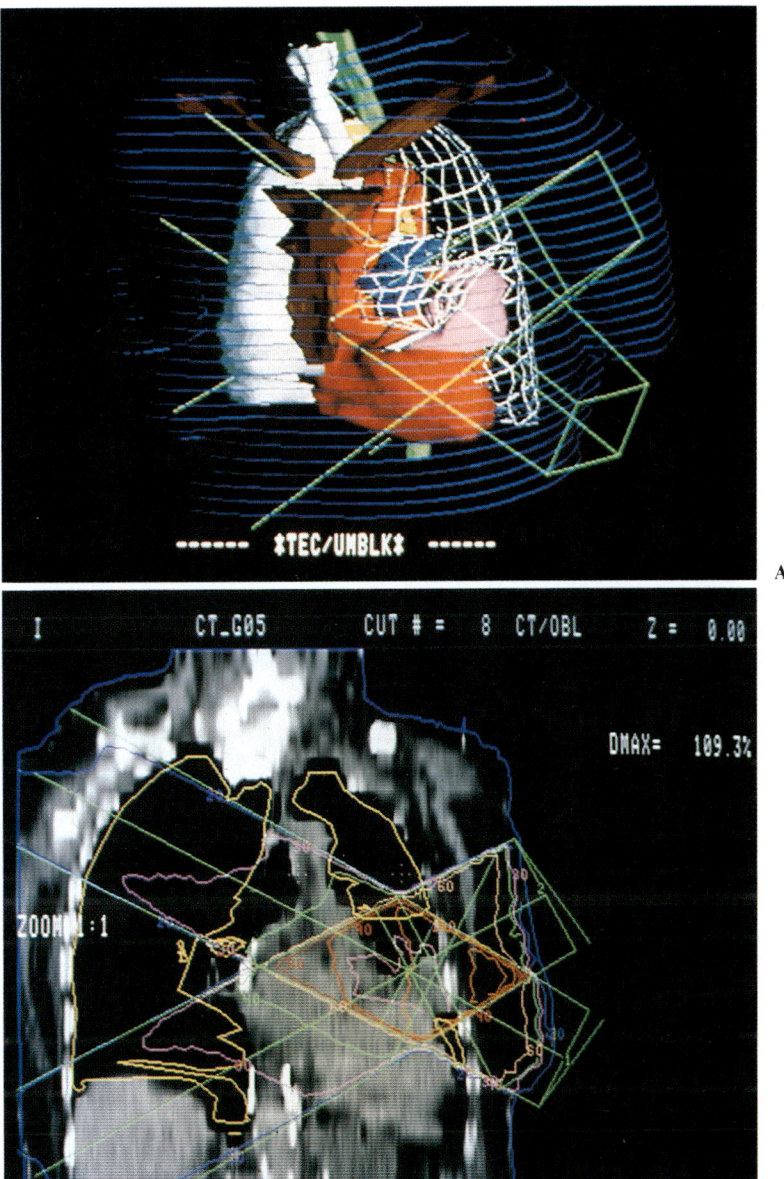

recently that it has been implemented for routine clinical use in a number of planning systems (GOITEIN et al. 1983; MOHAN et al. 1988; FRAASS et al. 1987b; MCSHAN et al. 1990; CHEN 1983; MCSHAN and GLICKSMAN 1984). Display of the patient anatomy with the BEV gives the planner a very effective tool for determining the best location and shape of the radiation field so that the target volume is treated to high dose, while minimizing dose delivered to critical normal structures. The design of field shape, and particularly shielding blocks and multileaf collimator shapes, is easily performed using the BEV

information. Finally, the anatomy displayed on the BEV can be used for detailed verification of portal and simulator images as one tries to make sure that the planned treatment plan is actually delivered correctly.

4.3.3 Block Design

The implementation of the routine use of the BEV display leads directly to the need to design field shapes with that display. Field shapes which are drawn on the simulator typically do not benefit from

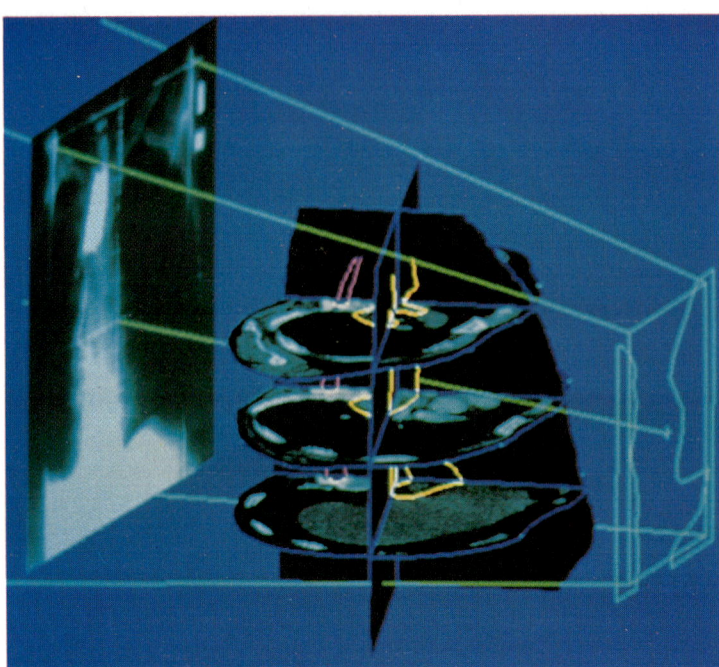

Fig. 4.10. Illustration of divergent beam geometry relative to axial CT slices and to a transmission image

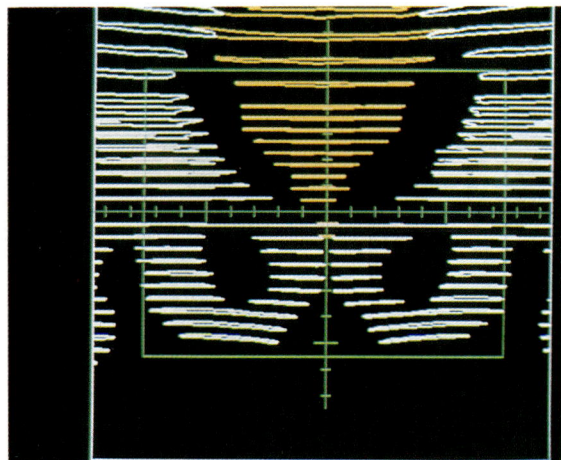

Fig. 4.11. BEV display showing projected outlines of pelvic anatomy

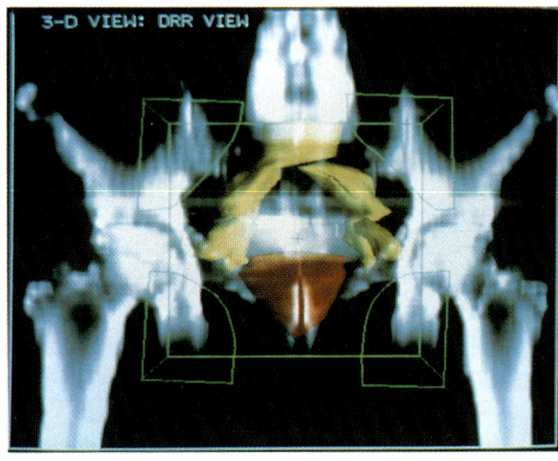

Fig. 4.12. Digitally reconstructed radiograph for same case shown in Fig. 11. Beam and blocks are shown as *green lines*, and nodal target volumes (*yellow*) and other anatomical structures are also projected on top of DRR

a detailed study of the CT-based target and normal tissue volumes, so when such blocks are viewed with respect to the defined anatomy, areas which are inappropriately blocked, or where not enough normal tissue is shielded, are very obvious. Once the decision is made to base treatments on CT and MRI-based target volumes, most initial field shaping will be performed using the BEV display.

Field shape design is performed in several ways. Blocks drawn elsewhere (on the simulator) can be digitized off the simulator films and then displayed. A mouse or joystick can be used to draw block shapes directly on top of the BEV display. This allows the block designer to use all of the information displayed in the BEV to help in the design. The use of a cursor of specified radius is very helpful in assisting the planner to draw blocks with a particular margin around certain structures. Any block design method will include at least the ability to use the mouse/joystick to edit block shapes on the BEV display.

A third method for field shape which is used for both blocks and multileaf collimator shaping is automatically designed blocks ("autoblocks") (MCSHAN et al. 1990). With this method, a structure or structures are selected (typically the target volume) to be the basis for the field shape. A desired margin is selected, and the computer then designs field shaping which leaves the specified margin about the target structure (Fig. 4.13). This is accomplished by projecting all of the selected structures to the isocentric BEV plane (a plane perpendicular to the central axis of the beam, and going through the isocenter), and then using the surface normals to the composite surface to define a shape for the blocks. Fields shaped in this fashion are often called "conformal," in that the field shape is specifically designed to conform to the shape of the target. The use of these autoblocks requires some additional effort, however:

1. Many of these blocks will need some editing in order to properly account for situations where normal structures are very near to the defined target volume, and there may be a need for some changing of the block margin in some particular parts of the field.
2. Patient setup errors, organ motion (breathing, for example), and other uncertainties may lead to the need for nonisotropic margins for many target volumes.
3. Most importantly, the geometric block margin is not the critical issue, since the most important thing is to correctly set the margin of the high-dose region with respect to the target. Therefore, block margins which give an appropriate dose margin about the target when a four-field axial technique is used will not be appropriate when a six-field noncoplanar technique is used.

Therefore, the block margins used with autoblock are typically used as a guide, and a number of different kinds of editing of the block shapes will occur as the dose distribution is optimized. When the field shaping is being performed for a multileaf collimator (MLC) field, additional considerations also come into play. For example, the autoshape algorithm must include the effects of the special behavior of the MLC penumbra and must be able to rotate the field so the leaf motion is oriented correctly toward the target shape (Fig. 4.14).

The block or MLC shape must then be transferred from the planning system to the machine or block cutter. Since the MLC shape will be used by a the computer-control system of the machine, MLC parameters will likely be sent directly to the machine. A number of different architectures for use with computer-controlled machines are presently in the developmental stage (MOHAN et al. 1990; FRAASS et al. 1994a; MCSHAN et al. 1994; KESSLER et al. 1994). Computer-controlled block cutters have been studied and used routinely in the clinic (WEEKS et al. 1989), but most block fabrication is done using manual styrofoam cutters. One way to make computer-generated blocks is to output a BEV plot, magnified to an appropriate source-film distance, and to use this plot as the basis for the block cutting,

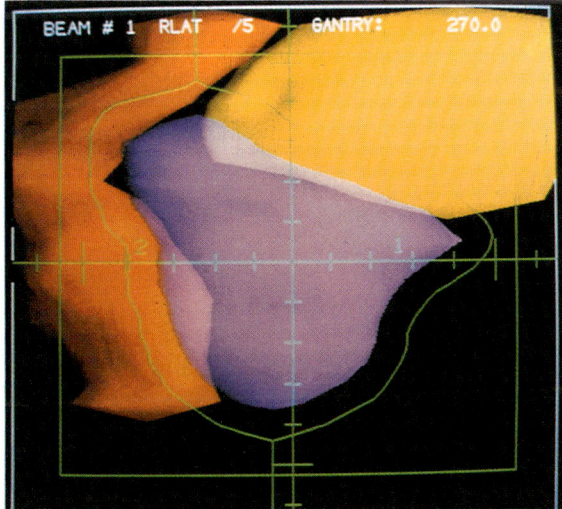

Fig. 4.13. Conformally shaped block (*green lines*) for prostate target volume (*solid purple*). Bladder (*transparent yellow*) and rectum (*transparent brown*) shown

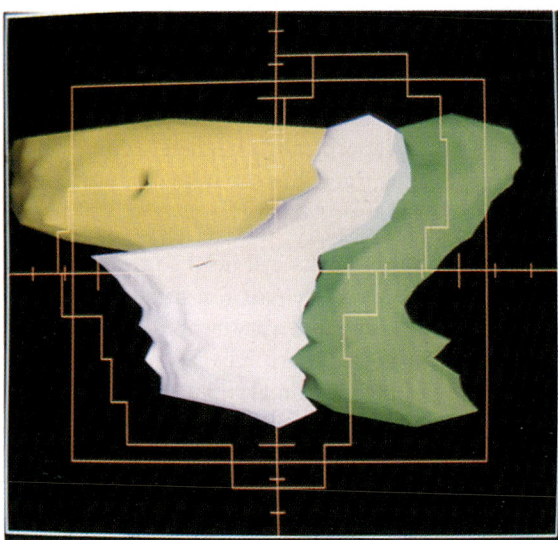

Fig. 4.14. Multi-leaf collimation with conformal shapping for prostate target volume

rather than the film (Fraass 1990). The block shapes can also be transferred from the BEV plots to simulator films.

4.4 Dose Calculations and Treatment Planning

After the patient anatomy and beam arrangement have been determined, the next step in the process is the calculation of the dose distribution. Discussion of dose-calculation algorithms is contained in another chapter of this text, and is not mentioned further here. However, there are many operational, organizational, and other facets of the use of any algorithm for treatment planning which are described in the following section.

One of the most important changes that the 3-D planning system has caused in the routine use of the treatment planning computer has been the change in the emphasis of the work of the treatment planners from "dose calculation and display" to "treatment planning." Routine 2-D treatment planning often consists of performing dose calculations for beam arrangements which have been determined in the simulator, or by the kind of tumor or site, before any information is entered into the planning system. The only kind of optimization and "planning" which may occur is the investigation of beam weights, wedges, and minor field size changes.

Three-dimensional planning system capabilities, and the ability to actually treat the patient with 3-D treatment techniques, open another dimension of possibilities for beam positions and shapes. Our

collective lack of experience with the possibilities, constraints, and usefulness of considering various 3-D plans forces us to evaluate several different plans for most cases, rather than knowing from the outset that only one type of beam arrangement is worth investigating. In order to support the iterative and comparative treatment planning which is often used with a 3-D system, one must have the ability to (a) work on, analyze, and compare many different beam arrangements concurrently, (b) easily iterate through different plan techniques, and make incremental improvements in specific parts of a complex plan, (c) have access to all the dose, technique, and anatomic display features from any point in the system, and (d) quantitatively compare the plan evaluation and analysis results for a number of different plans.

These kinds of requirements make the overall organization of a 3-D planning system more complex in organization than a 2-D system, and make constant access to many of the features of the system a virtual requirement. Since all data are accessible and can be changed from many places in the system, a sophisticated logic design is important to maintain the validity of specific calculations and to determine when a particular change has made the dose calculation results invalid. The logic must decide whether new calculations are required, or whether the doses from individual beams just need to be resumed into a new total dose distribution, or whether the doses are valid as currently maintained inside the system. This kind of logic is necessary to minimize the amount of recalculation that occurs in the system, since the dose-calculation time is an important part (perhaps a limiting part) of the system. Discussion of automated or semiautomated plan optimization techniques, including compensators and other dose optimization, and the use of biologic modeling (see Sect. 5) for plan scoring and optimization, is included in Chap. 11 by Brahme.

The number of calculational options available to a 3-D system is large, so the way the calculation parameters are set up is important. Since accurate 3-D calculations still are time consuming, especially when performed on a large grid or volume, it may be very useful to perform some of the iterative calculations (which are used to optimize the plan) on just a few 2-D cuts, rather than through the entire volume. The use of an axial, sagittal, and coronal cut through the target volume can give a very good impression of the three-dimensional dose distribution, and it takes much less calculational time. Another method used to speed up calculations is to use a large grid size. Typically, while photon dose calculations are performed on grid with spacing of approximately 0.5 cm (0.2 cm may be more typical for electrons and/or MLC-based plans), optimizing the plan on a much larger grid (1.0–1.5 cm) may be appropriate. If dose-volume histograms are the dose evaluation tool which will be used to compare and/or rate plans, then dose-calculation points which are specifically [or randomly (NIEMIERKO and GOITEIN 1990)] assigned inside the specific structures which are being histogrammed may allow very fast iterations. Dose-volume histogram results are very insensitive to grid-size type effects (DRYZMALA et al. 1991).

4.5 Dose Display and Plan Evaluation Techniques

The evaluation of the dose distribution is one of the areas where 3-D planning has forced the largest amount of change. For the standard 2-D planning system, dose display and plan evaluation were very straightforward: isodose lines were superimposed on whatever kind of anatomic display was available, and the results were then plotted out. Since the field shapes and sizes were typically planned on the simulator, the only thing to monitor on the dose distribution was whether the dose inside the target was fairly uniform. If not, then a wedge or beam weight would be changed to make the dose uniform. Accounting for tolerance of normal tissues would typically consist in checking the dose to the normal tissues which had been identified in the contour.

With 3-D planning, this simple approach to plan evaluation is unworkable for a number of reasons. Dose display is a much more complicated operation, since there can be as many as 150 CT and MR cuts which might be used for display of doses. In addition, one must provide the ability to display, and to inspect in detail, the 3-D dose distribution with respect to the 3-D anatomy and beam arrangements which have been used. Secondly, the factors which should be used to compare one plan against another, in order to pick the better of the two plans, include the dose distribution delivered to each normal tissue, as well as the uniformity and coverage of the target volume.

A very effective method for summarizing the dose which is delivered to each organ, the dose-volume histogram (DVH), has quickly become an essential part of the evaluation of virtually all 3-D treatment plans. However, since many more factors are used for the comparison, the relative importance of each of the factors (for example, the importance of each organ's DVH) must be considered. The use of normal tissue complication probabilities (NTCP)

and tumor control probabilities (TCP) as a way to attempt to combine the biologic tolerances of various tissues to irradiation with the known dose distributions throughout the volume of the area of interest was pioneered by the NCI-funded work of several cooperative working groups. These methods may eventually describe in somewhat analytical terms which plans are acceptable or unacceptable, and which plans are better than others.

4.5.1 Dose Display Tools

Display and evaluation of the results of the dose calculations are done with both graphic and analytical techniques. The graphic methods allow the display of anatomic information and the dose distribution in several formats, including planar 2-D displays and full 3-D displays. These kinds of displays can be used for special dose distributions obtained from compositing a number of plans, or from plan subtraction, as well as for display of normal dose distributions.

4.5.1.1 2-D Dose Display Techniques

Once the dose has been calculated for any 2-D cut, the dose distribution can be displayed overlaid on the gray-scale image which corresponds to that cut (Fig. 4.15A). This essentially 2-D dose display technique is still the most often-used technique, since it lets the planner and physician to analyze, in detail, the dose distribution with respect to the anatomy which is displayed by the gray-scale image. If datasets based on MRI or other imaging modalities have been registered with the CT dataset, then dose may be displayed on a sagittal image (for example) just as easily as on an axial CT image (Fig. 4.15). The standard display mode is color-coded isodose lines overlaid on the gray-scale image. Point dose displays are useful for optimization of treatment plans, as dose is displayed at those points and is updated as the beam techniques are changed. The use of a mouse- or joystick-driven cursor to interrogate and display the dose at specific points is often used. This technique is particularly helpful for dose verification/quality assurance testing if the point readout contains 3-D coordinate readout (in a number of different coordinate reference systems, if available). A "color-wash" display, which assigns color values corresponding to dose to each pixel on the display (or every other pixel for a transparent color wash), paints the whole dose distribution in transparent bands of color on top of the image (Fig. 4.17B, C).

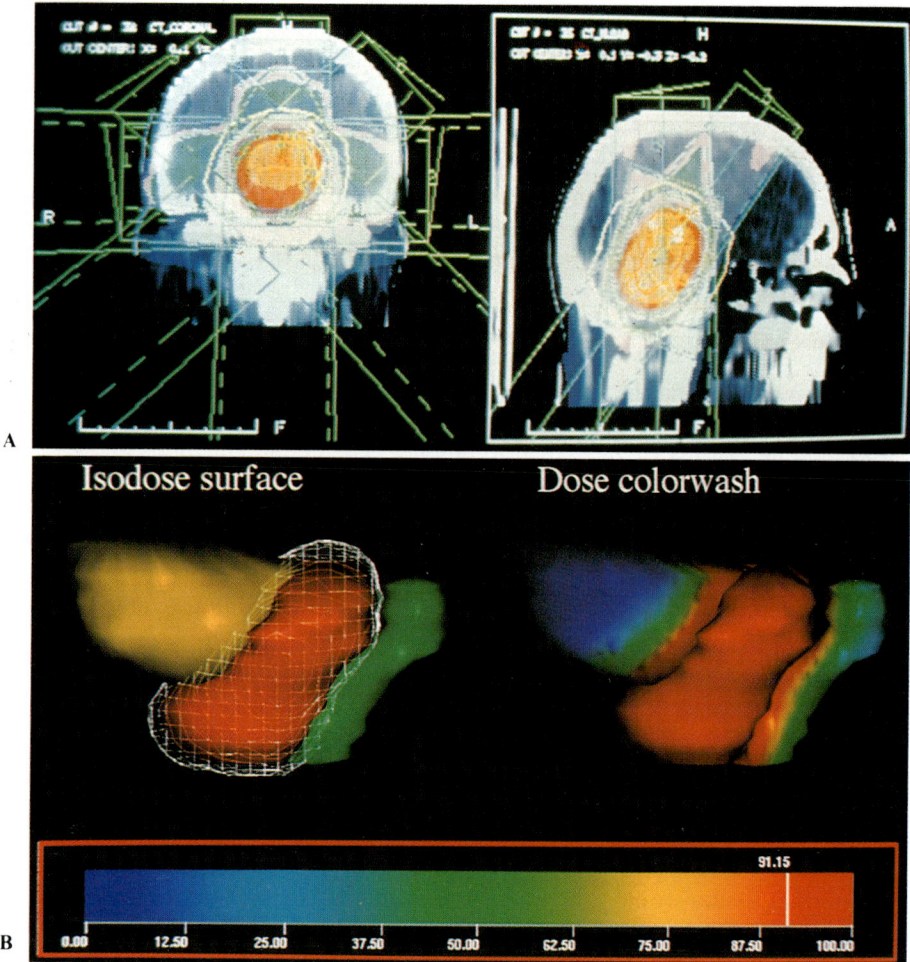

Fig. 4.15. Dose display techniques: **A** Isodose lines and colorwash on coronal and sagittal CT slices. **B** Isodose surface (wire frame) and colorwash (right panel) showing surface dose to displayed anatomy

By changing the dose to color assignment in the display look-up table (LUT), the color wash can be used to interactively scan a band of color (dose) over the image interactively.

4.5.1.2 3-D Dose Display Techniques

As with axial cuts, orthogonal and oblique cuts can be examined individually in order to assess dose distributions. However, because of the need to coalesce the individual results into a mental 3-D picture of the overall distribution, other techniques are also available. The multiframe display is often used so that multiple 2-D planar images overlaid with dose can be placed side by side. Another useful technique is to produce an axonometric display of multiple 2-D images (cuts). The user picks selected cuts to be displayed in this perspective view and can also decide to display the gray-scale images corresponding to those cuts. Since it can often be helpful to save these 3-D views to illustrate a particular issue, the orientation, perspective, scale, and other display parameters should be stored, so that each view can be recreated and/or edited as needed. Another useful capability is the ability to save the entire screen image to disk, for recovery whenever desired.

Several kinds of techniques are available which display both dose and anatomic information using 3-D solid surfaces as well as gray-scale images. If a 3-D volumetric dose grid has been calculated, then 3-D isodose surfaces can be generated. Display of these isodose surfaces, compared to solid surface displays of the target volume and critical normal structures, can be used to display the full 3-D extent of the high dose volume (Fig. 4.15B). When the target volume is shown sticking through the isodose surface, for example, one can easily see where the regions of target volume underdose are located. In

Fig. 4.16. Dose difference display colorwash for comparison of plans

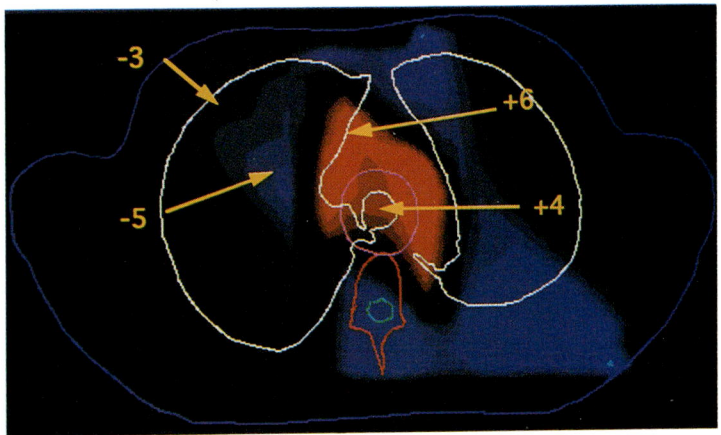

some situations, it may also be useful to display the dose on a particular object's surface by interrogating the 3-D dose grid and using bilinear interpolation to fill in dose color values instead of the shading color values.

4.5.1.3 Displays for Plan Comparisons

Several techniques for display of dosimetric plan comparisons can be useful in certain situations. Side-by-side comparisons of the doses from different plans can be displayed on a multiple window display. However, these displays quickly become fairly complicated to use for direct point-to-point comparisons. Quantitative comparisons can be performed by creating a dose difference display: this method involves the subtraction of one plan's dose distribution from the distribution of the other plan. Where the two distributions are the same, the resulting difference is 0. A color-wash display can then highlight the areas in which one distribution is hotter (higher dose) than the other, and vice versa (Fig. 4.16). Any of the 2-D and 3-D dose evaluation tools can then be applied to evaluate the difference display. This dose difference display is a special case of the more general composite plan capability which can be used for a number of different situations. In its most general form, a composite plan's function allows the planner to (a) define a collection of plans which are to be composited, (b) define the prescription dose and prescription isodose line chosen for each plan, and (c) evaluate the composite dose distribution. This feature can be used to generate the total dose actually delivered to the patient, to plan how much dose should be delivered by a boost treatment plan, or even to add brachytherapy and external beam doses.

4.5.2 Dose-Volume Histograms

The 2-D and 3-D displays of the dose distributions discussed above are very useful ways to transmit the dosimetric results of the plan, but there is also a need for a way to summarize the dose being received by each of the structures of interest. This summary is provided by the dose-volume histogram (DVH), which shows how much of the volume of each structure receives how much dose. On this one line graph, the planner can quickly see how the dose is distributed over each structure. All spatial information is lost, but a good summary of the situation is plotted. The use of DVHs for analysis of treatment planning dose distributions (NCI 1987; AUSTIN-SEYMOUR et al. 1986; DRYZMALA et al. 1991; LAWRENCE et al. 1990) has been one of the most important improvements in the technology used for treatment planning in recent years.

The most basic form of the DVH is the frequency distribution plotted in Fig. 4.17B. To make this plot, the volume of the patient is divided into voxels, as shown in Fig. 4.17A. Each voxel is flagged with the 3-D structures that it is included in. A number of dose bins are created, from zero dose up to a bin which includes the maximum dose calculated for the plan. Then, for the structure being histogrammed, the number of voxels which are part of each particular dose bin are counted. Plotting the number of voxels per bin on the vertical axis and the dose for each bin on the horizontal axis gives the frequency plot shown in Fig. 4.17B. If the volume of each voxel is known, then the scale on the vertical axis can be absolute volume, or the percent of the volume of the structure, rather than just the number of bins.

Two other forms of DVH are more commonly used in treatment planning, the true differential DVH and the cumulative DVH. The main difference

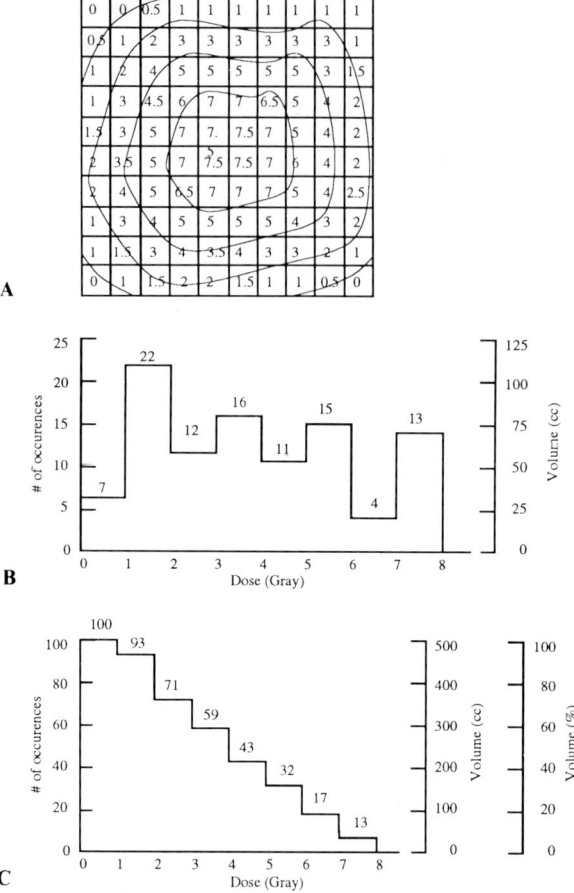

A

B

C

Fig. 4.17. Dose volume histograms. **A** Dose distribution. **B** Frequency plot. **C** Cummulative (integral) DVH

pixels which contribute certain doses to the plan. In fact, use of the differential DVH with a dose display tool which highlights specific ranges of dose can help one understand which dose on the plan actually causes the shape of the DVH to look as it does (Kessler et al. 1994b).

What should these DVHs look like? Figure 4.18A,B illustrates two ways of looking at a simple plan which does a good job of treating the target volume with a relatively uniform dose, while at the same time keeping the dose to a minimum for the spinal cord. The cumulative histograms shown in Fig. 4.18 show the same information as the differential DVHs in Fig. 4.18B. Compare these histograms to a similar set of histograms for a plan which does not completely treat the target volume, and does worse with the spinal cord also (Fig. 4.19). Note in particular that the differential histo-grams show much more clearly exactly how much of the target is underdosed by what dose.

4.5.3 Normal Tissue Complication Probability

The section above on DVHs shows how the DVH can help the planner and physician summarize the results of the volumetric dose calculation with respect to the target volume and critical structures. However, it does not tell the planner what to do when the comparison of two plans yields DVHs for the critical normal tissue which cross each other, as illustrated in Fig. 4.20. It is not clear from first principles whether (for this tissue) more dose to less tissue is more or less likely to cause a complication than less dose to more tissue. The DVH illustrates the issue, but says nothing about which of the two situations is better. When this kind of complicated comparison must be performed for a number of different tissues simultaneously, it becomes very difficult to know how to rate or score different treatment plans.

The concept of using a model to try and predict the probability of complications from a certain plan was first published by John Lyman in 1985 (LYMAN 1985). The normal tissue complication probability (NTCP) model has been discussed in detail by a number of authors (LYMAN and WOLBARST 1987; BURMAN et al. 1991; KUTCHER et al. 1991; LAWRENCE et al. 1992; NIEMIERKO and GOITEIN 1991; DRITSCHILO et al. 1987). This model, or variations of it, is in quite common use for assisting with plan evaluation and/or comparison. Although this subject is discussed in more detail elsewhere in this

between the frequency plot and the differential histogram is that for the differential DVH, the volume plotted on the vertical axis is divided by the volume of the bin size, so that this form of the DVH is independent of the voxel and dose bin sizes. This is a detail which is critical if one wants to compare DVHs created for different plans with different bin or vixel sizes. The third type of DVH is the "cumulative" DVH, shown in Fig. 4.17C. On the vertical axis of this plot are plotted all of the voxels which contain a dose higher than or equal to the bin dose. This plot typically starts at 100% on the volume axis and decreases to 0% from left to right.

Which of these kinds of presentations should be used depends on the situation being analyzed. If one is interested in comparing the overall behavior of a number of different structures for one particular plan, the cumulative histogram may be the most useful. If, on the other hand, one wants to try to understand in detail the shape of the histogram, then the differential DVH shows the actual groups of

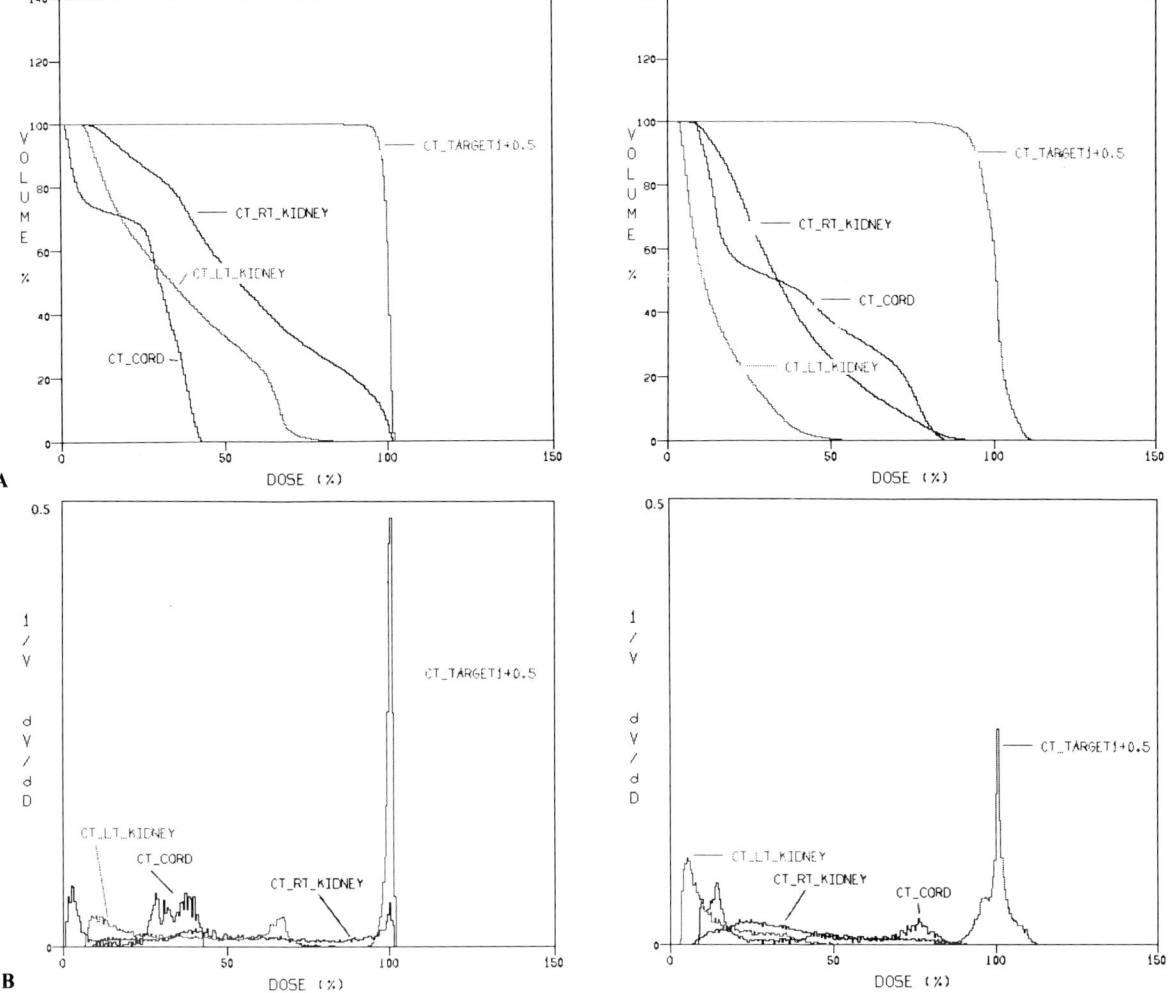

Fig. 4.18. Dose volume histogram plots for target and normal anatomy **A** Cumulative **B** Differential

Fig. 4.19. Histograms for different plan to same site as Figure 4.18 but resulting in higher dose to spinal cord. **A** Plan #2 Cumulative DVH. **B** Plan #2 Differential DVH

volume, a brief description of the main points is included here so that the effect this analysis has on treatment planning is understood.

The NTCP model of Lyman assumes a very simple power law relationship between the probability of complication in a normal tissue and the volume of the organ which is treated to a particular dose:

$$\text{NTCP} = (2\pi)^{1/2} \times \int_{-\infty}^{t} \exp(-t^2/2) \, dt,$$

where

$$t = (D - TD_{50}(v))/(m \times TD_{50}(v))$$
$$v = V/V_{ref}$$
$$TD(v) = TD(1) \times v^{-n}$$

$TD(v)$ is the tolerance dose (for a particular complication) as a function of the fraction of the volume of

the organ v which is uniformly irradiated. There are four adjustable parameters, $TD50(1)$, m, n, and vref (vref is often taken to be the volume of the whole organ).

The dose-volume numbers used in this equation depend on the shape of the DVH for the organ. Although there are a number of different schemes which are used to reduce a complicated clinical DVH to the single-step DVH used for this formula (LYMAN and WOLBARST 1989; KUTCHER and BURMAN 1989), eventually the DVH is reduced to a single dose and effective volume number. Fractionation and other corrections have been incorporated in the model as well (LYMAN, unpublished work, 1989; 1991).

The NTCP formula above is fit with parameters generated from input clinical tolerance data, eventually creating a dose-volume-complication

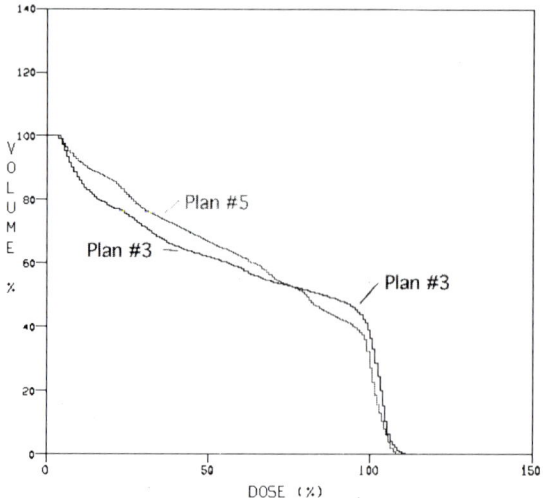

Fig. 4.20. Comparison of cummulative liver DVH curves for plans shown in Figs. 4.18 and 4.19

function which is shown in Fig. 4.21. The parameters used for this model depend on the tolerance of the organ when certain fixed percentages of the organ are irradiated to certain doses. While this is a somewhat artificial construct, it has allowed the generation of a first set of parameters to be used for a series of normal tissues of interest in radiation therapy by the NCI Photon Treatment Planning Working Group (EMAMI et al. 1991). Although the dose tolerances used in this work were the result of a survey of physicians rather than hard data, at least these numbers give some idea of the current state of knowledge of partial organ tolerance doses.

The use of the NTCP model with some chosen set of input parameters and a chosen method for histo-

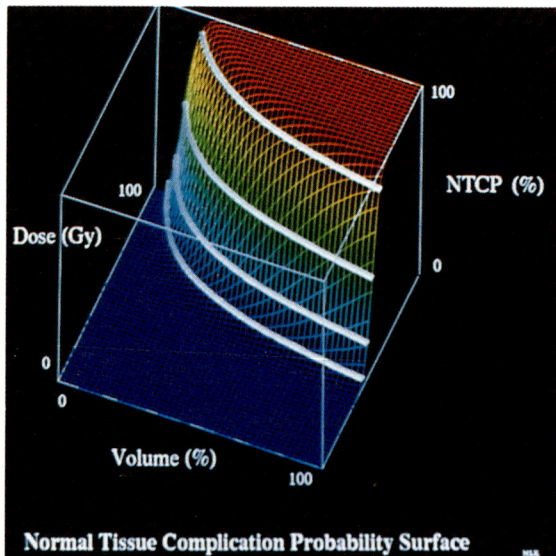

Fig. 4.21. Dose-volume-complication plot of *NTCP* for some normal organ

gram reduction allows the generation of a complication probability for each normal tissue in each plan. This kind of analysis can be valuable for many reasons. First, even if the NTCP values generated from this model are far from the real complication probabilities, it is reasonable to expect that the NTCP values would be a good way to order plans as one tries to choose the best plan from a group of candidate plans. Second, in a similar way, the NTCP values may be helpful (at least in a qualitative way) for the design of dose-escalation trials, in which the acceptable complication rates will be the main factor deciding the limits for the escalation. Finally, the model can be a good way to analyze the actual dose-volume-complication data which will be generated in coming years from the use of full 3-D dose-volume analysis of patients clinically treated and followed for complications. In fact, in a recent study of the dose, volume, and complications in irradiation of the liver (LAWRENCE et al. 1992), the NTCP model could be parameterized to describe the clinical data. This study is the first of a whole series of studies which are needed to determine quantitatively, and for the first time, the real dose-volume-fractionation-complication relationships for each organ which might limit the dose which can be delivered to the tumor.

4.5.4 Tumor Control Probability

The NTCP concepts discussed above help determine how to evaluate the effects of various normal tissues which are relevant to the plan, but do not help analyze the response of the tumor to the treatment which is being considered. However, similar model-based approaches to tumor control probability (TCP) have also been created by a number of workers (GOITEIN and SCHULTHEISS 1985; GOITEIN 1986; WOLBARST et al. 1982; THAMES et al. 1991; BRAHME and AGREN 1987; ZAGARS et al. 1987). These models allow plan evaluation to include a more sophisticated result than "Is the target volume dose uniform?" This work is discussed in much more detail in Chap. 16 by Schultheiss.

4.5.5 Probability of Uncomplicated Control

Several of the cooperative NCI-funded treatment planning working groups have established functions which were used to combine the NTCP values for each organ at risk, and the estimated TCP values, into a "probability of uncomplicated control" that could be used to compare competing treatment plans

(MUNZENRIDER et al. 1991). This number was obtained by taking each NTCP, multiplying by a weight factor (w) that took into account how significant the complication would be, and then combining those factors with the TCP value. This probability thus became:

$$P = TCP \times w_1(1\text{-}NTCP)_1$$
$$\times w_2(1\text{-}NTCP)_2 \times \ldots w_n(1\text{-}NTCP)_n.$$

Although these schemes have been used to analyze some comparative treatment planning studies (KUTCHER and BURMAN 1989), they have not yet been used to determine which treatment plans to use clinically.

4.6 Plan Verification

Sophisticated treatment plans generated using BEV techniques, with computer-designed blocks and 3-D dose calculations, and with calculated values for the NTCP and TCP for the proposed plan, could be less effective than the most simple AP-PA rectangular fields if they are not properly treated on the patient each day. Development and use of plan verification techniques are therefore a critical part of treatment planning, especially for treatment planning which uses conformal techniques. Important parts of the plan verification process include the use of patient immobilization techniques, patient localization consistency checks, plan verification techniques using BEV and DRR displays, and treatment verification using portal imagers.

4.6.1 Patient Immobilization

One of the most obvious ways to improve the consistency and accuracy with which patients are planned and treated is to position the patient in the same way each day. Patient positioning has always been critical. However, with the improved imaging that is available with CT and MRI, as well as the technology of 3-D treatment planning which makes possible the design of conformal field shapes that have margins about the target volume of less than a centimeter, the positioning of the patient is ever more critical. Several reports have shown that without special immobilization devices, daily uncertainties and motion in patient setup are typically of the 5–8 mm range (RABINOWITZ et al. 1985; BOYER 1987). Special immobilization systems including thermoplastic masks (THORNTON et al. 1991a), use of markers, masks, and diagnostic x-rays (VERHEY et al.

1982), and body casts and molds (JAKOBSEN et al. 1987) have shown that attention to these details can decrease uncertainties to 2 mm (THORNTON et al. 1991b) or even less (VERHEY et al. 1982).

4.6.2 Use of BEV/DRR Plan Verification Checks

Comparison of the location of anatomic landmarks as described in the treatment planning system, as opposed to the location determined on the patient during simulation or treatment, is the major plan verification method. In 2-D planning, due to the very limited amount of anatomic information contained in the planning system, one is limited to checking the distance from an obvious structure (the vertebral bodies, for example) to an edge of the radiation field. In many situations, there was either not enough anatomic detail to make a comparison, or the comparisons were not near the areas of most interest to the physician.

With the full 3-D capabilities of modern 3-D planning systems, there can be a great deal of anatomic information contained in the patient's description inside the treatment planning system. The display of this information, particularly through the use of the BEV display, can now be the basis of sophisticated, and in the future automatic, alignment and verification techniques. The key to this verification method is that BEV displays, showing bony or other anatomy which should be imaged using simulator (or megavoltage) films of the radiation portals, can be directly compared to the films themselves. The use of this technique to perform a so-called plan verification simulation is described here. Similar techniques are used for treatment machine verifications, which are discussed in the next section.

In modern 3-D treatment planning, the "simulation" which was performed on the simulator has been replaced by treatment planning on the planning computer [this has been called "virtual simulation" (SHEROUSE et al. 1987)]. Some workers start the treatment planning process with an initial procedure on the simulator which defines the patient treatment position, immobilization device, and a coordinate system for later planning (this has been called a localization simulation) (FRAASS 1992). When the virtual simulation paradigm is used, this localization procedure actually occurs on the CT scanner, before the treatment planning CT scan is acquired.

In any event, most workers prescribe that a "plan verification" procedure take place after computerized treatment planning has been completed. At this procedure, the simulator is used to simulate all facets of the treatment plan which has been developed. Each step of this procedure should be verified with the data inside the planning system to make sure that the plan is implemented on the patient as accurately as possible. The three basic steps are:

1. Verify the patient setup using the orthogonal pair of localization films obtained at the original localization simulation.
2. Verify the location of the new plan isocenter(s) determined in the planning system, using orthogonal films obtained at the new isocenter(s).
3. Verify the shape and location of each of the treatment fields designed in the planning system.

In order to perform all three of the above verification checks, BEV displays, with or without the appropriate DRRs, are compared to the simulator films which are obtained during the procedure. One straightforward method for the comparison is to plot out the BEV display at a magnification equal to a specified source-film distance (SFD) and direct comparison of that BEV plot to the film shot at the same SFD. Care must be taken to outline and plot structures which will image on the simulator film (bones, for example), and the definition of the edge which is used for the bone outlines must be consistent and tested to compare accurately with the projections of the bone on the film (Fig. 4.22).

If a digital imaging system is available for image acquisition on the simulator, direct comparison of images on the planning system may be the best way to do the comparison. In that case, a digital image from the simulator is acquired by the planning system (by laser digitization of the simulator film or by direct digital acquisition from the image intensifier). The BEV graphics information can then be overlaid directly over the simulator image and compared. Registration techniques can then be used to determine the translations and rotations which are necessary to align the simulator images with the planning system. A more sophisticated method adds the use of DRR images to the graphics BEV data. The use of the DRR may be very important where the projection of the anatomic information may be difficult to visualize from simple BEV contours. However, the use of DRRs also requires more time for DRR calculation and more sophisticated image registration and display techniques than the simple BEV graphics method discussed above. There are clearly situations where each of the described methods is advantageous.

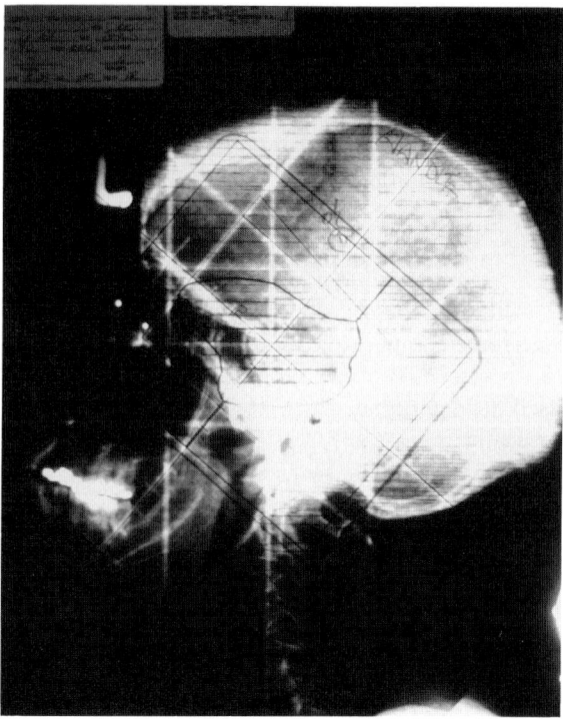

Fig. 4.22. Simulator film (*lateral head*) underlaid with contours from BEV

4.6.3 Treatment Verification and Portal Imagers

The verification of the positioning and alignment of the patient, as well as the accuracy of the radiation delivered to the patient by the therapy beam, can be monitored by portal films and/or portal imaging devices. This verification is a critical step in the treatment planning process, as this is the accuracy which can determine the error margins which must be included in the decisions made during treatment planning. The techniques used to actually compare the treatment planning information with the portal images obtained on the patient are similar to those described above. The treatment verification situation is much more difficult, however:

1. Megavoltage portal images contain much less anatomic information due to the very low contrast realized in this imaging situation.
2. Treatment fields should be verified each day, if possible, not just in a simulator-based plan verification procedure. In order to do these checks practically, the speed with which the alignment between the plan and actual field displays can be made is critical.
3. The alignment between the two studies must be done in a quantitative and accurate way that is

reproducible each day. This tends to make automatic methods necessary, yet automatic methods which can be used to identify anatomic information on very low contrast portal images are not yet readily available.

4. To prevent treatments which are not optimally delivered, the image acquisition and all the comparisons and decision-making must be performed with a few seconds, to allow the Go–No Go decision to be made before all the dose from the field is delivered to the patient.

More information on portal imagers and their use can be found in Chap. 8.

4.7 Treatment Plan Normalization and Monitor Unit Calculations

Dose distributions generated in a treatment planning system may be displayed in two different ways: (a) in absolute dose and (b) in relative dose, referred to some defined point of normalization. Since the purpose of treatment planning is eventually to help the user deliver the correct absolute dose to the patient, understanding how to interpret the doses displayed by the planning system, and how to implement them (i.e., calculate monitor units to be used by the treatment machine), is a critical issue. The calculation of the actual machine monitor unit setting (i.e, the calculation of the amount of radiation to be delivered by the machine) to be used for each radiation field is a critical issue, and should be the subject of numerous and redundant checks which assure that the actual dose delivered to the patient is correct. One of the most important hazards which is present in treatment planning is the possibility of incorrectly calculating the dose to be delivered by each field.

Monitor unit (MU) calculations, therefore, are of primary importance in the delivery of the correct dose to the patient. There are many different philosophies about how MU calculations should be performed, who should perform them, how checks should be performed, etc. Many institutions require the MU calculation step to be a manual calculation, so that a hand calculation is used to confirm that the dose delivered, or at least the dose delivered to one point inside the patient for each field, is as correct as any non-computer-based calculation can be. It is therefore appropriate to consider dose inside the treatment planning systems in two fashions: (a) how relative dose calculations inside the treatment planning system are performed and (b) how those relative doses are converted to absolute doses and MUs (whether inside or outside the "treatment planning system").

4.7.1 Basic Concepts

In the sections following, the various ways in which beams and plans can be normalized are described. This flexibility is important to the full use of the 3-D capabilities of the planning system, since very complex plans may require different kinds of normalization techniques than are required for simple AP-PA fields. As always, the treatment planning system user should have a good understanding of the way in which plans are normalized before using the various options. In fact, the user should always confirm that the local method of MU calculation is consistent with the kinds of plan normalization used in the planning system.

There are a number of basic normalization-related issues which must be defined and explained carefully:

1. *Beam normalization*: the (relative) dose delivered by each beam. This is often called the "beam weight" in 2-D planning systems. It is more general to call this value the "relative dose delivered to the defined beam normalization point," since this specifies exactly what the term implies. The individual beam normalizations thus determine the relative weights of the contributions of each beam to the total dose distribution.

2. *Plan normalization*: the way the relative total dose distribution is normalized after the dose distribution is added up by the planning system. Typically, the user picks one point inside the patient (here called the "isodose reference point") and then picks the value of the dose to be displayed at that point (the "isodose reference dose"). The planning system then rescales the total dose distribution so that the chosen value is displayed at that point.

3. *Prescription isodose surface* (isodose line is a 2-D concept): the isodose surface (for the plan presented to the physician) which is chosen as the volume to which the prescription dose should be delivered.

4. *MU calculations*: the calculation of the MUs to be delivered for each field depend on the prescription dose, the prescription isodose line, the plan normalization method, the beam normalization point and method, and details of the dosimetric settings for the MU counter for each machine and energy.

In addition, there are a number of different ways to calculate doses to be used for beam or plan normalization. One can assume that these values are calculated relative to a standard situation (flat surface water phantom at 100 cm SSD), or relative to the patient assuming no tissue inhomogeneity. This complicating factor has major implications for how dose is displayed and how one calculates MUs for the fields which are involved.

4.7.2 Beam Normalization

Beam normalization is the general term used here to describe the process which is often described as "beam weights." Both terms refer to the method of describing the relative dose contribution of each beam to the total dose distribution. Typically, the beam weight is the relative dose delivered to the standard normalization point. Since the usual choices for this point are on the beam central axis, at a depth of d_{max}, or at the isocenter, for simple situations the meaning of the beam weights will be easily understood. However, during 3-D planning, one is much more likely to end up with the beam central axis out in air rather than inside the patient, or underneath a shielding block, or inside an inhomogeneity. Therefore, in 3-D planning, it is critical to clearly define how the beam normalization is performed.

One solution to the difficulties of beam normalization is to allow the planner to define the point at which each beam is normalized. This "beam normalization point" can be placed wherever is appropriate for the situation. Then the "beam weight" will be the relative dose which is to be delivered to the beam normalization point. This means that internally to the system's calculations, the raw dose at each point due to the beam being considered will be multiplied by the ratio of the desired beam weight divided by the raw dose calculated at the beam normalization point. These rescaled doses will then be added to the individually rescaled doses from each of the other beams to give the total dose distribution.

The dose used for rescaling the dose to the beam normalization point may be calculated in a number of different ways, including (a) with a flat surface, unit density material, and no blocks, (b) with flat surface and unit density material, but including the effect of blocks, and (c) with a full density-corrected type calculation. Although any of these different methods can be used while maintaining a consistent

system for dosimetry and MU calculations, probably the most straightforward choice is to use the open field (unblocked), unit density calculation as the reference calculation for the beam normalization point. This choice makes sense since the point of this normalization is to make contact with the usual method of calculating MUs. Most MU calculations, particularly those performed by hand, assume the use of rectangular (equivalent square) fields, incident on a flat water tank, with no density effects taken into account. With the use of this method, the beam weight gives the relative contribution a simple rectangular field would give to the plan, and so it scales with the MU for each field. If the beam normalization dose calculation took into account the effect of blocks, for example, then the effect of a field with a block would change dramatically as the beam normalization point became closer to the edge of the block. The location, SSD, depth, and other parameters at the beam normalization point are important for MU calculations, since the MU for each field will be calculated by using the depth, SSD, and other parameters to deliver a specified dose to the geometric location of the beam normalization point. These characteristics must be carefully checked for each user's planning system and monitor unit calculation method.

4.7.3 Overall Plan Normalization

The second (and distinct) step in the plan normalization process is the definition of the way the overall dose distribution is to be normalized and displayed. After the total dose distribution is created using the beam normalizations (weight factors) as described above, the entire dose distribution is renormalized to give the desired dose at a particular point inside the patient. Although this overall plan normalization is not essential if all the beam normalization points and the beam normalization dose values (beam weights) are consistently and rigidly maintained, there are several advantages to the use of the plan normalization:

1. The planner can choose the value of the dose displayed on the plan (it can be absolute or relative dose, for an entire treatment, for a daily treatment, or as a relative percentage only).
2. One can easily redefine the dose to this normalization point, to display different effects.
3. The effects of blocked field shapes, inhomogeneity, and other corrections can be shown by

displaying the magnitude of these corrections for the planner and physician.

4. Use of the chosen normalization point is the connection between the essentially relative dose distributions that the planning system generates and the actual delivered doses and MUs required to fulfill the plan's prescription.

The point chosen for this plan normalization (the isodose reference point), the value which would be displayed at that point (the isodose reference value), and the different methods of normalization are described below.

4.7.3.1 Isodose Reference Point

The point inside the patient which is used as the point at which the whole plan will be normalized can be called the isodose reference point. This must be a well-defined coordinate inside the patient which is chosen as a point that will determine the value of the overall dose distribution which will be displayed. Note that (as described below) the physician may pick a particular isodose line to use to modify the prescription dose which will be delivered to this point. The choice of the prescription point, and the careful determination of the dose delivered to that prescription point, is of course critical to accurate treatment and to the documentation of dose delivered to the patient, particularly for patients involved in research studies.

The isodose reference point for a plan is typically taken to be at some point of special interest or symmetry. In a 2-D plan, the point is often at the isocenter (for isocentric treatments), at the center of the target volume, or perhaps at the point of maximum dose (or at d_{max} on the central axis of one of the beams). For 3-D plans, more flexibility is needed to handle the complex situations which develop. Some typical locations for the isodose reference point include:

1. The beam normalization point for one of the beams
2. The isocenter of one of the beams
3. The point at d_{max} on the central axis of one of the beams
4. The intersection point of the central axes of the beams (the point which is typically used to define the target absorbed dose).

If there is no normalization point defined for the overall plan, then the total dose which is displayed on the plan is totally dependent on the way the individual beam weights and beam normalization points

are defined. This option should be rarely used, since for most clinical treatments the user wants to be able to prescribe and calculate the dose to some particular point inside the patient.

4.7.3.2 Isodose Reference Dose

The isodose reference dose is the dose value that the planner would like to have appear at the isodose reference point. Typical values for this parameter would include 100%, the daily prescription dose (e.g., 2 Gy or 200 cGy), or the total dose prescription for the plan (e.g., 45 Gy or 4500 cGy). Changing this value, or the units of the dose, should require only a rescaling of the displayed dose values, which is typically a very fast calculation.

4.7.3.3 Type of Plan Normalization

Different institutions, dosimetric systems, and clinical situations require different kinds of overall normalization of each treatment plan. In some institutions, the physician is used to viewing dose distributions with the "100%" isodose line running through the isodose reference point, for fully density corrected plans, and would like to prescribe "density-corrected" doses. In other institutions, the physicians would like the 100% isodose line to document the uncorrected dose calculation, since that is what their clinical experience is based on. In order to handle the variations in attitude that are involved in this issue, a number of different types of plan normalization are used for 3-D plans.

Four different methods for isodose reference point dose calculation are discussed in this section, but there are certainly others. This issue is critical to understand, since each method will require a different kind of MU calculation for that plan to be actually implemented on the patient. These four situations should give the reader an idea of the issues involved, so that any perturbation of these systems could be understood and safely utilized. Three of the methods described here correspond to choosing a different kind of dose-calculation method for the calculation of the dose (at the isodose reference point) which is being renormalized to the desired isodose reference dose. Before displaying the dose distribution, the system renormalizes the dose distribution to the desired value by using one of these three calculations of the dose to the reference point. The fourth method relies on no overall plan normalization, but directly uses the beam normalizations.

Unit Density, Open Field Normalization The reference point dose which is used for renormalization is calculated using unblocked fields incident on a flat block of unit density material. In other words, the reference point dose calculation is done analogously to the typical hand dose calculation which is done for MU calculations. With this type of choice, the dose to the reference point which is displayed shows the dose relative to the unblocked, unit density, flat phantom situation. Therefore, all effects of blocks, patient contours, inhomogeneities, etc. are reflected in the displayed dose, relative to the chosen desired dose. The dose displayed at the reference point is typically several percent less than the chosen value, generally due to block effects.

Monitor units are then calculated using the open (unblocked) field, and without worrying about density corrections. All effects of blocking and/or density corrections are included by the planning system in the isodose distribution.

Unit Density, Blocked Field Normalization The reference point dose is calculated using the blocked field, but with a unit density patient. In this event, the dose displayed at the reference point will be the chosen value, even with heavily blocked fields, if density corrections have little effect. If there are density correction effects, then the displayed dose will demonstrate the entire magnitude of the dose changes caused by the correction method.

With this option, MUs are calculated using the blocked field, without worrying about density corrections. All effects of density corrections are included in the isodose distribution, but the effects of blocking on the MU calculation must be put into the hand MU calculation by using the equivalent square of the blocked field.

Blocked Field and Density-Corrected Normaization The reference point dose calculation includes all corrections, including density corrections. Therefore, the chosen dose value will always be displayed at the chosen isodose reference point. With this option, MUs are calculated using the blocked field, and density effects must be included by the planning system when it displays the dose to the beam normalization points which is used to determine the dose required from each field for the MU calculations. The effects of blocks and density corrections are shown by the values of the beam and plan normalization numbers in the plan, or, if they are not, then both block and density correction effects must be taken into account by the hand MU calculation.

No Plan Normalization This type of normalization does not include an overall plan renormalization. Therefore, the dose values (or weights) entered into the plan to determine how much dose is delivered from each beam and the beam normalization point choices completely determine how the doses will be displayed on the plan. This kind of normalization system is somewhat less flexible than the use of the other three methods discussed above; however, it has the advantage that the dose (or weight) factor entered for each beam directly determines the dose displayed, without any more renormalizations.

Figure 4.23 illustrates the same plan and dose distribution, with each of the three methods of normalization which are discussed above.

4.7.4 Prescription Dose and Prescription Isodose Surface (Line)

Before MU calculations can be discussed, the ways that doses are prescribed on treatment plans must be briefly reviewed.

1. The easiest method to understand is based on prescribing the dose to the isodose reference point of the plan (typically the isocenter or pseudo-isocenter for a multifield plan). Most commonly, the isodose reference point is chosen to be the isocenter of the isocentric plan, and the isodose reference value is chosen to be 100%. Then the prescription dose (say 200 cGy/fraction) is prescribed to the isocenter. Thus, the 100% on the plan corresponds to 200 cGy. The dose at this point is often called the target absorbed dose and is a very important reference dose for many cooperative protocols. The main advantage of this kind of plan normalization is that it clearly defines the dose which is delivered to a particular point in the patient, which is typically in the center of the target and tumor volumes. In this way, it is reasonable to easily check the dose to this point by hand, to record this dose in the chart. It is useful for studies, since the dose to the center of the tumor and/or target may give a much better indication of the dose delivered to the bulk of the disease, compared to documenting just the minimum dose delivered to the edge of the target volume.

2. The second method is perhaps the most common: the physician looks at the isodose lines (IDLs) on the treatment plan and picks an IDL to be used as the prescription IDL. In this case, the 100% which is typically defined to be the dose at the isocenter (or isodose reference point) does not correspond to the

Fig. 4.23. Wedge pair isodose plots with three different normalization techniques

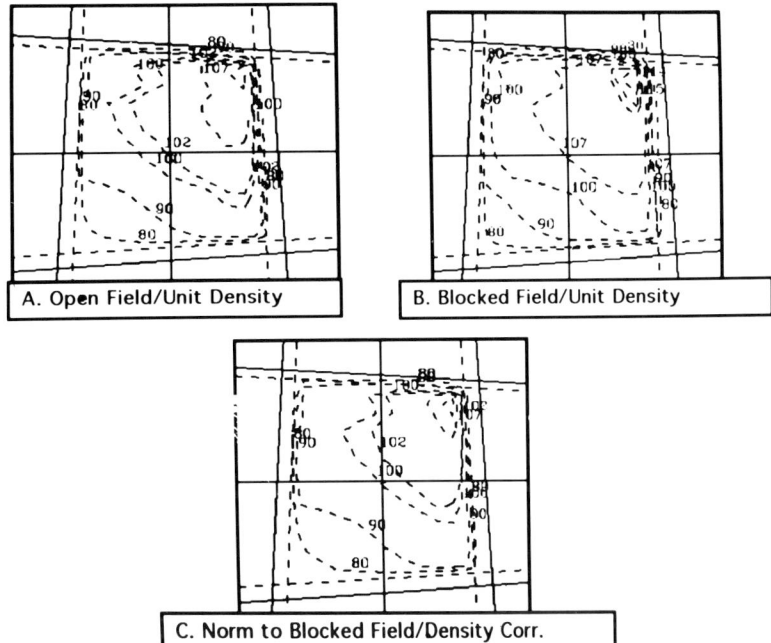

Fig. 4.23. Wedge pair isodose plots with three different normalization techniques

treatment prescription (200 cGy, for example). If the 97% IDL was chosen, then the dose to be delivered to the plan isodose prescription point (the isocenter, typically) would be 200 cGy/0.97, or 206.2 cGy. This prescription method is typically used to prescribe to the isodose line which goes around the target volume, which in effect is an attempt to prescribe the minimum dose to the target volume. While some protocols are written in this fashion, it is often difficult to know exactly what the dose actually delivered to the bulk of the target (or tumor) was unless one also carefully describes the target volume dose uniformity and other details of the plan.

3. The third method is the most straightforward, but is probably the least used. The physician simply prescribes the dose to be delivered by each field. Then, the weight factor of each beam is used as the dose from that field, and the doses are simply added up and displayed. When dose is prescribed in this way, very little information about the relationships of the prescribed dose and the dose which was delivered to the tumor and/or target is available without detailed inspection of the treatment plan.

4.7.5 Monitor Unit Calculations

The type of MU calculations which are required to deliver a specific prescribed dose to any treatment plan is dependent on all of the information discussed

above, on the planning system definitions of these concepts, and on the particular plan, beam arrangement, and normalization options which have been chosen for use in a particular institution. Some of the factors which determine the kind of MU calculations preferred include:

1. Tissue phantom ratio (TPR) vs fractional depth dose (FDD)-based MU calculations. Typically, for 3-D treatment planning which almost always is based on the use of isocentric and pseudo-isocentric techniques, a TPR-based approach is preferred.

2. Is the planning system used to take into account all block-related effects, or does the institution choose to base their MU calculations on the equivalent square method of MU calculations? With the latter well-known method, the equivalent square of the blocked field is used for the field size look-up of factors like the TPR and phantom scatter factor. This is the traditional way of accounting for the changes caused by adding blocks to rectangular fields. However, as the treatment planning system becomes sophisticated enough to accurately take into account the more complex behavior of fields shaped with blocks or multileaf collimators, this method will likely be replaced. If the planning system doses (beam weight factors, as described above) are determined relative to the unshaped, unit density, flat surface situation, the planning system will account for all of the complex shape, density, and

surface effects, while a very simple MU calculation can be used.

3. A third issue is whether physicians prescribe density-corrected doses or uncorrected doses. If the physician has experience with dose values which have been corrected for inhomogeneities, the former option is probably most reasonable. However, since most of the clinical and published data on tumor control and normal tissue complications are based on uncorrected (unit density) doses, it is quite reasonable to prescribe uncorrected doses. If density corrections are available in the treatment planning system, the density-corrected plan should always be displayed also, so that eventually most people in the field will have the experience to allow the "real" (density-corrected) doses to be used for all prescriptions and research study dosimetry.

Given the complexity of the many options described above, as well as the complexity of modern 3-D treatment plans, the most critical issue for MU calculations is the verification of the methodology used at each institution. The entire prescription – dose calculation – monitor unit calculation sequence must be verified to confirm that the entire process is consistent and that it gives the appropriate answers over a wide range of different situations.

4.8 Clinical 3-D Treatment Planning

Now that all of the parts of treatment planning have been discussed, it is appropriate to attempt to describe some of the methods which have proven useful for actually planning patient treatments. How one actually goes about creating good treatment plans is fairly well known for 2-D planning and has been described in detail (BENTEL 1992). 3-D treatment planning is much less well understood, and the techniques which are routinely used in the clinic are less well publicized. This discussion will attempt only to give a very brief overview of some of the useful concepts which have been applied to this problem.

4.8.1 Definition of Patient Anatomy

One of the most time-consuming parts of the 3-D treatment planning process is currently the definition of patient anatomic structures, since this is usually performed by drawing contours around each of the structures as visualized on serial images (typically CT). Until more automated structure definition software tools are developed, the planner must make careful decisions about which structures must be defined and which can be ignored.

Anatomic structures are outlined for three major reasons:

1. Some organs, for example the prostate gland when the patient has prostate cancer, are outlined to form the basis of the tumor volume, since the entire organ must be treated. In many sites, however, the whole organ is not irradiated to high dose, and the actual macroscopically visualized tumor is the basis of the tumor and target volumes.

2. A second reason for organ definition is that the organ is considered to be at risk for radiation-related complications, and so the dose to that organ must be considered as part of the plan optimization criteria. A number of organs are nearly always outlined, as the dose to those structures is nearly always important. These structures include the eyes, spinal cord, kidneys, rectum, and bladder. If dose-volume histo-gram (DVH) analysis will be used in the plan evaluation, it is important to scan and then contour all of the appropriate structures, so that the DVH shows the relationship of the dose to the entire organ.

3. A final reason for organ delineation is that many organs can be imaged on simulator and/or portal images which are used to verify patient alignment. Assuring that the patient is aligned correctly with the treatment plan is critical to accurate plan implementation. Objects which are typically outlined for use in BEV-based alignment checks include the inner table of the skull, vertebral bodies, pelvic bones, and long bones like the femur and humerus. If gradient-tracking automatic contouring options are used to define the contours of these structures, especially the skull, then the user should investigate the effects of the autotrack algorithm parameters (gradient limits, etc.) to make sure that the algorithm is tracking the same bony boundaries that the user outlines by hand, and to understand how the gradient-defined edges relate to the projection x-rays and/or DRRs which are used for the alignment check.

4.8.2 Determination of the Beam Arrangement

Two-dimensional treatment plans very often consist of very well-known beam arrangements: the four-field box, three-field AP and lateral plans, and the wedged pair, for example. With the implementation and use of 3-D planning, however, the space which is available for investigation has been broadened considerably, and there are many other kinds of beam arrangements to be investigated. How does the planner make the transition from these simple axial plans to full 3-D plans?

One of the keys to a quick transition to 3-D planning is the ability to make use of the tried and true 2-D planning techniques, with some 3-D modifications. For example, plans for brain lesions often are noncoplanar many-field plans. Many of these plans, however, really just consist of either two or three sets of wedge pair fields, with the 3-D difference being that the plane of the wedge pairs is not axial, but coronal (vertex plan), sagittal, and/or oblique.

A way to make creation of these kinds of plans easier is to visualize the target volume and normal structures in three dimensions and to look for planes of symmetry, either in the target volume shape/location or in the location of the critical normal structures. Creating oblique CT reconstructions along these planes of symmetry allows the planner to use 2-D planning knowledge to generate the plan. For example, investigation of a midline frontal lobe lesion in the brain may appear to be a complex situation. However, by creating an angled coronal-axial plane which contains the bulk of the target, but misses the eyes and other normal structures, one may find that a simple three-field plan, in the oblique plane, will satisfy all the constraints. Planning in the oblique slice makes creation of nonaxial and noncoplanar beam plans fairly straightforward.

To make use of the full power of 3-D planning, however, one should not be limited to the use of 2-D concepts such as the wedged pair. As described in more detail elsewhere (McSHAN 1990), some of the well-known 2-D planning concepts can be expanded into their 3-D analogs. For example, if one imagines beams entering the patient along the edges of a cone, one finds that the cone is really a "3-D wedged pair," since the wedged pair is simply the cone collapsed onto a 2-D plane. The purpose of the cone is to irradiate the target with a number of different beam directions, each of which enters and exits the patient from a different direction. A cone made up of four beams would look like two orthogonal wedge pairs, while a cone made up of only three beams, or five beams, would look like a much more complex noncoplanar field arrangement. As reported by McSHAN (1990), a number of workers are investigating the use of these 3-D beam constructs for use in computer-controlled conformal therapy. However, the current 3-D planner can use concepts like cones and multiple orthogonal arc segments as an aid in visualizing beam arrangements which may be of use. All of these beam arrangements attempt to deliver the dose to the target from a number of different directions while

trying to assure that the entrance and exit regions of the various beams do not overlap, so that the dose in those normal tissues is kept to a minimum.

Once the beam arrangement has been determined and optimized, a number of practical issues should be considered. For example, it is very useful to create oblique CT reconstructions in the plane of wedged pair fields, even if they were not created while the plan was being determined. The display of the beams and wedges in this plane can be an excellent check that the directions and size of the wedges are appropriate. Even in more complex noncoplanar beam plans, where the collimator of some of the fields may be turned, with wedges in place, so that the wedges can be used as simple dose compensators for the noncoplanar beam arrangement, these kinds of display checks are useful to help the planner (and the treatment therapist, as well) understand the complex beam arrangements that have been designed. One other key to 3-D planning is the determination, for the particular set of immobilization devices, simulation, and treatment equipment which is used at each institution, of the practical limits on nonaxial treatments. These limitations involve more than simple collision avoidance between the different types of machinery; they also involve institutional policies on simulation, plan verification, and other issues.

4.8.3 Field Shape Design

Design of field shapes for 3-D and conformal therapy is an important part of 3-D planning. As described above, conforming the shape of the field to that of the target volume can now be performed automatically by features like Autoblock (McSHAN et al. 1990). However, the actual need is to create a certain margin between a particular isodose surface (typically the 95% isodose surface) and the target volume (expanded to include setup and patient/organ motion uncertainty). Creation of this "dose margin" requires significantly more effort that simply pushing the "Autoblock button."

Several effects must be considered in selecting the block margins. For example, one often must edit the block shape to allow a particular margin around a critical structure. The margin which is required to account for organ motion is often anisotropic: for example, we use a 2-cm superior and inferior margin for liver motion due to respiration, while maintaining only a 1-cm margin in the other directions (LAWRENCE et al. 1992). Another important effect involves the dosimetric results of the other fields which make up the plan. For example, to maintain a uniform dose margin about a target volume using a simple axial four-field plan, one must use larger block margins at the superior and inferior edges of the target than in the lateral or anterior/posterior edges. This is due to the fact that in the axial plane, the 95% isodose line is shifted toward the block edges owing to the addition of dose from the other fields, while in the coronal plane, for example, at the superior aspect of the target volume, the 95% line is much farther from the block edges. With experience, the specific block margins which are required in the various circumstances are learned, and the field shapes for similar plans are designed with those modified parameters. Of course, after the dose calculations are performed, the field shapes can again be modified to ensure that the desired dose distribution is attained.

4.8.4 Optimizing Plans

Another of the time-intensive parts of 3-D treatment planning is the plan optimization procedure. Until automated optimization techniques become routinely available, however, this kind of effort will be required. Currently, most plan optimization is still performed interactively, that is, the planner designs a plan, calculates the dose distribution, analyzes it with respect to the stated goals for the plan, then makes iterative changes in the plan until reaching an acceptable plan. How close this plan actually is to the "optimal" depends on the planner and physician time and resources available, knowledge and creativity of the planner, reasonableness of all the stated goals for the plan, and many other factors.

Several techniques or planning system features can speed the interactive optimization process. Since in general, dose-calculation times are still too long, anything that will decrease calculation time can be used as part of the optimization process. This includes the use of coarse dose-calculation grids (> 1 cm resolution), calculation on a few orthogonal cuts rather than throughout the full 3-D volume, use of small numbers of randomly or pseudo-randomly placed calculation points, and use of simplified dose-calculation algorithms which ignore scatter or other complex parts of the dose deposition process. Each user should investigate the effectiveness of these techniques, and also determine the limits of the technique. Obviously, any of these time-reduction techniques can be applied too strenuously, increas-

ing the inaccuracy in the resulting dose distribution so much that the "optimization" is actually invalid. Many of these techniques can be extremely useful for the fast calculation of DVHs for particular organs. If all of the plan evaluation criteria can be related to organ-based DVHs, then this approach can be fairly simple. However, most clinical treatment plan evaluations still contain some amount of physician input and qualitative judgement, and depend on DVH, dose distribution, and other criteria.

4.9 Quality Assurance in Treatment Planning

One of the most important jobs of the clinical medical physicist is associated with quality assurance (QA) for all aspects of the technical part of radiation therapy. QA for treatment planning is a very important aspect of that work and is critical for the appropriate and accurate use of any treatment planning for patient care. The routine use of tomographic and multimodality imaging, 3-D treatment planning systems, image-based targeting, full 3-D dose calculations, integrated plan verification techniques, and computerized plan evaluation and analysis techniques as part of the clinical routine has dramatically altered the scope of QA in treatment planning.

In most 2-D treatment planning, many of the important decisions were made externally to the computerized treatment planning system. For example, block design and block shapes were rarely even input into the planning system, so verification and QA of this important part of the therapy process was a mechanical and operational issue, essentially unrelated to "treatment planning." In many clinical operations, the main decisions made as part of so-called treatment planning involved decisions about whether wedges were needed, beam weight changes to make the dose over the target uniform, and perhaps some investigation of different gantry angles or field sizes for particular plans. Most publications and work on quality assurance for 2-D treatment planning have thus concentrated on the most obvious issue, the accuracy of the dose calculations (McCullough and Krueger 1980; Westmann et al. 1984; Sauer et al. 1987; Rosenow et al. 1987), although there are recent discussions of some of the broader issues (Curran and Starkschall 1991; Ten Haken et al. 1991b; van Dyk et al. 1993).

The situation is much different now, as one considers QA for the entire 3-D treatment planning process. In particular, many more of the treatment-related decisions may now be based on the computerized planning system. This can include any use of images, all tumor and target identification, all use of normal anatomy, definition of beam parameters, block design, dose calculations, dose display, and plan analysis tools, computation of the normal tissue and tumor control probabilities, and the calculation of the monitor units to be used. In order to analyze the QA program which is needed to address this new situation, a number of separate areas must be discussed:

1. QA testing of the computerized treatment planning system, including its software, procedures, training, and other facets of its use
2. Measurement, testing, and verification of the dosimetric aspects of the planning system
3. QA of the clinical use of treatment planning throughout the entire treatment planning and treatment processes.

Although not entirely unrelated to each other, the three major topics can each be dealt with by using a fairly independent approach. The detailed information in this section should be taken as part of the more comprehensive discussion on QA which is presented in Chap. 18 by Rosen.

4.9.1 Quality Assurance of the Treatment Planning Software

Quality assurance of the treatment planning system software, and of all the other treatment planning system-related components, is a critical part of QA for the treatment planning process. However, as with other software and computer-dependent equipment like computer-controlled treatment machines, the software QA of the system is usually not under the direct control of the user. Rather, the QA process is the responsibility of the vendor of the system.

The user of the commercial products is able to affect the situation, however. As suggested in a recent AAPM Task Group report on Accelerator Safety for Computer-Controlled Medical Accelerators (AAPM 1992a), the user should require enough documentation from the vendor that the user can be convinced that the system design, implementation, and QA program are robust enough for the clinical use that the user intends. There are numerous published descriptions of the standard software engineering tools and procedures which are used to meet the quality standards (BEZIER 1984; MEYERS 1979), and the user can exert pressure for as much information as can reasonably be provided by the vendor or assimilated by the user. The user should always expect that there are unreported errors in large

software systems such as treatment planning systems (JACKY and WHITE 1990).

4.9.2 Dosimetric Quality Assurance

Basic verification of the accuracy of dose calculations has been discussed by a number of workers, although most have concentrated on 2-D dose calculations (McCULLOUGH and KRUEGER 1980; WESTMANN et al. 1984; SAUER et al. 1987; ROSENOW et al. 1987; AAPM 1992b; WITTKAMPER et al. 1988; LEPINOY et al. 1984). In addition, most users of radiation therapy treatment planning systems have themselves performed at least some tests of their systems to convince themselves that the dose calculations presented by the system agree with the data measured at that institution. Although QA of a treatment planning system includes much more than just dose calculations, it is in this area that nearly all QA work on treatment planning has concentrated. Considering the importance of dose calculations, this is not unreasonable, even though there is much more to consider. As discussed in the sections below, dosimetric checks of a planning system involve a number of different operations.

4.9.2.1 Measured Dataset

The measurement of a complete and self-consistent dataset characterizing the dose distributions obtained on a particular machine is one of the most important activities needed in order to perform reasonable testing on the system. With older 2-D systems, the only data needed were a few profile measurements, and perhaps some depth-dose curves, for a number of different square field sizes. As long as these few profiles could be renormalized to the values given by the depth doses, the consistency of the data was not too important, since there were few checks of that consistency.

The situation is much different with a 3-D planning system. The dose calculations are used to predict the dose throughout an entire volume of the patient. Therefore, the dataset should be obtained throughout the volume, and the comparison should also occur throughout the volume of interest. Since it is typically difficult to make measurements directly on a volumetric basis [although there have been some efforts (THOMAS et al. 1992)], measurements are still often made with 1-D curves (depth-dose and profile measurements). When this kind of data is obtained, one curve at a time, the difficulty of obtaining a 3-D set of data curves that is all consistent becomes very difficult. When the dose on the sagittal plane does not agree with the dose at the same coordinates, but when measured in an axial set of data, then knowledge of which value is correct at that point is extremely difficult.

One method which can be used to generate a complete 3-D dose distribution from measured data has recently been reported by STERN et al. (1992). This method uses film dosimetry to measure the dose distribution in a number of 2-D planes perpendicular to the beam central axis ("BEV planes") and a depth-dose curve (measured with ion chamber or other appropriate dosimeter). The depth-dose curve is used to generate a nonlinear interpolation between the BEV plane data, along divergent ray lines. This method, which uses the symmetry and physics of the divergent geometry generally followed by the photon beam, generates 3-D dose distributions which can be directly compared to the calculated 3-D dose distributions obtained from the 3-D planning system. These kind of data are essential if the full 3-D dose distribution calculated by a 3-D system is to be fully characterized and verified.

4.9.2.2 Techniques for Dose Distribution Comparison and Verification Checks

Verification checks of 2-D dose distributions are often performed simply by overlaying two plots of profiles, depth doses, or an isodose chart measured on the central axis axial contour. For checks of the entire 3-D dose distribution, however, more sophisticated techniques can make the analysis of the results much easier to obtain, while also being more quantitative. Many standard techniques for comparisons are useful for any treatment planning system. For example, graphic plots comparing measured and calculated lines of data have been widely used for both depth-dose and profile behavior (Fig. 4.24A). For specific inspection of the planning system depth dose-related behavior, one of the most useful techniques has been the creation of tables of FDD (or TPR) as a function of field size and depth (VAN DE GEIJN and FRAASS 1984). Comparison of these tables, by taking the difference between the calculations and the data (Fig. 4.24B) can highlight the overall quality of the agreement, as well as trends of errors which may lead to understanding how to improve the calculations. The third standard kind of display comparison technique is the direct comparison of isodose charts generated from the data and by the calculations. Comparisons of these isodose charts by overlaying one chart on another, or by

two different kinds of curves on the same image, are typically used to compare doses on the axial contours which are always used in 2-D planning (Fig. 4.24C).

For 3-D dose distribution comparisons, much more sophisticated tools are often used so that the comparison can be done more efficiently and quantitatively. Just as the display of the actual dose distribution is significantly improved by the use of orthogonal and axonometric displays, use of isodose curve overlays on sagittal, coronal, and axonometric displays can be a very effective way to show the comparison in three dimensions. Coronal (Fig. 4.25A) cuts are very useful, since they show the beam dosimetry in the same fashion as the BEV display which is so critical to how the fields are designed. The axonometric display is one of the easiest ways to show the actual comparisons, while showing the 3-D distributions as well (Fig. 4.25B).

In order to show more clearly areas of relatively small differences, one can use a dose difference display (in 2-D cuts or in a full 3-D volume). As illustrated for a 3-D case in Fig. 4.25C, the dose distribution generated by calculation can be subtracted from the measured dose distribution. Using isodose lines or colorwash dose displays can be a very effective way to highlight the differences. The dose difference display can then be histogrammed (in 3-D in particular) to illustrate the agreement which is found throughout the entire volume of interest. This dose (difference) volume histogram can be used as a quantitative measure of the overall agreement between calculations and data (Fig. 4.25D) (FRAAS et al. 1994b). Another method which is useful in particular for analyzing electron beam dose distributions uses not only the dose difference, but the distance from the measured isodose line to the calculated isodose line. This distance difference is often more important than the actual dose difference, particularly in the penumbral regions and the region behind the 90% isodose line in electron beams, where the depth dose is falling off very rapidly (NCI 1991).

4.9.2.3 Verification of Input Test Data

Once the user has acquired a consistent set of data to be used for planning system modeling, testing, and verification, the next step is to input the necessary data into the dose-calculation model, and to verify that the model adequately reproduces the input data. In many 2-D calculation models, which are directly based on the data or even on beam libraries which are

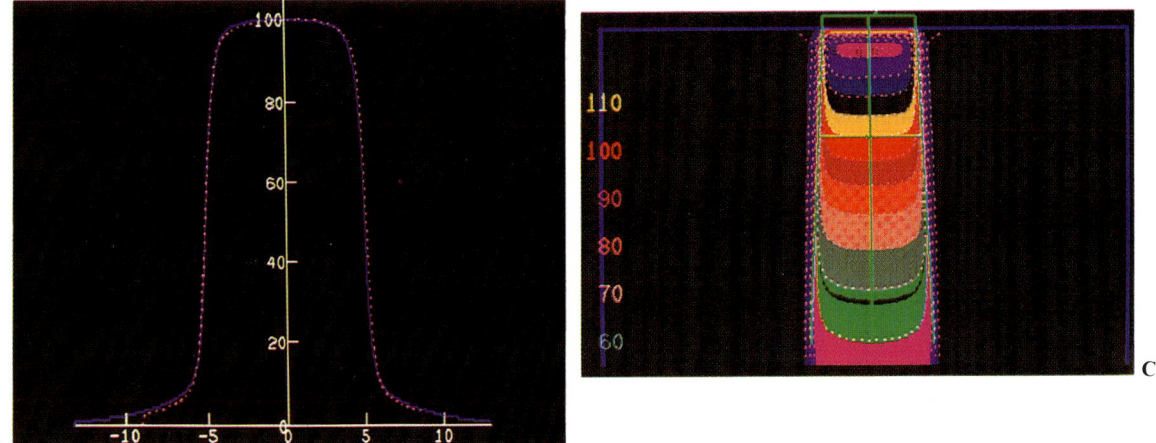

```
FILENAME: X10_D10_4_FDD.DIF     DATE:30-JUL-1993 08:43:50.43

DIFFERENCE: CALC(X10_D10_4.FDD) - DATA(X10_D10_DAT.FDD)
```

	3	4	5	6	7	8	10	12	14	17	20	30	31
	3	4	5	6	7	8	10	12	14	17	20	30	40
2.0	24	35	34	29	28	26	18	18	7	2	-6	-12	-9
3.0	-6	-1	2	3	2	4	3	3	0	-1	-4	-9	-9
4.0	-12	-8	-6	-4	-3	-1	0	1	-1	0	-1	-5	-4
5.0	-8	-5	-3	-4	-2	0	0	2	0	2	0	-2	-1
6.0	-7	-5	-3	-2	-1	0	0	3	1	2	2	0	0
7.0	-2	-1	-1	0	0	3	3	3	2	3	2	0	1
8.0	-3	1	0	0	0	2	1	4	3	3	2	1	1
9.0	-1	1	0	1	1	1	1	1	0	2	1	0	2
10.0	0	1	0	0	0	0	0	0	0	0	0	0	0
11.0	-4	2	-1	-1	0	0	-1	-3	-1	-1	-1	0	-1
12.0	-4	0	-1	0	0	1	-1	-2	-1	-1	-2	-1	0
13.0	-2	1	-2	0	-1	-1	-1	-3	-2	-2	-2	-4	-3
14.0	-2	1	-1	0	0	1	0	-3	-2	-2	-2	-4	-3
15.0	-3	1	0	2	1	-1	-1	-2	-3	-4	-4	-5	-4
16.0	-3	0	0	1	1	0	0	-2	-3	-3	-3	-5	-5
17.0	-6	1	-1	2	0	1	-1	-3	-2	-3	-4	-7	-6
18.0	-4	1	0	2	2	1	0	-2	-2	-2	-4	-7	-6
19.0	-6	0	-1	1	1	2	0	-2	-2	-4	-4	-7	-7
20.0	-6	0	0	2	1	1	1	-3	-2	-4	-5	-8	-8
21.0	-5	0	1	1	2	2	1	-1	-1	-3	-5	-8	-8
22.0	-6	0	1	1	2	2	1	-2	-1	-3	-4	-6	-10
23.0	-6	0	0	2	3	3	1	-2	-1	-3	-4	-8	-8
24.0	-7	-1	0	4	3	3	2	-2	0	-2	-4	-7	-9
25.0	-4	1	0	5	3	4	2	0	0	-2	-3	-7	-8
26.0	-6	1	2	2	3	3	3	1	1	-1	-2	-5	-9
27.0	-6	1	4	5	4	5	4	2	2	0	-1	-6	-9
28.0	-5	2	1	3	4	6	4	2	1	1	-1	-7	-8
29.0	-2	1	4	6	6	6	6	4	4	1	0	-6	-7
30.0	-4	1	3	7	7	6	7	2	4	2	1	-6	-8

Fig. 4.24. 2-D dose verification methods: **A** 1-D profile comparison: blue line is calculation, purple dotted lines is data. **B** FDD table of differences, obtained by subtracting measured FDD curves from calculated FDD curves, displayed in units of 0.1% of the dose at dmax. **C** Axial contour comparing measured and calculated dose distributions: *dotted lines* are measured isodose lines, *color wash* is the calculated dose distribution.

just compilations of measured data, this is a very straightforward step which is mostly a check on the software and/or the consistency of the measured data and the entry of that data. However, the situation is somewhat different for the generally more complex calculation algorithms which are used in 3-D planning systems. It is much more difficult to build a completely data-based calculation algorithm in 3-D than in 2-D. Far more data are required, and the importance of field shaping devices, wedges, oblique angulations, and nonaxial beams, compensators, and other devices and situations are much more important when the full 3-D dose distribution is being calculated. Therefore, these algorithms (which are discussed in much more detail in Chap. 3 by Bloch and Altschuler) contain more modeling of the real physics of the situation, so that they can be used in situations in which direct measurements are unavailable or difficult to obtain. Since a number of these models are based on first principles theories and parameters [for example the Monte Carlo-based kernels used for the convolution techniques developed by MACKIE et al. (1988) and AHNESJO (1989)], often the reproduction of the usual input data (field flatness and details of the depth-dose behavior) must be carefully checked. These results must be well characterized and understood, because they will form the basis for some fraction of the differences between calculations and data in all the other comparisons which are discussed below.

4.9.2.4 Verification of Applicability and Limits of the Dose-Calculation Algorithm

A very important point is made in an excellent paper by JACKY and WHITE (1990) on QA testing for treatment planning dose calculations. All current dose-calculation algorithms for electrons, photons, and brachytherapy sources contain approximations and limitations. The testing of the algorithm must be

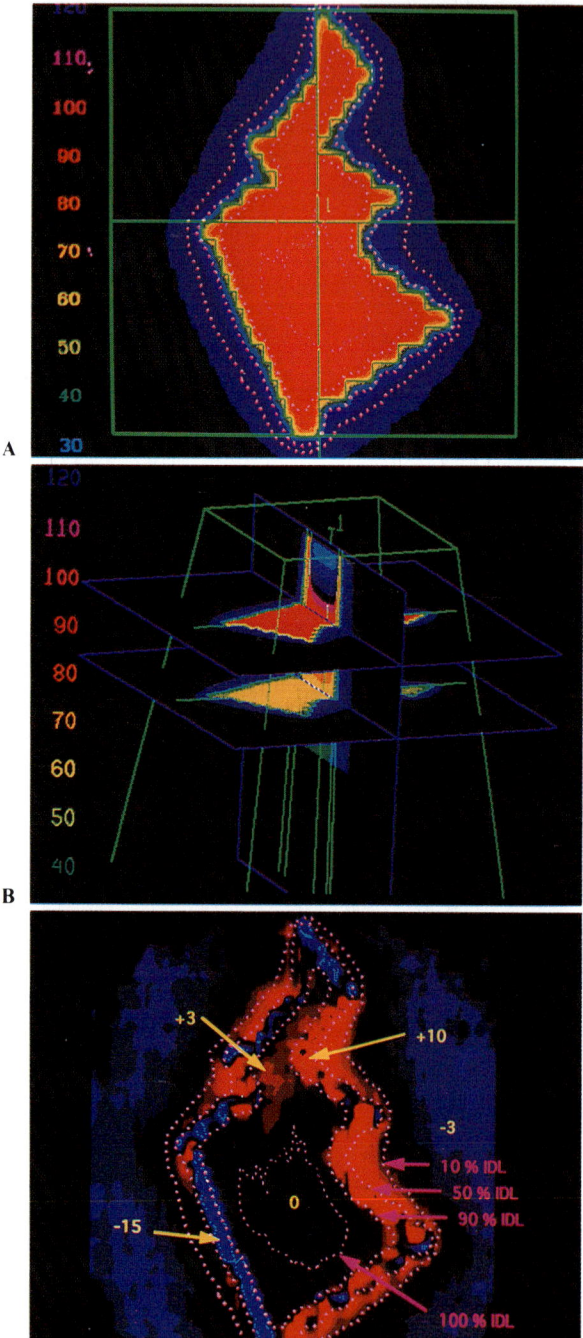

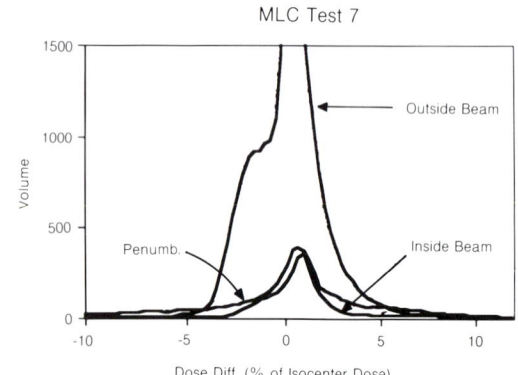

Fig. 4.25. Dose verification techniques used for 3-D systems. **A** Coronal cut is used for dose comparisons with an isodose line display. **B** Axonometric display of axial, sagittal and coronal cuts with isodose line display. **C** Dose Difference display colorwash on axonometric image. **D** Dose Volume histogram of 3-D dose difference display

performed to confirm that it behaves correctly, as the algorithm is designed to work. That may mean that if the algorithm is a simple one with many approximations, the agreement between calculations and data may be very poor, even if the algorithm is doing exactly what it is supposed to do. Therefore, this fact points out the need for adequate information and understanding about what the algorithm does, how it works, and what kind of results one should expect. These tests may be very different in scope, operation, or results than the tests in the next section, which determine whether the algorithm predicts the dose well enough to be used for clinical planning. The design and operation of tests of the algorithm are, of course, individualized to the particular dose calculation algorithm in question.

4.9.2.5 Dose Verification over the Range of Clinical Usage

One of the most traditional kinds of verification check is comparison of the dose-calculation results with measured data over the range of applicability of the calculations. These comparisons are the most relevant relative dose comparisons because they compare the planning system results with the measured data. This comparison will show the overall differences between the predictions and the measurements. Errors in input data, fitting, algorithm coding, and/or design and various other kinds of errors are all incorporated in the results. Therefore, this is a critical kind of test since it shows the overall precision with which particular kinds of calculations may be performed. However, if there is a discrepancy, these tests will likely not help explain the reasons for the problem. Appendix 1 gives an example of some of the considerations that might be involved in the design of this set of verification tests.

4.9.2.6 Verification of Absolute Dose Output and Plan Normalization

All of the tests above have concentrated on verification of the relative dose distribution calculated by the planning system. However, perhaps the most critical issue is the absolute dose which will be delivered to the prescription point in the patient. The verification that the absolute dose is appropriately handled throughout the entire treatment planning process, and delivered to the patient, is thus very important.

Important parts of the process which are related to the absolute dose include:

1. Plan prescription methods
2. Prescription dose
3. Prescription point
4. Plan normalization method inside the planning system
5. Beam normalization method inside the planning system
6. Machine monitor unit calibration method
7. Monitor Unit calculation method

Since there are numerous ways to handle each of the factors above, a detailed discussion of test procedures is clearly beyond the scope of this work. However, it is also clear that careful confirmation of the accuracy of the dose prescription–plan normalization–monitor unit calculation process is critically important (CUNNINGHAM 1984).

4.9.3 Tests of Nondosimetric Functions

Modern planning systems, especially 3-D systems, are involved in many more parts of the treatment planning process than simply dose calculations. These nondosimetric parts of the planning process also need to be verified and quality assured. Although there have been some preliminary reports on nondosimetric QA tests (BURMAN et al. 1989), a comprehensive set of tests has not been described. Appendix 2 lists an example of the kinds of tests which may be needed to confirm that all of the nondosimetric features of a planning system are adequately documented and verified. The scope of this test is just like the dosimetric tests described in Appendix 1: the amount of testing which is required depends directly on the activities and features of the system which are used clinically.

4.9.4 QA for the Entire Clinical Planning and Treatment Process

After all treatment planning system verifications have been performed, there is still another major segment of the QA process which must be implemented. Modern treatment planning systems contain a large degree of flexibility, and that means that the process requires checks of the way it is routinely (or not so routinely) used. The best way to make sure that the entire planning-treatment process is working correctly is to build redundant and backup checks into the normal routine of the process. Some examples of procedures which can fulfill this aim are listed below.

1. *Routine physics checks.* Inspection of the dosimetric aspects of each plan and its prescription and monitor unit calculation are, of course, essential.

2. *Verification of image registration.* The use of multiple imaging modalities is becoming much more common in treatment planning. While this typically means the use of CT and MRI, it can also include much simpler issues, such as registering the treatment planning CT scan with the coordinate system placed on the patient with orthogonal films and skin marks made to align with the lasers used at the initial localization simulation. Careful checks of the accuracy of the alignment between the different datasets is essential for accurate implementation of the CT-based plan on the real patient.

3. *Use of sagittal and coronal CT reconstructions.* The use of CT reconstructions made in sagittal and coronal planes can highlight numerous potential problems, while at the same time being a useful way to show many of the 3-D characteristics of the patient and plan. If the reconstructions are constructed from all of the CT data available, then any problems with inconsistencies in the CT data (due to table motion, patient movement, field-of-view changes, mislabeling or misreading of slice location information, etc.) will be demonstrated by problems or artifacts in the reconstructed images.

4. *3-D surface displays.* To create the 3-D structures which are used for planning, such as the tumor and target volume, as well as normal structures, contours are typically drawn on serial axial CT slices. It is extremely difficult to envision the 3-D extent of these structures as one draws contours on one slice at a time, and there are volume averaging and other effects which sometimes make it difficult to accurately contour particular structures on some images, anyway. Viewing the 3-D surfaces created from these contours is essential to verifying that the contours are all consistent and realistic. Viewing a solid surface display of the target volume will quite often show several places where the surface looks unrealistic, and where inspection and perhaps editing of the contours used on a particular slice will make the volume more realistic and self-consistent Fig. 4.26. Sometimes this display can also be formed by "cutting" a contour from the surface onto orthogonal image(s) (CT sagittal and/or coronal). This review of the 3-D surface, and the "sculpting" of that surface which must be performed if a problem is identified, is one of the things that distinguish 2-D planning from 3-D planning.

5. As one learns to plan in three dimensions, there are several tricks that can be used to help verify that plans are being performed correctly. In particular, it is often somewhat difficult to determine how wedges should be oriented when the fields are nonaxial. Since many nonaxial plans are just relatively standard wedge pair or three-field plans, generation of an oblique CT reconstruction in the plane of the beams can make the determination and/or checking of wedge orientations very straightforward.

6. *Patient alignment consistency checks.* Verification that the patient is correctly positioned with respect to the treatment plan should be performed whenever possible. In particular, this should be performed carefully at the verification simulation, and when the patient begins treatment, at least. One way to do this check is to use a comparison of BEV plots (with bony anatomy, as discussed in Sect. 4.6) with the orthogonal simulator (or treatment) films taken to verify the isocenter of the plan, before any of the fields are checked. The field ports should also then be verified. Accurate checks of this kind require a good understanding of which anatomy to draw into the planning system, and how that anatomy projects onto films and BEV displays. The use of DRR displays rather than simple BEV plots can make this technique somewhat more straightforward, although it requires more technology to implement quantitatively.

7. *Target volume checks.* Especially as institutions begin to use CT target volume definition rather than targets drawn on simulator films, the projection of the CT target volume onto BEV displays, and then onto the simulator films themselves, may help the physicians and other staff to make the logical connection between the way targets are drawn with both the old and the new techniques, improving the confidence of the staff in the newer techniques.

8. *Point dose calculations.* Where possible, it is very helpful if the point to which the dose is prescribed, or the plan normalization point, is situated so that a hand calculation of the dose delivered to that point is possible. If that is the case, then one can verify that the total dose prescribed by the plan is actually being delivered, at least to one point inside the patient. This will help remove the possibility of any major dosimetric problems involving the wrong total dose to the patient.

9. *SSD checks.* It is possible, in the simulator and/or treatment machine, to measure the SSD to the central axis of each treatment field. This routine check measurement, when compared to the SSDs expected by the planning system, is a reasonably

Fig. 4.26. Nasal tumor volume with sharply irregular shape possibly requiring editing

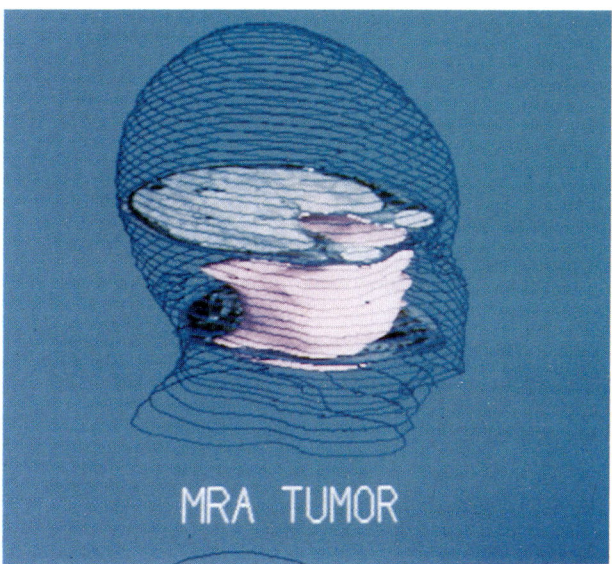

good check for large setup errors or other inconsistency somewhere in the planning process. In a similar fashion, checks of the table coordinates which are required to set the isocenter at the correct spot inside the patient are a similar candidate for a consistency check that will show possible errors in SSD indicator or other machine-related problems that might affect the correctness of the treatment.

4.10 Summary

The treatment planning and treatment process has changed dramatically in recent years. The introduction of the routine and comprehensive use of CT (and more recently MRI) imaging for definition of tumor, target, and normal anatomy has made a complete 3-D description of the patient possible. Recent developments of new computer and graphics display hardware and software have made possible the development of fully 3-D treatment planning systems which have applications in all phases of the planning and treatment process. These new tools have greatly widened the possibilities for treatment: no longer are we constrained to treating patients with a maximum of three or four fields, all in the axial plane of the patient, with simple field shapes guided by standard anatomic rules. Current 3-D planning technology is quickly leading toward fully conformal treatment planning, with numerous fields, all shaped by the computer to fully conform the high dose to the target volume, with the multiple fields delivered by computer-controlled treatment machines.

The purpose of all this new technology should not be lost amid all the computer graphics. Conformal treatment planning has the promise of two different kinds of improvements for the patient:

1. Since the high dose volume is more confined to the target volume, it should be possible to deliver more dose to the tumor, hopefully resulting in a better local control rate, while at the same maintaining the dose to normal tissues, or at least the complication rate, at the same level as with current treatment techniques.
2. If the dose to the target is not escalated, then the dose to normal tissues can be decreased, thereby decreasing complications.

Either of these two situations can significantly improve the clinical results of radiotherapy. These questions are beginning to be studied with the technology of 3-D planning which has recently become available. Careful quantitative studies must be performed to delineate the tolerance of normal tissues to partial volume irradiation and the limits of dose escalation for various kinds of tumors. Then, it will be possible to perform the studies which will finally show whether the 3-D technologies discussed here are of direct benefit to patients.

Acknowledgments. The authors would like to thank Randy Ten Haken, Ph.D., Mary Martel, Ph.D., and Marc Kessler Ph.D. for their helpful comments and suggestions. We would also like to thank Lon Marsh, CMD, for preparation of many of the figures and photographs included here.

Appendix A. Clinical Dose Calculation Verification Testing

The following set of suggested test cases is an example of the kind of organized testing that can be implemented to verify the behavior of the treatment planning system dose calculations over a wide range of clinically used situations. The tests suggested here are a graded series of cases which range from basic checks of depth-dose curves to much more sophisticated dose-calculation situations including heavily blocked fields and inhomogeneous phantom situations.

The test techniques which are used for calculation verification have already been discussed (Sect. 4.8.2.2).

1 Basic Depth-Dose/Output Data

1.1 Fractional Depth-Dose Curves (FDD)

One of the most critical and basic tests for the planning system is the ability to predict accurately the depth dose for standard field size situations. In this section, FDD curves are measured for a number of field sizes, all at a standard SSD for the machine in question. Although typically the FDD data are obtained at 100 cm SSD (for modern linear accelerators), it is quite reasonable to use an SSD like 90 cm, since it is much closer to the average condition under which patients are treated if all plans are performed isocentrically. Field sizes of 3×3, 4×4, 5×5, 6×6, 7×7, 8×8, 10×10, 12×12, 14×14, 17×17, 20×20, 25×25, 30×30, 35×35, and 40×40 are often used. The FDD table accuracy should be checked at other SSDs which cover the clinical range which is used (80–110 cm is reasonable).

1.2 Tissue Phantom Ratio (TPR)

Perform TPR measurements for a number of field sizes and depths. Since these measurements are quite time intensive, field sizes might be limited to 4×4, 6×6, 10×10, 20×20, 30×30, and 40×40 at depths of nominal d_{max}, 5, 10, and 20 cm.

1.3 Phantom Scatter Factor (PSF)

These data are typically obtained at the same field sizes as are used for the FDD data.

1.4 Collimator Scatter Factor

Same as 1.3.

1.5 Other Factors (Tray, Wedge, etc.)

As required and/or used by the planning system.

2 Open Fields

The basic starting condition for any dose calculation modeling and/or verification is open square fields. The checks of these situations can be made with 2-D isodose curves and charts, or with full 3-D comparisons if the data and the analysis tools are available.

2.1 Field Size Dependence, Standard SSD

The basic calculation verification tests listed below should be obtained at the standard SSD used for all the basic data. A subset of the field sizes used to measure the depth-dose table can be used here, for example 3×3, 5×5, 10×10, 20×20, 30×30, and 40×40.

2.2 Extended SSD

A subset of the field sizes above should be tested at smaller and larger SSDs.

2.3 Rectangular Fields

The behavior of the depth dose as rectangular fields are used must be tested. A zeroth order check can just verify that the equivalent square concept is reproduced. However, real measurements of fairly extreme rectangular situations will give different results than the equivalent square would suggest, so some measurements and checks are appropriate.

2.4 Oblique Incidence

2.4.1 30×30 Field, $30°$ Oblique

The oblique incidence data should be obtained at the largest angle possible for the large field, which is typically something like a 30×30 field with a $30°$ oblique gantry angle.

2.4.2 Surface Irregularity

Use a step phantom or some other easily reproducible shape to look at the effects of nonflat surface contours. Use a 30 × 30 field incident on a large (5 cm) step in the surface of the phantom. This is probably the most extreme test of the photon algorithm for homogeneous phantoms.

2.4.3 Nonaxial Beam Entry

This test is similar to 3.3.1, but the table angle is set to 90° so that the beam is oblique to the axial plane.

2.5 Arc Field Calculation

2.5.1 110° Arc, 6 × 6 Field, Square Phantom

Create a square phantom using soild water, and measure the dose distribution from a 180° arc calculation for a 6 × 6 field (arc centered about the normal to the phantom surface). Measure the dose at least in a coronal plane and in an axial plane.

3 Wedged Fields

Dose calculation verification measurements must be made and verified for each physical wedge used for each photon beam. Since the dose calculations are used three-dimensionally, the dose distribution must be checked at a minimum in both the axial and the sagittal plane, and preferably in a three-dimensional fashion. If the dose calculation algorithm is sensitive enough to predict changes in the wedge-related dose contributions due to changing SSD, then a series of SSDs should be measured. The wedge modifies the behavior of the beam in the sagittal direction as well as the axial, so data and comparisons in the sagittal or coronal planes, where the difference can actually be noted, must be studied. In addition, depending on how the wedges are taken into account by the calculation algorithm, the behavior may change as a function of field size, so measurements at a number of field sizes will likely be appropriate.

4 Blocks

The use of blocks in treatment planning varies widely among institutions. While some institutions routinely use fairly simple hand-placed blocks, others routinely use computer-generated conformal blocks on nearly every field. The number of verification checks that should be done on the accuracy of the blocked field calculations thus depends on the use of blocking that is made at a particular institution, and, in particular, on the kind of clinical blocking decisions

that are based on the output of the treatment planning system.

The list of test cases below thus should be used appropriately by each institution. If very sophisticated blocking is used routinely, the institution should attempt to complete all of the relevant parts of the test below. If only simple situations are going to be used inside the planning system, then some of the testing may not be immediately required.

4.1 Normalization of Data and Plans

The blocked field dose calculations are used in two ways in the planning system: (a) to predict the relative dose distribution (i.e., isodose curves) and (b) to calculate the change in the dose to the plan normalization point due to the blocking. In order to perform dose verification checks of both features simultaneously, the test measurements and calculations must be normalized as described below.

For each test case, the data should be normalized to the value obtained at the normalization point without the block (but including the tray), so that the dose at the normalization point reflects the effect of the block. These normalization conditions thus require that ion chamber normalization measurements be made for each blocked field case, with and without the block in place (but including the tray), so that the absolute dose difference due to the blocks is known.

4.2 Straight Edge Block

4.3 30 × 30 Field Blocked to 25 × 25

4.4 Same as Above, but at Different SSD

4.5 Mantle Field

4.6 Superclavicular or Pelvic Field (Corner Blocks)

4.7 Conformal (Computer-Designed) Fields

4.8 Simple Island Block

4.9 15 × 15 Blocked to 4 × 15

4.10 Transmission Blocks

If clinical treatments involving transmission blocks are designed or planned using the system, then these situations must be checked. The method of determination of block thickness versus "transmission" must be measured, and then the corresponding "block transmission factors" must be determined and checked.

4.11 Straight Versus Focused Blocks

5 Patient Shape Effects

The effect of the shape of the patient is studied with a number of simple phantom studies in which the specific effects caused by the shape differences are easy to study. The first test employs a geometry similar to that of a tangential breast field, while second test looks at the situation when the field "flashes" off the edge of a square phantom.

6 Inhomogeneity Correction Tests

There are very few 2-D or 3-D sets of inhomogeneity correction data for photon beams, and many of the basic data on which verification checks are based are for simple square phantoms with rectangular inhomogeneities. There is clearly a need for 2-D and 3-D datasets which can be used for verification checks of modern 3-D algorithms.

6.1 Verification Tests for Correction Method

Straightforward square phantoms with various inhomogeneities are used for these tests.

6.2 Use of a Measured 1-D Dataset

To document the accuracy of the correction method in a number of basic but clinically relevant geometries, the dataset measured and reported by RICE et al., 1988 can be used.

6.3 Profile Tests for Individual Energies

Aside from changes in the depth dose in the center of the radiation field, the profile of the dose in and behind inhomogeneities is very important.

6.4 Coronal Plane Dose Distributions

The use of coronal plans can give much more information than even the single 1-D lines of data which were discussed above.

7 Compensator Tests

The kinds of tests which are used for compensators depends a great deal on the kind of compensation that is performed.

7.1 Tests of Missing Tissue Compensation

For missing tissue compensation, in which the change in patient shape is used directly to create the compensator for that particular field, the algorithms which (a) find the surface shape, (b) create the compensator, and (c) create a flat surface in the dose calculation to mimic the supposed behavior of the compensator can be relatively easily verified with a few simple phantom tests.

7.1.1 Lateral Head/Neck Field

7.1.2 Anterior Mantle Field with Lung Blocks

7.2 Test Cases for Dose Compensation

Tests of dose compensation techniques will be much more detailed. In this case, many more kinds of geometries of patient and compensator need to be checked, since the algorithms used likely attempt to do a much better job in predicting the dosimetric behavior of the compensator than does the missing tissue algorithm. In addition, dose compensation is likely used with a much more sophisticated dose optimization scheme which may optimize the dose for several beams at the same time. The complexity of these algorithms should be the main guide in designing the tests.

7.2.1 Lateral Head/Neck Field

7.2.2 Anterior Mantle Field with Lung Blocks

7.2.3 Noncoplanar Brain Plan, Three Fields

7.2.4 Nonaxial Abdomen Plan, Three Fields

8 Anthropomorphic Phantom Tests

It can be quite useful to generate some standard test data and plans to verify the accuracy of the calculations in anthropomorphic phantoms. These test cases should be similar to treatment techniques used in the clinic, and of interest to the physicians and physicists who are responsible. Some examples of clinical situations which could be verified if desired are:

8.1 Mantle Field: TLD/Film, Coronal Midline Plane of Humanoid Phantom

8.2 Breast Field, Tangential Breast with Lung. Axial Plane

9 Multileaf Collimator

Testing of the MLC is similar in principle to the testing used for blocked fields. One must ensure that both the absolute and the relative doses from all the test plans are checked, since the behavior of output, phantom scatter, and collimator scatter factors for multileaf collimators have not yet been widely studied.

10 Segmental Field Calculations

Testing of segmental treatment fields (fields made up of more than one fixed segment or beam portal) is similar to testing of rotational arc calculations. One is testing the ability of the planning system to handle the additional bookkeeping involved, just as one does with an arc calculation check. One additional complexity here is that the conformal calculation could be generated in two ways: (a) where all the segment definitions are done earlier, and the calculation is just run, or (b) where the conformal MLC settings are generated "on the fly" during the calculation. This second method, of course, requires more rigorous checking.

Appendix B. Nondosimetric QA Tests

The extent of tests necessary for 3-D treatment planning is extremely large. The intent of this discussion is not to provide a comprehensive or complete list of things which should be verified, but rather to provide an example of the kinds of functions which should at least be considered when planning the clinical testing for the planning system. People with simple 2-D planning systems will find this list nearly extraneous, while those with a complete 3-D system may find it very incomplete. As with other test procedures, only those features which will be used clinically need be tested initially. However, one should be aware that some of these features may be important to understand, even if no explicit use of the feature is intended, due to mistakes, exploration, or design of the system.

1 Image conversion
  1.1 CT
  1.2 MRI
  1.3 Digitized film images
  1.4 Video digitizer
      1.4.1 Laser digitizer
  1.5 Other imaging modalities
2 Anatomic Structures
  2.1 Mechanical contours
      2.1.1 Digitizer checks
      2.1.2 Keyboard entry
      2.1.3 Mouse/joystick entry
  2.2 Contours on axial images (CT, MRI, etc.)
  2.3 Autotracking contours
  2.4 Generation of 3-D structures
  2.5 Capping (how structures based on axial contours are closed)
  2.6 Use of nonaxial contours for surface generation
  2.7 Extraction of contours from surfaces
  2.8 Surface expansion
  2.9 Bolus
  2.10 Contours drawn on projection images
  2.11 3-D points
3 Dataset registration
  3.1 Coordinate system transforms
  3.2 Reference point-based registration
  3.3 Interactive structure-based registration
  3.4 Automatic registration (head/hat)
  3.5 Goodness of fit checks
  3.6 Structure-dataset dependence
4 Density representation
  4.1 Density matrix operations
  4.2 Use of bulk density

This test protocol should include a number of routine clinical cases which are used to run through the entire treatment planning process as a check on the entire process, including dose prescription, monitor unit calculations, and measurement of doses in phantom for final verification of the entire system. Geometric and dosimetric tests of all information (displays and hard-copy output) will be used to verify the consistency of the process.

References

AAPM (1993) AAPM Task Group 35 Report Medical accelerator safety considerations. Med Phys 2014: 1261–1276

AAPM (1992) AAPM Task Group 23 Report Radiation treatment planning dosimetry verification

Ahnesjo A (1989) Collapsed cone convolution of radiant energy for photon dose calculation in heterogeneous media. Med Phys 16: 577–592

Austin-Seymour MM, Chen GTY et al. (1986) Dose volume histogram analysis of liver radiation tolerance. Int J Radiat Oncol Biol Phys 12: 31–35

Bentel G (1992) Radiation therapy planning. Macmillan, New York

Bezier B (1984) Software system testing and quality assurance. Van Nostrand Reinhold, New York

Boyer AL (1987) Patient positioning and immobilization devices. In: Kereiakes JG, Elson HR, Born CT (eds) Radiation oncology Physics – 1986. American Institute of Physics, New York (Medical Physics Monograph No. 15)

Brahme A, Agren AK (1987) Optimal dose distribution for eradication of heterogeneous tumors. Acta Oncol 26: 1–9

Burman C, Kutcher GJ, Hunt M, Brewster L (1989) Acceptance testing criteria for a CT based 3D treatment planning system (abstract). Med Phys 16: 465

Burman C, Kutcher GJ, Emami B, Goitein M (1991) Fitting of normal tissue tolerance data to an analytical function. Int J Radiat Oncol Biol Phys 21: 123–135

Chen GTY (1983) Computed tomography in high LET radiotherapy treatment planning. In: Ling CC, Rogers CC, Morton RS, (eds) Computer tomography in radiation therapy. Raven, New York, pp 221–227

Chen GTY, Pelizzari CA (1989) Image correlation techniques in radiation therapy treatment planning. Comp Med Imaging Graphics 13: 235–240

Coffey CW, Hines HC, Wang PC, Smith SL (1984) The early applications and potential usefulness of NMR in radiation therapy treatment planning. Proceedings of the Eighth International Conference on the Use of Computers in Radiation Therapy. IEEE Computer Society, Toronto, Canada, pp 173–180

Cunningham JR (1972) Scatter-air ratios. Phys Med Biol 17: 42–51

Cunningham JR (1984) Quality assurance in dosimetry and treatment planning. Int J Radiat Oncol Biol Phys 10 (Suppl 1): 105–109

Curran B, Starkschall G (1991) A program for quality assurance of dose planning computers. In: Starkschall G, Horton JL (eds) Quality assurance in radiotherapy physics. Med Phys Publishing, Madison, Wisc., pp 207–228

Dritschilo A, Chaffey JT, Bloomer WD, March A (1987) The complication probability factor: a method for selection of radiation treatment plans. Br J Radiol 51: 37

Drzymala RE, Mohan R, Brewster L, Chu J, Goitein M, Harms W, Urie M (1991) Dose volume histograms. Int J Radiat Oncol Biol Phys 21: 71–78

Emami B, Lyman J, Brown A, et al. (1991) Tolerance of normal tissue to therapeutic irradiation. Int J Radiat Oncol Biol Phys 21: 109–122

Fraass BA, McShan DL (1987) 3-D treatment planning. I. Overview of a clinical planning system. In: Bruinvis IAD, van der Giessen PH, van Kleffens HJ, Wittkamper FW (eds) The use of computers in radiation therapy. Elsevier Science, North-Holland, pp 273–276

Fraass BA, McShan DL, Diaz RF, et al. (1987a) Integration of MRI into radiation therapy treatment planning. Int J Radiat Oncol Biol Phys 13: 1897–1908

Fraass BA, McShan DL, Weeks KJ (1987b) 3-D treatment planning: III. Complete beam's-eye-view planning capabilities. In: Bruinvis IAD, van der Giessen PH, van Kleffens HJ, Wittkamper FW (eds) The use of computers in radiation therapy. Elsevier Science, North-Holland, pp 193–196

Fraass BA, McShan DL, Weeks KJ (1990) Computerized beam shaping. In: Benedetto AR, Huang HK, Ragan DP (eds) Computers in medical physics. Amer Inst Physics, Woodbury, N.Y. (Medical Physics Monograph 17, pp 333–340)

Fraass BA (1992) Clinical application of 3-D treatment planning. In: Purdy JA (ed) Advances in radiation oncology physics. Dosimetry, Treatment planning and Brachytherapy. American Institute of Physics, Woodburg, NY, pp 967–997

Fraass BA, Matrone GM, McShan DL (1994a) An electronic chart for computer-controlled conformal therapy. In: Hounsell AR, Wilkinson JM, Williams PC (eds) Proceedings of the XIth International conference on the use of computers in radiation therapy. Medical Physics Publishing, Madison, Wisc., pp 218–219

Fraass BA, Martel MK, McShan DL (1994b) Tools for dose calculation verification and QA for conformal therapy treatment techniques. In: Hounsell AR, Wilkinson JM, Williams PC (eds) Proceedings of the XIth International conference on the use of computers in radiation therapy. Medical Physics Publishing, Madison, Wisc., pp 256–257

Glatstein E, Lichter AS, Fraass BA, van de Geijn J (1985) The imaging revolution and radiation oncology: use of CT, ultrasound and NMR for localization, treatment planning and treatment delivery. Int J Radiat Oncol Biol Phys 11: 1299–1311

Goitein M (1986) Causes and consequences of inhomogeneous dose distributions in radiation therapy. Int J Radiat Oncol Biol Phys 12: 701–704

Goitein M, Abrams M (1983) Multi-dimensional treatment planning. I. Delineation of anatomy. Int J Radiat Oncol Biol Phys 9: 777–787

Goitein M, Schultheiss TE (1985) Strategies for treating possible tumor extension: some theoretical considerations. Int J Radiat Oncol Biol Phys 11: 1519

Goitein M, Wittenberg J, Mendiondo M, et al. (1979) The value of CT scanning in radiation therapy treatment planning: a prospective study. Int J Radiat Oncol Biol Phys 5: 1787–1798

Goitein M, Abrams M, Rowell D, Pollari H, Wiles J (1983) Multidimensional treatment planning. II. Beam's eye-view, back projection, and projection through CT sections. Int J Radiat Oncol Biol Phys 9: 789–797

Hall E (1987) Computer image processing and recognition. Academic Press, New York pp 76–78

Henkelman RM, Poon PY, Bronskill MJ (1984) Is magnetic resonance imaging useful for radiation therapy planning. In: Proceedings of the Eighth International Conference on the Use of Computers in Radiation Therapy. IEEE Computer Society, Toronto, Canada, pp 181–185

Hunt M, Kutcher G, Burman C, Fass D, Harrison L, Leibel S, Fuks Z (1989) Effect of positional uncertainties on the treatment of nasopharynx cancer. Med Phys 16: 456 (Abstract)

ICRU (1987) ICRU Report 29. Dose specification for reporting external beam therapy with photons and electrons. International Commission on Radiation Units and Measurements, Washington, D.C.

ICRU (1993) ICRU Report 50. Prescribing, recording and reporting photon beam therapy. International commission on radiation units and Measurements. Washington, DC

IEEE Computer Society Press (1990) Proceedings of the First Conference on Visualization in Biomedical Computing, May, 1990, Atlanta, Ga.

Jacky J, White CP (1990) Testing a 3-D radiation therapy planning program. Int J Radiat Oncol Biol Phys 17: 253–261

Jakobsen A, Iversen P, Gadeberg C, Hansen JL, Hjelm-Hansen M (1987) A new system for patient fixation in radiotherapy. Radiother Oncol 8: 145–151

Kessler ML (1987) Computer techniques for correlating NMR x-ray CT imaging for radiotherapy treatment planning. In: Bruinvis IAD, van der Giessen PH, van Kleffens HJ, Wittkamper FW (eds) The use of computers in radiation therapy. Elsevier Science, North-Holland, pp 441–444

Kessler ML, Pitluck S, Petti P, Castro JR (1991) Integration of multimodality imaging data for radiotherapy treatment planning. Int J Radiat Oncol Biol Phys 21: 1653–1667

Kessler ML, McShan DL, Fraass BA (1992) Displays for 3-D treatment planning. Semin Radiat Oncol 2(4): 226–234

Kessler ML, McShan DL, Fraass BA (1994a) A graphical simulator for design and verification of computer-controlled treatment in delivery. In: Hounsell AR, Wilkinson JM, Williams PC (eds) Proceedings of the XIth International conference on the use of computers in radiation therapy. Medical Physics Publishing, Madison, Wisc., pp 80–81

Kessler ML, Ten Haken RK, Fraass BA, McShan DL (1994b) Expanding the use and effectiveness of dose-volume histograms for 3-D treatment planning I: integration of 3-D dose-display. Int J Rad Oncol Bio Phys (in press)

Koral KF, Ten Haken RK, McShan DL (1990) Superimposition of SPECT and CT images and transfer of RoI for quantification. 37th Annual Meeting of the Society of Nuclear Medicine, June 19–22, 1990, Washington, D.C.

Kutcher GJ, Burman C (1989) Calculation of complication probability factors for non-uniform normal tissue irradiation: the effective volume method. Int J Radiat Oncol Biol Phys 16: 1623–1630

Kutcher GJ, Burman C, Brewster L, Goitein M, Mohan R (1991) Histogram reduction method for calculating complication probabilities for 3D treatment planning evaluations. Int J Radiat Oncol Biol Phys 21: 137–146

Lawrence TS, Tesser RJ, Ten Haken RK (1990) An application of dose volume histograms to treatment of intrahepatic malignancies with radiation therapy. Int J Radiat Oncol Biol Phys 19: 1041–1047

Lawrence TS, Ten Haken RK, Kessler ML, et al. (1992) The use of 3D dose-volume analysis to predict radiation hepatitis. Int J Radiat Oncol Biol Phys 23: 781–788

Lepinoy D, Aletti P, Boisserie G, et al. (SFPH) (1984) Quality assurance program for computers in radiotherapy: progress report. IEEE, pp 322–327

Lichter AS, Fraass BA, van de Geijn J, Fredrickson HA, Glatstein E (1983) An overview of clinical requirements and clinical utility of computer tomography (CT)-based radiotherapy treatment planning. In: Ling CC, Rogers CC, Morton RS (eds) Computer tomography in radiation therapy. Raven, New York, pp 1–21

Lifshitz LM, Pizer SM (1990) A multiresolution hierarchial approach to image segmentation based on intensity extrema. IEEE Transactions on Pattern Analysis and Machine Intelligence 12 (6): 29–540

Ling CC, Rogers CC, Morton RS (eds) (1983) Computer tomography in radiation therapy. Raven, New York

Lyman JT (1985) Complication probability as assessed from dose volume histograms. Radiat Res 104: S-13–S-19

Lyman JT (1991) Normal tissue complication probabilities: variable dose per fraction. Int J Radiat Oncol Biol Phys 22: 247–250

Lyman JT, Wolbarst AB (1987) Optimization of radiation therapy. III. A method of assessing complication probabilities from dose volume histograms. Int J Radiat Oncol Biol Phys 13: 103–109

Lyman JT, Wolbarst AB (1989) Optimization of radiation therapy. IV. A dose volume histogram reduction method. Int J Radiat Oncol Biol Phys 17: 433–436

Mackie TR, Bielajew AF, Rogers DWO, Battista JJ (1988) Generation of photon energy deposition kernels using the EGS Monte Carlo code. Phys Med Biol 33: 1–20

Mageras GS, Podmaniczky KC, Mohan R (1992) A model for computer-controlled delivery of 3-D conformal treatments. Med Phys 19: 945–953

McCullough EC (1987) Potentials of computed tomography in radiation treatment planning. Radiology 129: 765–768

McCullough EC, Krueger AM (1980) Performance evaluation of computerized treatment planning systems for radiotherapy: external photon beams. Int J Radiat Oncol Biol Phys 6: 1599–1605

McShan DL (1990) Conformal treatment planning. Med Phys Bull Assoc Med Phys India 15: 190–199

McShan DL, Fraass BA (1987) Integration of multi-modality imaging for use in radiation therapy treatment planning. In: Lemke HU, Rhodes ML, Jaffee CC, Felix R (eds) Computer assisted radiology. Springer, Berlin Heidelberg New York, pp 300–304

McShan DL, Glicksman AS (1984) Graphical simulation and design of beam portal blocking. In: Proceedings of the Eighth International Conference on the Use of Computers in Radiation Therapy. IEEE Computer Society, Toronto, Canada, pp 114–118

McShan DL, Silverman A, Lanza D, Reinstein LE, Glicksman AS (1979) A computerized three-dimensional treatment planning system utilizing interactive color graphics. Br J Radiol 52: 478–481

McShan DL, Fraass BA, Lichter AS (1990) Full integration of the beam's eye view concept into computerized treatment planning. Int J Radiat Oncol Biol Phys 18: 1485–1494

McShan DL, Matrone G, Fraass BA, Lichter AS (1993) A large screen digitizer system for radiation therapy treatment planning. Int J Radiat Oncol Biol Phys 26: 681–684

McShan DL, Fraass BA (1994) UM-CCRS/SP: sequence processor for computer controlled radiotherapy treatment delivery. In: Hounsell AR, Wilkinson JM, Williams PC (eds) Proceedings of the XIth International conference on the use of computers in radiation therapy. Medical Physics Publishing, Madison, Wisc., pp 210–211

Meyers GJ (1979) The art of software testing. John Wiley, New York

Mohan R, Barest G, Brewster LJ, Chui CS, Kutcher GJ, Laughlin JS, Fuks Z (1988) A comprehensive three-dimensional radiation treatment planning system. Int J Radiat Oncol Biol Phys 15: 481–495

Munzenrider JE, Brown AP, Chu JC, et al. (1991) Numerical scoring of treatment plans. Int J Radiat Oncol Biol Phys 21: 147–163

Mustafa AA, Jackson DF (1983) Phys Med Biol 28(2): 169–176

NCI (1987) Evaluation of treatment planning for particle beam radiotherapy. Radiotherapy Development Branch, Radiation Research Program, Division of Cancer Treatment, National Cancer Institute, Bethesda, M.D

NCI Photon Treatment Planning Working Group (1991) Three dimensional dose calculations for radiation therapy treatment planning. Int J Radiat Oncol Biol Phys 21: 25–36

Niemierko A, Goitein M (1990) Random sampling for evaluating treatment plans. Med Phys 17: 753–762

Niemierko A, Goitein M (1991) Calculation of normal tissue complication probability and dose volume histogram reduction schemes for tissues with critical element architecture. Radiother Oncol 20: 166–176

Pelizzari CA, Chen GTY (1987) Registration of multiple diagnostic imaging scans using surface fitting. In: Bruinvis IAD, van der Giessen PH, van Kleffens HJ, Wittkamper FW (eds) The use of computers in radiation therapy. Elsevier Science, North-Holland, pp 437–440

Rabinowitz I, Broomberg J, Goitein M, McCarthy K, Leong J (1985) Accuracy of radiation field alignment in clinical practice. Int J Radiat Oncol Biol Phys 11: 1857–1867

Rice RK, Mijnheer BJ, Chin LM (1988) Benchmark Measurements for lung dose correction for x-ray beams. Int J Radiat Oncol Brol Phys 15: 399–409

Rosenow UF, Dannhausen H-W, Lübbert K, et al. (1987) Quality assurance in treatment planning. Report from the German Task Group. In: Bruinvis IAD, van der Giessen PH, van Kleffens HJ, Wittkamper FW, (eds) The use of computers in radiation therapy. Elsevier Science, North-Holland, pp 45–58

Sauer O, Nowak G, Richter J (1987) Accuracy of dose calculations of the Philips treatment planning system OSS for blocked fields. In: Bruinvis IAD, van der Giessen PH, van Kleffens HJ, Wittkamper FW, (eds) The use of computers in radiation therapy. Elsevier Science, North-Holland, pp 57–60

Schad LR, Boesecke R, Schlegel W, Hartmann GH, Sturm V, Strauss LG, Lorenz WJ (1987) 3-D image correlation of CT, MR and PET studies in radiotherapy treatment planning of brain tumors. J Comput Assis Tomogr 11: 948–954

Schwade JG, Houdek PV, Landy HJ (1990) Small-field stereotactic external-beam radiation therapy of intracranial lesions: fractionated treatment with a fixed-halo immobilization device. Radiology 176: 563–565

Sewchand W, Aygun C, Nicholson G, Salazar OM (1986) Patient immobilization during CT for treatment planning of head and neck cancer. Radiology 158: 251–252

Sherouse GW, Mosher CE, Novins K, Rosenman J, Chaney EL (1987) Virtual simulation: concept and implementation. In: Bruinvis IAD, van der Giessen PH, van Kleffens HJ, Wittkamper FW (eds) The use of computers in radiation therapy. Elsevier Science, North-Holland, pp 433–436

Sherouse GW, Bourland JD, Reynolds K, McMurry HL, Mitchell TP, Chaney EL (1990) Virtual simulation in the clinical setting: some practical considerations. Int J Radiat Oncol Biol Phys 19: 1059–1065

Sontag MR, Galvin JM, Axel L, Bloch P (1984) The use of NMR images for radiation therapy treatment planning. Proceedings of the Eighth International Conference on the Use of Computers in Radiation Therapy. IEEE Computer Society, Toronto, Canada, pp 168–172

Sterling TD, Glicksman AS, Knowlton K, Weinkam J (1971) Three-dimensional treatment plan display on computer-produced films. In: Glicksman AS, Cohen M, Cunningham JR (eds) Computers in radiotherapy. (Proceedings of the 3rd International Conference on Computer in Radiotherapy, Glasgow, September 1970.) Br J Radiol (Special Report) 5

Stern RL, Fraass BA, Gerhardsson A, McShan DL, Lam KL (1992) Generation and use of measurement-based 3-D dose distributions for 3-D dose calculation verification. Med Phys 19: 165–174

Ten Haken RK, Forman JD, Heimburger DK, et al. (1991a) Treatment planning issues related to prostate movement in response to differential filling of the rectum and bladder. Int J Radiat Oncol Biol Phys 20: 1317–1324

Ten Haken RK, Kessler ML, Stern RL, Ellis JH, Niklason LT (1991b) Quality assurance of CT and MRI for radiation therapy treatment planning. In: Starkschall G, Horton JL (eds) Quality assurance in radiotherapy physics. Med Phys Publishing, Madison, Wisc., pp 73–103

Ten Haken RK, Thornton AF, Sandler HM, et al. (1992) A quantitative assessment of the addition of MRI to CT-based, 3-D treatment planning of brain tumors. Radiother Oncol 25: 121–133

Thames HD, Schultheiss TE, Hendry JH, Tucker SL, Dubray BM, Brock WA (1991) Can modest escalations of dose be detected as increased tumor control? Int J Radiat Oncol Bio Phys 22: 241–246

Thomas SJ, Wilkinson ID, Dixon AK, Dendy PP (1992) Magnetic resonance imaging of Fricke-doped agarose gels for the visualization of radiotherapy dose distributions in a lung phantom. Br J Radiol 65: 167–169

Thornton AF Jr, Ten Haken RK, Weeks KJ, Gerhardsson A, Correll M, Lash KA (1991a) A head immobilization system for radiation simulation, CT, MRI, and PET imaging. Med Dosim 16: 51–56

Thornton AF Jr, Ten Haken RK, Gerhardsson A, Correll M (1991b) Three-dimensional motion analysis of an improved head immobilization system for simulation, CT, MRI, and PET imaging." Radiother Oncol 20: 224–228

Thornton AS Jr, Sandler HM, Ten Haken RK, et al. (1992) The clinical utility of MRI in 3-D treatment planning of brain neoplasms. Int J Radiat Oncol Biol Phys 24: 767–775

Tsien KC (1955) The application of automatic computing machines to radiation treatment planning. Br J Radiol 28: 432–439

Urie MM, Goitein M, Doppke K, et al. (1991) The role of uncertainty analysis in treatment planning. Int J Radiat Oncol Biol Phys 21: 91–107

van de Geijn J (1965) The computation of two and three dimensional dose distributions in cobalt-60 teletherapy. Br J Radiol 38: 369–377

van de Geijn J, Harrington FS, Lichter AS, Glatstein E (1983) Simplified bite-block immobilization of the head. Radiology 149: 851

van de Geijn J, Fraass BA (1984) The net fractional depth dose: a basis for a unified analytical description of FDD, TAR, TMR, and TPR. Med Phys 11: 784–793

Van Dyk J, Barnett RB, Cygler JE, Shragge PC (1993) Commissioning and quality assurance of treatment planning computers. Int J Rad Oncol Biol Phys 26: 261–273

Verhey LJ, Goitein M, McNulty P, Munzenrider JE, Suit HD (1982) Precise positioning of patients for radiation therapy. Int J Radiat Oncol Biol Phys 8: 289–294

Weeks KJ, Fraass BA, McShan DL, Hardybala SS, Hargreaves EA, Lichter AS (1989) Comparison of automated and manual shielding block fabrication. Int J Radiat Oncol Biol Phys 16: 501–504

Westmann CF, Mijnheer BJ, van Kleffens HJ (1984) Determination of the accuracy of different computer planning systems for treatment with external photon beams. Radiother Oncol 1: 339–347

Wittkamper FW, Mijnheer BJ, van Kleffens HJ (1988) Dose intercomparison at the radiotherapy centers in The Nether-lands. 2. Accuracy of locally applied computer planning systems for external photon beams. Radiother Oncol 11: 405–414

Wolbarst AB, Chin LM, Svensson GK (1982) Optimization of radiation therapy: integral-response of a model biological system. Int J Radiat Oncol Biol Phys 8: 1761–1769

Yanke BR, Ten Haken RK, Aisen A, Fraass BA, Thornton AF (1991) Design of MRI scan protocols for use in 3-D, CT-based treatment planning. Med Dosim 16: 205–211

Zagars GK, Schultheiss TE, Peters LJ (1987) Inter-tumor heterogeneity and radiation dose-control curves. Radiol Oncol 8: 353–361

5 Electron Beam Dose Calculations*

DAVID JETTE

CONTENTS

5.1 Introduction

Electron beams are now in widespread use in the radiation treatment of cancer because their rapid dose falloff minimizes irradiation of critical healthy tissues beyond the treatment volume. Electron dose-calculation algorithms are required for clinical treatment planning, and in conjunction with computed tomography (CT) scan data they can potentially be made quite accurate. Unfortunately, it is not yet possible to routinely calculate the absorbed dose distribution accurately (to within 5% or, preferably, 3%) in many clinical situations, particularly in the presence of tissue inhomogeneities and body curvature. Much progress has been made in tackling this difficult problem, however, and in this chapter we shall examine current electron dose-calculation methods and identify the problems requiring further research.

We shall start, in Sect. 5.2, by looking at what we call "classical approaches" to the problem of electron dose calculation, by which we mean those which are not based upon, or extend, Fermi-Eyges multiple-scattering theory. Particularly in modeling broad beams irradiating a homogeneous medium, such approaches have through the years proven to be quite useful.

* This work was performed while the author was Director of Therapeutic Radiological Physics at the Institute of Applied Physiology and Medicine in Seattle

DAVID JETTE, PhD, Executive Director, The Lawrence H. Lanzl Institute of Medical Physics, 3876 Bridge Way N, Suite 300 Seattle, WA 98103-7951, USA and Professor of Medical Physics, Rush-Presbyterian-St. Luke's Medical Center, Chicago, IL 60612, USA

In the first years of use of high-energy electron beams for cancer therapy, the effect of such tissue inhomogeneities as air cavities, lung, and bone was described by elementary theories of equivalent thickness, or by empirical formulas. Such an approach was necessary in order to achieve at least qualitative understanding of the dose distribution before the advent of powerful computing facilities. Subsequently, investigators realized that a pencil electron beam ("ray") spreads out laterally approximately as a Gaussian function, with the amount of spread increasing with depth, and it became increasingly possible to achieve quantitative agreement between theory and reality. Thus a useful method of electron beam dose calculation computes the dose at a particular point by adding up the contributions of spreading pencil beams, the summation being taken over the whole field at the surface.

However, the crucial advance in electron dose calculation came with the application of Fermi-Eyges multiple-scattering theory, which describes the variation with penetration depth of the parameters of a pencil beam of electrons. The fundamental importance of multiple-scattering theory results from the fact that in penetrating the body, high-energy electrons suffer a great many collisions and follow a tortuous, although mainly forward, path. Thus multiple scattering is the dominant physical process involved, and multiple-scattering theory provides a reasonable first approximation for electron dose calculation. In Sect. 5.3 we shall explain the Fermi-Eyges theory in some detail, including some proposed modifications of it, and in Sect. 5.4 we shall examine the most recent (and accurate) models for dealing with localized inhomogeneities, most of which start from the Fermi-Eyges theory.

Finally, in Sect. 5.5 we shall review particular problems which need to be addressed by practical dose-calculation algorithms. One such problem, which is totally ignored by the (small-angle) Fermi-Eyges theory and its derivatives, is that of dose enhancement from electron backscatter at a tissue interface, such as from bone back to soft tissue.

Oblique incidence of the beam, resulting from surface curvature of the patient, poses another basic problem for an accurate dose-calculation algorithm, as does the effect of bremsstrahlung photons and high-energy secondary electrons arising from the collimation system. These contaminant particles can strongly influence output factors for various field sizes, and within the patient there is also redistribution of beam energy by bremsstrahlung photons and high-energy secondary electrons created here. In the more advanced algorithms it is necessary to be able to calculate scattering power and stopping power accurately from CT scan information, and there can be difficulties caused by end-of-range effects (range- and energy-straggling, diffusion of electrons rather than their being essentially forward-directed). Finally, for accurate dose calculation it is necessary to be able to model the effects of beam divergence and the location of the (virtual) source of the beam electrons.

Before starting our review of methods of electron beam dose calculation, we should mention some general references. The meaning of absorbed dose, and its relationship to fluence, are explained by ALM CARLSSON (1981), BRAHME (1982), and ANDREO and BRAHME (1983). The interactions undergone by electrons in penetrating matter are presented in detail by ANDERSON (1984). And useful surveys of the problems involved in electron dose calculation are included in works of the ICRU (1984a), KLEVENHAGEN (1985, 1988), and the AAPM Radiation Therapy Committee Task Group No. 25 (KHAN et al. 1991).

5.2 Classical Approaches

The earliest methods of electron dose calculation were empirical: isodose distributions were determined from measurements in water phantoms for particular treatment configurations [source-to-surface distance (SSD), field size, the particular treatment machine used]. One could then extrapolate to different SSDs using the inverse-square law and appropriate geometric shifting of the isodose curves, extrapolate (hopefully accurately) to field sizes different from the standard ones used in the measurements, and compensate for patient surface curvature through shifting of the isodose curves along beam rays. This provided the radiotherapist with a reasonably correct understanding of the dose to be expected in irradiating a homogeneous patient.

Unfortunately, real people are not very homogeneous, especially in regions of interest for electron radiotherapy such as the head and neck and the chest wall. It was still necessary to take into account the effects of tissue inhomogeneities (air cavities, bone, and lung), and this problem was first approached by shifting the isodose curves along beam rays in accord with the material encountered along such rays. In analogy with the decrease in primary photons along photon beam rays, the electron dose along a beam ray was considered to be the same as that given by a broad-beam depth-dose curve for water, but with this curve contracted or expanded by factors (perhaps empirically determined, but approximately the ratio of electron density of the tissue encountered to that of water) along the ray. Thus the absorption equivalent thickness (AET) factor method was pioneered by LAUGHLIN (1965a), who also introduced the ray concept and the correction for the polarization density effect. The AET factor method was refined with the coefficient of equivalent thickness method (BOONE et al. 1967; ALMOND et al. 1967) and the modified absorption coefficient method (BAGNE 1976). Clear reviews of these methods have been given by HOLT et al. (1978), STERNICK (1978), ANDREO (1985a), and KLEVENHAGEN (1985).

These "equivalent thickness" methods have little to do with the physics of electron transport, although historically they have provided useful parameterizations of electron dose distributions in the central region of broad beams for horizontally layered media (with the beam coming down from above along the z-axis, as usually depicted). In the central region of a broad beam, the main physical processes determining the dose are the loss of electrons (particularly those scattered at wide angles) with increasing depth z, the increasing skewness of the electrons' trajectories with increasing z, and the redistribution of beam energy by high-energy secondary electrons. If one considers only the dose deposited directly by the primary electrons, then the depth-dose curve is determined by two competing processes: loss of electrons, which tends to decrease the curve as z increases; and increasing skewness, which tends to *increase* the curve as z increases, since the dose deposited by the electrons is proportional to the secant of the angle between their direction and the z-axis, for a given planar fluence (taken normal to the z-axis). This is seen in Fig. 5.1, which gives the primary-electron depth-dose curve for an (infinitely) broad beam of 10 MeV electrons in water, as determined by EGS4 Monte Carlo calculations. Up to a depth of 3 cm, increasing skewness of the electrons' trajectories is dominant, resulting in a steadily rising depth-dose curve, but beyond 3 cm loss of electrons takes over.

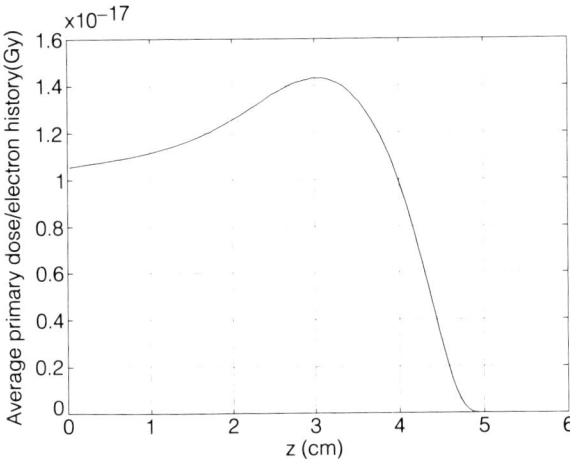

Fig. 5.1. The average primary dose per electron history in water as a function of depth z, for a broad beam of 10 MeV electrons. For this calculation, 991 000 electron histories were run, using an EGS4 Monte Carlo program which deposited locally all of the energy conveyed to secondary electrons. (Work carried out by Suzan Walker)

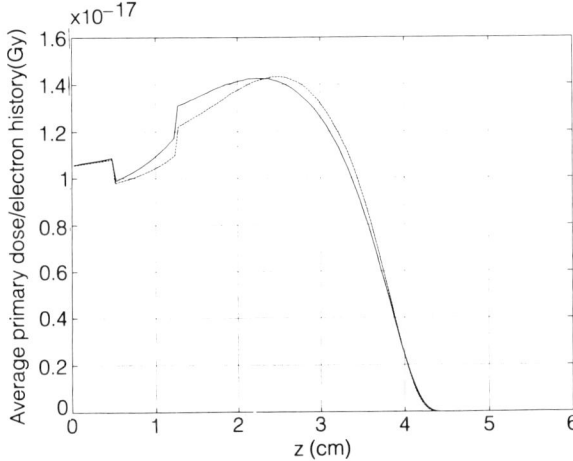

Fig. 5.2. Same as Fig. 5.1, but for a configuration with a slab of cortical bone inserted in the water between depths of 0.5 cm and 1.25 cm; 707 000 histories were run, and the results are plotted as the *solid curve*. The *dashed curve* represents an equivalent thickness calculation for the same configuration, as described in the text. (Work carried out by Suzan Walker)

Figure 5.2 (solid line) provides the primary-electron depth-dose curve for a similar configuration, in which a cortical bone slab has been inserted between $z = 0.5$ cm and $z = 1.25$ cm. Physically, these are both fictitious curves, for the Monte Carlo program used (JETTE and BIELAJEW 1989) ignores the transport of secondary particles (high-energy secondary electrons and bremsstrahlung photons) and rather considers the energy which would have been carried off by the secondary electrons to be deposited locally.

(The energy carried off by the bremsstrahlung photons is simply discarded, although it is deducted from the primary electrons' energy.) Nonetheless, these Monte Carlo calculations allow us to investigate the accuracy of an equivalent thickness approach to modeling the dose in the central part of the beam in a (horizontally) layered material. The dashed curve in Fig. 5.2 is the curve of Fig. 5.1 scaled in the following way. The portion from 0 to 0.5 cm was of course unchanged, but for 0.5 cm $< z <$ 1.25 cm the water curve was contracted in depth by a factor of 0.577 (the ratio of the total linear stopping powers for water and cortical bone at an energy of 8 MeV, as used in this computer program) and its value was multiplied by 0.905 (the ratio of the unrestricted mass colision stopping powers for cortical bone and water). For z >1.25 cm, the water curve was simply shifted toward $z = 0$ by 0.55 cm, since the energy lost in the 0.75 cm of bone was equal to that lost in 0.75/0.577=1.30 cm of water. Although such scaling seems to make sense physically, we do in fact see clinically significant discrepancies between the two depth-dose curves of Fig. 5.2. This is because energy is deposited more rapidly with depth in the real bone configuration than in its water-scaled equivalent, since bone scatters the primary electrons much more than would water of the same (mass) density as bone.

Thus, in understanding the distribution of energy deposited by electron beams, it is important to disabuse oneself of the concept of equivalent thickness as applied to scaling isodose or depth-dose curves. Indeed, fairly recent studies (MORI 1985; OGAWA et al. 1987) have demonstrated the inaccuracy of the AET method. One really cannot understand electron beam dose distributions without starting from multiple-scattering theory, for multiple scattering is in fact the dominant physical process involved. As we shall see in Sect. 5.4, it is quite possible to scale depth-dose curves accurately for layered media, so long as the energy and the state of angular dispersion (due mainly to multiple scattering) of the primary electrons is matched at the layer interfaces (LAX and BRAHME 1985). Also as described in Sect. 5.4, a theoretical treatment of the increase in dose due to increasing skewness of the primary electrons' trajectories is provided by "second-order multiple-scattering theory" (JETTE 1985b; JETTE and BIELAJEW 1989).

These equivalent thickness methods provide simple analytic forms for fitting depth-dose curves to experimental data through least-squares analysis (JETTE 1982), and analytic forms of various complexity are still being proposed for this purpose (JETTE et al. 1981; STRYDOM 1984, 1991; MEIGOONI and DAS

1987; CHEN 1988a; KOVÁŘ 1989; TABATA et al. 1991).
Broad-beam depth-dose distributions are highly
machine-dependent, being highly sensitive not only
to the design of the treatment head (which produces
"contamination" of the beam in the form of
bremsstrahlung photons and high-energy secondary
electrons), but even to the dispersion in energy of the
primary electrons (JOHNSEN et al. 1983). However, by
taking into account the state of angular dispersion of
the primary electrons at depth, WERNER (1983) and
WERNER et al. (1983) have been able to construct universal depth-dose curves. Any method of accurate
representation of broad-beam depth dose can be a
useful first step in constructing broad-beam dose distributions for a homogeneous medium (e.g., C.E.
NELSON et al. 1984). Nonetheless, as pointed out
above, one cannnot accurately obtain corresponding
dose distributions even for a horizontally layered
medium (the next simplest case) through scaling,
without taking the state of angular dispersion into
account. The physics of electron transport has got to
be involved in such modeling. (However, by including backscattering corrections in addition to the use
of scaling, TABATA and ITO (1991 and 1992) have constructed a semi-empirical model for electron depth-
dose curves which provides moderately good results,
as reviewed by TABATA (1993).)

One classical approach which has evoked considerable interest has been the use of "age-diffusion
theory," which goes back to the work of BETHE et al.
(1938). ROESCH (1954) early applied this theory to the
calculation of dose from certain distributions of beta
emitters, and KAWACHI (1975) later generalized from
age-diffusion theory as used in nuclear reactor
physics, to provide a model for electron dose distribution for a rectangular beam penetrating a homogeneous medium. In two dimensions (x,z), Kawachi
obtained

$$D(x,z,\tau) = A\left[erf\left(\frac{x_0 - x}{2\sqrt{\kappa\tau}}\right) + erf\left(\frac{x_0 + x}{2\sqrt{\kappa\tau}}\right)\right]$$
$$\cdot \cos\left(\frac{2\pi}{3R_e}z - \frac{\pi}{6}\right)e^{-\left(\frac{2\pi}{3R_e}\sqrt{\kappa\tau}\right)^2}.$$

In KAWACHI's equation, the dose $D(x,z,\tau)$ depends
also on the "age" τ of the electrons, but this quantity
is taken to be a function of depth z:

$$\sqrt{\kappa\tau} = \left(\frac{cz}{R_e} + \rho\right)^n$$

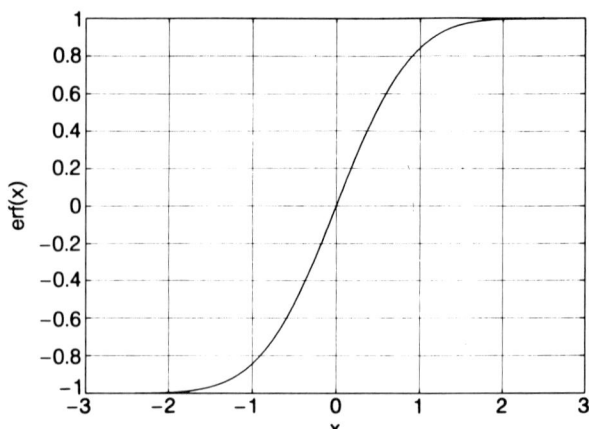

Fig. 5.3. The error function $erf(x)$

A is a normalization constant, R_e is the extrapolated electron range, κ is the diffusion coefficient of
the medium, and c, n, and ρ are arbitrary parameters
used to obtain a best fit with experimental data. X_0 is
one-half the field width at the surface, but it can be
modified through use of the inverse-square law for
divergent beams. The "error function" erf is defined
by

$$erf(x) = \frac{2}{\sqrt{\pi}}\int_0^x e^{-t^2}dt,$$

and is plotted in Fig. 5.3

The Kawachi formula provides a three-dimensional representation of the dose when another transverse factor with error functions in y is inserted, and
an overall inverse-square-law factor is included. Its
dependence on the transverse variable x is given by
the sum of the error functions, and from Fig. 5.3 we
see that this sum is zero well outside the beam, but
essentially constant within the central portion of the
beam if X_0 is large enough. However, to fit this representation closely to experimental data, it turned out
to be necessary to modify the argument of the cosine
factor by using an arbitrary polynomial in z. STEBEN
et al. (1979) used a polynomial of degree 2, while
MILLAN et al. (1979) used one of degree 6; their fits to
data for 10 MeV electrons in water are given in Figs.
5.4 and 5.5.

Starting from the Kawachi formula, other
researchers have proposed models for three-dimensional dose distributions in homogeneous media
(NÜSSLIN 1980; Shackleton 1984). And the work of
Steben et al. (1979) has provided a three-dimensional pencil-beam model, obtained by dividing the field
into small square areas and calculating the dose contributed from each subbeam (SUNTHARALINGAM and

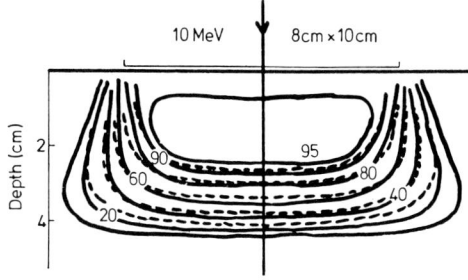

Fig. 5.4. Central-plane isodose distribution for a 10-MeV electron beam of 8 cm width by 10 cm, as calculated using an age-diffusion method. *Dashed lines* denote measurements, while *solid lines* denote calculations. (From STEBEN et al. 1979)

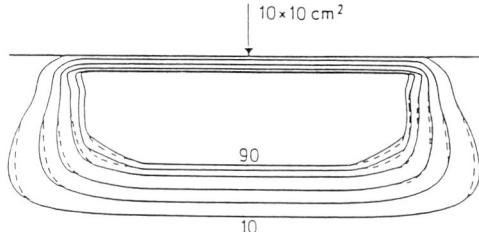

Fig. 5.5. Isodose distribution in central plane for 10 MeV electrons, as calculated using another age-diffusion method. *Dashed lines* denote measurements, while *solid lines* denote calculations. (From MILLAN et al. 1979)

AYYANGAR 1984). This procedure can give reasonably accurate dose calculations for a large inhomogeneity such as the lung (CYGLER and ROSS 1988), but it is not known how well it handles localized inhomogeneities. Age-diffusion theory can be useful in analytically modeling the dose deposited in a homogeneous medium, but it is not applicable to the theoretical description of electron beams except near the end of their range, where the diffusion process becomes dominant. Indeed, its inapplicability has been demonstrated by the need to modify the argument of the cosine term in Kawachi's formula: the resulting formula no longer satisfies the fundamental age equation.

These broad-beam models have been reviewed by several authors (NÜSSLIN 1979; ANDREO 1985a; JETTE 1985a). Dose-calculation models for homogeneous media have also been proposed using scatter functions, in analogy to photon dose calculation (DUTRIEX and BRIOT 1985; VAN DE GEIJN et al. 1987). But what is really necessary, in developing accurate dose-calculation algorithms applicable to localized inhomogeneities, is to start with the dominant physical process involved: multiple scattering of the primary electrons. Thus we turn next to understanding Fermi-Eyges multiple-scattering theory.

5.3 Multiple-Scattering Theory

Logically, the next step in developing an electron dose-calculation algorithm capable of dealing accurately with localized inhomogeneities is to consider a broad beam as being made up of many individual pencil beams, as was done for age-diffusion models. Then the effect of an inhomogeneity on each pencil beam can be analyzed and taken into account, and it is straightforward to allow for varying intensity across the field and for patient surface curvature. Pencil-beam summation also provides immediate calculation of dose distributions for irregular fields, assuming that the effect of collimator scatter is properly included. Initially, investigators used experimental pencil-beam data directly for this purpose (ROZENFELD et al. 1969; LILLICRAP et al. 1975; NATH et al. 1980).

An important advance in pencil-beam summation was realizing that the radial dose profile of pencil beams is approximately Gaussian. GOITEIN et al. (1978) and GOITEIN (1978) first used Gaussian representations of pencil-beam profiles to explain quantitatively the occurrence of "hot and cold spots" beyond the vertical edge of an inhomogeneity. The mechanism for the creation of such hot and cold spots is seen in Fig. 5.6, in which a dense inhomogeneity such as bone is located at depth $z = z_0$, for $x \geq 0$. The beam is passing through scattering material, such as muscle tissue, which continuously increases the angular dispersion of the beam electrons with increasing depth; this angular dispersion is indicated by the pairs of arrows in the figure. The inhomogeneity increases the angular dispersion

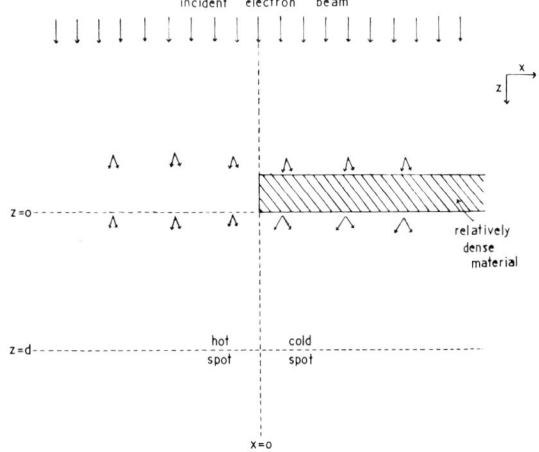

Fig. 5.6. Formation of hot and cold spots beneath the sharp edge of a dense inhomogeneity. (From JETTE et al. 1989)

drastically as the right part of the beam passes through it. The result is an increase in electrons for x slightly less than zero at depth $z = d$, since additional electrons are being scattered to there by the inhomogeneity, over the number of electrons from the right part of the beam which would reach that place in the absence of the inhomogeneity. Thus there is a "hot spot" there, and, conversely, a "cold spot" at $z = d$ for x slightly greater than zero.

Mathematically, this mechanism of hot- and cold-spot creation can be understood in terms of Gaussian pencil beams G of r.m.s. width $\sigma(z)$:

$$G\bigl(x, z; \sigma(z)\bigr) = \frac{e^{-\left(\frac{x^2}{2\sigma(z)}\right)}}{\sqrt{2\pi\sigma^2(z)}}.$$

If the half-slab is considered to be (infinitely) thin and no primary electrons are lost within it, it can be shown (JETTE et al. 1989) that physically it is as if Gaussian pencil beams of certain initial angular dispersion were propagated from the level $z = 0$. (This is the concept of "beam redefinition", which we shall examine in the next section as a way to deal accurately with localized inhomogeneities.) These Gaussian pencil beams on the right side (for $x > 0$) have larger initial angular dispersion than those on the left side, so at depth $z = d$ their r.m.s. width $(D_+/2)^{1/2}$ is greater than the r.m.s. width $(D_-/2)^{1/2}$ of those pencil beams emanating from the left side. For a fixed calculation point (x, d) we integrate the contributions of all these pencil beams (i.e., over x at $z = 0$) to obtain the (normalized) dose in the central region given by Eq. 62 of JETTE et al. (1989):

$$L(x, z) = 1 - \frac{1}{2}\left[erf\left(\frac{x}{\sqrt{D_-}}\right) - erf\left(\frac{x}{\sqrt{D_+}}\right)\right].$$

With $D_+ > D_-$, inspection of Fig. 5.3 shows that there will be a cold spot for x somewhat greater than zero, and a hot spot for x somewhat less than zero. This is the basic way in which multiple-scattering theory models the effects of localized tissue inhomogeneities.

Gaussian pencil beams were used in dose-calculation models by ABOU MANDOUR and HARDER (1978) and KOZLOV and SHISHOV (1982). But what is necessary is to keep track not only of the lateral dispersion of a pencil beam, but also of its angular dispersion; this is the only way that one can accurately take into account the effects of inhomogeneities. This is done through Fermi-Eyges multiple-scattering theory,

which is based upon the usual small-angle approximation:

$$\sin \Theta \approx \Theta$$
$$\cos \Theta \approx 1.$$

Let us use the coordinate system depicted in Fig. 5.7, which defines the *projections* Θ_x and Θ_y (in the small-angle approximation) of the polar angle Θ upon the x–z and y–z planes respectively. In two dimensions, we deal with the distribution function $P(x, \Theta_x, z)$ defined in the following way: $P(x, \Theta_x, z) \Delta x \Delta \Theta_x$ is the number of electrons whose location is between x and $x + \Delta x$, and whose direction is between Θ_x and $\Theta_x + \Delta \Theta_y$, when they reach depth z. Then P must satisfy the Fermi equation (ROSSI and GREISEN 1941):

$$\frac{\partial P}{\partial z} + \Theta_x \frac{\partial P}{\partial x} - \frac{T}{4}\frac{\partial^2 P}{\partial \Theta_x^2} = 0.$$

In this equation T is the (linear) scattering power, considered to be a function of position; at a point

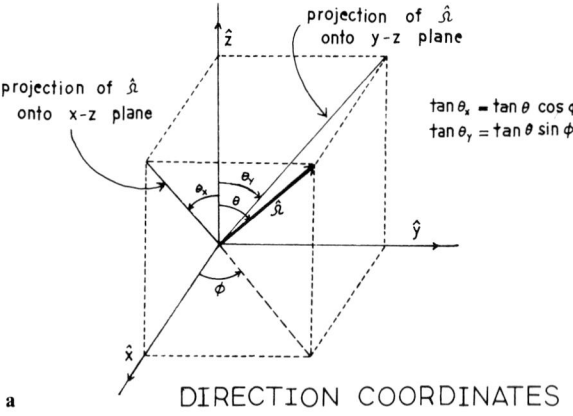

a DIRECTION COORDINATES

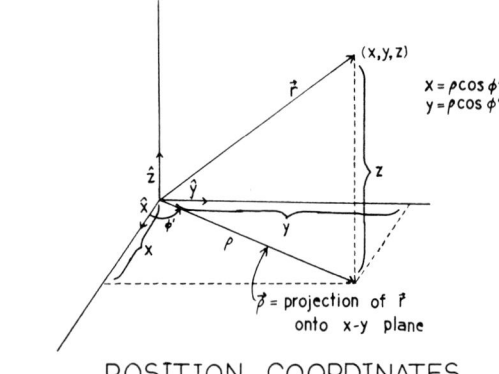

b POSITION COORDINATES

Fig. 5.7 a,b. Definition of coordinates used for Fermi-Eyges multiple-scattering theory. (From JETTE 1988b)

(x, z), it is essentially determined by the energy of the primary electrons and the electron density of the material there.

Fermi originally solved this equation for constant T, which physically means both that the energy loss within the medium is negligible, and that the medium itself is homogeneous. This case is appropriate for the study of electron beams in air before reaching the patient (BRAHME 1971; HUIZENGA and STORCHI 1987; POLMAN and VAN DER LINDEN 1987a, b; SANDISON 1987; SANDISON and HUDA 1988). And even when dealing with a dense medium, the approximation of constant scattering power can be useful as a means of getting a handle on what is going on (PERRY and HOLT 1980). But for accurate calculation of dose in a homogeneous or horizontally layered medium, what is needed is the solution to the Fermi equation for scattering power T dependent upon depth z, and this was provided by EYGES (1948).

As explained in previous reviews (JETTE et al. 1983; JETTE 1984a, 1988b), for a ray (or pencil beam) initially directed along the z-axis at $x = y = z = 0$, the Eyges solution to the Fermi equation for $T = T(z)$ is

$$P\left(x, \theta_x z\right) = \frac{e^{\left(-\frac{A_0 x^2 - 2A_1 x\theta + A_2 \theta_x^2}{A_0 A_2 - A_1^2}\right)}}{\pi\sqrt{A_0 A_2 - A_1^2}},$$

where the A_n are functions of z:

$$A_0(z) \equiv \int_0^z T(\xi)\,d\xi \approx Tz$$

$$A_1(z) \equiv \int_0^z T(\xi)(z - \xi)\,d\xi \approx \frac{1}{2}Tz^2$$

$$A_2(z) \equiv \int_0^z T(\xi)(z - \xi)^2\,d\xi \approx \frac{1}{3}Tz^3.$$

In these expressions for the $A_n(z)$ we have also given their approximations for constant scattering power T, which hold for a homogeneous medium when the energy loss is neglected. Such approximations can be useful in helping to understand what all this mathematics means physically.

In these expressions for $A_n(z)$ it is necessary to re-express the scattering power, which depends on the electrons' energy at a point, as a function of the depth. This is accomplished by first calculating an average electron energy at depth through use of the total stopping power, and then assuming a power-law dependence of T on electron energy, as explained by JETTE (1988b). Figure 5.8 gives the $A_n(z)$ for 10 MeV electrons in water. [Note that Table I of JETTE (1988b), which gives parameters of the power-law fit for T, has been superseded by Table I of JETTE and

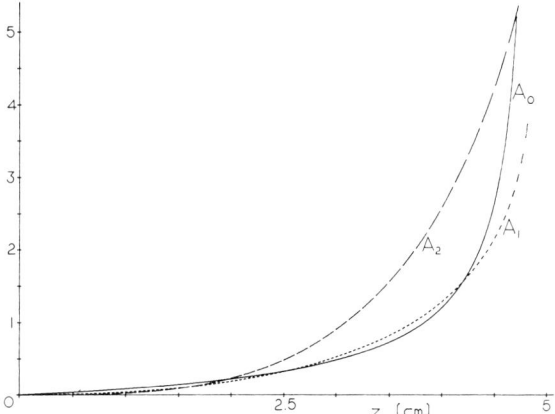

Fig. 5.8. The functions $A_0(z)$, $A_1(z)$, and $A_2(z)$ for 10 MeV electrons in water. The units of $A_n(z)$ are $(cm)^n$. (From JETTE et al. 1988b)

WALKER (1992a), and that before 1992 the Jette group used k rather than T to represent the scattering power.]

In three dimensions, P is simply the product of two two-dimensional P's: $P(x, \Theta_x; y, \Theta_y; z) = P(x, \Theta_x, z)P(y, \Theta_y, z)$. But in order to find the dose at a point (x, y, z), we need the planar fluence (or "location distribution") $L(x, y, z)$ defined as the integral of P over all angles Θ_x and Θ_y. (These integrals are taken as going from $-\infty$ to ∞ for mathematical simplicity – in the small-angle approximation this is valid to do.) Thus

$$L(x, y, z) = \frac{e^{\left(-\frac{x^2 + y^2}{A_2(z)}\right)}}{\pi A_2(z)},$$

and the dose D is then

$$D(x, y, z) = \frac{S}{\rho}(x, y, z)L(x, y, z),$$

where $(S/\rho)(x, y, z)$ is the restricted mass collision stopping power at (x, y, z), a function of the energy of the primary electrons and the material there. (We are now calculating the dose directly deposited by the primary electrons – the dose deposited by secondary particles must be determined separately.) "Iso-probability curves" of constant $L(x, y, z)$ are shown in Fig. 5.9, for a pencil beam of 10 MeV electrons in water.

Dose distributions for broad beams, of possibly varying intensity across the field, can be built up simply by summing up contributions from these pencil beams. In particular, for a collimated uniform parallel beam normally incident upon the flat surface

102

David Jette

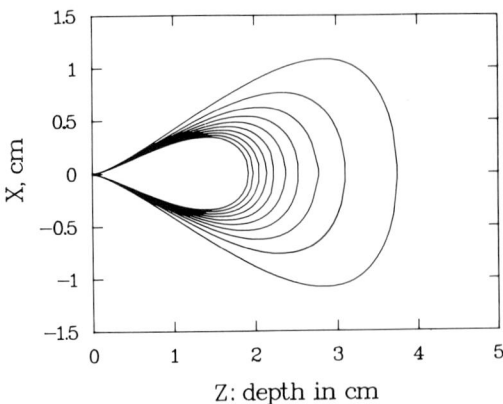

Fig. 5.9. Isoprobability curves in the x–z plane for a narrow beam of 10 MeV electrons in water, initially at the origin and traveling along the z-axis. (From JETTE et al. 1983)

of a layered medium, one obtains (HOGSTROM et al. 1981; WERNER et al. 1982; ABOU MANDOUR et al. 1983; JETTE et al. 1983)

$$D(x,y,z) = C(z)\left[erf\left(\frac{a+x}{\sqrt{A_2(z)}}\right) + erf\left(\frac{a-x}{\sqrt{A_2(z)}}\right)\right]$$
$$\cdot \left[erf\left(\frac{b+y}{\sqrt{A_2(z)}}\right) + erf\left(\frac{b-y}{\sqrt{A_2(z)}}\right)\right],$$

where $C(z)$ is constant in the transverse variables x and y, and the field size is $2a \times 2b$. As it happens, the transverse dependence here is of the same form as that for the age-diffusion models of the previous section, which explains their success for broad beams in homogeneous media. EDWARDS and COFFEY II (1979) have studied this "cumulative normal" distribution for collimated electron beam profiles.

This, in a nutshell, is the Fermi-Eyges multiple-scattering theory. It has been discussed in texts on the scattering of electrons (ROSSI 1952; SCOTT 1963; ZERBY and KELLER 1967; LUO and BRAHME 1993b). In 1967 WR Nelson provided a thorough derivation of it, for application to accelerator shielding problems (unpublished notes, Stanford University). Only very recently has the Fermi equation been derived rigorously from the Boltzmann equation: WV Prestwich at McMaster University, in unpublished lecture notes, has shown how to do this in a straightforward fashion; and Larsen (1993) has provided the full derivation, examining the various approximations entering into it. The application of the Fermi-Eyges theory to electron-beam dose calculation was first suggested by BRAHME (1975), but unfortunately that

early work was not readily available in North America. BRAHME et al. (1981) subsequently published a dose-calculation model using the Fermi-Eyges theory, at around the same time that HOGSTROM et al. (1981) and WERNER et al. (1982) published such models. All three models were applicable to (horizontally) layered media and arbitrary field shapes, but the Hogstrom group's work was already explicitly in the form of an algorithm which could deal practically with localized tissue inhomogeneities, and that of the Brahme group was also developed (particularly by Lax) into a practical dose-calculation algorithm, as we shall see in the next section. In any case, all three groups share credit for introducing Fermi-Eyges multiple-scattering theory for use in calculating electron beam dose.

The Fermi-Eyges theory has been applied to a number of problems concerning the effects of angular dispersion. It has been used effectively for energy and spatial scaling with horizontally layered media and in going from one homogeneous medium to another (BRAHME and LAX 1983; WERNER et al. 1983; LAX and BRAHME 1985; NAHUM and BRAHME 1985), for which the state of angular dispersion must be taken into account. JETTE (1984b) and JETTE et al. (1989) have used the Fermi-Eyges theory to predict "focused hot spots" without accompanying cold spots under the tips of wedges. The relationship between the initial angular variance of the beam electrons and penumbra width in air has been studied by a number of investigators (HUIZENGA and STORCHI 1987; SANDISON 1987; MOHAN et al. 1988; SABBAS et al. 1987; SANDISON and HUDA 1988; DEASY et al. 1992; HUIZENGA and VAN BATTUM 1992), and SABBAS et al. (1987) have also shown that a divergent beam is expected to have a markedly *narrower* penumbra within the patient than the equivalent parallel beam would have.

Fermi-Eyges multiple-scattering theory suffers from a number of limitations which come into play toward the end of the electron range. One obvious problem is that, as a small-angle theory, it is oblivious to backscattering of the primary electrons. For that matter, the Fermi-Eyges theory completely breaks down toward the end of the range of the beam electrons, where the dominant process is diffusion rather than transport of the electrons in a more-or-less forward direction. Another problem can be seen by integrating the planar fluence $L(x,y,z)$ over all x and y, to obtain the number of beam electrons at depth z (or, in the case of a single electron as was modeled above, the probability that the electron will reach depth z). Regardless of depth, this integrated planar fluence is constant with z: the beam electrons think they can continue merrily forever, and it is necessary to externally impose a finite range on them.

A great strength of the Fermi-Eyges theory is that, even for a horizontally layered medium, it predicts the r.m.s. width $\sigma_r(z)$ of a pencil-beam profile; multiplying the expression above for $L(x,y,z)$ by $r^2(=x^2+y^2)$ and integrating over all x and y, one finds that $\sigma_r(z) = [A_2(z)]^{1/2}$. This expression for σ_r is reasonably accurate to perhaps 70% of the electron range, but, as seen in Fig. 5.8, $A_2(z)$ becomes infinite as the end of the range is approached. In fact, σ_r goes to zero toward the end of the range as widely scattered electrons run out of energy and stop. ABOU MANDOUR et al. (1983) and NÜSSLIN (1984) have studied this problem, and VAN GASTEREN (1984) and VAN DER LINDEN et al. (1984) have had to use measured values of $\sigma_r(z)$ in constructing an electron pencil-beam algorithm (in place of values of σ_r given by age-diffusion models). WERNER et al. (1982) and LAX et al. (1983) have suggested modifications to the Fermi-Eyges theory to make σ_r go to zero at the end of the range. However, although both of these modifications provide dramatic improvement over the prediction of σ_r by the Fermi-Eyges theory towards the end of the range, they have still been found to be inaccurate there (SANDISON 1987; SANDISON et al. 1989).

The problem of modeling $\sigma_r(z)$ accurately is not yet solved, but new efforts have been made in addressing it. VAN GASTEREN (1987) has used decrement lines to model primary electron loss, and McPARLAND et al. (1988) have suggested a parameterization of σ_r particularly for use in modeling the dose distribution of small fields. The addition of a term to the right-hand side of the Fermi equation has been proposed (PAPIEŻ and SANDISON 1990; SANDISON and PAPIEŻ 1990) in order to account for electron loss, but this model erroneously predicts the buildup of the electron depth-dose curve as resulting (only) from the energy deposited by stopped primary electrons. Another model (McLELLAN et al. 1991) also proposes that a term be added to the Fermi equation without any physical justification for doing so, this time in order to oppose the scattering of electrons to large angles, and it remains to be seen whether this method will be proven to be physically valid.

A major shortcoming of the Fermi-Eyges theory is that it does not model large-angle single scattering, which produces a tail on the otherwise Gaussian radial dose profile of a pencil beam. To accurately model radial dose profiles, LAX et al. (1983) have constructed these profiles from three Gaussian functions, with the six parameters obtained by fitting Monte Carlo data. Ideally, however, one would like to integrate Molière scattering theory (BETHE 1953; SCOTT 1963; BRAHME 1971; ANDREO 1985b; BERGER and WANG 1988) into the formalism of the Fermi-Eyges theory, for the Molière theory as modified by BETHE (1953) does include large-angle single scattering. This has just been accomplished by JETTE and WALKER (1992b), with the addition of a convolution term of the form $G(\Theta_x, \Theta_y, z) \circ P(x, \Theta_x; y, \Theta_y; z)$ to the right-hand side of the Fermi equation. (The practical implications of this proposed new theory are still under investigation.)

While it is necessary to use multiple-scattering theory to model electron-beam dose distributions accurately, one need not be restricted to small-angle theory, but rather one can start more fundamentally from the Boltzmann transport equation (BRAHME 1985c). This approach, using a "moments" expansion and the continuous slowing-down approximation, was taken early by KESSARIS (1966) in computing flux, current, energy deposition, charge accumulation, and angular distributions for a broad parallel beam penetrating water; his calculations predicted both positive and negative charge accumulation, as found experimentally by LAUGHLIN (1965b). Subsequently the approach was used by OZDEMIR et al. (1985) in constructing a multigroup method for modeling electron transport, and by PERRY (1988) in pursuing (with great mathematical complexity) the multiple-scattering analysis of YANG (1951). A "Phase Space Time Evolution" model is presently being developed (HUIZENGA and STORCHI 1989; HUIZENGA et al. 1992; MORAWSKA-KACZYŃSKA and HUIZENGA 1992) incorporating fundamental electron and (bremsstrahlung) photon processes directly. Thus far it is applicable only to broad beams penetrating homogeneous or layered media, but its very high accuracy and reasonable calculation time may well make it a viable alternative to other analytic or Monte Carlo calculation of electron dose distributions.

Another approach is to divide the transport of the primary electrons into two parts: forward-directed electrons serve as the source for diffusing electrons. LUO (1985a,b) has carried out this analysis, again with great mathematical complexity, and LUO and BRAHME (1990, 1993a) have extended this "bipartition model" to include energy straggling and the redistribution of beam energy by secondary particles. Independently, starting from the diffusion theory of BETHE et al. (1938), WERNER (1985) has developed a similar model, mathematically much simpler and directed toward calculating dose perturbations at tissue interfaces. This latter model has been applied to the calculation of dose from distributions of beta-particle emitters (WERNER 1987, 1991; WERNER and DAS 1987; WERNER et al. 1988, 1991).

5.4 Localized Tissue Inhomogeneities

The existence of localized tissue inhomogeneities, such as air cavities, bone, and lung, presents the most difficult problem for electron beam dose calculation. In this respect, photon beam dose calculation is considerably simpler, for the (primary) photons travel through the irradiated material in straight lines until discrete interactions take place. Electrons, on the other hand, suffer innumerable minor collisions as they penetrate the material, spreading out from their original line of travel in a highly complicated way when inhomogeneities lie along their path. Thus an electron dose-calculation algorithm which accurately takes into account the effects of localized tissue inhomogeneities must be based on multiple-scattering theory in some form. Even for a homogeneous medium, the multiple scattering of electrons makes the design of bolus far more complicated than for photon beams, and usually impossible for obtaining a uniform dose distribution conforming to a treatment volume (Low et al. 1992).

Practically speaking, the big boost for use of Fermi-Eyges multiple-scattering theory in electron dose calculation has been provided by HOGSTROM et al. (1981), in developing a pencil-beam dose-calculation algorithm which incorporates directly the detailed information available from CT scans. This "Hogstrom algorithm" has been reviewed in some detail by NAHUM (1985a), and it has been extended to use in electron arc therapy (HOGSTROM and KURUP 1987; HOGSTROM et al. 1989). In principle it provides a full three-dimensional calculation of dose distributions, but as originally implemented, the calculation was only two-dimensional in order to be computationally practical. (Mathematically this means that to find the dose deposited at a particular point, one looks at the CT slice containing that point and considers the material found in that slice to extend indefinitely in the third dimension.) However, recently the computer implementation of this model has been improved to the point at which a full three-dimension calculation takes only four times as long as a two-dimensional one (STARKSCHALL et al. 1991).

Figure 5.10 and 5.11 show the very good accuracy of the Hogstrom algorithm for inhomogeneous phantom configurations simulating the lung and the

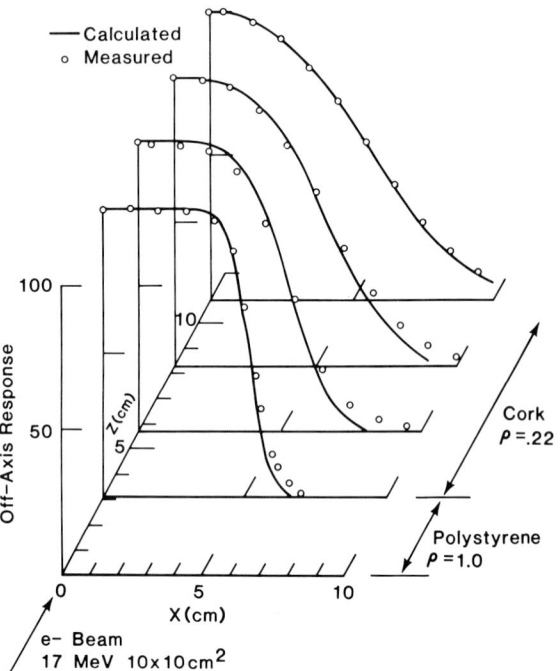

Fig. 5.10. Comparison of measured with calculated (Hogstrom algorithm) beam profiles in a polystryrene/cork/polystyrene slab phantom for a 10×10 cm field size, using 17 MeV electrons from a Therac 20 linear accelerator. (From HOGSTROM et al. 1981)

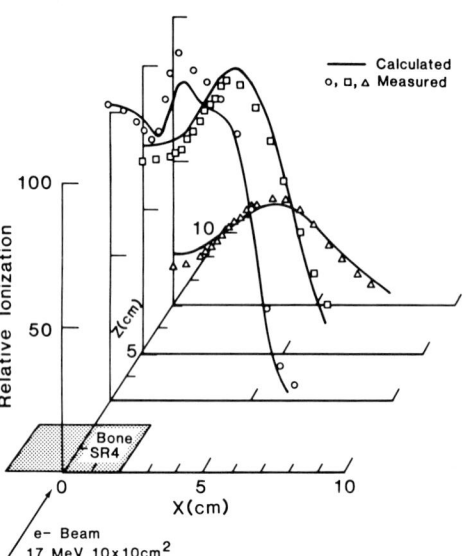

Fig. 5.11. Comparison of measured with calculated (Hogstrom algorithm) ionization profiles at various depths in a water phantom behind a 4 cm wide by 2 cm thick bone substitute block in a 10×10 cm field size, using 17 MeV electrons from a Therac 20 linear accelerator. The data have been normalized to the calculation at the central axis of the first profile. (From HOGSTROM et al. 1981)

sternum, respectively. The key to the Hogstrom algorithm's success is a clever mathematical device which allows the effects of the scattering of the beam (in air and off the collimator walls) above the final collimation level to be ignored in calculating the effects of localized tissue inhomogeneities. Suitably weighted pencil beams of zero initial angular dispersion are propagated from the level of (final) collimation, first through the remaining air to the patient surface and then through the patient. The Fermi-Eyges theory is applicable only to a layered medium, of course, so in the Hogstrom algorithm the material encountered along the axis of each pencil beam is considered to extend indefinitely in the lateral directions, for the purpose of calculating the dose contribution from that pencil beam to a particular point in the patient.

The Hogstrom algorithm uses the Fermi-Eyges theory only for its lateral distribution factor, taking its axial factor from measured data in a water phantom for each configuration of beam energy and electron applicator. The strength of this approach is that secondary processes, such as large-angle single scattering, production of high-energy secondary electrons, and scattering off the collimation system, are largely compensated for, so that in many cases the algorithm is able to provide high accuracy deep into the material, well beyond the limit (perhaps 50% of the primary electron range) of good accuracy of the Fermi-Eyges theory. The weakness of the approach is that these secondary process are not modeled from physical laws, so that the high accuracy may be lost for more complicated configurations.

The M.D. Anderson Cancer Center group early provided detailed evaluations of their algorithm (HOGSTROM and ALMOND 1983; HOGSTROM et al. 1984; HOGSTROM 1986). MORI (1985) compared the predictions of the Hogstrom algorithm and of earlier "effective thickness" algorithms with experimental data for a cork layer simulating lung, and KARLSSON (1985) conducted a similar study in the vicinity of bolus edges; both investigators found significant improvement in accuracy using the Hogstrom algorithm. However, BIELAJEW et al. (1987) found some discrepancies between the predictions of the Hogstrom algorithm and EGS4 Monte Carlo calculations for layered phantoms, and SHIU and HOGSTROM (1991a) have found the algorithm to underestimate by as much as 6% the increased dose in bone near a bone–tissue interface. Even for homogeneous phantoms, KURUP et al. (1992) reported discrepancies as large as 5% for isodose contours below 10%; while these discrepancies were judged not to be

clinically significant for fixed-beam therapy, they were thought to be important for arced beams.

Lack of accuracy of the Hogstrom algorithm in dealing with localized tissue inhomogeneities is pronounced for its two-dimensional implementation. ROGERS et al. (1984) and NAHUM (1985a) found such problems when using a short cylinder of air as the inhomogeneity, and the extensive comparisons with experimental data carried out by CYGLER et al. (1987) confirmed lack of accuracy of the Hogstrom algorithm for localized inhomogeneities of regular shape. Detailed studies by LAX (1987) using an anthropomorphic phantom also demonstrated significant inaccuracy of the Hogstrom algorithm, and KIRSNER et al. (1987) have concluded that a three-dimensional inhomogeneity correction is required instead for accurate dose calculation for treatment of retinoblastoma.

The need for a full three-dimensional inhomogeneity correction in certain cases was also concluded by MAH et al. (1989) and by STARKSCHALL et al. (1991), but significant inaccuracy of the Hogstrom algorithm actually goes beyond the dimensionality of its implementation. As we shall see later in this section, in Fig. 5.14, the Hogstrom algorithm is not able to handle deep localized inhomogeneities accurately. The basic problem is that when a pencil beam gets deep into the irradiated material, it has already spread out so much that it is relatively insensitive to the sharp lateral edges of the inhomogeneity, so that the inhomogeneity's effect tends to be washed out. Thus, in comparison with dose measurements for a spinal phantom, discrepancies as large as 14% were found for the Hogstrom algorithm (DOMINIAK 1991; DOMINIAK et al. 1991).

The Hogstrom algorithm has been refined and implemented by a number of investigators. BLOCH et al. (1984) implemented it in three dimensions using tissue-phantom ratios to calculate the dose on the axis of the pencil beams. KOOY and KIJEWSKI (1988) and KOOY and RASHID (1989) have worked out a fast three-dimensional implementation by breaking up the field into a set of rectangular subbeams through use of a quadtree data structure (depending upon the field shape and the distribution of underlying inhomogeneities) and propagating those subbeams through the irradiated material instead of uniformly distributed pencil beams. This approach provides an efficient way to calculate dose for an irregular field, but for deep inhomogeneities it suffers from the same limitations in accuracy as the (three-dimensional) Hogstrom algorithm. Finally, WILSON and ALMOND (1991) have used the Hogstrom algorithm to calculate the dose distribution for a laterally translated electron beam in a homogeneous medium, for application to a proposed treatment method in which the therapy couch is moved at varying speed to produce a nonuniform dose distribution.

As explained in Sect. 5.2, a problem with the Fermi-Eyges theory is that it does not include large-angle single scattering, so that it incorrectly represents the radial dose profile of a pencil beam as being Gaussian. LAX et al. (1983) addressed this problem by replacing the Gaussian description of a pencil beam by the sum of three Gaussians, the six parameters of which were fit to Monte Carlo or experimental data for homogeneous media, and scaled for layered media by the method described by LAX and BRAHME (1985). In this way, LAX (1985, 1986a) obtained his Generalized Gaussian Model, a pencil-beam model which is quite like the Hogstrom algorithm in considering only the material lying along the axis of a pencil beam in calculating that pencil beam's contributions to dose throughout the volume. Although also suffering from this "semi-infinite slab approximation" (as Lax terms it), the Generalized Gaussian Model does appear to outperform the Hogstrom algorithm in terms of accuracy (LAX 1986b, 1987). The crucial finding by Lax in making this comparison for a head phantom is that the Hogstrom algorithm tends to make local overestimations of dose, whereas his own model tends to make local underestimations: the former error can lead to local recurrence of the tumor, while the latter error may not be clinically significant. HYÖDYNMAA et al. (1987) have implemented the Generalized Gaussian Model on an array processor, thereby achieving a tenfold saving of computer time and opening the way to a full three-dimensional implementation of this algorithm.

These dose-calculation algorithms were critically reviewed by BRAHME (1985a,b), and BRAHME and NILSSON (1984), NAHUM (1985a), and LAX (1986b, 1987) all emphasized the inaccuracy inherent in using the "semi-infinite slab approximation", by both the Hogstrom algorithm and the Generalized Gaussian Model. The problem is that the Fermi-Eyges theory is applicable only to a layered medium, so that treating the material lying along a pencil beam's axis as extending laterally throughout the volume is a natural assumption to make, in evaluating its dose contributions. But if one is going to make such an

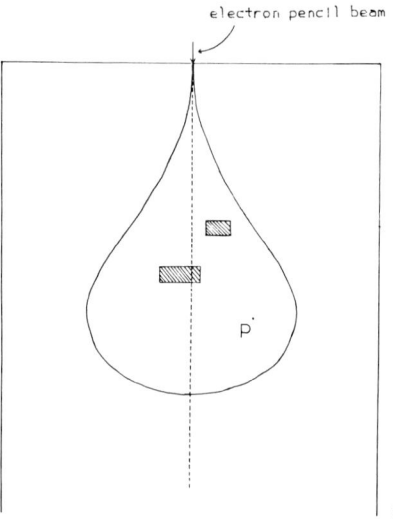

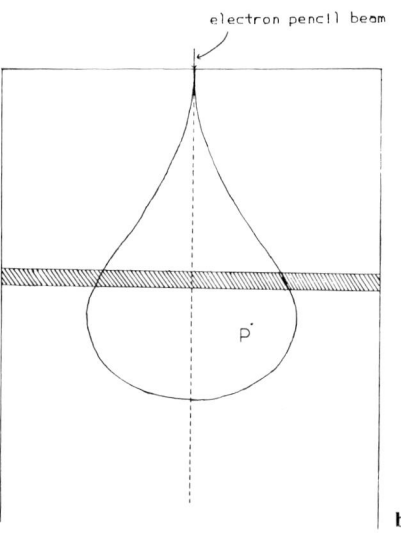

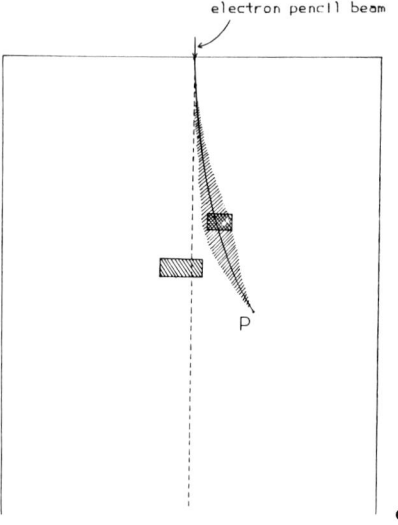

Fig. 5.12 a–c. The representative path approach. **a** "Fermi-Eyges bubble" of high concentration, for a pencil beam. **b** What Fermi-Eyges theory "sees" in calculating the dose at P. **c** "High-concentration bubble" for those pencil-beam electrons actually reaching *P*. (From JETTE 1986)

approximation, one may as well be more selective about the material chosen for lateral extension. Specifically, given a pencil beam and a point of calculation P off the central axis of the pencil beam, one can consider only the pencil-beam electrons which actually reach P, and look at the material through which most of *those* electrons pass. Thus, in calculating the dose contribution from the pencil beam to a point P, one looks at the material along a "representative path" from the pencil beam to P, rather than along the pencil beam's axis. This idea is illustrated in Fig. 5.12.

The representative path method was first put forward by PERRY and HOLT (1980). Their model did not include electron energy loss, but JETTE (1984b, 1986) subsequently provided representative path formulas for the Fermi-Eyges theory, which includes energy loss. Using this method, C.X. YU et al. (1988) have developed a Multi-Ray Model of electron dose calculation which treats localized inhomogeneities more realistically than the Hogstrom algorithm does. They found good agreement between the results of their calculations and Monte Carlo simulations. On the other hand, LAX (1986b) has found the original version of the representative path method (not including energy loss) to provide little improvement in accuracy over the Hogstrom algorithm, for a thick half-slab configuration. Futher comparison between the two algorithms is clearly needed, but it may be that while the representative path method conceptually seems advantageous to use in place of the Hogstrom algorithm, this advantage in dealing with localized inhomogeneities gets washed out for broad beams because of the many incident rays. Furthermore, the Multi-Ray Model appears to be quite impractical for routine treatment planning, for YU et al. give a typical time for the calculation of dose throughout an inhomogeneous medium as 6 h, and that is only for one of the many rays which constitute the incident field! (However, as we shall see in the next section, the representative path approach does have application to the problem of range straggling.)

In order to improve the accuracy of electron dose calculation for inhomogeneous configurations, researchers have used the technique of *beam redefinition:* the electron beam, whose fluence distribution is already known (approximately) at one layer of CT voxels (perpendicular to the beam), is propagated to the next layer taking into account the electron density (as determined by CT scanning) of irradiated material there, and this process is continued until the electrons run out of energy. This "moment method," as they termed it, was first introduced by STORCHI and HUIZENGA (1985), and in order to achieve sufficient

accuracy STORCHI et al. (1987) and STORCHI and VAN DER LINDEN (1989) have subsequently refined their method through use of a rather complicated sum of Hermite polynomials. [This technique of beam redefinition is *not* what was used in the Hogstrom algorithm (HOGSTROM et al. 1981), for in that work what is mathematically propagated from the collimation level is physically fictitious pencil beams with no initial angular dispersion.]

Independently, SHIU and HOSTROM (1991b) have developed a "pencil-beam redefinition algorithm" which is essentially the same as the original moment

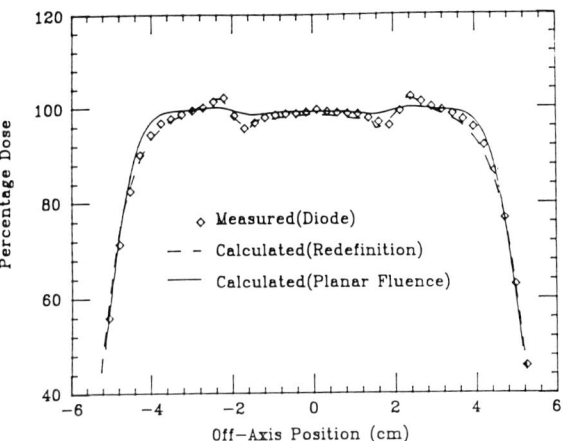

Fig. 5.13. Measured vs calculated (pencil–beam redefinition model) cross-beam profiles, for a 0.5-cm-thick bone at a 2.0-cm depth of water situated in the center of 10×10 cm field at 100-cm SSD. The depth of the profiles is 3.0 cm, which is 0.5 cm beneath the bone. (From SHIU and HOGSTROM 1991b)

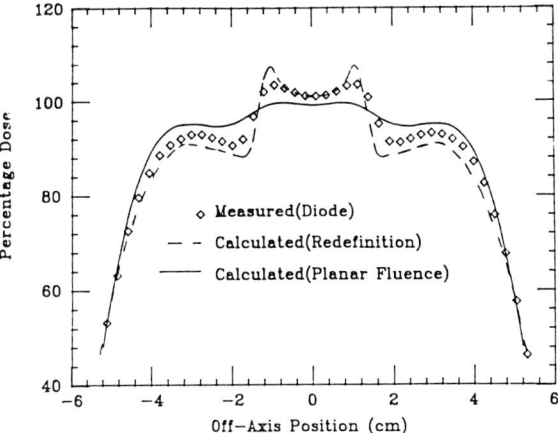

Fig. 5.14. Measured vs calculated (pencil beam redefinition model) cross-beam profiles, for a 1.0-cm-thick air cavity at a 2.0-cm depth of water situated in the center of a 10×10 cm field at 100-cm SSD. The depth of the profiles is 4.0 cm, which is 1.0 cm beneath the air cavity. (From SHIU and HOGSTROM 1991b).

method, except for keeping track of the electron energy distribution through energy bins. This algorithm does indeed give high accuracy even for deep inhomogeneities, and in this respect it provides major improvement over the Hogstrom algorithm, as seen in Figs. 5.13 and 5.14 (in which the predictions of the Hogstrom algorithm are also plotted, as "planar fluence"). However, as Shiu and Hogstrom have pointed out, the calculation time of their algorithm is presently too long (some 6 h on a Digital VAXstation 3500 computer) for it to be suitable for clinical use. McLELLAN et al. (1992a) have just proposed a general beam-redefinition formalism which can incorporate any theoretical model describing the angular scattering of electrons with depth. However, although they obtain reasonable calculation times, their treatment of the problem is only two-dimensional, and evidently a full three-dimensional algorithm would result in prohibitively large calculation times.

Meanwhile, Jette and co-workers have been approaching the problem of electron-beam dose calculation by improving the accuracy of the multiple-scattering theory used. This was already seen in the previous section, with the proposed incorporation of Molière scattering theory into the Fermi-Eyges theory (JETTE and WALKER 1992b). This systematic effort commenced with the generalization of a number of extant scattering theories into a single "Gaussian multiple-scattering theory" (JETTE 1988b). Under the usual small-angle approximation ($\sin\Theta \simeq \Theta$, $\cos\Theta \simeq 1$) this theory gives the Fermi equation (examined in the previous section), but it is also straightforward to use a more accurate small-angle approximation:

$$\sin\Theta \simeq -\Theta^3/6$$
$$\cos\Theta \simeq -\Theta^2/2$$

This latter approximation is indeed more accurate than the usual one, for fairly large angles: for example, for $\Theta = 60°$ it gives $\sin\Theta \simeq 0.856$ instead of 1.047 (correct value: 0.866), and $\cos\Theta \simeq 0.452$ instead of 1 (correct value: 0.500).

Under this more accurate approximation a "second-order multiple-scattering theory" has been derived (JETTE 1985b; JETTE and BIELAJEW 1989), and comparisons with Monte Carlo calculations have demonstrated its increased accuracy over the Fermi-Eyges theory. Indeed, since in this theory $\sec\Theta$ is not set equal to 1, it is able to model the component of the buildup in electron depth-dose curves resulting from the increasing skewness with increasing depth of the beam electrons' trajectories. Its formula for the dose deposited by a pencil beam is simply that of Fermi-

Eyges theory multiplied by second-degree polynomial (with coefficients dependent upon z) in r^2, so going from the (first-order) Fermi-Eyges theory to the second-order theory requires little additional calculation time. (The main calculation time results from the exponential factor, which is already in the Fermi-Eyges formula.) An example of the increased accuracy of the second-order theory is seen in Fig. 5.17 of the next section, concerning oblique incidence.

Another theoretical advance has been the solution of the Fermi equation for scattering power T as an arbitary function of position (x, y, z) (JETTE 1991), thereby removing the inaccuracy inherent in the "semi-infinite slab approximation." This solution takes the form of a perturbation series about an "average configuration" (a fictional layered medium), and only the first two or three terms of this perturbation series are required for highly accurate dose calculations for localized tissue inhomogeneities (JETTE and WALKER 1992a), as seen in Fig. 5.15. (Figure 5.15 also gives the predictions of the Hogstrom algorithm for this configuration; as already explained, the latter does poorly for deep localized inhomogeneities.) Besides having high accuracy, this new dose-calculation method provides short calculation times (currently, of about a minute for the inhomogeneity correction), for its formulas are much simpler when expressed in Fourier space; thus most of the dose calculation can be carried out there and then transformed to physical space using the Fast Fourier Transform, as suggested by BOYER (1984).

Haneman et al. (1993) have demonstrated the high accuracy and practicality of this new electron dose-calculation method for a clinically realistic situation involving a complex structure of localized inhomogeneities. What was considered was a spine phantom using a rehydrated human spine embedded in muscle substitute material, created at M.D. Anderson Cancer Center. There, CT scans were taken of the phantom, and the dose was measured throughout the phantom for comparison with predictions of the (three-dimensional) Hogstrom algorithm, which gave errors up to 14% (DOMINIAK 1991, DOMINIAK et al. 1991). Using the same CT data, Haneman et al. calculated the dose distribution using the new algorithm, as seen in Fig. 16a. The agreement of this new dose calculation with Dominiak's experimental data was much better, being within several percent (Fig. 16b). Calculation times were also quite satisfactory, being 37 seconds for a 128 × 128 × 64 dose-calculation matrix and 68 seconds for a 256 × 256 × 64 matrix.

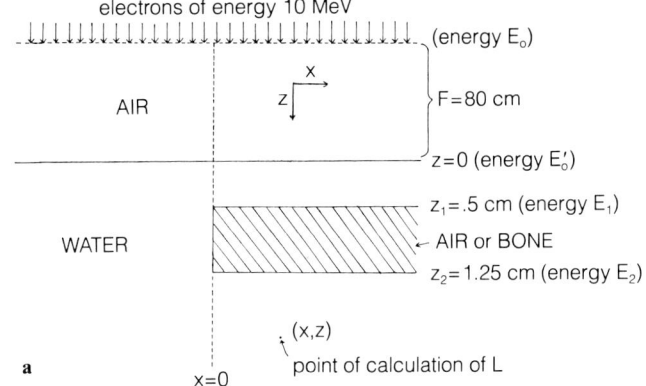

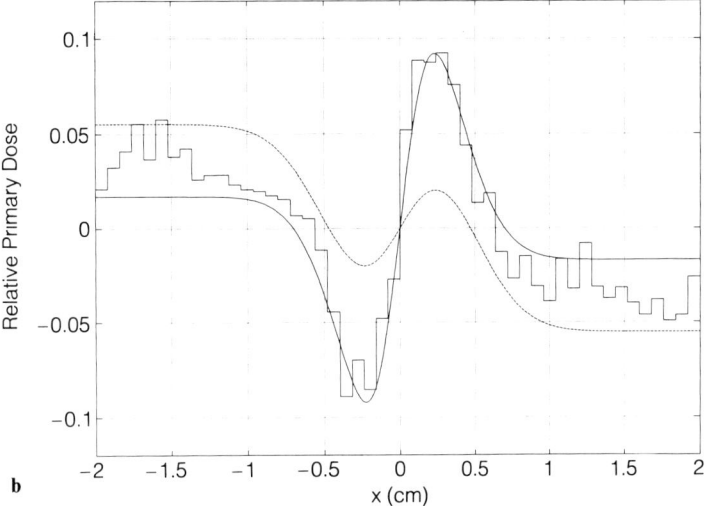

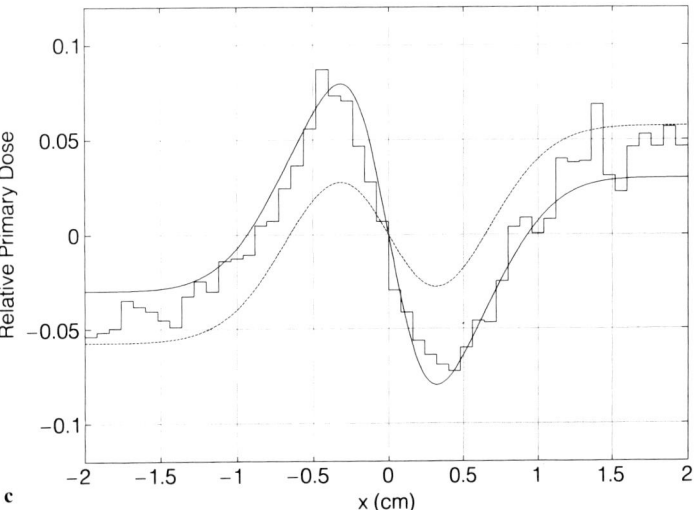

Fig. 5.15 a–c. Dose calculations for a broad, normally incident parallel beam of 10 MeV electrons. **a** Configuration for a half-slab of air or bone; the configuration is independent of y. **b** EGS4 Monte Carlo calculation (*steps*), perturbation-series method (*solid curve*), and Hogstrom algorithm for a 10-cm collimator-surface distance (*dashed curve*), at depth $z = 2.0$ cm and for an air half-slab. **c** Same as **b**, but for a bone half-slab. (From JETTE and WALKER 1992a)

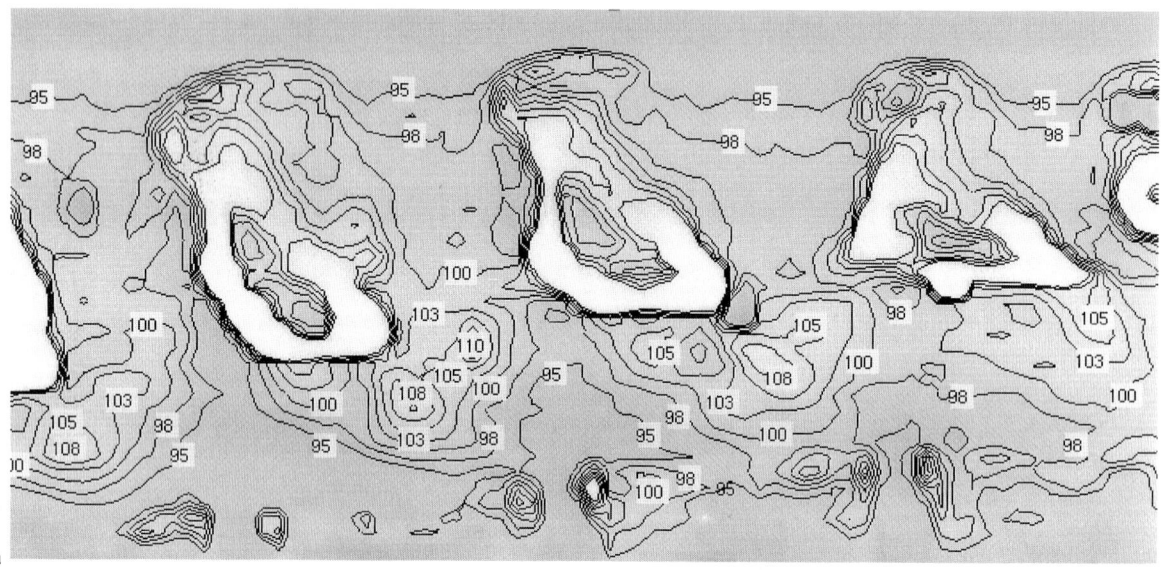

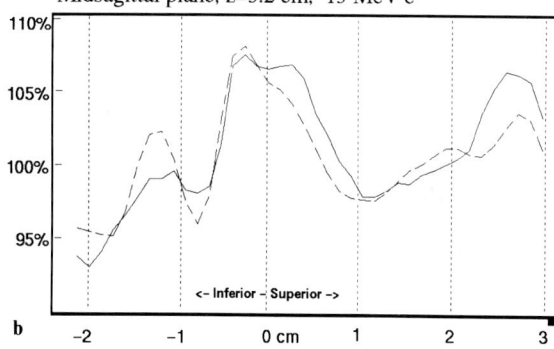

Calculated (– – ·) vs. measured (——) dose profiles
Midsagittal plane, z=3.2 cm; 15 MeV e⁻

Fig. 5.16 a. Calculated isodoses for a 15 MeV electron irradiation of a spine phantom. The area of the cord is shown; spinous process contours are visible in the dose images due to differences in material stopping powers. (From HANEMAN et al. 1993). **b** Calculated (*dashed line*) vs measured (*solid line*) dose profiles for the spine phantom of **a** (15 MeV electrons). The dose profiles were taken at a depth of 3.2 cm in the midsagittal plane. (From HANEMAN et al. 1993)

The most accurate way to calculate electron beam dose distribution, of course, is not analytically (even using advanced multiple-scattering theories), but through the Monte Carlo method (NAHUM 1985b, 1988, MACKIE 1990; ROGERS and BIELAJEW 1990; ANDREO 1991; ROGERS 1991). Monte Carlo provides great accuracy in examining the effects of the treatment head upon the electron beam actually arriving at the patient surface (UDALE 1988, UDALE-SMITH 1992). Furthermore, it allows the investigators to carry out "impossible experiments" to check the accuracy of theories describing individual physical processes, for example the dose deposited directly by the primary electrons without regard for the redistribution of beam energy by secondary particles, as seen in Figs. 5.1 and 5.2. One important use for Monte Carlo is the calculation of accurate databases with which to compare the predictions of proffered dose-calculation models, using standard inhomogeneous configurations (BRAHME 1983b; ROGERS et al. 1984; SHORTT et al. 1986), although experimental data should also be used for this purpose (CYGLER et al. 1987; MAH et al. 1989; SHIU et al. 1992).

The Monte Carlo code most widely used in medical physics research is EGS4 (W.R. NELSON et al. 1985; NELSON and ROGERS 1988) with its PRESTA electron-step enhancement (BIELAJEW and ROGERS 1987, 1988a) because of its flexibility, alterability, and user-friendliness, as well as its well-established accuracy (SHORTT et al. 1986). This code is readily available in both UNIX (BIELAJEW and ROGERS 1992) and PC microcomputer (WALKER et al. 1992) versions. Other general-purpose Monte Carlo codes for electron-photon transport are ETRAN (SELTZER 1988) and the Integrated Tiger Series (HALBLEIB 1988). Also AL-BETERI and RAESIDE (1992a, b) have developed their own Monte Carlo code for use in electron dose calculation.

But besides being an important tool in developing and testing analytic models of electron beam dose calculation, Monte Carlo can be used directly in treatment-planning algorithms, and its very high accuracy makes such use extremely attractive. One such use for Monte Carlo is the precalculation of dose kernels arising from pencil beams (MACKIE 1987, ALTSCHULER et al. 1992), although such kernels can also be extracted from measured broad-beam profiles (CHUI and MOHAN 1988). Unfortunately, this method is not very accurate in the vicinity of localized inhomogeneities, through which the propagation of the kernels is not accurately modeled. The big problem with Monte Carlo as a routine dose-calculation method is its very long calculation time, and one approach is to greatly reduce this time by transporting the electrons in large-scale macroscopic steps through the irradiated material (MACKIE and BATTISTA 1984; NEUENSCHWANDER and BORN 1992), but this method also suffers from inaccuracy in the vicinity of localized inhomogeneities. However, with the advent of ever faster and less expensive computer workstations, and with various statistical techniques for reduction of calculation time (BIELAJEW and ROGERS 1988b), it may yet be possible to use Monte Carlo for routine clinical dose calculations; this is the goal of the OMEGA Project (MACKIE et al. 1990; ROGERS et al. 1990; HOLMES et al. 1991a, 1992).

5.5 Particular Problems

There is much more to electron-beam dose calculation than simply modeling the tranport of the primary electrons through the patient under multiple-scattering theory. In this concluding section we examine theoretical and practical problems which must be faced in order to arrive at accurate, usable dose-calculation algorithms for clinical treatment.

One major problem involves the multiple-scattering theories which we have thus far been considering, for the Fermi-Eyges theory and its drivatives totally ignore backscattering of the (primary) electrons. This is seen in Fig. 5.7, in the relationship given there between the polar angles (Θ, ϕ) and the "projected angles" (Θ_x, Θ_y) which we actually use: since, for example, $\tan\Theta_x = \tan\Theta \cos\phi$, Θ is necessarily restricted to the range $0 \leq \Theta < \pi/2$. Thus these theories transport electrons only in the forward direction – the distribution of electrons at depth z is entirely determined by the material at depths less than z, without regard for the underlying material.

It is not clear, however, how important backscattering of the *primary* electrons is to dose in the vicinity of a (horizontal) interface between two kinds of tissue. SHIU and HOGSTROM (1991a) have carried out careful measurements of dose in the vicinity of bone–soft tissue interfaces, and have found maximum dose enhancements in bone of 7%. However, they consider this enhanced dose to be due mainly to the transport of secondary electrons across the interface, which makes sense since these "delta rays" tend to emerge from the beam in a sideways (although somewhat forward) direction. Considerable effort has been put toward measuring electron backscatter factors and fitting them to empirical formulas (BAILY 1980; GAGNON and CUNDIFF 1980; KLEVENHAGEN et al. 1982; LAMBERT and KLEVENHAGEN 1982; KLEVENHAGEN 1985, 1991; HUNT et al. 1988), and TABATA and ITO (1991, 1992) have studied the problem using a semi-empirical depth-dose computer code. It will take Monte Carlo investigations to determine the relative importance of primary and secondary backscattered electrons, and to verify the accuracy of theories developed to account for such dose enhancement. For primary electrons, it will probably be necessary to use a diffusion theory such as that being developed by WERNER (1985) and WERNER and DAS (1987) to account for dose perturbations at tissue interfaces, as discussed at the end of Sect. 5.3.

Another problem is that of oblique incidence of the electron beam. Unlike most experimental phantoms, real patients annoyingly present curved surfaces to the beam, thereby distorting the dose distribution within them. MCKENZIE (1979) and RITENOUR et al. (1983) have pointed out that the usual shifting of isodose curves, as used in photon dose calculation, is unsatisfactory for oblique angles of incidence, and EKSTRAND and DIXON (1982) analyzed this problem theoretically, using the strip-beam approach of WERNER et al. (1982). Invoking Fermi-Eyges multiple-scattering theory, Ekstrand and Dixon showed why an obliquely incident beam should have an elevated dose maximum at a shallower depth, and this is seen in Fig. 5.17 in the Monte Carlo calculations for dose directly deposited by the primary electrons. If, instead of regarding the beam as being obliquely incident, one considers the surface to be oriented obliquely, then the dose along the beam's central axis is reduced, and often this must be taken into account in intraoperative and intraoral electron therapy (BIGGS 1984).

In order to handle oblique incidence, KHAN et al. (1985) have proposed using experimentally determined

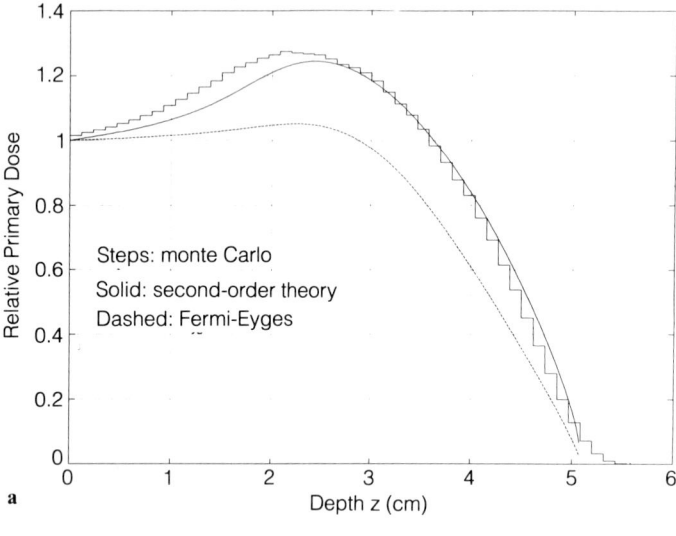

a

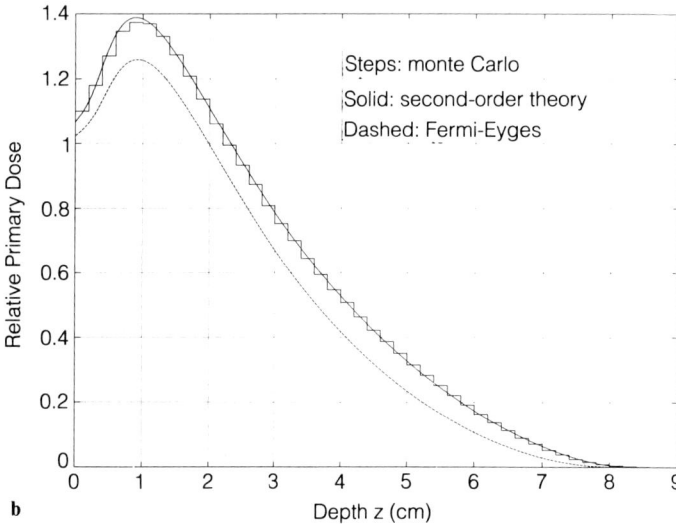

b

Fig. 5.17 a,b. Relative primary dose for a broad parallel beam of 10 MeV electrons obliquely incident upon a flat water surface: **a** 30° angle of obliquity; **b** 60° angle of obliquity. (From JETTE and WALKER 1991)

correction factors, while ULIN and STERNICK (1989) have suggested using experimentally determined isodose shift factors. But a pencil-beam approach based in multiple-scattering theory ought to be more satisfactory for this purpose, for each pencil beam does not start expanding significantly until the patient surface is encountered. This of course is how the Hogstrom algorithm (HOGSTROM et al. 1981) works, but as seen in Fig. 5.17, the Fermi-Eyges theory alone only partially models the elevated primary dose buildup due to oblique incidence. [Presumably, however, the remainder of this buildup is calculated with the Hogstrom algorithm through its experimentally determined depth-dose factor $g(z)$, which includes the effect of increasing skewness with increasing depth of the primary electrons' trajectories.] JETTE (1988a) and JETTE and WALKER (1991) have studied

this problem on the basis of multiple-scattering theory, and have found that second-order multiple-scattering theory (JETTE 1985b; JETTE and BIELAJEW 1989) models the elevated primary dose buildup fairly well, as seen in Fig. 5.17.

The dose distribution of electron beams is far more sensitive to collimation than that of photon beams, because of scatter of the primary electrons off the walls of the collimation system, as well as the creation of contamination electrons there. In dealing with rectangular fields of various sizes there is immediately the problem of calculating the dose distribution in water-like tissue for those fields in terms of measured dose distributions for standard field sizes, particularly their output factors. (The output factor for a field, possibly of irregular shape, is defined as the ratio of the maximum dose on the central axis of that field to that of a reference field, often 10×10 cm.) One simple way to calculate output factors for rectangular beams is through the method of equivalent squares, using the ratio of area to perimeter as the defining parameter. This technique has been used successfully with photon beams, and H. YU (1983) has proposed using it for electrons. However, MCPARLAND (1992) has concluded that the use of equivalent fields is either immaterial or inaccurate for electron-beam dose calculations.

MILLS et al. (1982) have taken a more theoretical approach to the problem of calculating output factors, using Fermi-Eyges multiple-scattering theory to derive two possible formulas. The square-root method uses $OF(X, Y) = [OF(X, X) \cdot OF(Y, Y)]^{1/2}$, where $OF(A, B)$ is the output factor for a field of size $A \times B$; the one-dimensional method takes $OF(X, Y) = OF(X, R) \cdot OF(R, Y)$ where R is the side of a square reference field. The second method requires more fields for its database, but it is more accurate because it takes into account asymmetry between the x- and y-collimation. The accuracy of these three methods has been evaluated for small elongated fields, but with conflicting results: S.C. SHARMA and WILSON (1985) found the square-root method to be satisfactory, and NIROOMAND-RAD (1989) determined both the equivalent-square and the square-root methods to be accurate, but RASHID et al. (1990) found a clear preference for the one-dimensional method over the other two. MILLS et al. (1985) subsequently have had to use a correction factor for the one-dimensional method, and MCPARLAND (1987, 1989c) and CHEN (1988b) have each proposed two-dimensional parameterizations of the output factor for increased accuracy.

For irregularly shaped fields, accurate calculation of output factors is considerably more difficult, although a number of elementary methods have recently been proposed (WU et al. 1987; MCPARLAND 1989b; JONES et al. 1990; MULLER-RUNKEL 1993). Somewhat surprisingly, PLANE and TREVOR (1992) have found experimentally that the output factor for an open applicator is valid for any irregular shape within a given range. Nonetheless, for accurate calculation of dose throughout the patient for an irregularly shaped field, a pencil-beam method based in multiple-scattering theory is necessary (unless Monte Carlo is used), and the quadtree method of representing an irregular field (KOOY and KIJEWSKI 1988; KOOY and RASHID 1989) is particularly well suited to this task. However, electron scatter off the (final) beam-defining collimator ("applicator," "cone," "cutout," or "trimmer") can be quite

significant, as NYERICK et al. 1991 and HUIZENGA and VAN BATTUM (1992) have emphasized. LAX and BRAHME (1980) early studied this problem, and WERNER (1983) and WERNER et al. (1983) were able to incorporate the effect of collimator scatter into their "universal depth-dose curves" for broad beams. But the basic theoretical treatment of this problem has been carried out (BRUINVIS et al. 1983a, 1984; BRUINVIS 1987) using the Fermi-Eyges theory to model the scattering off the electron applicator separately. Essentially, Bruinvis et al. have broken up the electron beam into two components, calculating the contributions of each with different experimentally determined parameters $\sigma_r(z)$: the main component is the usual one for pencil-beam algorithms, and includes scatter off the main collimator walls, while the second component treats the electron applicator with field-defining frame as an additional source of scattered electrons. BRUINVIS and MATHOL (1988) have found that neglecting this additional source results in differences of up to 8% between measured and calculated dose.

Besides scattered primary electrons and high-energy secondary electrons coming from the treatment head, there is another external source of dose to the patient other than the idealized primary beam. Bremsstrahlung from the treatment head, particularly from the scattering foil (if the beam is not magnetically scanned) and the transmission ion chamber, can provide as much as 7% of the dose, and can be quite significant for arc therapy (EL-KHATIB et al. 1991). GUR et al. (1979) and RUSTGI and RODGERS (1987) have studied such photon contamination of the electron beam, and KUBO (1990) has found substantial bremsstrahlung dose under electron cutouts which are supposed to be blocking the irradiation of critical healthy tissue. Electron beam dose-calculation algorithms usually take account of bremsstrahlung by postulating a wide uniform "photon beam" which is either constant or exponentially decaying within the patient; its parameters can be determined from dose measurements of the "bremsstrahlung tail" deep within a water phantom, beyond the practical range of the primary electrons.

High-energy secondary electrons ("delta rays") created within the patient by the primary electrons are important in redistributing the beam energy from that which would be calculated simply by using multiple-scattering theory (which gives the dose directly deposited by the primary electrons). This redistribution of energy can account for as much as 30% of the dose (ICRU 1984a; WERNER et al. 1987). In practical dose-calculation algorithms (e.g., HOGSTROM et al.

1981), such redistribution of beam energy is subsumed into a broad-beam depth-dose factor measured for a homogeneous phantom, but the effect of high-energy secondary electrons for inhomogeneous configurations really needs to be studied. An early theoretical investigation of secondary electron transport was contained in the "moments method" work of KESSARIS (1966) mentioned in Sect 5.3. WERNER et al. (1987) have provided a simple model for the production of secondary electrons, and KRITHIVAS and RAO (1989) have carried out an experimental study of this problem (but their "secondary electrons" include laterally scattered primary electrons). Another source of energy redistribution is bremsstrahlung production within the patient, but RUSTGI and RODGERS (1987) have found this process not to be very significant. (However, bremsstrahlung photons can carry off a significant portion of the primary electrons' energy, escaping from the patient, so this process must be modeled accurately.) Monte-Carlo-based investigations should be carried out on these processes of beam energy redistribution by secondary particles.

For calculating electron dose distributions using multiple-scattering theory, two physical quantities are of fundamental importance: the scattering power T, which enters into the formulas for the spreading out of a pencil beam [e.g., the $A_n(z)$ of Sect. 5.2], and the stopping power S/ρ, which is used both to calculate the dose from the fluence and to determine the energy of the primary electrons throughout the irradiated material (which in turn is used to calculate the scattering power there). ICRU Report 35 (1984a) provides tables of these quantities, and ICRU Report 37 (1984b) gives much more extensive tables of stopping power. Applying the power-law approximation of WERNER et al. (1982) to the ICRU Report 35 data, JETTE (1988b) and JETTE and WALKER (1992b) have parameterized the scattering power for use in pencil-beam algorithms. However, further investigation of accepted scattering-power values is warranted, for MCPARLAND (1989a) has derived a formula for the scattering power giving values which are some 6% less than those of ICRU Report 35. ANDREO and FRANSSON (1989) and ANDREO et al. (1989) have recently studied the effects of primary electron energy and angular spread, and of electron and photon contamination of the beam, on the values of stopping-power ratio. In a practical dose-calculation algorithm it is necessary to convert CT number to values of scattering power and stopping power (by way of the electron density of the material), and a number of investigators have shown how to do this

(HOGSTROM et al. 1981; MUSTAFA and JACKSON 1983; HENSON and FOX 1984; HUIZENGA and STORCHI 1985; MCCULLOUGH and HOLMES 1985; KNÖÖS et al. (1986).

Another important quantity in electron dose calculation is the range (total distance traveled) of the primary electrons. ICRU report 35 (1984a) gives tables of range versus energy for electrons, and the relationship between range and energy has recently been debated (ANDERSON and ST. GEORGE 1985; MARKUS 1986; ANDERSON 1986). Also, GROSSWENDT and ROOS (1989) have investigated energy–range relations and depth scaling among various materials. For a broad electron beam penetrating a (horizontally) layered medium, the dose finally drops off with depth quickly but not completely abruptly, partly because some of the electrons are scattered obliquely and run out of energy at smaller depths than do those electrons which keep going essentially forward, and partly because of catastrophic process (creation of high-energy secondary electrons and of bremsstrahlung photons) which take large portions of their energy. The first of these processes is still multiple scattering under the continuous slowing-down approximation, and in initial work JETTE (1986, 1987) has modeled it both by using the representative path method described in Sect. 5.4 and by using a more accurate expression for actual pathlength than that given by YANG (1951). BRUINVIS (1987) and BRUINVIS et al. (1989) have modeled this first process more completely, using the representative path method and calculating the mean energy along a representative path accurately; agreement of the predictions of their model with Monte Carlo data is quite good. The "Phase Space Time Evolution" model presently being developed (HUIZENGA and STORCHI 1989; HUIZENGA et al. 1992; MORAWSKA-KACZYŃSKA and HUIZENGA 1992) explicitly includes the catastrophic collisions of the second process, and MCLELLAN et al. (1992b) have improved the Landau theory (BERGER and WANG 1988) for electron energy straggling (for the second process).

Finally, the effects of beam divergence must be incorporated into a practical algorithm for electron beam dose calculation. JETTE et al. (1989) have investigated the modifications necessary in the parallel-beam formulas of the Fermi-Eyges theory, and have found that somewhat more is necessary than a simple geometric shifting (keeping the same material along beam rays, and not laterally shifting the material at the level of dose calculation) combined with use of the inverse-square law; however, if $\sigma_r(z)$ is determined from experimental data (for that divergent beam), then only these usual modifications are needed. Also, the position of the source of a divergent beam must be specified, and this is greatly complicated by scatter off the collimation system. An effective extended source has been suggested (BRAHME 1983a; ICRU 1984a), derived from the Fermi-Eyges theory. Measurements of the position of an effective source point of the electron beam have been contradictory and dependent upon the field size (CECATTI et al. 1983; JAMSHIDI et al. 1986; THOMAS 1988; S.C. SHARMA and JOHNSON 1991; PLANE and TREVOR 1992; A.K. SHARMA et al. 1992), and SANDISON and HUDA (1989) have concluded that the use of a fictitious virtual source concept for a magnetically scanned electron beam is unnecessary.

References

Abou Mandour M, Harder D (1978) Calculation of the dose distribution of high energy electrons within and behind tissue inhomogeneities of any width. II. Influence of multiple scattering. Strahlentherapie 154: 546–553

Abou Mandour M, Nüsslin F, Harder D (1983) Characteristic functions of point monodirectional electron beams. Acta Radiol Suppl (Stockh) 364: 43–48

Al-Beteri AA, Raeside DE (1992a) A Monte Carlo electron transport code for the desktop computer. Comput Phys 6: 633–642

Al-Beteri AA, Raeside DE (1992b) Optimal electron-beam treatment planning for retinoblastoma using a new three-dimensional Monte Carlo-based treatment planning system. Med Phys 19: 125–135

Alm Carlsson G (1981) Absorbed dose equations: on the derivation of a general absorbed dose equation and equations valid for different kinds of radiation equilibrium. Radiat Res 85: 219–237

Almond PR, Wright AE, Boone MLM (1967) High energy electron dose perturbations in regions of tissue heterogeneity. Part II. Physical models of tissue heterogeneities. Radiology 88: 1146–1153

Altschuler MD, Bloch P, Buhle EL, Ayyalasomayajula S (1992) 3d dose calculations for electron and photon beams. Phys Med Biol 37: 391–411

Anderson DW (1984) Absorption of ionizing radiation. University Park Press, Baltimore, Chaps. 3 and 4

Anderson DW (1986) Comments on range energy relationships for a scanning beam electron linear accelerator. Phys Med Biol 31: 683–685

Anderson DW, St. George F (1985) Range energy relationship for a scanning beam electron linear accelerator. Phys Med Biol 30: 461–465

Andreo P (1985a) Broad beam approaches to dose computation and their limitations. In: Nahum AE (ed) The computation of dose distributions in electron beam radiotherapy. Umeå University, Sweden, pp 128–150

Andreo P (1985b) The interaction of electrons with matter. II. Scattering. In: Nahum AE (ed) The computation of dose distributions in electron beam radiotherapy. Umeå University, Sweden, pp 56–71

Andreo P (1988) Electron pencil-beam calculations. In: Jenkins TM, Nelson WR, Rindi A (eds) Monte Carlo transport of electrons and photons. Plenum, New York, pp 437–452

Andreo P (1991) Monte Carlo techniques in medical radiation physics. Phys Med Biol 36: 861–920

Andreo P, Brahme A (1983) Fluence and absorbed dose in high energy electron beams. In: Brahme A (ed) Computed electron beam dose planning. Acta Radiol Suppl (Stockh) 364: 25–33

Andreo P, Fransson A (1989) Stopping-power ratios and their uncertainties for clinical electron beam dosimetry. Phys Med Biol 34: 1847–1861

Andreo P, Brahme A, Nahum A, Mattsson O (1989) Influence of energy and angular spread on stopping-power ratios for electron beams. Phys Med Biol 34: 751–768

Bagne F (1976) Electron beam treatment planning system. Med Phys 3: 31–38

Baily NA (1980) Electron backscattering. Med Phys 7: 514–519

Berger MU, Wang R (1988) Multiple-scattering angular deflections and energy-loss straggling. In: Jenkins TM, Nelson WR, Rindi A (eds) Monte Carlo transport of electrons and photons. Plenum, New York, pp 21–56

Bethe HA (1953) Molière's theory of multiple scattering. Phys Rev 89: 1256–1266

Bethe HA, Rose ME, Smith LP (1938) The multiple scattering of electrons. Proc Am Philos Soc 78: 573–585

Bielajew AF, Rogers DWO (1987) PRESTA: The `parameter reduced electron-step transport algorithm' for electron Monte Carlo transport. Nucl Instr Meth B18: 165–181

Bielajew AF, Rogers DWO (1988a) Electron step-size artefacts and PRESTA. In: Jenkins TM, Nelson WR, Rindi A (eds) Monte Carlo transport of electrons and photons. Plenum, New York, pp 115–137

Bielajew AF, Rogers DWO (1988b) Variance-reduction techniques. In: Jenkins TM, Nelson WR, Rindi A (eds) Monte Carlo transport of electrons and photons. Plenum, New York, pp 407–419

Bielajew AF, Rogers DWO (1992) A standard timing benchmark for EGS4 Monte Carlo calculations. Med Phys 19: 303–304

Bielajew AF, Rogers DWO, Cygler J, Battista JJ (1987) A comparison of electron pencil beam and Monte Carlo calculational methods. In: Bruinvis IAD, van der Giessen PH, van Kleffens HJ, Wittkämper FW (eds) The use of computers in radiation therapy. Elsevier Science, Amsterdam, pp 65–68

Biggs PJ (1984) The effect of beam angulation on central axis per cent depth dose for 4-29 MeV electrons. Phys Med Biol 29: 1089–1096

Bloch P, Altschuler MD, Wallace RE, Baren J (1984) Three dimensional electron beam dose calculations. In: Cunningham JR, Ragan D, Van Dyk J (eds) Proceedings of the eighth international conference on the use of computers in radiation therapy. IEEE Computer Society Press, New York, pp 132–136

Boone MLM, Jardine JH, Wright AE, Tapley ND (1967) High energy electron dose perturbations in regions of tissue heterogeneity. Part I. In vivo dosimetry. Radiology 88: 1136–1145

Boyer AL (1984) Shortening the calculation time of photon dose distributions in an inhomogeneous medium. Med Phys 11: 552–554

Brahme A (1971) Multiple scattering of relativistic electrons in air. The Royal Institute of Technology, Stockholm, Sweden (TRITA-EPP 71–22)

Brahme A (1975) Simple relations for the penetration of high energy electron beams in matter. National Institute of Radiation Protection, Stockholm, Sweden (SSI: 1975–011)

Brahme A (1982) Physics of electron beam penetration: fluence and absorbed dose. In: Paliwal B (ed) Proceedings of the symposium on electron dosimetry and arc therapy, University of Wisconsin, U.S.A. American Institute of Physics, New York, pp 45–68

Brahme A (1983a) Geometric parameters of clinical electron beams. Acta Radiol Suppl (Stockh) 364: 11–19

Brahme A (1983b) Standard geometry for comparison of dosimetry and dose planning programs for electron beams. Acta Radiol Suppl (Stockh) 364: 101–102

Brahme A (1985a) Brief review of current algorithms for electron beam dose planning. In: Nashum AE (ed) The computation of dose distributions in electron beam radiotherapy. Umeå University, Sweden, pp 271–290

Brahme A (1985b) Current algorithms for computed electron beam dose planning. Radiother Oncol 3: 347–362

Brahme A (1985c) Elements of electron transport theory. In: Nahum AE (ed) The computation of dose distributions in electron beam radiotherapy. Umeå University, Sweden, pp 72–79

Brahme A, Lax I (1983) Absorbed dose distribution of electron beams in uniform and inhomogeneous media. Acta Radiol Suppl (Stockh) 364: 61–72

Brahme A, Nilsson B (1984) Limitations of pencil beam algorithms in electron beam dose planning. In: Cunningham JR, Ragan D, Van Dyk J (eds) Proceedings of the eighth international conference on the use of computers in radiation therapy. IEEE Computer Society Press, New York, pp 157–160

Brahme A, Lax I, Andreo P (1981) Electron beam dose planning using discrete Gaussian beams: mathematical background. Acta Radiol Oncol 20: 147–158

Bruinvis IAD (1987) Electron beams in radiation therapy: collimation, dosimetry and treatment planning. Doctoral dissertation, Drukkerij Elinkwijk BV, Utrecht, The Netherlands

Bruinvis IAD, Mathol WAF (1988) Calculation of electron beam depth-dose curves and output factors for arbitrary field shapes. Radiother Oncol 11: 395–404

Bruinvis IAD, Van Amstel A, Elevelt AJ, Van der Laarse R (1983a) Calculation of electron beam dose distributions for arbitrarily shaped fields. Phys Med Biol 28: 667–683

Bruinvis IAD, Van Amstel A, Elevelt AJ, Van der Laarse R (1983b) Dose calculations for arbitrarily shaped electron beams. Acta Radiol Suppl (Stockh) 364: 73–79

Bruinvis IAD, Van der Laarse R, Mathol WAF, Nooman MF (1984) An electron beam dose planning method for arbitrary field shapes. In: Proceedings of the Eighth International Conference on the Use of Computers in Radiation Therapy. IEEE Computer Society Press, New York, pp 152–156

Bruinvis IAD, Mathol WAF, Andreo P (1989) Inclusion of electron range straggling in the Fermi-Eyges multiple-scattering theory. Phys Med Biol 34: 491–507

Cecatti ER, Gonçalves JF, Cecatti SGP, da Penha Silva M (1983) Effect of the accelerator design on the position of the effective electron source. Med Phys 10: 683–686

Chen FS (1988a) An analytical equation of electron beams percentage depth ionization curve along the central axis. Med Phys 15: 407–409

Chen FS (1988b) An empirical formula for calculating the output factors of electron beams from a Therac 20 linear accelerator. Med Phys 15: 348–350

Chui CS, Mohan R (1988) Extraction of pencil beam kernels by the deconvolution method. Med Phys 15: 138–144

Cygler J, Ross J (1988) Electron dose distributions in an anthropomorphic phantom—verification of Theraplan treatment planning algorithm. Med Dosim 13: 155–158

Cygler J, Battista JJ, Scrimger JW, Mah E, Antolak J (1987) Electron dose distributions in experimental phantoms: a comparison with 2D pencil beam calculations. Phys Med Biol 32: 1073–1086

Deasy JO, Wilson D, Almond PR (1992) Direct measurements of electron flux angular distributions. Med Phys 19: 792 (abstract #G25)

Dominiak GS (1991) Dose in spinal cord following electron irradiation. MSc thesis, University of Texas, Houston

Dominiak GS, Starkschall G, Shiu AS, Hogstrom KR (1991) Dose in spinal cord following electron irradiation. Med Phys 18: 848 (abstract #WP2-5)

Dutreix A, Briot E (1985) The development of a pencil-beam algorithm for clinical use at the Institut Gustave Roussy. In: Nahum AE (ed) The computation of dose distributions in electron beam radiotherapy. Umeå University, Sweden, pp 242–270

Edwards FH, Coffey II CW (1979) A cumulative normal distribution model for simulation of electron beam profiles. Int J Radiat Oncol Biol Phys 5: 127–133

Ekstrand KE, Dixon RL (1982) The problem of obliquely incident beams in electron-beam treatment planning. Med Phys 9: 276–278

El-Khatib EE, Scrimger J, Murray B (1991) Reduction of the bremsstrahlung component of clinical electron beams: implications for electron arc therapy and total skin electron iradiation. Phys Med Biol 36: 111–118

Eyges L (1948) Multiple scattering with energy loss. Phys Rev 74: 1534–1535

Gagnon WF, Cundiff JH (1980) Dose enhancement from backscattered radiation at tissue-metal interfaces irradiated with high energy electrons. Br J Radiol 53: 466–470

Goitein M (1978) A technique for calculating the influence of thin inhomogeneities on charged particle beams. Med Phys 5: 258–264

Goitein M, Chen GTY, Ting JY, Schneider RJ, Sisterson JM (1978) Measurements and calculations of the influence of thin inhomogeneities on charged particle beams. Med Phys 5: 265–273

Grosswendt B, Roos M (1989) Electron beam absorption in solid and in water phantoms: depth scaling and energy-range relations. Phys Med Biol 34: 509–518

Gur D, Bukovitz AG, Serago C (1979) Photon contamination in 8-20 MeV electron beams from a linear accelerator. Med Phys 6: 145–146

Halbleib J (1988) Structure and operation of the ITS code system. In: Jenkins TM, Nelson WR, Rindi A (eds) Monte Carlo transport of electrons and photons. Plenum, New York, pp 249–262

Haneman B, Jette D, Walker S (1993) Electron dose calculations in a spine phantom using Gaussian multiple-scattering theory. Med Phys 20: 934 (abstract #HH3)

Henson PW, Fox RA (1984) The electron density of bone for inhomogeneity correction in radiotherapy planning using CT numbers. Phys Med Biol 29: 351–359

Hogstrom KR (1986) Evaluation of electron pencil beam dose calculation. In: Kereiakes JE, Elson HR, Born CG (eds) Radiation oncology physics. American Institute of Physics, New York, pp. 532–561 (Medical physics monograph no. 15)

Hogstrom KR, Almond PR (1983) Comparison of experimental and calculated dose distributions. Acta Radiol Suppl (Stockh) 364: 89–99

Hogstrom KR, Kurup RG (1987) A pencil-beam algorithm for arc electron dose distributions. In: Bruinvis IAD, van der Giessen PH, van Kleffens HJ, Wittkämper FW (eds) The use of computers in radiation therapy. Elsevier Science, Amsterdam, pp 73–77

Hogstrom KR, Mills MD, Almond PR (1981) Electron beam dose calculations. Phys Med Biol 26: 445–459

Hogstrom KR, Mills MD, Meyer JA, Palta JR, Mellenberg DE, Meoz RT, Fields RS (1984) Dosimetric evaluation of a pencil-beam algorithm for electrons employing a two-dimensional heterogeneity correction. Int J Radiat Oncol Biol Phys 10: 561–569

Hogstrom KR, Kurup RG, Shiu AS, Starkschall G (1989) A two- dimensional pencil-beam algorithm for calculation of arc electron dose distributions. Phys Med Biol 34: 315–341

Holmes M, Mackie TR, Sanders CA, Kubsad SS, Rogers DWO, Bielajew AF (1991a) The Omega Project: calculation of inverse cumulative probability distribution functions for electron transport for fast Monte Carlo simulations. Med Phys 18: 608 (abstract #F20)

Holmes M, Mackie TR, Sohn W, Bielajew AF, Rogers DWO (1991b) The Omega Project: variance reduction techniques for Monte Carlo simulations as applied to electron beam dosimetry. Med Phys 18: 642 (abstract #P25)

Holmes M, Mackie TR, Sohn W, Bielajew AF, Rogers DWO (1992). The Omega Project: the calculation of correction factors using correlated sampling and EGS4 Monte Carlo. Med Phys 19: 809 (abstract #P4)

Holt JG, Mohan R, Caley R, Buffa A, Reid A, Simpson LD, Laughlin JS (1978) Memorial electron beam AET treatment planning system. In: Orton CG, Bagne F (eds) Practical aspects of electron beam treatment planning. Am. Inst. Phys., New York, pp 70–79 (Medical Physics Monograph No. 2)

Huizenga H, Storchi PRM (1985) The use of computed tomography numbers in dose calculations for radiation therapy. Acta Radiol Oncol 24: 509–519

Huizenga H, Storchi PRM (1987) The in-air scattering of clinical electron beams as produced by accelerators with scanning beams and diaphragm collimators. Phys Med Biol 32: 1011–1029

Huizenga H, Storchi PRM (1989) Numerical calculation of energy deposition by broad high-energy electron beams. Phys Med Biol 34: 1371–1396

Huizenga H, van Battum LJ (1992) On the initial angular variance s_{0x}^2 of electron beams of various manufacturers' accelerators. Med Phys 19: 729–793 (abstract #G26)

Huizenga H, Morawska-Kaczyńska M, Riedeman D (1992) Feasibility of the phase-space-time-evolution model as 3D-electron beam dose calculation model in radiotherapy. Med Phys 19: 809 (abstract #P5)

Hunt MA, Kutcher GJ, Buffa A (1988) Electron backscatter corrections for parallel-plate chambers. Med Phys 15: 96–103

Hyödynmaa S, Lax I, Israelsson A (1987) Array processor application of the generalized Gaussian pencil beam algorithm for electron dose computation. In: Bruinvis IAD, van der Giessen PH, van Kleffens HJ, Wittkämper FW (eds) The use of computers in radiation therapy. Elsevier Science, Amsterdam, pp 79–81

ICRU (1984a) Radiation dosimetry: electron beams with energies between 1 and 50 MeV. International Commission on Radiation Units and Measurements Report 35, Bethesda

ICRU (1884b) Stopping powers for electrons and positrons. International Commission on Radiation Units and Measurements Report 37, Bethesda

Jamshidi A, Kuchnir FT, Reft CS (1986) Determination of the source position for the electron beams from a high-energy linear accelerator. Med Phys 13: 942–948

Jette D (1982) Approximation formulas for least-squares fitting of functions of the form $f[\mu(x-x_0)]$. Med Phys 9: 106–109

Jette D (1984a) The problem of electron dose calculation. I. Multiple-scattering methods. J Am Assoc Med Dosim 9: 6–13

Jette D (1984b) The problem of electron dose calculation. II. Inhomogeneities and beam shaping. J Am Assoc Med Dosim 9: 12–17

Jette D (1985a) The problem of electron dose calculation. III. Homogeneous configurations. J Am Assoc Med Dosim 10: 11–15

Jette D (1985b) Second-order multiple-scattering theory for charged-particle teletherapy beams. Med Phys 12: 178–182

Jette D (1986) Representative electron paths and straggling. Med Phys 13: 604 (abstract #R7)

Jette D (1987) Electron energy- and range-straggling resulting from multiple scattering. Med Phys 14: 464 (abstract #H7)

Jette D (1988a) The effect of oblique incidence upon electron-beam dose distributions. Phys Med Biol 33 (Suppl 1): 137 (abstract #MP34-25)

Jette D (1988b) Electron dose calculation using multiple-scattering theory: a. Gaussian multiple-scattering theory. Med Phys 15: 123–137. Erratum: Med Phys 16: 920 (1989)

Jette D (1991) Electron dose calculation using multiple-scattering theory: localized inhomogeneities – a new theory. Med Phys 18: 123–132

Jette D, Bielajew A (1989) Electron dose calculation using multiple-scattering theory: second-order multiple-scattering theory. Med Phys 16: 698–711

Jette D, Walker S (1991) Incorporation of oblique beam incidence into a developing model of electron dose calculation. Med Phys 18: 606–607 (abstract #F14)

Jette D, Walker S (1992a) Electron dose calculation using multiple-scattering theory: evaluation of a new model for inhomogeneities. Med Phys 19: 1241–1254

Jette D, Walker S (1992b) Incorporation of Molière scattering theory into the Fermi-Eyges theory for electron dose calculation. Med Phys 19: 792 (abstract #G22)

Jette D, Lanzl LH, Rozenfeld M, Pagnamenta A (1981) Analytic representation of electron central-axis depth dose data. Med Phys 8: 877–881

Jette D, Pagnamenta A, Lanzl LH, Rozenfeld M (1983) The application of multiple scattering theory to therapeutic electron dosimetry. Med Phys 10: 141–146

Jette D, Lanzl LH, Pagnamenta A, Rozenfeld M, Bernard D, Kao M, Sabbas AM (1989) Electron dose calculation using multiple-scattering theory: thin planar inhomogeneities. Med Phys 16: 712–725

Johnsen SW, LaRiviere PD, Tanabe E (1983) Electron depth-dose dependence on energy spectral quality. Phys Med Biol 28: 1401–1407

Jones D, Andre P, Washington JT, Hafermann MD (1990) A method for the assessment of the output of irregularly shaped electron fields. Br J Radiol 63: 59–64

Karlsson M (1985) Validity of dose calculation with the M.D.A.H. model at bolus edges. In: Nahum AE (ed) The computation of dose distribution in electron beam radiotherapy. Umeå University, Sweden, pp 185–190

Kawachi K (1975) Calculation of electron dose distribution for radiotherapy treatment planning. Phys Med Biol 20: 571–577

Kessaris ND (1966) Penetration of high-energy electron beams in water. Phys Rev 145: 164–178

Khan FM, Deibel FC, Soleimani-Meigooni A (1985)

Obliquity correction for electron beams. Med Phys 12: 749–753

Khan FM, Doppke KP, Hogstrom KR, et al. (1991) Clinical electron-beam dosimetry: report of AAPM Radiation Therapy Committee Task Group No. 25. Med Phys 18: 73–109

Kirsner SM, Hogstrom KR, Kurup RG, Moyers MF (1987) Dosimetric evaluation in heterogeneous tissue of anterior electron beam irradiation for treatment of retinoblastoma. Med Phys 14: 772–779

Klevenhagen SC (1985) Physics of electron beam therapy. Adam Hilger, Bristol Boston (Medical physics handbooks 13)

Klevenhagen SC (1988) Current status of electron therapy – clinical and physical aspects. Br J Radiol (Suppl) 22: 34–51

Klevenhagen SC (1991) Implication of electron backscattering for electron dosimetry. Phys Med Biol 36: 1013–1018

Klevenhagen SC, Lambert GD, Arbabi A (1982) Backscattering in electron beam therapy for energies between 3 and 35 MeV. Phys Med Biol 27: 363–373

Knöös T, Nilsson M, Ahlgren L (1986) A method for conversion of Hounsfield number to electron density and prediction of macroscopic pair production cross-sections. Radiother Oncol 5: 337–345

Kooy HM, Kijewski PK (1988) Quadtrees as a representation for irregularly shaped fields in radiotherapy applications. Int J Radiat Oncol Biol Phys 15: 1251–1256

Kooy HM, Rashid H (1989) A three-dimensional electron pencil-beam algorithm. Phys Med Biol 34: 229–243

Kovář I (1989) Analytical approximation of depth-dose curves for electron beams. Phys Med Biol 34: 939–948

Kozlov A, Shishov V (1982) Calculation of high-energy electron dose distributions in tissue-equivalent media. I. Determination of the dose function of point unidirectional sources. Strahlentherapie 158: 298–304

Krithivas G, Rao SN (1989) A study on the secondary electrons in a clinical electron beam. Phys Med Biol 34: 1021–1028

Kubo H (1990) Effects of electron cutouts on absorbed dose in and outside Varian-20 electron fields. Med Dosim 15: 61–66

Kurup RG, Hogstrom KR, Otte VA, Moyers MF, Tung S, Shiu AS (1992) Dosimetric evaluation of a two-dimensional, arc electron, pencil-beam algorithm in water and PMMA. Phys Med Biol 37: 127–144

Lambert GD, Klevenhagen SC (1982) Penetration of backscattered electrons in polystyrene for energies between 1 and 25 MeV. Phys Med Biol 27: 721–725

Larsen EW (1993) A mathematical derivation of Fermi theory and higher-order corrections in electron dose calculations. Med. Phys 20: 887 (abstract # L23)

Laughlin JS (1965a) High energy electron treatment planning for inhomogeneities. Br J Radiol 38: 143–147

Laughlin JS (1965b) Studies of absorption of high energy electron beams. In: Zuppinger A, Poretti G (eds) Symposium on high-energy electrons. Springer, Berlin Heidelberg New York, pp 11–16

Lax I (1985) A generalized Gaussian model for electron beam dose planning. In: Nahum AE (ed) The computation of dose distributions in electron beam radiotherapy. Umeå University, Sweden, pp 191–210

Lax I (1986a) Development of a generalized Gaussian model for absorbed dose calculation and dose planning in therapeutic electron beams. Doctoral dissertation, University of Stockholm, JINAB, Sweden

Lax I (1986b) Inhomogeneity corrections in electron-beam dose planning: limitations with the semi-infinite slab approximation. Phys Med Biol 31: 879–892

Lax I (1987) Accuracy in clinical electron beam dose planning using pencil beam algorithms. Radiother Oncol 10: 307–319

Lax I, Brahme A (1980) Collimation of high energy electron beams. Acta Radiol Oncol 19: 199–207

Lax I, Brahme A (1985) Electron beam dose planning using Gaussian beams: energy and spatial scaling with inhomogeneities. Acta Radiol Oncol 24: 75–85

Lax I, Brahme A, Andreo P (1983) Electron beam dose planning using Gausian beams: improved radial dose profiles. Acta Radiol Suppl (Stockh) 364: 49–59

Lillicrap SC, Wilson P, Boag JW (1975) Dose distributions in high energy electron beams: production of broad beam distributions from narrow beam data. Phys Med Biol 10: 30–38

Low DA, Starkschall G, Bujinowski SW, Wang LL, Hogstrom KR (1992) Electron bolus design for radiotherapy treatment planning: bolus design algorithms. Med Phys 19: 115–124

Luo ZM (1985a) Improved bipartition model of electron transport. I. A general formulation. Phys Rev B32: 812–823

Luo ZM (1985b) Improved bipartition model of electron transport. II. Applications to inhomogeneous media. Phys Rev B32: 824–836

Luo ZM, Brahme A (1990) MONKEY, a highly efficient microcomputer program for calculation of transport of high energy electrons in media. In: Proceedings of the tenth international conference on the use of computers in radiation therapy. IEEE Computer Society Press, New York, pp 29–32

Luo ZM, Brahme A (1993a) High energy electron transport. Phys Rev B46: 15739–15752

Luo ZM, Brahme A (1993b) An overview of the transport theory of charged particles. Radiat Phys Chem 41: 673–703

Mackie TR (1987) Calculating electron dose using a convolution/superposition method. In: Bruinvis IAD, van der Giessen PH, van Kleffens HJ, Wittkämper FW (eds) The use of computers in radiation therapy. Elsevier Science, Amsterdam, pp 445–448

Mackie TR (1990) Applications of the Monte Carlo method in radiotherapy. In: Kase KR, Bjärngard BE, Attix FH (eds) The dosimetry of ionizing radiation, vol III. Academic Press, San Diego, pp 541–620

Mackie TR, Battista JJ (1984) A macroscopic Monte Carlo method for electron beam dose calculation: a proposal. In: Cunningham JR, Ragan D, Van Dyk J (eds) Proceedings of the eighth international conference on the use of computers in radiation therapy. IEEE Computer Society Press, New York, pp 123–127

Mackie TR, Kubsad SS, Rogers DWO, Bielajew AF (1990) The Omega Project: electron dose planning using Monte Carlo simulation. Med Phys 17: 732 (abstract #BBI)

Mah E, Antolak J, Scrimger JW, Battista JJ (1989) Experimental evaluation of a 2D and 3D electron pencil beam algorithm. Phys Med Biol 34: 1179–1194

Markus B (1986) Range-energy relationship for a scanning beam electron linear accelerator. Phys Med Biol 31: 657–661

McCullough AC, Holmes TW (1985) Acceptance testing computerized radiation therapy treatment planning systems: direct utilization of CT scan data. Med Phys 12: 237–242

McKenzie AL (1979) Air-gap correction in electron treatment planning. Phys Med Biol 24: 628–635

McLellan J, Sandison GA, Papież L, Huda W (1991) A restricted angular scattering model for electron penetration in dense media. Med Phys 18: 1–6. Erratum: Med Phys 18: 328 (1991)

McLellan J, Papież L, Sandison GA, Huda W, Therrien P (1992a) A numerical method for electron transport calculations. Phys Med Biol 37: 1109–1124

McLellan J, Papież L, Sandison G, Sawchuk S, Battista J (1992b) An improved method for the calculation of electron energy straggling distributions. Med Phys 19: 810 (abstract #P6)

McParland BJ (1987) A parameterization of the electron beam output factors of a 25-MeV linear accelerator. Med Phys 14: 665–669

McParland BJ (1989a) A derivation of the electron mass scattering power for electron dose calculations. Nucl Instr Meth Phys Res A274: 592–596

McParland BJ (1989b) A method of calculating the output factors of arbitrarily shaped electron fields. Med Phys 16: 88–92

McParland BJ (1989c) Methods of calculating the output factors of rectangular electron fields. Med Dosim 14: 17–21

McParland BJ (1992) An analysis of equivalent fields for electron beam central-axis dose calculations. Med Phys 19: 901–906

McParland BJ, Cunningham JR, Woo MK (1988) The optimization of pencil beam widths for use in an electron pencil beam algorithm. Med Phys 15: 489–497

Meigooni AS, Das IJ (1987) Parametrisation of depth dose for electron beams. Phys Med Biol 32: 761–768

Millan PE, Millan S, Hernandez A, Andreo P (1979) Parametrisation of linear accelerator electron beam for computerised dosimetry calculations. Phys Med Biol 24: 825–827

Mills MD, Hogstrom KR, Almond PR (1982) Prediction of electron beam output factors. Med Phys 9: 60–68

Mills MD, Hogstrom KR, Fields RS (1985) Determination of electron beam output factors for a 20-MeV linear accelerator. Med Phys 12: 473–476

Mohan R, Chui CS, Fontenla D, Han K, Ballon D (1988) The effect of angular spread on the intensity distribution of arbitrarily shaped electron beams. Med Phys 15: 204–210

Morawska-Kaczyńska M, Huizenga H (1992) Numerical calculation of energy deposition by broad high-energy electron beam. II. Multi-layered geometry. Phys Med Biol 37:2103-2116

Mori T (1985) Tissue heterogeneity corrections for high energy electron treatment planning. Jpn J Radiol Technol 4: 282–286 (English translation from Jpn J Radiol Technol 40, 1984)

Muller-Runkel R (1993) Dosimetry of shaped electron fields using a radial integration method. Med Dosim 17: 207–211

Mustafa AA, Jackson DF (1983) The relation between x-ray CT numbers and charged particle stopping powers and its significance for radiotherapy treatment planning. Phys Med Biol 28: 169–176

Nahum AE (1985a) The M.D.A.H. pencil-beam algorithm. In: Nahum AE (ed) The computation of dose distributions in electron beam radiotherapy. Umeå University, Sweden, pp 151–184

Nahum AE (1985b) Monte-Carlo electron transport simulation. II. Application to dose planning. In: Nahum AE (ed) The computation of dose distributions in electron beam radiotherapy. Umeå University, Sweden, pp 319–340

Nahum AE (1988) Overview of photon and electron Monte Carlo. In: Jenkins TM, Nelson WR, Rindi A (eds) Monte Carlo transport of electrons and photons. Plenum, New York, pp 3–20

Nahum AE, Brahme A (1985) The computational of dose distribution in electron depth-dose distributions in uniform and non-uniform media. In: Nahum AE (ed) Electron beam radiotherapy. Umeå University, Sweden, pp 98–127

Nath R, Gignac CE, Agostinelli AG, Rothberg S, Schulz RJ (1980) A semi-empirical model for the generation of dose distributions produced by a scanning electron beam. Int J Radiat Oncol Biol Phys 6: 67–73

Nelson CE, Haneman W, Young K, O'Foghludha F (1984) Analytic calculation of electron beam isodose distributions. Med Phys 11: 242–246

Nelson WR, Rogers DWO (1988) Structure and operation of the EGS4 code system. In: Jenkins TM, Nelson WR, Rindi A (ed) Monte Carlo transport of electrons and photons. Plenum, New York, pp 287–305

Nelson WR, Hirayama H, Rogers DWO (1985) The EGS4 code system. Stanford Linear Accelerator Center Report 265, Stanford University, Stanford, Calif.

Neuenschwander H, Born EJ (1992) A macro Monte Carlo method for electron beam dose calculations. Phys Med Biol 37: 107–125

Niroomand-Rad A (1989) Film dosimetry of small elongated electron beams for treatment planning. Med Phys 16: 655–662

Nüsslin F (1979) Computerized treatment planning in therapy with fast electrons: a review of procedures for calculating dose distribution. Medicamundi 24: 112–118

Nüsslin F (1980) A simple model for calculating dose distributions in high-energy electron therapy. J Eur Radiother 1: 193–197

Nüsslin F (1984) Characterization of elementary electron beams. In: Cunningham JR. Ragan D, Van Dyk J (eds) Proceedings of the eighth international conference on the use of computers in radiation therapy. IEEE Computer Society Press, New York, pp 137–139

Nyerick CE, Ochran TG, Boyer AL, Hogstrom KR (1991) Dosimetry characteristics of metallic cones for intra-operative radiotherapy. Int J Radiat Oncol Biol Phys 21: 501–510

Ogawa K, Nohara H, Yukawa Y, Okada T (1987) Absorption equivalent thickness (AET) method for electron beam treatment planning. Jpn J Radiol Technol 6: 90–93

Ozdemir A, Teng SP, Lindstrom DG, Anderson DW (1985) Calculations for distributions in water for 10-MeV electrons. Radiat Res 101: 213–224

Papież L, Sandison GA (1990) A diffusion model with loss of particles. Adv Appl Prob 22: 533–547

Perry DJ (1988) On the penetration of fast charged particles. Radiat Res 115: 26–43

Perry DJ, Holt JG (1980) A model for calculating the effects of small inhomogeneities on electron beam dose distributions. Med Phys 7: 207–215

Plane JH, Trevor MM (1992) Virtual source distances and field geometry independent output factors for 5-14 MeV electron beams from a Siemens Mevatron M7145. Br J Radiol 65: 717–719

Polman HLA, van der Linden PM (1987a) Description of the uncollimated electron beam in air by means of directional pencil beam model. Phys Med Biol 32: 355–363

Polman HLA, van der Linden PM (1987b) Determination of three parameters describing the uncollimated electron beam in air. Phys Med Biol 32: 345–353

Rashid H, Islam MK, Gaballa H, Rosenow UF, Ting JY (1990) Small-field electron dosimetry for the Philips SL25 linear accelerator. Med Phys 17: 710–714

Ritenour ER, Cacak RK, Hendee WR (1983) Ionization produced by electron beams beneath curved surfaces. Med Phys 10: 669–671

Roesch WC (1954) Age-diffusion theory for beta ray problems. General Electric Company, Hanford Atomic Products Operation, Richland, Washington, Report HW-32121

Rogers DWO (1991) The role of Monte Carlo simulation of electron transport in radiation dosimetry. Appl Radiat Isot 42: 965–974

Rogers DWO, Bielajew AF (1990) Monte Carlo techniques of electron and photon transport for radiation dosimetry. In: Kase KR, Bjärngard BE, Attix FH (eds) The dosimetry of ionizing radiation, vol III. Academic Press, San Diego, pp 427–539

Rogers DWO, Bielajew AF, Nahum AE (1984) Monte Carlo calculations of electron beams in standard dose planning geometries. In: Cunningham JR, Ragan D, van Dyk J (eds) Proceedings of the eighth international conference on the use of computers in radiation therapy. IEEE Computer Society Press, New York, pp 140–144

Rogers DWO, Bielajew AF, Mackie TR, Kubsad SS (1990) The Omega Project: treatment planning for electron-beam radiotherapy using Monte Carlo techniques. Phys Med Biol 35: 285–286 (abstract)

Rossi B (1952) High-energy particles. Prentice-Hall, New York, pp 62–77

Rossi B, Greisen K (1941) Cosmic-ray theory. Rev Mod Phys 13: 240–309 (Fermi's work is given on pp 265–268)

Rozenfeld M, Lanzl LH, Newton CM, Skaggs LS (1969) Computation of distribution of absorbed dose and absorbed dose rate from a scanning electron beam. Strahlentherapie 138: 651–659

Rustgi SN, Rodgers JE (1987) Analysis of the bremsstrahlung component in 6–18 MeV electron beams. Med Phys 14: 884–888

Sabbas AM, Jette D, Rozenfeld M, Pagnamenta A, Lanzl LH (1987) Collimated electron beams and their associated penumbra widths. Med Phys 14: 996–1006

Sandison GA (1987) Application of Fermi-Eyges scattering theory to magnetically scanned therapeutic electron beams. Doctoral dissertation, University of Manitoba, Winnipeg, Canada

Sandison GA, Huda W (1988) Application of Fermi scattering theory to a magnetically scanned electron linear accelerator. Med Phys 15: 498–510

Sandison GA, Huda W (1989) Is the 'fictitious' virtual source a redundant concept for scanned therapeutic electron beams? Phys Med Biol 34: 369–378

Sandison GA, Papież L (1990) Dose computation applications of the electron loss model. Phys Med Biol 35: 979–997

Sandison GA, Huda W, Savoie D (1989) Comparison of methods to determine electron pencil beam spread in tissue-equivalent media. Med Phys 16: 881–888

Scott W (1963) The theory of small-angle multiple scattering of fast charged particles. Rev Mod Phys 35: 231–313

Seltzer SM (1988) An overview of ETRAN Monte Carlo methods. In: Jenkins TM, Nelson WR, Rindi A (eds) Monte Carlo transport of electrons and photons. Plenum, New York, pp 153–181

Shackleton D (1984) A new shape-fitting method for computerised electron beam radiotherapy planning. In: Cunningham JR, Ragan D, Van Dyk J (eds) Proceedings of the eighth international conference on the use of computers in radiation therapy. IEEE Computer Society Press, New York, pp 130–131

Sharma AK, Supe SS, Sathiya Narayanan VK, Subbarangaiah K (1992) Determination of virtual SSDs for electron beams from a dual energy linear accelerator. Strahlenther Onkol 168: 402–405

Sharma SC, Johnson MW (1991) Electron beam effective source surface distances for a high energy linear accelerator. Med Dosim 16: 65–70

Sharma SC, Wilson DL (1985) Depth dose characteristics of elongated fields for electron beams from a 20-MeV accelerator. Med Phys 12: 419–423

Shiu AS, Hogstrom KR (1991a) Dose in bone and tissue near bone-tissue interface from electron beam. Int J Radiat Oncol Biol Phys 21: 695–702

Shiu AS, Hogstrom KR (1991b) Pencil-beam redefinition algorithm for electron dose distributions. Med Phys 18: 7–18

Shiu AS, Tung S, Hogstrom KR, et al. (1992) Verification data for electron beam dose algorithms. Med Phys 19: 623–636

Shortt KR, Ross CK, Bielajew AF, Rogers DWO (1986) Electron beam dose distributions near standard inhomogeneities. Phys Med Biol 31: 235–249

Starkschall G, Shiu AS, Bujinowski SW, Wang LL, Low DA, Hogstrom KR (1991) Effect of dimensionality of heterogeneity corrections on the implementation of a three-dimensional electron pencil-beam algorithm. Phys Med Biol 36: 207–227

Steben JD, Ayyangar K, Suntharalingam N (1979) Betatron electron beam characterisation for dosimetry calculations. Phys Med Biol 24: 299–309

Sternick ES (1978) Algorithms for computerized treatment planning. In: Orton CG, Bagne F (eds) Practical aspects of electron beam treatment planning. Am Inst Phys, New York, pp 81–110 (Medical physics monograph no. 2)

Storchi PRM, Huizenga H (1985) On a numerical approach of the pencil beam model. Phys Med Biol 30: 467–473

Storchi P, van der Linden RJ (1989) A numerical method for the calculation of the diffusion of high energy electrons in a heterogeneous medium. J Computat Phys 85: 417–433

Storchi P, van der Linden R, Huizenga H (1987) Mathematical generalization of the moment method for clinical electron beam dose calculation. In: Bruinvis IAD, van der Giessen PH, van Kleffens HJ, Wittkämper FW (eds) The use of computers in radiation therapy. Elsevier Science, Amsterdam, pp 145–148

Strydom WJ (1984) An analytical expression for central axis depth-dose of electrons. Phys Med Biol 29: 267–269. Erratum: Phys Med Biol 29: 605 (1984)

Strydom WJ (1991) Central axis depth dose curve for electron beams. Med Phys 18: 1254–1255

Suntharalingam N, Ayyangar K (1984) 3-dimensional dose calculations. In: Bagne F (ed) Computerized treatment planning systems. U.S. DHHS, Bethesda, MD (FDA 84-8223), pp 124–135

Tabata T (1993) Semiempirical models for depth-dose curves of electrons in matter: An introductory review. Bulletin of University of Osaka Prefecture 41: 103–118

Tabata T, Ito R (1991) Electron-beam backscattering at media interfaces and a depth-dose algorithm. Proc. RadTech Asia '91, Osaka, 15-18 April 1991. RadTech Jpn, Tokyo, pp 528–533

Tabata T, Ito R (1992) Simple calculation of the electron-backscatter factor. Med Phys 19: 1423–1426

Tabata T, Andreo P, Ito R (1991) Analytic fits to Monte Carlo calculated depth-dose curves of 1- to 50-MeV electrons in water. Nucl Instr Meth Phys Res B58: 205–210

Thomas SJ (1988) Virtual source distances for electron beams between 5 and 20 MeV. Phys Med Biol 33: 1325–1328

Udale M (1988) A Monte Carlo investigation of surface doses for broad electron beams. Phys Med Biol 33: 939–953.

Udale-Smith M (1992) Monte Carlo calculations of electron beam parameters for three Philips linear accelerators. Phys Med Biol 37: 85–105

Ulin K, Sternick ES (1989) An isodose shift technique for obliquely incident electron beams. Med Phys 16: 905–910

van de Geijn J, Chin B, Pochobradsky J, Miller RW (1987) A new model for computerized clinical electron beam dosimetry. Med Phys 14: 577–584

van der Linden P, Bouwer W, van Gasteren H (1984) Implementation of an electron pencil beam algorithm in the TP-11 planning system. In: Cunningham JR, Ragan D, Van Dyk J (eds) Proceedings of the eighth international conference on the use of computers in radiation therapy. IEEE Computer Society Press, New York, pp 119–122

van Gasteren JM (1984) Pencilbeam parameters of clinical electron beams. In: Cunningham JR, Ragan D, Van Dyk (eds) Proceedings of the eighth international conference on the use of computers in radiation therapy. IEEE Computer Society Press, New York, pp 161–166

van Gasteren JM (1987) The influence of primary electron loss on the pencilbeam width in the calculation of electron beam dose distributions. In: Bruinvis IAD, van der Giessen PH, van Kleffens HJ, Wittkämper FW (eds) The use of computers in radiation therapy. Elsevier Science, Amsterdam, pp 137–140

Walker S, Bielajew A, Hale ME, Jette D (1992) Installation of EGS4 Monte Carlo code on an 80386-based microcomputer. Med Phys 19: 305–306

Werner BL (1983) Comparison of broad beam central axis depth dose curves from different accelerators using the universal depth dose curve model. Acta Radiol Suppl (Stockh) 364: 35–41

Werner BL (1985) The perturbation of electron beam dose distributions at medium interfaces. Med Phys 12: 754–763. Erratum: Med Phys 13: 966 (1986)

Werner BL (1987) Dose distributions in regions containing beta sources: small scale nonuniformities. Med Phys 14: 807–808

Werner BL (1991) Dose distributions in regions containing beta sources: irregularly shaped source distributions in homogeneous media. Med Phys 18: 1192–1194

Werner BL, Das IJ (1987) Dose distributions in regions containing beta sources: plane interface in a homogeneous medium. Med Phys 14: 797–806

Werner BL, Khan FM, Deibel FC (1982) A model for calculating electron beam scattering in treatment planning. Med Phys 9: 180–187. Erratum: Med Phys 9: 784 (1982)

Werner BL, Khan FM, Deibel FC (1987) Model for calculating depth dose distributions for broad electron beams. Med Phys 10: 582–588

Werner BL, Das IJ, Khan FM (1987) The production of secondary electrons in an electron beam. Med Phys 14: 992–995

Werner BL, Kwok CS, Das IJ (1988) Dose distributions in regions containing beta sources: large spherical source regions in a homogeneous medium. Med Phys 15: 358–363

Werner BL, Rahman M, Salk WN, Kwok CS (1991) Dose distributions in regions containing beta sources: uniform spherical source region in homogeneous media. Med Phys 18: 1181–1191

Wilson DL, Almond PR (1991) Dose calculations for a laterally translated electron beam. Med Phys 18: 614 (abstract #15)

Wu RK, Wang W, El-Mahdi AM (1987) Irregular field output factors for electron beams. In: Bruinvis IAD, van der Giessen PH, van Kleffens HJ, Wittkämper FW (eds) The use of computers in radiation therapy. Elsevier Science, Amsterdam, pp 73–77

Yang C (1951) Actual path length of electrons in foils. Phys Rev 84: 599–600

Yu CX, Ge WS, Wong JW (1988) A multiray model for calculating electron pencil beam distribution. Med Phys 15: 662–671

Yu H (1983) The applicability of the method of equivalent squares for photon and electron beams. Phys Med Biol 28: 1279–1287

Zerby CD, Keller FL (1967) Electron transport theory, calculations, and experiments. Nucl Sci Eng 27: 190–218

6 Clinical Electron Beam Physics

FRANCIS J. BOVA

CONTENTS

6.1 Introduction

As in all treatment planning, the aim of the procedure is to derive a plan for irradiating the target tissues while ensuring optimum sparing of nontarget tissues. With photon beam planning, the beam is exponentially absorbed by the tissues beyond the target depth, whereas the finite range of electrons provides a powerful tool for limiting the dose to deep-seated tissues. As always, with any advantage comes disadvantage. With electrons, these disadvantages involve the rapid decrease in field uniformity as one moves farther away from the point of final collimation. With electrons, one also experiences rapid and significant changes in the depth–dose curves for very small field sizes, and the rapid change in depth of dose penetration when traversing inhomogeneities. There are also difficulties in accurately predicting a virtual source position and subsequently predicting output at extended source-to-surface distances (SSDs). Furthermore, output for irregularly shaped fields is difficult to predict.

There are some basic rules for dealing with electron beams in a clinical setting. One is to avoid abrupt irregularities; an effort should be made to

FRANCIS J. BOVA, PhD, University of Florida, Department of Radiation Oncology, P.O. Box 100385, Gainesville, FL 32610-0385, USA

provide smooth, regular surfaces. This dictum applies to the surface contour of the patient as well as to the shape of the electron portal. Another rule is to set up a system where electron fields are monitored and changes in important beam parameters, such as depth dose and output, can easily be identified. In certain situations, monitoring can be handled through computer algorithms that have been rigorously verified to predict the changes in dose distributions caused by irregular surfaces, changes in tissue homogeneities, and changes in beam profile and output factors resulting from changes in collimator-to-surface distance. In other situations, monitoring of changes can be achieved only by routine clinical measurement.

The purpose of this text is to indicate the critical parameters that affect the dose distributions and output factors in clinical electron beam treatment planning, which include methods for production of electron beams, characteristics of electron beam isodose distributions, the shielding of electron beams, the modification of electron beams for special clinical applications, and the measuring of electron beam dose distributions.

6.2 Production of Electron Beams

Historically, the initial source of clinical electron beams was the medical betatron (KERST 1943; SKAGGS et al.1948), a device that accelerated electrons in circular orbits of increasing radius. The electron beam could be extracted from the device through the use of a deflecting coil. While electron beam currents within many betatrons were relatively modest, when compared with those necessary for photon beam production, they were more than adequate for clinical electron beam applications.

By the late 1940s, development of the medical linac had begun. On 19 August 1953, the first patient received treatment by a 2-MV linac at Hammersmith Hospital (MILLER 1954). The size and complexity of betatrons proved to be disadvantageous when compared with that of the smaller and simpler linacs, and by the mid-1960s, the linac had become the unit of choice. As with any such device, the initial linacs of the late 1950s and early 1960s have undergone significant improvement in the areas of beam stability and beam quality. The improvement in electron

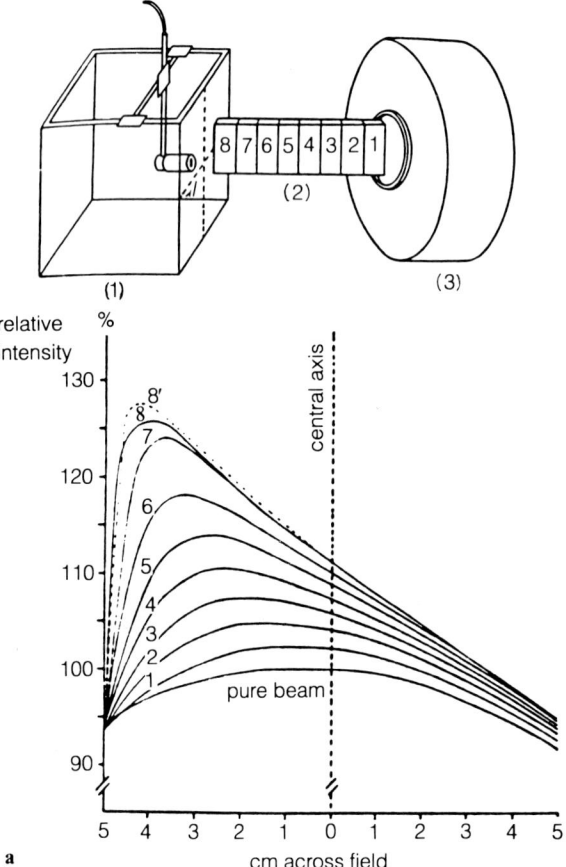

beam performance can be demonstrated by examining the depth–dose curves of the betatrons, early linacs, and current generation linacs. The combination of improved beam acceleration and electron optics, as well as the development of dual scattering foil systems, has provided the clinician with electron beams having sharper falloff, increased field uniformity, and less x-ray contamination.

Early linac designs employed a single scattering foil to spread the relatively small electron beam that emerges from the accelerator's electron window. These foils increase in thickness and atomic number with beam energy. While these single foils can effectively broaden the electron beam, the dose distribution available at the linac's isocenter lacks homogeneity and is unsatisfactory for clinical use. To achieve the required homogeneity, single-foil systems employ solid-walled applicators. A beam size sufficient to allow electrons to interact with the applicator walls is employed. The walls of the applicators scatter electrons back into the useful beam area. The applicator also incorporates a fixed-beam collimator, usually positioned 4–6 cm from the patient's surface.

The effects of the applicator walls can be seen in Fig. 6.1a. As the walls are extended, the amount of scatter increases. The beam profile is enhanced by the addition of a wall parallel to the direction of the beam profile, as well as walls perpendicular to the profile (Fig. 6.1b).

The electrons scattered from the walls of the applicator, as well as those scattered from the x-ray jaws, result in an overall increase in angular and energy spread of the beam as it enters the patient. This, in turn, results in a more rapid increase in dose in the buildup region (Fig. 6.2). While the scatter from the

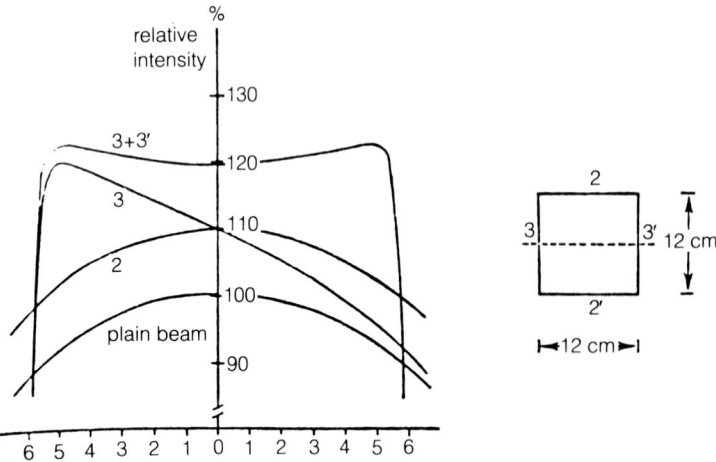

Fig. 6.1. a Experimental setup for the measurement of electron-scatter contribution from the walls of an electron cone from a single scattering foil system and the intensity of the 12-MeV electron beam as each section of the cone wall is added. b The scatter contribution from the walls for a 12-MeV electron beam. The beam profile is measured along the *dashed line* at a depth of 0.8 cm. *2*, Profile for wall 2 present. *3*, Profile with wall 3 present. *3+3'*, Profile for walls 3 and 3' present. (Redrawn from VAN DER LAARSE et al. 1978)

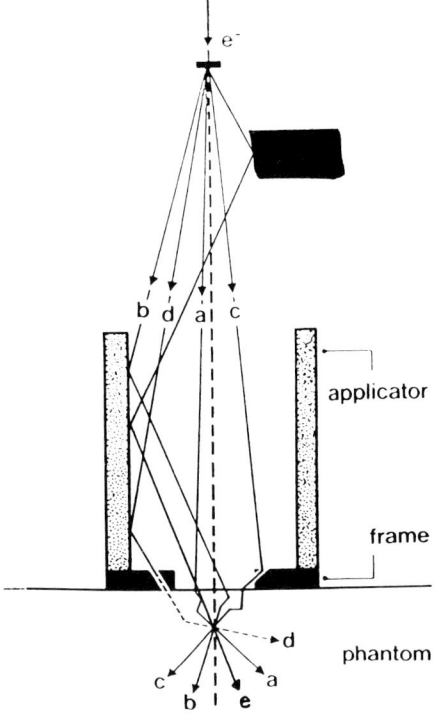

Fig. 6.2. The different contributions to the dose. *a*, Directly incident electrons; *b*, applicator wall-scattered electrons; *c*, frame-scattered electrons; *d*, electron beam which contributes from the applicator field but is eliminated when frames with smaller apertures are used; *e*, electrons scattered from the photon collimators. (Redrawn from BRUINVIS and MATHOL 1988)

collimator walls adds to the dose uniformity, this contribution is negligible beyond the point of dose maximum (LAX and BRAHME 1980). It has been shown also that this contribution to the buildup region changes rapidly with changes in cone-to-surface distance, as can be seen from Fig. 6.3.

In order to reduce energy and angular spread of the electron beam and to reduce the bremsstrahlung production, it is advantageous to reduce the amount of material used to scatter an electron beam. One approach to achieve these goals was suggested by BJARNGARD et al. (1976) for the optimization of a 12-MeV linac. With this method, two foils separated by a few centimeters are used. The first scatterer, which is very much smaller than the width of the electron beam, is used to scatter electrons from the beam's center to its outer edges. A second, and thinner, scatterer is used to fill in the dose depression in the center of the field created by the initial scattering foil.

A similar dual-foil system was suggested by ABOU MANDOUR and HARDER (1978) (Fig. 6.4). In their system, the first foil is used to widen the beam through the use of multiple scattering. The second foil is employed as a compensator to flatten the beam and scatter electrons to the field periphery. In both

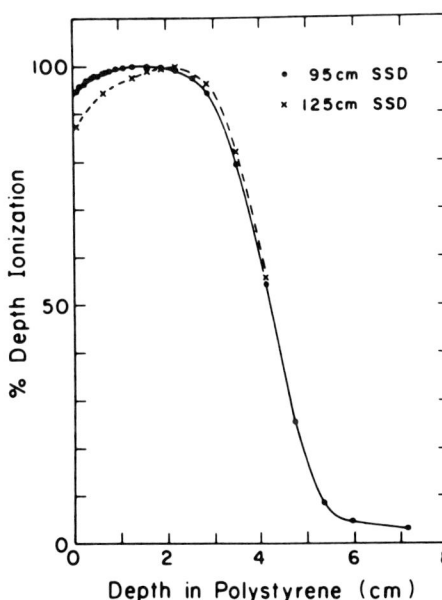

Fig. 6.3. Percent depth ionization for 10-MeV electrons $10 \times 10\,\text{cm}^2$ cone at 95 and 125 cm from the x-ray target. (From SWEENEY et al. 1981)

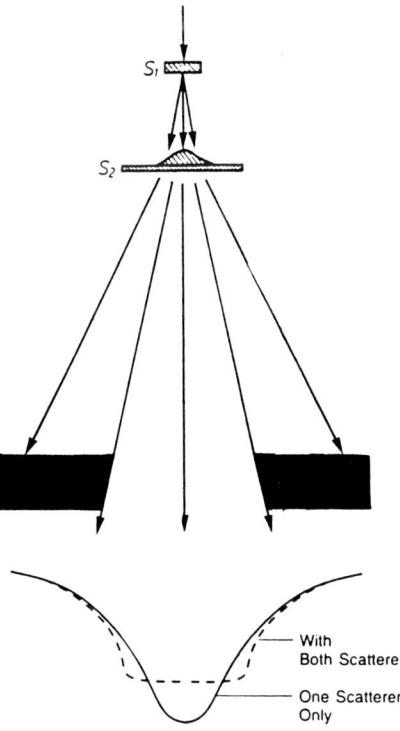

Fig. 6.4. Principle of the double scattering system. The initial electron beam exits the accelerating structure and is widened by S_1 and then again by scatterer S_2. The profiles below indicate the beam profile if S_1 were the only scatterer present and if both S_1 and S_2 were present. (From ABOU MANDOUR and HARDER 1978)

systems, the solid-wall applicator is abandoned, and only a field-defining collimator at the approximate location of those used in solid-wall applicators is used to define the final beam.

Electron beams that entirely avoid the use of scattering foil systems have also been employed. To produce a uniform dose distribution, these units use magnetic fields to sweep the relatively narrow unscattered electron beam across the field to be irradiated. Such scanning electron beam systems have been in use since the 1970s. These scanned beams traverse less material after emerging from the accelerating structure, which results in a beam with less spread both in energy and in angular direction. Because scanned beams lose less energy between the exit of the accelerating structure and entering the patient's surface, they also provide a beam with a narrower mean angular distribution of electrons. This, in turn, provides a deeper D_{max} and a steeper dose gradient in both buildup and falloff regions. Because they do not traverse a relatively high z scatterer, they have less x-ray contamination than most single scattering foil systems. The advantages provided by scanned beams are more pronounced when compared with beams from a single scattering foil system. The current generation of linacs utilizing dual-foil systems, however, produce uniform beams that are comparable to those produced by scanning technology.

It should be mentioned that the size of the scanned beam is not that of the 1- to 2-mm beam that exists at the high vacuum side of the electron window. The passage through the electron window, as well as the passage through the dosimetry system and the traversing of the approximately 90 cm of air prior to reaching the patient's surface, can produce a beam approaching 10 cm in diameter. While the scanning of such a large beam can produce a uniform field, these early generation scanning systems do not have the ability to produce compensated dose distributions.

The energy ranges of the clinical dual-foil systems start at approximately 4 MeV and continue to approximately 25 MeV. Because the thickness of the high atomic number scatter increases as the square of the energy of the electron beam and the square of the maximum field size for beams beyond 25 MeV, scanning systems still provide advantages in beam central-axis dose distribution, beam uniformity, and reduced x-ray contamination (ABOU MANDOUR and HARDER 1978).

It is interesting to note that the early betatrons provided, effectively, a dual scattering foil system. These early units had relatively thick donut windows

that acted much like the primary scatterer in a modern dual-foil system. As the betatron technology progressed, manufacturers were able to design thinner electron windows, and as thicknesses decreased, they ceased performing the function of primary scatterers. The single scattering foil, which had previously acted as the secondary or compensating foil, was then thickened to provide more uniform coverage of the electron field. These single scattering foil system designs were then carried into the early linac designs.

6.3 Electron Beam Central-Axis Depth–Dose Characteristics

The characteristics of the central-axis depth–dose distribution for electron beam are affected by beam energy, beam direction, and absorbing material. For a given clinical beam, many of these parameters remain relatively fixed. It is, however, important to understand how they affect the final dose distribution.

The characteristic electron beam depth-versus-dose curve is shown in Fig. 6.5. The curve shown is for a 14-MeV electron beam from a dual scattering foil system. The clinically important portions of the curve are the surface dose, D_s, defined as the dose at $R(0.5\,mm)$, the dose maximum range, R_{100}, the therapeutic range, R_t, the practical range, R_p, and the normalized dose gradient, G_0.

The surface dose for clinical electron beams ranges from approximately 75% depth dose for low-energy beams in the 4- to 6-MeV range to above 90% at the 20- to 25-MeV range (Fig. 6.6). Unlike photons, which experience a dose buildup due to the

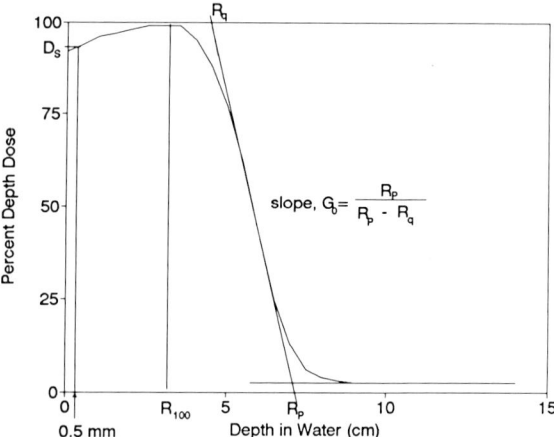

Fig. 6.5. Depth versus dose for a 14-MeV electron beam. D_s is the surface dose at 0.5-mm depth, R_{100} is the range of the depth of dose maximum, and R_p is the extrapolated range

Fig. 6.6. The relative surface dose for 6-, 9-, 12-, 15-, and 18-MeV electron beams for a 10×10 cm^2 cone. (From Turner 1980)

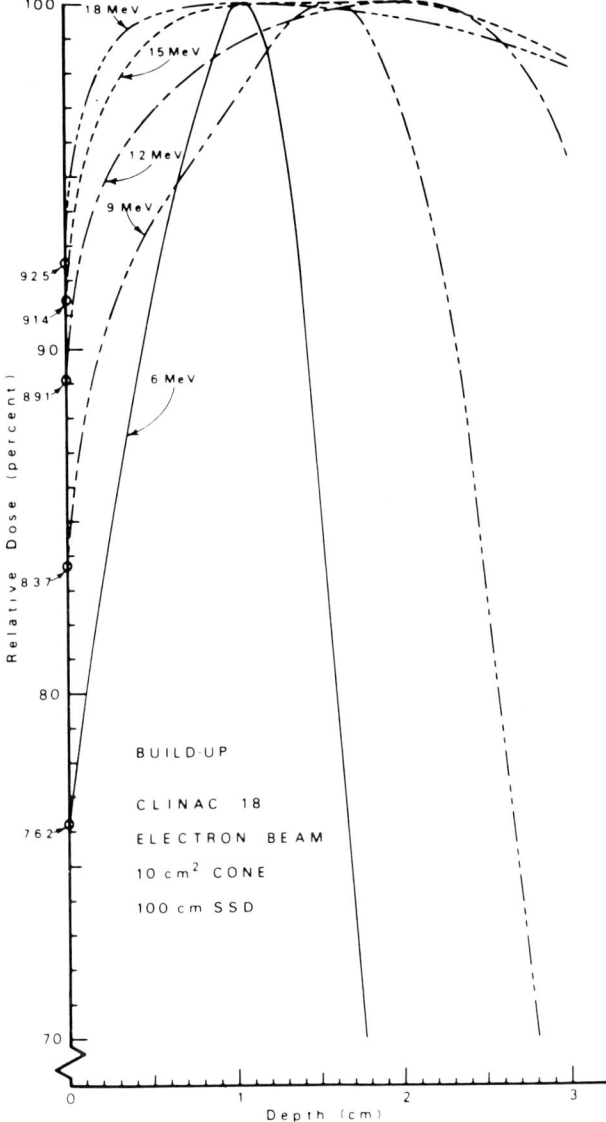

increase in fluence of forward-moving secondary electrons, the incident fast electron in a clinical electron beam loses a relatively constant amount of energy per unit path length. As the electrons penetrate the surface, they undergo multiple scattering. The path of the electrons becomes more oblique to the direction of the central axis of the beam. This change in direction increases the path length of travel of the electron per path length along the central ray (Boag 1972). As demonstrated (Berger and Seltzer 1981), the increase in obliquity of the electron beam is more pronounced for low-energy than for high-energy electrons. This increase results in a more pronounced deposition of dose at D_{max}, which results in a relatively lower, normalized surface dose (Fig. 6.7). Beyond the depth of D_{max}, the electrons have lost reference to

the initial beam direction and approach a state of full diffusion.

The surface dose for an individual beam is, however, a complex interaction of the beam scattering foil system, the collimation–applicator system, field size, and beam energy. The design of the scattering and applicator systems affects the buildup region of the central-axis depth–dose distribution by influencing the angular divergence and energy of the beam. An increase in angular distribution creates a situation where fewer scattering events are required to achieve an effective state of full diffusion. As mentioned, teletherapy units that utilize single scattering foil systems employ solid-walled applicators. The removal of applicator walls has been shown to reduce surface doses on the order of 7% (Udale 1988). While

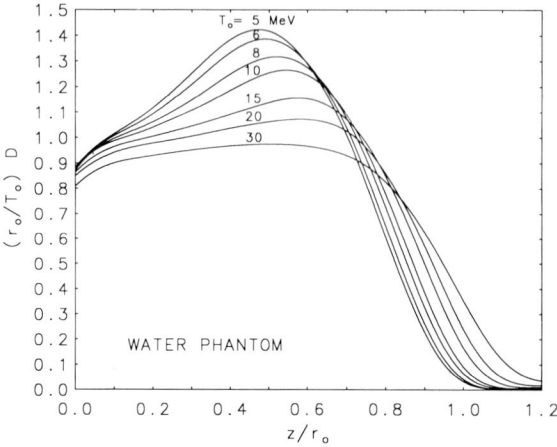

Fig.6.7. Fraction of energy deposited per MeV of incident energy versus scaled depth. (From BERGER and SELTZER 1981)

the applicators act to degrade the beam, by contributing scattered electrons into the field, narrow applicators degrade the beam less because they stop low-energy particles that have been scattered at large angles. It is the filtering effect of the side walls that causes a change in surface dose with applicator field size when solid-walled applicators are used.

For dual scattering foil systems, not only is the amount of scattering material traversed by the beam reduced, but also applicator scatter is not needed to increase beam homogeneity. This results in beams that have smaller angular spread as they enter the patient's surface, which, in turn, produces steeper dose gradients in both the buildup and falloff regions.

The effect of field size on depth dose is related to the range of the scattered electrons in phantom. In general, the field sizes with a diameter greater than one-half of the extrapolated range show minimal change in depth dose with further increases in field size. Clinically, little change in depth dose is realized for field sizes beyond 10 cm in diameter. For a given

field, the range of scattered electrons is also the distance that any point must be from all edges of the beam to receive full scatter. The effective distances required for full scatter increase with increasing energy. This is further illustrated in Fig. 6.8, which shows the relationship of depth of the central axis, 80% isodose vs equivalent field area.

Care should be taken in the design of irregular portals and the point of dose specification for all electron beam therapy. The desire to shield sensitive structures may lead a clinician to design a field that is sufficiently narrow to reduce the depth–dose coverage of the prescribed beam. Figures 6.9a and b show an electron beam portal and the resultant dose distributions along the plane indicated for 6-, 10-, and 20-MeV electron beams. As the electron portal narrows, the higher energies show a reduction of dose at depth. Figure 6.9c shows the dose distribution perpendicular to the central ray at the depth of D_{max} for each of the beams. This demonstrates the effect of lack of scatter into the corners of the portal.

Another difficult clinical situation involves the use of low energy and small beam diameters. In this situation, very rapid changes in depth dose can occur. This effect is demonstrated in Fig. 6.10. Changes include the surface dose, depth of D_{max}, a shift of the falloff region towards the surface, and a decrease in output calibration. These changes not only create a problem with depth dose, but the shift in D_{max} can also affect the depth at which the beams should be calibrated. For example, if one were to recalibrate the beam in Fig. 6.10 for a 1-cm-diameter field, one would first have to obtain a percent depth–dose relationship to ensure that the calibration was carried out at the actual depth of D_{max}. If the dose was prescribed to the 90% isodose depth for the open cone, a depth of 1.6 cm, the actual percent depth dose for the 1-cm field to that depth would be only 50%. This would result in a net underdose of 29%. If, on the other

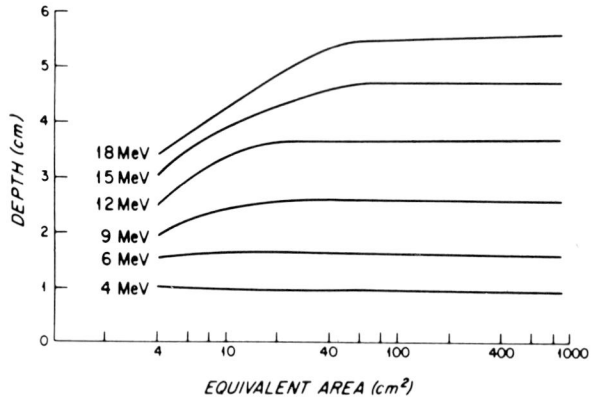

Fig. 6.8. The variation in the depth of the central-axis 80% isodose value with equivalent cone area. (From BIGGS et al. 1979)

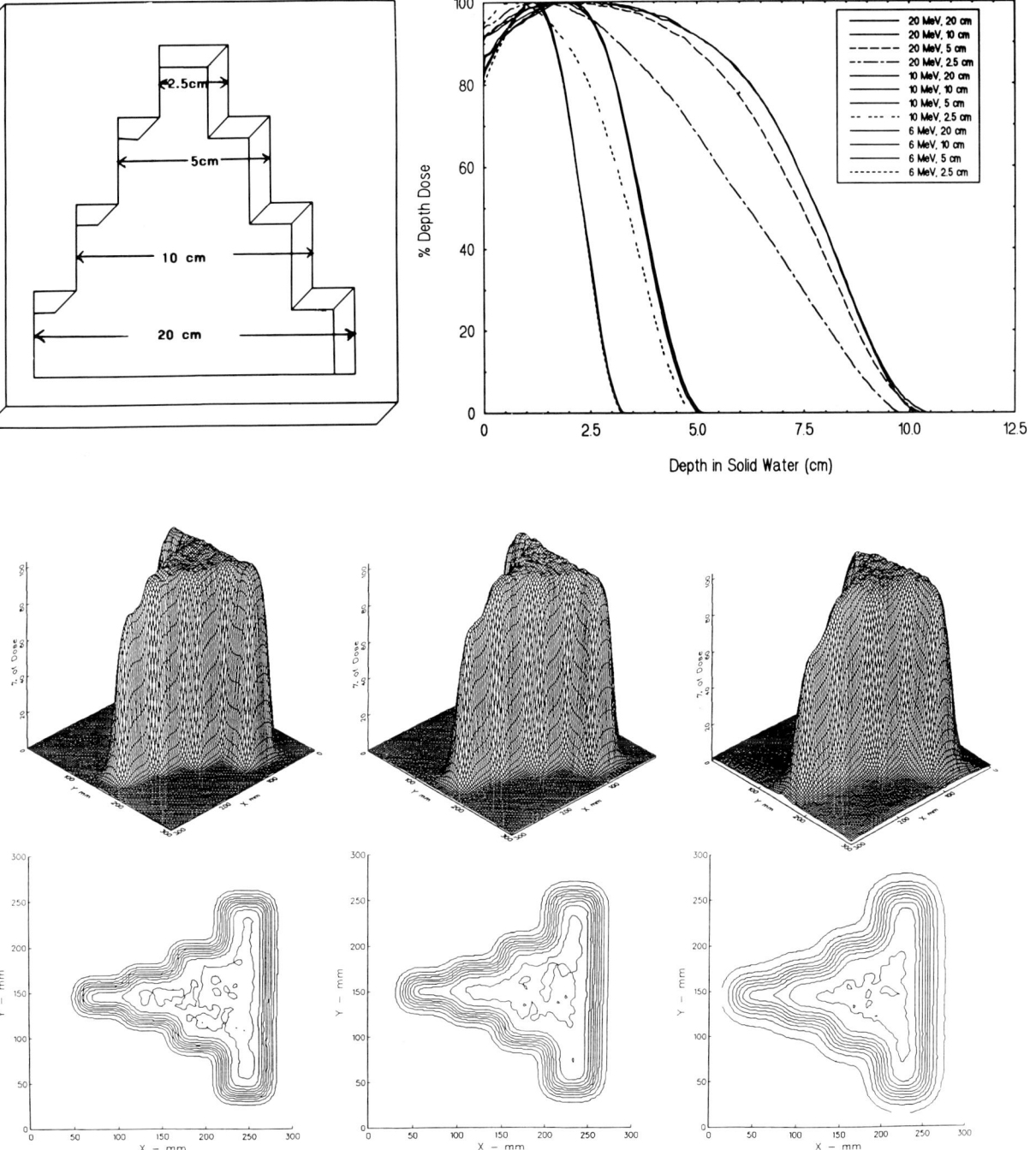

Fig. 6.9. a An electron block for a 25 × 25 cm cone. **b** The central-axis depth dose along the center of each step in beam width for 6-, 10-, and 20-MeV electrons from a dual scattering foil system. As can be seen, the 6-MeV C.A. depth dose is relatively unchanged. The effect is greater as the beam's energy is increased. **c** The dose distributions taken at the depth of the distal central-axis 90% isodose value perpendicular to the central ray. The upper surfaces represent the 90% isosurface at the plane of dose measurement. This corresponds to the 81% isodose level. The lower curves are the 100%, 90%, 80%, 70%, 60%, 50%, 40%, 30%, 20%, and 10% isodose lines normalized at the plane of dose measurement

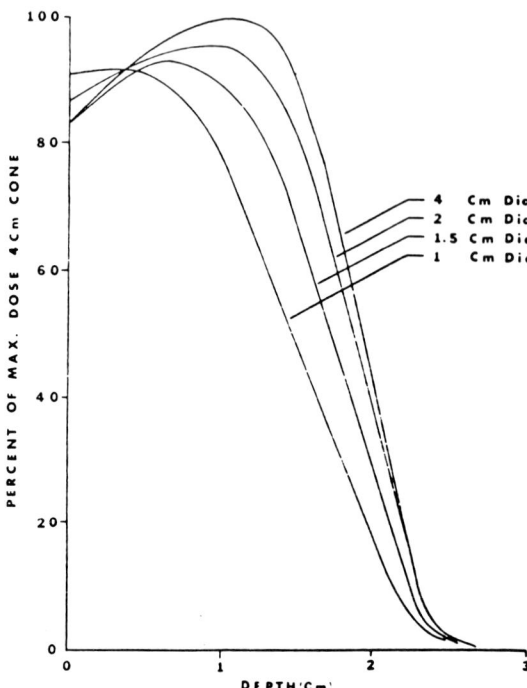

Fig. 6.10. Depth–dose curves as a percent of the peak dose for the 4-cm-diameter cone. (From McGinley et al. 1979)

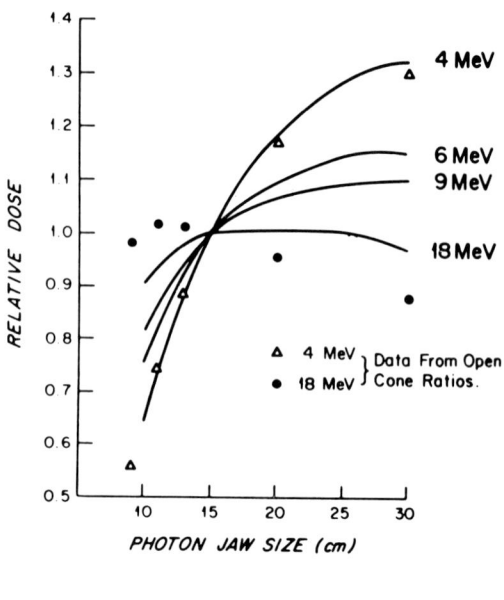

Fig. 6.11. Variation of the relative dose delivered through a 10 × 10-cm open cone to the flat ionization chamber at D_{max} with change in jaw settings. (From Biggs et al. 1979)

hand, the point of calibration, the point of D_{max}, were to remain at the depth of D_{max} for the open cone, a depth of 1.2 cm, this would correlate to a central-axis percent depth–dose value of 70% for the 1-cm-diameter beam. Calibrating the field at this depth would result in the actual depth of D_{max} receiving 140% of the dose at 1.2 cm, a net overdose of 40%. The determination of the extrapolated range is of little clinical value, since it is not significantly affected by field size.

6.3.1 Output Factors

The calibration, cGy per monitor unit, for a given electron beam is dependent upon the beam energy, the field (or cone) size, the size and shape of the irregularly shaped field insert, and numerous other accelerator parameters. Clinically, one of the most important linac parameters is the setting of the x-ray jaws (Biggs et al. 1979). Fig. 6.11 shows the effect of x-ray jaw position on beam output. The vast majority of linacs in clinical use employ fixed electron cones. The settings of the variable x-ray jaws for a given energy–cone combination are usually fixed. As previously mentioned, the setting of the x-ray jaws has an effect on beam flatness and symmetry. The consistent setting of the x-ray jaws is such an important beam parameter that it is incorporated into the

linac's dosimetry interlock chain. It is the recommendation of the American Association of Physicists in Medicine (AAPM) (Khan et al. 1991) that for accelerators that utilize applicators, a separate calibration point must be obtained for each energy–cone combination. For a standard linac with six energies and four fixed cone sizes, this requires a minimum of 24 calibration points. For daily clinical dosimetry, this issue is further complicated by the perturbation of the cone output factors with custom electron cutouts and varying source-to-surface distances.

As previously mentioned, in order for electrons to reach a point in an irradiated volume, they undergo a significant number of scattering events. These begin with the accelerated electron beam traversing the output window of the accelerating wave guide and then interacting with either a single or dual scattering foil system. The electron beam then undergoes collimation by the x-ray jaws. Some electrons are scattered off of the x-ray jaw surfaces while others progress directly towards the patient's surface. For single scattering foil systems, the electron beam then enters a walled applicator. Again a portion of the beam proceeds directly to the surface of the patient, while another portion is directed towards the patient's surface after scattering off of the applicator walls. The electron beam then undergoes final collimation by the end frame of the electron applicator. Here again, some of the electrons pass directly through to the patient's surface, while a portion is absorbed by the end collimator, and a portion of the beam is scattered off the edges of the collimator in the direction of the patient's surface. Once the electron beam enters the patient, it begins to scatter, continually deviating from the beam's original central-axis direction until it reaches a point of near random diffusion. The calibration factor of the electron beam is dependent upon all of these interactions. The output depends not only on the number of electrons reaching the patient's surface but also on the direction at which they enter the surface. It is, therefore, not surprising that individual calibration points are considered necessary for each energy–cone combination.

For most clinical situations, the square or rectangular fields defined by the manufacturer's applicator will not adequately shield all normal tissues. Collimating blocks fabricated from low-melting-point alloy are routinely inserted into the end of the electron applicators. When the blocking for a given cone is minimal, the change in output is usually negligible (KHAN et al. 1991; PALTA et al. 1990). For small cones and for fields that block an appreciable portion of the cone's open area, significant changes in output can occur. These changes in output are primarily a result of loss of side-scatter equilibrium. Lack of side-scatter equilibrium occurs when the width of the field is insufficient to provide full scatter to the point of interest, effectively placing the point in the beam's penumbra. Because the angular scattering power for electrons is inversely proportional to the square of the energy, this effect is most pronounced for the low-energy electron beam. As can be seen in Table 6.1, the effect of blocking a 6 × 6-cm cone to a 3 × 3-cm field can have a significant effect on the output factor. This effect is significantly reduced if a 6 × 14-cm cone is reduced to a 3 × 14-cm field and again reduced further if an 8 × 16-cm field is reduced to a 4 × 16-cm field. As can be seen from these data, the size of the final collimation significantly affects the output factor. While a trend across energies is demonstrated, it is difficult to draw any predictive conclusions from most measured data.

Table 6.1. Change in output factors with field cutouts in standard applicators (from PALTA et al. 1990)

Cone size	6 ×6	6 ×14	8 ×16
Insert block	3 ×3	3 ×14	4 ×16
6 MeV	0.826	0.947	0.985
10 MeV	0.816	0.939	0.972
20 MeV	0.893	0.969	0.985

The change in output factor with low-melting-point alloy inserts also varies across machine types. For a given energy, the output factor for a 6 × 6-cm field will vary depending upon the cone size used to define the field. Table 6.2 gives the output factor, cGy/mu, for a 6 × 6-cm field defined from a variety of cone sizes. While a trend with energy is apparent, this trend will vary depending upon the design of the linac's electron scattering system.

For linacs, which utilize variable electron collimators, the changes in output with variation in field size are more predictable. For scanning electron beam systems, relationships have been derived that characterize and predict these effects through the use of pencil-beam calculational algorithms (MILLS et al. 1982), as well as the parameterization of measured output factors (McPARLAND 1987). These models can predict the relative output to within 1.0%–1.5% of experimental results.

Dual scattering foil systems have demonstrated output characteristics that are consistent with those of the single-foil systems (PALTA et al. 1990; PURDY, personal communication, September 1993). As can be seen in Fig. 6.12, the behaviors for the dual-foil systems from different manufacturers are vastly different not only in magnitude of the output factors but also in the direction of these changes. Unlike the relatively narrow range of output factors that are experienced with photon beams, the range of output factors for electrons can change 20% as one progresses from a 6 × 6-cm to a 20 × 20-cm field.

The behavior of irregular electron beams for dual-foil systems has been investigated (BIGGS et al. 1979). For *a given electron cone*, a method of correlating the output of a rectangular field and a square field of the same area has been demonstrated. It was found that

Table 6.2. Relative output for a 6 × 6-cm insert in 6 × 6-, 10 × 10-, and 15 × 15-cm cones for 4-, 6-, 10-, 15-, and 18-MeV electrons normalized to a 10 × 10-cm open cone (adapted from BIGGS et al. 1979)

Cone size	4 MeV	6 MeV	9 MeV	10 MeV	15 MeV	18 MeV
6 × 6	0.74	0.84	0.9	0.94	1.0	1.01
10 × 10	0.98	0.97	0.96	0.96	0.94	0.91
15 × 15	1.15	1.03	0.99	0.95	0.93	0.91

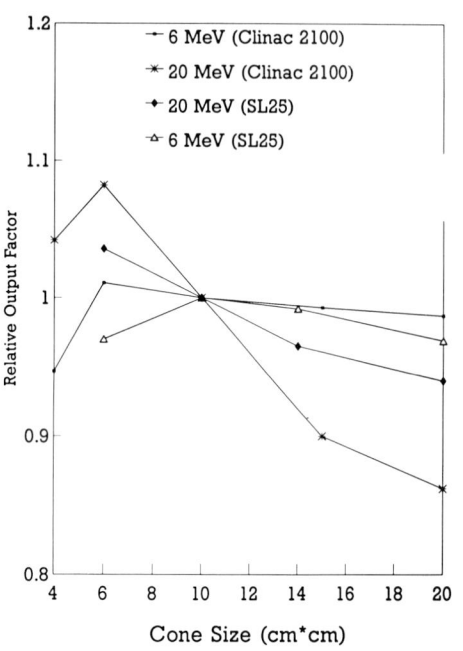

Fig. 6.12. Relative output factors versus cone size for a Clinac 2100 and an SL25

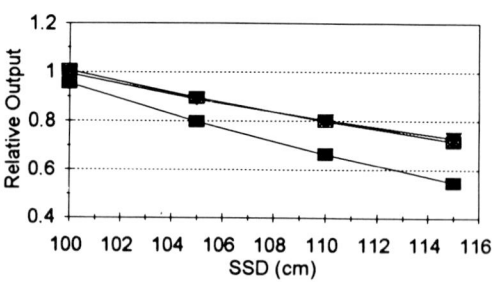

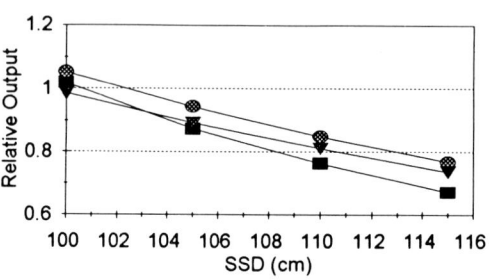

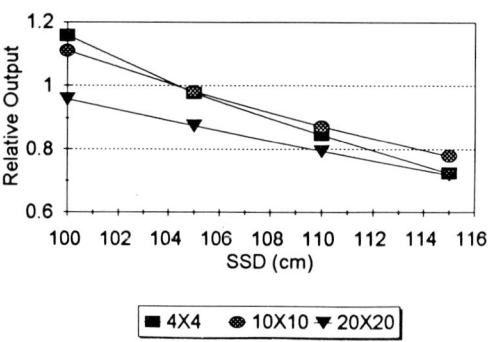

Fig. 6.13. Relative output for a Clinac 2500 for 6-, 12-, and 20-MeV from 100- to 115-cm SSD

for a given energy–cone combination, the output for a square field was almost identical to that of a rectangular field of the same area. With the equivalent area method, it is possible to predict rectangular field outputs to within 1% as long as the aspect ratio remains no greater than 2:1.

6.3.2 Source-to-Surface Distances

As with output, the scattering of electrons from multiple surfaces within the accelerator and collimating structures gives rise to multiple apparent source positions, or virtual sources. Two methods for determination of virtual source position have been suggested in the literature (ABOLGHASSEM et al. 1986). These are: (a) the inverse-square law and (b) the back-projection of the full width at half maximum (FWHM). For large field sizes across all energies, the two methods have been found to give consistent results (PALTA et al. 1990). It has, however, been reported that the position of the virtual source can vary significantly with changes in electron energies as well as in field sizes.

It has been suggested that in order to satisfy clinical needs it may be necessary to compile a table of effective SSDs across energy–cone combinations (KHAN 1984). Fig. 6.13 shows the change in output from SSD 100 to 115 for a range of electron energies and cone sizes. As can be seen from these data, the

variation in output with distance is a smooth function, and one method of dealing with this phenomenon is simply to create a table of output factors. One can then use any of several fixed cone-to-surface offsets or can linearly interpolate to the cone-to-surface offset needed.

Another method of dealing with this problem is to create a table of air gap correction factors (MEYER et al. 1984). With this technique, the nominal SSD is used to calculate the variation in output with varying distance. The difference between a set of measured and calculated data is then tabulated. Figure 6.14a shows the air gap correction factor for a 7-MeV beam across three cone sizes, while Fig. 6.14b shows the gap correction factor across four energies for a single

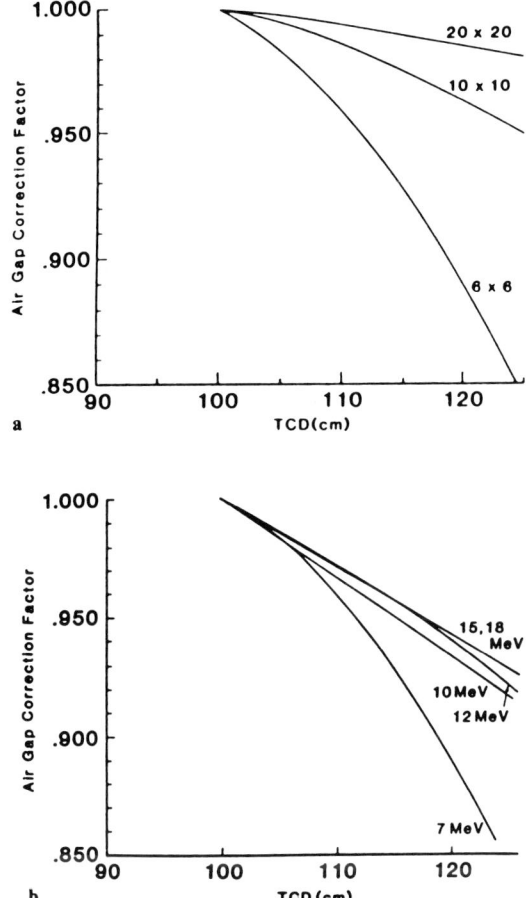

a

b

Fig. 6.14. a Air gap correction factor for 6 × 6-, 10 × 10-, and 20 × 20-cm cones for a 7-MeV beam. **b** Air gap correction factor for a 6 × 6-cm cone for 7-, 10-, 12-, 15-, and 18-MeV electrons. (From MEYER et al. 1984)

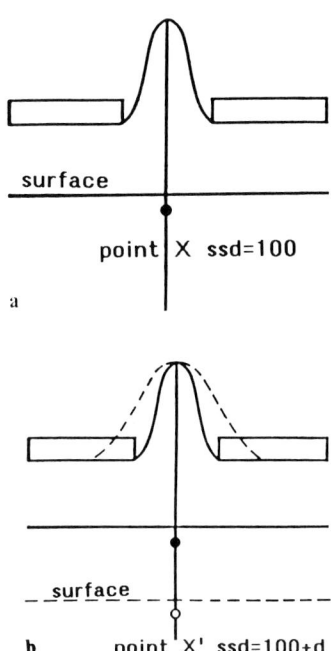

a

b

Fig. 6.15. a The *solid line* represents the gaussian-shaped profile which illustrates the probability distribution at the collimator of electrons reaching a point X on the central axis. **b** The same as above except the *dashed line* represents the probability distribution for the point X on the central axis, at the same depth, except at an extended distance from the surface. (From KHAN et al. 1991)

cone. Although these data are for a single scattering foil system, similar data have been presented for dual-foil systems. It can be seen from both sets of data that the correction factors are larger at low electron energies and small field sizes and smaller for high electron energies and large field sizes.

This phenomenon is due largely to the lack of side-scatter equilibrium for small fields at low energies. As previously discussed, electrons reach the patient surface through several different paths. Once the beam has reached the patient's surface, it continues to scatter. Figure 6.15a shows the probability distribution (gaussian in shape) at the final electron collimator of all electrons reaching point x on the central axis. As one moves away from point x, which is at SSD + x from the nominal source, the probability distribution increases in width (Fig. 6.15b). If the final collimator opening is small, then insufficient electrons will pass through the collimator to establish full side-scatter equilibrium. The width of the gaussian

distribution necessary for full side-scatter equilibrium is inversely proportional to electron energy. This explains the change in air gap correction factor with both energy and field size.

6.4 Dose per Monitor Unit Calibration

Dosimetry can be broken down into absolute dose determination and relative dose mapping. For absolute calibration, the current recommendations of the AAPM are detailed in their protocol for high-energy electron and photon calibration known as the TG21 protocol (Task Group 21, 1983). Absolute calibration of any treatment units should always be directly traceable to the facility's calibrated ion chamber.

When determining dose through the use of an ion chamber, the quantity measured is the charge that results from the separation of the gas which fills the volume of the chamber. The TG21 protocol details a methodology of relating the charge collected to a quantity of energy that would have been deposited in the medium if the chamber had not been present. This methodology first relates the amount of charge collected to the amount of energy required to produce that charge. Then through the use of the Spencer-Attix modification of the Bragg-Gray cavity theory, the protocol details a methodology of transferring the work performed in the gas to the work that would have been deposited in the medium if it had fully displaced the ion chamber and its gas cavity. The relationship is:

$$\text{Dose}_{\text{in the medium}}/\text{mu}$$
$$= (M/U)N_{\text{gas}}\,(\bar{L}/\mathcal{P})_{\text{air}}^{\text{medium}}\,P_{\text{ion}}P_{\text{repl}}.$$

where M is the electrometer reading normalized to 22°C and one standard atmosphere, U is the accelerator monitor units, N_{gas} is the cavity-gas calibration factor, $\bar{L}/\mathcal{P}$ is the mean restricted collisional mass stopping power ratio for the medium to air at the beam energy being calibrated, P_{ion} is the ion-recombination correction factor, and P_{repl} is a factor that corrects for the displacement of the medium by the ion chamber.

Unlike calibrations for photon beams, the energy of the electron beam cannot be considered constant as the beam penetrates through absorbing material. The energy of the electron beam continually decreases at a rate of 2 MeV per cm. Because $\bar{L}/\mathcal{P}$ and P_{repl} are dependent upon beam energy, the normalized depth ionization readings are not identical to percent depth–dose values, as is the case for photon dosimetry. These corrections are well known, and many current dose-scanning systems have algorithms incorporated for the conversion of ionization data to dose. It is, however, the responsibility of the physicist to understand these corrections and ensure their proper use.

While the AAPM's TG21 protocol is the standard for both electron and photon dosimetry, the special considerations regarding electron dosimetry are more explicitly covered in the subsequent TG25 report (KHAN et at. 1991).

6.5 Film Dosimetry

One of the most convenient methods of obtaining relative dose distributions for electron beams is through the use of film dosimetry. Film is a high-resolution detector which provides a permanent record of the dose distribution. It requires relatively little beam on-time and is applicable to both scattered- and scanned-beam systems. The data can be measured in the laboratory and not at the linac, decreasing the use of valuable machine time.

The primary disadvantage of film is its nonlinear optical density-to-dose relationship, known as the sensitometric curve. While some films, such as Kodak's XV2 film, show linearity for clinical electron beams between 0 cGy and 40 cGy, it is always best to map out the optical density-to-dose relationship for every batch of film and every specific phantom material being used. Although the optical density is dependent upon film processing parameters, most modern-day automatic film processors maintain a nearly constant processing environment. Often the sensitometric curve can be verified by the use of a single calibration film. If the processing conditions have altered, yielding a slight increase or decrease in optical density for a given dose, then the sensitometric curve may be globally adjusted to coincide with the new processing conditions.

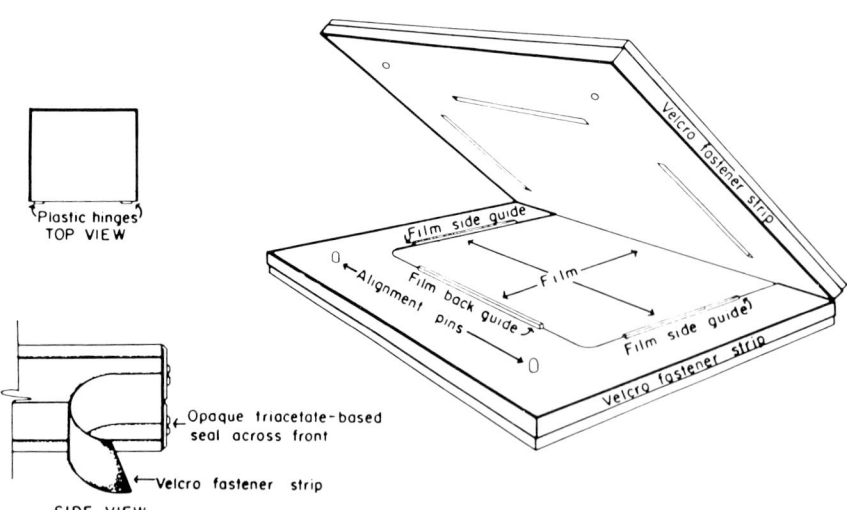

Fig. 6.16. Solid Water film phantom. (From BOVA 1990)

The most important consideration in film dosimetry is the film–phantom alignment. As noted (DUTREIX and DUTREIX 1969), the alignment of the film in the phantom, as well as the good phantom–film contact, is necessary for artifact-free film dosimetry. The design of a light-tight film–phantom cassette is given in Fig. 6.16 (BOVA 1990). This cassette provides for good front-edge alignment as well as good film–phantom contact. The use of Velcro fasteners enables quick and easy loading in the darkroom. If water-equivalent phantom materials, such as Solid Water (WHITE et al. 1977), are used in the construction of film phantoms, then problems associated with charge storage (GALBRAITH et al. 1984) and corrections for phantom density (KHAN et al. 1991; Task Group 21, 1983) can be eliminated.

6.6 Shielding for Electron Beams

For electrons, the best shielding technique is to place the final collimation in contact with the patient's surface. As the final collimation moves away from the surface, the width of the field's penumbra increases (Fig. 6.17). In practice, however, it is very often impractical to construct lead shielding that lies directly on the patient. The most practical and time-efficient technique involves the creation of low-melting point alloy shields, which attach to the linac's electron-cone applicators. Most manufacturers provide molding kits that produce blocks that can be mounted to the linac's electron applicators. These systems also allow encoding of individual blocks. The lead thickness, in millimeters, required for shielding of a specific electron beam can usually be

obtained by dividing the beam's energy in MeV by two. For most low-melting point alloys, the thickness necessary to adequately shield a beam is usually 20% greater than that necessary for lead (KHAN et al. 1981, 1991; GIARRATANO et al. 1975).

As previously mentioned, it may be advantageous to apply shielding directly to the patient's surface. This can be accomplished through the use of several different techniques. The oldest technique involves obtaining an impression of the surface through the use of plaster-impregnated gauze strips. It is often a good idea to outline the radiation portal first with a water-soluble ink. The strips can then be applied to the patient's surface. After the plaster has set, not only has a surface impression been obtained, but the outline of the radiation portal has also transferred to the mold. Once this mold is fully cured, sides are attached, and a liquid compound, which will harden in relatively short order, is then poured in to create a positive of the area to be shielded. Such compounds include dental stone, concrete, and low-melting-point alloys. As soon as the hard surface has been obtained, lead can be applied and molded to the surface. Once a good fit has been achieved, the radiation portal can be cut out of the shield. The application of custom surface shields, especially in the head and neck region, usually involves small electron portals. As previously mentioned, in these situations the physicist should be careful to verify not only the output of the final beam but also the depth–dose relationship.

6.6.1 Intracavitary Shielding

It is often necessary to use a high-density shield to prevent a beam from penetrating past a specific tissue boundary. Targets involving the lip and eyelid often require such shielding considerations. In these cases, two important shielding criteria must be met. The first is providing a shield that is thick enough to provide adequate shielding of underlying tissues. An example of such a criterion is the use of eye shields when treating a lesion confined to the lid or canthus. In this case, the shield is usually chosen to properly fit the patient's eye. The thickness of the shield is usually only sufficient to shield beams of 8 MeV or lower. The results of using high-energy beams on such a shield can be seen in Fig. 6.18. While the thickness of the target tissues usually does not require energies above 8 MeV, the relatively low surface dose provided by many low-energy electron beams may tempt clinicians to use higher beam energies.

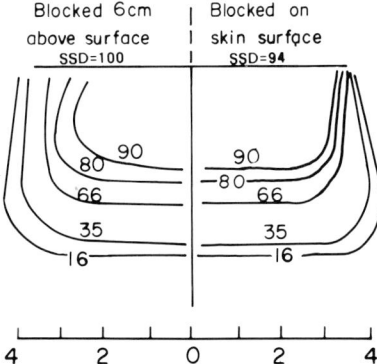

Fig. 6.17. Isodose distribution for a 10-MeV electron beam at 94-cm SSD, cone in contact with surface, and 100-cm SSD with a 6-cm gap between the end of the cone and the surface. (From BOVA 1994)

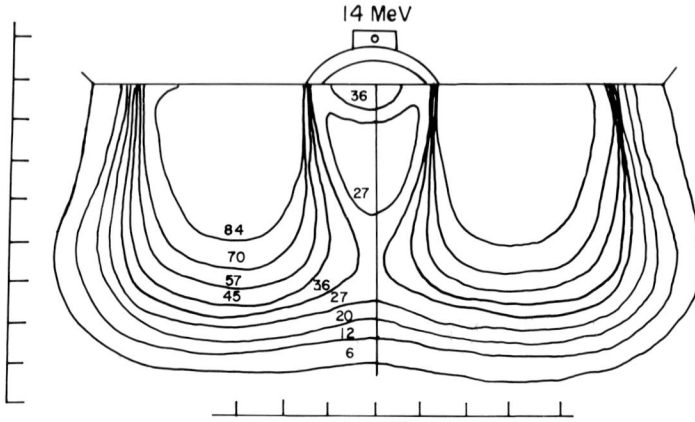

Fig. 6.18. The dose distribution which results from a 14- MeV electron beam being incident upon a standard gold-plated lead eye shield. (From Bova 1994)

In fact, the surface dose in such a clinical situation is usually much higher than anticipated. The enhancement of dose from the backscatter of electrons from high-density, high atomic number absorbers has been reported (Klevenhagen et al. 1982). This increase in surface dose at the tissue–shield interface can be negated through the use of lower atomic number and lower density absorbers (Bova 1984). Table 6.3 shows the effects of such absorbers on lead shields for both photons and electrons. Whenever an intracavitary shield is employed, extreme care should be taken to document and eliminate these effects whenever necessary.

6.7 Surface Contours

The effects of electron scatter at abrupt irregularities has been well documented (ICRU 1972; Sternick 1978). The clinical implications of these scattering effects can be seen in Fig. 6.19. Although the contour shown contains hot spots that are 20% greater than the central-axis maximum dose, planes just 1–2-cm

superior are, in general, flatter and exhibit much reduced dose inhomogeneities. While computational techniques have been developed to help predict these effects (Hogstrom and Almond 1983), it is often more desirable to eliminate the effects and produce a uniform distribution. This can be accomplished by the addition of surface bolus, as shown in Fig. 6.20. In this case, the bolus is added throughout the length of the field to produce a flat entrance surface for the electron portal. Reduced beam penumbra can also be provided, if the final beam collimation is on the surface of the bolus with the electron cone diaphragm defining a field that is 1 cm larger at all edges.

6.8 Oblique Incidence

The dose prescription for most electron portals is at or near the 90% isodose line. To allow for clinical uncertainties in target definitions, the prescribed depth is usally determined by adding a margin to the target depth. This margin does not usually allow for

Table 6.3. Backscatter measurements: increase above homogeneous dose (%). (From Bova 1994)

Millimeters of absorber placed over 3 cm of lead	250^{60} kVp[a]	Co[a]	8 MV[a]	17 MV[a]	5 MeV[b]	10 MeV[b]	20 MeV[b]
None	170	110	100	90	70	50	30
0.4 mm wax	0	50	60	70	No data	No data	No data
3.0 mm wax	0	0	20	30	No data	No data	No data
3.0 mm Al	No data	No data	20	30	10	10	20
3.0 mm Al + 1 mm Wax	0	0	10	10	10	10	20

[a]Relative ionization
[b]Relative TLD response (Hershaw TLD-100 chips), measurements at a depth equal to D_{max}

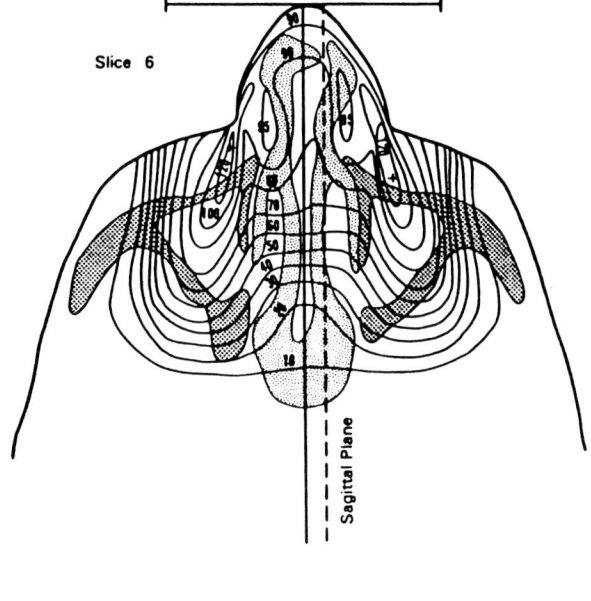

Fig. 6.19. Calculated isodose distribution in the transverse plane with a 7.9 × 7.9-cm 13-MeV electron beam at a SSD of 100-cm. (From HOGSTROM and ALMOND 1983)

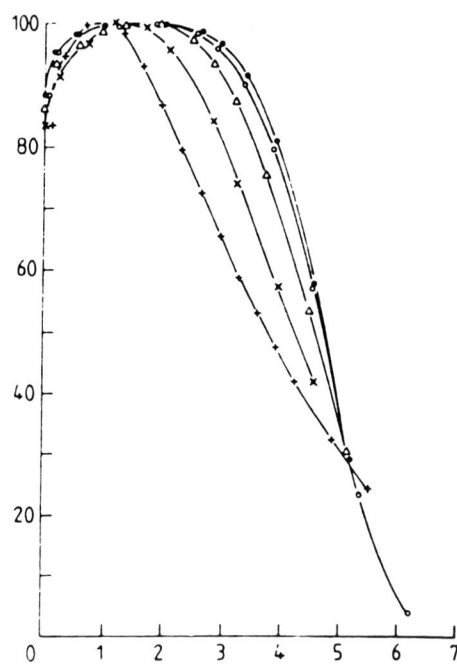

Fig. 6.21. The central-axis percent ionization curves for a 12-MeV electron beam at angles of: •, 0°; o, 15°; Δ,30°; ×,45°; +,60°. (From BIGGS 1984)

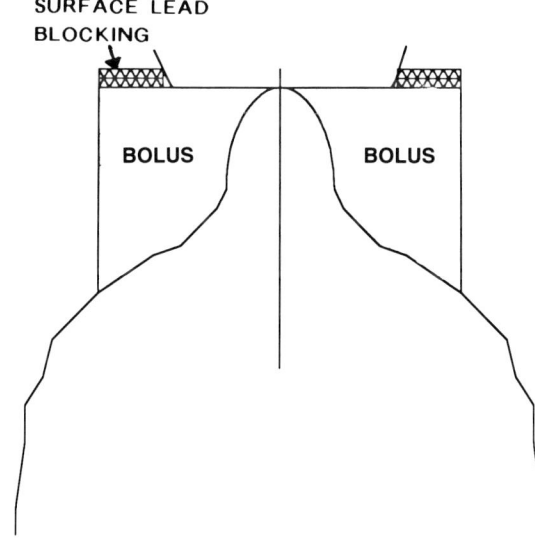

Fig. 6.20. The addition of unit density tissue equivalent bolus to eliminate surface irregularities. The final lead collimation has been applied directly to the surface of the bolus

shifts in the depth-dose curve. The effects of oblique incidence of electron beams onto the patient's surface is, therefore, of general clinical concern.

It has been demonstrated (BIGGS 1984) that as the obliquity of the beam increases, the entire isodose curve shifts towards the surface (Fig. 6.21). For low-energy beams, this shift can decrease the 80% ionization depth by a factor of 2, effectively reducing its coverage by half. The shift of dose towards the surface is greater for larger angles of obliquity and also larger for lower electron energies. This shift is due to the increase in attenuation from the side of the beam that reaches the patient's surface, that is, the side of increased attenuation, and the reduction of scatter from the side of the beam that traverses air instead of tissue. It is interesting to note that, as with other phenomena in which the central-axis isodose relationship is shifted towards the surface, here again the range of the oblique electron beam is relatively unaffected. From the clinical standpoint, it is best to try to maintain orthogonality for electron portals.

References

Abolghassem J, Kuchnir FT, Reft CS (1986) Determination of the source position for the electron beams from a high-energy linear accelerator. Med Phys 13: 942–948

Abou Mandour M, Harder D (1978) Systematic optimization of the double-scatter system for electrons beam field-flattening. Strahlentherapie 154: 328–332

Berger MJ, Seltzer SM (1981) Fundamental aspects of absorbed dose measurement. Proceedings of the Symposium on Electron Dosimetry and Arc Therapy. pp 1–19

Biggs P (1984) The effect of beam angulation on central axis percent depth dose for 4–29 MeV electrons. Phys Med Biol 29: 1089–1096

Biggs PJ, Boyer AL, Doppke KP (1979) Electron dosimetry of irregular fields on the Clinac 18. Int J Radiat Oncol Biol Phys 5: 433–440

Bjarngard BF, Piontek RW, Svensson GK (1976) Electron scattering and collimation system for a 12 MeV linear accelerator. Med Phys 3: 153–158

Boag JW (1972) Surface ionization ratio for electrons in the energy range 3 to 11 MeV. Br J Radiol 45: 229

Bova FJ (1990) A film phanton for routine film dosimetry in the clinical environment. Med Dosim 15: 83–85

Bova FJ (1994) Treatment planning for irradiation of head and neck cancer. In Million RR, Cassisi NJ (eds) Cancer of the head and neck : a multidisciplinary approach. Philadelphia, Lippincott, pp 209–230

Bruinvis IAD, Mathol WAF (1988) Calculation of election beam depth-dose curves and output factors for arbitrary field shapes. Radiother Oncol 11: 395–404

Dutreix J, Dutreix A (1969) Film dosimetry of high energy electrons. Ann NY Acad Sci 161: 33–42

Galbraith DM , Rawlinson JA, Munro P (1984) Dose errors due to charge storage in electron irradiated plastic phantoms. Med Phys 11: 197–203

Giarratano JC, Duerkes RJ, Almond PR (1975) Lead shielding thickness for dose reduction of 7- to 28-MeV electrons. Med Phys 2: 336–338

Hogstrom KR, Almond PR (1983) Comparison of experimental and calculated dose distributions. Acta Radiol Supl (Stockh) 364: 89–99

International Commission on Radiation Units and Measurements: ICRU Report 21 (1972) Radiation dosimetry: electrons with initial energies between 1–50 MeV. Washington, DC

Kerst DW (1943) The betatron. Radiology 40: 115–119

Khan FM (1984) The physics of radiation therapy. Williams & Wilkins, Baltimore, pp 320 and 322

Khan FM, Werner BL, Deibel FC Jr (1981) Lead shielding for electrons. Med Phys 8: 712–713

Khan FM, Doppke KP, Hogstrom KR et al. (1991) Clinical electron-beam dosimetry: Report of AAPM Radiation Therapy Committee Task Group No. 25. Med Phys 18: 73–109

Klevenhagen SC, Lambert GD, Arbabi A (1982) Backscattering in electron beam therapy for energies between 3 and 35 MeV. Phys Med Biol 27: 363–373

Lax I, Brahme A (1980) Collimation of high energy electron beams. Acta Radiol (Oncol) 19: 119–207

McGinley PH, McLaren JR, Barnett BR (1979) Small electron beams in radiation therapy. Radiology 131: 231–234

McParland BJ (1987) A parametrization of the electron beam output factors of a 25-MeV linear accelerator. Med Phys 14: 665–669

Meyer JA Palta JR Hogstrom KR (1984) Demonstration of relatively new electron dosimetry measurement techniques on the Mevatron 80. Med Phys 11: 670–677

Miller CW (1954) An 8 MeV linear accelerator for x-ray therapy. Proc IEE 101: 207–222

Mills MD, Hogstrom KR, Almond PR (1982) Prediction of electron beam output factors. Med Phys 9: 60–68

Palta JR, Daftari IK, Ayyangar KM, Suntharalingam N (1990) Electron beam characteristics on a Philips SC25. Med Phys 17: 27–34

Skaggs LS, Almy GM, Kerst DW, Lanzl LH (1948) Development of the betatron for electron therapy. Radiology 50: 167–173

Sternick E (1978) Algorithms for computerized treatment planning. In: Orton CG, Bagne F (eds) Practical aspects of electron beam treatment planning. Medical physics monograph No. 2, p 52. American Institute of Physics, New York

Sweeney LE, Gur D, Bukovitz AG (1981) Scatter component and its effect on virtual source and electron beam quality. Int J Radiat Oncol Biol Phys 7: 967–971

Task Group 21, Radiation Therapy Committee, American Association of Physicists in Medicine (1983) A protocol for the determination of absorbed dose from high-energy photon and electron beams. Med Phys 10: 741–771

Turner AP (1980) Surface dose measurements clinac 18 electron beams. Proceedings Eight Varian Clinac Users Meeting, January 31–February 2

Udale M (1988) A Monte Carlo investigation of surface doses for broad electron beams. Phys Med Biol 33: 939–954

van der Laarse R, Bruinvis IAD, Nooman MF (1978) Wall-scattering effects in electron beam collimation. Acta Radiol Oncol 17: 113–124

White DR, Martin RJ, Darlison R (1977) Epoxy resin based tissue substitutes. Br J Radiol 50: 814–821

7 Physics of Electron Arc Therapy

Dennis D. Leavitt

7.1 Introduction

Electron arc therapy is the application of a long, narrow strip beam of electrons to a superficial treatment volume as the gantry of the linear accelerator is rotated through an arc about the patient. Like photon arc therapy, the dose to a point within the treatment volume is the summation of dose contributions from the electron beam during each degree of arc; however, unlike photon arc therapy, the treatment volume is not irradiated continuously during the entire arc. In this respect, electron arc therapy is similar to moving strip photon therapy in that the irradiating strip is moved sequentially across the treatment surface. Electron arc therapy is ideally suited to the treatment of large superficial volumes in portions of the body having a cylindrical shape such as the thorax, abdomen, arms, or legs, and is less well suited to spherical or ellipsoidal surfaces such as the head. Several innovative techniques have been developed to improve the dose uniformity throughout the treatment volume and to conform the dose to the prescribed treatment volume. These techniques include variable-shaped electron arc apertures (BLACKBURN 1981; LEAVITT 1987, 1988; LEAVITT et al. 1989a,b), custom field shaping on the patient's surface using tertiary blocking (LEAVITT et al. 1990a; THOMADSEN 1981), application of different electron energies across different arc segments (LEAVITT et al. 1985), application of multiple electron energies within a single arc segment (LEAVITT et al. 1990b), variation of dose rate with angle (LEAVITT 1978), and use of cutomized tissue-equivalent bolus (LEAVITT et al. 1990a). Creative use of these techniques enables the delivery of a uniform dose to a large curved surface and to the underlying superficial volume, while sparing critical underlying organs and avoiding the problems of dose inhomogeneity at matchlines of abutted fixed electron fields.

Prior to the widespread use of electrons from clinical linear accelerators, electron arc therapy was investigated using betatrons (BECKER and WEITZEL 1956; HUDEPOHL and RASSOW 1973; RASSOW 1970a,b, 1972; RUEGSEGGER et al. 1979). Because of the greater range of electron energies available with the betatron, techniques were investigated to treat deeper-seated tumor volumes as well as the superficial treatment volumes accessible by lower energies. The capacity to produce electrons of up to 50 MeV using modern microtrons may rekindle an interest in electron arc techniques using higher energy beams (BRAHME et al. 1988); however, linear accelerator-based electron arc techniques will remain limited to the energy range from 4 MeV to about 20 MeV.

A common site for electron arc therapy is the post mastectomy chest wall. Figure 7.1 schematically illustrates the locations of the primary photon collimators, the secondary electron aperture, and the tertiary field-defining cast relative to the patient contour. The primary photon collimators and the secondary electron arc aperture define the shape and intensity of the electron arc beam incident on the patient, while the tertiary cast defines the patient's surface area to be irradiated. The integration of the instantaneous does to a point within the patient as the gantry arcs through the prescribed angle defines the total dose to that point. The implementation of this technique is demonstrated in the patient setup of Fig. 7.2. The patient lies supine on the treatment table; the tertiary field-defining cast is placed

DENNIS D. LEAVITT, PhD, Division of Radiation Oncology, University of Utah Medical Center, 50 N. Medical Drive, Salt Lake City, UT 84132, USA

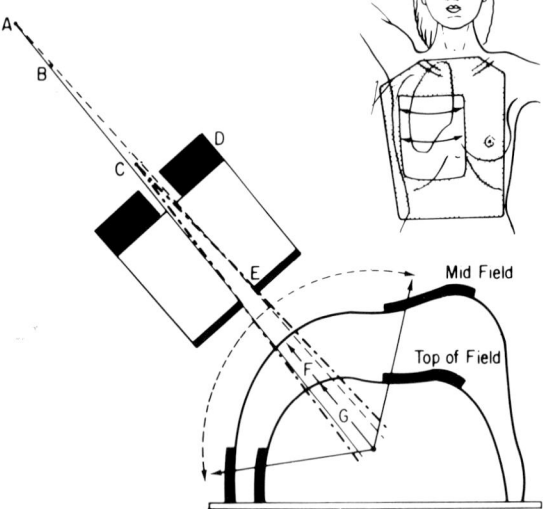

Fig. 7.1. Schematic representation of electron arc therapy setup. *A*, Location of x-ray target (100 cm from mechanical isocenter) (the x-ray target is not in the beam during electron arc; its position is shown for reference only); *B*, location of electron scattering foil; *C*, effective source for 6-MeV electrons using secondary cerrobend collimator; *D*, primary photon collimator; *E*, secondary cerrobend aperture used for electron arc; *F*, depth of isocenter in "mid field" central plane ($=r_0$ in text and equations); *G*, depth of isocenter in "top of field" plane ($=r$ in text and equations). Location of tertiary collimation breastplate is noted in insert. The two arrowed lines superimposed on the inset demonstrate the location of the midfield and top of field contours. (From LEAVITT et al. 1985)

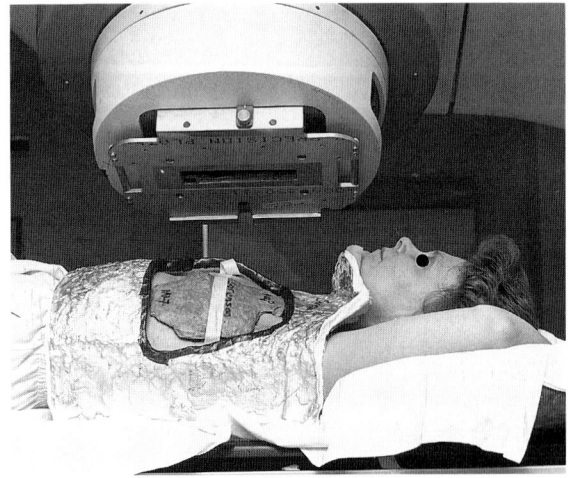

Fig. 7.2. Actual electron arc patient setup, showing the linear accelerator positioned for treatment, starting at the medial field edge. The secondary electron arc collimator and the tertiary field shaping cast are in place, and the wax compensator rests on the patient's skin surface. The tertiary field shaping cast is supported to the left and right of the patient by the treatment couch. The tertiary cast allows shallow breathing during treatment. (From LEAVITT et al. 1990a)

immediately over the patient and is fully supported by the treatment table; the secondary electron arc aperture is placed in the accelerator accessory mount; and customized paraffin bolus to limit the depth of penetration of electrons into the chest wall is placed immediately on the patient's chest. These two illustrations demonstrate the fundamental features of electron arc therapy. The dosimetric characteristics of the electron arc beams, dose calculation techniques, and dose delivery techniques necessary to implement electron arc therapy will be described in the text.

7.2 Electron Arc Beam Characteristics (Fixed Field)

Accurate dose delivery in electron arc therapy requires a complete description of the instantaneous electron arc beam. Thus, for each electron energy, the dose per monitor unit, depth dose including surface dose and buildup dose, and beam profiles in the narrow and long axis of the electron arc aperture must be measured for the range of aperture widths and lengths to be applied and for the range of isocenter depths to be expected in clinical use (LEAVITT 1992). It is especially important that the electron beam modelling applied must reproduce the measured data, since small errors in the calculated width of the electron profile or errors in representation of the electron penumbra will produce larger errors as the entire profile is integrated. For example, a change in profile width of only 2 mm produces a change of 2.6% in dose calculated for an electron arc using 6-MeV electrons and 3.8% for 20-MeV electrons. Similarly, failure to model the penumbra beyond the beam edge can introduce errors in calculated dose. HOGSTROM and LEAVITT (1986) recommend adjusting the effective width of the beam such that the area under the calculated and measured profiles agree, thereby preserving integral dose in the plane of rotation.

Figure 7.3 shows the depth-dose curves for electron energies of 6-, 9-, 12-, 16-, and 20-MeV electrons incident on a cylindrical phantom with radius of curvature of 15 cm. Fixed-field and 180° arcs are compared for an isocenter depth of 15 cm. The fixed-field depth doses using the electron arc aperture are similar to those using the standard electron applicators. Figure 7.4 shows the electron beam profiles in the long axis of a 10-cm-wide by 40-cm-long field defined by the primary photon collimators. This demonstrates that the electron intensity

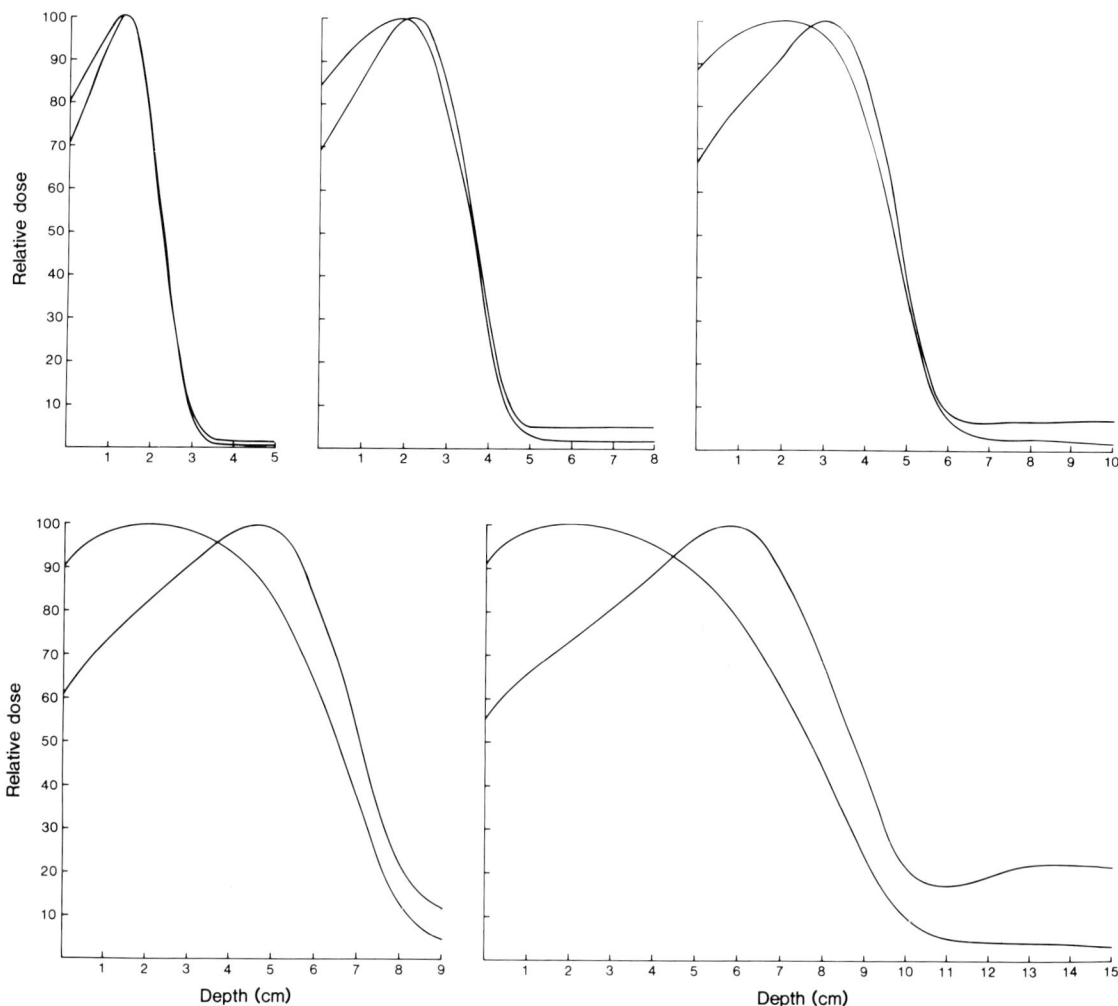

Fig. 7.3. Comparison of static and arc electron depth doses using the secondary electron arc aperture projecting a 5-cm-wide field at isocenter. The isocenter depth is held constant at 15 cm. Comparisons are for 6-, 9-, 12-, 16-, and 20-MeV electrons. Major differences as discussed in the text are: reduced surface dose, greater depth of maximum dose, faster dose, falloff beyond depth of maximum, and increased bremsstrahlung dose beyond range of electrons

15 cm superior or inferior to the central axis is reduced by as much as 20% for 6- MeV electrons and 50% for 20-MeV electrons. Although this decrease in dose can be compensated for by adjusting the shape of the secondary electron arc aperture, this effectively limits the superior-to-inferior length of the volume that can be treated using electron arc. Figure 7.5 compares the 20-MeV electron beam profiles defined by the primary photon collimators set to 10 cm wide by 40 cm long (solid lines) with the electron beam profiles defined by the rectangular electron arc aperture set to 5 cm wide by 30 cm long (dashed lines). In the long axis of the field the profile is modified only by the superior and inferior limits of the electron arc aperture; in the narrow axis of the field, the 10-cm-wide field defined by the photon collimators is reduced to 5 cm wide by the electron aperture. Similar effects are seen for the other electron energies. Unlike photon beams and standard-applicator beams, the measured width of the electron arc beam is consistently larger than the projected light field. Figure 7.6 illustrates the change in long-axis electron dose profile with shape of the electron aperture. A trapezoidal-shaped aperture reduces the intensity as the width is reduced superiorly. This trapezoidal shape is approximated by a three-vane step shape and a nine-vane step, and the dose profiles are seen to vary accordingly. Thus, a reduction in intensity of 20% can be easily achieved over part of the field length by reducing the field width

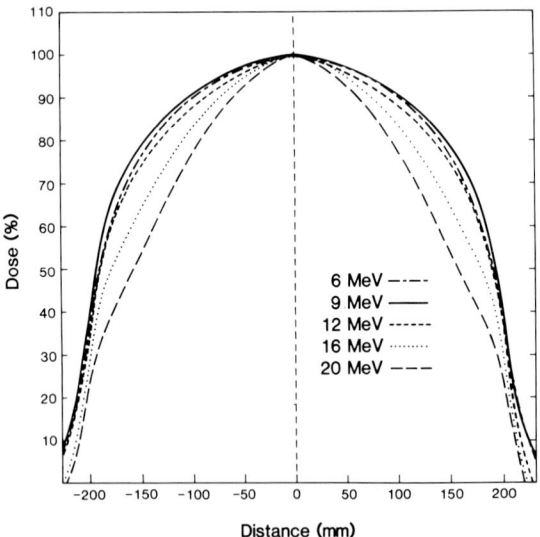

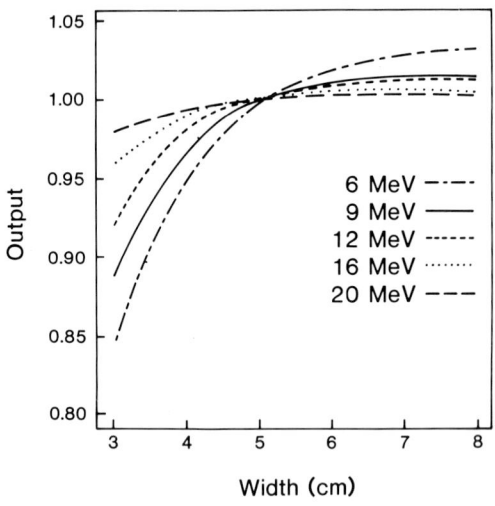

Fig. 7.4. Electron beam profiles in long axis of rectangular secondary electron arc collimator. Note the increased falloff with energy. These changes in intensity in the long axis of the collimator must be compensated for by changing the width of the electron arc collimator superior and inferior to the central axis. These profiles may vary with the scattering foil configuration used in the linear accelerator

Fig. 7.7. Dependence of electron intensity on secondary electron arc collimator width. Curves are normalized to 1.0 for the 5-cm-wide field for all energies. The variation in intensity is greatest for the lower energy beams. (From LEAVITT et al. 1992)

over that segment. The application of electron arc aperture shaping to dose optimization will be discussed later. Figure 7.7 illustrates the change is dose as the width of the electron arc aperture is varied from 3 cm wide to 8 cm wide (projected light field at isocenter). These measurements, normalized to unity for the 5-cm-wide light field at isocenter, show a reduced effect of aperture width as the electron energy is increased.

Bremsstrahlung contamination within the electron arc beam is a source of concern in all potential electron arc treatments, since the patient volume immediately surrounding the isocenter is irradiated during the entire arc by the contaminant photons (EL-KHATIB et al. 1991; KASE and BJARNGARD et al. 1979; LEAVITT et al. 1986; PLA et al. 1989). Figure 7.8 shows the bremsstrahlung dose profile in the long and narrow axes of the field defined by the primary

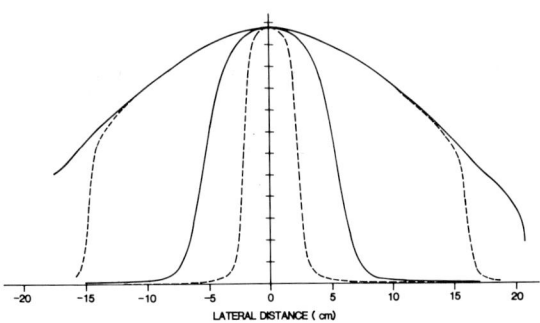

Fig. 7.5. 20-MeV electron beam profiles defined by the primary photon collimator set to 10 cm wide by 40 cm long (*solid lines*.) Electron profiles as modified by insertion of the 5-cm-wide by 30-cm-long secondary electron arc collimator are shown as *dashed lines*

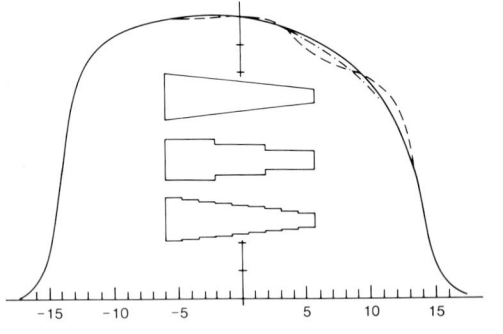

Fig. 7.6. For 6-MeV electrons, the effect of secondary electron arc collimator shape on profile intensity in the long axis of the field (inferior to superior) is demonstrated by these three shapes. All three shapes project a field 5 cm wide on the central axis at isocenter. The trapezoidal shape (*top*) is approximated by a three-vane shape in which the segments project widths of 3 cm, 5 cm, and 7 cm at isocenter (*middle*) and by a nine-vane shape (*bottom*). The intensity profile of the trapezoidal shape (*solid line*) is closely approximated by the nine-vane aperture (*dash-dot line*), while the three-vane shape shows a stair-step intensity pattern (*dashed line*). (From LEAVITT et al. 1992)

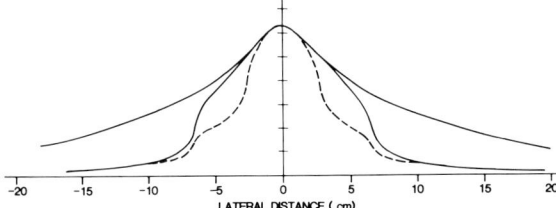

Fig. 7.8. Bremsstrahlung dose profiles measured at isocenter depth of 15 cm. *Solid lines* show profile defined by primary photon collimators. *Dashed line* shows modification of profile by secondary electron arc collimator set to 5 cm wide at isocenter. This modified profile shows a photon attenuation of approximately 50% through the electron arc collimator. This effect can be accounted for in dose calculations by including beam profiles measured beyond the range of the electrons for use in the dose calculation algorithm

7.3 Electron Arc Field Characteristics (Arced Field)

Electron arc isodose distributions differ significantly from the isodose distributions resulting from static application of the electron field as defined by the electron arc aperture. The major differences are: (a) reduced surface dose, (b) increased depth of maximum dose, (c) steeper dose gradient beyond the depth of maximum dose, and (d) increased bremsstrahlung dose due to focusing effect. These differences are clearly noted in Fig. 7.3. Although the depth of maximum dose and the dose gradient in the depths beyond this maximum are increased in the electron arc mode, the range of the electrons remains unchanged. The reduced surface dose, increased depth of maximum dose, and increased dose gradient are due simply to the focusing effect of electron arc, which means that a point closer to isocenter will remain in the electron beam for a longer time than a point further from isocenter (more superficially located). Similarly, the increased bremsstrahlung dose results since the isocenter is continuously irradiated during the entire arc. The dose in the penumbra region of the electron arc treatment volume is similar to that of standard electron fields, since tertiary collimation placed immediately on the patient's contour sharply defines the electron field edge. In the plane of gantry rotation, the electron arc beam is swept beyond the treatment field edge defined by the tertiary collimation and continued until the entire beam has been integrated at the field edge. In the cephalocaudal plane, the electron arc beam edge is extended several centimeters beyond the tertiary collimation in order to achieve a sharper penumbra.

photon collimators set to 10 cm wide by 40 cm long for 20-MeV electrons. This bremsstrahlung profile is modified by insertion of the electron arc aperture, which attenuates the photon beam by approximately 50%, resulting in a modified photon profile defined by the dashed lines. Measurement of depth doses should extend to a depth of 15 cm or greater so that accurate dose information can be calculated for the summed electron arc dose distributions. The relative fraction of bremsstrahlung dose in an electron arc field depends on the width of the electron arc aperture, on the setting of the primary photon collimators, and on the scattering foil configuration of the linear accelerator. Table 7.1 shows the relative photon dose at 15 cm depth versus electron energy and electron arc aperture width for a Varian Clinac 2100C. These measurements were limited to a fixed primary photon collimator setting of 10 cm wide by 40 cm long, since this is the standard photon collimator setting for clinical treatments, while the electron arc aperture width was varied across the clinically implemented range.

The dependence of electron arc dose on depth of isocenter differs from the simple distance effects of standard fixed-field electron dose calculation. A simplified derivation can demonstrate this dependence: referring back to Fig. 7.1, the ratio of doses at a depth

Table 7.1. Bremsstrahlung dose vs field width at 15 cm depth isocenter (static field and 90° electron arc)

MeV	Static field	Dose at isocenter per 100 per cGy at D_{max}					
		3 cm	4 cm	5 cm	6 cm	7 cm	8 cm
6	0.5	2.8	2.4	2.0	1.8	1.5	1.4
9	1.3	7.4	5.6	4.8	4.1	3.6	3.2
12	1.7	9.8	7.8	6.2	5.2	4.5	4.1
16	2.8	15.4	11.8	9.7	8.3	7.4	6.6
20	4.3	18.0	14.1	11.6	9.7	8.4	7.5

d below the surface of the two contours referenced by radii F (r_0 in following equations) and G (r in following equations) will be calculated assuming a uniform intensity profile of width w at isocenter which converges to the effective source location C (SAD in following equations). In this simplified illustration, the intensities at the two points distance *d* below the surface of each contour are related as:

$$I_G/I_F = ((SAD + d - r_0)/(SAD + d - r))^2. \qquad (7.1)$$

The widths of the profiles at these two positions are:

$$W_F = W_E (SAD + d - r_0)/(SAD - SCD) \qquad (7.2)$$

and

$$W_G = W_E (SAD + d - r)/(SAD - SCD). \qquad (7.3)$$

The total dose at each point as the field sweeps across the two contours is:

$$D_F = I_F \times t_F = I_F \times W_F/(\Omega \times (r_0 - d)) \qquad (7.4)$$

and

$$D_G = I_G \times t_G = I_G \times W_G/(\Omega \times (r - d)), \qquad (7.5)$$

where Ω = angular velocity of gantry rotation. The ratio of doses is then:

$$D_G/D_F = (I_G \times W_G/(r - d))/(I_F \times W_F/(r_0 - d)), \qquad (7.6)$$

which after substitution reduces to:

$$D_G/D_F = (SAD + d - r_0)/(SAD + d - r)$$
$$\times (r_0 - d)/(r - d). \qquad (7.7)$$

Alternative derivations addressing different profile shapes have been developed (BOYER et al. 1982; HOGSTROM and LEAVITT 1986; KHAN et al. 1977; LEAVITT et al. 1985). This derivation is valid as long as the arc limits are sufficiently large to allow integration of the entire electron arc beam profile. Electron arc doses were calculated for 180° arcs for each of the five electron energies available on a Clinac 2100C. The relative doses, normalized to the dose, for a 15-cm isocenter depth, are plotted in Fig. 7.9. Ratios calculated from the above formula were compared with the values Fig 7.9 and agree very well at larger isocenter depths, but diverge for smaller isocenter depths. This effect has been discussed by HOGSTROM and LEAVITT (1986), who noted that the approximation is only valid as long as the radius of curvature is greater than the electron arc field width. The slope of these lines differ with energy since the depth of maximum dose at which the normalization is done increases with electron energy.

For a fixed isocenter depth, electron arc dose varies approximately linearly with aperture width.

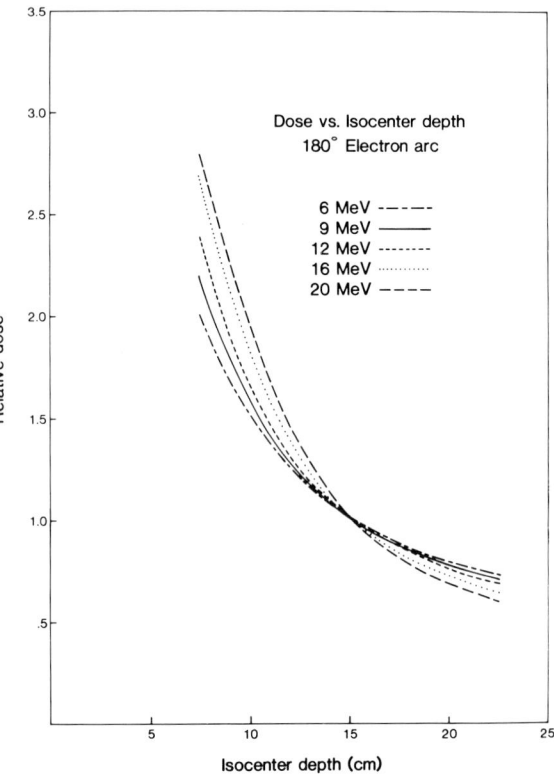

Fig. 7.9. Dependence of electron arc dose on depth of isocenter. Curves are normalized to 1.0 for the 15-cm isocenter depth for all energies. For all curves, the maximum dose resulting from an electron arc through 180°, using the same monitor units per degree (same total monitor units for every arc), was determined. The depth of maximum dose at which the comparison was made varied from 1.5 cm for 6-MeV electrons to 6 cm for 20-MeV electrons. The greatest variation is seen with the highest energy electrons, corresponding to the deepest reference depth. This is in agreement with Eq. 7 in the text. (From LEAVITT et al. 1992)

This relationship has been derived for different profile shapes (LEAVITT et al. 1985; HOGSTROM and LEAVITT 1986) and is valid as long as the depth of isocenter is greater than the projected aperture width. However, the slope of the curve changes with electron energy and with depth of isocenter. This makes it somewhat difficult to simply adjust the width of the aperture to compensate for a change in dose due to change in isocenter depth across the patient contour. Figure 7.10 illustrates the dependence of electron arc dose on projected light field width for a fixed isocenter depth of 15 cm. These curves can be fitted to a two-parameter model of the form:

$$D_{arc}(d, r, w, L)/(D_{arc}(d, r, w_0, L) = a + b \times (w/w_0),$$
$$(7.8)$$

where w = width of arc aperture light field projected

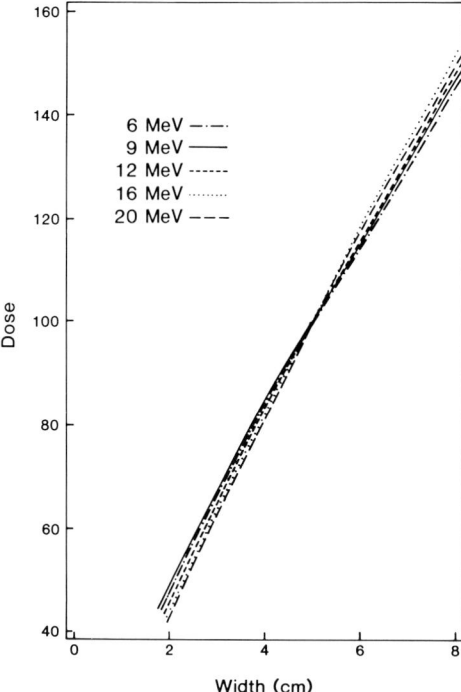

Fig. 7.10. Dependence of electron arc dose on field width. The electron arc maximum dose was calculated for 180° arc at 15 cm isocenter depth. Monitor units per degree were held fixed, while the electron arc collimator width projected to isocenter was varied from 2 cm to 8 cm wide. All curves were normalized to the value for the 5-cm-wide collimator for each energy. (From LEAVITT et al. 1992)

to isocenter, w_0 = width of standard arc aperture light field projected to isocenter (typically 5 cm), r = depth of isocenter, and L = distance above or below the central axis plane.

Using Eqs. 7.7 and 7.8, the width of the arc field aperture projected to isocenter can be calculated. Out of the central axis plane the depth of isocenter may change relative to the central axis plane. Thus, the ratio of doses can be estimated using Eq. 7.7. However, as seen in Fig. 7.4, the relative intensity of the beam in the long axis of the collimator decreases with distance away from the central axis. This effect is included as an off-axis factor in the long dimension of the collimator, OAF (L/L_0), where L/L_0 is the fractional distance from the center of the aperture along the long axis. The ratio of electron arc doses, Eq. 7.7, is multiplied by the off-axis factor corresponding to the proper distance out of the central axis plane and is set equal to the ratio of electron arc doses defined by Eq. 7.8. This yields:

$$(SAD + d - r)/(SAD + d - r_0) \times (r - d)/$$
$$(r_0 - d) \times OAF(L/L_0) = a + b \times (w/w_0), \quad (7.9)$$

which is solved for w to give:

$$w = w_0 \times \{(SAD + d - r)/(SAD + d - r_0)$$
$$\times (r - d)/(r_0 - d) \times OAF(L/L_0) - a\}/b. \quad (7.10)$$

This equation can be used as a manual check of aperture widths determined by more complex computerized treatment planning. However, it is not recommended as a substitute to the exclusion of computerized treatment planning, since visual evaluation of the computer-generated isodose distribution is an integral part of the decision process in treatment planning.

The above formulas for calculating the change in dose with isocenter depth and the corresponding required change in aperture width to deliver the same dose to a contour of different isocenter depth depend upon the complete integration of the dose profile within the electron arc. In many clinical circumstances this requirement is simply not satisfied. Although the entire electron arc may be of sufficient length to integrate the dose profile, the arc is frequently divided into three or more segments having different energy, aperture shape, or monitor units per degree. This complicates the use of simple formulas for estimation of dose or shape of the electron arc aperture. Figure 7.11 illustrates the integration of dose during an arc of 180°. The five graphs show the integration of dose for each electron energy (6, 9, 12, 16, and 20 MeV) for isocenter depths of 10, 15, 20, and 25 cm. In each of these curves, the electron arc aperture was constrained to project a 5-cm-wide field at isocenter, and the dose rate was constant at one monitor unit per degree of arc. thus, points on each curve corresponding to 20° of arc received an electron exposure corresponding to 20 monitor units; similarly, points corresponding to 180° of arc were exposed during an irradiation of 180 monitor units. This demonstrates that the arc length needed to fully integrate the electron arc dose profile depends strongly on the isocenter depth and on the electron energy. As a general clinical rule, an arc of 90° is required to fully integrate the dose profile for isocenter depths used in actual treatments. This is satisfactory except for the highest energy electrons, which are seldom if ever used for an entire arc. In actual use, graphs such as these which tabulate the dose versus arc length for fixed monitor units per degree can be constructed for the full range of isocenter depths encountered in clinical use. The expected dose to a reference or calculation point can then be estimated graphically, allowing rapid verification of computer-calculated treatment plans.

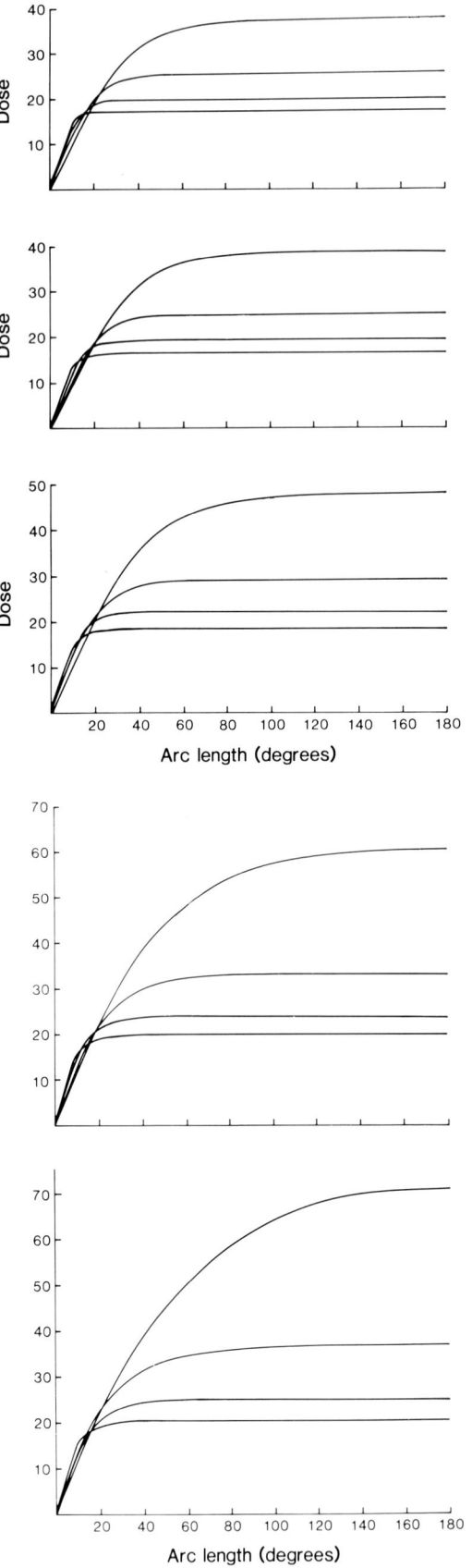

One advantage of electron arc therapy techniques is that on large curved surfaces such as the chest wall the isocenter can be placed such that, in the plane of rotation, the electron beam has perpendicular incidence to the surface across the entire arc. In the cephalocaudal plane, however, this advantage cannot be claimed. Thus, a large change in the patient dimensions superior-to-inferior may introduce additional complications to accurate estimation of dose distributions resulting from electron arc due to oblique incidence of the electron beam. The effects of oblique incidence of the electron beam include increased dose at shallow depths and decreased dose at normal treatment depths (BIGGS 1984; EKSTRAND and DIXON 1982). Ekstrand and Dixon reported increases in dose as high as 11% for 6-MeV and 18% for 9-MeV electrons for 60° oblique incidence, and showed that the depth of maximum dose moves toward the surface with increasing obliquity. These effects can complicate the process by which the shape of the electron arc aperture is determined, in that the simplified formulas described above cannot be used, nor can the two-dimensional pencil beam calculations be relied upon. In cases of extreme change in the chest wall dimensions superior-to-inferior, dose calculations must be accompanied by careful phantom measurements to determine the magnitude of these effects, which may be inadequately addressed by the simplified treatment planning models. Film and TLD measurements in specially constructed phantoms simulating patient chest walls with large cephalocaudal variation showed differences of 10% and more between measured and calculated electron arc doses in regions of oblique incidence (LEAVITT et al. 1989a). Clinical response, exhibited as small regions of moist desquamation near the upper border of the electron arc treatment field, has been observed in patients having large superior-to-inferior variation in chest wall dimensions (MCNEELY et al. 1988, STEWART et al. 1991; PEACOCK et al. 1984). The problem of depth-dose changes in regions of extreme obliquity may limit the

Fig. 7.11. Dependence of electron arc dose on length of arc. Electron arc doses for energies of 6, 9, 12, 16, and 20 MeV were calculated for isocenter depths of 10, 15, 20, and 25 cm using a fixed electron arc collimator projecting a field 5 cm wide at isocenter. Monitor units per degree were held fixed. Maximum dose was determined as the arc length was increased from 1° through 180°. Except for the shallowest isocenter depth for the two highest energy electron arcs, the entire profile is integrated within 90° of arc. *Upper left*: 6 MeV; *upper center left*: 9 MeV; *center left*: 12 MeV; *lower center left*: 16 MeV; *lower right*: 20 MeV

application of electron arc therapy in treatment of the intact breast and postmastectomy implants.

7.4 Electron Arc Treatment Techniques

Three common requirements in electron arc therapy planning are: (a) to adjust the depth of penetration of the electrons into the patient in order to minimize dose to critical underlying organs such as heart and lung; (b) to shape the treatment surface to conform to surgical scars, drainage sites, and excursions of tissue involvement beyond simple rectangular fields; and (c) to deliver a uniform dose across the entire treatment volume. These needs have led to the development of specialized electron arc therapy treatment techniques.

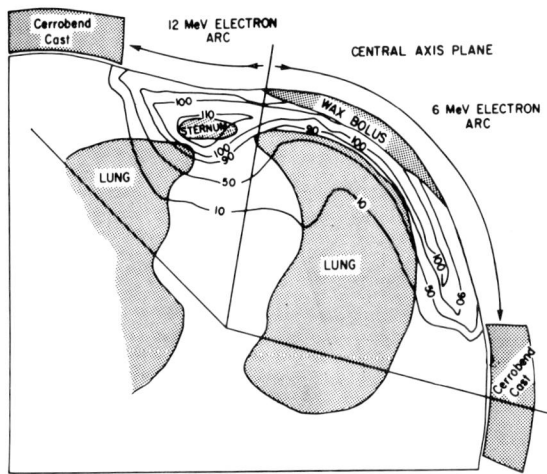

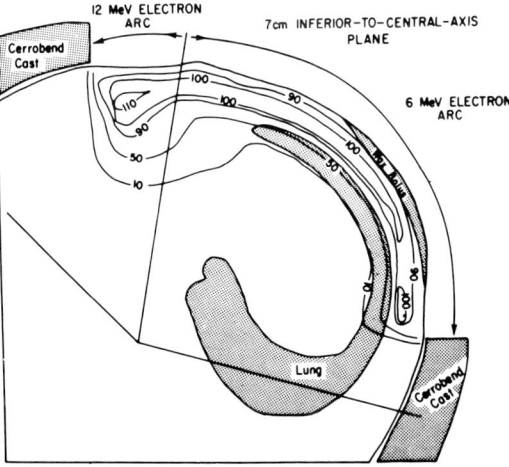

Fig. 7.12. Illustration of use of customized wax bolus as a missing tissue compensator to minimize dose to underlying lung. The shape of the bolus is determined from CT displays in multiple planes across the treatment volume. Placement of the bolus on the patient is demonstrated in Fig. 7.2 (From McNeely et al. 1988)

The desired depth of penetration of the electron field in treatment of the postmastectomy chest wall may change dramatically with location on the patient's surface as the gantry is rotated around the patient. Thus, a lower electron energy such as 6 MeV or 9 MeV may be applied across the fraction of the arc, where underlying lung constrains the treatment depth, while a higher electron energy such as 12–20 MeV may be applied across the mediastinum, where a deeper penetration is desired. Figure 7.12 illustrates the isodose distribution in the central axis plane and 7 cm inferior plane resulting from the combination of 6-MeV and 12-MeV electrons within a single treatment. In this illustration, the three lines radiating from isocenter correspond to the start/stop positions of the central axis of the electron beam. Thus, there is an overlap of high-and low-energy electrons at the matchline between the 12-MeV electron arc segment and the 6-MeV electron arc segment. At this matchline, the dose contribution from each arc segment is 50% of the maximum dose within that segment. Depending on electron energy, the dose falloff extends another 13°–30° beyond the stop angle before the 5% dose level is reached. This corresponds to a linear distance of 3–4 cm at the depth of maximum extension of the isodose line. This diffuse edge to the electron arc segment dose distribution is ideal for matching arc segments. However, it may be necessary to minimize the dose to lung from the high-energy mediastinal field. This can be accomplished by adding tissue-equivalent bolus of uniform thickness coincident on the patient's surface with the stop-angle of the high-energy electron arc segment. This bolus should be thick enough to stop the high-energy electrons from penetrating into the lung. This bolus would be placed on the patient only during the high-energy electron arc, then removed for the low-energy arcs. The dose received by the patient then represents the composite using different treatment modifiers during the same treatment. This technique has been discussed by Hogstrom and Leavitt (1986).

A problem commonly encountered in electron arc therapy is the need to adjust the depth of penetration across only part of an arc segment. This may be due to a tissue deficit caused by surgical procedures which excise tissue nearly to the rib cage across part of the chest wall while leaving overlying muscle intact across the remainder. In this situation, customized bolus can be constructed to compensate for the missing tissue. The shape and thickness of the bolus are determined by evaluating the chest wall thickness as determined by multiple CT slices equally spaced

1 cm apart across the region of potential bolus. The customized bolus is then formed from paraffin using hot blade techniques (LEAVITT et al. 1990). Such an application of bolus is illustrated in Fig. 7.12, showing the change in bolus thickness across the arc, as well as the extent of the bolus inferior-to-superior. A typical bolus, in treatment position, is demonstrated in Fig. 7.2. A second application of bolus is to increase the surface dose to the patient. Figure 7.3 shows that the surface dose in the absence of bolus is approximately 70% of maximum dose. Superposition of 1.5-cm bolus over the entire treatment surface raises the surface dose to 100% for 6-MeV electrons and to greater than 90% for 9-MeV electrons. However, the depth of maximum dose and the range of electrons are also decreased by 1.5 cm. An example of the use of bolus to raise the surface dose would be to substitute 9-MeV electrons plus 1.5 cm bolus for 6-MeV electrons in an arc segment. This raises the surface dose to approximately 90% while nearly matching the depth of penetration of the 6-MeV electron arc. Alternatively, dose uniformity in the region from the surface to the depth of maximum dose can be achieved by combining several electron energies within the same arc segment (LEAVITT et al. 1990). Figure 7.13 shows the radial depth doses for 90° electron arcs for the five standard electron energies plus a "modified depth dose" consisting of

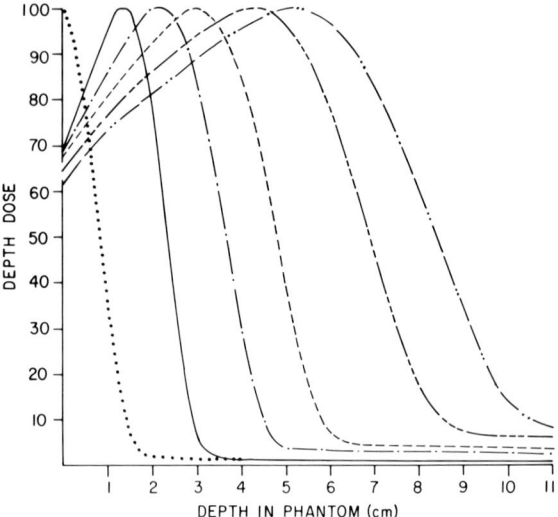

Fig. 7.13. Radial depth doses for electron arc energies. Shown in increasing depth of penetration: 6 MeV on patient contour modified by inclusion of 1.5-cm bolus; 6 MeV; 9 MeV; 12 MeV; 16 MeV; 20 MeV. Doses determined for 90° of arc. (From LEAVITT 1990b)

6-MeV electron arc applied to the contour supplemented by 1.5-cm bolus, thereby positioning the maximum dose immediately on the patient's surface. Using a least-squares optimization routine to determine the relative weightings of each energy within the arc, the radial depth dose can be made

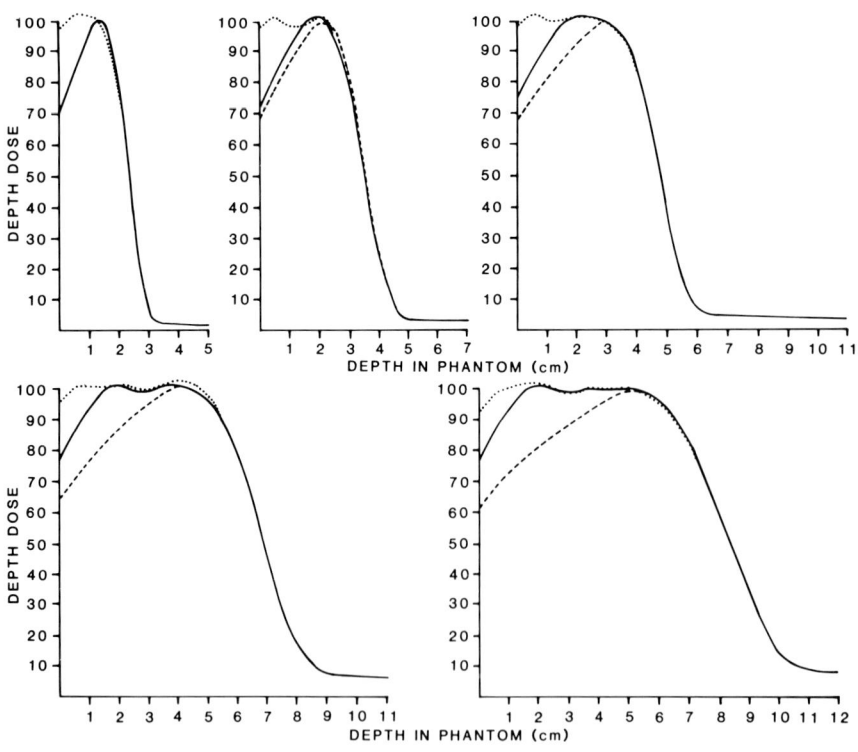

Fig. 7.14. Radial depth doses illustrating the effect of combining bolus and one or more electron energies by multiple arcs across a single arc segment. *Dashed lines*: depth dose for single electron energy without additional passes for bolus or lower electron energies; *solid lines*: depth dose for multiple passes combining several electron energies without bolus; *dotted lines*: depth dose for multiple passes combining several electron energies including 6 MeV electron arc modified by superposition of 1.5 cm bolus over patient contour. a = 6 MeV; b = 9 MeV; c = 12 MeV; d = 16 MeV; e = 20 MeV. (From LEAVITT et al. 1990b)

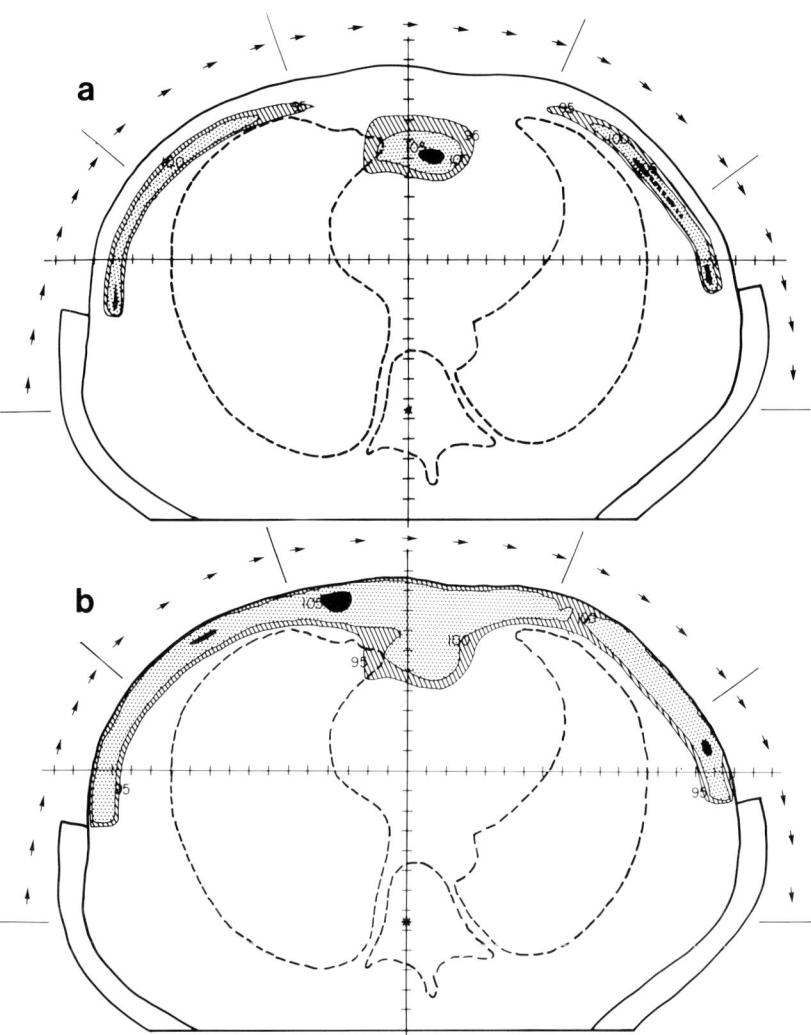

Fig. 7.15. a Isodose distribution resulting when only one electron energy is applied to each arc segment with the constraint to minimize the amount of treatment volume receiving more than 105% of prescription dose. *Arrows* outside contour represent direction of arc. *Dividing lines between arrows* represent border between arc segments, at which electron energy or monitor units per degree are changed. Cross-section of tertiary field defining cast is shown at left and right of patient contour. Center segment irradiated with 16-MeV electron arc; outer segments irradiated using 6-MeV electron arc. **b** Isodose distribution resulting from addition of multiple electron energies, including "bolused" 6 MeV, within each arc segment. The improvement in dose uniformity across the target volume is dramatic. (From LEAVITT 1990b)

uniform to within a few percent from the surface to the depth of maximum dose of the unmodified electron arc distribution. Thus, the 6-MeV electron arc can be modified by inclusion of a second 6-MeV arc in which the patient contour is modified by super-position of 1.5-cm bolus ("bolused 6 MeV");

similarly, the 20-MeV electron arc distribution can be modified by inclusion of fractional doses contributed by the five lower energy electron arcs. The summed relative weights for this technique range from 1.25 to 1.43, compared to 1.0 for the single-energy technique, reflecting the additional dose being deposited in the buildup region. The change in the radial depth doses is demonstrated in Fig. 7.14. Note that radial depth-dose uniformity is achieved without reducing the range of the maximum-energy arc used in each combination. This dose uniformity cannot be achieved through use of bolus alone. The efficacy of this technique can be seen in the clinical example of Fig. 7.15. The post-bilateral mastectomy chest wall is to be treated using an electron arc extending through 180°. The arc is divided into five segments, with the medial segment containing the lymph nodes in the internal mammary chain to be treated using 16-MeV electrons, while the remaining segments are to be treated using 6-MeV electrons. The treatment plan is constrained to minimize the volume receiving greater than 105% of prescription dose, while attempting to optimize the dose uniformity throughout the treatment volume. Figure 7.15a shows a cross-section of the volume receiving 95%–105% of the prescription dose, as achieved under the constraint that only one electron energy be applied to each arc segment. This distribution was judged to be unsatisfactory. Figure 7.15b shows the improved dose uniformity across the treatment volume when optimized weights of all six available electron arc treatments are combined. In examples such as this, the increased time necessary to deliver multiple electron energies within a single arc is clearly justified by the dramatically improved dose uniformity.

A clear advantage of electron arc therapy is the opportunity to customize the shape of the treatment surface to conform to the clinically determined superficial target volume, and to avoid field abutments through scars or gross disease. Unlike tangent photon irradiation techniques which require a straight match line, the upper margin of the electron arc treatment field can be extended to include the chest wall over the upper portion of the lung, thereby minimizing dose to the lung. Simultaneously, the matching supraclavicular photon field can be shaped to avoid the lung and can be angled obliquely to avoid the spinal cord and esophagus. The production of customized field shaping devices for electron arc therapy requires some forethought, as the weight of casting materials is too great to be supported by the patient. one proven technique (LEAVITT et al. 1990) creates a plaster of paris shell which conforms to the patient's chest and which is self-supporting by legs to the left and right of the patient. The desired treatment field is marked on the shell, a defining edge is masked around the treatment field, and cerrobend metal is poured onto the shell in the area surrounding the treatment field. The shell overlying the treatment field is then removed from the cast, leaving the treatment surface open to the incident electron arc beam, while the surrounding normal tissue is shielded by the cerrobend. This casting system, combined with reproducible patient postioning ensured by conforming an Alpha Cradle foam mattress to the patient's supine treatment contours, enables the consistent day-to-day treatment. Other field-shaping techniques that have been described include the use of rubberized lead cutouts or lead sheet strips (THOMADSEN 1981).

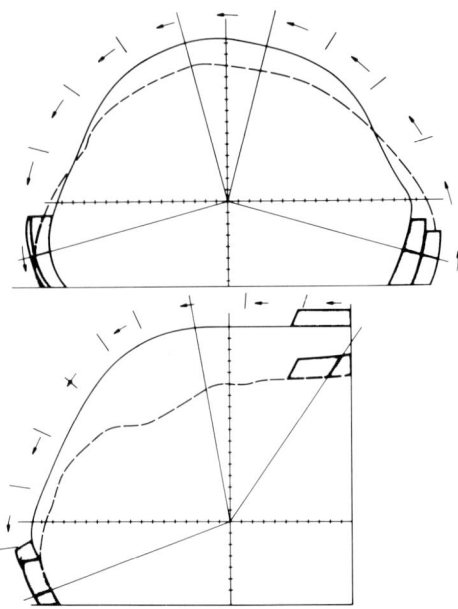

Fig. 7.16. Illustration of change in patient contours with treatment plane. *Upper section*: First clinical case demonstrating the contours at the central axis plane (*solid line*) and at the 8-cm superior plane (*dashed line*). Treatment included bilateral chest wall and nodes. *Lower section*: Second clinical case for treatment of a right-sided chest wall and nodes, showing contours at the central plane and at the 8-cm superior plane. The radial *solid lines* extending outward from isocenter indicate the limits of the three arc segments used in the clinical treatment plans. The disconnected radial lines indicate the limits of the ten arc segments used in the optimized plans. (From LEAVITT et al. 1989a)

The basic problem of delivering a uniform dose across the entire treatment volume is addressed through customized shaping of the secondary electron arc aperture. As illustrated in Figs. 7.9 and 7.10, changes in dose due to variation in the depth of isocenter across the patient contour can be compensated by adjusting the width of the electron arc aperture across segments of the arc. Figure 7.16 illustrates two clinical examples which compare the patient contour in the central axis plane with the contour 8 cm superior. Both cases illustrate that the patient contours do not change symmetrically from plane to plane, thereby making necessary individual adjustment of the secondary electron arc aperture with gantry angle as well as with distance superior or inferior to the central axis. Figure 7.17 illustrates the change in depth of isocenter with gantry angle for the superior plane and the central axis plane, and charts the change in calculated field width in the superior plane for each patient. Figure 7.18 shows the change in dose with gantry angle for fixed aperture width

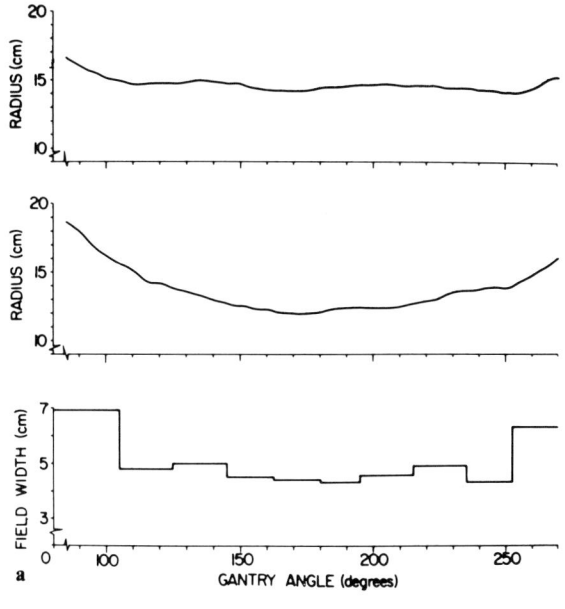

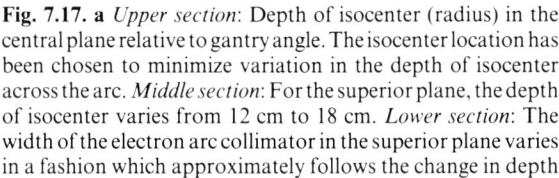

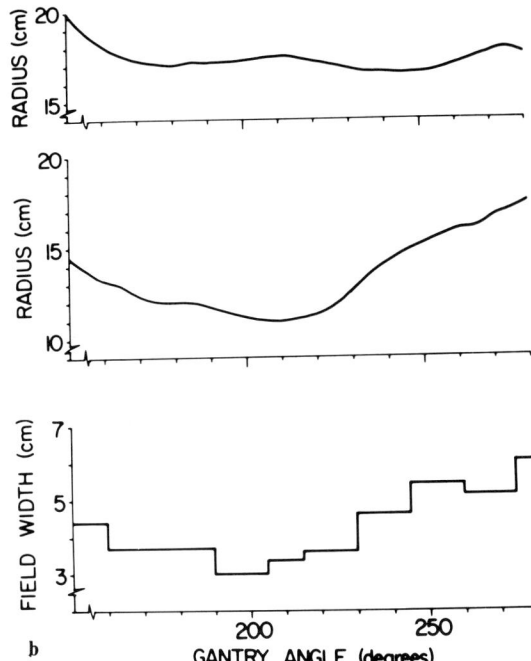

Fig. 7.17. a *Upper section*: Depth of isocenter (radius) in the central plane relative to gantry angle. The isocenter location has been chosen to minimize variation in the depth of isocenter across the arc. *Middle section*: For the superior plane, the depth of isocenter varies from 12 cm to 18 cm. *Lower section*: The width of the electron arc collimator in the superior plane varies in a fashion which approximately follows the change in depth

of isocenter with gantry angle. **b** *Upper section*: Depth of isocenter (radius) in the central plane relative to gantry angle for the right-sided electron arc. *Middle section*: For the superior plane, the depth of isocenter (radius) varies from 11 cm to 17 cm. *Lower section*: The change in patient radius is matched approximately by change in the superior collimator width. (From LEAVITT 1989a)

across the entire arc (top), distinct aperture widths for three arc segments (center), and individualized aperture widths for each of ten arc segments (bottom). Clearly, dynamic adjustment of electron arc aperture leads to improved dose distributions. In order to achieve the desired aperture shapes across the entire field without the interruption of multiple entries into the treatment room to physically remove one aperture and replace it with a modified shape aperture, a multivane computer-controlled electron arc collimator has been constructed (LEAVITT et al. 1989b). This is illustrated in Fig. 7.19.

7.5 Treatment Planning and Dose Calculations

Dose calculation techniques treat the electron arc dose distribution as the summation of a series of static beams equally spaced across the arc. The accuracy of the dose calculations depends upon the accuracy of modeling of the individual static beams. The critical parameters which must be accurately modeled were described earlier. Several techniques have been developed to model the static electron dose

distribution. The simplest of these are extensions of photon fan-line models which interpolate doses from measured tabular data (MILAN and BENTLEY 1974). For uniform density phantoms, these simple algorithms calculate dose distributions in good agreement with measurement, as long as the limits of the calculations remain within the domain of measured data. Thus, beam profiles may be measured and tabulated for electron fields from 2 cm to 8 cm wide, at three or more isocenter depths from 5 to 25 cm, at up to six depths below the phantom surface, for each electron energy. These data, when tabulated, then serve as an extensive base for interpolation. Alternatively, newer electron pencil beam dose calculation algorithms have been applied to electron arc therapy treatment planning (HOGSTROM et al. 1989; KOOY and RASHID 1989). With these algorithms, the change in electron dose profile with distance from the electron arc aperture can be explicitly calculated with inclusion of effects such as electron scatter in air. The amount of data that must be tabulated is greatly reduced through use of pencil beam dose calculation algorithms. However, the requirements for agreement with measured data

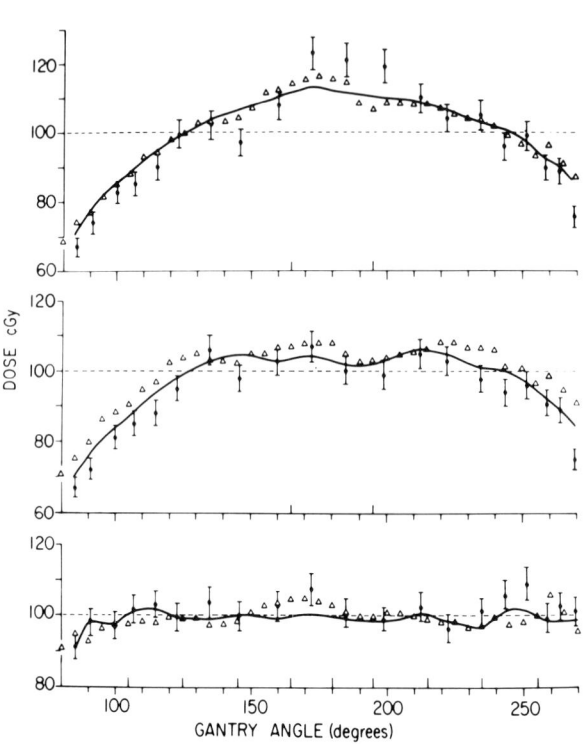

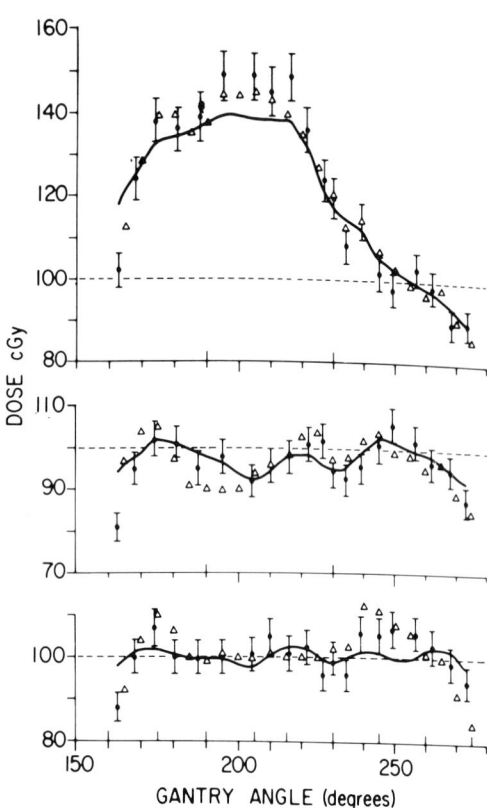

Fig. 7.18. Comparison of measured and calculated doses for two electron arc cases in the plane 8 cm superior to the central axis. *Top frame* shows doses when electron arc collimator is held fixed at 5 cm wide (projected to isocenter) for entire arc; *center frame* shows doses when collimator width is adjusted for each of three arc segments; *lower frame* shows doses when collimator width is adjusted for each of ten arc segments.

The *vertical tic marks* extending above the axis indicate the limits of the individual arc segments used in each study. Calculated doses are shown by *solid line*; film densitometry measurements are shown by *triangles*; TLD measurements are shown by *solid dots* with error bars. Idealized dose is shown by *horizontal dashed line*. (From LEAVITT et al. 1989a)

Fig. 7.19. Multivane electron arc collimator assembly with bottom plate removed. Eighteen independent vanes are shown from fully extended (*top*) to fully retracted (*bottom*). The six three-axis controllers (numbered 10–15) are shown with ribbon connectors to three motor assemblies each. This device can be operated from battery power or from direct power plug-in. (From LEAVITT et al. 1989b)

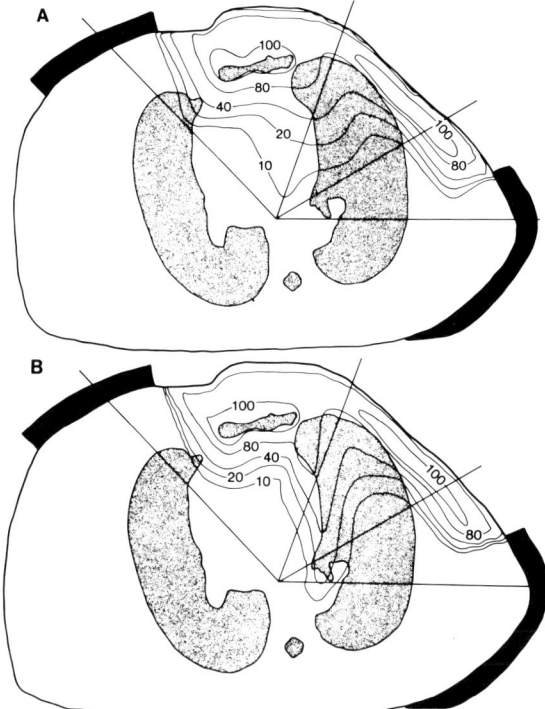

Fig. 7.20. Comparison of electron arc dose distribution calculated using pencil beam (**A**) and tabular data (**B**). The major difference between the two calculation techniques is seen in the greater predicted penetration of the high-energy electrons into the lung by the simple heterogeneity correction applied by the tabular data calculation

remain important (EL-KHATIB et al. 1992; LAM et al.1987; PLA et al. 1988; SHIU et al. 1989).

Although good agreement can be achieved between calculations using tabular data and pencil beam calculations for uniform density phantoms, significant deferences arise in the presence of heterogeneities. The tabular data approach allows only one-dimensional corrections for the presence of heterogeneities, while the pencil beam approach imcorporates a two-dimensional correction which includes the influence of lateral scatter into and from interal heterogeneities. The significance of this effect can be seen in Fig. 7.20. Although the relative dose distributions are similar in the regions superficial to the lungs, the dose distribution within and beyond the lungs is visibly different. The lung dose distribution due to the 12-MeV electron component over the mediastinum is greatly overpredicted using the tabular one-dimensional heterogeneity correction, compared to the pencil beam calculated distribution. Thus, any evaluation of dose within heterogeneities requires use of pencil beam type algorithms which account for lateral scatter into and from these structures.

The problem of oblique incidence of the electron arc beam superior to the central axis plane is not adequately addressed by either one-or two-dimensional calculation techniques. As employed currently the tabular and pencil beam calculations are performed plane-by-plane, without regard for oblique incidence to the external superior-to-inferior patient contour or for change in the size and shape of internal structures from plane to plane. The implementation of full three-dimensional dose calculations for electron arc therapy remains to be completed. The imminent arrival of a new generation of fast computer workstations may hasten the implementation of full 3D electron arc treatment planning capabilities.

7.6 Summary

The technique of electron arc therapy has enabled the treatment of the superficial volume immediately below large curved surfaces with a dose uniformity superior to that available through other treatment techniques. Electron arc therapy is a complex treatment technique which requires development of new treatment accessories such as secondary electron arc collimators and tertiary field shaping casts, as well as measurement of electron dose profiles and dose characteristics for nonstandard electron collimators. Dose calculations and measurements show the dependence of electron arc on energy, arc length, field width, and beam modifiers to be different from both fixed electron beam and photon arc characteristics. The effort required to implement electron arc therapy is equivalent to that required to commission the electron energies of a clinical linear accelerator. Once the effort has been committed, however, electron arc therapy can become a "routine treatment procedure" understood and applied by physician, physicist, and radiation therapist.

References

Becker J, Weitzel G (1956) New forms of moving radiation from a 15 MeV Siemens betatron. Strahlentherapie 101: 180–190
Biggs P (1984) The change in percentage depth dose of electrons due to beam angulation. Med Dosim 9: 25–28
Blackburn BE (1981) A practical system for electron arc therapy. In: Paliwal B (ed) Proceedings of the symposium on electron dosimetry and arc therapy. American Institute of physics, New York, pp 295–314
Boyer AL, Fullerton GD, Mira JG (1982) An electron beam pseudo arc technique for irradiation of large areas of chest wall and other curved surfaces. Int J Radiat Oncol Biol Phys 8: 1969–1974

Brahme A, Lind B, Kallman P (1988) Clinical possibilities with a new generation of radiation therapy equipment. Dosimetry in Radiotherapy, IAEA-SM-298/67, IAEA, Vienna

Ekstrand KE, Dixon RL (1982) The problem of obliquely incident beams in electron-beam treatment planning. Med Phys 9: 276–278

El-Khatib E, Scrimger J, Murray B (1991) Reduction of the bremsstrahlung component of clinical electron beams: implications for electron arc therapy and total skin irradiation. Phys Med Biol 36: 111–118

El-Khatib E, Antolak J, Scrimger J (1992) Radiation dose distributions for electron arc therapy using electrons of 6–20 MeV. Phys Med Biol 37: 1375–1384

Gerber RL, Purdy JA (1984) Dosimetry and treatment planning considerations in electron arc therapy. In: Proceedings Varian's Fourth European Clinac Users Meeting, Varian AG, Zug, Switzerland, pp. 92–97

Hogstrom KR, Leavitt DD (1986) Dosimetry of arc electron therapy. In: Kereiakes JG, Elson HR, Born CG (eds) Proceedings of the AAPM Summer School. Radiation Oncology Physics, Medical Physics Monograph #15, pp 265–295

Hogstrom KR, Kurup RG, Shiu AS, Starkshall G (1989) A two-dimensional pencil-beam algorithm for calculation of arc electron dose distributions. Phys Med Biol 34: 315–341

Hudepohl G, Rassow J (1973) Contribution to the deep electron therapy by pendulum irradiation. 7th communication: Dosimetry and example for the application of telecentric pendulum irradiation with electrons of 5–10 MeV under changing focus-skin-distance. Strahlentherapie 146: 546–558

Kase KR, Bjarngard BE (1979) Bremsstrahlung dose to patients in rotational electron therapy. Radiology 133: 531–532

Khan FM, Fullerton GD, Lee JMF, Moore VC, Levitt SH (1977) Physical aspects of electron beam arc therapy. Radiology 124: 497–500

Kooy HM, Rashid H (1989) A 3-dimensional electron pencil beam algorithm. Phys Med Biol 34: 229–243

Lam KS, Lam WC, O'Neill MJ, Lee DJ, Zinreich E (1987) Electron arc therapy: Beam data requirements and treatment planning. Clin Radiol 38: 379–383

Leavitt DD (1978) A technique for optimization of dose distributions in electron rotational therapy. Med Phys (Abstract 5: 347

Leavitt DD (1987) Optimization of electron arc therapy doses by dynamic collimator control. In: Bruinvis IAD, van der Giessen PH, van Kle Vens HJ, Wittkamper FW (eds) Proceedings of the 9th International Conference on the Use of Computers in Radiation Therapy. Elsevier, Amsterdam, pp 149–152

Leavitt DD (1988) Multileaf collimation in electron arc therapy. In: Proceedings of the Twelfth Varian Users' Meeting. Varian Associates, Palo Alto, CA, USA pp 63–67

Leavitt DD, Peacock LM, Gibbs FA Jr, Stewart JR (1985) Electron arc therapy: physical measurement and treatment planning techniques. Int J Radiat Oncol Biol Phys 11: 987–999

Leavitt DD, Gibbs FA, Moeller JH (1986) Electron arc therapy: influence of heterogeneities on dose to blood-forming organs. Radiology 161(P): 248

Leavitt DD, Stewart JR, Moeller JH, Earley L (1989a) Optimization of electron arc therapy doses by multi-vane collimator control. Int J Radiat Oncol Biol Phys 16: 489–496

Leavitt DD, Stewart JR, Moeller JH, Lee WL, Takach GA (1989b) electron arc therapy: design, implementation and evaluation of a dynamic multi-vane collimator system. Int J Radiat Oncol Biol Phys 17: 1089–1094

Leavitt DD, Earley L, Stewart JR (1990a) Design and production of customized field shaping devices for electron arc therapy. Med Dosim 15: 25–31

Leavitt DD, Stewart JR, Earley L (1990b) Improved dose homogeneity in electron arc therapy achieved by a multiple-energy techinque. Int J Radiat Oncol Biol Phys 19: 159–165

Leavitt DD, Stewart JR, Moeller JH, Earley L (1992) Electron beam arc therapy. In: Purdy J (ed) Advances in radiation oncology Physics. Medical Physics Monograph # 19, pp 430–465

McNeely L, Jacobson G, Leavitt DD, Stewart JR 1988 Electron arc therapy: chest wall irradiation of breast cancer patients. Int J Radiat Oncol Biol Phys 14: 1287–1294

McNeely L, Leavitt DD, Stewart JR, Eggar M (1991) Dose volume histogram analysis of lung radiation from chest wall treatment: comparison of electron arc and tangential photon beam techniques. Int J Radiat Oncol Biol Phys 21: 515–520

Milan J, Bentley RE (1974) The storage and manipulation of radiation dose data in a small digital computer. Br J Radiol 47: 115–121

Peacock LM, Leavitt DD, Gibbs FA Jr, Stewart JR (1984) Electron arc therapy: clinical experience with chest wall irradiation. Int J Radiat Oncol Biol Phys 10: 2149–2153

Pla M, Pla C, Podgorsak EB (1988) The influence of beam parameters on percentage depth dose in electron arc therapy. Med Phys 15: 49–55

Pla M, Podgorsak EB, Pla C (1989) Electron dose rate and photon contamination in electron arc therapy. Med Phys 16: 692–697

Rassow J (1970a) Contribution to the deep electron therapy by pendulum irradiation. 3rd communication: basics of a new dosage integration technique for calculation of dose distribution during pendulum therapy. Strahlentherapie 139: 116–138

Rassow J (1970b) Contribution to the deep electron therapy by pendulum irradiation. 4th communication: about a new, specific telecentric small angle pendulum technique for primary unscattered electrons. Strahlentherapie 140: 156–172

Rassow J (1972) On the telecentric small-angle pendulum therapy with high electron energies. Electromedica 1: 1–6

Ruegsegger DR, Lerude SD, Lyle D (1979) Electron beam arc therapy using a high energy betatron. Radiology 133: 483–489

Shiu AS, Otte VA, Hogstrom KR (1989) Measurement of dose distribution using film in therapeutic electron beams. Med Phys 16: 911–915

Stewart JR, Leavitt DD, Prows J (1991) Electron arc therapy of the chest wall for breast cancer rationale, dosimetry and clinical aspects. Front Radiat Ther Oncol 25: 134–150

Thomadsen B (1981) Tertiary collimation of moving electron beams. In: Paliwal B (ed) Proceedings of the Symposium on Electron Dosimetry and Arc Therapy. American Institute of Physics, New York, pp 315–326

8 Treatment Verification Using Digital Imaging

Shlomo Shalev

CONTENTS

8.1 Treatment Verification

Treatment verification plays a vital role in the management of patients receiving radiation treatment. As part of the overall quality assurance program, it is designed to detect a variety of mistakes, errors, and inaccuracies that can occur during a protracted course of fractionated therapy. Mistakes are due to human misjudgments or performance failures, and may have serious implications if not detected and corrected promptly. Treatment errors arise when calculations or procedures are less accurate than expected and can occur in all the stages of the planning and treatment process. They can be divided into two main categories: those associated with the prescription, computation, and specification of target

---

Shlomo Shalev, PhD, Department of Medical Physics, Manitoba Cancer Treatment and Research Foundation, 100 Olivia Street, Winnipeg, Manitoba R3E OV9, Canada

doses and dose distributions, and those that occur during the delivery phase and involve the treatment machine or the patient. Portal verification, in which a transmission image of the patient is acquired just prior to or during teletherapy, is used to verify that such errors are within acceptable limits. While portal films have been the conventional methodology for many years, on-line portal imaging techniques are now coming into use and offer a variety of new capabilities for the detection, quantification, and correction of treatment errors.

8.1.1 Dose Specification Errors

The accuracy required in specifying and delivering the prescribed target dose has been estimated from the change in tumor control probability with dose. As early as 1976 a limit of $\pm 5\%$ was proposed (ICRU 1976), and subsequently Svensson (1984) showed that various contributing factors in the calculation and delivery of dose to a homogeneous phantom combine to give an uncertainty of this magnitude at the 95% confidence level. However, for tumors with a high dose response gradient, a limit of $\pm 3\%$ may be more appropriate (Brahme 1984, 1988), in which case even higher accuracy is required of the dose calculation algorithm, including corrections for heterogeneity, blocks, wedges, and missing tissue compensators. While in vivo dosimetry using diodes (Heukelom et al. 1992) has a role to play in verifying delivered dose, the emerging methodology of portal exit dosimetry may provide a means to verify both absolute dose and relative dose distributions under routine clinical conditions (Wong et al. 1990b, Ying et al. 1990, van Dam et al. 1992).

8.1.2 Machine Parameter Errors

Treatment errors can arise from a variety of causes related to the therapy unit and its accessories. Some of these will be due to inherent mechanical tolerances of the system, which affect gantry rotation, couch height, light field/radiation field alignment, back-pointer placement, and so on. A good quality control program will ensure that such uncertainties remain within the manufacturer's specifications, and tolerance limits have been proposed (BRAHME 1988). However, larger machine parameter errors may be caused by incorrectly installed field-shaping accessories, mechanical or electrical malfunctions, or operator error in the interpretation or implementation of treatment parameters. For example, one study showed that accidental errors of more than 30 s in timer settings for a cobalt unit occurred with a frequency of 3%, although this was reduced to 1.4% for a linac fitted with a digital timer (KARTHA et al. 1977). Many errors can be found by careful double-checking of the field parameters before each treatment, albeit at a considerable cost in time and personnel resources. Many therapy machines can now be fitted with computerized record and verify (RV) systems, which have been shown to dramatically reduce major errors, particularly those involving the presence or orientation of wedges and incorrect dose settings (MULLER-RUNKEL and WATKINS 1991). However, RV systems cannot check the presence or location of unmounted shielding blocks, or in most cases the position of the treatment couch. Furthermore, it has been shown that their use as an uncontrolled setup system is prone to human mistakes in data transfer, leading to significant errors in 26% of the treatments (LEUNENS et al. 1992).

8.1.3 Field Positioning Errors

This category relates to errors in the positioning of a treatment field relative to the patient, or more specifically relative to the prescribed target volume. They are also referred to as localization, geometric or field placement errors. It is convenient to group field positioning errors (FPEs) into two categories: systematic and random errors. Systematic FPEs occur consistently in successive fractionated treatments, and should be correctable once they have been identified. They can arise from incorrect data transfer between the computed tomography (CT) scanner, simulator, planning computer, and treatment machine, so that the prescribed treatment parameters are incorrect (such as field size, collimator or gantry angle, and block locations). Alternatively, they may occur due to improper design, marking or positioning of treatment accessories such as compensator plates, shielding blocks and immobilization

devices. *Random* FPEs are due to difficulties in repeatedly setting up the patient for each treatment fraction, and are highly dependent on the skill and experience of the therapist.

Field positioning errors are defined as a discrepancy between the observed portal image and a reference image on which the prescribed field has been marked. The reference image may be a simulator film, or a digitally reconstructed radiograph (DRR), or even the first portal image. Each of these approaches is subject to specific errors. For example, the DRR is computed from a set of CT images, usually obtained some time before the start of therapy, and one cannot assume that the patient's anatomy will remain unaltered over a treatment course lasting several weeks. The simulation procedure is necessary for placing marks on the patient's skin or cast, which are then used for setup on the treatment couch, usually with the aid of lateral and sagittal laser beams. The simulator isocenter, gantry angle, and field wires are subject to mechanical tolerances; patient skin marks can migrate relative to internal anatomical landmarks; casts are subject to deformation and permit a certain amount of independent patient mobility; patients may get thinner, tissue swelling may decrease, and tumors may shrink during a course of treatment; and organ movement can occur due to breathing, bladder or rectal filling, and so on. All of these effects can lead to FPEs.

8.1.4 Patient Immobilization

The use of skin marks has been shown to be unsatisfactory, at least for pelvic and mantle treatments (GRIFFITHS et al. 1987, 1991), and patient immobilization is essential if accurate treatment is to be assured. Using thermoplastic casts for head and neck treatments, one study found the average field displacement to be 5 mm, comprising the summation in quadrature of a systematic error of 3 mm in translating from the simulator film to the portal film, and an average random day-to-day variation in field alignment of 3 mm (HUIZENGA et al. 1988).

Movement of patients immobilized in a plastic mask was evaluated in three dimensions by THORNTON et al. (1991), and was found to average 3.8 mm (standard deviation 1.3 mm) over a period of eight weeks. These authors suggested that since marks on the cast are used for alignment, patient movement within the cast may account for part of these errors. MARKS and HAUS (1976) showed that the use of a bite block reduced the incidence of errors greater than 10 mm from 16% to 1%. Foam-cast techniques have been used by SHEROUSE et al. (1990) and by SOFFEN et al. (1991) who found a reduction in average range of movement from 8 mm to 3.3 mm when casts were used for pelvic treatments. For breast irradiations, errors greater than 10 mm were found in 3.4% of the treatments using foam-cast techniques, but in only 0.9% with a vacuum bag (JAKOBSON et al. 1987). Many centers use lasers for patient set-up, which have been shown to have an accuracy of 2.2 ± 1.4 mm (VERHEY et al. 1982), so that 5% of treatments set up in this way may have errors greater than 5 mm. Repeated laser setups for a tumor in the lower abdomen demonstrate an even broader spatial distribution, with a standard deviation of 3.9 mm (SVENSSON 1984).

8.2 Portal Radiography

Portal radiography is the conventional method of verifying the accuracy of megavoltage radiation treatments. A film is placed on the beam exit side of the patient and records a two-dimensional projection of those parts of the patient's anatomy which are situated within the field boundaries. Three types of portal radiographs can be used: A *localization radiograph* is acquired with a brief exposure of a few monitor units and is then processed and examined before the remainder of the treatment is delivered. While this permits the identification and correction of gross FPEs, the procedure is time consuming and unsuitable for routine implementation of every treatment fraction. A *verification radiograph* uses a less sensitive film and is exposed over the entire treatment, causing minimal interference with the clinical work-load. However, errors can only be corrected on the following treatment, and image quality may be reduced if there is significant patient movement. The third type is a *double exposure-radiograph,* in which a short exposure of a large unblocked field is followed by a second exposure to the actual treatment field. This has the advantage of showing anatomical features outside the field boundaries, which is often very useful in identifying patient positioning, but frequent use is limited by the radiation tolerance of the exposed normal tissues.

8.2.1 Contrast in Portal Images

In comparison to diagnostic films, the quality of portal films is very poor with regard to both contrast and spatial resolution. Figure 8.1a shows a diagnostic

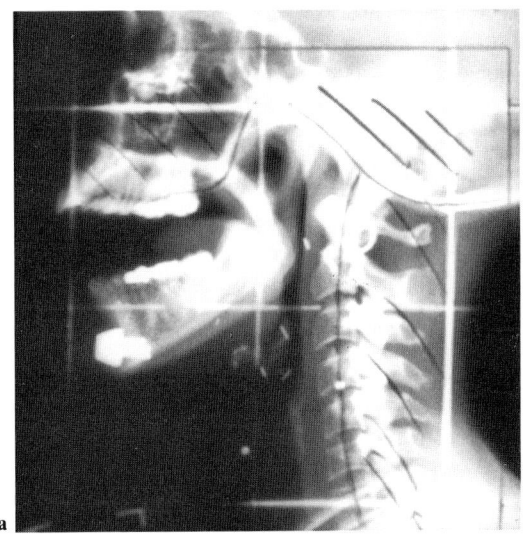

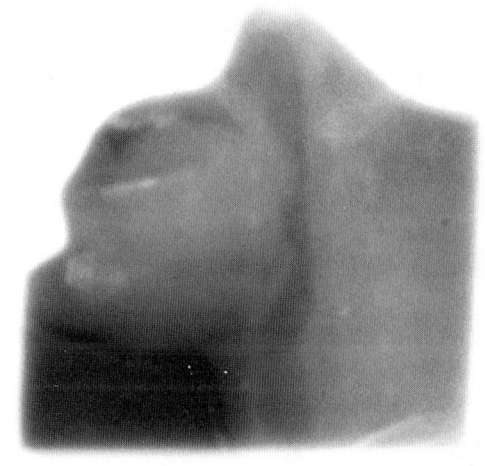

Fig. 8.1 a,b. 23-year old patient being treated for carcinoma of the tongue. **a** Digitized simulator film showing prescribed treatment field. **b** Digitized portal radiograph acquired during treatment with 4-MV photons. (See Fig. 8.14c for enhanced image)

radiograph of a 23-year old patient treated for cancer of the tongue and Fig. 8.1b shows the portal film, acquired during treatment with 4-MV photons, after digitization into a 512×480 pixel frame buffer to permit subsequent processing (see Fig. 8.14c). The reduced contrast between bone and soft tissue at megavoltage energies is apparent, and the rather blurry edges of the shielding blocks attests to the poor spatial resolution.

Following WEBB (1988) we define subject contrast for the simple model shown in Fig. 8.2, comprising a uniform block of tissue with attenuation coefficient μ_1 in which is embedded a small anatomical structure of thickness x and attenuation coefficient μ_2. If the

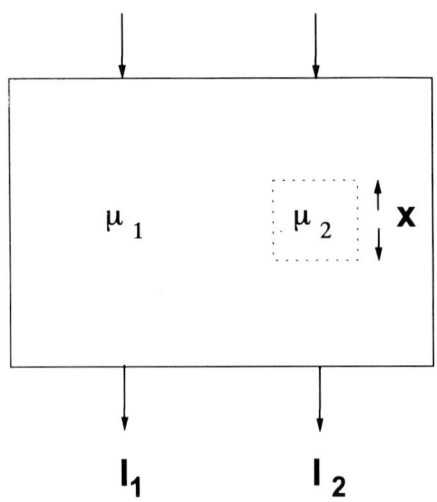

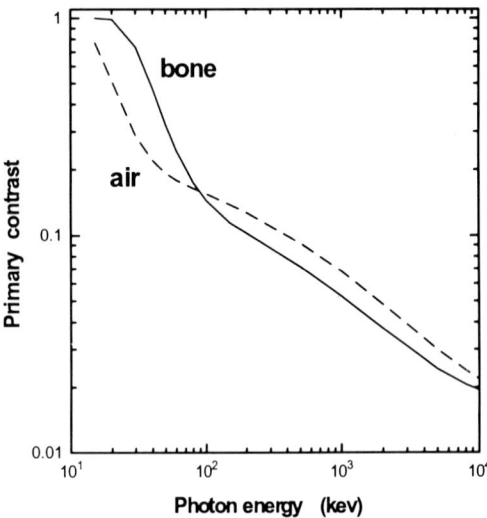

Fig 8.2. Simple model for definition of subject contrast. A small object of thickness x and attenuation coefficient μ_2 is embedded in a uniform block of tissue with attenuation coefficient μ_1. I_1 and I_2 are the recorded exit fluxes

Fig. 8.3. Primary contrast for bone and air 1 cm thick embedded in a uniform block of tissue

recorded exit fluxes behind the uniform block and the embedded structure are I_1 and I_2 respectively, the primary (unscattered photon) contrast is:

$$C_p = \frac{|I_1 - I_2|}{I_1} = 1 - \exp\left[-\left(\mu_2 - \mu_1\right)x\right] \qquad (8.1)$$

This is plotted in Fig. 8.3 for $x = 1$ cm, using data from JOHNS and CUNNINGHAM (1983) for monoenergetic photons. At diagnostic energies (30–50 keV) contrast is high, with bone dominant, but at megavoltage energies the primary contrast is much

reduced, and a 1-cm void is more visible than an equal thickness of bone. Figure 8.4a shows the simulation film of an 87-year-old woman prior to treatment for squamous cell carcinoma of the mandible. The target area and regions to be shielded have been marked. Figure 8.4b shows the right and left lateral portal images acquired at 6 MV and 23 MV respectively. The degradation of image contrast with increasing energy is apparent. Contrast will also be degraded by scattered photons, since an antiscatter grid is not practical at megavoltage energies, and the

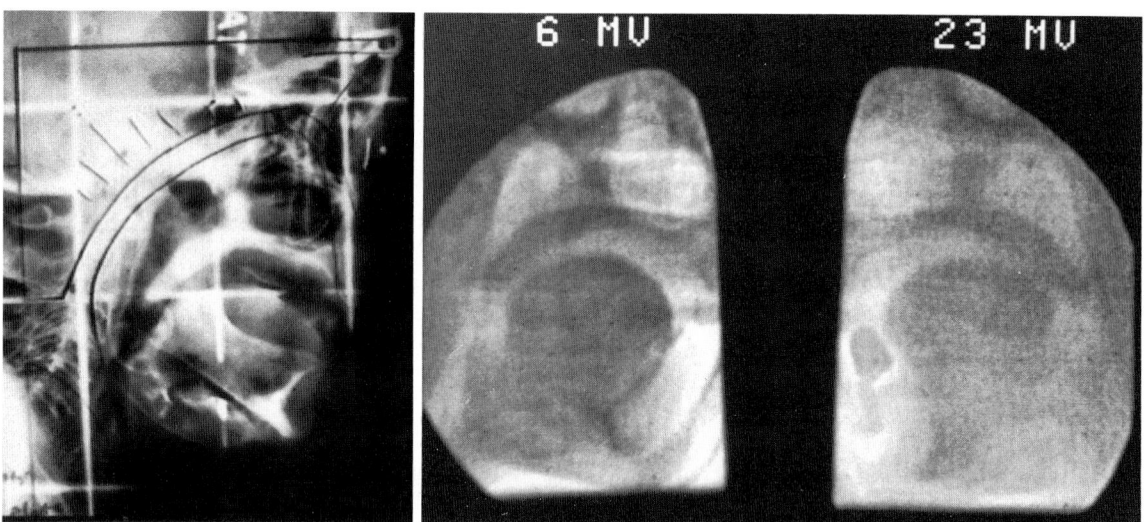

Fig. 8.4a,b. 87-year old patient being treated for carcinoma of the mandible. **a** Digitized simulator film showing prescribed target area and shielding block. **b** Right and left lateral portal images acquired with 6-MV and 23-MV photons

overall subject contrast C is a function of the scatter-to-primary ratio S/P:

$$C = C_P(1 + S/P)^{-1}. \qquad (8.2)$$

The effect of scatter can be reduced by having an air gap between the patient and the detector for 4- and 6-MV photons, but at higher energies scattering is highly forward directed and the air gap has little effect (AMOLS et al. 1986).

Contrast enhancement of portal films has been demonstrated by "gamma multiplication," which is achieved by photographic duplication onto a second high contrast film (REINSTEIN and ORTON 1979). REINSTEIN et al. (1987a) conducted an observer study which indicated some improvement in object detectability, at least under the restriction of limited viewing times. However, this technique has not met with widespread acceptance, since more convenient digital enhancement techniques have since become available.

Since image contrast decreases rapidly with increasing photon energy, improved image quality can be attained by using low-energy photon beams for the verification of patient setup, prior to the high-energy treatment. BIGGS et al. (1985) mounted a diagnostic x-ray source on a 10-MV linear accelerator for this purpose, while SHIU et al. (1987) applied the same technique to a cobalt therapy unit and obtained a portal image superimposed on a diagnostic quality x-ray image. An alternative approach suggested by GALBRAITH (1989) involves replacing the high-Z electron target with a thin low-Z target, so that sufficient low-energy photons will be produced to provide high contrast portal images. More widely used is a simple technique proposed by GALKIN et al. (1978) and GOER (1983), in which the linac is set to the lowest possible photon energy, and a preset short exposure is made to acquire a localization film. Some modern computer-controlled linacs can be programmed to open the collimators in this mode of operation, known as "port film mode", and reset them before the actual treatment, although this is impractical if shielding blocks are used.

8.2.2 Port Film/Screen Combinations

Early development of megavoltage radiography was based on a slow diagnostic film used for mammography, which was later marketed by Kodak as a therapy localization (TL) film. Haus and colleagues introduced an even slower film (XV) which could be used to verify a complete treatment (HAUS et al. 1970, 1973). It was suitable for automatic processing and

was marketed in a "ready-pack" light-tight envelope so that a cassette was unnecessary. Since then a considerable number of publications have dealt with the desirability of using heavy metal front and rear screens. Certainly a front screen is very important to remove electrons scattered from the patient, since they would degrade the image contrast, and for this purpose a sufficient screen thickness is about 1 g/cm² and 1.5 g/cm² for 4-MV and 8-MV photons respectively (DROEGE and BJARNGARD 1979a).

The front metal screen acts as a converter of photons to electrons by the Compton and (at high energies) the pair production processes, which in turn expose the film. This increase in sensitivity is obtained at the expense of spatial resolution, since the electrons scatter in the screen before reaching the film. Measurement of the spatial resolutions of metal screen/film combinations showed that lead front screens consistently gave a higher modulation transfer function (MTF) than copper owing to the lower electron path length (DROEGE and BJARNGARD 1979b). These authors also found that a rear lead screen reduces MTF by more than 40% for spatial frequencies above 2 cycles/mm due to the broad angular distribution of back-scattered electrons. On the other hand, HAMMOUDAH and HENSCHKE (1977) recommend using both front and rear lead screens with regular diagnostic film in order to improve contrast, and claim that image sharpness is better than for the TL film.

Fluorescent screens are widely used in diagnostic imaging, but they are seldom employed in portal radiography. One reason is that they must be used together with a front metal screen to exclude incident electrons, and it has been claimed that the additional spread of light in the phosphor will cause unacceptable loss of spatial resolution (DROEGE and BJARNGARD 1979a, REINSTEIN et al. 1987b). However, SEPHTON et al. (1989) have developed a new system consisting of a front lead screen and a rear fluorescent screen, with a single-emulsion high-contrast graphics film sandwiched between them. Tests over 2 years using 20 such cassettes on 6-MV linacs have confirmed superior contrast and spatial resolution compared to commercial portal radiography metal screen/film combinations. A disadvantage of fluorescent screens is their higher sensitivity, so that very low monitor unit settings are required and control of film density is difficult. On the other hand, KIHLEN et al. (1991) have found their high sensitivity useful for verifying lung shielding in total body irradiations, where the source-detector distance may exceed 4 m and the incident dose rate is low.

Clearly there are a number of parameters which must be taken into account in the optimization of port film techniques. The source size and energy, patient thickness, air gap, screen/film combination, cassette design, and weight will all affect the image contrast and resolution. A comprehensive review of port film methodology was carried out by the AAPM Task Group No. 28, in which these and other technical considerations were discussed, including screen design, viewing conditions, and film processing (REINSTEIN 1986; REINSTEIN et al. 1987b). The report includes a number of useful recommendations regarding the design and use of portal films in order to obtain good quality images.

8.2.3 Digital Enhancement of Portal Radiographs

There are numerous problems associated with the handling, viewing and storage of portal radiographs. They are of inherently poor contrast and low spatial resolution, and one study has shown that in 11 out of 23 institutions, over half the portal films were judged to be of poor quality (REINSTEIN et al. 1984). Frequently they are under- or overexposed, since the required number of monitor units is difficult to assess for different sites, patient separations, wedges, compensators, and so on, although the development of technique charts is useful in this regard (DROEGE and STEFANAKOS 1985; FAERMANN and KRUTMAN 1992). While too dense films can be viewed under a "hot light," it is difficult to examine films with very low optical densities. Additional difficulties arise when a comparison is required between successive portal films, or between a portal film and a simulator film, unless the magnification is identical for both. These problems can be ameliorated by converting the radiograph to digital format, and using the same image processing, display and storage techniques that are used for other digital radiographic images.

For digitizing portal films, a number of different types of scanning densitometers are available. One of the simplest approaches is to transilluminate the portal film on a viewbox and use a monochrome video camera to acquire a digital image, usually with an 8-bit analog-to-digital converter. Inexpensive "frame-grabber" boards are available which will acquire a complete video image of (typically) 512×480 8-bit pixels at the video scan rate, or about 1/30 s. The availability of a 16-bit frame buffer will allow up to 256 frames to be summed and the average stored as an 8-bit image for subsequent processing and display. If all noise sources are random (i.e., no

significant fixed-pattern noise from the camera or electronics), frame averaging will improve the signal-to-noise ratio (SNR) by up to a factor of 16. A correction image, acquired without the film present, can be subtracted to reduce nonuniform response due to lens vignetting, camera shading, and luminance variations across the viewbox.

Once it is available in digital format, the image can be stored, retrieved, processed, and displayed in the same way as other digital images, such as those acquired in computed radiography, on-line portal imaging, digital subtraction angiography, CT, magnetic resonance imaging, and so on. Comparisons can be made between successive portal images, and correlations performed with digitized radiographs from a therapy simulator, with suitable allowance made for unequal magnifications.

The portal image can be manipulated in a number of ways to improve the perceptibility of anatomical structures. During digitization the camera lens f-stop is adjusted to allow for different film densities, and the digitizer gain and offset can be adjusted to ensure that the maximum dynamic range is achieved. The stored digital image can be processed in a number of ways to enhance contrast or resolution and to permit the analysis of field size, shape, and position, and these topics are discussed below in the context of on-line portal image processing.

A number of authors have proposed that portal film digitization and enhancement be carried out on simple inexpensive microcomputers (SHALEV et al. 1984, 1985; AMOLS and LOWINGER 1987; MOK and FELDMEIER 1988; TAIT and HANSON 1989; CROOKS and FALLONE 1991). Such a device, used as an "electronic view box" for portal films, would considerably reduce the effort and strain of reviewing portal films, and provide improved visualization of anatomical structures, as well as the capability of comparing portal images with simulator radiographs. It will be quite some time before on-line digital portal imaging becomes routine practice on all megavoltage treatment units, and meanwhile portal radiographs will continue to be acquired and reviewed on a regular basis. The introduction of a low-cost, simple, semiautomatic viewing station using digital enhancement and analysis techniques would be a welcome addition to the techniques currently available for treatment verification.

8.2.4 Computed Radiography

Computed radiography (CR) is a relatively new technology in which radiographic film is replaced by a reusable photostimulable phosphor plate. After exposure in a conventional cassette, the latent image is developed by a scanning laser beam and the emitted luminescence is detected, digitized, and stored as a two-dimensional digital image. Prescanning is used to ensure correct normalization, and the dose-response function is linear over four orders of magnitude. Contrast and edge enhancement can be applied, and hard copy produced on film within a few minutes of the exposure.

Computed radiography was initially evaluated for megavoltage portal imaging by WILENZICK et al. (1987), who compared a Fuji system with conventional portal films. Other groups evaluated the Kodak CR system (GUR et al. 1989; WEISER et al. 1990) and the Toshiba system (ROEHRIG et al. 1990). The consensus seems to be that the digitally enhanced CR images show superior visualization of soft tissue and bone in portal images when compared to conventional film. Other advantages are the ease of storage and retrieval, the wide dynamic range (eliminating repeat exposures due to error in exposure), and the possibility of digital processing and analysis. However, CR is not a real-time modality, and access to CR systems is still difficult for many radiation therapy centers.

8.3 Electronic Portal Imaging Systems

The first megavoltage electronic portal imaging devices (EPIDs) were used to monitor 2-MV (ANDREWS et al. 1958) and 30-MV (BENNER et al. 1962) photon beams. Image contrast was poor, especially at the higher energy, but these studies demonstrated the potential of on-line verification of radiation treatments. Later improvements in photon beam intensities, phosphor design and video camera technology have led to dramatic improvements in the quality of video-based EPID image quality. Recent developments in solid-state and liquid ionization detectors have also been exploited in the design of EPIDs suitable for megavoltage portal imaging, several of which were reviewed by BOYER et al. (1992).

8.3.1 Solid State Detectors

LAM et al. (1986) constructed a linear array of 256 silicon diodes, which scanned across the radiation field in 2-mm increments. A 1.1 mm thick lead converter plate was used for imaging a 4-MV beam.

Another solid-state scanning array developed by MORTON et al. (1991) uses 128 zinc tungstate scintillating crystals arranged in two overlapping rows of 64 detectors, each crystal being coupled to a photodiode. A full scan over the 19 × 19 cm field of view (FOV) takes 4 s.

Scanning arrays are not efficient photon detectors and impose conflicting requirements for high spatial and contrast resolution (slow scanning speed) versus rapid image acquisition (fast scan). These disadvantages can be overcome with an area detector based on a multielement amorphous silicon detector array (MASDA) under development at the University of Michigan (ANTONUK et al. 1991, 1992). A matrix of 256 × 240 photodiodes, each coupled to a field effect transistor, forms a plane optical detector 10 × 11 cm in area, which is placed in contact with a metal/phosphor converter plate. Initial studies show that this concept has the potential to provide large area, efficient, and high-resolution portal imaging systems.

8.3.2 Liquid Ionization Chambers

A scanning liquid ionization chamber (SLIC) developed at the Netherlands Cancer Institute consists of a 256 × 256 matrix ionization chamber filled with isooctane and covered by a 1-mm-thick steel converter plate (VAN HERK and MEERTENS 1988). The signal is read out by sequentially switching the polarizing voltage applied to rows of chambers, and reading the ionization current of each column. The 32.5 × 32.5 cm FOV is scanned in 5.9 s, although faster image acquisition is possible with reduced spatial resolution (VAN HERK et al. 1992). The detector, and its associated read-out electronics, is contained in a 52 × 52 × 4 cm box and is suitable for mounting on a linear accelerator fitted with a beam stopper.

8.3.3 Video-Based Imaging Systems

An early demonstration of fluoroscopic portal imaging used a large flat metal/phosphor screen viewed through a 45° mirror by a silicon intensified target (SIT) video camera (BAILY et al. 1980). A schematic representation of this system, and later developments on the same principles, is shown in Fig. 8.5. The system was too bulky to attach to the gantry, but demonstration images were obtained with ^{60}Co and 6-MV beams. LEONG (1986) constructed a system

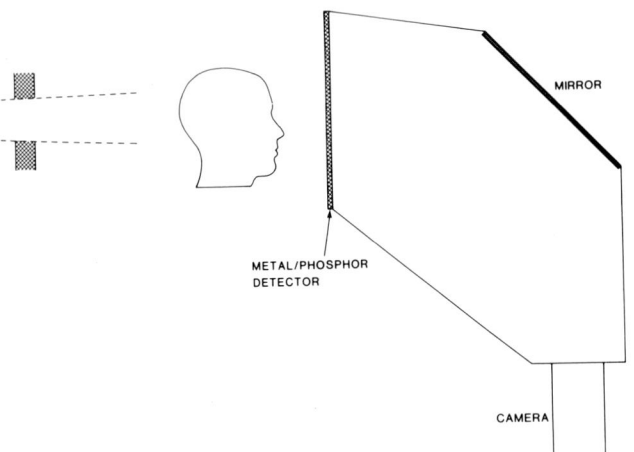

Fig. 8.5. Schematic diagram of a video-based portal imaging system

similar to Baily's but with the additional capability of digital processing for image enhancement. A partial solution to the size problem was proposed by WONG et al. (1990a) at the Mallinckrodt Institute of Radiology, who developed a plastic fiber-optic reducer to transmit the image from a metal/phosphor screen to a video camera. A matrix of 256×256 fiber bundles covered a FOV of 40×40 cm, and since the device was only 12 cm thick it could be mounted on a beam stopper.

Several other groups have also developed and evaluated video-based EPIDs. Using a structural design similar to Fig. 8.5, they differed mainly in the details of the metal/phosphor screen and in the selection of video camera. MUNRO et al. (1990) used a plumbicon camera with on-target signal integration to reduce electronic noise. VISSER et al. (1990) in Rotterdam selected a CCD camera, which

provides some advantages but has only a 512×256 pixel matrix. The group in Winnipeg, Canada originally opted for a SIT camera (SHALEV et al. 1989; LESZCZYNSKI et al. 1990) but later changed to a Newvicon for reasons which are discussed below. Commercial versions of a SLIC and three video-based EPIDs are currently available and their main characteristics are given in Table 8.1.

The selection of a suitable video camera depends on the luminance (brightness) L_s of the phosphor during a clinical treatment, and on the design parameters of the imaging system. The camera faceplate illuminance E_c is the product of the luminance L_c and the subtended solid angle Ω_c at the lens exit pupil (RCA 1974). Reduction in luminance in the optical system is expressed as T_r, the lens transmission factor, for which $T_r = 0.8$ is a reasonable value for a multielement lens.

Table 8.1 Comparison of commercial available EPIDs

Type:	SLIC		Video	
R&D group:	NKI	Rotterdam	London, Ontario	Winnipeg
Supplier:	Varian	Philips	Infimed	Siemens
Plate	Steel		Copper	Brass
Thickness (mm)	1.0		1.0	1.36
Size (cm)	32.5×32.5	40×30	40×40	44×35
Detector	Ion chamber		Gd_2O_2S	Gd_2O_2S
mg/cm^2		411		~400
Camera		CCD	Plumbicon	SIT, Newvicon
Lens f/			0.95	1.1
Matrix	256×256	512×256	512×512	512×480
Pixal (mm)	1.27×1.27	0.78×1.17	0.78×0.78	0.64×0.81
SDD (cm)	Variable	160	Variable	140
FOV (cm)	Variable	25×119	Variable	33×26

Fig. 8.6. Response curves for video cameras. SIT, silicon-intensified target; IA, image amplifier; ST, silicon target

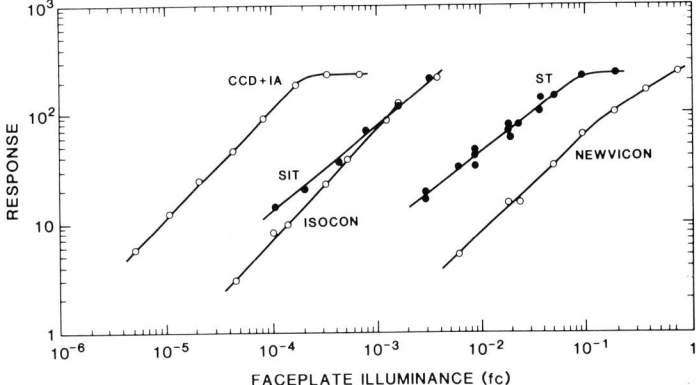

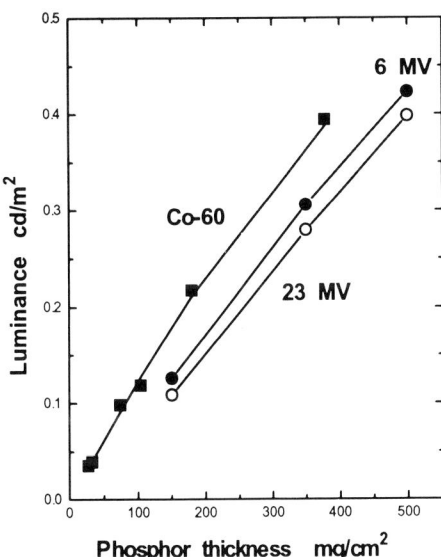

Fig. 8.7. Luminance of Gd_2O_2S screens irradiated by cobalt-60, 6-, and 23-MV photon beams. The data are normalized to an incident dose rate of 100 cGy/min

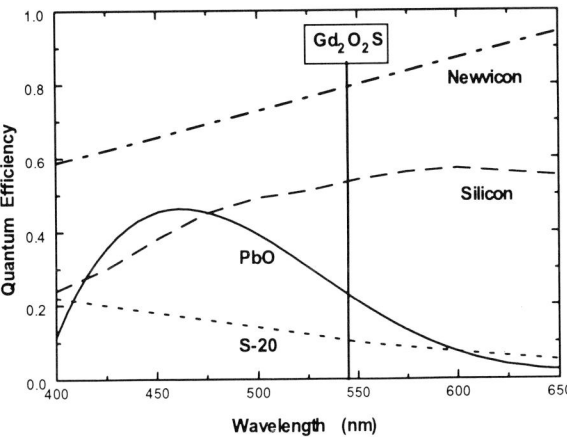

Fig. 8.8. Quantum efficiency of Newvicon, CCD (Silicon), Plumbicon (PbO), and SIT (S-20) video cameras. The light emitted by a Gd_2O_2S screen is predominantly at 545 nm

For an f-stop F and magnification (image/object) m, we have:

$$E_c = L_c\Omega_c$$
$$= L_sT_r\Omega_c$$
$$= L_sT_r\frac{\Pi}{4F^2}\left(\frac{1}{1+m}\right)^2. \qquad (8.3)$$

For a 30×40 cm screen viewed by a nominal 1" camera (9.6×12.8 mm) the magnification m=0.032. Experimental response measurements are shown in Fig. 8.6 for a CCD camera fitted with an image amplifier, SIT and Isocon cameras, a silicon target tube, and a Newvicon.

In a prototype design (SHALEV et al. 1989) for a 4-MV beam using a commercial screen (100 mg/cm²), screen luminance was found to be 0.16 cd/m² (0.046 fL), which gives a faceplate illuminance of 4.7×10^{-2} lx (4×10^{-3} fc) for a f/1.4 lens. This is an upper limit, since under clinical conditions the brightness would be reduced due to attenuation of the beam by the patient, compensator plates and wedges. Based on the response curves shown in Fig. 8.6 a SIT camera was selected (LESZCZYNSKI et al. 1990) although the more expensive Isocon would also have been appropriate. Thicker phosphor screens provide a brighter image, as shown in Fig. 8.7 for measurements with a ^{60}Co beam (BUCHANAN et al. 1974) and 6- and 23-MV photon beams (WOWK et al. 1994). Using a 500 mg/cm² Gd_2O_2S screen, an incident dose rate of 100 cGy/min and a f/0.85 lens, the camera faceplate illuminance can be increased to 3.3×10^{-2} fc, which is in the range of a Newvicon camera.

Figure 8.8 shows quantum efficiency plotted against wavelength. The light emitted from Gd_2O_2S screen is predominantly at 545 nm, and the Newvicon has an eight-fold advantage over the SIT camera (which has an S-20 response). Figure 8.9

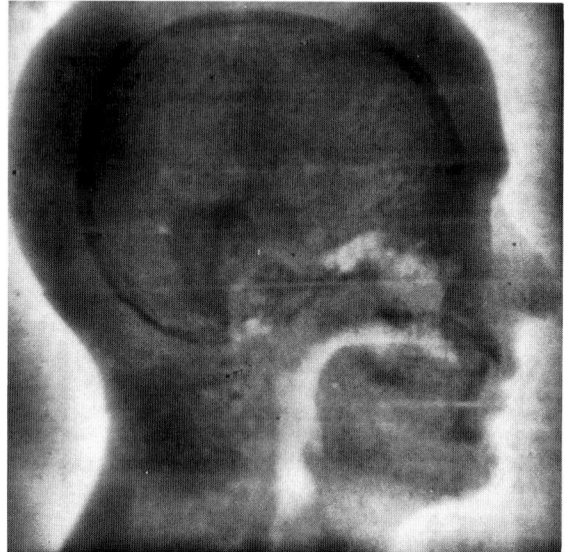

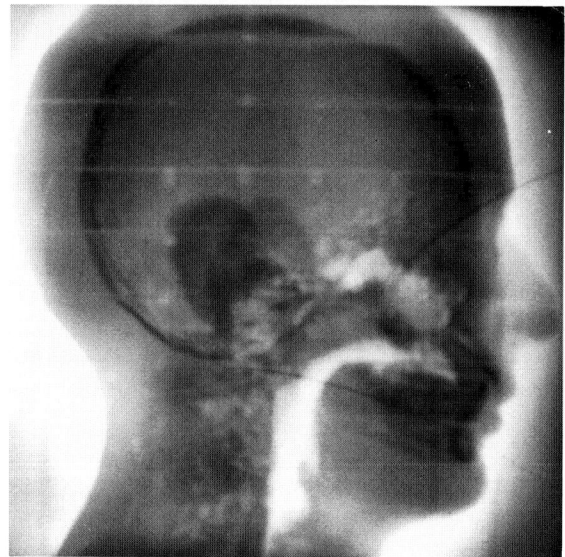

Fig. 8.9a,b. Anthromorphic head phantom irradiated with 6-MV photons. **a** Portal image acquired with a SIT camera. **b** Portal image acquired with a Newvicon camera. Both images exposed for 8.5 s (256 video frames)

shows two processed portal images of an anthropomorphic head phantom, irradiated at 6 MV and acquired with SIT and Newvicon cameras with an 8.5 sec exposure (256 video frames). The superior contrast of the Newvicon camera is clearly seen.

8.3.4 Comparison of EPID Systems

The various commercial EPIDs, and systems under development in research laboratories, are based on different design parameters. Optimization of the variables, such as contrast and spatial resolution, FOV, display matrix, and detection efficiency, is a subjective exercise, and various solutions have been proposed. We will examine some of the design parameters involved in this process.

8.3.4.1 Spatial Resolution

Spatial resolution depends on the focal spot size, source-detector geometry, and display matrix. Other factors can also be of influence, such as video camera readout sequence and lag, detector and/or patient movement, and screen thickness. Consider the geometry shown in Fig. 8.10, where for simplicity the focal spot is assumed to be of gaussian shape defined by σ_s, and the detector pixel size is denoted by σ_d. The magnification M is defined as the ratio of the source-detector distance (SDD) to the source-axis distance (SAD). The line spread function (LSF) at the isocenter will consist of contributions of unsharpness from both the source and the detector (BARRETT and SWINDELL 1981):

$$\sigma_{LSF}^2 = \frac{\sigma_d^2}{M^2} + \sigma_s^2\left(\frac{M-1}{M}\right)^2, \tag{8.4}$$

which is minimized for an optimal magnification

$$M_{opt} = 1 + \left(\frac{\sigma_d}{\sigma_s}\right)^2 \tag{8.5}$$

The effective focal spot size for most therapeutic linear accelerators is in the range 0.9–1.6 mm, although values as low as 0.5 mm for the Siemens KD-2 and as high as 3 mm for the Therac 6 have been measured (MUNRO et al. 1988; LOEWENTHAL et al. 1992; JAFFRAY et al. 1993). Equation 8.4 is plotted in Fig. 8.11 for three source sizes and for two typical detector matrix sizes. The curve marked 512 × 480 represents the Siemens and Infimed video systems

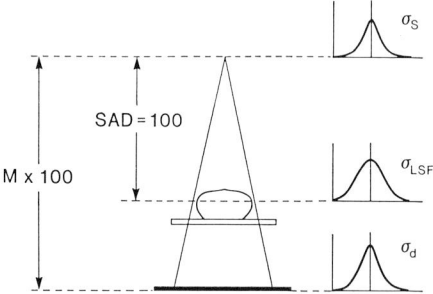

Fig. 8.10. Contributions to degradation of spatial resolution. The focal spot is assumed to have a Gaussian shape defined by σ_s and the detector pixel size is σ_d, which results in a line spread function at isocenter characterized by σ_{LSF}

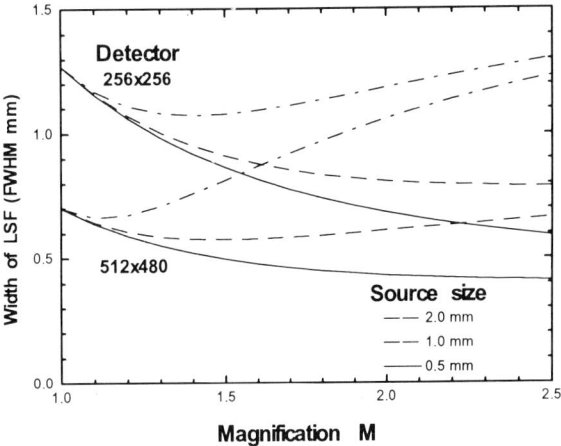

Fig. 8.11. Width of line spread function at isocenter for three focal spot sizes and two detector matrix sizes

with average pixel size σ_d = 0.7 mm, and the curve marked 256 × 256 represents the Varian SLIC detector with σ_d = 1.27 mm. Clinically relevant values of M are 1.4–1.6, as larger values result in small FOV and low photon detection efficiency. Except for very large source sizes, the curves are relatively flat, and there is no significant advantage in using the M_{opt} values predicted from Eq. 8.5. Indeed, the recommended magnification is less than M_{opt} if one takes into account the inverse square reduction in photon flux with increasing magnification (BISSONNETTE et al. 1992). From Fig. 8.11 one

concludes that a significant loss of spatial resolution occurs with the 256 × 256 detector matrix, and that such a detector should be used at as high a magnification as possible. On the other hand, the 512 × 480 pixel video systems should be used in the range M = 1.4 – 1.6, except for very large sources, when increasing magnification has a very deleterious effect on the spatial resolution. An experimental measurement of the MTF of such a system gave FWHM = 0.35 mm⁻¹ (BISSONNETTE et al. 1992).

8.3.4.2 Contrast Resolution

Defining the contrast resolution of an imaging system is difficult, as many factors will affect the result, and contrast-detail analysis is an accepted approach (COHEN et al. 1981; KOTRE et al. 1992). Various phantoms have been proposed (MUNRO et al. 1990; VISSER et al. 1990), and in the interest of standardization a contrast-detail phantom was distributed by this author to participants of a workshop in Las Vegas in 1989. Constructed from aluminum with holes of various diameters and depths, the "Las Vegas" phantom has been used to define system resolution in an objective way (MORTON et al. 1991). Figure 8.12 shows a portal image of the phantom acquired at 4 MV with an exposure of 18 cGy. A contrast-detail curve is obtained by delineating the smallest object visible at each contrast level, and

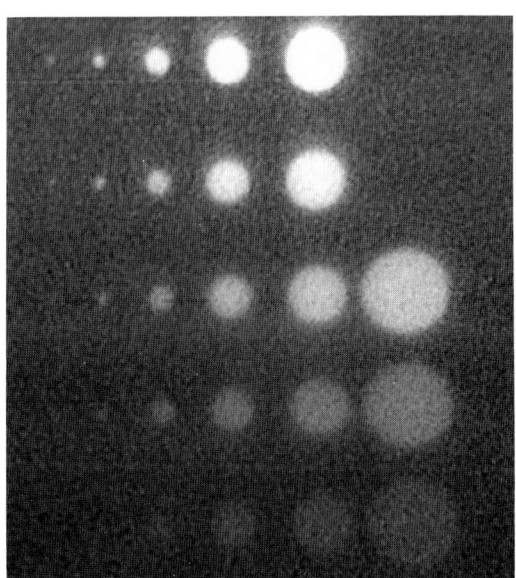

Fig. 8.12. Portal image of the "Las Vegas" phantom acquired with 4-MV photons and a dose of 18 cGy

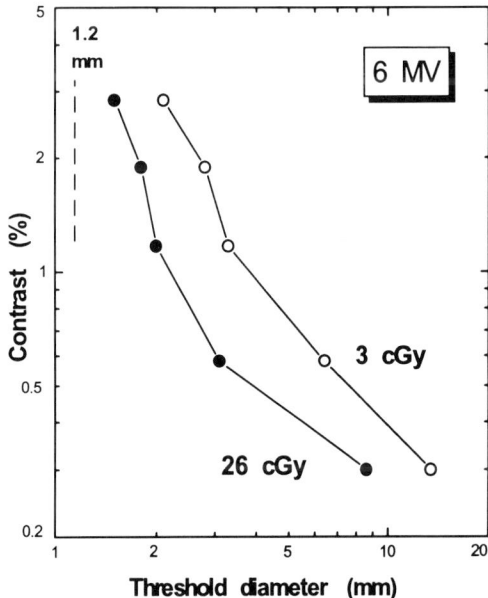

Fig. 8.13. Contrast-detail plot for the Siemens BEAMVIEW portal imaging system for 6-MV photons, using the "Las Vegas" phantom

Fig. 8.13 shows the results of such a test on the Siemens BEAMVIEW system at 6 MV. For increasing contrast the curves approach the asymptotic limit of 1.12 mm determined by the Nyquist frequency for pixels of 0.81 mm width at M=1.4.

8.3.5 Image Processing

Most of the digital image processing techniques used on portal images fall under three headings: contrast enhancement, noise reduction, and edge crispening. These are documented elsewhere (e.g., GONZALEZ and WINTZ 1987; PRATT 1991), and only brief mention will be made of their application to EPID images. The simplest method to enhance contrast is by manipulating the output look-up-table, so as to map the stored pixel values to different gray levels. Windowing is an example of global amplitude scaling, in which the transfer function is linear with adjustable slope and end limits. LEONG (1984), SMITH (1987) and MEERTENS et al. (1988) have applied various forms of global histogram modifications and frequency domain filters to digitized port film images. Global contrast enhancement is limited in effectiveness where there is a significant low frequency component, and contrast limited adaptive histogram equalization (CLAHE) has been found to be better suited to portal films and to x-ray localization films used to verify the location of brachytherapy seeds (SHEROUSE et al. 1987; PIZER et al. 1987). However, this technique does cause some blurring of the field edges, which is undesirable since the field location cannot be determined accurately. A technique called selective contrast limited adaptive histogram equalization (SCLAHE) has been developed to overcome this problem, in which the image is first segmented into regions inside and outside the field, and then CLAHE is applied to the inner region only (LESZCZYNSKI et al. 1992a). In regions of low contrast, artifacts can be introduced due to the regional bilinear interpolation inherent in the technique. Moving Histogram Equalization (LESZCZYNSKI and SHALEV 1989) and matched Fourier filtering (MCGEE et al. 1993) can be used to reduce their effect.

Figure 8.14a shows an on-line portal image for the same patient as Fig. 8.1, being treated for cancer of the tongue with 4-MV photons. The same image after processing by SCLAHE is shown in Fig. 8.14b. Digital processing is equally effective for portal radiographs, and Fig. 8.14c shows the corresponding portal film from Fig. 8.1b after processing by high-pass filtering. At higher energies digital

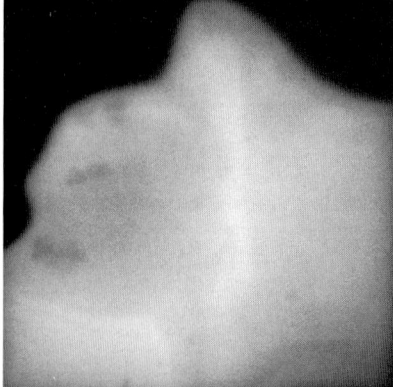

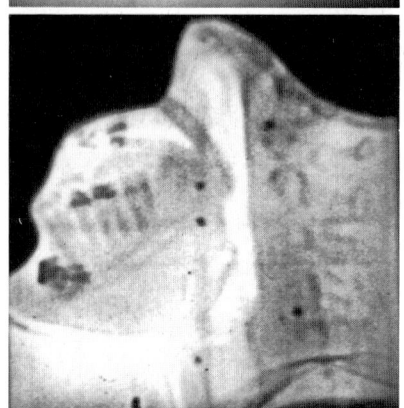

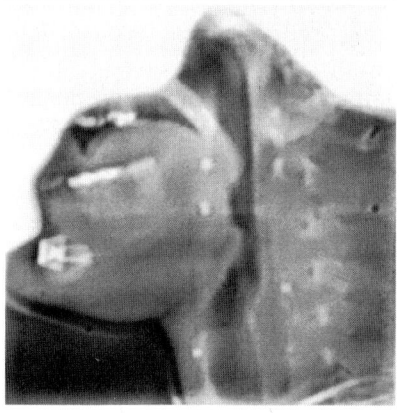

Fig. 8.14a–c. Portal images of same patient as Fig. 8.1. **a** Unprocessed on-line portal image. **b** Same image as **a** after processing by SCLAHE. **c** Portal radiograph (Fig.8.1b) after processing by high-pass filtering

processing is even more useful, since the unprocessed images demonstrate very poor contrast. Figure 8.15 shows a processed pelvic on-line image for treatment of cancer of the ovary at 23-MV. The two images are identical, except that they are displayed with reversed look-up tables for "black-bone" and "white-bone" displays, respectively.

A review of image processing techniques by CUMBERLIN et al (1989) interjects a word of caution, that while contrast enhancement of portal films is

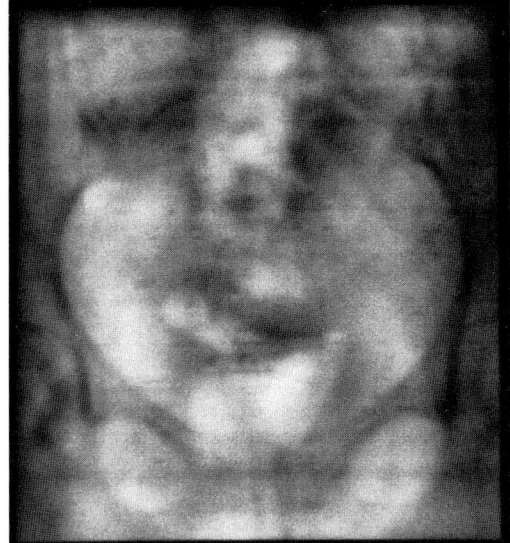

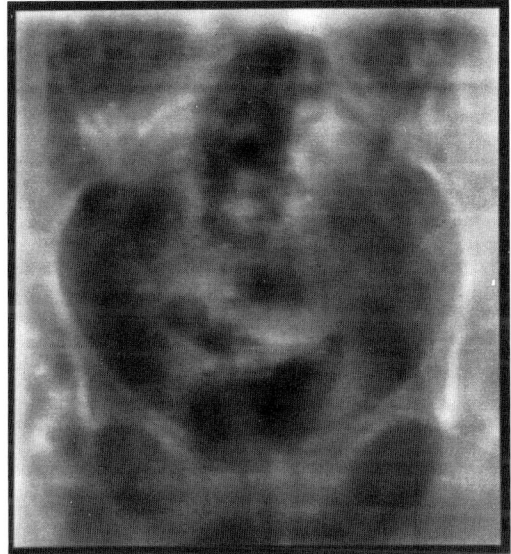

Fig. 8.15a,b. On-line portal image of a 40-year-old woman treated for cancer of the ovary with 23-MV photons. The images are identical except for display on reversed gray scale: **a** "black-bone" display; **b** "white-bone" display. Processing was by SCLAHE. Image acquisition with a Newvicon camera and 0.5 s exposure

feasible, its clinical utility still needs to be evaluated in an objective manner. One approach is to evaluate the accuracy with which observers locate landmarks on unprocessed and enhanced images (LESZCZYNSKI et al. 1992b). A study of image processing algorithms (SHALEV and McGEE 1992) suggests that the previous experience and skill of the observer may be more important than image processing for the perception of small objects in on-line portal images.

8.4 Clinical Applications

Conventional port films are used quite infrequently, with only 40% of the institutions in one survey taking films on even a weekly basis (REINSTEIN et al. 1987b). Consequently radiation treatment verification is based on a very sparse sample, which may not be indicative of normal practice if extra care is taken on days on which port films are acquired. The time and trouble involved in examining localization films and adjusting patient setup before continuing treatment discourages their more general use. On the other hand, the low cost and simplicity of on-line portal imaging (OPI) permits the verification of every field and every fraction, as well as providing display of sequential images during a single treatment (intra-treatment movies) or inter-treatment time-lapse sequences throughout a fractionated course of treatment. How, then, will OPI be used for treatment verification, and in what manner should the new technology be implemented in routine clinical practice?

8.4.1 Gross Setup Errors

An important function of OPI is the detection of gross setup errors early on in the treatment, so that the beam can be shut off and the error corrected before too large a dose has been delivered. Gross errors occur when incorrect shielding is used, or shielding is placed incorrectly in the field (e.g., left-right reversal, wrong shadow tray), or when a shielding block is missing. For treatment machines not fitted with automatic RV systems, field size errors are possible, as well as incorrect collimator rotations and similar problems. Such errors in the size, shape and rotation of the field are independent of the patient's position within the beam.

The simplest method of detecting gross setup errors is for the operator to view the on-line portal images at the start of every treatment fraction, and shut off the beam if necessary. A less stressful approach is to give an initial dose of a few monitor units, and examine the portal image at leisure before continuing the treatment. Several centers are developing techniques to automatically detect the field edge and compare the field size and shape with the prescribed parameters, permitting completely automatic shutdown of the treatment machine if a gross error is detected (BIJHOLD et al. 1992a; LESZCZYNSKI and SHALEV 1993).

8.4.2 *Image Alignment*

In order to determine whether the beam is targeted correctly relative to the target volume, it is necessary to identify both the field edge and a set of anatomical features in the portal image. A reference image is also required, which may be a previous (correctly located) portal image, or a digitized simulator film, or a digital reconstructed radiograph on which the prescribed field edge has been marked. A variety of approaches have been proposed to determine whether the relationship of the patient (anatomical features) to the beam (field edge) is the same in both the portal and the reference images. Interactive registration of colored (WEINHOUS 1990; GRAHAM et al. 1991) or interlaced images (EVANS et al. 1992) provides a visual indication of field placement accuracy. Where image quality is poor, graphic templates drawn over anatomical features have been proposed (BIJHOLD et al. 1991). Corresponding sets of fiducial reference points in both images can be used (BOESECKE et al. 1990; MEERTENS et al. 1990; HALVERSON et al. 1991; LAM et al. 1991), although care must be taken to ensure that the landmarks are actual anatomical structures and not the projections of overlying features which are at different depths in the patient, since these would be very sensitive to small changes in the viewing angle. More reliable is the use of features such as open curve segments (BALTER et al. 1992) or correlation methods (JONES and BOYER 1991). Radiopaque markers placed on the immobilization shell or on the patient (LAM and TEN HAKEN 1991) provide easily visible registration points, but implantation of markers into the tumor volume may prove to be the most accurate approach. Once the anatomical features have been identified, and the field edges delineated, one of two approaches can be followed: (a) align the field edges and compare any displacement in the anatomical features; or (b) align the anatomical features and compare the locations of the field edges.

8.4.2.1 Alignment by Field Edges

After detection of the field edges on both the reference and the portal image, they are aligned by a rigid-body two-dimensional transformation according to the least squares criterion or the Procrustes algorithm (SCHONEMANN and CARROLL 1970). Care must be taken to evaluate the goodness of fit, since an error in field size or shape will lead to a poor fit. The transformation parameters are then used to align the portal image fiducial points or features onto the reference image, and any discrepancy is interpreted as a field positioning error (FPE).

8.4.2.2 Alignment by Anatomical Features

In this approach the fiducial points or anatomical features are aligned, and the transformation parameters are used to register the portal field edge onto the reference image. Any discrepancy between the portal field edge and the reference field edge is interpreted as an FPE. It is important, of course, that the alignment procedure should not introduce errors comparable to the FPEs which are to be measured. In many treatment sites there are few anatomical landmarks, and if they are closely grouped, even small inaccuracies in delineating their positions can lead to large errors in the transformation parameters. Distortion of the reference or portal image and out-of-plane rotation by the patient will also lead to erroneous results. However, FPEs greater than 2–3 mm or rotations > 5° can probably be determined with a high degree of confidence by image alignment techniques.

8.4.3 *Quantification of Field Positioning Errors*

Once the reference and portal images are correctly aligned, any FPEs can be described in terms of lateral displacement vectors and rotations, either of the field perimeter or center of gravity of the field area, or of anatomical features within the target volume. Other useful parameters are the total field area, target coverage (the fractional area of the prescribed target covered by the actual portal field), and the area of normal tissue which should have been outside the irradiated area but was in fact within the treatment portal.

Following BIJHOLD et al. (1992b), the displacement vector $\mathbf{m}_{pf}$ is defined for patient p and fraction f over a treatment series. The actual displacement vector of the target volume relative to the treatment beam consists of a systematic component Δ_p throughout the whole treatment, and a random component δ_{pf} for each fraction. Errors can also arise in the alignment process, which may have a systematic component E_p and a random component ε_{pf}, so that

$$\mathbf{m}_{pf} = \Delta_p + \delta_{pf} + E_p + \varepsilon_{pf}$$

or

$$\mathbf{m}_{pf} = M_p + \mu_{pf} \tag{8.6}$$

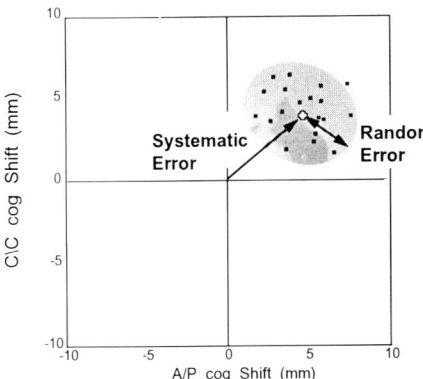

Fig. 8.16. Example of plotting the displacement vector $\mathbf{m}_{pf}$ for shifts in the center of gravity of the treatment field edge. Shifts in the anterior/posterior (A/P) and cranial/caudal (C/C) directions for a lateral treatment. Each black dot represents the displacement during one fraction. The white cross represents the mean shift

where $M_p = \Delta_p + E_p$ and $\mu_{pf} = \delta_{pf} + \varepsilon_{pf}$ are the vector components due to systematic and random errors, respectively. A typical example is shown in Fig. 8.16 for displacements of the center of gravity of the field edge in the anterior/posterior and cranial/caudal directions during a lateral treatment. Each individual dot represents the displacement $\mathbf{m}_{pf}$ during one fraction. The ellipse represents the 95% confidence region, calculated from all the data points at the completion of treatment. The spread of $\mathbf{m}_{pf}$ about the mean $\overline{\mathbf{m}}_{pf}$, assuming normal distribution functions for the random errors, is represented by the variance σ_p. The displacement of $\overline{\mathbf{m}}_{pf}$ from the intended (zero-shift) location is the systematic error Σ_p. RABINOWITZ et al. (1985) studied a variety of treatment sites, and concluded that systematic errors were dominant, but these results were not consistently supported by other studies of the chest (GRIFFITHS et al. 1987), head and neck (HUIZENGA et al. 1988), breast (VAN TIENHOVEN et al. 1991), and mantle (CREUTZBERG et al. 1992). In general one can expect both σ_p and Σ_p to be dependent on treatment site, immobilization technique, patient size, and other factors related to the simulation and treatment procedures.

8.4.4 Dynamic Portal Imaging

From the earliest demonstrations of OPI, there has been an interest in viewing a live video image of the target volume during a treatment (ANDREWS et al. 1958, BAILY et al. 1980, LEONG and STRACHER 1987). The presence and magnitude of relative movement is more readily appreciated on movies than on a sequence of static images. Day-to-day movements can also be visualized by acquiring portal images during consecutive treatments and subsequently viewing them in rapid succession.

8.4.4.1 Intratreatment Movies

During a typical daily treatment, the beam may be on for 20–30 s, and during this time a number of individual portal images can be acquired. Viewing them after the completion of treatment as a "movie-loop" shows any organ movement that occurred due to breathing, cardiac motion, etc. Such megavoltage movies have been shown at several conferences in the form of video presentations (SHALEV et al. 1988, 1990; REINSTEIN et al. 1988; SHALEV 1992). Surprisingly, even in vacuum-molded head shells, one can see considerable motion of the skull, while in lung treatments the effect of breathing is clearly seen. Digital subtraction of successive images provides dramatic enhancement of motion, which can be displayed as phase-colored overlays on a static portal image (SHALEV et al. 1988). Such studies of patient motion can be used to evaluate the efficacy of immobilization devices and to acquire data on patient and organ motion under treatment conditions, but it is doubtful whether they will be used routinely in a busy clinical environment.

8.4.4.2 Intertreatment Time-lapse Sequences

The verification portal images acquired for each daily treatment can be displayed as a time-lapse sequence, so that daily variation in patient positioning and shielding block locations can be visualized. However, since most commercial EPIDs are not precisely registered from one treatment to the next, the images tend to "jump around," and viewing time-lapse sequences is best done with images that have been realigned to a common coordinate system.

Nevertheless, time-lapse display is a very rapid method of reviewing patient setup reproducibility throughout a course of treatment, and one can expect it to become an accepted part of routine quality assurance procedures.

8.4.5 Interventional Treatment Verification

The ultimate goal of digital portal imaging is to improve the accuracy of treatment delivery. The installation and operation of EPIDs, and the acquisition and analysis of numerous portal images, is of no avail unless it leads to some action which will have a positive impact on the final dose distribution delivered to the patient. Clearly the discovery and correction of gross setup errors is advantageous, but these are (hopefully) infrequent, and will become even more so as RV systems gain in popularity. In the final analysis, the impact of OPI on radiation therapy will depend on its capability of providing rapid, accurate and reliable information on what is wrong and what must be done to correct the situation. This can be done in two ways.

8.4.5.1 Intertreatment Correction

Intertreatment correction is the conventional approach used with portal films. The verification portal image is examined at the end of the daily treatment, and if necessary a correction in patient setup is made at the following treatment session. From Fig. 8.16 it is clear that a single measurement at the first fraction is not an effective approach to treatment verification. The predictive value of a single measurement depends on the (usually unknown) ratio of systematic to random errors (MITINE et al. 1991; LEBESQUE et al. 1992; DUTREIX et al. 1992). If $\Sigma_p \gg \sigma_p$, that is, if systematic errors dominate, then a single correction to the patient's setup will be advantageous. However, if $\Sigma_p \ll \sigma_p$, any displacement found on a single fraction is only one sample from a random distribution, and changes to the setup are just as likely to reduce the accuracy of the following fraction as they are to improve it. A series of measurements without corrections permits an estimate of both Σ_p and σ_p, so that a decision can be made on the advisability of changing the setup.

DALLY et al. (1992) have calculated the number of measurements required in order to estimate (and correct for) the systematic error. For example, if the standard deviation of the random error is 6 mm, eight measurements are necessary in order to estimate the position of the field center to within 3 mm at the 95% confidence level, and hence to discover a systematic error large than 3 mm. Two measurements are sufficient to correct for systematic errors greater than 6 mm under these conditions. An alternative approach is to make intertreatment corrections according to a shrinking action level, determined from consecutive measurements of the FPEs (BIJHOLD et al. 1992b; BEL et al. 1992).

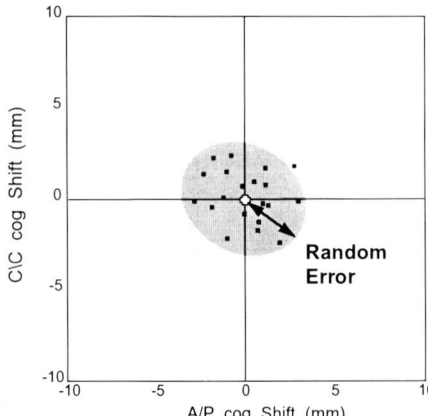

Fig. 8.17. Plot of displacement vector m_{pf} after application of intertreatment corrections. Only random errors remain. Abbreviations as in Fig. 8.16

Once all the initial systematic errors have been corrected, subsequent fractions should exhibit only random errors with variance σ_p, as shown in Fig. 8.17. This is the best that can be achieved with intertreatment corrections. However, measurements should continue to be made throughout the course of treatment, since new systematic errors could be introduced and there is some evidence for a time trend of setup errors (DENEVE et al. 1992; ELGAYED et al. 1992), although this effect was not seen in an earlier study of infradiaphragmatic fields (GRIFFITHS et al. 1987).

8.4.5.2 Intratreatment Correction

Intratreatment correction is highly interactive and interventional. A short localization portal image is acquired at the start of treatment, using a small "localization dose" and the beam is shut off. The localization image is examined and if no FPE is detected, the remainder of the treatment dose is delivered. If a significant FPE is found, a correction is made to the setup, and treatment is continued, with the option of acquiring another localization image if required. While this approach is time consuming and expensive in terms of resources and patient throughput, it does have the potential to correct for both systematic and random errors, and to ensure that every fraction is delivered to the target volume with the highest possible precision. A full course of treatment using intratreatment correction on every fraction should yield a center of gravity shift distribution as shown in Fig. 8.18, where the residual error ε_{pf} is due to point placement and other random

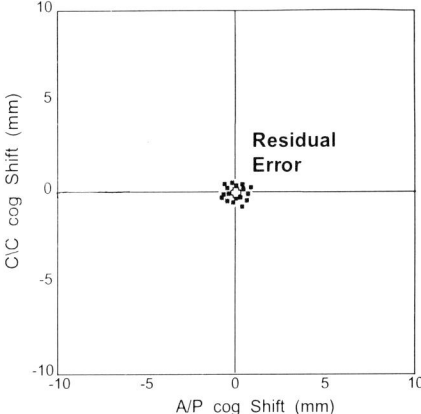

Fig 8.18. Plot of displacement vector m_{pf} after application of intratreatment corrections. Only small residual errors remain. See text for explanation; abbreviations as in Fig. 8.16

technique-related inaccuracies. Several trials are in progress (Ezz et al. 1992; GILDERSLEVE et al. 1992) which indicate that FPEs are significantly reduced by intratreatment correction. A major study was carried out by DE NEVE et al. (1992) involving 883 fields on 21 patients receiving treatment in the head and neck, thorax, and pelvic regions. EPID localization images were compared to simulation films, and the patient setup was corrected for FPEs greater than a predefined limit in the range 0.5–1.0 cm. Less than 4% of the thoracic and head and neck setups required correction, but over half of the pelvic setups were found to be in error greater than the prescribed tolerance, and were corrected before continuing treatment. While intratreatment correction is time consuming, it appears to improve considerably the accuracy of treatment delivery, especially for patients who are not treated in an immobilization device.

References

Amols H, Lowinger T (1987) An inexpensive microcomputer based portal film image enhancement system. Med Phys 14: 483

Amols H, Reinstein L, Lagueux B (1986) A quantitative assessment of portal film contrast as a function of beam energy. Med Phys 13: 711–716

Andrews J, Swain R, Rubin P (1958) Continuous visual monitoring of 2 MEV Roentgen therapy. Am J Roentgenol 73: 74–78

Antonuk L, Yorkston J, Boudry J (1991) Large area amorphous silicon photodiode arrays for radiotherapy and diagnostic imaging. Nucl Instr Meth Phys Res A310: 460–464

Antonuk L, Boudry J, Huang W et al. (1992) Demonstration of megavoltage and diagnostic x-ray imaging with hydrogenated amorphous silicon arrays. Med Phys 19: 1455–1466

Baily N, Horn R, Kampp T (1980) Fluoroscopic visualization of megavoltage therapeutic x-ray beams. Int J Radiat Oncol Biol Phys 6: 935–939

Balter J, Pelizzari C, Chen T (1992) Correlation of projection radiographs in radiation therapy using open curve segments and points. Med Phys 19: 329–334

Barrett H, Swindell W (1981) Radiological imaging, vol 1. Academic press, New York, p 137

Bel A, Bartelink H, El-Gayed A, van Herk M, Vijlbrief R, Lebesque J (1992) Portal imaging and decision rules for correcting patient setups. Radiother Oncol 24 (Suppl):S33

Benner S, Rosengren B, Wallman H, Netteland O (1962) Television monitoring of a 30 MV X-ray beam. Phys Med Biol 7: 29–33

Biggs P, Goitein M, Russell M (1985) A diagnostic X-ray field verification device for a 10 MV linear accelerator. Int J Radiat Oncol Biol Phys 11: 635–643

Bijhold J, van Herk M, Vijlbrief R, Lebesque J (1991) Fast evaluation of patient set-up during radiotherapy by aligning features in portal and simulator images. Phys Med Biol 36: 1665–1679

Bijhold J, Gilhuijs K, van Herk M (1992a) Automatic verification of radiation field shape using digital portal images. Med Phys 19: 1007–1014

Bijhold J, Lebesque J, Hart A, Vijlbrief R (1992b) Maximizing setup accuracy using portal images as applied to a conformal boost technique for prostatic cancer. Radiother Oncol 24: 261–271

Bissonnette J-P, Jaffray D, Fenster A, Munro P (1992) Physical characterization and optimal magnification of a portal imaging system. SPIE 1651: 182–188

Boesecke R, Bruckner T, Ende G (1990) Landmark based correlation of medical images. Phys Med Biol 35: 121–126

Boyer A, Antonuk L, Fenster A, et al. (1992) A review of electronic portal imaging devices (EPIDs). Med Phys 19: 1–16

Brahme A (1984) Dosimetric precision requirements in radiation therapy. Acta Radiol Oncol 23: 379–391

Brahme A (1988) Accuracy requirements and quality assurance of external beam therapy with photons and electrons. Acta Oncol (Suppl 1)

Buchanan R, Sklensky A, Maple T, Bailey H (1947) Metal-phosphor intensifying screens for high energy imaging applications. IEEE Trans NS-21: 692–694

Cohen G, Wagner L, Amtey S (1981) Contrast-detail-dose and dose efficiency analysis of a scanning digital and a screen-film-grid radiographic system. Med Phys 8: 358–367

Creutzberg C, Visser A, De Porre P, Meerwaldt J, Althof V, Levendag P (1992) Accuracy of patient positioning in mantle field irradiation. Radiother Oncol 23: 257–264

Crooks I, Fallone B (1991) PC-based selective histogram equalization for contrast enhancement of portal films. Med Phys 18: 618

Cumberlin R, Rodgers J, Fahey F (1989) Digital image processing of radiation therapy portal films. Comput Med Imaging Graph 13: 227–233

Dally M, Hunter K, Wheat J, Leslie G, Fahey P, Hamilton C, Denham J (1992) Objective decision making following a portal film: the results of a pilot study. Radiother Oncol 24 (Suppl): S45

De Neve W, Van den Heuvel F, DeBeukeleer M et al. (1992) Routine clinical on-line portal imaging followed by immediate field adjustment using a tele-controlled patient couch. Radiother Oncol 24: 45–54

Droege R, Bjarngard B (1979a) Influence of metal screens on contrast in megavoltage X-ray imaging. Med Phys 6: 487–493

Droege R, Bjarngard B (1979b) Metal screen-film detector MTF at megavoltage X-ray energies. Med Phys 6: 515–518

Droege RT, Stefanakos TK (1985) Portal film technique charts. Int J Radiat Oncol Biol Phys 11: 2027–2031

Dutreix A, van der Schueren E, Leunens L (1992) Quality control at the patient level: action or retrospective introspection? Radiother Oncol 25: 146–147

El-Gayed A, Bartelink H, Bel A, Vijlbrief R, Lebesque J (1992) Evaluation of the time trend of setup deviations during the course of pelvic irradiation using an electronic portal imaging device. Radiother Oncol 24 (Suppl): S45

Evans P, Gildersleve J, Morton E et al. (1992) Image comparison techniques for use with megavoltage imaging systems. Br J Radiol 65: 701–709

Ezz A, Munro P, Porter A, Battista J, Jaffray D, Fenster A, Osborne S (1992) Daily monitoring and correction of radiation field placement using a video-based portal imaging system: a pilot study. Int J Radiat Oncol Biol Phys 22: 159–165

Faermann S, Krutman Y (1992) Generation of portal film charts for 10-MV x-rays. Med Phys 19: 351–353

Galbraith D (1989) Low-energy imaging with high-energy bremsstrahlung beams. Med Phys 16: 734–746

Galkin B, Wu R, Suntharalingam N (1978) Improved technique for obtaining teletherapy portal radiographs with high-energy photons. Radiology 127: 828–830

Gildersleve J, Dearnaley D, Evans P, Law M, Rawlings C, Swindell W (1992) A randomised trial of patient repositioning during pelvic radiotherapy. Presented at Megavoltage Portal Imaging, London

Goer D (1983) Radiation therapy treatment: the role of treatment aids and accessories IEEE Trans NS-30: 1784–1787

Gonzalez RC, Wintz P (1987) Digital image processing, 2nd edn. Addison-Wesley, Reading

Graham ML, Cheng AY, Geer LY, Binns WR, Vannier MW, Wong JW (1991) A method to analyze 2-dimensional daily radiotherapy portal images from an on-line fiber-optic imaging system. Int J Radiat Oncol Biol Phys 20: 613–619

Griffiths S, Pearcey R, Thorogood J (1987) Quality control in radiotherapy: The reduction of field placement errors. Int J Radiat Oncol Biol Phys 13: 1583–1588

Griffiths S, Khoury G, Eddy A (1991) Quality control of radiotherapy during pelvic irradiation. Radiother Oncol 20: 203–206

Gur D, Deutsch M, Fuhrman C, Clayton P, Weiser J, Rosenthal M, Bukovitz A (1989) The use of storage phosphors for portal imaging in radiation therapy: therapists' perception of image quality. Med Phys 16: 132–136

Halverson K, Leung T, Pellet J, Gerber R, Weinhous M, Wong J (1991) Study of treatment variation in the radiotherapy of head and neck tumors using a fiber-optic on-line radiotherapy imaging system. Int J Radiat Oncol Biol Phys 21: 1327–1336

Hammoudah M, Henschke U (1977) Supervoltage beam films. Int J Radiat Oncol Biol Phys 2: 571–577

Haus A, Pinsky S, Marks J (1970) A technique for imaging patient treatment area during a therapeutic radiation exposure. Radiology 97: 653

Haus A, Marks J, Griem M (1973) Evaluation of an automatic rapid- processable film for imaging during the complete radiotherapeutic exposure. Radiology 107: 697–698

Heukelom S, Lanson J, Mijnheer B (1992) In vivo dosimetry during pelvic treatment. Radiother Oncol 25: 111–120

Huizenga H, Levendag P, De Porre P, Visser A (1988) Accuracy in radiation field alignment in head and neck cancer: A prospective study. Radiother Oncol 11: 181–187

ICRU (1976) Errors in clinical dosimetry. Report 24: 45

Jaffray D, Battista J, Fenster A, Munro P (1993) X-ray sources of medical linear accelerators: Focal and extra-focal radiation. Med Phys 20: 1417–1427

Jakobsen A, Iversen P, Gadeberg C, Lindberg Hansen J, Hjelm-Hansen M (1987) A new system for patient fixation in radiotherapy. Radiother Oncol 8: 145–151

Johns HE, Cunningham JR (1983) The physics of radiology, Charles C. Thomas, Springfield, Ill.

Jones S, Boyer A (1991) Investigation of an FFT-based correlation technique for verification of treatment setup. Med Phys 18: 1116–1125

Kartha P, Chung-Bin A, Wachtor T, Hendrickson F (1977) Accuracy in radiotherapy treatment. Int J Radiat Oncol Biol Phys 2: 797–799

Kihlen B, Cederlund T, Lagergren C, Nordell B, Ruden B (1991) Improved portal film image quality in radiation therapy with high energy photons. Acta Oncol 30: 1–5

Kotre C, Marshall N, Faulkner K (1992) An alternative approach to contrast-detail testing of X-ray image intensifier systems. Br J Radiol 65: 686–690

Lam K, Ten Haken R (1991) Improvement of precision in spatial localization of radio-opaque markers using the two-film technique. Med Phys 18: 1126–1131

Lam K, Partowmah M, Lam W (1986) An on-line electronic portal imaging system for external beam radiotherapy. Br J Radiol 59: 1007–1013

Lam W, Herman M, Lam K, Lee D (1991) On-line portal imaging: computer-assisted error measurements. Radiology 179: 871–873

Lebesque J, Bijhold J, Hart A (1992) Detection of systematic patient set-up errors by portal film analysis. Radiother Oncol 23: 198

Leong J (1984) A digital image processing system for high energy X-ray portal images. Phys Med Biol 29: 1527–1535

Leong J (1986) Use of digital fluoroscopy as an on-line verification device in radiation therapy. Phys Med Biol 31: 985–992

Leong J, Stracher M (1987) Visualization of internal motion within a treatment portal during a radiation therapy treatment. Radiother Oncol 9: 153–156

Leszczynski K, Shalev S (1989) A robust algorithm for contrast enhancement by local histogram modification. Image and Vision Computing 7: 205–209

Leszczynski K, Shalev S (1993) Verification of radiotherapy treatments: Computerized analysis of the size and shape of radiation fields. Med Phys 20: 687–694

Leszczynski K, Shalev S, Cosby S (1990) A digital video system for on-line portal verification. SPIE 1231: 401–405

Leszczynski K, Shalev S, Cosby S (1992a) The enhancement of radiotherapy verification images by an automated edge detection technique. Med Phys 19: 611–622

Leszczynski K, Shalev S, Ryder S (1992b) A study on the efficacy of digital enhancement of on-line portal images. Med Phys 19: 999–1005

Leunens G, Verstraete J, Van den Bogaert W, Van Dam J, Dutreix A, van der Schueren E (1992) Human errors in data transfer during the preparation and delivery of radiation treatment affecting the final result: "garbage in, garbage out". Radiother Oncol 23: 217–222

Loewenthal E, Loewinger E, Bar-Avraham E, Barnea G (1992) Measurement of the source size of a 6- and 18-MV radiotherapy linac. Med Phys 19: 687–690

Marks J, Haus A (1976) The effect of immobilisation on localisation error in the radiotherapy of head and neck cancer. Clin Radiol 27: 175–177

McGee K, Shalev S (1993) Reduction of interpolation artifacts introduced by MHE in digital on-line portal images. Phys Med Biol 38: 601–614

Meertens H, van Herk M, Weeda J (1988) An inverse filter for digital restoration of portal images. Phys Med Biol 33: 687–702

Meertens H, Bijhold J, Strackee J (1990) A method for the measurement of field placement errors in digital portal images. Phys Med Biol 35: 299–323

Mitine C, Leunens G, Verstraete J, Blanckaert N, Van Dam J, Dutreix A, van der Schueren E (1991) Is is necessary to repeat quality control procedures for head and neck patients? Radiother Oncol 21: 202–210

Mok E, Feldmeier J (1988) Digital enhancement of treatment verification films using a low cost video digitizer with a personal computer. Phys Med Biol 33 (Suppl 1): 47

Morton E, Swindell W, Lewis D, Evans P (1991) A linear array, scintillation crystal-photodiode detector for megavoltage imaging. Med Phys 18: 681–691

Muller-Runkel R, Watkins S (1991) Introducing a computerized record and verify system: Its impact on treatment errors. Med Dosim 16: 19–22

Munro P, Rawlinson J, Fenster A (1988) Therapy imaging: source sizes of radiotherapy beams. Med Phys 15: 517–524

Munro P, Rawlinson J, Fenster A (1990) A digital fluoroscopic imaging device for radiotherapy localization. Int J Radiat Oncol Biol Phys 18: 641–649

Pizer SM, Amburn EP, Austin JD, et al. (1987) Adaptive histogram equalization and its variations. Comput Vision Graphics Image Process 9: 355–368

Pratt WK (1991) Digital image processing. Wiley, New York

Rabinowitz I, Broomberg J, Goitein M, McCarthy K, Leong J (1985) Accuracy of radiation field alignment in clinical practice. Int J Radiat Oncol Biol Phys 11: 1857–1867

RCA (1974) Electro-optics handbook. RCA, New Jersey, pp 216–217

Reinstein L (1986) Radiotherapy portal imaging quality. In: Kereiakes J, Elson H, Born C (eds) Radiation oncology physics. American Institute of Physics, New York, p 627 (AAPM medical physics monograph no. 15)

Reinstein LE, Orton CG (1979) Contrast enhancement of high-energy radiotherapy films. Brit J Radiol 52: 880–887

Reinstein LE, Durham M, Tefft M, Yu A, Glicksman AS (1984) Portal film quality: A multiple institutional study. Med Phys 11: 555–557

Reinstein L, Alquist L, Amols H, Lagueux B (1987a) Quantitative evaluation of a portal film contrast enhancement technique. Med Phys 14: 309–313

Reinstein L, Amols H, Biggs P, Droege R, Filimonov A, Lutz W, Shalev S (1987b) Radiotherapy portal imaging quality. American Institute of Physics, New York (AAPM report no.24)

Reinstein L, Shalev S, Leszczynski K, Cosby S, Meek A (1988) Megavoltage movies. Int J Radiat Oncol Biol Phys 15 (Suppl 1): 200

Roehrig H, Lutz W, Barnea G, Pond G, Dallas W (1990) Use of computed radiography for portal imaging. Proc SPIE 1231: 492–497

Schonemann PH, Carroll RM (1970) Fitting one matrix to another under choice of a central dilation and a rigid motion. Psychometrika 35: 247–255

Sephton R, Green M, Fitzpatrick C (1989) A new system for port films. Int J Radiat Oncol Biol Phys 16: 251–258

Shalev S (1992) The design and clinical application of digital portal imaging systems. Int J Radiat Oncol Biol Phys 24 (Suppl 1): 104

Shalev S, McGee K (1992) Evaluation of image processing techniques by ROC analysis. Med Phys 19: 822

Shalev S, Arenson J, Stewart M (1984) Digital enhancement of treatment verification films. Radiology 153(P):154

Shalev S, Cheng CW, Arenson J (1985) Port film enhancement by digital processing. SPIE 555: 103–108

Shalev S, Leszczynski K, Cosby S, Reinstein L, Meek A (1988) On-line verification of radiation treatment portals. Proc IEEE/EMBS New Orleans: 382–383

Shalev S, Lee T, Leszczynski K, Cosby S, Chu T, Reinstein L, Meek A (1989) Video techniques for on-line portal imaging. Comput Med Imaging Graph 13: 217–226

Shalev S, Leszczynski K, Cosby S, Agbay H (1990) Clinical implementation of on-line portal imaging. Med Phys 17: 554

Sherouse GW, Rosenman J, McMurry HL, Pizer SM, Chaney EL (1987) Automatic digital contrast enhancement of radiotherapy films. Int J Radiat Oncol Biol Phys 13: 801–806

Sherouse G, Bourland D, Reynolds K, McMurry H, Mitchell T, Chaney E (1990) Virtual simulation in the clinical setting: Some practical considerations. Int J Radiat Oncol Biol Phys 19: 1059–1065

Shiu A, Hogstrom K, Janjan N, Fields R, Peters L (1987) Technique for verifying treatment fields using portal images with diagnostic quality. Int J Radiat Oncol Biol Phys 13: 1589–1594

Smith V (1987) Routines for enhancement for radiation therapy images. In: Bruinvis I (ed) The use of computers in radiation therapy. Elsevier Science, North Holland, pp 37–40

Soffen E, Hanks G, Hwang C, Chu J (1991) Conformal static field therapy for low volume low grade prostate cancer with rigid immobilization. Int J Radiat Oncol Biol Phys 20: 141–146

Svensson G (1984) Quality assurance in radiation therapy: physics efforts. Int J Radiat Oncol Biol Phys 10 (Suppl 1): 23–29

Tait W, Hanson M (1989) Microcomputer processing of film radiographs. Br J Radiol 62: 613–619

Thornton A, Ten Haken R, Gerhardsson A, Correll M (1991) Three-dimensional motion analysis of an improved head immobilization system for simulation, CT, MRI, and PET imaging. Radiother Oncol 20: 224–228

van Dam J, Vaerman C, Blanckaert N, Leunens G, Dutreix A, van der Schueren E (1992) Are port films reliable for in vivo exit dose measurements? Radiother Oncol 25: 67–72

van Herk M, Meertens H (1988) A matrix ionisation chamber imaging device for on-line patient setup verification during radiotherapy. Radiother Oncol 11: 369–378

van Herk M, Bijhold J, Hoogervorst B, Meertens H (1992) Sampling methods for a matrix ionization chamber system. Med Phys 19: 409–418

van Tienhoven G, Lanson J, Crabeels D, Heukclom S, Mijnheer B (1991) Accuracy in tangential breast treatment set-up: a portal imaging study. Radiother Oncol 22: 317–322

Verhey L, Goitein M, McNulty P, Munzenrider J, Suit H (1982) Precise positioning of patients for radiation therapy. Int J Radiat Oncol Biol Phys 8: 289–294

Visser A, Huizenga H, Althof V, Swanenburg B (1990) Performance of a prototype fluoroscopic radiotherapy imaging system. Int J Radiat Oncol Biol Phys 18: 43–50

Webb S (1988) The physics of medical imaging. Adam Hilger, Bristol, pp 26–32

Weinhous M (1990) Treatment verification using a computer workstation. Int J Radiat Oncol Biol Phys 19:1549–1554

Weiser J, Gur D, Gennari R (1990) Evaluation of analog contrast enhancement and digital unsharp masking in low-contrast portal images. Med Phys 17: 122–125

Wilenzick R, Merritt C (1987) Megavoltage portal films using computer radiographic imaging with photostimulable phosphors. Med Phys 14: 389–392

Wong J, Binns R, Cheng A, Greer L, Epstein J, Purdy J (1990a) On-line radiotherapy imaging with an arrary of fiber-optic image reducers. Int J Radiat Oncol Biol Phys 18: 1477–1484

Wong J, Slessinger E, Hermes R, Offutt C, Roy T, Vannier M (1990b) Portal dose images. I. Quantitative treatment plan verification. Int J Radiat Oncol Biol Phys 18: 1455–1463

Wowk B, Radcliffe T, Shalev S, Rajapakshe R (1994) Optimization of metal/phosphor screens for on-line portal imaging. Med Phys 21: 227–235

Ying X, Geer L, Wong J (1990) Portal dose images. II. Patient dose estimation. Int J Radiat Oncol Biol Phys 18: 1465–1475

9 Computer-Controlled 3D Conformal Radiation Therapy

GERALD J. KUTCHER, RADHE MOHAN, STEVEN A. LEIBEL, ZVI FUKS, and C. CLIFTON LING

CONTENTS

---

[1] "Computer-controlled" usually implies that all phases of the treatment (except initiation) are implemented without human intervention, while "computer-aided" usually implies human intervention at different points in the treatment. We do not distinguish between the two in this paper and often use the term computer-controlled when either could apply

GERALD J. KUTCHER, PhD, RADHE MOHAN, PhD, C. CLIFTON LING, PhD, Department of Medical Physics, Memorial Sloan-Kettering Cancer Center, 1275 York Avenue, New York, NY 10021, USA; STEVEN A. LEIBEL MD, ZVI FUKS, MD, Department of Radiation Oncology, Memorial Sloan-Kettering Cancer Center, 1275 York Avenue, New York, NY 10021, USA

9.1 Introduction

Computer-controlled or -aided[1] radiation therapy is a technique with the potential to safely deliver radiation treatments more expeditiously and accurately than traditional methods. One of the most important applications is in the delivery of conformal treatments for which the goal of maximizing the dose to the tumor and minimizing the dose to normal structures often leads to treatment plans of complex design. Although conformal therapy has always been a goal of radiotherapy, until recently the type of treatment plans that could practically be implemented was quite restrictive. With recent advances in three-dimensional (3D) treatment planning, and (as we will attempt to demonstrate) with emerging techniques for computer-aided delivery of treatments, it appears likely that full-scale studies may be realized to test whether conformal therapy improves tumor control, decreases tissue toxicity, and ultimately enhances survival. In fact, recent work suggests that 3D conformal radiation therapy (3DCRT) should make it possible to increase tumor doses to levels beyond those feasible with traditional two-dimensional (2D) techniques (LEIBEL et al. 1992). Although improved tumor coverage and higher prescribed doses are likely to enhance local tumor control, this remains to be demonstrated.

Three-dimensional conformal radiation therapy tends to be more complex than traditional treatments and may consist of a large number of coplanar and non-coplanar fields, virtually all of irregular shape, and possibly with their intensity distributions modulated. Until recently, 3DCRT has been implemented with custom-fabricated cerrobend blocks and mechanical wedges and compensators, which makes treatments highly labor intensive. Consequently, conformal treatment plans have been restricted to a relatively small number of fields per fraction. To overcome this problem, computer-controlled treatment delivery systems are being developed which achieve field shaping with a multileaf collimator (MLC) and which will control all moving components of the treatment machine, as well as dose and dose rate.

Neither the delivery of treatments under computer control nor the MLC are new concepts. For example, Bjarngard and co-workers (BJARNGARD and KIJEWSKI 1976, 1978; LEVENE et al. 1978; CHIN et al. 1981) implemented computer-controlled therapy on a customized linear accelerator (Siemens Mevatron XII) in which gantry rotation, four independent jaws, collimator rotation, couch translations, turntable rotation, and dose rate were computer controlled. Each treatment was divided into groups of trajectories of the mechanical machine components in order to scan the radiation over selected volumes of the patient. The velocity of each component could be set precisely to ensure delivery of a requisite number of monitor units for each trajectory. The authors demonstrated the versatility of this system by delivering conformal dose distributions to highly complex target volumes. Although this scanning technique could be designed to modulate dose for almost any target shape, dose delivery often required a substantial amount of time. An alternative approach, which can potentially deliver conformal dose distributions in times comparable to traditional treatments, is to shape the dose distribution with MLCs. Several investigators in Japan fabricated and employed MLCs for treatments in the 1960s and 1970s (see for example TAKAHASHI 1965; UEDA 1969; MATSUDA et al. 1980), although their value was limited due to the large leaf widths and the small field sizes. Moreover, the trajectories of the MLC leaves were cam driven rather than controlled by digital computers.

To some extent the methods and technology developed by these early investigators were ahead of their time since 3D radiation treatment planning and computer-controlled radiation treatment delivery are essential for proper implementation of conformal therapy. Until recently, 3D treatment planning systems had not evolved to levels where they could be used to design treatments to take advantage of computer-controlled treatment machines. Over the last decade, the explosive growth in computer technology, coupled with the wider availability of CT and MR imaging devices, has enabled several institutions to apply 3D treatment planning. In addition, there have been a number of significant advances in the field of engineering – for example, the development of powerful miniaturized stepping motors have made it practical to design and fabricate MLCs with small leaf widths.

The automated delivery of 3DCRT may be classified into two types. In one, dynamic therapy, irradiation takes place while the components of the treatment machine are in motion. In the second, multisegment therapy, the treatment session is divided into a sequence of fixed fields, called "segments," which are set up and treated in succession under computer control, without human intervention (an offshoot and simpler version would be MLC segmented therapy with human intervention between segments). From a practical point of view, dynamic therapy can be adequately approximated by a sequence of segments. Furthermore, segmented therapy is technically simpler to implement since it is not necessary to synchronize the mechanical motions with radiation nor with one another – as we shall see, the latter is important for collision avoidance. It should be noted that intensity modulation can be achieved with both approaches.

We first discuss in this paper the biologic and clinical rationale of 3DCRT. After discussing the rationale, we attempt to demonstrate that 3DCRT is not practical without computer control. In the central sections we review in some detail one approach for computer-controlled dose delivery, namely multisegment therapy. This is followed by a description of the characteristics and potential of MLCs for computer-controlled 3DCRT. We then discuss the role and importance of on-line imaging for monitoring and correcting computer-controlled 3DCRT. In the last sections we discuss safety issues: for computer-controlled accelerators per se and for multisegment therapy.

9.2 Biologic Rationale for 3DCRT

The ultimate aim of 3DCRT is to improve tumor control probability by improving the spatial distribution of dose relative to that achievable with traditional techniques. This could lead to a more homogeneous target dose distribution[2] with less normal tissue irradiation; consequently, the prescribed dose can be increased, and local control improved. However, it has been argued that modification of the dose distribution and/or increased tumor dose may not be sufficient to increase local control due to inherent tumor radioresistance and patient population heterogeneity (DEACON et al.

---

[2] 3DCRT is also more effective when an inhomogeneous target dose distribution is desired (e.g., when there are critical structures abutting the target volume)

1984; PETERS et al. 1981). These factors may significantly flatten the dose-response curves such that a significant escalation in dose would be required for a modest improvement in local control (THAMES et al. 1991). On the other hand, the shallowness of the clinically observed dose-response curves may be related to underdosed regions within the tumor or marginal misses. If these deficiencies in treatment can be reduced, it may be more likely that an effect of increased dose on local control will be observed.

It has also been argued that even if 3DCRT improves local control, this improvement would be limited by the subsequent appearance of metastatic disease. While this may be true in some cases, in others it has been argued with the support of clinical and laboratory data that improved local control leads to increased survival. One possibility is that the probability of local control and metastatic disease are independent so that survival probability would increase in proportion to local control (SUIT 1982, 1992). Another suggestion, based upon retrospective clinical studies on the patterns of failure after curative local therapy, is that distant metastasis may arise from the regrowth of the "un-eradicated" tumor (FUKS et al. 1991a,b; LEIBEL et al. 1991a). Thus improved local control will decrease the rate of metastatic spread and improve survival.

The preceding discussion emphasizes the need for the complete eradication of the primary tumor and suggests a testable hypotheses: using 3DCRT to escalate the dose to the tumor, one would observe increased local control and improvement in metastatic-free survival.

9.3 Clinical Studies of 3DCRT

To address the above hypothesis in clinical studies, appropriate disease sites should be chosen in which, ideally, local control is inadequate with conventional radiotherapeutic methods. Also, a tumor dose-response relationship should be anticipated such that the impact of improved local control on outcome can be tested (LEIBEL et al. 1991a). Studies have been performed to assess whether 3DCRT leads to improved dose distributions with the potential of dose escalation for locally advanced nasopharynx and prostate tumors and non-small cell lung carcinoma (LEIBEL et al. 1991b; ARMSTRONG et al. 1993; LEIBEL et al. 1993; TEN HAKEN et al. 1989). In these studies 3DCRT has been shown to lead to improved target coverage and reduced normal tissue doses when compared to traditional treatments.

9.3.1 Nasopharynx

The advantages of 3DCRT in patients with carcinoma of the nasopharynx has recently been demonstrated for ten newly diagnosed and five previously irradiated patients with locally recurrent disease (LEIBEL et al. 1991b). New patients received 50.4 Gy via parallel opposed fields to the primary tumor and the cervical lymph nodes followed by 19.8 Gy via 3DCRT to the region of the primary tumor. Patients with locally recurrent disease received their entire treatment with 3DCRT. Two teams planned the patient independently: one developed 2D plans, while the other designed 3D plans with beam's eye view (BEV) techniques (MOHAN et al. 1988) using generally five to nine posterior, lateral, and posterior-oblique fields. The 2D and 3D plans were ranked for each patient, according to 3D dose distributions, dose-volume histograms (DVHs), estimated tumor control probabilities (TCPs) (GOITEN 1987), and normal tissue complication probabilities (NTCPs) (KUTCHER and BURMAN 1989; KUTCHER et al. 1991a).

With few exceptions, the 3D plans had better target coverage. For example, the volume of the target receiving less than 95% of the prescribed dose was reduced on the average by 15%, and similar results were demonstrated for the 90% level (LEIBEL et al. 1991b). Moreover, the mean tumor dose for the 3D plans was increased on the average by 13%: this was the result of less underdosing of the target and a redistribution of the high-dose regions from normal tissues (for 2D plans) to within the target volume (for 3D plans). Consequently, the dose to lateral tissues (mandible, temporomandibular joints, and parotid glands) was significantly improved, and the volume of normal tissues receiving greater than 80% of the prescribed dose was reduced to one-half of that for the 2D technique. Using the probability of uncomplicated control as a criterion (KUTCHER 1992), Fig. 9.1 demonstrates that 3DCRT yielded a significant improvement in therapeutic effect for each of the ten patients tested.

9.3.2 Prostate

LEIBEL et al. (1993) have reported on a study to explore the feasibility of increasing the target dose in patients with carcinoma of the prostate. Ninety patients were treated with 3DCRT with minimum tumor doses of 66.6 Gy, 70.2 Gy, and 75.6 Gy in 40, 34, and 16 patients, respectively.

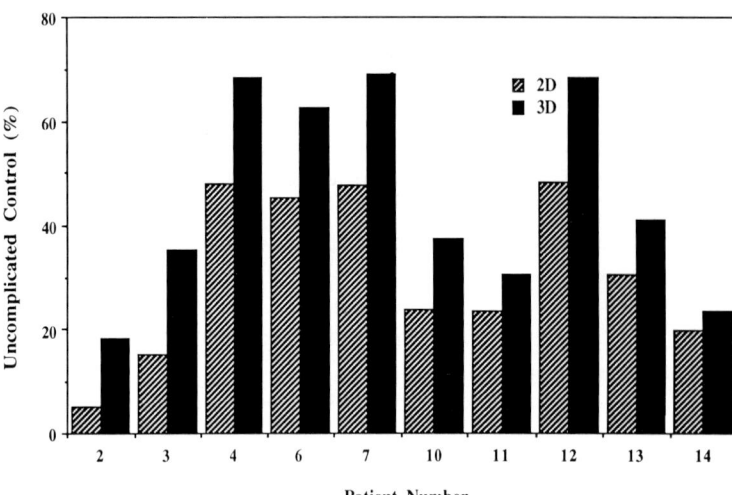

Fig. 9.1. The probability of uncomplicated control [the product of TCP for the nasopharynx and (1-NTCP) for all normal organs] for 3D and traditional treatment plans of primary tumors of the nasopharynx for ten patients. The parotid gland is excluded from the calculation since NTCP is near unity. (Redrawn from LEIBEL et al. 1991b, where further details can be found)

A six-field technique (two lateral and four oblique coplanar fields) was used in a majority of the patients (78/90) for dose escalation based upon a dosimetric analysis of target and normal tissue DVHs. Although four-, six-, and eight-field techniques were capable of uniformly irradiating the prostate, there were differences in rectal and femoral head dose distributions. While the four-field technique irradiated the least volume of the rectum, the femoral head dose would be unacceptably high if the prescribed dose were increased to 80 Gy. Consequently, the six-field plan was chosen to balance rectal and femoral head irradiation. Other techniques, including shrinking fields and non-coplanar methods, will need to be explored as the limit of dose escalation is approached.

The patients were assessed for acute morbidity (according to criteria established by RTOG) with a median follow-up time of 12 months (range 3–38 months). Only 27% of these patients had grade 2 acute morbidity, requiring medication for symptomatic relief. Two patients had higher grade acute genitourinary toxicity. The incidence of grade 2 or higher acute toxicity is about one-half of that expected with conventional approaches (SOFFEN et al. 1991). No late complications have been observed.

Complete tumor regression as evaluated by digital rectal examinations was observed in all the patients. Post-treatment serum prostate specific antigen (SPA) concentrations normalized in 61% of patients (55/90) and progressively decreased in an additional 27% (24/90) and did not normalize in 8/90 patients. The median time to normalization was 2 months (range 1–16 months). Normalized PSA concentrations, which subsequently rose to abnormal levels, were observed in three of 90 patients. Six of the 11 patients with elevated PSA concentrations developed bone metastases.

9.3.3 Lung

ARMSTRONG et al. (1993) have compared 3D and 2D treatment techniques in nine patients with non-small cell lung cancer treated to a prescribed dose of 70.2 Gy for local disease and 50.4 Gy for elective nodal volumes. 3DCRT was shown to yield reduced doses to normal tissues while target volume coverage was comparable to traditional techniques. For example, the average volume of lung receiving 25 Gy or more was reduced by 11% and 51% for the ipsilateral and contralateral lungs, respectively. And for some patients, the volume of contralateral lung receiving 25 Gy or more was reduced from 25% to 5% or less. In addition, the average volume of the esophagus receiving 60 Gy was reduced by 25% and NTCPs (BURMAN et al. 1991; KUTCHER and BURMAN 1989) were reduced by 11%. These results suggest that dose can conceivably be increased by approximately 25%.

9.3.4 Dose-Escalation Studies

Based upon these preliminary investigations, LEIBEL et al. (1992) have reported on studies whose aim is to establish the maximum tolerable doses that can be delivered to patients with T3 and T4 carcinoma of the nasopharynx, T1–T4, N0–N2 non-small cell lung carcinoma, and stage B2 or C carcinoma of the prostate. These studies (presently in progress) are designed to escalate the dose in 5.4 Gy increments provided the incidence and severity of treatment-related morbidity are no worse than with conventional radiotherapeutic techniques. The primary endpoints are late toxicity for nasopharynx and prostate and late and acute toxicity (since acute pneumonitis is life threatening) for the lung. As previously discussed, these dose-escalation studies represent a first step in a program to establish the relationship between high-dose 3DCRT, local control, metastatic outcome, and survival.

9.4 Computer-Controlled Multisegment Therapy

9.4.1 Rationale

It should be apparent, even for the preliminary studies described in Sects. 9.1–9.3, that 3DCRT requires more fields than traditional treatments. Future developments are likely to lead to treatment plans with an even larger number of fields and with an increased emphasis on non-coplanar beam arrangements. For example, we have been exploring the use of a ten-field non-coplanar technique to treat the prostate (BURMAN et al. 1992) which consists of four anterior-oblique fields arranged on a conical arc centered on the prostate, four posterior-oblique fields similarly arranged, and two lateral fields.

Clearly, the delivery of conformal plans using traditional static fields, cerrobend blocks, and mechanical field modifiers is not really practical: the treatments take too long, are more error prone, and are ultimately too costly. Moreover, non-coplanar techniques will almost certainly require that the source to tumor distance exceeds 100 cm to permit clearance between the patient and the head of the accelerator. Consequently, it will be necessary to change the height and lateral position of the couch for different gantry positions, leading to more complex dose delivery.

The time necessary to implement a treatment session is an especially important consideration. For example, if it takes a therapist 2 min to enter the treatment room, change the blocks, wedges, and gantry angle and return to the treatment console and if the initial patient setup takes 10 min, then a ten-field treatment plan would be delivered with traditional technology in approximately 30 min. If the treatment also included multiple couch motions for non-coplanar fields, and collimator rotations, it is likely that the time would be significantly increased. While it is possible to use traditional methods to deliver 3DCRT when the number of patients and fields are limited (KUTCHER et al. 1993), this is not realistic for large-scale studies, much less as a routine form of therapy.

One approach is to treat a limited number of fields per fraction (e.g., four per fraction in a 12-field plan). While this could reduce the time per fraction substantially, it creates problems of treatment design and delivery. The distribution of dose to the tumor would depend upon which subset of fields were treated, and the dose per fraction to normal tissues would be significantly altered. Consequently, it would be necessary to develop treatment planning techniques to minimize the change in the target dose distribution when using a limited number of fields per fraction. Moreover, the increased dose per fraction to nontarget structures would become an additional treatment variable. In addition, the difficulty of delivering non-coplanar techniques would not be directly addressed. It would appear that this approach might prove more viable, for example, in proton therapy, where the number of fields could be more limited and where intensity modulation is usually mandatory.

An alternative approach is to use computer-control to deliver 3DCRT. While a number of solutions are possible, we will describe one method, multisegment therapy. It has the potential of delivering treatments in times comparable to traditional techniques. For example, the ten-field plan previously described could be delivered in approximately 15 min–10 min for initial patient setup, 2–3 min to confirm that it is collision free (virtual treatment), and another 2–3 min for treatment. Moreover, the treatments can be implemented precisely as designed and prescribed[3] to deliver such treatments safely. For example, there are methods, as will be described, of preventing collisions between treatment machine components as well as between the treatment machine and the patient.

Multisegment 3DCRT can be implemented using different approaches. We will describe a model designed and implemented on the Scanditronix MM50 (MAGERAS et al. 1992), although it can potentially be applied with some modifications to other accelerators. In this model there are two primary computers. One controls all operations of the machine in

---

[3] The term "prescribed" or "prescription" is used in a general sense in this paper. It includes all intended settings of a treatment machine as specified by a physician and/or indicated by a treatment plan, rather than just the dose or monitor units

conjunction with numerous microprocessors internal to the treatment machine: we call this the control computer (CC). This computer communicates with a second computer, called the external computer or host computer (HC), which has access to patient data, results of treatment planning, and the corresponding treatment prescription data. The software running on the HC utilizes the results of 3D treatment planning to compose the treatment prescription for all segments of the multisegment treatment and "downloads" it onto the CC. For each patient there may be one or more single-segment or multisegment treatment. The operator then sets up the patient and the treatment machine under the control of the CC, stepping through each segment using a specially designed hand pendant. If necessary, the settings of any segment may be adjusted or any parameters not already defined in the downloaded prescription may be set up manually. Once this is accomplished a "virtual treatment" is performed, in which the machine components are moved through all segments of the actual treatment but without irradiation, thereby allowing the operator to check for possible collisions.

Upon the completion of patient setup in the treatment room, the CC sends to the HC a list of any parameter values that were modified or manually set during setup. The HC verifies these values against the corresponding prescribed values. If the discrepancy exceeds a predefined tolerance, it is reported to the user, who specifies whether the changed value should be used for this treatment only, or for subsequent treatments as well. The CC delivers treatment, segment by segment, pausing momentarily (after the segment is set up and before the beam goes on) for an authorization from the HC. During this pause the HC checks whether the current machine settings are consistent with the values at the time of initial verification. Irradiation commences if the consistency check is successful. The treatment is automatically aborted (terminated) if the consistency check fails. The CC notifies the HC every time radiation begins and ends and the HC records the monitor units delivered and the reason for termination of radiation. Special provisions are available for recovery from faults and prematurely terminated treatments.

9.4.2 Treatment Planning for 3DCRT

Planning of 3DCRT requires the availability of functions not generally found in standard 3D radiation treatment planning (3DRTP) systems. At a minimum, the 3DRTP system must incorporate MLCs and produce output data suitable for communication to the treatment machine. We describe some desirable features that we have developed.

For each beam, one or more apertures are drawn. For example, one of the apertures may frame the entire target volume while others may also shield specific normal structures. Each aperture is formed automatically by including the tumor volume and a margin and, optionally, excluding an overriding normal structure and its margin, if any (BREWSTER et al. 1991, 1992a). Automatic aperture definition is desirable for treatment plans involving large numbers of beams.

Each aperture shape is then converted to leaf settings of the MLC with leaves positioned relative to the aperture boundary in one of three ways: the entire edge of each leaf may be just inside the aperture boundary, the entire edge of each leaf may be just outside the aperture boundary, or each leaf may intersect the aperture at its midline. Initial dosimetric studies indicate that the last of these schemes may be the most suitable one (MOHAN 1992; LOSASSO et al. 1992). We have implemented an improvement of this scheme in which the area of overlap of each leaf and the aperture is exactly equal to the area of the exposed normal tissue between the leaf and the aperture.

Certain convoluted apertures cannot be approximated by an MLC. Therefore, it may be necessary to divide the aperture into two or more smaller apertures or, alternatively, to accept exposure to larger regions of normal tissues outside the aperture. In addition, depending upon the shape of the aperture, optimization of collimator angle and couch translational positions and jaw positions may be required to minimize the volume of the normal tissue exposed.

Once the treatment plan is approved and the target dose prescribed, the treatment plan is converted into a format suitable for transmittal to the treatment machine. Each aperture corresponds to a segment of an automated treatment. Each segment consists of a set of parameter values including gantry angle, collimator angle, jaw settings, couch positions, leaf positions, wedge or electron applicator, modality, energy, dose rate, and monitor units. The values of all parameters for the first segment are specified in absolute units. The absolute settings of some of the parameters (e.g., couch parameters and possibly gantry and collimator angles) for the first segment may vary somewhat from day-to-day due to differences in patient positioning. However, their values for the subsequent segments relative to the first segment remain constant and are, therefore, specified in relative units. Note also that for first-time treatment, the values of couch settings for the first segment are, in general, undefined. They are defined after the completion of the first day's setup and are recorded in the prescription. The couch positions for the first segment of subsequent treatments is known accurately only if the patient is immobilized and the immobilization device is securely and reproducibly fastened to the couch.

9.4.3 Treatment Setup and Virtual Treatment

The downloading of the treatment plan is initiated at the treatment machine terminal attached to the CC. Following downloading, the HC waits for the treatment setup to be completed and to receive any modifications to the prescribed values.

Treatment setup is carried out entirely under the control of the CC using a specially designed hand pendant interacting with software resident on the CC (shown in Fig. 9.2 is the MM50 hand pendant). A monitor in the treatment room displays current machine settings and status (e.g., whether computer-controlled setup is in progress or is complete). For safety reasons, since the first segment positions of the couch and gantry are designated as "manually set," the operator is required to move the couch and gantry without computer control to their initial locations. Automatic setup of the other parameters for the first segment (e.g., leaves, collimator angle, jaws, etc.) is initiated by depressing the appropriate button on the hand pendant.

To set up the next segment, the operator presses another button and all parameters change simultaneously. If necessary, individual parameters (including leaf positions) may be adjusted using buttons and switches. After one or more parameters have been modified, the current values may be saved or the settings restored to the values stored in the CC. In this fashion the operator may step through the setup segment by segment to check and modify the treatment parameters. After the last segment the operator returns the treatment machine to the first segment under computer control. At any time during the setup, the operator may cancel the treatment and load another treatment prescription.

To check for a possible collision, a virtual treatment is required: the exposed components of the treatment machine, that is, couch, gantry, and head,

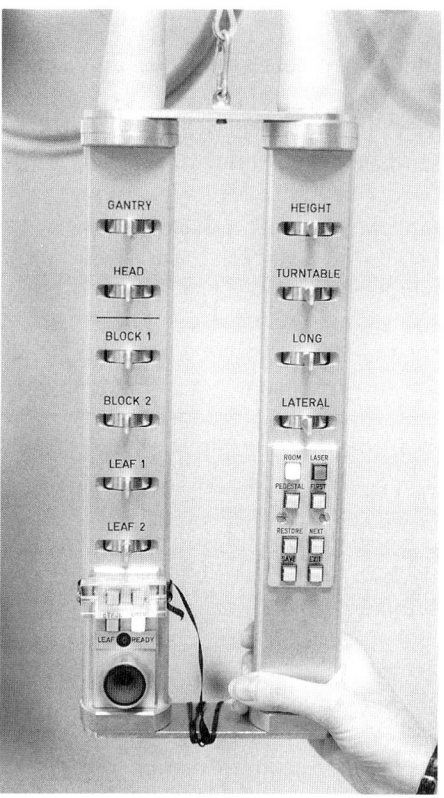

Fig. 9.2. Specially designed hand pendant on the Scanditronix MM50, which is used for multisegment therapy

undergo changes as in a regular treatment – but without radiation. To initiate a virtual treatment, the operator presses the appropriate button on the hand pendant while keeping the dead-man's switch depressed. The machine moves from one segment to the next, pausing momentarily between segments. For most fractions, except for the first, where parameter adjustments are necessary, the operator (after positioning the patient on the couch) proceeds directly to the virtual treatment.

After the virtual treatment is complete, the operator returns the treatment machine to the first segment and indicates that setup is complete by pressing the associated button on the hand pendant. The CC checks that the treatment machine is set up according to the first segment values stored on the CC. It then sends a list of any parameter values that were modified or defined during setup to the HC for verification and waits for permission to proceed with treatment.

9.4.4 Treatment Verification and Consistency Check

The HC first verifies that the treatment machine settings are in agreement with the prescribed settings (within tolerance limits) that were determined either manually or were modified during setup. In effect, verification requires the operator to confirm any modified or manually set values that lie outside their tolerances. Verification is not performed for those treatment parameters whose prescribed values are downloaded from the HC and are set up automatically without modification.

If a verification failure occurs, the operator must take a corrective action by: (a) overriding the failure if the tolerance is too tight or if this is a one-time only change in the treatment; (b) changing the setting back to the prescribed one; or (c) modifying the prescription if this setting is to be used for the current and subsequent fractions. The operator may also cancel the treatment and download another treatment.

If corrective actions include any changes in the treatment settings, the operator must perform another virtual treatment. This is followed by another verification attempt. If further verification failures occurs, the process is repeated.

Once verification is complete, treatment is initiated by pressing a button on the treatment machine console. Before radiation commences for the first segment, the CC seeks the permission of the HC again. The HC performs a consistency check by comparing the current machine settings with those stored at the time of the most recent verification (which, as noted above, may not be the same as the prescribed settings). The tolerances for consistency check, which are determined by the reproducibility of the machine parameters, are much tighter than for verification. As a further precaution the CC additionally performs its own consistency check.

This consistency check is repeated prior to radiation for each segment. It also occurs if there is a pause during irradiation due to a machine malfunction, as a consequence of patient motion, or another emergency. In this case, the settings at the resumption of treatment are compard with those at the beginning of the pause. If a consistency check failures occurs, the treatment should be aborted and resumed as a separate treatment.

9.4.5 Treatment Delivery, Termination, and Recording

Each time radiation starts and stops, the CC notifies the HC, which uses this information to determine the number of segments completed, the current segment, and the next segment. When the treatment terminates (as opposed to pausing temporarily) either normally at the completion of all segments, or abnormally because of treatment termination by the therapist or a machine malfunction, the HC composes and logs a treatment record, termination status (normal or abnormal), and the number of monitor units delivered. This information is used for resuming a prematurely terminated treatment and to prevent accidental retreatment.

9.4.6 Resumption of Prematurely Terminated Treatments

To complete a prematurely terminated treatment, the HC composes a temporary prescription in which the segments already treated are set to zero monitor units, and the partially treated segment is set to the remaining monitor units. This scheme simplifies the patient setup and ensures that every treatment always starts with the first segment setup and traverses through all the segments sequentially, according to the previously designed collision-free path.

9.4.7 Manual Recording of Treatments

Manual recording of treatments (or beam film exposures) is necessary any time a failure prevents the automatic recording of a completed or a prematurely terminated treatment due to, for example, a breakdown of the HC or CC. In the event that the HC becomes unavailable during treatment, the CC completes the current segment, detects the failure, and asks the operator whether the remaining segments (if any) should be treated. After the treatment of segments is ended (either by canceling the remaining segments or by allowing them to finish), the operator can manually record the treatment on the HC when it becomes available. The only entry required is the total number of monitor units delivered. From this information the HC software computes whether the treatment was completed normally, or the segment in which termination occurred.

In the event of a CC breakdown during treatment, the HC, which is waiting for a signal from the CC that radiation has terminated, receives no response. When the HC, either through an operator action or automatically, determines that the CC is not responding, it informs the operator and requests the monitor units that were delivered at the time treatment stopped. The HC provides the last recorded number of monitor units delivered as the default value.

9.4.8 Portal Films and Portal Images

Segments for portal films or portal images (when using an on-line imager) are processed essentially like treatment segments with the following special features: images are always taken with x-rays (usually with the lowest energy photons available); electron applicators (with arbitrary collimator settings) are allowed; the number of monitor units must be less than a predefined limit. Typically, there are two exposures for each portal image, one with the treatment MLC and jaw settings, the other with the MLC and jaws retracted to a large rectangular field. The other geometric parameters (gantry, couch, collimator angle) are the same as for the corresponding treatment segment. When the operator selects a segment for imaging, a two-segment portal image prescription is automatically generated corresponding to the two exposures. For portal films, there is a pause between film segments to allow the operator to remove or install blocks, change films, etc.

Alternatively, portal images may be taken with geometric settings that are different from any of the treatment segments. For example, a portal image may be taken from two different (orthogonal) gantry angles. In this case, there is a single portal image segment for each gantry angle.

9.5 Multileaf Collimation

As should be evident from the previous discussion, one of the essential components of multisegment therapy is automated field shaping with MLCs. The general features of MLCs of four manufacturers is reviewed in Table 9.1. Of these commercially available MLCs, all consist of 50 or more leaves each driven by an independent motor. Another important design feature pertains to the alignment of the leaf edge to the radiation source: two are singly focused

Table. 9.1 Features of MLCs of four manufacturers

	Scanditronix	Varian	Siemens	Phillips
X-ray jaws	1 pair (upper)	2 pairs	2 pairs	2 pairs
No. of leaves	32 × 2	26 × 2	27 × 2	40 × 2
Leaf width at isocenter	1.25 cm	1.00 cm	1.00 cm	1.00 cm
Max. travel over center	5 cm	16 cm	10 cm	12.5 cm
Max. field size	32 × 40 cm^2	26 × 40 cm^2	16 × 16 cm^2	40 × 40 cm^2
Leaf focusing	Double	Single	Double	Single
Transmission				
Through leaves	< 1.5%	< 2%		< 2%
Between leaf sides	< 3%	< 2%		< 3.1%
Between leaf ends	< 1.5%	25%–30%		

with respect to the radiation source, i.e., in a direction perpendicular to leaf motion, and two are doubly focused, i.e., in directions perpendicular and parallel to leaf motion. For a singly focused system, the leaves move in a plane perpendicular to the central ray and therefore it is necessary to shape the end of the leaves to reduce the penumbra. In principle, this solution produces a penumbra which is smaller than an unfocused straight edge, but larger than a focused design. In practice, the difference appears to be small since other factors come into play (e.g., flattening filter design, leaf transmission, distance of leaves from the target). For example, for 6-MV x-rays the "true" penumbra of the Varian MLC is approximately 1 mm larger than the lower x-ray jaws (LOSASSO et al. 1992). The focusing also appears to affect the transmission through the ends of the leaves. [Measurements for the Scanditronix and Varian units (CHUI 1992) show higher transmission for the single focus design when the leaves are abutted at the center of the field. Presumably, the higher transmission through the Varian MLC is also due to the leaves stopping short, perhaps to prevent collission. The transmission can be reduced by abutting the MLCs off the center of the field.]

Another important parameter to consider is the width of the leaves, which vary from 1.0 cm to 1.25 cm (in the plane of the isocenter) for the MLCs listed in the table. when the ends of the leaves are staggered (as would occur in a typically shaped irregular field) the finite leaf width produces a staircase pattern of radiation in the region of the 50% isodose curve. This situation may be described by defining an "effective penumbra" for an MLC which would be larger than

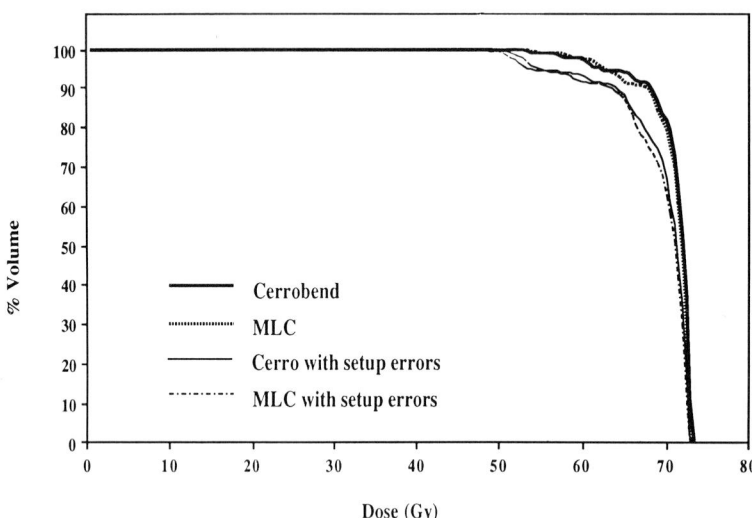

Fig. 9.3. DVH for the prostate for a six-field conformal plan prescribed to a minimum dose of 70 Gy for MLC and cerrobend treatment plans. The solid lines represent the nominal treatment plan without setup errors. The dotted lines represent the average DVH when random setup errors with a standard deviation of 0.5 cm are included. (Redrawn from LoSasso et al. 1992)

that produced by cerrobend or an MLC with a very small leaf width. Consequently, there is a possibility that treatment plans with MLCs may be degraded in comparison to plans using continuous field apertures. Losasso et al. (1992) have investigated this problem for multiple field treatments. After verifying with film that MLC dose distributions could be accurately predicted, treatment plans were generated for both MLC and cerrobend apertures for a prostate and the nasopharynx patient using the field arrangements described in Sects. 9.3.1 and 9.3.2. For MLC fields, each leaf was positioned to transect the intended field aperture such that an equal area of each leaf was inside and outside the aperture. Treatment plans with MLC and cerrobend apertures were compared using 3D dose distributions, DVHs, TCPs, and NTCPs. In addition, dose distributions which incorporated the effects of patient setup uncertainty were generated to provide a potentially more realistic representation of a full course of therapy. For these calculations it was assumed that the setup errors were randomly distributed with a standard deviation of 0.5 cm.

Figure 9.3 shows DVHs of a prostate target volume for a six-field plan using either a cerrobend aperture or the MLC of the Scanditronix MM50. It is evident for the six-field conformal plan that there is little or no perceptible difference in the DVHs between MLC and cerrobend, although the spatial distributions (not shown) still exhibit some structure reflecting the finite leaf width, especially for the 50% isodose level. If setup errors are included, then the cerrobend and MLC isodose distributions are virtually identical. Moreover, it is also evident from the DVHs that setup errors have a significant impact

on the treatment (as will be described further in Sect 9.6), while the difference between MLC and cerrobend does not.

The role of MLC leaf width for treatments with a small numbers of fields (e.g., parallel-opposed) has yet to be adequately addressed. Although it is unlikely that smaller leaf widths (e.g., 0.5 cm) will prove necessary, further studies are warranted.

An additional consideration is the potential use of MLCs for field modulation where leaf width and leaf design may play an important role. In this respect, distance of travel of the leaves across the midline to the opposite side is important since this dictates the size of the target volume which can be modulated. Futhermore, to reduce transmission through adjacent leaves, which are focused, the leaves may be notched with a tongue and groove arrangement. In this circumstance, it has been suggested that because of the variation in transmitted radiation in the vicinity of the tongue and groove, the use of leaves for field modulation may be compromised (GALVIN and SMITH 1991). However, alternate strategies for the trajectory of the leaves than those used by Galvin should circumvent some of these problems (CONVERY and ROSENBLOOM 1992).

9.6 On-line Imaging of Computer-Controlled 3DCRT

While portal films are traditionally used for treatment verification, on-line imaging is more appropriate for computer-controlled 3DCRT. First with on-line systems, it is possible to image all segments during treatment (i.e., in verification mode) if the imager has been mounted and shown to be free of collision during the virtual treatment. With verification films, it is not possible to do this without stopping treatment and entering the room. Second, it is possible to perform double-exposure images prior to treatment by identifying the segments to be imaged, or adding additional segments if nontreatment segments are to be imaged (e.g., orthogonal open field images). An important consequence of this approach is that the overall treatment time would be only marginally increased. A double-exposure treatment field image takes the time to move the MLCs from the aperture settings to a rectangular setting and back and the exposure time.

Although MLC positions for each segment should be checked by independent and redundant systems, on-line verification images provide a further check on the position of the leaves. Moreover, the aperture can be detected by computer methods (MEERTENS et al. 1990). In addition to direct monitoring, there are other potential benefits of on-line systems for 3DCRT. For example, it is possible to systematically study setup errors and patient motion during treatment – and to use the measurements to assess the dosimetric consequences. And, it may also be possible to develop computer-assisted techniques which provide the means for technologists, after the identification of internal anatomic features on double-exposure on-line images, to correct patient setup errors prior to treatment.

9.6.1 Effects of Setup Errors on 3DCRT

Although there are a number of studies which attempt to assess the magnitude of setup errors (e.g.,

see BIJHOLD et al. 1992; EZZ et al. 1992; HUNT et al. 1992; MEERTENS et al. 1990; RABINOWITZ et al. 1985), data on their effects on conformal plans are quite sparse (BIJHOLD et al. 1992; HUNT et al. 1992). In our analysis of setup errors for the nasopharynx (HUNT et al. 1992), we found that 3D conformal plans were more sensitive to setup errors than traditional plans, and that nominal treatment plans (i.e., those generated without taking setup errors into consideration) are an overly optimistic representation of the actual dose distributions. To measure setup errors, two double-exposure portal films, approximately 90° apart, were taken before each fraction for six patients with nasopharyngeal tumors. Five to seven landmarks were identified on each film, and the translation and rotation of the patient were calculated using a least squares algorithm (HUNT et al. 1992). For each fraction, 3D dose distributions were calculated for each fraction using the measured position of the radiation isocenter relative to the patient.

Both systematic and random setup errors were found, and each had an average uncertainty in the position of the isocenter of approximately 2.5 mm. The dosimetric effect of these errors is demonstrated in Fig. 9.4, which shows DVHs for the brainstem of a conformal multifield treatment plan for one of the six patients. Included are the nominal DVH (i.e., the planned treatment without accounting for setup errors), the mean DVH for all treatments, and the range of DVHs for each fraction (i.e., the maximum and minimum DVHs).

One striking feature is that the nominal plan is an overly optimistic estimate of the intended treatment. As indicated by the range of daily DVHs and the mean DVH, almost all daily treatments irradiate a

Fig. 9.4. DVHs for the brainstem with and without setup errors for a conformal multifield treatment plan of the nasopharynx. The setup error for each day of the boost phase was measured and the resulting dose distribution for the conformal plan calculated (HUNT et al. 1992). Shown are the nominal DVH (without setup errors), the range of measured daily dose distributions, i.e., the minimum and maximum DVHs, and the average DVH for all treatments. (Redrawn from HUNT et al. 1992)

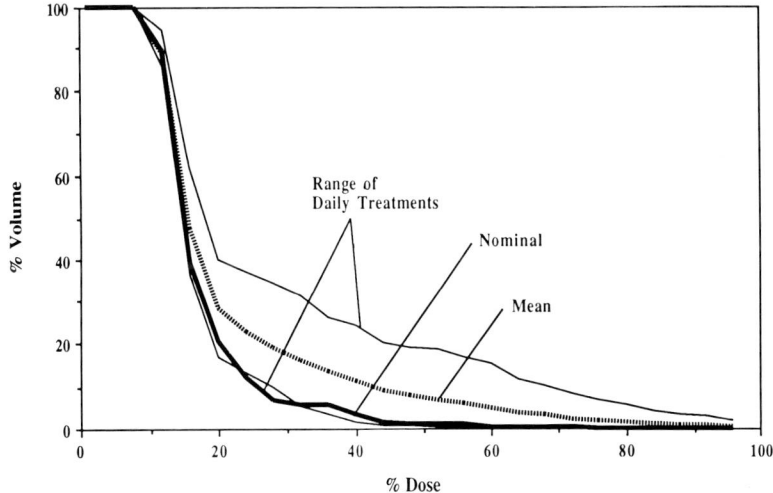

larger volume of the brainstem (at each dose level) than the planned treatment. This result appears to be a consequence of the fact that almost any setup error leads to an increase in dose to the brainstem since the dose distribution in this case surrounds the brainstem on three sides. In contrast, if the patient were treated with a traditional parallel opposed plan (not shown), the range of daily dose distributions would be closer to the nominal or intended plan. Similar results and conclusions follow when DVHs for target volume are analyzed (HUNT et al. 1992).

9.6.2 Monitoring and Correction of Setup Errors in Multisegment Therapy

An extension of the technique described above might be used to correct setup errors prior to treatment. First, at least two segments are identified for double-exposure imaging; these may be open field orthogonal images or two or more of the treatment segment. After virtual treatment and under computer control images are obtained. While the remainder of the session (the treatment per se) is halted, the therapist identifies pre-specified anatomic structures (points and curves) on the images, or these are determined automatically, or some combination of both is employed. Using computer techniques, the structures are compared to the planned configuration. If the deviation between the intended and measured positions of the identified structures exceeds some tolerance limits, the computer calculates the gantry, collimator, and couch positions needed to rectify the error. To be practical, this process should take no more than 2 or 3 min.

If the corrections are within safety tolerance limits, the gantry, collimator, and couch positions are modified under computer control. With the model of multisegment therapy described here, it is appropriate to modify the couch, gantry, and collimator positions for the first segment in order to correct the setup error since the coordinates of all subsequent segments are relative to the first segment. Tolerance limits on what constitutes a safe modification in the initial treatment parameters (i.e., that would not lead to a collision) would need to be determined.

Such an approach might be used to correct setup errors throughout treatment (irrespective of whether they are random or systematic) or to correct only systematic errors during the initial phase of the treatment, and then to use daily imaging to assure that the expected random errors are not exceeded.

The dosimetric consequences of these different approaches as well as the effects of other uncertainties (e.g, organ motion) may be studied analytically. We have developed two methods to incorporate uncertainties into treatment plans. For rapid calculations, the mean dose distribution can be obtained via convolution (CHUI et al. 1992). This approach is suited to interactive planning, but the average DVH does not represent the range of the dosimetric uncertainty as demonstrated in Fig. 9.4. Consequently, some additional measures of uncertainty are required, for example confidence limits in dose. To incorporate confidence limits in treatment plans we have developed a method in which the spatial distribution of setup errors is sampled, and for each sample point a dose distribution is calculated and both the mean and percentiles in dose are generated (KUTCHER et al. 1991b). For example, assuming a gaussian distribution in setup errors for the prostate, it is possible to determine the range in dosimetric uncertainties (e.g., 10% and 90% confidence limits in dose) as a function of the frequency of imaging (KUTCHER et al. 1991b). From such results it is possible to determine an optimum frequency of imaging. This is important since double-exposure images add unwanted dose to nontarget structures.

9.7 Safety Issues

9.7.1 General Safety Considerations for Computer-Controlled Accelerators and Computer-Controlled Therapy

The use of digital logic, microprocessors, and software for the monitoring and control of accelerators and automated treatments (e.g., multisegment 3DCRT) raises a number of safety concerns. For example, on multimodality accelerators there have been incidents in which patients were overdosed when these units were unintentionally operated at dose rates in the neighborhood of 3000 Gy/min (AAPM 1993). Faults of this type can occur if the electron gun produces the high electron currents required for x-ray therapy and the electron scattering foil, rather than x-ray target and flattening filter, is positioned in the beam. Such errors may be due to mechanical malfunction, software design errors, or software bugs arising from operating the software in an unanticipated environment.

It has been recommended (AAPM 1993) that medical accelerators should be designed so that the likelihood of a failure which could lead to serious medical consequences (e.g., death or injury due to overdose or collisions) is less than 5×10^{-6} per patient per treatment. Unfortunately, it is difficult to estimate failure rates: unanticipated errors do occur even when presumably large additional "safety factors" are included in the design of complex systems. It therefore behooves the manufacturer and the user to follow accepted procedures for the design, documentation, testing, commissioning, training, and quality assurance of computer-controlled systems. While there are guidelines for the design and testing of computer-controlled medical accelerators for manufacturers (ANSI/IEEE 1984; IEC 1981; FDA 1987, 1988), the number of published procedures for acceptance testing and quality assurance of such systems by medical physicists is quite limited (AAPM 1993; WEINHOUS et al. 1990). The FDA appears to be the major source of standards for manufacturers who want to market computer-controlled accelerators in the United States. Briefly, for computer-controlled systems these may be presented as follows (WEINHOUS 1991): software programming standards, which include structure, documentation, coding, and testing; life cycle, which includes design, implementation, testing, installation, operation, and maintenance; quality assurance, which includes items like auditing; and software test methodologies, which cover items like module testing, systems testing, and security.

In this section we briefly review (AAPM 1993) some aspects of this problem, and in the next section we describe some unique safety issues apropos of multisegment computer-controlled therapy.

1. *Manufacturer's testing and documentation.* The design and testing of software for computer-controlled accelerators and treatment systems should include verification of performance, quality assurance, and hazards analyses. In addition, the systems should be designed so that failure, due to built-in errors or bugs, or introduced by the environment, will always lead the system to a safe state. The software interface should include isolation between the clinical and service modes. Moreover, the manufacturers should produce documentation for the user on the overall design philosophy and provide detailed descriptions of all subsystems and input–output functions. In addition, the manufacturer should provide detailed procedures for verifying specifications, safety testing, recommended testing for upgrades, and suggested quality assurance procedures.

2. *Acceptance testing and commissioning.* The user should verify that the system performs according to specifications. It is important to thoroughly test the user interface and try to cause the system to go into an unsafe state by extensively exercising the interface. Interlock testing should be performed during acceptance testing, since exercising certain interlocks may require the manufacturer's participation. In the event of a computer or computer-hardware-related failure, the user should confirm that the system returns to a safe state. Unfortunately, the number of defects in computer-controlled software[4] that may be resident in the software – even after undergoing extensive tests by the manufacturer and user – can still be significant (JACKY and WHITE 1990; SCHMUHL 1985). However, the number of such errors is expected to decline in updated versions following extensive field use.

3. *Quality assurance.* There is a necessity to test the software, over and above hardware quality assurance, for computer-controlled accelerators and computer-controlled treatment delivery systems on a scheduled basis. It is possible that latent software bugs could arise as a consequence of the aging of mechanical machine components, or that corruption of data sets or code could occur with time. Checksums or other methods of verifying that the code and data files have not changed should be reviewed regularly.

The testing of computer-controlled systems after software or hardware upgrades is problematic since the extent of the tests required to assure safe operation may not be clear. The manufacturer should recommend which subsystems are to be tested after an upgrade and whether (or to what extent) the interlock system also should be retested.

4. *Therapist training.* Training of therapists (and other personnel) is important since computer-controlled systems are generally more complex to operate than electromechanical systems, and the likelihood of mishaps is expected to decline as the therapist's understanding of the system grows. Therapists should have: an overview of the computer-controlled accelerator (and computer-controlled delivery system, if available) and an understanding of all control operations that are invoked from the console and keyboard outside the room and from the

---

[4] The number of bugs are a consequence of many issues including, the number of lines of source code (which has been estimated for computer-controlled acelerators to be as high as 3×10^6) and the number of logical brances in the code – assuming 1500 two-way control-transfer statements, this would produce 3.5×10^{451} possible execution pathways (WEINHOUS 1991)

switches, hand pendants, keyboards, and other hardware residing inside the treatment room. Furthermore, the therapists should be fully versed in emergency and safety procedures. The importance of keen participation and awareness during treatment (e.g., viewing the patient on a video monitor during treatment delivery) is even more important for automated and semiautomated treatments, where there may be a greater potential for collision.

9.7.2. Safety Issues in Multisegment Therapy

Computer-controlled treatment delivery itself introduces additional concerns due to the fact that treatment machine motions and irradiation occur under computer control without active human intervention. Consequently, it is important to have safeguards to ensure data and software security, integrity and validity of the treatment, and to protect against possible collisions. The following is a list of some of the approaches which would improve the safety of multisegment therapy, particularly with respect to collision avoidance.

1. *Access by individuals.* Access to treatment planning and delivery software should be restricted to qualified and authorized individuals.

2. *Restricted access to the control computer (CC) and the host computer (HC) from other computers or terminals on the network.* All communication between the CC and HC must be initiated by the CC, with the exception of the queries for independent monitoring of the machine. Similarly, communication over the network must be initiated only by the CC. This requirement reduces the chance that unauthorized instructions or other data will be successfully transferred to the CC.

3. *Independent prescriptions.* To reduce the chance of treating a patient with corrupted data, a second independent copy of all treatment prescriptions is maintained on the CC, and these are checked against each other continually (approximately every 2 s) throughout the course of setup and treatment.

4. *Patient immobilization.* The patient should be fixed with respect to an immobilization device which itself is fixed to the couch. This maintains the position of the patient relative to all machine components between and during treatments.

5. *First segment setup.* The couch and the gantry for the first segment must be manually setup by the therapist at their prescribed positions within specified tolerances. This requirement is imposed to eliminate the possibility of a collision between a

potentially arbitrary starting position and the first segment.

6. *Virtual treatment.* A virtual treatment should be carried out before each treatment to assure that the technique is safe; for example, to verify that a planned or inadvertent adjustment in any treatment parameter would not lead to collision.

7. *Independent and redundant readouts and monitoring of all movable components.* On the Scanditronix MM50, for example, hardware circuits utilize the primary readouts to control motions, whereas the CC and a second computer independently monitor the primary and secondary readouts.

8. *Sequential, rather than simultaneous, motion of couch and gantry.* The couch and gantry motion are always assigned to separate segments; therefore, one of them must correctly reach their final prescribed position before the other begins to move. Although a collision could still occur if there was an error in reading or transmitting the final position of the couch or gantry, this possibility is substantially reduced if there is redundant monitoring of mechanical positions.

The use of separate segments for couch and gantry has other desirable features. Since the trajectory is only defined by its endpoints, it is not necessary to monitor intermediate positions. This is a fundamentally difference from dynamic therapy, where all parameters may change simultaneously and a knowledge of the trajectory in time of all components is required. Consequently with multisegment therapy, the requirement that the treatment machine hardware have the capability to guarantee reproducible trajectories can be somewhat relaxed. Finally, patients may be less nervous and disoriented if couch and gantry motions do not occur simultaneously. One disadvantage of separate couch segments is that the treatment time is longer, especially if each segment requires couch motions.

9. *Computer simulation.* 3D graphical simulation software may be useful for detecting and eliminating potential collisions during the planning process, and for ensuring the feasibility of the setup – particularly for non-coplanar treatments. While a potential collision would be discovered during the initial virtual treatment, it should be eliminated in the planning process in the interest of efficiency and safety.

10. *Collision Sensors.* Mechanical or other type (e.g., optical) of sensors attached to parts of the treatment machine can be used to monitor and stop potential collisions between the machine and patient. Although it is unlikely that any practical

arrangement of mechanical devices could prevent all possible collisions, strategic placement of the sensors (e.g., on the lower face of the collimator and applicators) could prevent certain likely encounters between machine and patient.

11. *Collision avoidance software*. It is be possible to develop computer models of the treatment machine and patient. During simulation and treatment machine parameters are monitored and a distance of closest approach between the envelopes of the patient and machine calculated. If these are less than some tolerance, machine motion is disabled. As above, this approach cannot prevent all collisions since, for example, the readouts may fail and transmit false information of the position of the machine. Yet in combination with other strategies, collision avoidance software presents another layer of protection.

12. *Therapist involvement*. It is important to design computer-controlled systems so that the therapists are fully involved in all phases of the treatment (e.g., manual setup of the first segment, virtual treatment) since human intervention at key steps is an important safety feature.

9.8 Conclusions

We have reasoned that 3DCRT may lead to improved dose distributions to the target volume and normal tissues, and that dose escalation may be initiated with the expectation that higher prescribed doses can be achieved. However, because of the increased number of fields and the potentially more complex field arrangements required, more sophisticated dose delivery systems are needed. We have outlined one approach, computer-controlled multisegment therapy, which is expected to reproducibly deliver treatments in times comparable to traditional techniques. We have also discussed the importance of accounting for and correcting setup and other uncertainties in order to better realize the potential of 3DCRT. We have also described some of the safety issues. We believe that the multisegment approach for treatments without couch motion is at least as safe as traditional rotational treatments. With couch motions there are further concerns. One important avenue towards reducing the likelihood of collision is to assign couch and gantry motions to separate segments. Further gains are possible if hardware and software collision avoidance systems are added to these treatments.

Multisegment 3DCRT is a tool that could be used to test whether 3DCRT will improve local control and survival. At the least, it already appears that 3DCRT leads to improved dose distributions and reduces the toxic effects of treatment. Furthermore, computer-controlled 3DCRT will have the practical consequence of reducing the treatment time per fraction even for simpler traditional treatments (although to a lesser extent than multifield complex treatments). The latter reason alone may indeed prove the driving force for the implementation on a broad scale of computer-controlled or assisted therapy with MLCs

Acknowledgments. We would like to thank Dr. G. Mageras for helpful discussions on computer-controlled treatment delivery. This work was supported in part by grant CA 54749 from the National Cancer Institute, Department of Health and Human Services, Bethesda, Maryland.

References

AAPM (1993) Medical accelerator safety considerations. Report of Task Group 35, Radiation Therapy Committee of AAPM. Med Phys 20: 1261–1275

ANSI/IEEE (1984) Software quality assurance plans. Standard 730–1984. IEEE, New York

Armstrong JG, Burman C, Leibel SA, et al.(1993) Conformal three dimensional treatment planning may improve the therapeutic ratio of high dose radiation therapy for lung cancer. Int J Radiat Oncol Biol Phys 26: 685–689

Bijhold J, Lebesque JV, Hart AM, Vijlbrief RE (1992) Maximizing setup accuracy using portal images as applied to a conformal boost technique for prostatic cancer. Radiother Oncol (in press)

Bjarngard BE, Kijewski PK (1976) The potential of computer control to improve dose distributions in radiation therapy. In: Sternick ES (ed) Computer applications in radiation oncology. The University Press of New England, Hanover, N.H.

Bjarngard BE, Kijewski PK (1978) Computer controlled radiation therapy. In: Orthner FH (ed) Proceedings of the 2nd annual symposium on computer applications in medical care. IEEE Comp Soc, Long Beach, California, pp 86–92

Brewster L, Mageras G, Mohan R (1991) Automatic generation of beam aperture shapes. Med Phys 18: 610

Brewster L, Mageras G, Mohan R (1992a) Automatic generation of beam aperture shapes. Med Phys 20: 1337–1342

Brewster L, Mohan R, Mageras G, Burman C, Leibel S, Fuks Z (1992b) Incorporation of multileaf collimation in clinical 3D conformal treatment planning (abstract). Int J Radiat oncol Biol Phys 24. Supplement 1: 157

Burman C, Kutcher GJ, Emami B, Goitein M (1991) Fitting of normal tissue tolerance data to an analytic function. Int J Radiat Oncol Biol Phys 21: 123–135

Burman C, Kutcher GJ, Zalefsky M, Leibel SA (1992) 3-dimensional treatment planning of the prostate with non-coplanar field arrangements (abstract). Med Phys 19: 842

Castro JR, Chen GT, Blakely EA (1985) current considerations in heavy charged particle radiotherapy. Radiat Res (Suppl 8): PS 227–234

Chin LM, Kijewski PK, Svensson GK, Chaffey JT, Levene MB, Bjarngard BE (1981) A computer -controlled radiation therapy machine for pelvic and paraaortic nodal areas. Int J Radiat Oncol Biol Phys 7: 61

Chui CC, Kutcher GJ, LoSasso T (1992) A convolution method for incorporating uncertainties in dose calculation (abstract). Med Phys 19: 814

Convey DJ, Rosenbloom ME (1992) The generation of intensity-modulated fields for conformal radiotherapy by dynamic collimation. Phys Med Biol 37: 1359–1374

Deacon JM, Peckham MJ, Steel GG (1984) The radio-responsiveness of human tumours and the initial slope of the cell survival curve. Radiother Oncol 2: 317–323

Ezz A, Munro P, Porter AT et al. (1992) Daily monitoring and correction of radiation field placement using a video-based portal imaging system: a pilot study. Int J Radiat Oncol Biol Phys 22: 159–165

FDA (1987) Technical reference on software developmental activities. United States Food and Drug Administration

FDA (1988) Draft: reviewer guidance for computer-controlled medical devices. United States Food and Drug Administration

Fuks Z, Leibel SA, Wallner KE, et al. (1991a) The effect of local control on metastatic dissemination in carcinoma of the prostate: long term results in patients treated with $_{125}$I implantation. Int J Radiate Oncol Biol Phys 21: 549–566

Fuks Z, Leibel SA, Kutcher GJ, et al. (1991b) Three dimensional conformal treatment: a new frontier in radiation therapy. In: DeVita VT Jr, Hellman S, Rosenberg SA (eds) Important advances in oncology. Lippincott, Philadelphia, pp 151–172

Galvin J, Smith AR (1991) Dosimetric characteristics of a multileaf collimator system (abstract). Int J Radiat Oncol Biol Phys 21 (Suppl): 184

Goitein M (1985) Calculation of the uncertainty in dose delivered during radiation therapy. Med Phys 12: 608–612

Goitein M (1987) The probability of controlling an inhomogeneously irradiated tumor. In Zink S (ed) Evaluation of treatment planning for particle beam radiotherapy. National Cancer Institute, Bethesda, pp 5.8.1–5.8.17

Hunt MA, Kutcher GJ, Burman C, et al. (1993) The effect of positional uncertainties on the treatment of nasopharynx cancer. Int J Radiat Oncol Biol Phys 27: 437–447

IEC (1981) Particular requirements for the safey of medical electron accelerators in the range 1 MeV to 50 MeV. Revision of publication 601–2–2: Medical electrical equipment part 1

Jacky J, White CP (1990) Testing a 3-D radiation therapy planning program. Int J Radiat Oncol Biol Phys 18: 253-261

Kutcher GJ (1992) Quantitative plan evaluation In: Purdy JA (ed) Advances in radiation oncology physics. Proceedings of the 1990 AAPM Summer School, Lawrence, Kansas, Amer Inst Phys, New York, pp 998–1021

Kutcher GJ, Burman C (1989) Calculation of complication probability factors for non-uniform normal tissue irradiation: the effective volume method. In J Radiat Oncol Biol Phys 16: 1623–1630

Kutcher GJ, Burman C, Brewster L, Goitein M, Mohan R (1991a) Histogram reduction method for calculating complication probabilities for 3D treatment planning evaluations. Int J Radiat Oncol Biol Phys 21: 137–146

Kutcher GJ, Chui C, LoSasso T (1991b) Incorporation of set-up uncertainties in treatment plan calculations (abstract). Int J Radiat Oncol Biol Phys 21: 123

Kutcher GJ, Leibel SA, Mohan R (1993) Advances in precision treatment: some aspects of 3D conformal radiation therapy.

In: Meyer JA (ed) Frontiers in radiation therapy and oncology. Karger, Basel, pp 209–226

Leibel SA, Ling CC, Kutcher GJ, et al. (1991a) The biological basis of conformal three-dimensional radiation therapy. Int J Radiat Oncol Biol Phys 21: 805–811

Leibel SA, Kutcher GJ, Harrison LB, et al. (1991b) Improved dose distributions for 3D conformal boost treatments in carcinoma of the nasopharynx. Int J Radiat Oncol Biol Phys 20: 823–833

Leibel SA, Kutcher GJ, Mohan R et al. (1992) Three dimensional conformal radiation therapy at the Memorial Sloan-Kettering Cancer Center. Semin Radiat Oncol 2: 274–289

Leibel SA, Heinmann R, Kutcher GJ, et al. (1994) Three-dimensional conformal radiation therapy in locally advanced carcinoma of the prostate: preliminary results of a phase I dose-escalation study. Int J Radiat Oncol Biol Phys 28: 55–65

Levene MB, Kijewski PK, Chin LM, Bjarngard BE, Hellman S (1978) Computer-controlled radiation therapy. Radiology 129: 769

LoSasso TJ, Chui CS, Rebo I, et al. (1992) The use of multi-leaf collimator for conformal radiotherapy in carcinomas of the prostate and nasopharynx. Int J Radiat Oncol Biol Phys 25: 161–170

Mageras GS, Podmaniczky KC, Mohan R (1992) A model for computer- controlled delivery of 3D conformal treatments. Med Phys 19: 945–954

Matsuda T, Matsuoka A, Inamura K (1981) Computer controlled multi-leaf conformation radiotherapy. In: Umegaki Y (ed) proceedings of the 7th international conference on computers in radiation therapy 1980. Japan Radiological Society, Tokyo, Japan. pp 298–302

Meertens H, Bijhold J, Strackee J (1990) A method for the measurement of field placement errors in digital portal images. Phys Med Biol 35: 299–323

Mohan R (1992) Secondary field shaping, asymmetrical collimators and multileaf collimators. In: Purdy JA (ed) Advances in radiation oncology physics. Proceedings of the 1990 AAPM Summer School, Lawrence, Kansas, Amer Inst Phys, New York, pp 307–345

Mohan R, Podmaniczky KC, Caley R, Lapidus A, Laughlin JS (1984) A computerized record and verify system for radiation treatment. Int J Radiat Oncol Biol Phys 10: 1975–1985

Mohan R, Barest G, Brewster LJ, et al. (1988) A comprehensive three-dimensional radiation treatment planning system. Int J Radiat Oncol Biol Phys 15: 481–495

Peters LJ, Withers HR, Thames HD, et al. (1981) Keynote address – The problem: Tumor radioresistance in clinical radiotherapy. Int J Radiat Oncol Biol Phys 8: 101–108

Podmaniczky KC, Mohan R, Kutcher GJ, Kestler C, Vikram B (1985) Clinical experience with a computerized record and verify system. Int J Radiat Oncol Biol Phys 11: 1529–1537

Rabinowitz I, Broomberg J, Goitein M, et al. (1985) Accuracy of radiation field alignment in clinical practice. Int J Radiat Oncol Biol Phys 11: 1857–1867

Schmuhl EH (1985) Software QA for critical care systems. In: Proceedings of the Seventh Annual Conference on the Engineering of Medicine and Biology Society. IEEE, New York, pp 171–174

SoVen EM, Epstein BE, Hunt MA, et al. (1991) Decreased acute morbidity with conformal static field radiation therapy treatment of early prostate cancer as compared to non-conformal techniques (abstract). Int J Radiat Oncol Biol Phys 21 (Suppl 1): 152

Suit HD (1982) Potential for improving survival rates for the cancer patient by increasing the efficacy of treatment of the primary lesion. Cancer 50: 1227–1234

Suit H (1992) Local control and patient survival. Int J Radiat Oncol Biol Phys 23: 653–660

Takahashi S (1965) Conformation radiotherapy rotation techniques as applied to radiography and radiotherapy of cancer. Acta Radiol (Suppl), p 242

Ten Haken RK, Perez-Tamayo C, Tesser RJ, McShan DL, Fraass BA, Lichter AS (1989) Boost treatment of the prostate using shaped fixed beams. Int J Radiat Oncol Biol Phys 16: 193–200

Thames HD, Schultheiss TE, Hendy JH, et al. (1991) Can modest escalations of dose be detected as increased tumor control? Int J Radiat Oncol Biol Phys 22: 241–246

Ueda T (1969) Conformation technique in linear acceleration therapy. I.S.R.R.T. 4th World Congress, Tokyo, Japan, p 69

Weinhous MS (1991) Quality assurance of computer-controlled accelerators. In: Smith A (moderator) Symposium: problems and promises of computer controlled accelerators. AAPM Annual Meeting, Med Phys 18: 672

Weinhous MS, Purdy JA, Granda CO (1990) Testing of a medical linear accelerator's computer-control systems. Med Phys 17: 95–102

10 External Beam Stereotactic Radiosurgery Physics

Michael C. Schell and Andrew Wu

CONTENTS

Michael C. Schell, PhD, University of Rochester Cancer Center, Department of Radiation Oncology, 601 Elmwood Avenue, Box 647, Rochester, NY 14642-8647, USA; Andrew Wu, PhD, Division of Radiation Oncology, Allegheny General Hospital, Medical Center of Pennsylvania, Pittsburgh, PA 15212, USA

10.1 Introduction

Radiosurgery or stereotactic radiosurgery (SRS) is defined as the irradiation of intracranial lesions with a single fraction of focused small ionizing radiation beams, such as x-rays or gamma rays, eliminating the need for conventional invasive surgery. Stereotactic radiation therapy (SRT) is the treatment of intracranial lesions with the stereotactic apparatus and multiple fractions. A stereotactic frame allows for rigid immobilization of the patient and accurate localization of the target. The goal of radiosurgery is to locate and define the intracranial lesion and deliver single or multiple doses to the target with small x-ray beams without exceeding the radiation tolerance of normal tissues adjacent to the target volume. Radiosurgery was initiated in 1950 by Lars Leksell to treat dysfunctional intracranial abnormalities originally using orthovoltage x-rays in conjunction with a stereotactic frame of his own design (Leksell 1951). He later used protons and developed the dedicated unit (gamma knife) of 201 cobalt-60 sources converging on a single point in space. Other institutions have employed a variety of beam modalities including helium ions, heavy charged particles, and x-rays (linac-based radiosurgery designs) to irradiate benign and malignant lesions (Barcia-Salario et al. 1985; Betti et al. 1989; Colombo et al. 1987; Fabrikant et al. 1984; Kamerer et al. 1988; Larson et al. 1990; Larson et al. 1974; Marin-Grez 1988; Nedzi et al. 1990; Rand et al. 1987; Schwade et al. 1990).

10.2 Rationale

Radiosurgery has been used to treat benign and malignant lesions and functional disorders. Kihlstrom (1986) reported 1311 gamma unit radiosurgical procedures between 1968 and 1986, the most frequent categories being arteriovenous malformations (AVMs) (41%), acoustic neuroma (14%), and functional radiosurgery (14%). Chierego et al.

(1988) listed 150 patients treated with a linac-based system, the most frequent causes of treatment being AVM (44%) and malignancy (33%). Inoperable AVMs may be favorably influenced by radiosurgery, as discussed below. The risk of hemorrhage is 3% per year and 5% immediately following hemorrhage.

Radiosurgery as the sole treatment modality, with its dose localization and single fraction characteristics, is contraindicated in the treatment of primary malignant intracranial lesions, where tumor cells are known to infiltrate beyond the borders of abnormalities seen on computed tomography (CT) or magnetic resonance imaging (MRI) (HALPERIN et al. 1989, HOCHBERG and PRUITT 1980, WALLNER et al. 1989). Implicit in the isoeffect curves of KJELLBERG (1986) as a function of dose and volume is the need for multiple fractionation. The single fraction iso-complication dose decreases as the tumor volume increases. Clearly, the tumor dose must increase to achieve the same control rate. Hence, the normal tissue response increases the need for fractionation as the tumor volume increases. The role of radiosurgery in radiation oncology may be analogous to that of interstitial brain implants as a high-dose boost following the standard course of external beam therapy (HALPERIN et al. 1989).

10.3 Physics of Radiosurgery

This chapter focuses on linac-based radiosurgery for several reasons. The cost of charged-particle therapy limits the application to a few facilities. The gamma knife is a dedicated unit which allows tertiary care institutions with large patient loads to accommodate up to ten patients per week. The gamma knife differs from the linac in photon energy, dose delivery, and collimator size (4, 8, 14, and 18 mm) (WU et al. 1990). Field diameters on the linac typically step from 10 mm to 40 mm in 2.5-mm steps. The dose distribution changes with the arc geometry on the linac and collimator plug pattern on the gamma knife (FLICKINGER et al. 1990). However, the dose gradients in normal tissue have been shown to be comparable (PODGORSAK et al. 1989). Linac-based applications are less costly for most radiation oncology departments and therefore more common (BETTI et al. 1989; COLOMBO et al. 1987; FREIDMAN and BOVA 1989; HARTMANN et al. 1985; LUTZ et al. 1988; PODGORSAK et al. 1988). Consequently, the emphasis is on the linear accelerator approach.

Linac-based radiosurgery was first suggested in 1974 by LARSSON et al. (1974), and initial treatments in the United States were commenced in 1984 by PATIL (1989). Subsequently many treatment techniques and geometries have been developed (HARTMANN et al. 1985; LUTZ et al. 1988; NEDZI et al. 1990; PODGORSAK et al. 1988). Most approaches employ tertiary collimators to produce a well-localized x-ray beam. The stereotactic frame can be either couch mounted or affixed to a floor-mounted pedestal. The typical arc geometry involves a gantry rotation for fixed table (patient support assembly) angles. Dynamic radiosurgery was developed at McGill University and consists of the simultaneous rotation of the gantry and patient support assembly (PSA) or couch during irradiation. The PSA rotates from −75° to 75° while the gantry rotates from −150° to 150°. Dynamic radiosurgery avoids the overlap of beam entry and exit paths. This constraint minimizes the dose to normal tissues. Other configurations include the continuous rotation of the patient in a revolving treatment chair (McGinley) and angulation of the treatment chair relative to the incident beam (BETTI et al. 1989).

10.3.1 Radiosurgery Apparatus

Figure 10.1 illustrates the hardware design of LUTZ et al. (1988). The tertiary collimator system consists of a coaxial tube which holds cylindrical collimators of fixed diameters from 12.5 mm to 40 mm at isocenter. A BRW pedestal is mounted to the base plate of the PSA and the origin of the frame system (pedestal) is aligned with the isocenter of the linac. When the target coordinates are set on the pedestal and the patient/frame is attached, the target is then located at the focus of the arc geometry. The principal advantage of the Lutz approach is the verification of the target alignment with the x-ray beam.

10.3.2 Tertiary Collimation

Fixed cylindrical collimators provide more accurate collimation of the beam, a spherical dose distribution from arcs about isocenter, and a sharper dose falloff relative to secondary collimators. The principal benefits are increased accuracy in beam collimation and a steeper penumbra (SMITH et al. 1993). The field size tolerances are ±1mm at isocenter with each of the upper and lower jaws; thus a 10-mm field can vary between 8 mm and 12 mm with secondary collimation. Secondly, the tertiary collimator yields a drastically improved dose gradient relative to arc geometries with secondary collimation. The penumbra (80%–20%) is reduced by 20% with the addition of tertiary collimation.

Fig. 10.1. The Joint Center hardware configuration, with the BRW pedestal mounted on the PSA base plate. The stereotactic frame is attached to the stand with the target positioned at the locus of the rotation axes of the gantry, table, and collimator

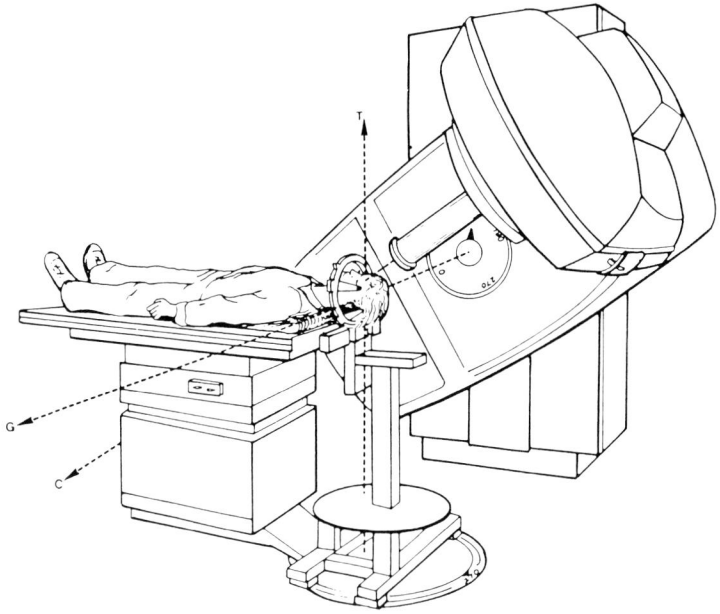

10.3.3 Stereotactic Frame Systems

Stereotactic frame systems are designed for rigid fixation to the skull of the subject. Several systems allow for repeat fixation and at least two commercial products are noninvasive. Frame systems were originally designed for sterotactic biopsies with an accuracy of ± 2 mm. The patient position during the imaging sequence is frequently different from that in surgery. Uncertainty in the target position is affected by two factors: flexion of the frame and movement of the brain within the cranium. The brain has slight negative buoyancy in the cerebrospinal fluid and can shift slightly relative to changes in the orientation of the skull. Consequently, the weight of the patient applies different stresses to the frame during the two procedures. GALLOWAY et al. (1991) have studied and reported on the uncertainties in localization and neurosurgery for four stereotactic frame systems. Unlike neurosurgery, the patient treatment and imaging positions are usually identical in radiosurgery. Hence, the frame deformations from the patient are similar in each procedure and the uncertainty in target localization is minimized.

10.4 Target Localization

The target is localized with a combination of the above imaging modalities. Dose delivery to the target requires that a minimum of three frames of reference be accurately correlated: the CT scan data set, the stereotactic frame system, and the linac (DE-SALLES et al. 1987; OLIVIER et al. 1987; PETERS et al.

1987). CT and positron emission tomography (PET) provide transverse scans of the patient by transporting the patient through a detector array. The scans are characterized by slice thickness, slice separation, and voxel size. The voxel size is normally used to estimate the uncertainty in the target localization. The uncertainty in the couch position is frequently overlooked and often exceeds the uncertainty introduced by the voxel dimensions. The impact of the couch reproducibility on target localization is negated if the scan-frame coordinate transformation is derived from the slice with the target center.

10.4.1 Tomographic Localization

The scan or image is a plane in both the image and frame space. The localizers are composed of vertical and oblique rods which enable the user to determine correspondence between the two sets of three points in each space. The derivation of the coordinate transformation from the image coordinate system to the BRW coordinate system has been previously published (SAW et al. 1987; WEAVER et al. 1990). The resolution of the rods in the scan (rod diameter relative to pixel dimensions) directly determines the uncertainty target coordinates in BRW space.

10.4.2 Angiographic Localization

A localizer box with four fiducial markers on each of the four acrylic panes is attached to the frame during angiography (Fig. 10.2). AVMs and fiducial markers

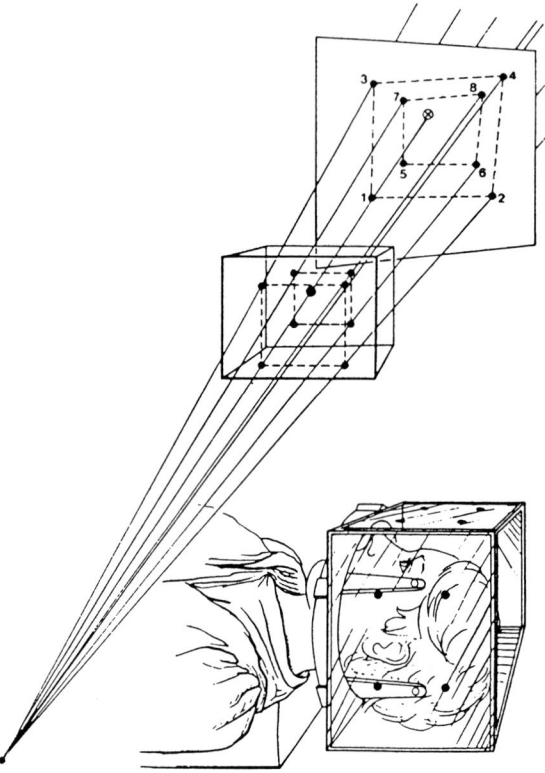

Fig. 10.2. The angiographic localizer box is attached to the BRW head ring. Knowledge of the BRW coordinates of the fiducial markers allows for target (AVM) localization from pairs of quasi-orthogonal radiographs.

This is in contrast to the mean deviation between the simulation and port films (RABINOWITZ et al. 1985) of 7 mm for the conventional irradiation of brain tumors. The uncertainty in the clinical definition of the targets (boundaries of AVM or tumor) exceeds the net uncertainty in the dose delivery.

10.5 Patient Safety

Pedestal-mounted frame systems should be modified to reduce the risk of patient injury from couch movement while the frame is fixed in space. Vertical movement should be constrained with a mechanical lock. The power to the couch should be switched off during treatment. Access to couch release switches should be limited to eliminate accidental use.

Couch-mounted frame systems and pedestal-mounted frame systems should both be modified to reduce the risk of gantry collision with the patient or ancillary equipment. Consideration should be given to gantry rotation limits (electrical or mechanical), contact interlocks which would disable gantry motion, and arc limits which maintain a safe distance from the pedestal. An interlock system has been designed and implemented which limits gantry rotation as a function of PSA angle, fixes the secondary collimators, and immobilizes the PSA (DEMAGRI et al. 1993). These safety features minimize the risk of gantry/patient collision and the risk of PSA movement during the SRS procedure. An interlock system should be implemented on linac-based SRS systems. The patient should be monitored visually, not only for accidental movements, but for accidental airway obstruction and health status as well.

must be visualized in two quasi-orthogonal angiograms. The BRW coordinates of the fiducial markers are known and allow the user to derive the co-ordinates of the AVM (SIDDON and BARTH 1987). Use of cut-film angiography avoids possible geometric distortion of the image intensifier of digital subtraction angiography.

10.4.3 Overall Uncertainty

The net uncertainty is the consequence of the uncertainty in each step of the procedure. The principal or dominating factors are the frame (1 mm), the imaging of the target (1.7 mm), the treatment machine (1 mm), and tissue motion between the target imaging sequence and dose delivery (1 mm), the net uncertainty is typically between 1.9 and 2.4 mm. The addition of uncertainty in the couch position raises the net uncertainty from 2.4 to 2.8 mm. The University of Florida system and the gamma knife have reduced tolerances relative to other linac configurations. The net uncertainty for the former two units are 1.3–2.0 mm without the CT couch uncertainty.

10.6 Prospective Risk Analysis and Quality Assurance Routines

Prospective risk analysis is the application of fault hazard analysis (FHA) to a system. Risk analysis techniques were introduced to anticipate and minimize malfunctions in aircraft and spacecraft systems (LAMBERT 1973). The purpose was to avoid trial and error approaches which had previously allowed for losses of aircraft during prototype tests. The application of FHA to the procedure consists in identifying steps in the procedure where risk or injury can occur. The probability of injury is assessed. The procedure is then redesigned with added procedural sequences to minimize risk when the initial risk probability is considered excessive.

An example of a quality assurance (QA) program was that designed by LUTZ et al. (1988) and TSAI et al. (1991) at the Joint Center For Radiation Therapy. This approach consists of hidden target verification for the entire SRS procedure, target coordinate verification prior to each patient irradiation, and the use of the arc assembly for ascertaining the frame fixation to the patient during the SRS procedure. Lutz tested the SRS system's ability to localize a target within a phantom and position the target at the isocenter of the linac. Lutz performed the evaluation for the CT localization and angiographic localization. The maximum localization errors were 1.5 mm and 0.6 mm for the two processes, respectively.

Verification of the patient position can be achieved with the phantom base and target simulator of the BRW system. The phantom base (Fig. 10.3) is adjusted to the tumor coordinates and the target simulator is then positioned at the target location in BRW space. The target coordinates are set on the BRW floor stand and the target simulator is attached to the floor stand. Radiographs are obtained to confirm the target alignment with the x-ray beam (Fig. 10.4).

QA procedures for radiation therapy can be found in AAPM report 13 (1984). QA procedures for linac-based radiosurgery have been developed for a variety of frame/linac configurations (LUTZ et al. 1988; PODGORSAK et al. 1989; DRYZMALA 1991). The principal features of the QA programs are (a) verification of the mechanical tolerances, (b) x-ray/light field alignment with isocenter, and (c) verification of the target/tumor with isocenter prior to treatment.

10.7 Dosimetry

10.7.1 Phantom Material

Phantoms of polystyrene (BJARNGARD et al. 1990; RICE et al. 1987), water (ARCOVITO et al. 1985; HOUDEK et al. 1983), and water-equivalent plastic (FRIEDMAN and BOVA 1989) have been used to measure the dose output and dose distributions of single stationary beams. Anthropomorphic phantoms and film have been used to obtain dose distributions from

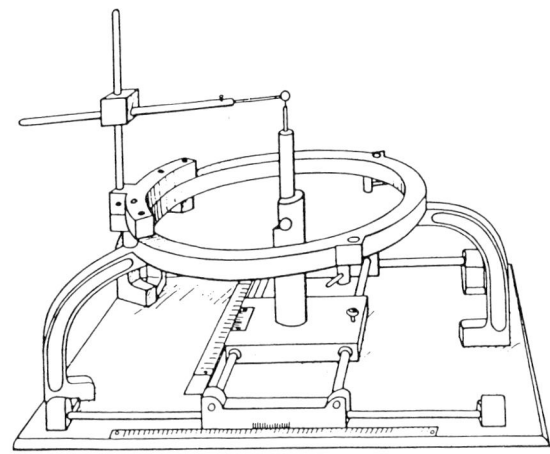

Fig. 10.3. The target simulator (steel ball) is attached to the BRW phantom. The phantom coordinate system duplicates the coordinates of the BRW stand. The pointer is set to position the target simulator at the lesion's coordinates. The BRW phantom is also used in QA procedures to verify the angiographic box localization in SRS

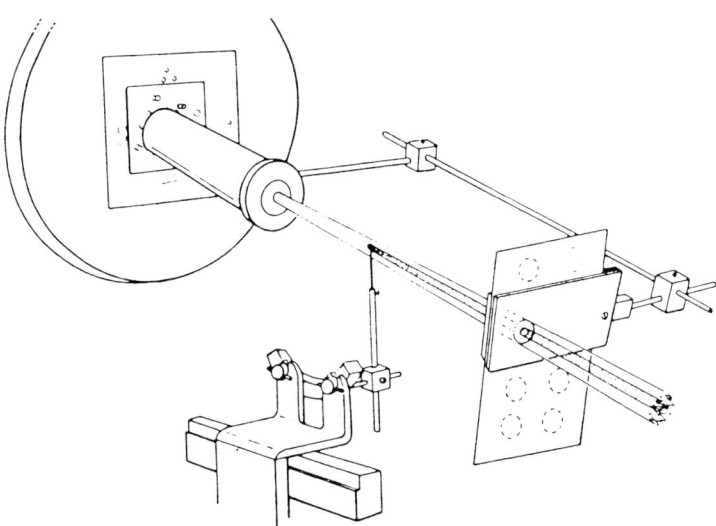

Fig. 10.4. Radiographs are obtained at various gantry/table angle combinations of the target simulator to verify the lesion's BRW coordinates prior to treatment. Target alignment with the x-ray field should be 1.0 mm or less prior to treatment

various arc geometries (CHIEREGO et al. 1988; SERAGO et al. 1989; SMITH et al. 1993). Existing anthropomorphic phantoms require modifications for use in the dosimetry of radiosurgery. Consequently, the gamma knife phantom design was altered to facilitate the larger field sizes encountered with linacs (AAPM TG-42 draft report). The water-equivalent phantom is 16 cm in diameter with cassettes for film sheets or TLDs; the film cassette accommodates film sheets 12 cm in diameter.

10.7.2 Depth Dose and Tissue-Maximum Ratio

Depth dose and tissue-maximum ratio (TMR) data have been reported in the literature for beam energies from 4 MV to 18 MV (ARCOVITO et al. 1985; HAIMSON 1963; HOUDEK et al. 1983; RICE et al. 1987). A compilation of SRS beam data by JANI (1993) covers the beam energy range from 4 MV TO 15 MV. A table of TMR values is listed as a function of field size for a 6-MV in Table 10.1. The TMR values exhibit a maximum difference of 6% as the field diameter increases from 12.5 to 30 mm. Thus, the TMR and depth dose data are weakly dependent on field size.

Differences due to the use of various beam energies have been reported by SERAGO et al. (1989) with dose-volume histogram analysis. The dose in tissue adjacent to the target is not strongly affected by beam energy. The reasons are twofold: (a) the target volume is small relative to the beam attenuation and (b) the arc geometry has the dominant impact on normal tissue dose.

10.7.3 Beam Profile

Tightly collimated megavoltage x-ray beams (12.5- to 40-mm diameters) pose unique dosimetry requirements. Figure 10.5 depicts the beam profiles at isocenter for 6-MV beam field diameters of 12.5 mm, 20.0 mm, and 30.0 mm. The beam profile for the 12.5-mm collimator slopes immediately with dis-

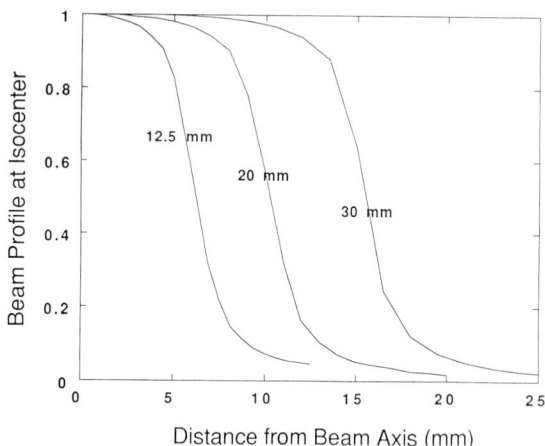

Fig. 10.5. 6-MV beam profiles of the 12.5-mm, 20.0-mm, and 30.0-mm cylindrical collimators are compared. Note that the 80% dose coincides with the nominal field diameter and that the 12.5-mm profile is dominated by the penumbra

tance from the beam axis. BJARNGARD et al. (1990) have shown that "practical lateral" electronic equilibrium exists for field diameters of 10 mm and larger. The active volume of conventional ionization chambers perturbs the radiation field. The lack of lateral electronic equilibrium can impair the absolute dose calibration. DAWSON et al. (1986) and RICE et al. (1987) investigated the effect of detector size on the measurement of beam profile.

Novel dosimetry techniques have been applied which ameliorate the difficulties presented by the lack of lateral electronic equilibrium. Radiochromic film is tissue equivalent and eliminates the film development process. The spatial resolution is sufficient for the measurement of the 4-mm beam profile of the gamma knife unit. Furthermore, the line spread function and the modulation transfer function of the radiochromic film/laser film digitizer system permit the accurate characterization of stereotactic beam profiles (MCLAUGHLIN et al. 1993). MRI of the Fricke-gel dosimeters is capable of measuring the dose distribution of small x-ray beams in three dimensions (OLSON et al. 1992). MRI is subject to the constraints of spatial and geometric resolution of 1 mm and 4 mm respectively (EHRICKE et al. 1992; SCHAD et al. 1992).

10.7.4 Output Factor

The output factor as a function of field diameter is shown in Fig. 10.6 for a 6-MV beam. The output factor decreases rapidly with field size below 20 mm.

Table 10.1 TMR vs Collimator Diameter (mm) for 6-MV x-rays

Depth (cm)	Collimator diameter (mm)		
	12.5	22.5	30.0
3.0	0.94	0.95	0.96
5.0	0.85	0.87	0.88
10.0	0.66	0.69	0.70
15.0	0.52	0.54	0.55

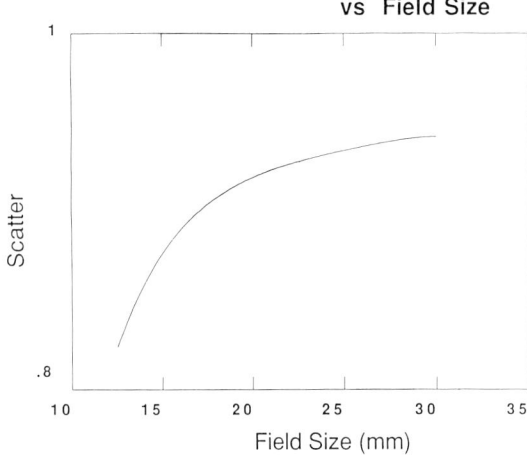

Fig. 10.6. The total scatter factors for a typical 6-MV beam as a function of field diameter monotonically increases from 0.84 (12.5 mm) to 0.92 (30.0 mm). The scatter factor increases rapidly from 12.5 mm to 17.5 mm and more gradually for the remaining field diameters

10.7.5 Machine Accuracy

The mechanical alignment of the axes of rotation of the gantry, collimator, and patient support assembly should coincide within a sphere of 1-mm radius in order to provide accurate beam delivery. The same requirement applies to the x-ray beam alignment with the mechanical isocenter. The verification of the mechanical and beam alignment is normally specified to ± 2 mm; however, most linac installations meet the 1-mm tolerance limit.

10.8 Calculation

Radiosurgery is performed with beam energies from cobalt-60 to 18 MV. The photon interactions with the skull and intracranial tissues are predominantly Compton. The narrowly collimated beams, simple geometry of the head, and small change in tissue densities simplify the requirements on the dose calculation algorithm. Several investigators employ TMR or percent depth dose tables in the algorithm (Arcovito et al. 1985; Rice et al. 1987; Pike et al. 1987). Rice et al. found that the narrow 6-MV beams could be accurately reproduced with tables of off-axis ratios, TMRs, and output factors. The dose delivered is accurate to 3% without accounting for obliquity, or tissue density. The dose at an arbitrary point is:

$$D(s,r,d) = \frac{D(10 \times 10, 0, d_{max})}{((100+d_{max})/(SSD+d))^2} \times TMR(s,d)\, S_t(s)\, OAR(s,r),$$

where s is the field size at depth d, and r is the distance from central axis. TMR is the tissue-maximum ratio, S_t is the total scatter factor, OAR is the off-axis ratio and $D(10x10,0, d_{max})$ is the reference dose.

In contrast to a semiempirical beam parameterization, Kubsad et al. (1990) simulated the collimation system and x-ray production with Electron-Gamma Shower-4 Monte Carlo code and calculated dose distributions for radiosurgery. The axial symmetry of cylindrical collimators allows for the use of a simple dose model based on TMRs, OARs, and the inverse-square dependence on dose. Complex collimation requires accurate accounting of the beam profile and differential scatter about the beam axis.

10.8.1 Software Requirements

Treatment planning software codes should produce 2D dose distributions in the plane of the image (tranverse cut or reconstructed in the plane of desired orientation) in a timely manner. Radiosurgery plans are typically performed interactively by a team consisting of the physicist, radiation oncologist, and neurosurgeon. The accuracy of dose delivery and complex neural structures necessitate the interdisciplinary planning effort.

10.8.2 2D Dose Distributions

Planar dose distributions overlaid on the corresponding 2D images are the primary plan evaluation tool in most systems. The pixel dimensions and slice thickness limit the resolution of the tumor or lesion in three dimensions. The grid size of the dose matrix must be sufficiently fine to faithfully reproduce the dose both in the treatment volume and in normal tissue within the limits of the scan resolution. The steep dose gradient in the penumbra region requires a mesh of one to two pixel widths. Figure 10.7 shows a comparison of the data and calculation of a 2D dose distribution in the coronal plane of a 16-cm-diameter sphere from a five-arc geometry with a 2.0-cm collimator. The isodose contour values are 80%, 60%, 40%, 20%, and 10% for the normalization of 100% at isocenter. The data were obtained with film and analyzed with an isodensitometer and an

TRANSVERSE   **CORONAL**   **SAGITTAL**

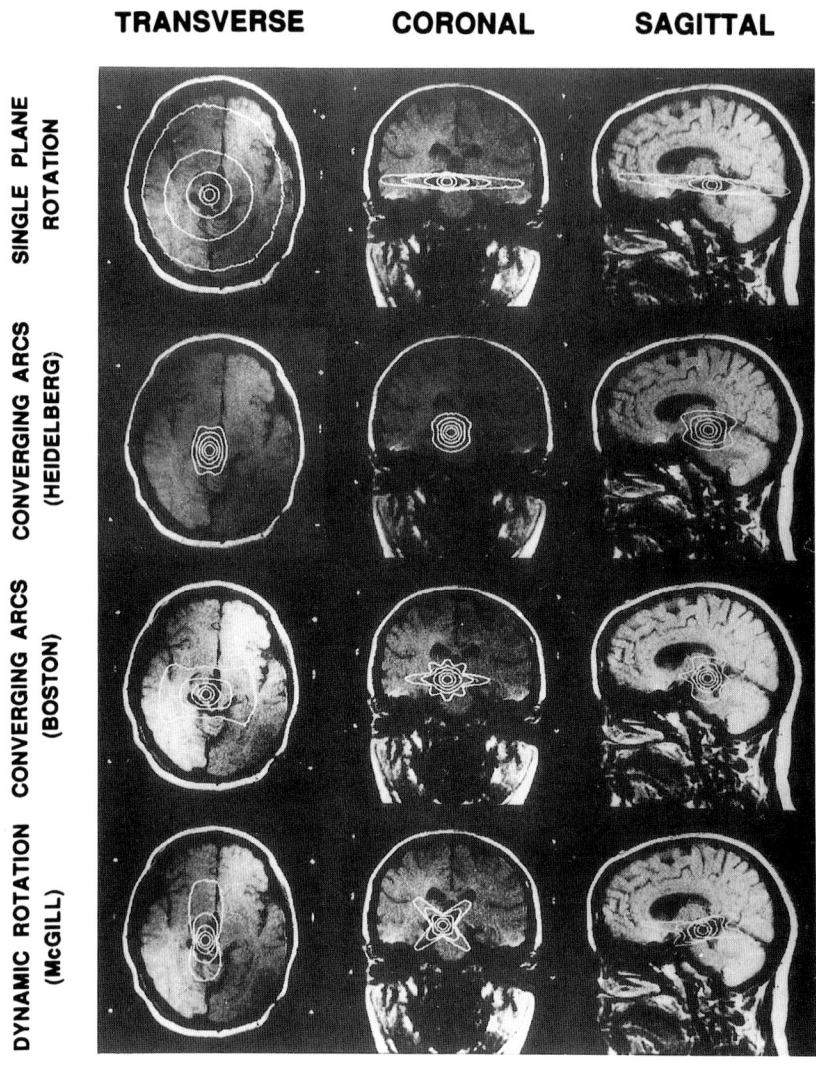

cm

Fig. 10.7. Planar dose distributions are shown for four arc techniques in the transverse, coronal, and sagittal planes. The calculations were performed for a 10-MV beam by Pike et al. for a 360° arc, the Heidelberg 11-arc approach, a four-arc design, and dynamic radiosurgery. The collimator diameter is 1.0 cm and the isodose values are 90%, 50%, 20%, 10%, and 5%

aperture diameter of 2.0 mm. The grid spacing and the dose calculation matrix are 1.3 mm.

10.8.3 3D Dose Distributions and Dose-Volume Histograms

The plan design is greatly facilitated with the beam's-eye view (BEV). The orientation of the arc plane and arc limits are positioned to avoid critical structures, while focusing on the lesion. Real-time manipulation of the patient anatomy in the BEV enables optimal plan design with minimal calculation iterations.

Several treatment planning systems are capable of simultaneous 2D and 3D dose calculations. Dose-volume histograms of the target and critical struc-

tures can then be used to evaluate the relative merits of individual plans. Cumulative dose-volume histograms are shown in Fig. 10.8 for the irradiation of a 16-cm-diameter sphere with a 12.5 mm cone size and one-, two-, and five-arc treatment geometries. The target volume occupies 1.0 cm^3 at the center of the sphere and is enveloped by the 80% dose. The isocenter dose is normalized to 100%. Note that the irradiated volume of normal tissue at a given dose decreases as the arc traversal increases.

Radiosurgery presents a unique opportunity in 3-D treatment planning. The requisite planning data are also sufficient for retrospective analysis of subsequent treatment complications. Archival storage of the complete 2D and 3D plans permits rapid analysis of tumor control and complication rates.

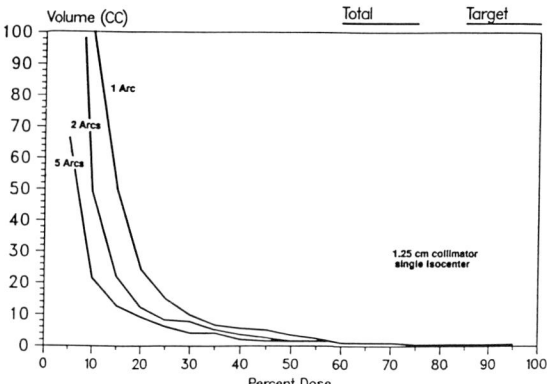

Fig. 10.8. The cumulative dose-volume histograms are shown for a 1.25-cm collimator irradiating a 16-cm sphere with one-, two-, and five-arc configurations. Note that the treatment volumes above 60% are identical, while significant differences appear between the techniques below 20% of the isocentric dose. Rano five, plan 6; UCSF radiosurgery

10.8.4 Software Quality Assurance

The software QA program must be designed to maintain the accuracy in the dose delivery, spatial representation of patient anatomy, and the coordinate transformation from the scan reference frame to the stereotactic reference frame. The QA procedure should be based on a prospective risk analysis, which provides a relative weight to the QA check systems steps based on the error risk estimates and the consequence of the error to the patient. Software QA recommendations for conventional treatment planning systems also apply to radiosurgery systems (CURRAN and STARKSCHALL 1991; VAN DYK et al. 1993).

10.9 Treatment Planning

10.9.1 Lesion Types

The treatment plan depends upon the abnormality. Benign lesions, such as AVMs and acoustic neuromas, were the focus of the initial investigations. These lesions were irradiated with a single high-dose fraction which obliterated the tissue in the treatment volume in 1–2 years. Malignancies occur as either a primary lesion in the neural tissues or as a metastasis from a distant site. Primary malignancies frequently have microscopic extensions into the normal tissue. Consequently, radiosurgery is normally used as a high-dose boost to the macroscopic/enhancing lesion on CT or MRI. Metastatic tumors are nor-

mally well localized and can be irradiated with a single fraction.

10.9.2 Treatment Volume/Target Volume

The treatment volume of tumors is defined as the volume of irradiated tissue as encompassed by a specified isodose surface. The target volume is normally encompassed by the treatment volume. The macroscopic image of a malignancy is defined by the voxels containing contrast enhancement and usually occupy an irregularly shaped volume. Consequently, the spherical dose distribution provided by cylindrical collimation encompasses not only the target but also normal tissue.

Target volume definition of the AVM is the nidus where the arteries shunt directly to the veins. Small AVMs usually present a well-defined nidus while the nidus of the larger AVMs is likely to be obscured by the feeding and draining vessels. Furthermore, volume definition by two orthogonal angiograms is insufficient for a unique solution (BOVA and FREIDMAN 1991).

10.9.3 Single Isocenter Distributions

Radiosurgery concentrates a large radiation dose in a small treatment volume and a minimal normal tissue dose. The ratio of target to normal tissue doses is maximized by arcing the x-ray source about the target. Thus, the integral dose in normal tissue is averaged over a larger volume for a constant target dose. Differences in dose distributions were demonstrated by Podgorsak et al. for the dynamic rotation, a four-arc approach, the Heidelberg technique, and a 360° arc in the transverse plane. Figure 10.7 shows dose distributions in the coronal plane. The distributions are normalized to 100% at isocenter for a 10-MV x ray beam collimated to 1.0 cm. The arc geometry has a dramatic impact on the shape of the dose distribution in normal tissue adjacent to the target. The 360° arc has the steepest dose falloff of all the distributions, but only perpendicular to the plane of rotation. The dose falloff is the least in the rotational plane. The treatment plan for a particular patient can be designed to minimize the dose to critical structures by optimizing the arc geometry. For a lesion that is adjacent and lateral to the brainstem, arc planes can be oriented towards the sagittal plane in order to minimize the dose to the brainstem.

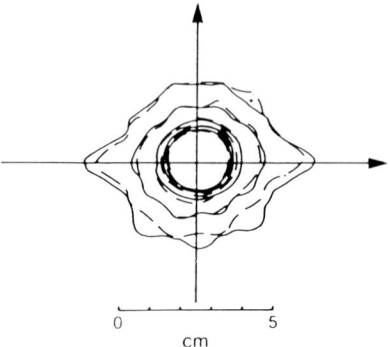

Fig. 10.9. Computer-calculated relative dose profiles are compared with film measurements in the vertical and lateral directions of the BRW reference frame for the five-arc geometry. The collimator diameter is 20 mm. 80, 60, 40, 20, 10% isodose contours

SCHELL et al.(1991) and SERAGO et al. (1989) have compared the dose-volume histograms for different arc configurations. It was shown that the differences between arc geometries occur at the 10% dose level when normalized to 100% at isocenter (Table 10.2). These differences are in the range of 2.5 Gy for most treatment plans. The reduction of the irradiated volume of normal tissue as a function of arc number is in large part accomplished with five arcs for a 1.25-cm-diameter lesion. A ten-arc geometry yields a 10% reduction in the 5% dose volume. The cumulative dose-volume histograms are shown in Fig.10.8 for one, two, and five arcs about the center of a 16-cm-diameter sphere using a 12.5-mm-diameter collimator. Above 60% of the isocentric dose, the volumes are equal. Large differences between the arc techniques occur below 30%. The reduction of normal tissue volume at the 5% dose level is at best 10% between five and ten arcs. Hence, in most cases, the

differences in beam geometry or technique have minimal clinical significance.

It was demonstrated that the normal tissue dose is minimized by several factors: (a) total arc traversals of more than 400° per isocenter, (b) equal angles between arc planes, (c) maintaining individual arc lengths less than 180° and (d) avoiding the overlap of arc planes of different isocenters. Finally, the differences between arc techniques are minimal when the field size is larger than 2.5 cm.

Isocenter is the locus of a set of arcs arranged in three dimensions. A target is irradiated with one or more isocenters, depending on the shape of the lesion. The initial isocenter is the center of mass of the target. Subsequent variation of isocenter position and collimator diameter modify the dose distribution and provide the desired minimal target dose. One-millimeter changes in the isocenter location result in large percentage changes in the penumbra dose. The dose volumes in the high-dose region are approximately spherical from converging non-coplanar arcs. Simple rules of thumb can be used to relate the collimator diameter with a given percent isodose diameter.

10.9.4 Effects of Arc Length and Location

Beam entry patterns for the gamma knife, a 360° arc, a four-arc technique, and dynamic radiosurgery are shown in Fig.10.10 (PODGORSAK et al. (1989). The gamma knife sources subtend an angle of ± 80° in the coronal plane and ± 48° in the sagittal plane. The four-arc plan traverses a total of 560° as compared to 300° for the dynamic radiosurgery approach. Figure 10.9 shows planar dose distributions in the coronal plane for the 360° rotation in the transverse plane, with a five-arc technique. The 360° dose distribution extends into normal tissue at the lower isodose values. This is a result of beam opposition, or beams entering through beam exit points. The same comment applies to the four-arc plan. The Heidelberg approach (Hartmann, Sturm) yields the smallest dose volume of irradiated normal tissue by arcing about the target with arc lengths between 1000° and 1100°.

Figure 10.8 and Table 10.2 depict the effect of additional arcs on the normal tissue dose for the 12.5-mm collimator irradiating a 16-cm-diameter sphere. From Table 10.2 it is clear that the ten-arc geometry reduces the 5% volume from 67 cm³ to 60 cm³. The reduction from one to five arcs is significant. Arc geometry is limited by the radiosurgery and linac

Table 10.2 Isodose volumes for four linac-based arc techniques

Technique	Field size	% Isodose volume (cm³)		
	(cm)	80%	20%	5%
Five arcs,	1.25	0.8	8.6	78
opposed	2.0	4.5	32	343
	3.0	18	104	875
Five arcs	1.25	0.8	8.0	67
non opposed	2.0	4.5	29	282
	3.0	18	98	846
Dynamic	1.25	0.6	7.3	124
radiosurgery	2.0	4.0	33	392
	3.0	14	102	817
Ten arcs	1.25	0.8	7.9	60
	2.0	4.5	28	236
	3.0	14	97	728

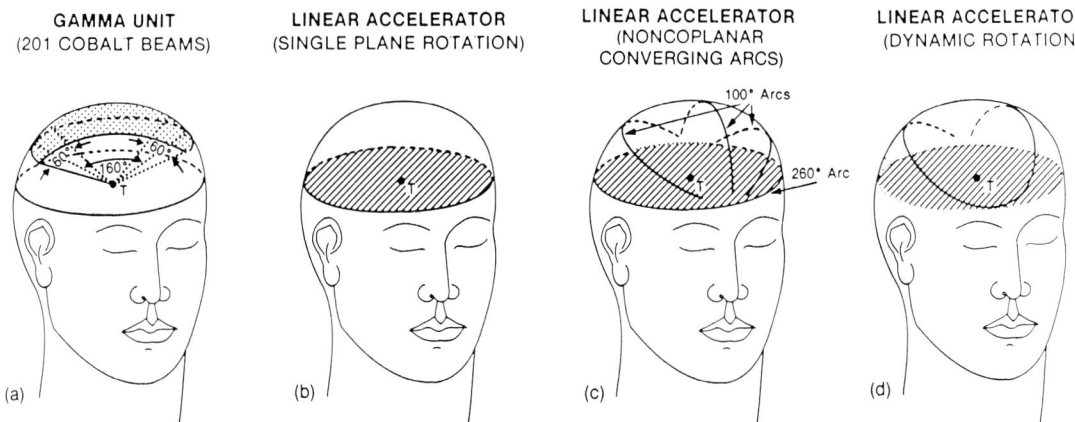

Fig. 10.10a–d. Four beam entry patterns are compared: **a** the gamma knife, **b** a 360° rotation in the transverse plane of the patient, **c** a four-arc configuration, and **d** the 300° sweep from dynamic radiosurgery brain implant

hardware of the particular installation. The normal tissue dose is minimized by evenly distributing the arcs (equal angles between the arc planes.) Ideally, this implies that the angles between the arc start and stop angles equal the angle separating the arc planes. However, the arc planes may be tightly grouped towards the sagittal plane to reduce the dose to the brainstem, for example. The BEV should be utilized to prevent arc traversal through critical structures. Individual arcs may contain gaps or truncation or be eliminated. The final arc geometry may differ substantially from the initial configuration.

Table 10.2 contains the 5% volumes for 12.5-mm, 20-mm, and 30-mm-diameter field sizes as a function of arc length and technique. The 5% volumes are strongly dependent on arc length for field diameters less than 20 mm. The dependence of the 5% dose volumes on field size is weak for diameters greater than 20 mm. This finite extent of the human brain becomes the limiting factor. The 5% dose volume becomes constrained by the surface of the head in the plane of the arc as the field size increases above 20 mm. Consequently, arc techniques are principally arranged to minimize the dose to the critical structures while delivering the tumor dose.

normal tissue volume that is approximately four times the volume from the multiple isocenter plan. Multiple isocenters produce an overlapping region and doses greater than 100% for contiguous target volumes. While an overlap within the target volume is clearly preferable, any overlap may result in an increased complication rate (NEDZI et al. 1990). In addition to the single isocenter planning guidelines, one must minimize the normal tissue dose by minimizing the overlap of the arc planes. In some cases, this can be achieved by a couch angle offset between isocenters.

10.9.6 Dose to Extracranial Critical Structures

The isocentric dose is typically 20 Gy with a rapid decrease in normal tissue proximal to the tumor. The 80% isodose surface is typically used to envelop the target. The dose estimates to extracranial organs are listed in Table in 10.3. The lens of the eye receives approximately 10 cGy for an isocentric dose of 20 Gy. Dose estimates for the other sites are proportionately less.

10.9.5 Multiple Isocenters

When the aspect ratio of a target is large, multiple isocenter geometries may reduce the normal tissue volume drastically. A peanut-shaped volume can be enveloped by two smaller collimator diameters or one large field. The larger field size would irradiate a

Table 10.3 Dose estimates to extracranial organs

Organ	Dose (%)
Eyes	0.5
Thyroid	0.4
Liver	0.07
Gonads	0.05

10.10 A Comparison of Three Treatment Techniques: Linac-Based Radiosurgery, Gamma Knife, and Brain Implants

A comparison of the normal and target (tumor) dose-volume histograms was performed for the three modalities. This comparison was based solely on the dosimetric characteristics of the techniques and did not account of the dose-rate effects on the irradiated tissues. Figures 10.11 and 10.12 illustrate the cumulative dose-volume histograms for the irradiation of normal tissue and an 18-mm-diameter target volume in a 16-cm-diameter sphere with the gamma knife, a five-arc geometry with a 6-MV linac, and an iodine-125 implant. The 18-mm-diameter treatment volume was selected as the ideal volume for the three techniques. The brain implant consisted of a single catheter with two seeds and a total activity of 30 mCi. The gamma knife employed the 18-mm helmet and 201 cobalt-60 sources. The arc geometry of the linac was five arcs with 40° between the arc planes. The total arc traversal was 550°.

The target volume was 3.0 cm³. The 80% isodose surface enveloped the target in both the linac plan and the gamma knife plan. The 45 cGy/h dose rate surface encompassed the target in the brachytherapy plan. The 80% isodose surfaces and the 45 cGy/h surface were chosen as the prescription point and normalized to unity. The target dose-volume histo-grams for the linac and gamma knife are identical. The maximum target dose is 25% greater than the prescription surface. The target dose-volume histogram for the brain implant demonstrates that one-third of the target volume receives a minimum of 178% of the prescribed dose. The large dose gradients within the target volume result from the r^2 dose distribution about each seed. Modest differences exist between the external beam and interstitial techniques. The 20% dose volume of the implant equals that of the linac and differs by 25% from that of the gamma knife. The differences are most likely clinically insignificant.

Interstitial brachytherapy of intracranial lesions has several advantages relative to radiosurgery. The seed activity and configuration can be arranged to conform the isodose surface to the tumor shape. The shape of the isodose surfaces is not inherently spherical as in the case of radiosurgery. Radiosurgery techniques must resort to multiple isocenter plans to irradiate irregularly shaped volumes. Hence, less normal tissue is irradiated with an implant. Secondly, the oxygen enhancement ratio is about 1.8 for implants compared to 3.0 for radiosurgery (LING et al. 1985). Hypoxic tumor tissue is less protected for implants than radiosurgery. Consequently, implants should have a higher therapeutic gain factor than single fraction treatments with radiosurgery.

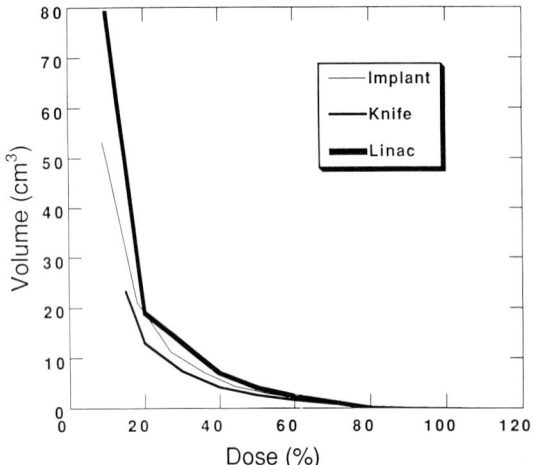

Fig. 10.11. The normal tissue cumulative dose-volumes are compared for treating an 18-mm-diameter spherical target with three modalities: the gamma knife, linac-based radiosurgery, and an iodine-125 brain implant. The linac plan consists of five arcs with 40° between the arc planes. The central arc is in the sagittal plane. Note the general agreement between techniques above 20%

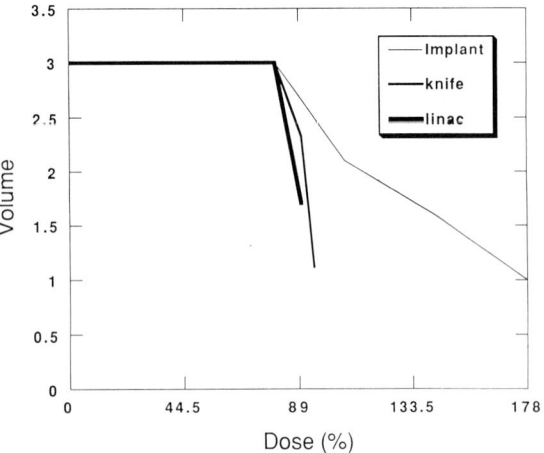

Fig. 10.12. The target dose-volume histograms for the three modalities are identical at and below the prescription dose value by definition. The linac and gamma knife plans are in general agreement with modest dose gradients within the target volume. The brain implant is a single catheter with two seeds. The iodine-125 sources produce a significant dose gradient with one-third of the target volume receiving twice the prescription dose

The advantages of radiosurgery are that the procedure is non-invasive and normal brain tissue is not directly damaged by neurosurgery; the outpatient procedure is complete in less than 1 day.

10.11 Conformal SRS

The primary limitation of Linac-based radiosurgery is the use of cylindrical collimation to irradiate intracranial lesions. Cylindrical collimation produces an approximately spherical volume of concentrated dose in the brain. Beam weighting techniques can change the shape into ellipsoidal volumes. However, most lesions are of irregular extent and the dose distributions do not conform to the shape of the tumor. The advantage of brain implants is that the placement of catheters and seed activity can produce dose distributions which do conform to the tumor shape. Efforts are under way to design collimator systems which produce dose distributions that spare neural tissues. One such system is the Peacock system introduced by Nomos. The Peacock system is a complete conformal SRS planning and treatment delivery system. The treatment planning software package is based on the inverse algorithm which consists of filtered back-projection techniques that generate the dose distribution by beam intensity modulation and the position of collimator jaws. The dose distribution is designed by the planner to conform to the tumor shape. The treatment planning algorithm then generates a configuration of beam geometries as a function of gantry angle to deliver the specified dose distribution. This system irradiates the target volume in transverse planes. Precise control of the PSA couch by special supplemental hardware produces a smooth dose distribution through the target. The treatment delivery hardware consists of pneumatically driven tertiary multileaf collimator jaws which move in or out of the beam as a function of gantry angle. The resulting dose distribution conforms to the tumor and delivers minimal dose-to-normal tissue.

10.12 Summary

Stereotactic radiosurgery is a treatment technique that has become widely accepted in North America. The procedure requires close collaboration between neurosurgery, neuroradiology, and radiation oncology. The initial results indicate increased survival time for patients with brain metastases, and complete obliteration of small and intermediate volume AVMs. Local control of astrocytomas and glioblastome multiforme tumors radiosurgery acting as a final high dose boost to conventional radiation therapy is competitive with invasive techniques. Treatment results for acoustic neuroma are limited to arrest of growth and modest hearing improvement. Patients receiving radiosurgery for any of the above disorders benefit from the non-invasive nature of the outpatient procedure (POLLOCK et al. 1993).

References

American Association of Physicists in Medicine (1984) Physical aspects of quality assurance in radiation therapy, American Institute of Physics, AAPM Report No. 13

American Association of Physicists in Medicine (1993) Task Group 42 Report. Stereotactic external beam irradiation

Arcovito G, Piermattei A, DAbramo G, Bassi FA (1985) Dose measurements and calculations of small radiation fields for 9-MV x-rays. Med Phys 12: 779–784

Barcia-Salorio JL, Roldan P, Hernandez G, Lopez GL (1985) Radiosurgical treatment of epilepsy. Appl Neurophysiol 48: 400–403

Betti OO, Munari C, Rosler R (1989) Stereotactic radiosurgery with the linear accelerator: treatment of arteriovenous malformations. Neurosurgery 24: 311–321

Bjarngard BE, Tsai JS, Rice RK (1990) Doses on central axis of narrow 6-MV X-ray beams. Med Phys 17: 794–799

Bova FJ, Friedman WA (1991) Stereotactic angiography: an inadequate database for radiosurgery? Int J Radiat Oncol Biol Phys 20: 891–895

Chierego G, Marchetti M, Avanzo RC (1988) Dosimetric considerations on multiple arc stereotaxic radiotherapy. Radiother Oncol 12: 141–152

Colombo F, Benedetti A, Casentini L, Zanusso M, Pozza F (1987) Linear accelerator radiosurgery of arteriovenous malformations. Appl Neurophysiol 50: 257–261

Curran BH, Starkschall G (1991) A program for quality assurance of dose planning computers. In: Starkschall G, Horton J (eds) Quality assurance in radiotherapy physics. Proceedings of an American College of Medical Physics Symposium. Medical Physics Publishing, Madison, Wisc

Dawson DJ, Schroeder NJ, Hoya JD (1986) Penumbral measurements in water for high-energy x-rays. Med Phys 13: 101–104

DeMagri CE, Smith V, Schell MC, Larson DA (1994) Interlock system for linear accelerator radiosurgery. Int J Radiat Oncol Biol Phys (in press)

De-Salles AA, Asfora WT, Abe M, Kjellberg RN (1987) Transposition of target information from the magnetic resonance and computed tomography scan images to conventional x-ray stereotactic space. Appl Neurophysiol 50: 23–32

Dryzmala RE (1991) Quality assurance for linac-based stereotactic radiosurgery. In: Starksschall G, Horton J (eds) Quality assurance in radiotherapy physics. Proceedings of an American College of Medical Physics Symposium. Medical Physics Publishing, Madison, Wisc

Dutreix A, Bridier A (1985) Dosimetry of photons and electrons. In: Kase KR, Bjarngard BE, Attix FH (eds) The dosimetry of ionizing radiation, vol I. Academic Press, New York

Ehricke H, Schad LR, Gademann G, Wowra B, Engenhart R,

Lorenz WJ (1992) Use of MR angiography for stereotactic planning. J Comput Assist Tomogr 16: 35–40

Fabrikant JI, Lyman JT, Hosobuchi Y (1984) Stereotactic heavy-ion Bragg peak radiosurgery for intra-cranial vascular disorders: method for treatment of deep arteriovenous malformations. Br J Radiol 57: 479–490

Flickinger JC, Schell MC, Larson DA (1990) Estimation of complications for linear accelerator radiosurgery with the integrated logistic formula. Int J Radiat Oncol Biol Phys 19: 143–148

Friedman WA, Bova FJ (1989) The University of Florida radiosurgery system. Surg Neurol 32: 334–342

Galloway RL, Maciunas R, Latimer JW (1991) The accuracies of four stereotactic frame systems: an independent assessment. Biomed Instrum Technol 25: 457–460

Haimson J (1963) Megavoltage x-ray dose distributions for millimetric field sizes. Radiology 80: 117–118

Halperin EC, Bentel G, Heinz ER, Burger PC (1989) Radiation therapy treatment planning in supratentorial glioblastoma multiforme: an analysis based in post mortem. Topographic anatomy with CT correlations. Int J Radiat Oncol Biol Phys 17: 1347–1350

Hartmann G, Schlegel W, Sturm V, Kober B, Pastyr O, Lorenz W (1985) Cerebral radiation surgery using moving field irradiation at a linear accelerator facility. Int J Radiat Oncol Biol Phys 11: 1185–1192

Hochberg FH, Pruitt A (1980) Assumptions in the radiotherapy of glioblastoma. Neurology 30: 907–911

Houdek PV, VanBuren JM, Fayos JV (1983) Dosimetry of small radiation fields for 10-MV x rays. Med Phys 10: 333–336

Jani SK (1993) Handbook of dosimetry data for radiotherapy. CRC Press, Boca Raton, Fl.

Kamerer DFB, Lunsford LD, Miller M (1988) Gamma knife: an alternative treatment for acoustic neurinomas. Ann Otol Rhinol Laryngol

Kelly PJ, Daumas-Dupport C, Kispert DB, Rall BA, Scheithauer BW, Illig JJ (1987) Image-based stereotaxic biopsies in untreated intracranial glial neoplasms. J Neurosurg 66: 865–874

Kihlstrom L (1986) Stereotactic radiosurgery – epidemiologic considerations. Karolinska Hospital, Stockholm

Kjellberg RN (1986) Stereotactic Bragg peak proton beam radiosurgery for cerebral arteriovenous malformations. Ann Clin Res 47: 17–19

Kooy HM, Nedzi LA, Loeffler JS et al. (1991) Treatment planning for stereotactic radiosurgery of intra-cranial lesions. Int J Radiat Oncol Biol Phys 21: 683–693

Kubsad SS, Mackie TR, Gehring MA, Misisco DJ, Paliwal BR, Mehta MP, Kinsella TJ (1990) Monte Carlo and convolution dosimetry for stereotactic radiosurgery. Int J Radiat Oncol Biol Phys 19: 1027–1035

Lambert HE (1973) Systems safety analysis and fault tree analysis. Lawrence Livermore Laboratory, UCID-16238

Larson DA, Gutin PH, Leibel SA, Phillips TL, Sneed PK, Wara WM (1990) Stereotaxic irradiation of brain tumors. Cancer 65: 792–799

Larsson B, Linden K, Sarby B (1974) Irradiation of small structures through the intact skull. Acta Radiol Ther Phys Biol 13: 512–534

Leksell DG (1987) Stereotactic radiosurgery. Present status and future trends. Neurol Res 9: 60–68

Leksell LT (1951) The stereotaxic method and radiosurgery of the brain. Acta Chir Scand 102: 316–319

Linq CC, Spiro IJ, Mitchell JM, Stickler R (1985) The variation of OER with dose rate. Int J Radiat Oncol Biol Phys 11: 1367 1373

Lunsford LD, Flickinger J, Lindner G, Maitz A (1989) Stereotactic radiosurgery of the brain using the first United States 201 cobalt-60 source gamma knife. Neurosurgery 24: 151–159

Lutz W, Winston KR, Maleki N (1988) A system for stereotactic radiosurgery with a linear accelerator. Int J Radiat Oncol Biol Phys 14: 373–381

Marin-Grez M (1988) High dose percutaneous stereotactic irradiation of solitary brain metastases using a 15-MeV linear accelerator (abstract). Int J Radiat Oncol Biol Phys 15(S1): 231

McGinley PH, Butker EK, Crocker JR, Landry JC (1990) A patient rotator for stereotactic radiosurgery. Phys Med Biol 35: 649–657

McLaughlin WL, Soares CG, Sayeg JA, McCullough EC, Kline RW, Wu A, Maitz A (1993) Chromic radiation detectors for gamma knife dose characterstics. Med Phys (to be published)

Nedzi LA, Kooy H, Alexander E, Loeffer JS (1990) Dose volume consideration in field shaping for stereotactic radiosurgery using a linear accelerator. Radiosurgery update Pine Manor College, Chestnut Hill, Mass., June 1990

Nedzi LA, Kooy H, Alexander E, Loeffler JS (1990) Variables associated with the development of complications from radiosurgery of intracranial tumors. Int J Radiat Oncol Biol Phys 19:149

Olivier A, de LA, Peters T, Pike B, Ethier R, Melanson D, Bertrand G, Podgorsak E (1987) Combined use of digital subtraction angiography and MRI for radiosurgery and stereoencephalography. Appl Neurophysiol 50: 92–99

Olson LE, Arndt J, Fransson A, Nordell B (1992) Three-dimensional dose mapping from gamma knife treatment using a dosimeter gel and MR-imaging. Radiother Oncol 24: 82–86

Patil AA (1989) Radiosurgery with the linear accelerator. Neurosurgery 25: 143

Peters TM, Clark J, Pike B, Drangova M, Olivier A (1987) Stereotactic surgical planning with magnetic resonance imaging, digital subtraction angiography and computed tomography. Appl Neurophysiol 50: 33–38

Pike B, Peters TM, Podgorsak E, Pla C, Oliver A de LA (1987) Stereotactic external beam calculations for radiosurgical treatment of brain lesions. Appl Neurophysiol 50: 269–273

Podgorsak EB, Olivier A, Pla M, Lefebvre PY, Hazel J (1988) Dynamic stereotactic radiosurgery. Int J Radiat Oncol Biol Phys 14: 115–126

Podgorsak EB, Pike, GB, Olivier, A, Pla, M, Souhami L (1989) Radiosurgery with high energy photon beams: a comparison among techniques. Int J Radiat Oncol Biol Phys 16: 857–865

Polock P, Lunsford LD, Konziolka DS, Maitz AH, Flickinger JC (1993) Patient outcomes after stereotactic radiosurgery for "operable" arteriovenous malformations

Rabinowitz I, Broomberg J, Goitein M, McCarthy K, Leong J (1985) Accuracy of radiation field alignment in clinical practice. Int J Radiat Oncol Biol Phys 11: 1857–1867

Rand RW, Khonsary A, Brown WJ, Winter J, Snow HD (1987) Leksell stereotactic radiosurgery in the treatment of eye melanoma. Neurol Res 9: 142–146

Rice RK, Hansen JL, Svensson GK, Sisson RL (1987) Measurements of dose distributions in small beams of 6-MV x-rays. Phys Med Biol 32: 1087–1099

Saw CB, Ayyangar K, Suntharalingam N (1987) Coordinate transformation and calculation of the angular and depth parameters for a stereotactic system. Med Phys 14: 1042–1044

Schad LR, Ehricke H, Wowra B et al. (1992) Correction of spatial distortion in magnetic resonance angiography for radiosurgical treatment planning of cerebral arteriovenous malformations. Magn Reson Imaging 10: 609–621

Schell MC, Smith V, Larson DA, Flickinger J, Wu A (1991) Evaluation of radiosurgery techniques with cumulative dose volume histograms in linac-based stereotactic external beam irradiation. Int J Radiat Oncol Biol Phys 20: 1325–1330

Schell MC, Evans JH, Martel MK, Wu A (1993) A methodology for the analysis of stereotactic radiosurgery beam data. Med Phys 20: 932

Schwade JG, Houdek PV, Landy HJ et al. (1990) Small-field stereotactic external beam radiation therapy of intracranial lesions: fractionated treatment with a fixed-halo immobilization device. Radiology 176: 563–565

Serago CF, Lewin AA, Houdek PV et al. (1992) Stereotactic radiosurgery: dose-volume analysis of linear accelerator techniques. Med Phys 19: 181–185

Siddon RL, Barth NH (1987) Stereotaxic localization of intracranial targets. Int J Radiat Oncol Biol Phys 13: 1241–1246

Smith V, Larson D, Schell MC (1993) Role of tertiary collimation used for linac-based radiosurgery. Radiat Oncol Invest 1: 71–75

Sturm V, Kober B, Hover KH et al. (1987) Stereotactic percutaneous single dose irradiation of brain metastases with a linear accelerator. Int J Radiat Oncol Biol Phys 13: 279–282

Tsai J, Buck BA, Svensson GK, Alexander E, Cheng C, Mannarino EG, Loeffler JS (1991) Quality assurance in stereotactic radiosurgery using a standard linear accelerator. Int J Radiat Oncol Biol Phys 21: 737–748

Van Dyk J, Barnett RB, Cygler JE, Shragge PC (1993) Commissioning quality assurance of treatment planning computers. Int J Radiat Oncol Biol Phys 26: 261–273

Wallner KE, Galcich JH, Malkin MG, Arbit E, Krol G, Rosenblum MK (1989a) Patterns of failure following treatment for glioblastoma multiforme and anaplastic astrocytoma. Int J Radiat Oncol Biol Phys 16: 1405–1409

Wallner KE, Galcich JH, Malkin MG, Arbit E, Krol G, Rosenblum MK (1989b) Inability of computed tomography appearance of recurrent malignant astrocytoma to predict survival following reoperation. J Clin Oncol 7: 1492–1496

Weaver KA, Smith V, Lewis J et al. (1990) A CT-based computerized treatment planning system for I-125 stereotactic brain implants. Int J Radiat Oncol Biol Phys 18: 445

Winston KR, Lutz W (1988) Linear accelerator as a neurosurgical tool for stereotactic radiosurgery. Neurosurgery 22: 454–464

Wu A, Maitz A, Kalend AM, Lunsford LD, Flickinger JC, Bloomer WD (1990) Physics of gamma knife approach on convergent beams in stereotactic radiosurgery. Int J Radiat Oncol Biol Phys 18: 941–949

11 Treatment Optimization Using Physical and Radiobiological Objective Functions

ANDERS BRAHME

CONTENTS

11.1 Introduction

Radiation therapy is a truly multidisciplinary field where the developments have taken place gradually and almost coherently in many different areas. This is fortunate because a chain is only as strong as its weakest link, and significant developments in one single area are not always sufficient for general improvements of the overall performance. The development of radiation therapy planning during the last decade has been enormous. We have witnessed an unprecedented improvement in three-dimensional (3D) diagnostic imaging through the advent of computed tomography (CT), magnetic resonance imaging, single-photon emission tomography, positron emission tomography, and ultrasound techniques. Simultaneously, we have seen a considerable improvement of dose-planning systems that now are capable of making use of this new diagnostic information and in some cases perform true 3D dose planning to improve the accuracy in the delivered dose distributions. Since the middle of the 1980s there has also been a considerable improvement in computational algorithms for both electron and photon beam calculations, as well as for treatment optimization in general. The continued development of the performance of computer hardware has been an equally important factor for the improved clinical performance.

Quite generally, the development can be expressed by the gradual change from two- to three-dimensional therapy planning in a large number of areas of fundamental importance (Table 11.1). This is particularly true in the diagnostic areas but also evident in the development of 3D biological response models (WOLBARST 1984; KÄLLMAN et al. 1992b), treatment simulation with CT-augmented simulators (WEBB 1990), real-time portal imaging and dose verification with CT capabilities (SWINDELL et al. 1983; BRAHME et al. 1987; BOYER et al. 1992), and, last but not least, computer graphics that make the display of the new images manageable and clear (MCSHAN and FRAASS 1987). The understanding of the value of fractionated radiotherapy has also increased considerably during the last decade partly because of increased knowledge about the time dependence and potential for repair in malignant and normal tissues (THAMES and HENDRY 1987).

Until the last few years, the missing link has been radiation therapy equipment capable of true 3D dose delivery. However, most companies are

ANDERS BRAHME, PhD, Department of Medical Radiation Physics, The Karolinska Institute and University of Stockholm, P.O. Box 260, S-17176 Stockholm, Sweden

Table 11.1. Developments towards 3D radiation therapy planning

Treatment decision:
3D diagnostic and patient data
3D target volume definitions
3D biologic response models
3D treatment objective functions

Dose planning:
3D energy deposition kernels
3D dose calculation algorithms
3D dose distribution measurements
3D dose optimization methods
3D diagnostic data and dose display

Patient setup:
3D treatment simulations
3D patient fixation aids
3D anatomic reference points

Treatment execution:
3D dose delivery
3D portal dose verification
3D treatment response monitoring and follow-up

working on various dynamic augmentations of their therapy equipment, developing dynamic asymmetric jaw movements, variable-wedge-angle devices, multileaf collimators, scanning beams, and conformation therapy capabilities, making true 3D dose delivery a reality in some cases (BRAHME 1987; KÄLLMAN et al. 1988; ISHIGAKI et al. 1990; WEBB 1993). At present, the least developed link in the therapy chain is probably methods capable of performing a true 3D optimization of the dose delivery and the selection of the best treatment technique.

From a mathematical point of view, classical radiation therapy planning is basically a forward process as it tries to answer the following question: How will the absorbed dose in the target volume and surrounding normal tissues be distributed for a given target volume, associated patient geometry, and suggested configuration of the incident beams? This is schematically illustrated in the upper panel of Fig. 11.1. Classical radiation therapy optimization is therefore generally a trial and error process, where gradually improved dose plans can be found by trying an increasing number of incident beam configurations (BRAHME 1992a).

However, in mathematical terminology, radiation therapy planning is fundamentally an inverse problem. This is so because what we really want to find is the optimum combination of incident beams for a given target volume (cf. also Table 11.2). More exactly, the planning process should answer the

Fig. 11.1. Schematic illustration of the difference between conventional forward radiation therapy planning and inverse planning. Dose optimization using forward planning is generally a trial and error process whereas inverse planning directly results in optimal beam profiles (lower left with physical objective function) and isodose distributions (lower right with biologic objective functions)

Forward Calculation:

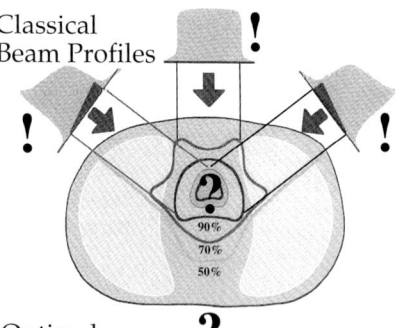

Classical
Beam Profiles

Inverse Calculations:

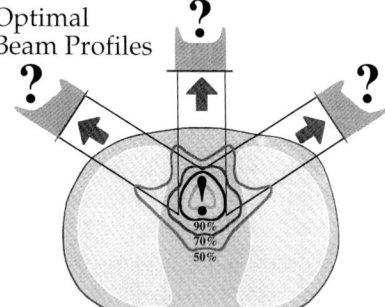

Optimal
Beam Profiles

Physical objective function

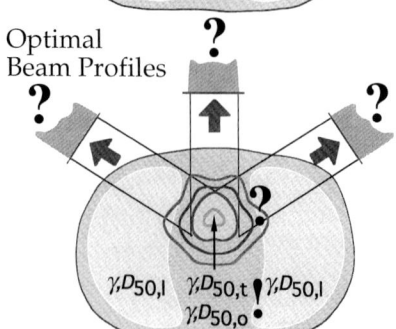

Optimal
Beam Profiles

Biological objective function

Table 11.2. Optimization problems in radiation therapy planning

Field	Objective	Kernel	Desired quantity	Relation
Brachytherapy	Generation of a desired dose distribution: $D(r)$	Point source dose distribution $d_p(r)$	Optimal source density distribution: $\varphi(r)$	$D(r) = \iiint d_p(r - \rho)\varphi(\rho)d^3\rho$
Arc and conformation therapy	Generation of a desired dose distribution: $D(r)$	Dose distribution in convergent pencil beam point irradiation: $d_c(r)$	Point irradiation density: $F(r)$, and incident beam profile	$D(r) = \iiint d_c(r,\rho)F(\rho)d^3\rho$ Fredholm eq. of first kind
External beam therapy	Generation of a desired depth-dose curve: $D(z)$	Monochromatic depth-dose curve: $d(E, z)$	Optimal incident particle spectrum: Ψ_E	$D(z) = \int d(E, z)\,\Psi_E dE$ Fredholm eq. of first kind
Photon therapy	Determination of photon beam spectrum from transmission: $\Psi(z)$	Energy dependence of attenuation coefficient: $\mu(E)$	Incident photon spectrum: $\Psi_E(0)$	$\Psi(z) = \int \Psi_E(0)e^{-\mu(E)z}dE$ Fredholm eq. of first kind
Electron therapy	Determination of dose distribution from incident electron fluence: Φ_θ	Scatter distribution in a thin layer: $\varphi_{\Delta r}(\theta)$	Spatial dose and fluence distribution: $\Phi_\theta(r)$	$\Phi_\theta(r + \Delta r) = \int \Phi_\theta(r)\varphi_{\Delta r}(\theta - \theta')d\theta'$
Beam flattening Beam compensation	Generation of a desired lateral dose distribution: $D(x,y)$	Elementary beam dose distribution: $d(x,y)$	Optimal scanning density distribution: $F(x,y)$	$D(x,y) = \iint d(x-\xi, y-\eta)F(\xi, \eta)d\xi d\eta$
Photon dose planning	Determination of dose distribution from incident photon fluence and resultant terma: $T(r)$	Mean energy imparted point spread function: $h(r)$	Resultant dose distribution: $D(r)$	$D(r) = \iiint h(r-\rho)T(\rho)d^3\rho$
Photon dose delivery	Generation of a desired dose profile in a patient: $D(x,y)$	Dose distribution in pencil beam: $d(x,y)$	Fluence profile in the incident beam: $\Phi(\xi, \eta)$	$D(x,y) = \iint d_1(x-\xi, y-\eta)\Phi(\xi, \eta)d\xi d\eta$
External beam therapy	Generation of a desired lateral dose distribution: $D(r)$	Collimated slit beam dose distribution: $H(r)$	Collimator opening density: $F(r)$	$D(r) = \iiint F(\rho)H(r-\rho)d^3\rho$
Dosimetry	Determination of true dose distribution from a measurement: $D_m(x)$	Detector response function: $S(x)$	True dose distribution: $D(x)$	$D_m(x) = \int S(x-u)D(u)du$
Tumor imaging in nuclear medicine	Determination of true uptake and improved resolution in measured distribution: $I(r)$	Point source response: $i_p(r)$	True uptake: $U(r)$	$I(r) = \iiint i(r-\rho)U(\rho)d^3\rho$

question of which configuration and shape of the incident beams are best for controlling the tumor growth with minimal damage to normal tissues. This question is illustrated in the lower half of Fig. 11.1. At least under the assumption that the desired dose to the target volume (lower left panel) or the geometric and radiobiologic properties of the tumor and normal tissues of the patient are known (lower right panel), it should be possible to find the optimal irradiation technique (BRAHME and ÅGREN 1987, KÄLLMAN et al. 1992a).

This conceptual difference between the classical forward calculation and the inverse approach is further clarified by comparing the three panels in Fig. 11.1. The exclamation marks indicate the known quantities whereas the question marks indicate the principal unknown quantity to be calculated, such as the optimal isodose distribution in the patient or the optimal incident beam profiles as further quantified in Table 11.2. Obviously the resultant absorbed dose distribution in the patient is also obtained by the inverse calculation either by an ordinary forward calculation or by the inversion method itself (lower right panel and BRAHME et al. 1990; HOLMES et al. 1991; KÄLLMAN et al. 1992a).

Strictly speaking, there is no such thing as a generally acceptable and truly optimal treatment plan, because there are so many clinical factors that are specific for each patient and for the responsible radiotherapeutic team and all of them have to be taken into account in the optimization. The principal problem is that most of these factors concern incompatible entities which cannot be compared on a common scale. In the general case, the number of degrees of freedom in the selection of incident beam directions and beam profiles is also very large. It is therefore difficult to integrate their respective advantageous and adverse radiation effects to maximize the clinical outcome, for example in terms of the quality of life of the patient. In a given clinical setting, such factors as the simplicity of the irradiation technique and the accuracy in dose delivery often have a strong impact and may sometimes disqualify very complex plans. In general, a strict mathematical optimization is therefore not possible, and several simplifying assumptions have to be made. Even if this is the case, considerable improvements of presently used forward treatment planning techniques can still be made by introducing new degrees of freedom in the dose delivery, so that only a marginal gain may be achieved by a very strict optimization.

Fig. 11.2. Comparison of the degrees of freedom of the classical and the generalized conformal therapy optimization methods. For photon beam dose optimization nonuniform dose delivery is a necessity when the target volume has concave sections. The few-field technique is the hardest to optimize as the entire phase space has to be checked before the true optimum can be found with certainty. The generalized conformal techniques quite often require non-coplanar dose delivery even if it is not always necessary. Classical conformation therapy according to TAKAHASHI (1965) is included for comparison

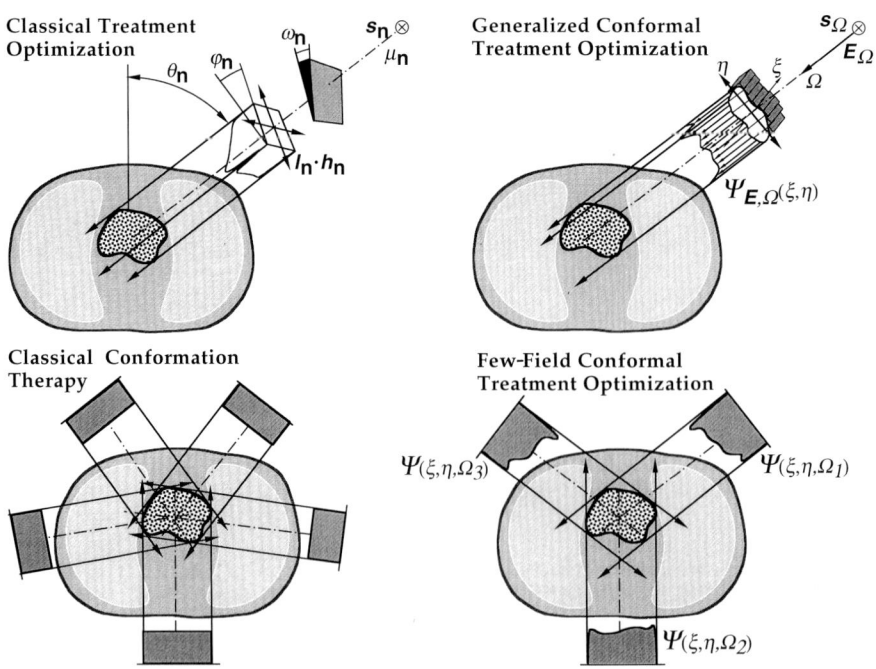

Fig. 11.3. Comparison of six different methods available for delivery of nonuniform therapeutic beams. T_0 is the standard treatment time of about 1 min for uniform dose delivery to the target volume. Only the lower three methods allow dynamic beam shaping but at greatly varying treatment times

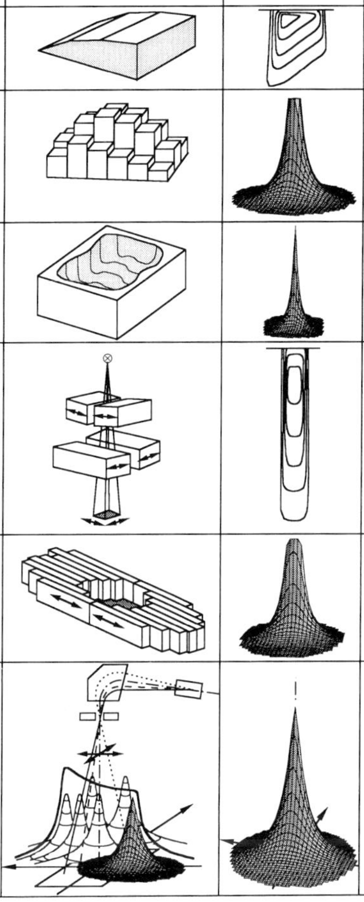

NONUNIFORM DOSE DELIVERY			
Method	Schematic	Kernel	Treatment Time
Wedge Filters			1.1-2.0 T_0
Compensating Filters or Bolus			1.0-1.5 T_0
Transmission Blocks			1.1-1.5 T_0
Dual Dynamic Asymmetric Jaw Pairs			> 20 T_0
Dynamic Multileaf Collimation			1.5-2 T_0
Scanned Elementary Beams			0.5-1.0 T_0

A more general and equally important criterion for treatment optimization is therefore that the dose plan should be as simple as possible to implement, in order to increase accuracy and safety in dose delivery. The properties and flexibility of the treatment unit are therefore an essential factor in treatment optimization. The parameters of the irradiation technique that are open for optimization of the dose delivery include the field size and shape, field entrance location and angulation, wedge angle and more generally the lateral dose distribution of the beam, and finally, the number of beam portals, their respective weights, and time dose fractionation as illustrated in Fig. 11.2. Optimization of classical isocentric beam delivery (left panel) using jaw collimators allows for the selection of arbitrary beam entrance angles, collimator rotation angles, and wedge angles in rectangular beams. Modern treatment units with 3D capabilities using beam blocks, transmission blocks, and more recently multileaf collimators allow almost perfect lateral confinement

of the incident beam to the target volume from multiple directions (Fig. 11.2, right panel, and BRAHME 1993). In addition, wedge filters, beam compensators, dynamic collimation, and, in the latest generation of treatment units, scanned beams facilitate the generation of arbitrary desired lateral dose distributions in the target volume for complex irradiation techniques (cf. Fig. 11.3). The aim of this chapter is to indicate some of the methods and results that can be achieved by optimization of the treatment using these new degrees of freedom of truly 3D radiation therapy.

11.2 Fundamentals of Treatment Optimization

11.2.1 The Integral Equations of Radiation Therapy Planning

The principal problem of radiation therapy planning can be formulated in the form of an integral equation that expresses the resultant dose distribution in the patient for a given incoming radiation field. The most elementary incident radiation beam is a point monodirectional beam $p^m(E, \boldsymbol{\Omega}, \boldsymbol{r}, \boldsymbol{\rho})$, that describes the energy deposition at $\boldsymbol{r}$ for a given energy E, point $\boldsymbol{\rho}$, and direction $\boldsymbol{\Omega}$ of incidence and modality or type of uncharged radiation beam such as photons, neutral pions, and neutrons (m). The absorbed dose at a point $\boldsymbol{r}$ in the patient is then given by an integral over the incident energy fluence $\boldsymbol{\Psi}^m_{E,\Omega}(\boldsymbol{\rho})$, differential in energy, and angle of such incident beams on points $\boldsymbol{\rho}$ on the patient surface:

$$D(\boldsymbol{r}) = \oiint_S \iiint \sum_m p^m(E, \boldsymbol{\Omega}, \boldsymbol{r}, \boldsymbol{\rho})\, \boldsymbol{\Psi}^m_{E,\Omega}(\boldsymbol{\rho}) dE d\boldsymbol{\Omega}\, d^2\rho,$$

(11.1a)

where the spatial integrals have to be performed over the relevant entrance surface, S, of the patient (cf. also Fig. 11.8 below and GUSTAFSSON et al. 1993). The most suitable unit for p is for neutral particles the mean specific energy $\bar{z}$ (J kg^{-1}) imparted per unit incident radiant energy R(J), and thus kg^{-1}. For Ψ the most suitable unit is the increment in incident

radiant energy dR (J) per unit area da (m^2), that is, J m^{-2}, (ICRU 1980).

For incident charged particles, like electrons, protons, and heavy ions, it is more natural to use the fluence of the incoming particles rather than the energy fluence as used in Eq. 11.1a. The pencil beam energy deposition equation then takes the form:

$$D(\boldsymbol{r}) = \oiint_S \iiint \sum_m \pi^m(E, \boldsymbol{\Omega}, \boldsymbol{r}, \boldsymbol{\rho})\, \Phi^m_{E,\Omega}(\boldsymbol{\rho}) dE d\boldsymbol{\Omega}\, d^2\rho,$$

(11.1b)

where $\boldsymbol{\Phi}^m_{E,\Omega}(\boldsymbol{r})$ is the incident particle fluence differential in energy and angle at a point $\boldsymbol{r}$ on the patient surface. The energy deposition kernel $\pi^m(E, \boldsymbol{\Omega}, \boldsymbol{r}, \boldsymbol{\rho})$ is the mean specific energy imparted at $\boldsymbol{r}$ per charged particle of energy E, direction, $\boldsymbol{\Omega}$, and particle type m incident at $\boldsymbol{\rho}$. For $\boldsymbol{\Phi}^m$ the most suitable unit is m^{-2} (ICRU 1980) and for π^m it is J kg^{-1}.

For mixed incident radiation fields of neutral and charged particles Eqs. 11.1a and 11.1b may be combined as a sum over all incident radiation modalities. When the time dependence is relevant the temporal variation of D, $\boldsymbol{\Phi}^m_{E,\Omega}$ and $\boldsymbol{\Psi}^m_{E,\Omega}$ also has to be

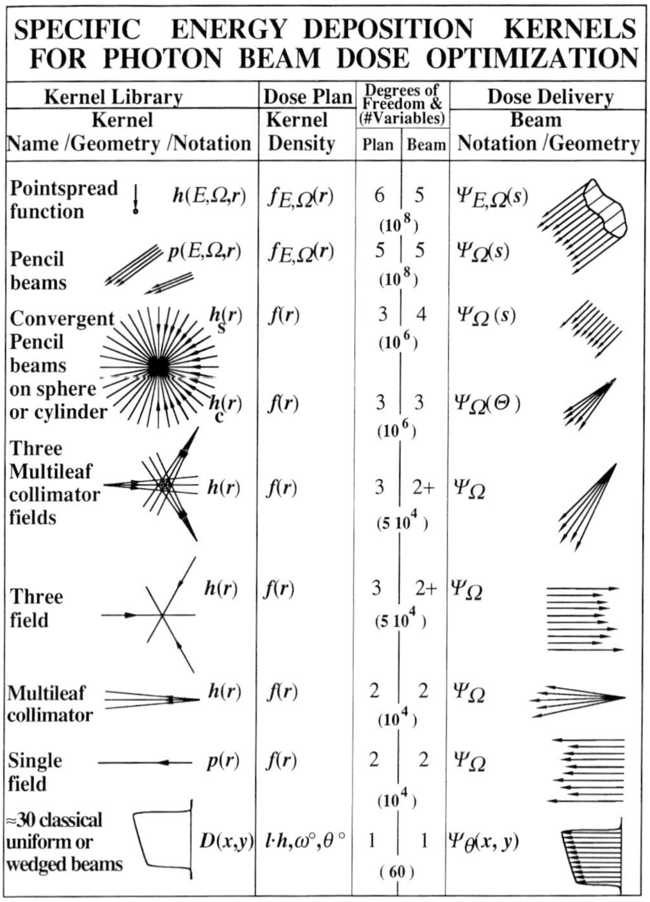

SPECIFIC ENERGY DEPOSITION KERNELS FOR PHOTON BEAM DOSE OPTIMIZATION

Kernel Library		Dose Plan	Degrees of Freedom & (#Variables)		Dose Delivery
Kernel		Kernel			Beam
Name /Geometry /Notation		Density	Plan	Beam	Notation /Geometry
Pointspread function	$h(E,\Omega,r)$	$f_{E,\Omega}(r)$	6 (10^8)	5	$\Psi_{E,\Omega}(s)$
Pencil beams	$p(E,\Omega,r)$	$f_{E,\Omega}(r)$	5 (10^8)	5	$\Psi_{\Omega}(s)$
Convergent Pencil beams on sphere	$h(r)\atop{s}$	$f(r)$	3 (10^6)	4	$\Psi_{\Omega}(s)$
or cylinder	$h(r)\atop{c}$	$f(r)$	3 (10^6)	3	$\Psi_{\Omega}(\Theta)$
Three Multileaf collimator fields	$h(r)$	$f(r)$	3 $(5\,10^4)$	2+	Ψ_{Ω}
Three field	$h(r)$	$f(r)$	3 $(5\,10^4)$	2+	Ψ_{Ω}
Multileaf collimator	$h(r)$	$f(r)$	2 (10^4)	2	Ψ_{Ω}
Single field	$p(r)$	$f(r)$	2 (10^4)	2	Ψ_{Ω}
≈30 classical uniform or wedged beams	$D(x,y)$	$l \cdot h, \omega°, \theta°$	1 (60)	1	$\Psi_{\theta}(x, y)$

Fig. 11.4. Overview of the free variables and the degrees of freedom in kernel-based radiation therapy planning. The flat or wedge-shaped beams of conventional treatment optimization are machine oriented. Most of the other kernels are instead tumor oriented and have the potential both to improve the accuracy and to optimize the dose delivery

included in this combined equation. The principal unknown quantities to be determined in the optimization of the treatment plan are the incident energy fluence $\Psi^m_{E,\Omega}(\rho)$ for neutral particles and the incident fluence $\Phi^m_{E,\Omega}(\rho)$ for charged particles (cf. ICRU 1980 and LIND and BRAHME 1992).

For indirectly ionizing particles like photon or neutron beams it is more natural to use Eq. 11.1a in a differential form along the incident rays and then p is replaced by the *point energy deposition kernel, h,* which describes the specific energy distribution by a primary interaction at a point r' in the patient (cf. Fig. 11.4 and MOHAN et al. 1986; AHNESJÖ et al. 1987). The resultant dose distribution then becomes:

$$D(r) = \oiiint_V \iiint h(E, \Omega, r, r') f_{E,\Omega}(r') \, dE \, d\Omega \, d^3 r',$$
(11.2)

where the spatial integral has to be evaluated over the whole irradiated volume where the kernel or irradiation density f is larger than zero. The unit for h is the same as for p but the kernel or irradiation density f is expressed in units of the increment in incident radiant energy dR per unit volume dV (cf. LIND and BRAHME 1992).

Classical radiation therapy planning is concerned with the evaluation of this type of integral for a given beam kernel p or h and incident energy fluence Ψ or kernel density f, and the resultant dose distribution may be obtained by a direct evaluation of the integrals. This classical approach is therefore a forward problem (Fig. 11.1, upper panel). In treatment optimization the problem is reversed as again p or h is known, but we want to find the optimal energy fluence Ψ, which generates a desired dose distribution D, or preferably the optimal dose distribution $\hat{D}$ as determined by the treatment objectives. A related well-posed problem is to determine the energy fluence Ψ for a known kernel p and measured dose distribution D (cf. Table 11.2 and AHNESJÖ and TREPP 1991). A third situation may also be of interest, namely when the dose distribution D and the irradiation density f or energy fluence Ψ are known and, in principle, it should be possible to determine the associated point energy deposition kernel h or the pencil beam kernel p (cf. AHNESJÖ 1984 and CHUI and MOHAN 1988, respectively). These two latter adjoint problems are in mathematical terminology often called inverse problems since the unknown is implicitly defined and an inversion of the integral is required to find the solution (Fig. 11.1, lower panels). In the mathematical literature Eqs. 11.1–11.4 are generally called Fredholm integral

equations of the first kind (cf. also Table 11.2 and MILLER 1974).

In a *monodirectional* photon or neutron beam the equation may for simplicity be integrated over all angles to obtain the fundamental equation of kernel-based radiation therapy planning with indirectly ionizing radiations. In this equation h is again the point energy deposition kernel but f is often given as the terma differential in energy T_E (total energy released per unit mass, to be distinguished from the kinetic energy released per unit mass, as specified by the kerma) and the resultant dose distribution becomes:

$$D(r) = \iiiint h(E, r, r') \rho(r') T_E(r') \, dE \, d^3 r'.$$
(11.3)

This equation is the base for modern forward treatment planning algorithms (cf. AHNESJÖ et al. 1987 and HOLMES et al. 1991, where the density ρ was included in h for simplicity). By comparison with Eq. 11.2 the irradiation density $f_{E,\Omega}$ can in unit density materials be identified as the terma differential both in energy and angle $T_{E,\Omega}$.

If we instead integrate Eq. 11.2 over both energy and angle and assume the incident energy fluence distribution to be spatially invariant in these variables at least over the target volume, the basic equation for inverse kernel-based therapy planning is obtained:

$$D(r) = \iiint h(r - r') f(r') \, d^3 r'.$$
(11.4)

Here the kernel h is the mean specific energy distribution per unit incident radiant energy of the elementary radiation field centered at r' and f is the irradiation or kernel density (for more details see LIND and BRAHME 1992, and a related equation for heavy ions formulated by LEEMANN et al. 1977). This latter convolution-type equation is much easier to handle mathematically than Eqs. 11.1–11.3 and it has been extensively used in kernel-based treatment optimization in recent years.

However, the inversion of Eq. 11.4 in radiation therapy is associated with serious mathematical and numerical difficulties. The principal problem is that both $D(r)$ and $f(r)$ are physical quantities (the absorbed dose and the irradiation density, respectively) which by necessity are larger than or equal to zero. The straightforward solution of the integral Eq. 11.4 using Fourier transformation (cf. BRAHME 1988a) will therefore only work under very special conditions, as was shown by BRAHME et al. (1982). In fact, for most dose distributions, D, of clinical interest, and for the associated energy deposition kernels, h, there does not even exist an exact solution with $f \geq 0$ which fulfills Eq. 11.4 over the entire volume of the patient. However, over a small volume such as the target volume, it may be possible to find an exact solution, provided the irradiation kernel h is appropriately chosen and well confined relative to the desired dose distribution D (see Sect. 11.5). These problems make it necessary to solve Eq. 11.4 by approximate methods or by reformulating the principal problem of radiation therapy planning, as will be discussed in more detail in the subsequent sections. For completeness some of the most important optimization problems of radiation therapy planning are summarized in Table 11.2. Most of them are linked to integral equations like Eqs. 11.1–11.4 and they can be handled by the methods discussed in this chapter.

11.2.2 Well-Posed and Ill-Posed Optimization Problems

From the above discussion of the integral equations it is clear that a direct exact solution of the inverse problem of radiation therapy planning as formulated by the Eqs. 11.1–11.4 is far from trivial and in general is an impossible task. To proceed there are two alternatives: either the clinical objective to generate a desired dose distribution has to be reformulated to get a more well-posed formulation or some numerical method has to be developed in order to solve the inverse problem as accurately as possible.

In either case it is very useful to transform the relevant integral equation into an algebraic form by discretizing the transport quantities along the coordinates of the free variables. In this discretized form it is natural to describe the dose distribution by a vector d where its components d_i run over the volume of interest of $D(r)$ in the 3D Cartesian space R^3. With a uniform m-fold sampling along each coordinate i will run from 1 to $n = m^3$. The convolution integral in Eq. 11.4 may then for example be described as a matrix multiplication between the convolution matrix, H, built up by the kernel vector h and the irradiation density vector f. The convolution matrix is a $m^3 \times m^3$, so-called Toeplitz matrix, where the kernel vector h is shifted one position for each new row in the matrix. With this notation Eq. 11.4 may be rewritten as:

$$d = Hf. \tag{11.5}$$

For the more complex case of a kernel that changes shape depending on where it is centered (cf. Eq. 11.2) the kernel vector of the convolution matrix will also change from row to row.

Whether we would like to reformulate the problem or find an approximate solution to Eq. 11.4, the best possible solution has to be found using some optimization procedure. In order to find the best possible solution to the integral equation the deviations between the desired dose d and the best achievable (Hf) should be minimized according to the most appropriate measure. This measure could either be physical in terms of the deviations in the absorbed dose delivered ($d-Hf$) or biologic when we are trying to quantify the influence of the dose deviations on tumor control and normal tissue reactions. To allow a strict optimization a scalar measure is required which accounts for all deviations of the components $d_i - (Hf)_i$. Most generally this weighting is mathematically described by a functional F, i.e., a mapping from R^n to R^1. In mathematical terms we thus have to minimize some functional F of the dose deviations (cf. Eqs. 11.15–11.23) according to:

$$\begin{cases} \min F(d-Hf) \\ f_i \geq 0 \end{cases}. \tag{11.6}$$

The systematic solution of this problem is discussed in Sect. 11.4 (cf. particularly Sect. 11.4.6).

The second and generally preferred alternative is to restate the treatment objectives in order to get a well-posed problem. Instead of trying to find the best way to produce a desirable dose distribution, this goal could be achieved by trying to find the dose distribution which cures the patient with minimal risk for treatment-related morbidity and severe adverse reactions in normal tissues. It is clear that one way to get a well-posed problem is to formulate the clinically relevant treatment objectives as precisely as possible. Due to the uncertainties in: (a) diagnoses, (b) tumor and target volume specification, (c) identification of organs at risk, and (d) treatment setup, we can at best express the *probability to achieve the treatment objectives* in n treatments by fractionated radiotherapy. If we, for the moment, call the associated functional F_+^n, the well-posed problem of radiation therapy may be formulated:

$$\begin{cases} \max F_+^n(Hf) \\ f_i \geq 0 \end{cases}. \tag{11.7}$$

In the subsequent sections we will give examples of the functionals F and F_+^n (cf. Eqs. 11.11–11.14) and discuss algorithms capable of finding the optimal f (cf. Eqs. 11.37, 11.41–11.45).

If we replace Hf by the more general integral from Eqs. 11.1–11.4, we can see that the optimization problem of radiation therapy can also be described as a variational problem where we want to find the incident beam combination which maximizes the functional that specifies the treatment objectives. This kind of formulation will be applied in the special case when the density of clonogenic tumor cells is approximately known, e.g., by quantitative diagnostic imaging (cf. Sect. 11.4.5).

11.2.3 Degrees of Freedom in Radiation Therapy Optimization

Since the beginning of radiation therapy a very extensive number of methods, beam qualities, and irradiation techniques have been developed. Throughout this period photon beams from external radiation sources have dominated the field closely followed by external electron beams and intracavitary and interstitial therapy with sealed sources. During recent decades particle therapy with neutrons, protons, and heavy ions have also been used extensively. The optimal use of these more exotic beam modalities has been discussed at length in the literature (RAJU 1980; FOWLER 1981; BRAHME 1982; BRAHME et al. 1991; KÄLLMAN 1992) and will therefore not be further discussed here. Similarly, brachytherapy will be left out due to the greatly differing irradiation techniques, even if most of the optimization methods that are discussed here apply to both these latter modalities (BRAHME 1988a; BRAHME et al. 1990; HOLMES et al. 1991).

In addition to the choice of radiation modality the degrees of freedom include dose fractionation schedule, beam energy, beam directions, beam collimation, beam profiles, and the irradiation technique in general as determined by the type of equipment used. Figure 11.2 compares the degrees of freedom in classical treatment optimization as it was developed in the mid 1960s (STERLING et al. 1965; HOPE and ORR 1965; BAHR et al. 1968; REDPATH et al. 1976; EBERT 1977) with present-day full-blown generalized conformal therapy (BRAHME 1988a; BORTFELD et al. 1990; BRAHME et al. 1990; WEBB 1990; HOLMES et al. 1991). The former used rectangular wedged fields from a large number, n, of beam directions even though for practical reasons the number was generally less than 20. Classical conformation therapy as developed by TAKAHASHI (1965) instead used a continuum of generally uniform or blocked beams conforming to the target volumes and the organs at risk as seen from the point of view of the beam source (lower left panel of Fig. 11.2). The ultimate step in the

therapy development is to allow full freedom in the shape of the delivered beams with regard to beam energy, beam direction, and beam profiles as illustrated by Eq. 11.1 by the upper right panel in Fig. 11.2, and by Fig. 11.4 (BRAHME 1987; LIND 1991; GUSTAFSSON et al. 1993). These new degrees of freedom of what might be called generalized conformal therapy will allow a full-blown optimization of the dose delivery also for very complex concave and heterogeneous target volumes. However, such treatments are quite complicated both to plan and to deliver, so a more practical treatment optimization as discussed above requires some treatment parameters or degrees of freedom to be locked to make planning and dose delivery practical and manageable. For example, the beam energy or directions of incidence may be preset to clinically obvious values for the target at hand (SÖDERSTRÖM and BRAHME 1993a,b; SÖDERSTRÖM et al. 1993).

For many simple target volumes few field techniques with uniform beams are quite sufficient whereas for more complex shapes nonuniform dose delivery is generally much more advantageous. In fact, it can be shown that the classical conformation therapy method with uniform beams is almost equal to the fully optimized generalized conformal method only for the special case of homogeneous circular symmetric target volumes. For most other target volumes nonuniform dose delivery will be clearly advantageous.

The number of degrees of freedom for different irradiation techniques are quantified and illustrated in Fig. 11.4. From this figure it is evident that the new treatment optimization methods allow several additional degrees of freedom and the number of free variables are several orders of magnitude higher than for classical treatment optimization. In principle, the most general treatment situation is to allow an arbitrary fluence in every direction, from the radiation source to the target volume, for all possible or desirable source positions. Alternatively, this may be expressed such that each point in the target can be irradiated by arbitrary ray intensities from all directions in space depending on the pattern of motion of the radiation source. This means in general five degrees of freedom: either three for the source position and two for the angle of emission or three for a point in the target volume and two for the angle of incidence. On top of this, the energy spectrum and fractionation can in principle vary from point to point. Disregarding these latter variables, this corresponds to about $100^3 \times 100^1 = 10^8$ free variables for coplanar irradiations and about 10^{10} for arbitrary beam directions assuming a spatial and angular grid of calculation with 100 rays along each coordinate. This explains the substantial improvements in the delivered dose distributions with some of the new treatment techniques.

The main methods for nonuniform dose delivery are reviewed in Fig. 11.3. From the figure it is seen that if full dynamic flexibility and reasonable treatment times are required in the clinical application of these new multidimensional optimization techniques, the best methods for nonuniform dose delivery are dynamic multileaf collimation and scanned elementary beams. The dynamic jaw collimation method in principle allows full modulation of the incident beam but at the cost of very extended treatment times. Furthermore, it requires that both the upper and the lower jaw pairs are fully asymmetric so that a narrow rectangular beam spot could be scanned arbitrarily across the entire target volume. If very high dose rates were available and the speed of motion of the collimator jaws was very fast, the time required could be reduced, but this is not a very realistic method with present accelerator systems. The classical filter and block techniques have the flexibility but they are probably not realistic for more than some three portals per patient. They could, for example, work with the few field techniques indicated in the lower panel of Fig. 11.3 either by manual change or with a filter revolver on the front end of the treatment head carrying three to five filters. In recent years several compensator optimization techniques have been developed (Brix et al. 1988; Miller 1988; Ulsø and Brahme 1988; Djordjevich et al. 1990; Weeks and Sontag 1991; Söderström et al. 1993) which are quite useful to handle few field techniques provided suitable beam directions can be identified. In reality the optimal choice of beam direction is one of the most difficult problems of treatment optimization since it involves a restriction on the phase space of feasible beam combinations. This cannot be achieved without having tested all possible beam combinations, which in practice is equal to a global optimization. It also accentuates a difficult radiobiological problem, in a way the Scylla and Charybdis of radiation therapy: With a single beam the small volumes of normal tissues in the entrance region receive a rather high local dose, whereas on the other extreme, with a continuum of arc beams, large volumes receive rather low doses (Popp et al. 1975). To allow a strict optimization realistic radiobiological objective functions capable of distinguishing between these extremes are needed (see Sect. 11.3.2).

11.3 Objective Functions for Treatment Optimization

11.3.1 General

The primary objective of radiation oncology, and for all medical care for that matter, is to make the quality of life for the patient as high as is reasonably achievable from the time of his or her first contact with the medical system. In principle some kind of integral measure of the *quality of life* would be a suitable objective function since a long, healthy, and comfortable life, accounting also for all steps of medical care, is most desirable. However, a very long life with severe loss of life quality is not generally desirable (Bush 1982), so some kind of weighted integral measure should therefore be most suitable. Obviously the quality of life of a person has to be a subjective quantity (Campbell et al. 1976) but from a medical point of view it also has a more strict and objective side as a general health index. This index can range from unity, corresponding to perfect health, to zero when the person in medical terms is dead or does not want to live anymore. There are several measures which are aimed at quantifying the condition of the patient, such as the Karnofsky status (Karnofsky et al. 1948), but all of them are far too complex and difficult to relate to therapeutic actions to be used directly for therapy optimization. We will therefore first discuss some simplifications that need to be introduced to obtain an objective function that is useful for computed radiation therapy optimization.

In practice, two main groups of objective functions have been used, physical and biological. The most commonly employed objective functions are physical and they often describe the properties of the delivered dose distribution in the *target volume* and affected normal tissues or *organs at risk*. In fact, the definition of these latter concepts is extremely important for how accurate and precise a treatment can be performed, as illustrated in Fig. 11.5. It is fundamental that the target volume and the organs at risk are defined with the narrowest possible safety margins in relation to the anatomical reference points that are used for setting up the patient for external beam therapy. In principle the target volume should include all the volumes and margins whose dimensions cannot be affected by changing or developing dose planning or treatment techniques (including fixation of the patient). Therefore the target volume is obtained by adding only an anatomical margin to the oncological volume to account for all internal

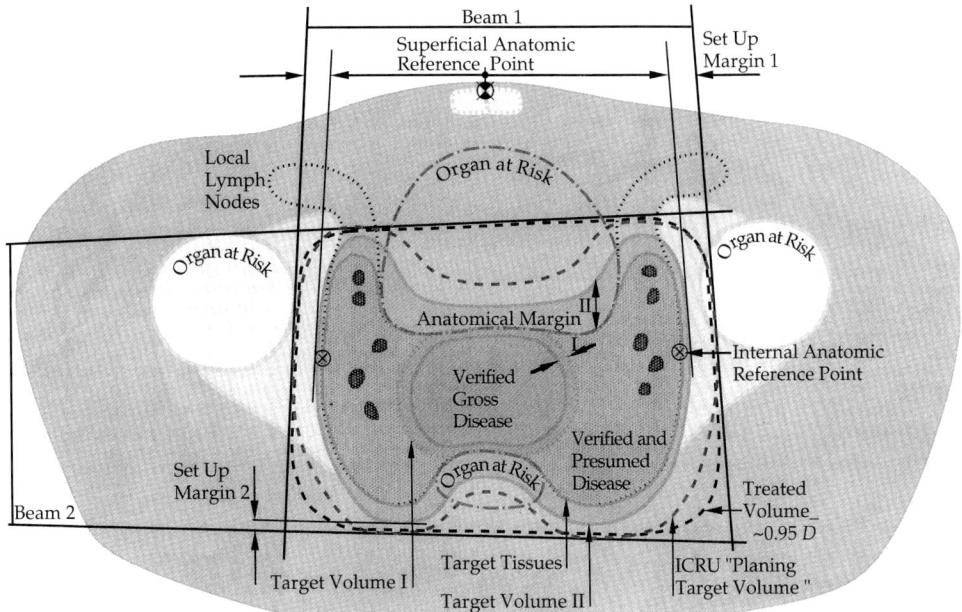

Fig. 11.5. Illustration of the target volume concept and its relation to the anatomic margin and the treatment setup margin. Different anatomic margins may be needed for the gross tumor and the microscopic disease. The anatomic reference points are fundamental for defining the target volume in the coordinate system of the patient and for the accurate setup of the therapeutic beams

shape changes of organs and their possible motions relative to the anatomical reference points (AALTONEN 1992, NACP 1993).

To this target volume, setup margins may later have to be added depending on the treatment technique and the quality of the treatment unit. It is fundamental that precisely this target volume should be used for dose prescription and reporting as it is the most relevant one for the treatment outcome. The biological objective functions aim at quantifying precisely this quantity, namely the probability that the patient will have a desirable treatment outcome. From this point of view the radiobiological objective functions therefore do quantify the quality of life of the patient after therapy, as will be discussed in more detail now.

11.3.2 Biological Objective Functions

There are a large number of factors that by necessity make the biological objective functions statistical quantities. First of all, the vital end point of killing all clonogenic tumor cells to eradicate the tumor makes the beneficial treatment outcome "tumor control" a stochastic quantity. We can only state the probability, P, and standard deviation, σ, for a certain treatment outcome due to the large uncertainty in hitting the very last tumor clonogen.

In the space of clonogenic tumor cells the treatment objective can be formulated: The maximum value of the tumor recurrence probability should be as low as possible throughout the target volume without causing severe damage to normal tissues. This formulation is quite interesting because it solves the inversion problem in a simple way, but it requires information about the density and sensitivity of clonogenic tumor cells (see Sect. 11.5 for further details). Since tumor eradication in a group of patients is a multidimensional truly binomial process (a given patient can either be cured or not), it is straightforward to calculate the relative standard deviation of the number of cured patients $\sigma_{P_B} = (1-P_B)P_B$ where P_B is the probability to cure each one of them (cf. KÄLLMAN 1992). The uncertainty thus has its highest value of 25% when P_B is 50%, as would be expected.

On top of this uncertainty there are uncertainties both in defining the target volume, so that all tumor clonogens really are included, and in dose delivery, so that the target volume is fully irradiated. Furthermore, for a given patient and tumor classification (e.g., according to the TNM system to minimize the influence of the extent of the disease), it is generally not known whether a given tumor is more sensitive or resistant than that of the "mean patient" for which established dose-response relations should be applicable. The same is true for the normal tissues of that patient and both these facts add to the total uncertainty.

For these reasons it is also generally assumed that the probability of severely injuring the patient P_I is statistically independent from the probability of a beneficial treatment outcome P_B, i.e., tumor control. The probability of a successful treatment, P_+, can then be expressed since the covariance of P_B and P_I is zero and the product law of statistics can be applied to determine the conditional probability of having tumor control without severe injury:

$$P_+ = P_B - P_B \cap P_I \approx P_B(1 - P_I), \qquad (11.8)$$

which is the traditional expression used by most workers (COHEN 1987; WOLBARST 1984; SCHULTHEISS et al. 1983 – the longer notations NTCP and TCP are sometimes used instead of P_I and P_B, respectively; the compact notation used here was introduced by COHEN 1960, MOORE and MENDELSOHN 1972, and ANDREW 1985). However, there are a number of factors that can alter this conclusion since it is strongly dependent on the type of dose delivery that is employed and also on whether there might be a true biological correlation between P_B and P_I.

There are several biological mechanisms that can introduce correlation, for example, if the patient happens to have an unusually efficient repair system for double-strand breaks (SCHWARTZ et al. 1988; ÅGREN et al. 1990). Since most oncogenes are different from the genes responsible for efficient repair of radiation damage, it is probable that both the tumor and the normal tissue will be more resistant to irradiation. Conversely, if the patient happens to have the ataxia telangectasia gene, which is one of several known genetic defects quite common in cancer patients (NORMAN et al. 1988; TIMME and MOSES 1988; ÅGREN et al. 1990), there would be an increased risk for normal tissue injury but also an increased probability of controlling the tumor. Furthermore, LEIBEL et al. (1991) pointed out that some tumors which show an increased risk for metastatic dissemination might be phenotypically distinct and their presence, like the radiation sensitivity, thus may be used as a predictive indicator which could influence the treatment decision.

However, there are also other genes like the tumor suppressor gene p53 which seem to have a regulatory effect on how DNA damage is handled by the cell. A mutant p53 gene is one of the most common defects found in tumor cells. This may be an important molecular biologic explanation why radiation therapy has a more severe effect on the tumor cells than on

healthy normal tissues with the wild-type p53, which is better capable of handling radiation-induced DNA damage (LANE 1992).

Depending on the type of dose delivery and the proportions of genetic variations for the tumor at hand, deviations from the simple expression 11.8 may therefore be seen. It is consequently advantageous to base the analysis as far as possible on clinical data. In a recent study of head and neck tumors by ÅGREN et al. (1990) a rather uniform dose distribution was used (parallel opposed beams). It was found that P_B and P_I were totally correlated at low doses, and only for large tumors at high doses did the uncorrelated portion δ reach a value of about 20%. The clinically observed P_+ value was then accurately described by:

$$P_+ = P_B - P_I + \delta (1 - P_B)P_I \qquad (11.9)$$

for clinically relevant doses as illustrated by the clinical data for large head and neck tumors in Fig. 11.6. A very convenient and radiobiologically coherent parametrization of the dose-response curves (solid curves) is obtained from:

$$P_{B,I} = 2^{-e^{\gamma e(1-D/D_{50})}}, \qquad (11.10)$$

where γ is the normalized dose-response gradient and D_{50} is the dose causing 50% probability of effect (tumor control or severe normal tissue reactions, cf. KÄLLMAN et al. 1992b for further details). Using Eqs. 11.9, 11.10, and 11.5, the associated functional may now be written:

$$F_+^0(f) = P_+(Hf). \qquad (11.11)$$

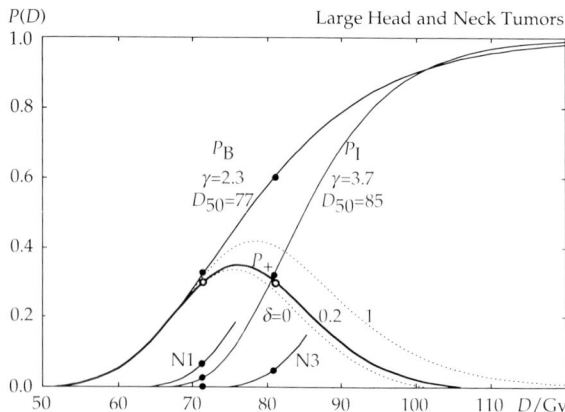

Fig. 11.6. Large head and neck tumors. Clinically established dose-response relations for tumor control (P_B) and fatal normal tissue reactions (P_I). The clinically established probabilities for complication-free tumor control (P_+) are indicated by *open circles* (cf. Eq. 11.9). The other clinical data points are indicated by *solid circles* and the curves are fitted using Eq. 11.10

To account for the possible variation in the alignment of the therapy beam with the target volume it may be even more desirable to maximize the expectation value $\langle P_+ \rangle$ rather than P_+ itself. It has recently been shown (LIND et al. 1992) that the functional $\langle P_+ \rangle$ is a very important quantity for optimization under the influence of stochastic processes. If the total dose distribution D_t is assumed to be a sum of the n individual statistically independent dose fractions, the expectation value of P_+ is given by

$$F_+^n = \langle P_+(D_t, \text{n}) \rangle$$
$$= \int_{-\infty}^{\infty} P_+ \left(\sum_{i=1}^{n} D(\boldsymbol{r} - t_i) \right) \prod_{i=1}^{n} \varphi_T(t_i) \mathrm{d}^{3n} t, \tag{11.12}$$

where φ_T is the probability density of the field displacement vector $\boldsymbol{t}$. This expression, which is an 3n-fold integral, is quite complex and very time consuming to evaluate in the general case. However, two cases of clinical interest can be handled more easily. For one single dose fraction (n = 1), Eq. 11.12 reduces to:

$$F_+^1 = \langle P_+(D_t, 1) \rangle$$
$$= \int_{-\infty}^{\infty} P_+(D(\boldsymbol{r} - t) \varphi_T(t) \mathrm{d}^3 t. \tag{11.13}$$

This equation offers a possibility to find the dose distribution $D(\boldsymbol{r})$ which in a single irradiation maximizes the expectation value of the probability of curing the patient without fatal complications (cf. Fig. 11.13).

For the case with a very large number of small dose fractions, $\varphi_T(t)$ will be very accurately sampled by $T_1 - T_n$. In the limit $n \to \infty$, the total dose is given by the sum of a series consisting of infinitesimal dose fractions, the limit value of which is a convolution integral with the probability density $\varphi_T(t)$. Since this limit value will be independent of t, it can be brought outside the outer multiple integral and the remaining integral is identical to unity according to:

$$F_+^\infty = \int_{-\infty}^{\infty} P_+ \left(\int_{-\infty}^{\infty} D(\boldsymbol{r} - t) \varphi_T(t) \mathrm{d}^3 t \right) \underbrace{\prod \varphi_T(t_i) \mathrm{d}^{3n} t}_{n \to \infty}$$
$$= P_+ \left(\int_{-\infty}^{\infty} D(\boldsymbol{r} - t) \varphi_T(t) \mathrm{d}^3 t \right) \tag{11.14}$$

In the limit with infinitely many treatments the optimal dose distribution under the influence of a stochastic process is well defined. The problem of finding the optimal beam profile $D(\boldsymbol{r})$ is equal to finding the resultant dose distribution $\hat{D}$ which by the influence of the stochastic process maximizes P_+. This simplifies the calculation since the biological

objective function does not need to be evaluated inside the integral (see Sect. 11.6.1).

11.3.3 Physical Objective Functions

To allow a strict optimization, the objective function should be a scalar quantity (more precisely a functional) as otherwise two different treatments cannot be compared on the same scale. This property is true for all the radiobiological objective functions presented above. However, to speed up calculations, further simplifications of the objective function are often desirable. This can be achieved by expressing the treatment objectives as simple functions of the dose distribution, but other parameters such as the total treatment time have also been used (EBERT 1977). The success of the dose optimization is directly linked to the ability to define such clinically relevant treatment objectives. The following physical features of importance for tumor control and normal tissue complications are probably the most significant ones and have been used extensively over the years:

1. Mean energy imparted to target volume and organs at risk ($\bar{\varepsilon}$, D)
2. Dose variance over target volume (σ_D)
3. Minimum absorbed dose in target volume (D_{min})
4. Peak dose to organs at risk (D_{max})
5. Conformity of treatment volume to target volume (cf. Figs. 11.5 and 11.21)

When the desired dose distribution, $D(\boldsymbol{r})$, in the target volume is known, it may also be possible to quantify the difference between it and the best achievable dose distribution, $\hat{D}(\boldsymbol{r})$. In principle, one could therefore apply the p:th norm on this difference to quantify deviations at all points $\boldsymbol{r}_i$ of interest according to:

$$\Delta_p = \left(\sum_i |D(\boldsymbol{r}_i) - \hat{D}(\boldsymbol{r}_i)|^p \right)^{1/p}. \tag{11.15}$$

The elliptic norm may also be of interest in this context since it can be regarded as a generalization of Δ_2:

$$\varepsilon_A^2 = \boldsymbol{d}^T A \boldsymbol{d}, \tag{11.16}$$

where A is a positively definite matrix. For the simple case when $A = \boldsymbol{I}$, the unity matrix, the elliptic norm is precisely equal to the 2nd norm. The 2nd norm is of special clinical interest since it can be shown to be related to the probability of achieving local control for a homogenous tumor (BRAHME 1984, 1992b):

$$P_B(D(\boldsymbol{r})) = P_B(\overline{D}) - \frac{\gamma^2}{2 P_B(\overline{D})} \left(\frac{\sigma_D}{\overline{D}} \right)^2 \cdots, \tag{11.17}$$

where

$$\overline{D} = \frac{\sum_{i=1}^{n} D(r_i)}{n} \text{ and } \sigma_D^2 = \frac{\sum_{i=1}^{n} (D(r_i) - \overline{D})^2}{n}. \quad (11.18)$$

Both the first two spatial moments of the dose distribution (and thus the 2nd norm) are therefore closely related to the clinical outcome as seen from the expression for the probability to control the tumor. In fact it can be shown that $\overline{D}$ is a very suitable quantity also for prediction of complication-free tumor control (BRAHME 1992b) and there are also clinical data supporting this (e.g., BURGERS et al. 1985). In addition, the use of $\overline{D}$ minimizes the second norm (Δ_2) which is a very valuable property according to Eq. 11.17. For tissues with a steep threshold-type response the infinite norm, Δ_∞, may also be of interest as it is a way of quantifying the maximum deviations.

In the above discussion some arguments are given for the frequent use of the least square deviation as a measure of the deviation from the desired dose distribution in treatment optimization (Δ_2^2 according to Eq. 11.15). The associated functional is given by:

$$F_2(f) = \frac{1}{2} f^T H^T H f - f^T H^T d$$

$$(= \frac{1}{2}(Hf - d)^T (Hf - d) - \frac{1}{2} d^T d). \quad (11.19)$$

However, this is only a first step towards treatment optimization since, for example, a narrow hot spot is generally better tolerated than a narrow cold spot provided the mean energy imparted to the tumor is the same (cf. BRAHME 1992b). The principal reason for this is that the lower dose inside the cold spot decreases the local overkill much more than an extended cold area, thus making the net recurrence probability with the cold spot significantly larger.

A way out of this clinical dilemma is to not even allow local underdosage but to minimize the local overdosage in all regions where a perfect match is not possible. The appropriate functional for this more clinically relevant objective function has recently been developed by LIND and co-workers (LIND 1990; LIND and BRAHME 1992):

$$F_1(f) = \frac{1}{2} f^T H f - f^T d. \quad (11.20)$$

The last functionals will be discussed in more detail below (Eqs. 11.38, 11.39).

In addition to the above functionals it may also be relevant to look at the entropy (cf. BOLTZMANN 1927;

SHANNON 1948) of the deviation from the desired plan using the definition:

$$S(d) = \sum \left[D(r_i) - \hat{D}(r_i) \right] \ln \frac{\left(D(r_i) - \hat{D}(r_i) \right)}{\overline{D}}. \quad (11.21)$$

This functional is particularly relevant when the microscopic extension of the tumor is uncertain (cf. Fig. 11.5) and it may be desirable to maximize the entropy of the deviation from some prior estimate of the uncertainties in the dose prescription. When such specifications are available they may even be accounted for using the maximum entropy method (JAYNES 1957, BUCK and MACAULAY 1991). To make comparisons with other measures and objective functions easier (cf. Fig. 11.18), it is sometimes useful to transform $S(d)$ linearly so it is always less than unity, corresponding to maximum information content (SÖDERSTRÖM and BRAHME 1993a), by putting:

$$S_-(d) = 1 + \frac{S_{min} - S(d)}{S_{max}}. \quad (11.22)$$

Another parameter which also may be useful to quantify the clinical value of a beam is its content of spatial frequencies, u. Generally it is fields with low spatial frequencies that can be most easily produced so an integral over the absolute value of the low frequency portion of the Fourier transform of the optimal dose distribution, $\tilde{D}$ may be a useful quantifier of gross structure in a beam:

$$Z = \int_0^{u_{max}} \left| \tilde{D}(u) \right| du. \quad (11.23)$$

Here it is natural to set the upper integration limit to correspond to the smallest structure of interest in the target volume, the resolution of the leaf collimator or the precision of patient setup, depending on the situation, and thus $u_{max} \approx 1/2\Delta r$ (cf. Fig. 11.14 and SÖDERSTRÖM and BRAHME 1993a).

Finally, it should be pointed out that it is sometimes of interest to use the difference in biological effect to quantify the deviations from the desired dose distribution rather than the dose difference as described by Eq. 11.15.

11.4 Mathematical Methods for Treatment Optimization

11.4.1 Overview

As discussed above, treatment optimization can follow two routes: either a restricted one where a *true inversion* of the integral equation is possible or a second more general one where a global numerical optimization has to be employed. Under simplifying assumptions an optimization sometimes can be performed by a direct mathematical inversion. We will first look closer at four examples of this technique. In all cases they work when the desired dose distribution is "well behaved." In the first case it is also assumed that the target volume poses rotational symmetry and in most cases the risk for damage to normal tissues is disregarded.

Almost all optimization methods use some kind of *iterative algorithm* which successively finds better and better solutions. Traditional "manual" or "visual" optimization belongs to this group, where the cumulated experience of the planner helps him to find better and better plans. The iterative algorithms may be either *stochastic* or *systematic* in their way of searching for the optimal solution whereas a human planner is generally a mixture of the two.

Traditional computer-aided visual treatment optimization is a quasistochastic iterative method where better and better treatment plans are found by a trial and error process in which the "adaptive experience" of the planner can help in speeding up the process. There are two slightly more systematic but still basically stochastic methods, namely: *inverse Monte Carlo* and *simulated annealing*. The inverse Monte Carlo approach is based on studying the adjoint transport equation in order to find the most suitable solution (DUNN 1981). So far this method has not been applied to radiation therapy planning problems, probably since it is associated with the same existence problems as Eqs. 11.4, 11.24, 11.26, and 11.27.

In recent years the simulated annealing technique has become quite popular for treatment optimization (WEBB 1989; MORRILL et al. 1990; MOHAN et al. 1992). The method, which in general is very time consuming, is based on a stochastic sampling of the total phase space of incident elementary beams (preferably pencil beams). The beam samples are accepted if they increase the objective function, and if not, they might still be accepted with a reduced probability given by the annealing expression, $e^{-\Delta F/kT}$. To start with the "annealing temperature" T is high and larger deviations are accepted, but as the iteration number increases, it is lowered to obtain a quasi-adiabatic annealing and to build up the fine structure of the optimal solution (cf. METROPOLIS et al. 1953; KIRKPATRICK et al. 1983). In principle, the functional F could be any of the objective functions of Sect. 11.3. Most logical is to use a biologic functional (MOHAN et al. 1992; MORRILL et al. 1990),

but the deviation from a desired dose distribution also may be useful, e.g., Δ_2 (WEBB 1989).

Most algorithms that have been developed over the last almost 30 years of treatment optimization belong to the *systematic iterative* group. In the present review there is not room to cover all of them in detail. Some of them have already been mentioned in Sect. 11.2.3 in connection with the degrees of freedom allowed in the optimization. We will briefly discuss them here, grouped in temporal order of introduction and according to the principal type of objective function and mathematical technique employed. The objective function could be linear or quadratic or generally nonlinear in the unknown variable and so could the associated constraining conditions.

The earliest techniques used various *score functions* in an attempt to quantify a good dose plan (HOPE and ORR 1965; HOPE et al. 1967; VAN DER LAARSE and STRACKEE 1976) mainly aiming at a good dose homogeneity in the tumor and little normal tissue damage. More recently biologic scores have also been used (GREMMEL et al. 1977; GOITEIN and NIEMIERKO 1988; NIEMIERKO 1993).

The *linear programming* technique has also been very popular over the years, with minimal mean energy imparted to normal tissue as the most common objective (BAHR et al. 1968; HODES 1974; EBERT 1977; MANTEL et al. 1977; BOLLMAN et al. 1981; LEGRAS et al. 1982; SONDERMAN and ABRAHAMSON 1985; LANGER and LEONG 1987; LANGER et al. 1991; and the recent review by ROSEN et al. 1991).

As soon as the least-square deviation from a desired dose distribution was sought, *quadratic* or *nonlinear methods* were needed (REDPATH et al. 1976; MCDONALD and RUBIN 1977; FRANKE 1980; STARKSCHALL 1984; DJORDJEVICH et al. 1990; BORTFELD et al. 1990).

A special class of algorithms has recently gained interest. They do not perform a strict optimization but rather a *feasibility search* for solutions within the limits of the treatment constraints such as minimal and maximal tumor and normal tissue doses. This method has been pioneered by CENSOR and co-workers (1988a,b) (cf. also POWLIS et al. 1989; VAN SANTVOORT and HUIZENGA 1991; STARKSCHALL and EIFEL 1992).

Due to their universal applicability the gradient and pencil beam methods are treated in some detail towards the end of this section. Finally a discussion of the similarities and differences of radiation therapy optimization and image reconstruction in computed tomography is given.

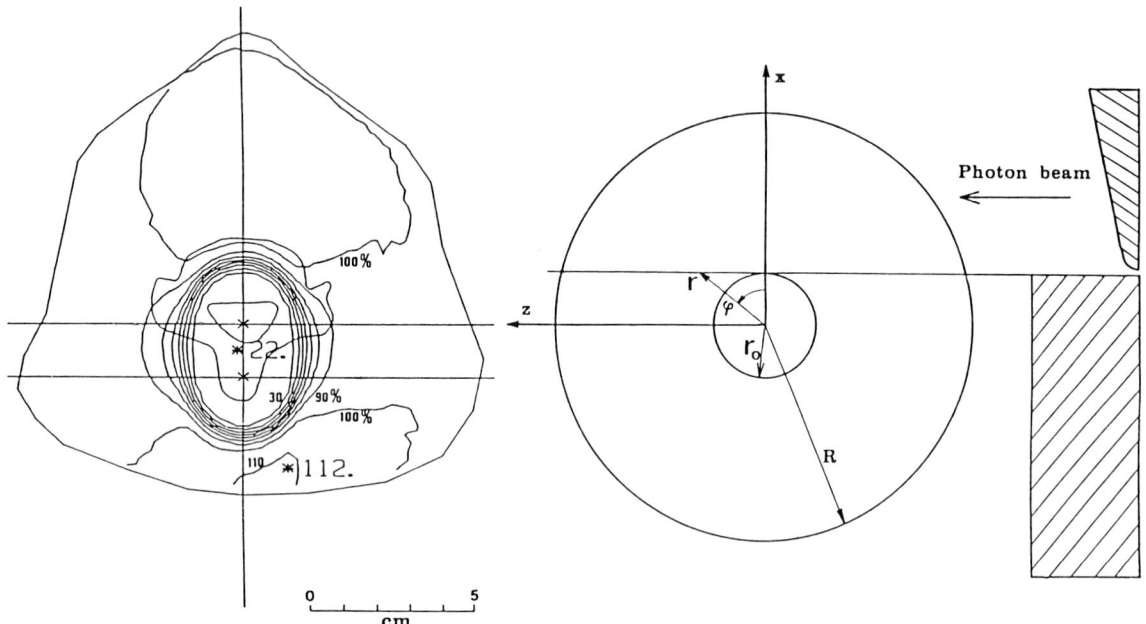

Fig. 11.7. The irradiation geometry and coordinate system used in the nonlinear wedge technique are shown to the *right*. The origin of the rectangular and polar coordinate system is located at the isocenter of the therapy machine. The location of the beam block and the nonlinear wedge-shaped filter is also indicated. The resultant dose distribution in a larynx cross-section is shown to the *left*. The gantry rotation starts with the gantry in a dorsal position, and the treatment is performed during a 180° arc using the ventral rotation center. Thereafter the treatment table is raised to a distance equal to that between the two rotation centers, and the treatment is continued for the remaining 180° (Lax and Brahme 1982)

11.4.2 Cylindrically Symmetric Target Volumes

One of the earliest applications of inversion methods in radiation therapy planning was the development of nonlinear wedge techniques for lymph node irradiation in the head and neck region, as shown in Fig. 11.7. Lax and Brahme (1982) developed a strongly nonlinear wedge filter to give a high uniform dose to the nodes and a low dose to the spinal cord using a double arc technique. First a numerical solution for the wedge profile was found by an iterative procedure but in addition the integral equation for rotationally symmetric arc therapy was also set up (cf. Fig. 11.7):

$$D(r) = \int d(x)\, e^{-\mu_p z}\, d\varphi / \pi \qquad (11.24)$$

By using Laplace transformation on this integral equation a general inversion formula was derived which allows the calculation of the required beam profile from a given desired dose distribution in the patient under cylindrical symmetry and full 360° rotation:

$$d(x) = \frac{d}{dx} \int_{r_o}^{x} \frac{\cos[\mu_p \sqrt{x^2 - r^2}]}{\sqrt{x^2 - r^2}} D(r) r\, dr. \qquad (11.25)$$

Here $D(r)$ is the desired radial dose profile in the patient, $d(x)$ is the required dose profile of the incident beam, and μ_p is the practical attenuation coefficient of the beam (Brahme et al. 1982). Unfortunately most patients are not rotationally symmetric so this general inversion equation is not generally applicable. However, a straightforward generalization of the theory is possible, namely to arbitrary body shapes, provided the target volume is still cylindrically symmetric. Because the derived dose profiles of the beam (cf. Eq. 11.25) pertain to a plane through the center of the target volume, it is just a matter of correcting for the nonuniform beam absorption in each CT slice of the body to obtain the optimal shape of the incident beam at each angle (cf. also Eq. 11.50 below). Obviously this results in angularly varying incident beam profiles in the general case even though the target volume is rotationally symmetric.

However, Eq. 11.25 does not result in physically possible incident dose distributions for arbitrary desired dose profiles, $D(r)$, in the patient since $d(x)$ may take both negative and infinite values. These problems were touched upon in Sect. 11.2.2 and are further discussed in Sect. 11.4.8 and in several recent publications (Brahme et al. 1982; Cormack 1987;

CORMACK and CORMACK 1987; BRAHME 1988a; CORMACK and QUINTO 1989; BARTH 1990; GUSTAFSSON et al. 1993). A closely related inversion problem, namely how to make an arbitrary picture just by drawing uniform straight lines, was solved by BIRKHOFF (1940). If the uniform lines were to be replaced by "pencil" beams and the solution could be generalized for this case, the inverse problem of radiation therapy would be solved. However, even if this solution is found it would in general be useless because for most desired dose distributions there is no nonnegative solution. Furthermore, the requirement given by Birkhoff for the existence of a solution in the drawing case is very hard to use since the coefficients of the resultant Fourier series have to be continuous and form a nonnegative continuous function. Such a requirement is of little help when trying to judge whether a desired dose distribution can be produced or not, and it is not useful for finding the best possible solution when no exact solution exists (cf. also BOCHNER 1948, and Secs. 11.4.3 and 11.4.8). The recent pencil beam algorithm of GUSTAFSSON et al. (1993) solves this difficult problem by a very efficient numerical algorithm (see Sec. 11.4.7).

11.4.3 Direct Fourier Transformation

Equations 11.3 and 11.4 are strict convolution integrals and can be solved exactly by direct Fourier transformation since the convolution in Fourier space is equal to the product of the Fourier transforms. Therefore, a strict mathematical inversion of Eq. 11.4 results in

$$f(r) = F^{-1}\left(\frac{\tilde{D}(u)}{\tilde{h}(u)}\right), \tag{11.26}$$

where $\tilde{D}(u)$ and $\tilde{h}(u)$ are the Fourier transforms of the desired dose distribution and the kernel respectively, and F^{-1} is the inverse Fourier transform operator (cf. BRAHME 1985, 1988a; HOLMES et al. 1991). Unfortunately, this method is not generally applicable for arbitrary D and h since f will often take negative values on some interval and may have severe oscillations at the zeros of $\tilde{h}$. This latter problem can sometimes be handled by a low pass filter in Fourier space according to:

$$f_\lambda(r) = F^{-1}\left(\frac{\tilde{D}(u)}{\tilde{h}(u)(1+\lambda|\tilde{h}(u)|^{-2})}\right). \tag{11.27}$$

However, in most difficult problems the inverse Fourier transformation method is not very useful

due to the negativity problem (DAVIS 1982; LIND and BRAHME 1985). The problem with negative f components is in principle handled by a theorem derived by BOCHNER (1948), but as in the case of the Birkhoff solution, it is very hard to use it in practice. It was pointed out by BIRKHOFF (1940) that if erasure or negative functions are allowed there always exists an exact solution of the drawing problem as in the radiotherapy case given by Eqs. 11.26 and 11.27 (see also Sect. 11.4.8).

One of the more interesting applications of Eq. 11.26 in recent years has been worked out by MACKIE and co-workers (cf. HOLMES et al. 1991) where they have replaced $\tilde{h}$ by a modified narrower kernel without zeros over the range of interest of $\tilde{D}$. The resultant f is only a crude first approximation to the optimal solution but it may serve as a suitable starting point for a more exact optimization similar to the case with Eq. 11.31 below (cf. HOLMES et al. 1991; KÄLLMAN et al. 1992a).

11.4.4 Inversion by Taylor Expansion

In cases where the function being studied is well behaved with multiple continuous derivatives an approximate inversion by Taylor expansion can sometimes be of value. We will illustrate this on the problem of optimization under uncertainties in beam patient alignment. According to Eq. 11.14 the total dose distribution under continuous irradiation with stochastic beam patient motions may be expressed by

$$D_t(r) = n\int D(r-t)\varphi_T(t)\mathrm{d}^3t. \tag{11.28}$$

By Taylor expansion of D around r and termwise integration, D_t is expressed in a series containing the derivatives of D. By assuming φ_T to be rather narrow compared to the fluctuations of D, this series may be inverted, giving:

$$nD(r) = D_t(r) - \frac{\sigma^2}{2}D_t^{ii}(r) + \frac{\sigma^4}{8}D_t^{iv}(r) - \frac{\sigma^6}{48}D_t^{vi}(r)..., \tag{11.29}$$

where σ is the standard deviation of φ_T (cf. BRAHME 1981). This method may be used, for example, to correct the desired target dose distribution D_t for the influence of known uncertainties in beam patient alignment (LIND et al. 1992).

11.4.5 Optimization as Variational Problem

The probability of controlling a tumor can be expressed mathematically using Poisson statistics when the initial density, $n_0(r)$, and sensitivity, $D_0(r)$, of clonogenic tumor cells are known. The associated integral equation becomes:

$$P_B = \exp(-\iiint_{V_t} n_0(r)\, e^{-D(r)/D_0(r)d^3r}).\qquad(11.30)$$

When the risk of causing normal tissue damage is negligible, this equation can be used to derive the optimal dose distribution $\hat{D}(r)$ which eradicates the tumor with a certain probability P_B and at the same time gives the lowest possible maximum value of the local recurrence probability in the target volume. This "Mini-Max" problem was treated independently by FISHER (1969) and BRAHME and ÅGREN (1987). The optimal dose distribution under this optimality criterion should obviously have a uniform recurrence density over the whole target volume, making the integrand in Eq. 11.30 constant and thus:

$$\hat{D}(r) = D_0(r)\, \ln\left\{\frac{n_0(r)\cdot V_t}{-\ln P_B}\right\},\qquad(11.31)$$

where V_t is the target volume. The optimal dose distribution in this meaning is thus proportional to the local D_0 value and the logarithm of the tumor cell density! This relation is very useful when $n_0(r)$ can be estimated by 3D diagnostic imaging and D_0 is known by experience or by a predictive assay. But it can also be used with approximate n_0 and D_0 values to find a first estimate of the optimal dose distribution as input to more complex algorithms (cf. KÄLLMAN et al. 1992a).

This method could be generalized to take normal tissue reactions into account using an equation similar to Eq. 11.30 for the tissue rescuing units of the normal tissues or preferably by using Eq. 11.10 and Eq. 11.8 or 11.9 to express the probability of achieving complication-free tumor control. Similar to the case with Eq. 11.7, this results in a variational problem where one wants to maximize P_+ by essentially maximizing the difference between integrals analogous to Eq. 11.30. If we, for simplicity, assume $\delta \approx 0$ and a uniform sensitivity and completely serial organization (cf. KÄLLMAN et al. 1992b) of surrounding normal tissues, the problem is:

$$\begin{cases} \max P_+ = \exp(-\iiint_{V_t} n_0(r)\, e^{-D(r)/D_0(r)}d^3r) \\ \qquad + \exp(\iiint_{V_n} v_0(r)\, \ln\left[1 - P_1\,(\hat{D}(r))\right]d^3r) - 1, \\ D(r) \geq 0 \end{cases}$$
$$(11.32)$$

where v_0 is the relative density of the normal tissues. The condition for Eq. 11.31 to be applicable is thus

that the P_1 term in Eq. 11.32 is negligible. In clinical practice the normal tissues are generally dose limiting and the full optimization problem of Eq. 11.32 has to be treated. However, this problem is outside the scope of this chapter.

11.4.6 Gradient Methods

Due to their universal applicability and general flexibility the gradient methods will be treated here in more detail. The gradient methods also have the advantage of being much faster than the stochastic methods, particularly for convex functionals. Successful applications of gradient methods include the work by COOPER (1978), LIND and co-workers (1987–1992), ULSØ and BRAHME (1988), BORTFELD et al. (1990), HOLMES et al. (1991), GUSTAFSSON et al. (1993), and SÖDERSTRÖM and BRAHME (1993). A natural starting point in the analysis is to make a second-order Taylor expansion of the objective function $F(f)$ around some point in the phase space of possible solutions f^k (cf. Sec. 11.2.2).

$$F(f) \cong F(f^k) + \nabla F^T(f^k)(f - f^k)$$
$$+ \tfrac{1}{2}(f - f^k)^T \nabla^2 F(f^k)(f - f^k),\quad(11.33)$$

where ∇F is the gradient vector of the objective function:

$$\nabla F^T = \left(\frac{\partial F}{\partial f_1}, \frac{\partial F}{\partial f_2}, \cdots \frac{\partial F}{\partial f_m}\right)\qquad(11.34)$$

and $\nabla^2 F$ is the associated Hessian matrix:

$$\nabla^2 F = \begin{cases} \dfrac{\partial^2 F}{\partial^2 f_1} & \dfrac{\partial^2 F}{\partial f_1\partial f_2} & \cdot\cdot & \dfrac{\partial^2 F}{\partial f_1\partial f_m} \\ \dfrac{\partial^2 F}{\partial f_2\partial f_1} & \cdot & \cdot\cdot & \cdot \\ \vdots & \vdots & \vdots\vdots & \vdots \\ \dfrac{\partial^2 F}{\partial f_m\partial f_1} & \cdot & \cdot\cdot & \dfrac{\partial^2 F}{\partial^2 f_m} \end{cases}\qquad(11.35)$$

From Eq. 11.33 it is clear that the simple quadratic form in f has its minimum value at f^{k+1} given by:

$$f^{k+1} = f^k - (\nabla^2 F)^{-1}\nabla F,\qquad(11.36)$$

which is Newton's method to find the minimum of the functional F by a systematic iterative search. According to the mean value theorem the exact minimum point may be shifted from Eq. 11.36 due to the influence of higher order terms in the expansion (Eq. 11.33). A large number of different methods have therefore been developed to simplify and speed up the basic Eq. 11.36 such as the *steepest descent* and the *conjugate gradient* methods. However, in clinical practice with very large vectors it is often too complex or time consuming to determine the inverse of the Hessian matrix and other corrections to the gradient term. It may even be faster to do an approximate calculation and instead repeat it several times to compensate for the loss in speed of convergence. Owing to the fundamental importance of the positivity constraint on the irradiation density f, Lind (1990) showed that the following simplified and modified gradient algorithm is ideally suited for fast calculations in radiation therapy:

$$f^{k+1} = C\left(f^k - aI\nabla F(f^k)\right). \tag{11.37}$$

Here, the inverse of the Hessian corresponds to a times the unity matrix, $\mathbf{I}$ (cf. also Bortfeld et al. 1990), and a positivity operator C has been added to ensure that f never falls below zero during the iterations. It can be shown that the iterative algorithm Eq. 11.37 converges with steepest descent in the quadratic norm $\varDelta_2$, since $A^{-1} = \mathbf{I}$ in this case (Ortega and Rheinboldt 1970). Equation 11.37 can now be combined with any of the objective functions from Sect. 11.3 above (F_1, F_2, F_+^n) to obtain a rapidly converging algorithm capable of optimizing different treatment aspects.

1. With F_1 (Eq. 11.20) it can be shown that f^∞ generates that dose distribution which is as close to d as possible but never below d. In fact, it can be shown that $Hf^\infty = d$ whenever $f_i^\infty \geq 0$ and $Hf^\infty \geq d$ when $f_i^\infty \geq 0$ and according to the Kuhn-Tucker theorem (cf. Luenberger 1973) F_m is minimized by the algorithm (Lind 1990). After insertion of F_m in Eq. 11.35 we obtain:

$$f^{k+1} = C(f^k - a(Hf^k - d)). \tag{11.38}$$

This type of algorithm has earlier been discussed by Schafer et al. (1981) in the context of image restoration.

2. With F_2 (Eq. 11.19) the algorithm will minimize the deviations between Hf and d in the least-square sense and according to Eq. 11.18 the variance of the dose distribution will be minimal and cause a high tumor control according to Eq. 11.17. This algorithm reduces to:

$$f^{k+1} = C\left(f^k - a(H^THf^k - H^Td)\right), \tag{11.39}$$

which is basically the same algorithm as Eq. 11.38 except for the factor H^T (cf. Lind 1991).

3. Finally, with $-F_+^n$ (Eq. 11.12) it is possible to maximize the treatment with respect to the probability of achieving complication-free tumor control. For this very important objective function we will only write out the full algorithm for the case of $n = 0$ and 1 by using the chain rule:

$$\nabla_f\left(-F_+^0(Hf)\right) = -\frac{\partial Hf}{\partial f}\nabla_{Hf}\left(F_+^0(Hf)\right)$$

$$= -H^T\nabla_{Hf}F_+^0(Hf) \tag{11.40}$$

Therefore for n = 0 Eq. 11.37 reduces to:

$$f^{k+1} = C\left(f^k + aH^T\nabla_{Hf}F_+^0(Hf^k)\right). \tag{11.41}$$

Similarly for n = 1

$$f^{k+1} = C\left(f^k + aH^T\boldsymbol{\Phi}\nabla_{Hf}F_+^0(Hf^k)\right), \tag{11.42}$$

where $\boldsymbol{\Phi}$ is the Toeplitz convolution matrix corresponding to φ_T. These last two algorithms differ from the two previous ones in that they are capable of determining the optimal dose distribution and dose level (Hf^∞) directly, whereas Eqs. 11.38 and 11.39 will generate the desired dose distribution d as truthfully as possible in terms of the treatment objectives (Källman et al. 1992a; Lind et al. 1992).

There is one remaining problem with all gradient algorithms which is associated with the initial estimate of f^0. If it is chosen outside the convex region of F, convergence is not ensured. A good starting point for Eqs. 11.38 and 11.39 is $f^0 = 0$ or $a \cdot d$, which will gain one iteration with Eq. 11.38. For Eqs. 11.41 and 11.42 the initial guess is even more important to ensure convergence and that the algorithm does not get trapped at local maxima. A very good method is to take Hf^0 from Eq. 11.31 and generally approach the optimal solution from above as complications are disregarded in this equation (Källman et al. 1992a). All the experience gained with these iterative algorithms is that trapping on local maxima is a very rare event, probably due to the interaction of the large number of variables which may collectively push f away from a local minimum, provided the convex region is not too large compared to the step length of the algorithm. This is particularly true for the pencil beam algorithms, to be discussed next, mainly because of the very large number of free variables (see Fig. 11.4 row 2).

11.4.7 Pencil Beam Methods

A very general optimization algorithm has recently been developed which can take into account almost

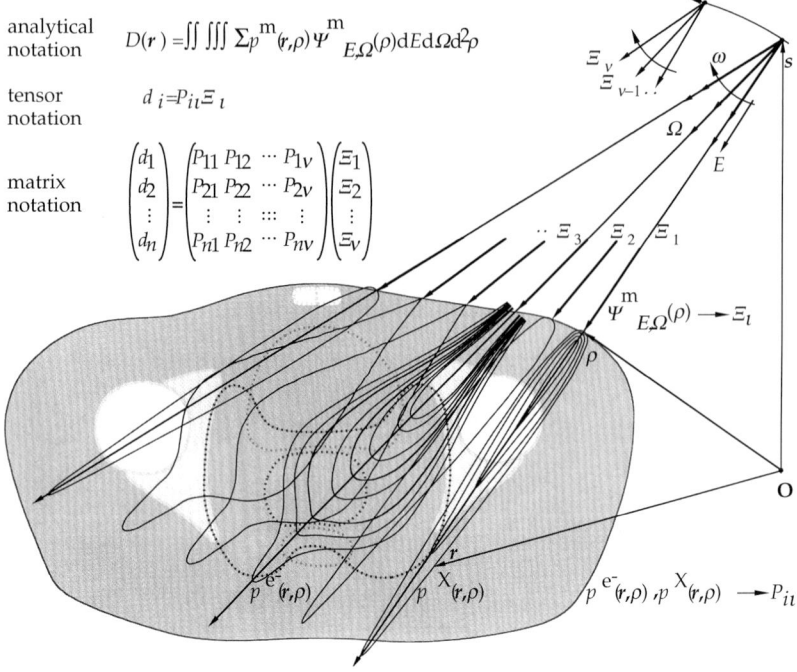

analytical
notation
$$D(r) = \iint \iiint \Sigma p^m{}_{(r,\rho)} \Psi^m{}_{E,\Omega}(\rho)\mathrm{d}E\mathrm{d}\Omega\mathrm{d}^2\rho$$

tensor
notation
$$d_i = P_{i\iota}\Xi_\iota$$

matrix
notation
$$\begin{pmatrix} d_1 \\ d_2 \\ \vdots \\ d_n \end{pmatrix} = \begin{pmatrix} P_{11} & P_{12} & \cdots & P_{1v} \\ P_{21} & P_{22} & \cdots & P_{2v} \\ \vdots & \vdots & \vdots\vdots\vdots & \vdots \\ P_{n1} & P_{n2} & \cdots & P_{nv} \end{pmatrix} \begin{pmatrix} \Xi_1 \\ \Xi_2 \\ \vdots \\ \Xi_v \end{pmatrix}$$

$$\Psi^m{}_{E,\Omega}(\rho) \longrightarrow \Xi_\iota$$

$$p^{e^-}{}_{(r,\rho)}, p^X{}_{(r,\rho)} \longrightarrow P_{i\iota}$$

Fig. 11.8. Illustration of irradiation geometry used in the optimization of the total dose distribution in the patient delivered by the fluence $\Phi^{e^-}_{E,\Omega}$ of electron pencil beams π^{e^-} and the energy fluence $\Psi^X_{E,\Omega}$ of photon pencil beams p^x. Through the use of accurately calculated pencil beams, even taking patient inhomogeneities into account, a very strict optimization is possible considering all major constraints on the dose delivery

all the degrees of freedom at an advanced radiotherapy department. As illustrated in Fig. 11.8, a very large number of beam modalities, beam energies, beam directions, and beam intensity distributions can be combined in a global optimization using pencil beams or more complex composite energy deposition kernels. By using the finest elemental radiation source, a pencil, or even a point monodirectional beam, different radiation modalities can be combined with a high geometrical resolution and computational accuracy on complex target volumes based on Eqs. 11.1a and 11.1b. Compared to the previously developed optimization techniques this new method allows direct determination of the incident energy fluence required by the different radiation modalities and accurate consideration of all possible degrees of freedom in the optimization.

In the numerical formalism the neutral particle pencil beams p^m and the charged particle pencil beams π^m are combined in a single pencil beam

matrix P, while the corresponding energy fluences $\Psi^m_{E,\Omega}$ and particle fluences $\Phi^m_{E,\Omega}$ can be represented in a single generalized fluence vector denoted Ξ. This generalized fluence could also include the source density of brachytherapy sources or other combined radiation fields such as from a multicobalt device or a Piotron with multiple convergent pi meson beams. The units of the different elements in the P matrix and the Ξ vector will thus depend on whether the particle referred to is charged or neutral and whether brachytherapy or external beam therapy is used. The mapping of Ξ and P on $D(r)$ is illustrated in Fig. 11.8 (GUSTAFSSON et al. 1993).

If the target volume is irradiated with v_m different beam modalities each with v_E different energy bins from v_Ω different source positions and the fluence profile for each source position is discretized into v_ω components, then the total number of components in the fluence vector Ξ is $v = v_m v_E v_\Omega v_\omega$. The pencil beam belonging to each fluence component is best discretized on the same grid as the dose distribution vector d. The pencil beam matrix P will therefore consist of $n \times v$ components, where n is the number of components in the vector representation of $D(r)$ as described in Sect. 11.2.2.

The dose distribution vector, the pencil beam matrix, and the fluence vector can respectively be denoted d_i, $P_{i\iota}$ and Ξ_ι where the range of i is 1, 2, ..., n and the range of ι is 1, 2, ..., v. Thus Latin indices are used for the volume quantities of dimensionality R^n and Greek indices for the surface quantities of dimensionality R^v.

The pencil beam equations, Eqs. 11.1a and 11.1b can now be directly implemented in Eq. 11.37, resulting in the algorithm:

$$\boldsymbol{\Xi}^{k+1} = C\left[\boldsymbol{\Xi}^k - a\nabla_{\boldsymbol{\Xi}} F(\boldsymbol{\Xi}^k)\right]. \qquad (11.43)$$

Unfortunately the functional F_1 of Eq. 11.20 cannot be used as a physical objective function in the fluence optimization, since $\boldsymbol{\Xi}^T$ is a v dimensional row vector while $\boldsymbol{P\Xi}$ and $\boldsymbol{d}$ are n dimensional column vectors. It is thus necessary to use other objective functions such as F_2 and P_+ that are compatible with the dimensions of the relevant vectors and matrices of the new problem (GUSTAFSSON et al. 1993). By analogy with Eq. 11.39 of the previous section one obtains for F_2:

$$\boldsymbol{\Xi}^{k+1} = C\left[\boldsymbol{\Xi}^k - a\boldsymbol{P}^T(\boldsymbol{P\Xi} - \boldsymbol{d})\right]. \qquad (11.44)$$

Similarly, for P_+ the iterative scheme can be written

$$\boldsymbol{\Xi}^{k+1} = C\left[\boldsymbol{\Xi}^k + a\boldsymbol{P}^T\nabla_{\boldsymbol{P\Xi}} P_+^0(\boldsymbol{P\Xi}^k)\right]. \qquad (11.45)$$

As discussed in more detail by GUSTAFSSON et al. (1993), both of these equations are similar to Eqs. 11.39 and 11.41 above, but differ in that they can consider a very large number of beam kernels as specified by $\boldsymbol{P}$. They therefore can take almost all important degrees of freedom of modern radiation therapy planning exactly into account (see Sect. 11.6.3).

11.4.8 Similarities and Differences in Treatment Optimization and Tomographic Imaging

In connection with the optimization of arc therapy, LAX and BRAHME (1982, see Sect. 11.4.2) found that there are fundamental similarities between the optimization method employed and the image reconstruction techniques used in CT. More specifically, they found that the optimal wedge filter for rotation therapy was reminiscent of the spatial filter used in filtered back-projection (cf. Fig. 11.9). Furthermore, the associated integral equation has similarities with that of CT (cf. Eq. 11.24 and BRAHME et al. 1982). This similarity is even more pronounced when comparing filtered back-projection with optimization of generalized conformal therapy, where angular dependent nonuniform incident beams are used to realize a uniform target dose (BRAHME 1985, 1988a). For example, the attenuated radon transform is linked to both these problems due to the influence of exponential photon absorption in both cases (cf. Eqs. 11.24 and 11.47). These similarities have recently been pursued by several authors for treatment

Fig. 11.9. Comparison of the similarities and differences of treatment optimization and image reconstruction. The ramp filter of the latter can unfortunately not be employed in treatment optimization owing to its negative beam portions. Instead a convergent planar kernel with essentially a $1/r$ dependence (*lower left*) has to be used in rotation therapy optimization (cf. Fig. 11.10 and the text)

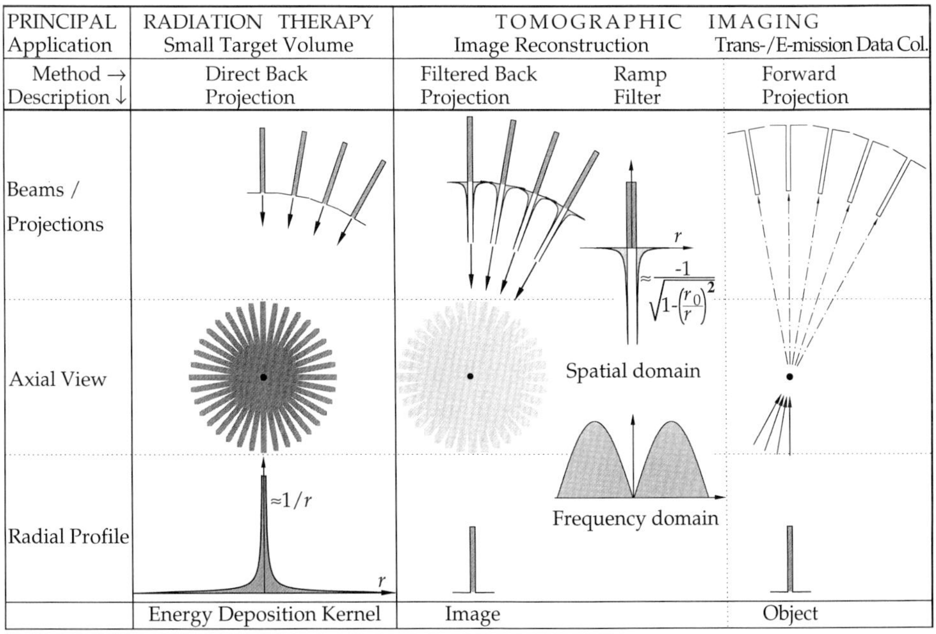

PRINCIPAL Application	RADIATION THERAPY Small Target Volume	TOMOGRAPHIC IMAGING		
		Image Reconstruction		Trans-/E-mission Data Col.
Method → Description ↓	Direct Back Projection	Filtered Back Projection	Ramp Filter	Forward Projection
Beams / Projections				
Axial View			Spatial domain	
			Frequency domain	
Radial Profile	$\approx 1/r$			
	Energy Deposition Kernel	Image		Object

optimization (CORMACK 1987; CORMACK and QUINTO 1989; BARTH 1990; BORTFELD et al. 1990; HOLMES et al. 1991). However, there are several serious differences which make a strict comparison impossible. First of all, in tomography we know that there always exists an exact solution from which the projections were collected, at least in the absence of projection noise, whereas for arbitrary desired dose distributions there does not generally exist an exact solution in the form of a set of incident external beam profiles containing nonnegative energy fluence.

An even more fundamental difference is that in tomography the physics of image production comes first and later the theory of image reconstruction is applied to restore the original image from its projections. In radiation therapy the causality of events is reversed since the desired dose distribution image is known or estimated from clinical knowledge. However, the beam profiles that will give the best conformation to this dose image are unknown. Therefore, in radiation therapy optimization the mathematical theory has to come first and the physics last, namely when the patient is irradiated by the determined optimal beam shapes. This considerable conceptual difference makes it possible in image restoration to use filter functions in the filtered back-projection, which have negative portions similar to the Abel kernel (cf. Fig. 11.9, right half, and ABEL 1823) capable of erasing the diffuse tail of each ray line during the back-projection. In radiation therapy this is impossible as a negative energy fluence or dose delivery is physically impossible unless there exists some kind of radiation that could stimulate or induce repair of radiation damage! This is also one way of explaining why any patient cross-section can be restored using filtered back-projection whereas most desired dose distributions cannot be realized in clinical practice (see also Sects. 11.4.2 and 11.4.3).

A further complication in the optimization of radiation therapy planning is that the desired absorbed dose distribution is not uniquely related to a certain energy fluence. The ray lines of the incident energy fluence, for example, have to be convolved with the energy deposition of the photon pencil beam to get the resultant energy deposition (Eq. 11.1). One way to get around this problem would be to first deconvolve the energy deposition kernel for isotropically convergent pencil beams from the desired dose distribution (Fig. 11.9, left half). The resultant kernel density would then represent the "sink density" of the incident energy fluence (LIND and BRAHME 1992) and could be used to determine the energy fluence of the photon radiation source during external rotation

therapy. This approach to get from the desired absorbed dose distribution to the required energy fluence, even if not strictly optimal, gives results which possess all the major advantages of true treatment optimization (cf. BRAHME et al. 1990; LIND 1991). In this context it should also be pointed out that the close $1/r$ dependence of this convergent energy deposition kernel in the plane is identical to the back-projection kernel of the radon transform (DEANS 1983; NATTERER 1986).

11.5 Treatment Execution

11.5.1 General

With the new kernel-based treatment planning methods presented here the required beams are often indirectly defined through the kernel or irradiation density. If we assume that a fully converged solution $f(r)$ or f^∞ of any of Eqs. 11.4, 11.6, 11.7, 11.25–11.27,

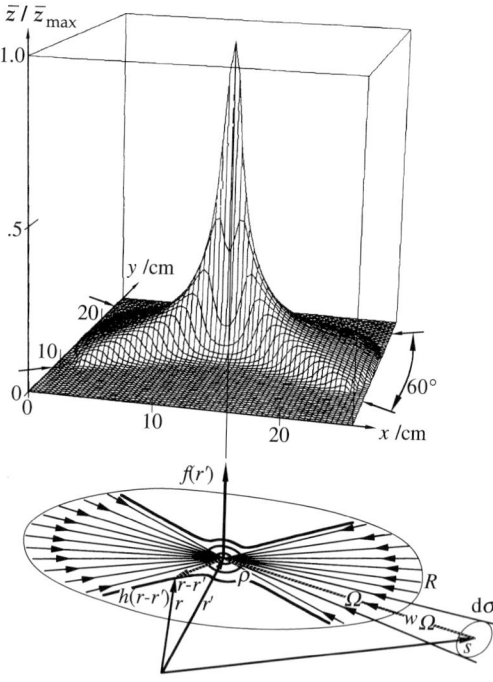

Fig. 11.10. The *upper panel* shows the energy deposition kernel resulting from the convergent irradiation of a cylindrical water phantom with convergent pencil beams in an angular interval of ±30° in a plane as symbolized by the *arrows* in the *lower half of the figure*. The coordinate systems used for the definition of the mean specific energy deposition kernel $h(r, r')$ are also shown, where r' is the center of the kernel and r is a point of interest where the mean specific energy deposition is specified. The vector s gives the position of a small spherical radiation source of cross-sectional area $d\sigma$, which is the effective photon source during dose delivery (cf. Eqs. 11.2 or 11.3 and CORMACK and QUINTO 1989)

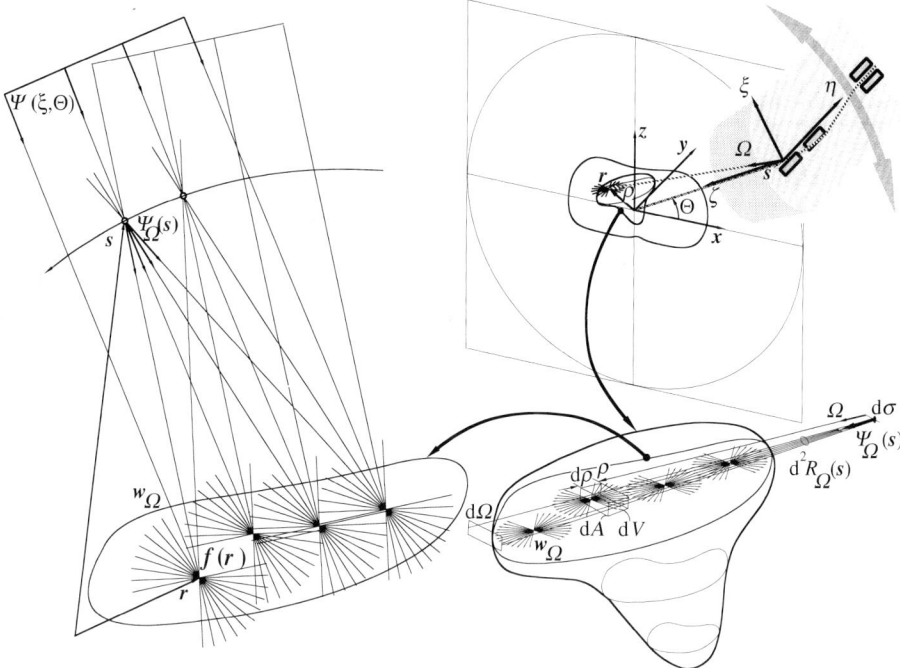

Fig. 11.11. Irradiation geometry and coordinate system of a radiation source located at a point s in the patient stationary coordinate system (*upper right panel*). The *shaded area* illustrates the treatment head of the therapy unit, showing the scanning magnets and the associated pathway of the deflected beam (*dotted line*). The determination of the energy fluence differential in angle $\Psi_\Omega(s)$ is illustrated as a ray-trace through the irradiation density along a ray ρ parallel to Ω (*lower right panel*). The angularly restricted point energy deposition kernels (w_Ω cf. Fig. 11.10) are used when determining the required incident beams (Ψ_Ω) from the kernel density f. The resampling of the mean specific energy distribution of the kernel into the required incident energy fluences from different source positions for plane parallel and point sources, respectively, is shown to the *left*

11.32, 11.36, 11.37, or 11.38–11.42 has been found; it expresses the local density or amplitude of kernels h centered at r. This means that a realization of the dose delivery in a patient requires the kernel to be reconstructed with the correct amplitude given by the irradiation density $f(r)$ and its assumed angular distribution ω_Ω (cf. Fig. 11.10) at each point r in the patient as described in detail in Fig. 11.11.

The most straightforward case would perhaps be the application of point source kernels in brachytherapy where the irradiation density would be proportional to the required local source strength. However, it could equally well describe the density of any of the kernels in Fig. 11.4. With external beams the realization of the kernel can be achieved by moving the patient in a stationary irradiation geometry which exactly reproduces the energy deposition

pattern of the kernel $h(r - r')$. Examples of such beam configurations are the movable intracavitary radionuclide source (cf. BRAHME 1988a; HOLMES et al. 1991), the multicobalt intracranial radiotherapy unit with about 200 equifocused cobalt sources (LARSSON et al. 1974), or the piotron with 72 convergent π^--meson beams (BLATTMAN 1979).

11.5.2 Back Projection of the Irradiation Density

However, with nonuniform external beam dose delivery a more generally applicable solution can be obtained by allowing the incoming fluence profile from a number of source positions to be suitably adjusted (Fig. 11.11 and BRAHME 1988a). In order to reproduce the irradiation density $f(r)$ during the execution of the treatment, the incoming photons that generate the dose, or more precisely, the mean specific energy distribution of the kernel, have to be back-projected (Fig. 11.11, lower right) and re-sampled to build up the incoming beam profiles. The resampling process can be made to fit either a rotary point source or a plane parallel source, as illustrated in the left half of Fig. 11.11. The required beam profiles in the coordinate system of the radiation source can then be determined. The required radiant energy $d^2R(s, \rho)$ that has to be emitted from a small but finite spherical radiation source located at s with cross-sectional area $d\sigma$ on to a volume element dV located at point ρ in the target volume (cf. Fig. 11.10,

11.11) to reproduce the local mean specific energy or dose distribution becomes

$$d^2 R(s, \boldsymbol{\rho}) = f(\boldsymbol{\rho}) w_{\boldsymbol{\Omega}} \frac{d\sigma}{\rho^2} dV, \qquad (11.46)$$

where $w_{\boldsymbol{\Omega}}\, d\sigma/\rho^2$ is the relative contribution to the kernel from photons traveling through the element of solid angle $d\sigma/\rho^2$ in direction $\boldsymbol{\Omega}$ as shown in the lower right corner of Fig. 11.11. The irradiated volume element dV in the target volume may be expressed in terms of the element of solid angle $d\boldsymbol{\Omega}$ from the source according to

$$dV = \rho^2 d\boldsymbol{\Omega}\, d\rho. \qquad (11.47)$$

Eq. 11.46 may thus be rewritten

$$d^2 R_{\boldsymbol{\Omega}}(s, \boldsymbol{\rho}) = f(\boldsymbol{\rho}) w_{\boldsymbol{\Omega}} d\sigma d\rho. \qquad (11.48)$$

And according to the definition of $\boldsymbol{\Psi}$ (cf. Eq. 11.1) the corresponding required energy fluence to be emitted in direction $\boldsymbol{\Omega}$ is given by

$$d\boldsymbol{\Psi_{\Omega}}(s, \boldsymbol{\rho}) = f(\boldsymbol{\rho}) w_{\boldsymbol{\Omega}} d\boldsymbol{\rho}. \qquad (11.49)$$

The principal quantity to be determined now is the required incident energy fluence differential in angle $\boldsymbol{\Psi_{\Omega}}(s)$ from the source at s per unit solid angle in direction $\boldsymbol{\Omega}$. By correcting for the difference in photon attenuation during the irradiation of the patient, and the reference situation in which the kernel was first calculated, $\boldsymbol{\Psi_{\Omega}}(s)$ can be expressed by

$$\boldsymbol{\Psi_{\Omega}}(s) = \int_0^{\infty} e^{-\mu_w r_0 \int_0^{\rho} \mu(t\boldsymbol{\Omega}) dt} w_{\Omega} f(\rho) d\rho \qquad (11.50)$$

The factor $e^{-\mu_w r_0}$ expresses the attenuation already present in the kernel (a water phantom of radius r_0). Equation 11.50 is a ray-trace through the kernel density $f(\boldsymbol{\rho})$ along $\boldsymbol{\rho}$, as expressed by the radial distance t along the unit vector $\boldsymbol{\Omega}$. It is clear that this integral corresponds to an "amplified" radon transform rather than an attenuated one (cf. DEANS 1983) but otherwise it is similar to the forward projection of CT (cf. Fig. 11.9). This fact illustrates that the projection order is reversed in therapy optimization as compared to tomographic imaging, cf. Sect. 11.4.8.

Two basic conditions have to be fulfilled to justify Eq. 11.4 and 11.50: naturally the kernel should be spatially invariant throughout the calculational volume in Eq. 11.4, and the delivered energy fluence differential in angle from all sources given by Eq. 11.50 should, at each point in the patient, reconstruct the kernel with the correct amplitude given by the irradiation density. The attenuation correction in Eq. 11.50 can, especially for low-energy photon beams, become quite large, but due to the purely exponential attenuation of a photon beam far from

entrance surfaces the local absorption gradient will be invariant at least for monoenergetic photons. This fortunate situation derives from the fact that an exponential function has a constant relative derivative. The assumption of a spatially invariant kernel in Eq. 11.4 implies infinite distance to the source, which is not necessarily true in Eq. 11.50. However, for the specially important case of generalized conformal therapy, where arbitrary beam profiles $\boldsymbol{\Psi_{\Omega}}(s)$ are delivered from any radiation source s on a circle (or sphere) around the patient, the kernel h can be resampled as illustrated to the left in Fig. 11.11. In this way the fluence delivered according to Eq. 11.50 exactly reproduces the desired local energy fluence (BRAHME 1988a; LIND and BRAHME 1992). It should be pointed out in this context that the pencil beam algorithm described in Sect. 11.4.7 has the advantage that it directly generates the required energy fluence without the need for a back-projection of the type described here.

11.6 Clinical Possibilities

We will now illustrate the potential of some of the new optimization methods discussed above. Rather than describing all methods we will concentrate on some of the most flexible techniques which have the greatest potential to improve the dose delivery with complex tumor shapes and large target volumes, where we have the most severe clinical problems today.

11.6.1 Nonuniform Dose Delivery Using Dynamic Multileaf Collimation

In most cases when multiple degrees of freedom are available nonuniform dose delivery will generally be advantageous or even a requirement. A very flexible technique for shaping of nonuniform beams is the use of dynamic multileaf collimation. The most elementary dose distribution that could be generated with a multileaf collimator is obtained when all leaves are closed except for a single leaf pair that is infinitesimally opened. This collimator setting will allow a very narrow line segment of photons, the length of which equals the leaf width (cf. Figs. 11.12, 11.13) to leave the treatment head and produce the elementary collimated slit beam in the phantom. The optimal setting of the collimator can thus be reduced to the problem of finding the opening density, f, of such narrow collimated slit beams, h, that generates the desired incident beam profile, D_i. This problem is analogous to many of the inverse problems discussed above (Table 11.2) and can be solved using

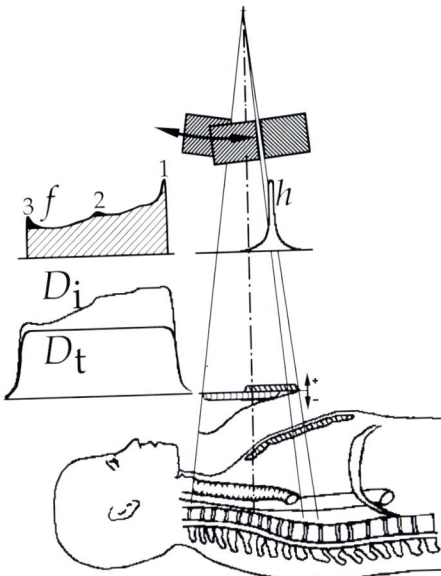

Fig. 11.12. Illustration of the use of dynamic multileaf collimation to generate a uniform dose, D_t, in the target volume (esophagus) by a nonuniform opening density, f, of the multileaf collimator, resulting in the required incident dose profile, D_i. The incident beam can in this way be compensated for the shape of the entrance surface and different heterogeneities such as the sternum and trachea but also for the jaggedness at the beam edge of the leaf collimator

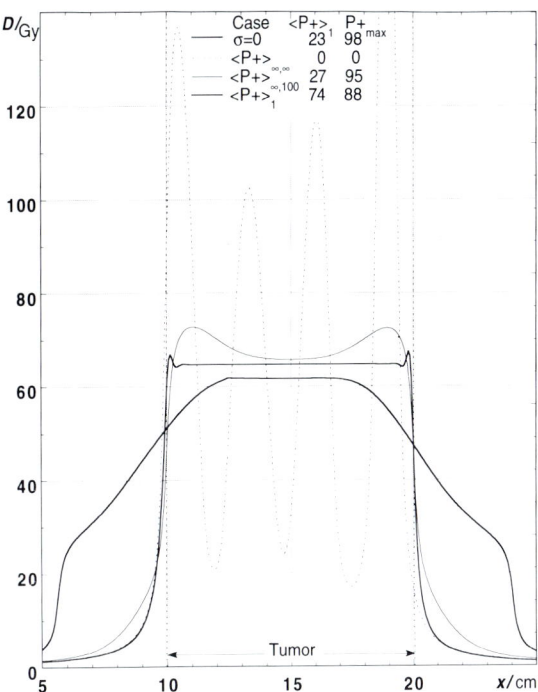

Fig. 11.13. The influence of the uncertainty in the alignment of the radiation beams with the tumor can be taken into account either by an extended setup margin or by an overcompensated beam. From a radiobiologic point of view the latter is advantageous since the dose to surrounding normal tissues is minimized in this way. The *thick rectangular solid line* gives the beam profile with the highest expectation value for P_+ with $\sigma = 0$. The oscillatory *thin dotted curve* represents the strict mathematical optimum with infinitely many irradiations and is obviously not realistic for clinical use, as seen from the associated maximum P_+ value and its expectation value. The overcompensated *solid lines* represent an intermediate result after 100 iterations, with a better $\langle P_+ \rangle$ value than the best field when $\sigma = 0$. The *lowest solid curve* shows the optimal field for a single high dose irradiation assuming exponential cell kill (Eq. 11.42 and $\alpha/\beta = \infty$)

any of the iterative algorithms, in Eqs. 11.37–11.45 (cf. also KÄLLMAN et al. 1988).

Dynamic multileaf collimation will be much faster than, for example, using asymmetric collimator jaws and will only prolong the irradiation time by about 50%–100% depending on the complexity of the field shape. It is therefore one of the most flexible techniques available for nonuniform dose delivery, especially since steep dose gradients may readily be generated. The technique is illustrated in Fig. 11.12 on one of the three fields for an esophageal tumor. The collimator opening density determined with the algorithm in Eq. 11.35 has generally small peaks near the field edges and at points with changes in the dose level to compensate for the loss of secondary electrons and scattered photons. Even if the calculation is performed with the elementary slit beam of the leaf collimator, h, the opening density can be realized by a repeated shrinking field technique on the three numbered peaks of the opening density as indicated in the figure. For complex opening densities it may be more suitable to move the leafs like the curtains of a camera shutter, as was recently shown by numerical optimization by CONVERY and ROSENBLOOM (1992) and by analytic optimization by SVENSSON et al. (1994).

We will also illustrate the influence of the uncertainty in beam alignment on the optimal shape of the incident beam in a simple one-dimensional case where uniform target dose is required, like in Fig. 11.12. (cf. Sect. 11.3.2). Let us assume that we want to deliver a uniform dose to a 10-cm-wide target volume, but we also want to take the 1-cm uncertainty (1σ) in beam patient alignment into account. This means that the delivered beam should be modified to allow for the known uncertainty in positioning (LIND et al. 1992). In the present example we have calculated the optimal dose distributions from the point of view of maximizing the expectation value of P_+ according to Eqs. 11.11–11.14 and Eqs. 11.41, 11.42 and 11.45. The results are summarized in Fig. 11.3. The dotted oscillatory curve is the strict mathematical result trying to optimize the expectation value of P_+ for the delivered dose distribution with infinitely many beams according to F_+^∞ from Eq. 11.14. The resultant beam is clearly unsuitable for therapy even if this strange beam when folded with a Gaussian with $\sigma = 1$ cm results in a suitable dose distribution. However, if the iteration process is allowed to continue only as long as no severe oscillations appear, e.g., 100 iterations, more clinically useful results are obtained, as seen by the thin solid line curve in Fig. 11.13. The thick solid bell-shaped curve, on the other hand, is optimized just for a single treatment (Eq. 11.42) and the oscillations are gone. In clinical practice it is clear that this type of optimization is applicable only for a very small number of irradiations such as in intraoperative radiotherapy or stereotatic radiotherapy. For the larger number of dose fractions in conventional radiation therapy the optimal beam according to Eq. 11.12 is extremely time consuming to calculate. However, from the limit cases of 1 or infinitely many beams as shown in Fig. 11.13 it can be concluded that the optimal beam shape would be wider than the tumor and slightly overcompensated. Such a beam would reduce the dose to surrounding normal tissues but still give a high uniform dose to the tumor. The solid thick rectangular-shaped curve is the optimal beam shape without setup uncertainty ($\sigma = 0$) and therefore gives a low expectation value when the uncertainty is considered.

11.6.2 Few-Field Generalized Conformal Therapy

As we now have a tool for optimized nonuniform dose delivery we will try it out on a target volume requiring a small number of beams. The first example is a case which is generally considered to be hard to treat well: a patient with a neck tumor with positive lymph nodes on both sides of the neck (Fig. 11.14a). The prescribed dose to the indicated target volume (++) would be of the order of 64 Gy in 32 × 2-Gy fractions, five fractions per week. A common technique is to irradiate the whole target volume with two lateral opposed beams which include the spinal cord up to about 45–50 Gy and then reduce these fields to exclude the spinal cord for the remainder of the treatment. The excluded lateral portions of the target volume are then treated with are techniques (cf fig. 11.7) or lateral electron beams. This is a quite laborious six-field technique with the problem of adjoining photon and electron beams. In Fig. 11.14a a very attractive dose distribution is shown which can be obtained with only three beams by using the ray line optimization algorithm of ULSØ

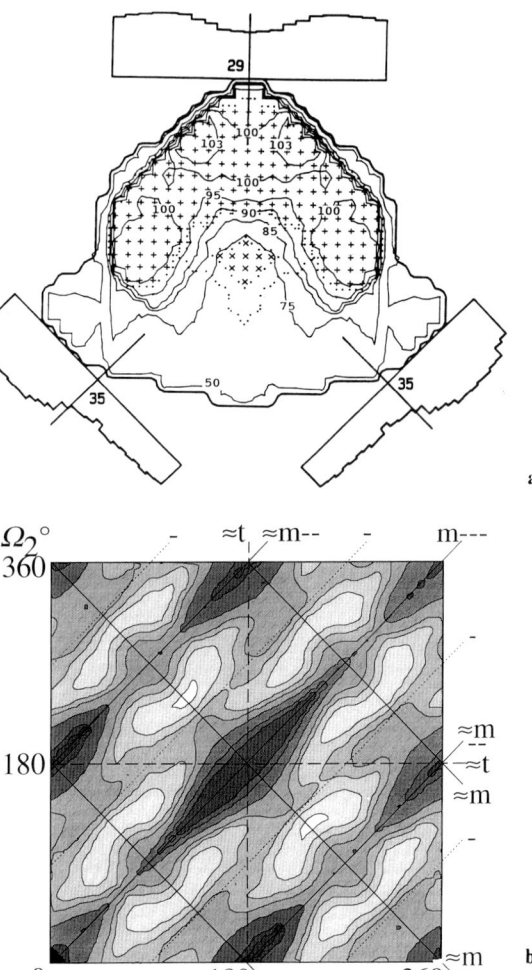

Fig. 11.14.a The use of three nonuniform fields for treating the lymph nodes on the neck. The ++ and × × indicate points of the calculation matrix in the target volume and the spinal cord respectively (ULSØ and BRAHME 1988). In **b** the phase space of P_+ for two-field techniques with angles of incidence Ω_1 and Ω_2 are shown. Areas with large P_+ values are *lighter shaded*. Exact lines of mirror symmetry are marked m and approximate ones ≈ m. *Dashed lines* with translation symmetry are marked t and the number of – signs indicates decreasing suitability in the angular combination (i.e., lower P_+, --- = parallel, -- = parallel opposed, - = perpendicular incidence). The peak P_+ value corresponds approximately to the two rear fields in **a**

(ULSØ and BRAHME 1988) to determine the optimal energy fluence distribution as indicated by the histograms at the beam portals. It is quite clear from the figure that a very suitable dose distribution is obtained with this intuitively selected three-field technique.

Recently SÖDERSTRÖM and BRAHME (1993b) tried out all possible two field combinations for this target volume to get a feeling for the shape of the distribution of P_+ as a function of the angle of incidence in 10° increments (Fig. 11.14b). It is interesting to see the considerable variation of P_+ for different pairs of nonuniform beams that have been optimized simultaneously. In some regions a very low value is seen corresponding to parallel (---) and parallel opposed (--) beams and in a regular meandering pattern very high P_+ values are obtained. It is also seen that an angular difference of 90° is not so advantageous either (dotted lines with one -). This is due to the fact that there is a better possibility to compensate for the attenuation in the opposed beam when the projected attenuation is reduced. However, as the angle approaches 180° and the attenuations compensate each other fully, the disadvantage of coinciding entrance and exit regions reduces P_+ even more (--), resulting in an optimal angular offset in the range 100–120°, depending on the size of the target volume and organs at risk. From this example it is evident that the best beam combinations are quite different

depending on whether uniform or nonuniform dose delivery is available.

When the number of beam portals increases, the ray line optimization technique gets slower and slower and faster methods are desirable. We will first look at the feasibility search method as used by POWLIS and co-workers (1989). The patient shown in Fig. 11.15a is postoperatively treated for a resected lower esophageal carcinoma with tumor extension into adjacent tissues and lymph nodes. The target volume and surrounding normal tissues are shown in the figure. The limiting doses were 54 Gy to the target, less than 44 Gy to the spinal cord, 20 and 30 Gy or less to the right and left lungs, respectively, and 60 Gy elsewhere. The dose distribution in Fig. 11.15a is based on six beam portals each with 150 rays optimized by the Agmon, Motzkin, and Schoenberg algorithm in accordance with the feasibility limits (cf. POWLIS et al. 1989).

In Fig. 11.15b the same target is treated with a three-field technique using the radiobiologic ray line optimization technique (SÖDERSTRÖM et al. 1993). It is seen that the different way of formulating the treatment objectives strongly alters the beam shapes. To maximize P_+ a uniform target dose is not necessary, especially with sensitive surrounding organs at risk, and therefore more rounded beam profiles are obtained. The P_+ values for the two plans are about 83% and 86.7% in a and b, respectively. The main advantage with the three-field technique is that parallel opposed beams were avoided in b, as pointed out above. Furthermore, the consideration of the serial (cord) and parallel (lungs) nature of the different organs at risk was also important for the choice of beam portals in b (cf. SÖDERSTRÖM et al. 1993; KÄLLMAN et al. 1992b).

11.6.3 Multiple-Field and Generalized Conformal Arc Therapy

When we go to an even larger number of fields and ultimately to continuous rotation therapy, different energy deposition kernels are needed and more efficient algorithms have to be used. In the previous two sections pencil beam and few-field kernels in principle were used (the lower half of Fig. 11.4). When the number of beam portals increases, a continuous isotropically rotated kernel is approached. This is based on the knowledge that as the number of fields increases, their optimal weighting tends to even out. For full 3D optimization a spherically convergent isotropic field of pencil beams

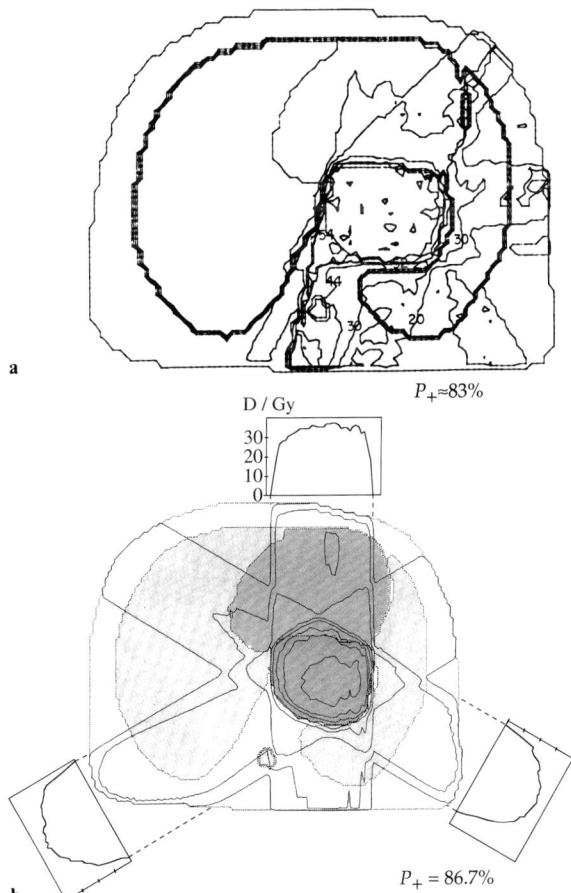

Fig. 11.15a,b. Comparison of dose plans for a resected lower tumor of the esophagus. **a** A six-field feasible solution according to POWLIS et al. (1989). **b** Radiobiologically optimized three-field plan according to SÖDERSTRÖM and BRAHME (1993b). The P_+ values for the two plans are also given

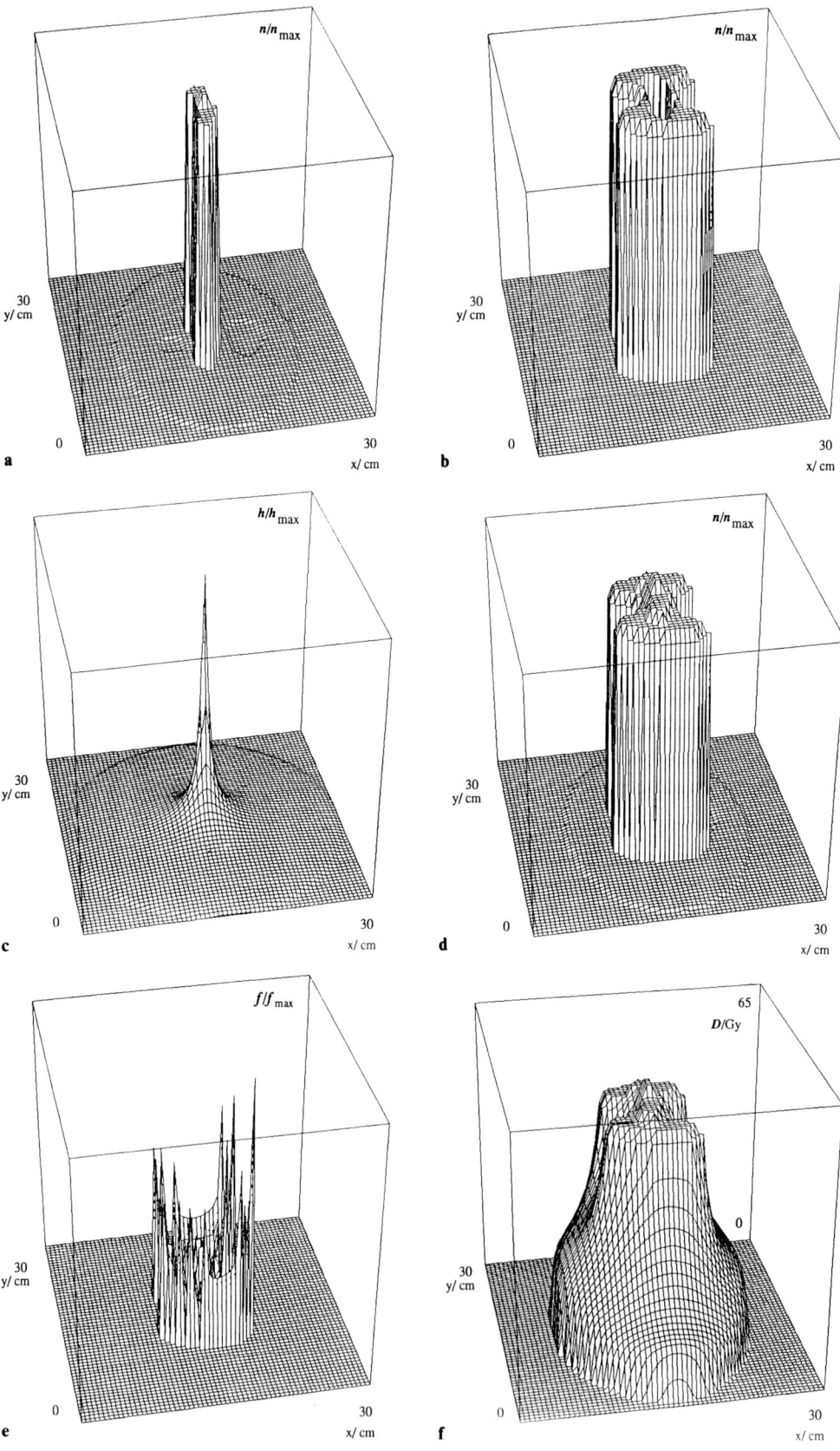

Fig. 11.16a–f

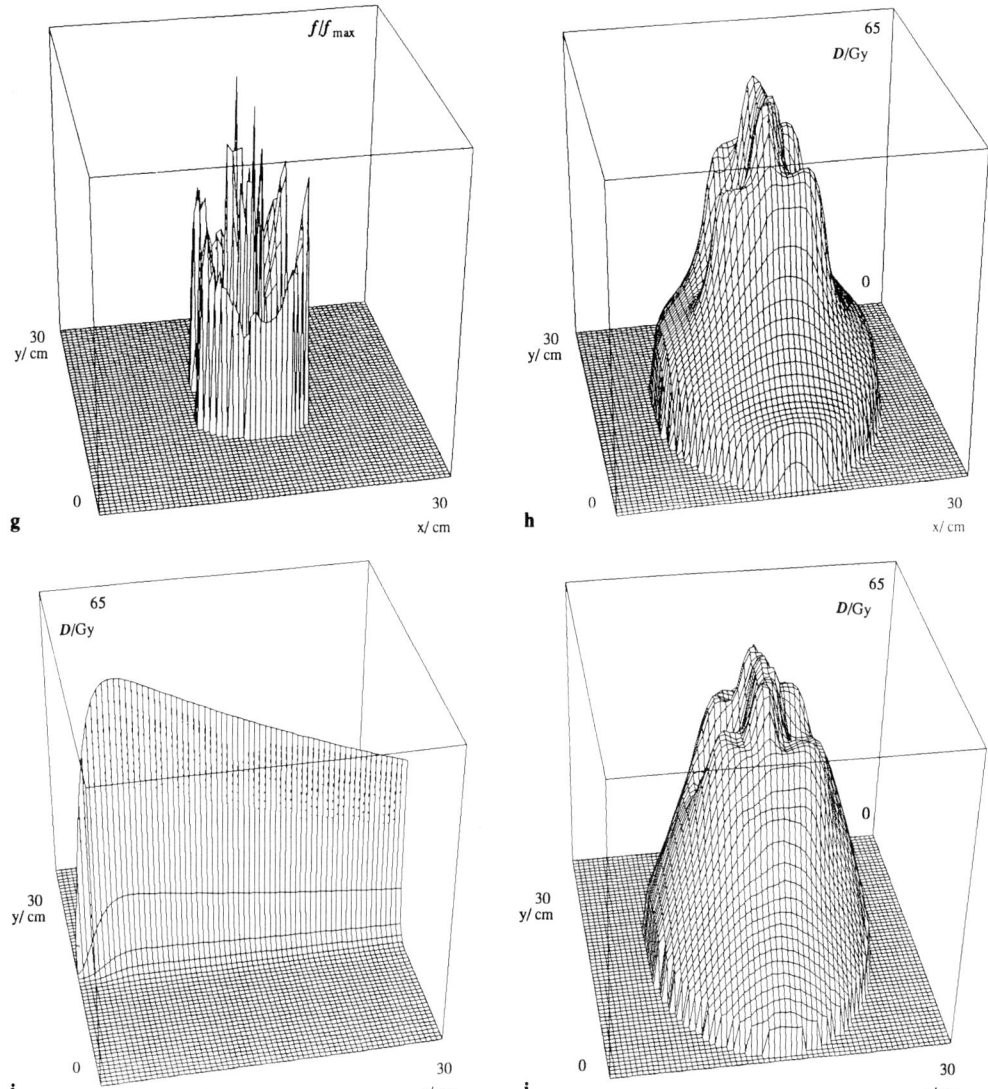

Fig. 11.16. **a** The gross target volume for an advanced cervical tumor. Also seen by relief effect are the rectum and bladder to the *left* and the *right* respectively. **b** The microscopic target volume including the locally involved lymph nodes. Everything outside the mentioned organs is from a radiation effect point of view regarded as small bowel. **c** The mean specific energy deposition kernel used to optimize the dose delivery to the cervical tumor by generalized conformal therapy. **d** The estimated optimal dose distribution for the cervical tumor and associated local nodes. It is used as prescribed dose for the algorithm in Eq. 11.38, cf. **e** and **f**. **e** The local irradiation density f^∞ for the prescribed dose distribution give in **d**. **f** The resultant optimal dose distribution $d^\infty = Hf^\infty$ for the prescribed dose distribution given in **d** and irradiation density in **e** ($P_+ = 87\%$, cf. BRAHME et al. 1990). **g** The local irradiation density f^∞ for the optimal treatment with the biologic algorithm, Eq. 11.41. **h** The resultant optimal dose distribution $d^\infty = Hf^\infty$. By comparison with **d** it is seen that the dose on the rectal side is lowered to reduce complications in the most sensitive organ ($P_+ = 91\%$). **i** One of two thousand photon pencil beam kernels used in the new pencil beam optimization technique (see Sect. 11.4.7 and GUSTAFSSON et al. 1993). **j** The final dose distribution using the powerful pencil beam algorithm with 72 beam directions ($P_+ = 92.6°$, Eq. 11.45)

would be desirable for general optimization with non-coplanar beams. It is interesting to note that irradiation densities determined with such kernels will also work well with multiple-field irradiation since the gross structure of the kernels will be quite similar (cf. BRAHME et al. 1990). Therefore, we will now turn to a somewhat more complex case, namely the convex–concave target volume in Fig. 11.5. We will employ both the biologic (Eq. 11.41) and the physical (Eq. 11.38) objective functions in the optimization to find suitable dose distributions.

Figures 11.16a and 11.16b illustrate the relevant target volumes for this quite complex case, an advanced cervical cancer; while a shows the gross target volume for the primary tumor, b shows the secondary target volume taking into account local lymphatic spread. By relief effect the organs at risk – rectum, bladder, and small bowel – are also shown in a. The dose-response relation of these tissues have been described in the calculations by the clinically based data set in Table 11.3. (cf. eq 11.10 and Källman 1992). The dose distribution kernel $h(r)$ is due to a 10 MeV photon pencil beam rotated isotropically 360° around the isocenter in the rotational plane of the photon target of the gantry as shown in c. The resultant optimal irradiation density f^∞ obtained from Eq. 11.38 is shown in g, and in h the associated optimal dose distribution $d^\infty = Hf^\infty$ is given. For comparison, e and f show the analogous dose distributions obtained with the physical dose prescription when the desired dose distribution $D(r)$ is chosen a priori according to the estimated optimal dose distribution shown in d (cf. Eq. 11.36 and BRAHME et al. 1990). By comparing e and f with g and h it is clear that the biological algorithm avoids irradiating the rectal side of the lymph nodes to a high dose to protect the rectum. Instead the bladder side is given a slightly higher dose. The improvement in complication-free tumor control P_+ is from about 87% to 91% with biologic optimization. This somewhat small improvement is partly due to the rather advanced treatment proposed in the reference case (Fig. 11.16d) with a desired nonuniform dose distribution, the dose to the primary tumor being approxi-

mately 10% higher than that to the surrounding lymph nodes. In Fig. 11.16i and j, finally, the results with the recent pencil beam algorithm (Eq. 11.45 and GUSTAFSSON et al. 1993) are illustrated. One out of about two thousand pencil beam kernels is shown in i, and in j the resultant optimal dose distribution with full freedom in dose delivery from 72 beam directions is given. It is clear from the higher P_+ of 92.6% and the radial streaks from the target, that a higher degree of freedom is really employed in this last plan (cf. Fig. 11.17c). With a more conventional reference case assuming uniform dose to the target by a four-field box technique the improvements would in general be much larger.

It is interesting to note that the distributions in both g and h have a smoother shape than those in e and f because of the restricting requirements on f that are imposed by the somewhat unnatural steps in the desired dose in d. This is even more clear in Fig. 11.17, where the required incident radiation fields have been plotted for every 5° of gantry rotation from 0 to 360°. The physical objective function with the estimated optimal dose distribution was used in a, the biologic algorithm in b, and the pencil beam algorithm in c. The dose profiles in a and b were obtained by projecting the irradiation density f in each direction as expressed by Eq. 11.47 (cf. BRAHME 1988a; LIND and BRAHME 1992). The dose profiles illustrate a new principle for treatment optimization. From directions where the beam crosses extended portions of the target volume, larger dose fractions should be delivered, since there are more tumor cells and at same time less normal tissues (BRAHME 1988a). This principle is the basis for protection of normal tissues in nonconvex targets like in Fig. 11.5.

The above optimal results for generalized conformal rotation therapy also may be used to find suitable beam portals for few-field irradiations, as has been described recently by SÖDERSTRÖM and BRAHME (1993a). In principle, the optimal beam profiles from Fig. 11.16 could be used to determine those beam directions where the largest control of the shape of the isodose distribution is achieved. For this reason we have in Fig. 11.18 plotted the angular variation of the information content, S (Eq. 11.22), the gross structure, Z (Eq. 11.23), and the probability to control the tumor without complications, P_+ (Eq. 11.9), for each single field from Fig. 11.17b. It is again clear that quite large variations in the importance of each field direction are seen (cf. Fig. 11.14b). Particularly the 0°–90°–180°–270° directions seem to have a larger internal structure and therefore may influence the isodose shapes more effectively, as also seen in the P_+ variation!

Table 11.3. Radiobiologic data for tumors and normal tissue

Organ	D_{50}	γ	s
Cervix	52	4	–
Lymph nodes	39	3	–
Rectum	55	3	0.69
Bladder	80	3	1.3
Small bowel	80	1.5	2.6

Fig. 11.17a–c. The angular dependence of the required incident radiation fields in **a**, **b**, and **c** corresponds to the resultant dose distributions in Fig. 11.16 f, h, and j respectively. The beam angle is increased in increments of 5° from 0° to 360° so each row corresponds to 45°

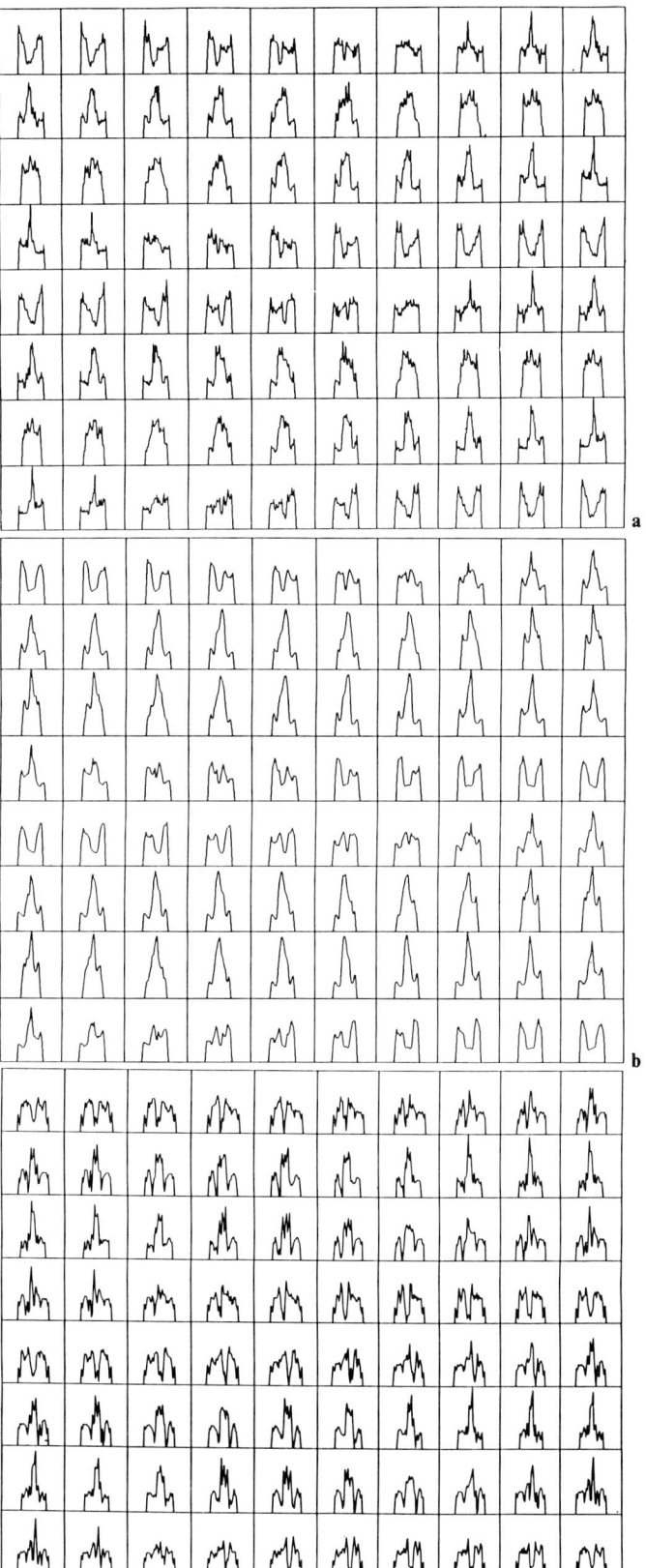

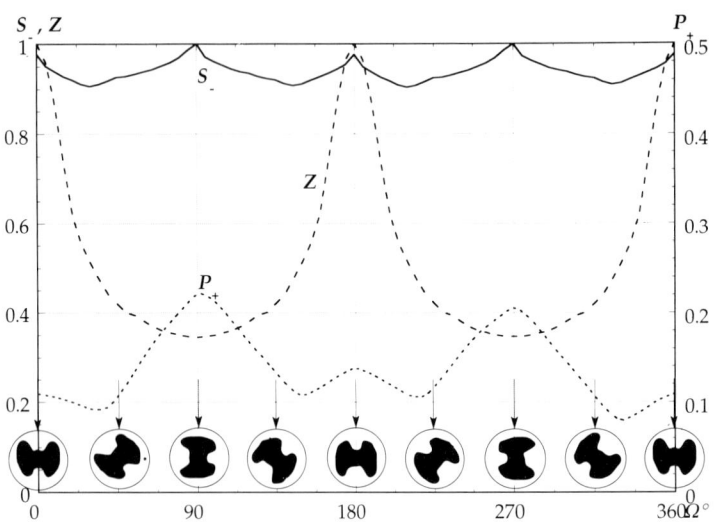

Fig. 11.18. The angular dependence of the information contents of the fields in Fig. 11.16b as given by the entropy measure (S_-, *solid line*), the low spatial frequency contents as given by the Fourier transform measure (Z, *dashed line*), and the biologic measure (P_+, *dotted line*). The orientation of the target volumes relative to the incoming radiation beam is presented along the Ω-axis. The good agreement between S_- and P_+ is particularly noticeable

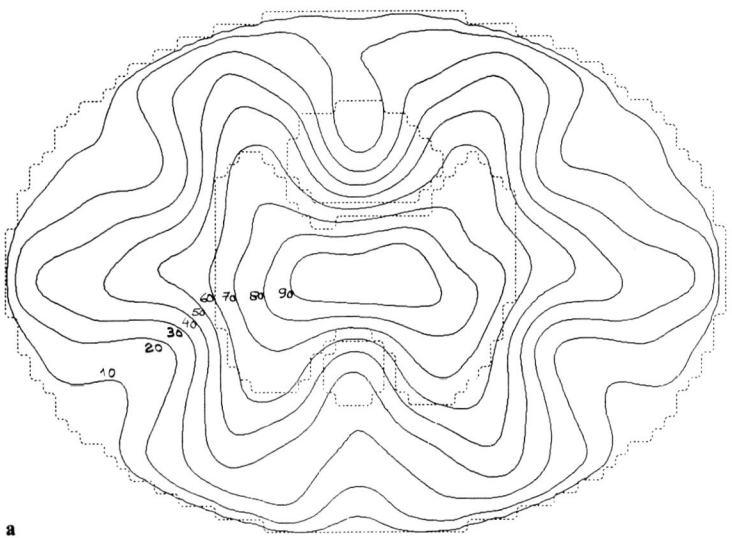

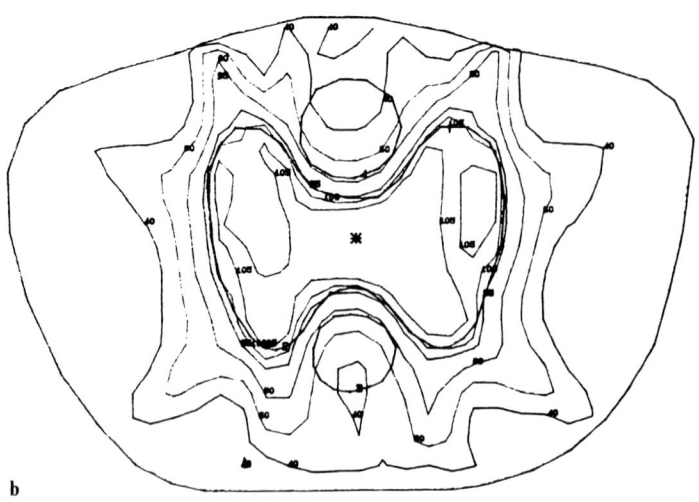

Fig. 11.19. a The resultant dose distribution with eight different 10-MV photon beam directions (≈ 30 MV) which are optimized for angles 0–$360°$ in $45°$ increments (cf. Fig. 11.18). The bladder and the rectum are the major organs at risk. The sparing of the rectum is clearly seen. None of the fields gives primary fluence in the rectum. **b** A generalized conformal high-energy electron beam treatment plan for the cervix target is shown using scanned 50-MeV pencil beams from nine portals. The large advantages with negligible exit dose and narrow electron kernel width (FWHM ≈ 15 mm) makes high-energy electrons the preferred modality for many deep-seated pelvic tumors

For this reason we show in Fig. 11.19a the result of a simplified eight-field technique where the main axial directions plus their diagonals were used with the biologic algorithm (Eq. 11.41, cf. KÄLLMAN 1992). Results for even fewer beam portals are given in the basic paper by SÖDERSTRÖM and BRAHME (1993), SÖDERSTRÖM et al. (1993a), and by BRAHME et al. (1990). It is clear from all these cases that few-field nonuniform dose delivery is the method of choice when practical aspects and safety aspects are of greater importance than the need for a truly optimal plan.

The last clinical example is also based on the cervical tumor and shows the very interesting poten-

tial of using scanned high-energy electron beams for deep targets (Fig. 11.19b). This dose distribution is produced by nine nonuniform beam portals delivered by a 15-mm-wide 50-MeV scanned electron beam from a Racetrack accelerator (VAN SANTVOORT and HUIZENGA 1991). These authors employed the Cimmino feasibility search algorithm which was adopted for radiation therapy by CENSOR et al. (1988b). The merits of high-quality scanned electron beams are very clear here since their low exit dose helps reduce the dose outside the target volume. This is clear by comparison with the photon plan in Fig. 11.19a (cf. also GUSTAFSSON et al. 1993 for further details). Owing to the high electron energy,

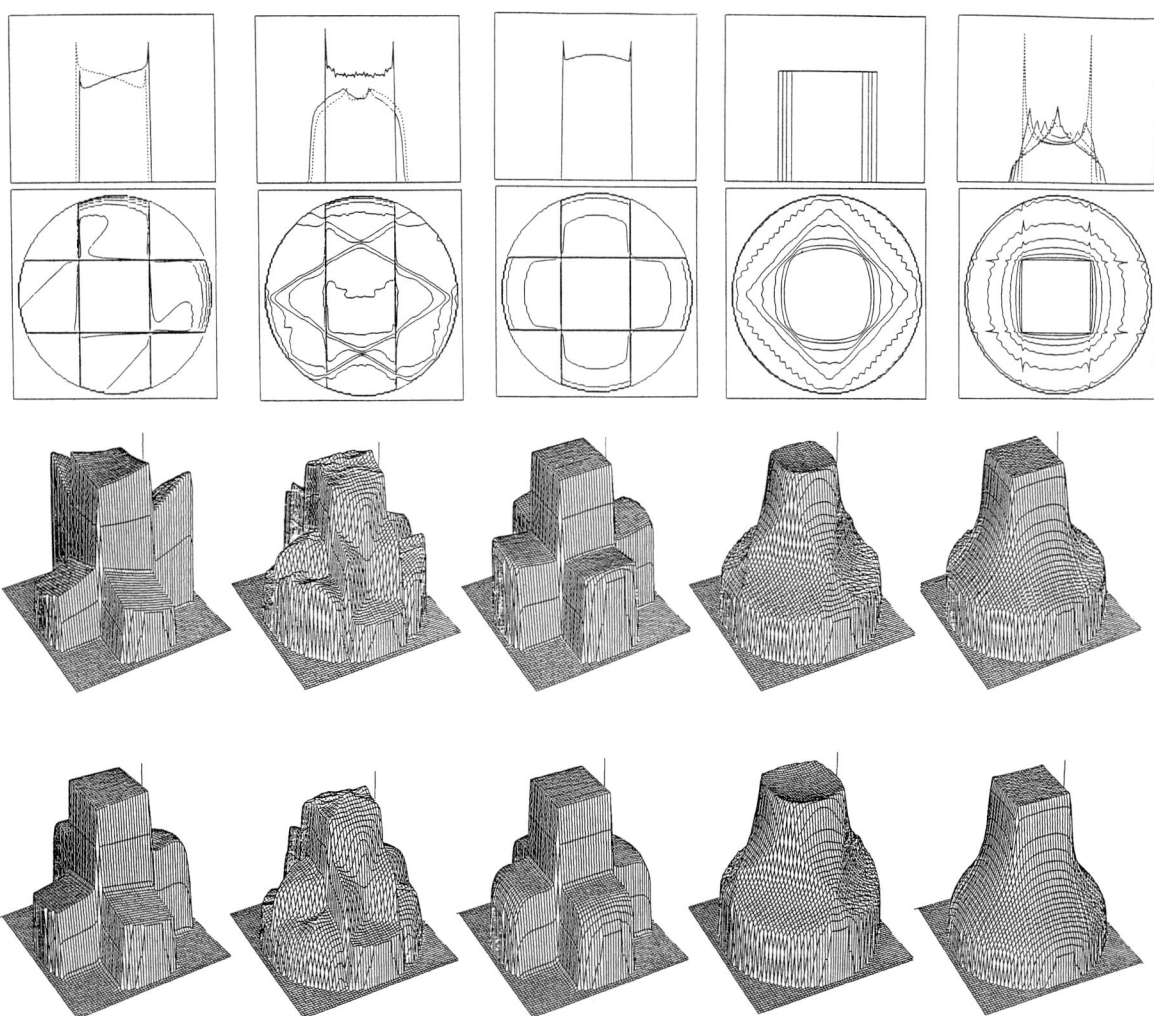

Fig. 11.20. Dose plans for treating a square target volume in a cylindrical phantom using five different irradiation techniques. The *top row* shows the incident beam profiles and the second row shows the corresponding isodose diagrams with 10%, 20%, 30%, 40%, 50%, 60%, 70%, 80%, and 90% isodoses indicated at 30 MV. The *lower two rows* show isometric plots of the resultant dose distributions at 6 and 30 MV respectively. The best dose distribution is generally obtained with the highest energy using the generalized conformation approach as shown by the *lower last diagram to the right*. Numerical dose distribution data are given in Table 11.4 for comparison

Table 11.4. Quantitative comparison of the treatment techniques for the square target in Fig. 11.19

Treatment technique	Acc. pot./MV	Target volume			Normal tissues		
		P_B	$\sigma_D/\bar{D}$	D_{min}/Gy	σ_D/Gy	D_{max}/Gy	$\bar{\varepsilon}$/J
2 wedge fields	6	95.7	1.64	58.0	8.7	58.1	23.7
2 wedge fields	30	95.9	1.03	59.5	15.9	43.2	21.9
3 generalized conformal fields	6	95.3	2.22	53.0	17.6	62.6	28.9
3 generalized conformal fields	30	93.8	3.84	49.8	16.4	63.6	26.9
4-field box	6	95.8	1.17	59.2	16.7	37.3	23.5
4-field box	30	95.9	0.97	60.0	15.5	38.4	21.8
Classical conformal field size = target	6	0	9.40	32.5	15.0	62.5	27.6
Classical conformal large field (+ 25%)	6	95.9	1.08	55.5	19.5	65.5	35.2
Generalized conformal	6	96.0	0.22	62.8	15.2	56.8	28.1
Generalized conformal	30	96.0	0.21	62.8	15.1	57.1	26.3

multiple scatter and inhomogeneities are minor problems in multiple high energy electron beam therapy (BRAHME 1988b).

To illustrate the power of inverse radiation therapy planning and its advantages and disadvantages over conventional uniform dose delivery, the incident beam profiles and resultant isodose diagrams for a simple square target volume are shown in the upper two rows of Fig. 11.20. The beam profiles for the two-field wedge and four-field techniques clearly show the characteristic overcompensation near the beam edges required to compensate for the loss of secondary electrons near the field edges. The strong nonuniformity of the required beams in inverse therapy planning is especially clear in the three-field and the generalized conformal therapy cases. In the two cases of classical and generalized conformal rotation therapy it is striking to see how much steeper the dose falloff is in all directions away from the target volume when the optimal strongly non-uniform beams of the generalized technique are used. In classical uniform beam conformation therapy the high-dose region is bulging out along the straight edges of the square target volume. This effect is eliminated using optimized conformation therapy with nonuniform dose delivery. For a more quantitative evaluation the most important dose distribution parameters for the cases in Fig. 11.20 are given in Table 11.4. Among the interesting conclusions are the facts that wedge techniques work better with higher photon energies and that the dose variance in the target volume is generally responsible for the loss in tumor control according to Eq. 11.7. It is also interesting to note that in a case like this, which is ideally suited for the four-field box technique (which gives the lowest mean energy imparted, $\bar{\varepsilon}$, to normal tissues), generalized conformal therapy still has some advantages, such as a lower standard deviation of the dose distribution in the target volume.

11.7 Summary and Conclusions

The multitude of classical approaches to dose optimization has improved the dose delivery in most cases but seldom succeeded in generating substantial improvements in the delivered absorbed dose distributions. Their resultant low clinical usage is not because of the shortcomings in the optimization techniques or lack of biologic data, but rather due to an insufficient degree of freedom in the optimization process since only conventional uniform or wedged fields have generally been employed. Also the lack of biologic models capable of choosing between the Scylla and Charybdis of radiation therapy, i.e., a limited high-dose volume or an extended volume irradiated to a low dose, may have contributed.

In Fig. 11.21 the principal properties of the treated volume (dashed lines) are schematically compared for different uniform and nonuniform external beam irradiation techniques on the cervical tumor and associated lymph nodes (shaded target volume). It is clear from the figure that with strongly concave target volumes none of the uniform beam irradiation techniques can produce a concave treated volume (except possibly for some fairly complex multicenter arc treatments). With nonuniform incident beams the flexibility is much greater and even a two- or three-field technique may do the job, as shown to the left in the lower row.

It is also evident from the clinical examples presented here that the new inverse and generalized conformal therapy approaches using radiobiologically optimized nonuniform beam profiles have important properties for a real optimization of radiation therapy. Furthermore, as improved radiobiologic data for tumors and normal tissue are continuously collected it will be possible to improve the models and increasingly trust in the optimal dose levels and dose distributions derived by the new algorithms. In addition the biologic models will allow the use of

CONVENTIONAL UNIFORM BEAM RADIOTHERAPY

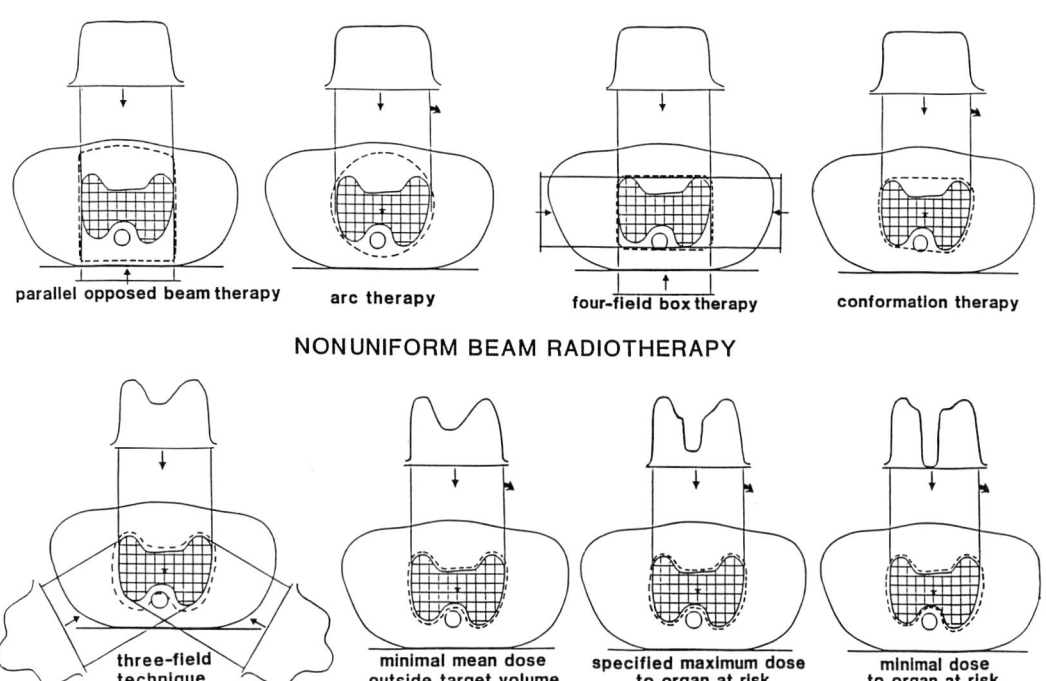

NONUNIFORM BEAM RADIOTHERAPY

Fig. 11.21. Schematic comparison of different external beam irradiation techniques. It is seen that nonuniform dose delivery using generalized conformal therapy and the inverse planning approach will in general allow a better matching of the treatment volume (*dashed line*) to the target volume (*shaded*)

clinical data such as tumor imaging and predictive assays.

From the discussions of the treatment optimization methods some basic principles can be recognized:

1. By a careful selection of the degrees of freedom that are most beneficial for a given treatment the number of free variables can be minimized and the optimization can be considerably simplified and speeded up with minimal loss in generality.
2. The use of composite elementary radiation field kernels is a very efficient and practical way of performing, this reduction of the degrees of freedom in treatment optimization.
3. By allowing nonuniform dose delivery most complex target volumes can effectively be irradiated by few-field techniques with a resultant increase in accuracy and safety.

Among the new principles for dose optimization that hold when nonuniform dose delivery is available are:

1. The largest dose to the tumor should be given from directions where the target volume has its largest extension along the beam.
2. Two-field techniques with parallel opposed or perpendicular beam combinations are generally less suitable than beams angled at about 100–120°.
3. The use of non-coplanar fields is another straightforward way of avoiding the coincidence of the entrance and exit regions of the beams.

In very general terms the function of nonuniform dose delivery is to protect normal tissues in front of, inside, or beyond the target volume whereas organs at risk outside or lateral to the target volume are spared by irregular field crossections (cf. Fig. 11.2). From the beam's point of view the function of the beam compensator or the scanned beam is therefore to save organs at risk *longitudinal* to the target volume. The block or multileaf collimation system, on the other hand, saves normal tissues in the *transversal* plane *lateral* to the target volume. To conclude, it should be pointed out that nonuniform dose delivery is the single remaining degree of freedom that still has been left largely unused in radiation therapy opti-mization. Strangely enough, at the same time it also seems to be one of the most powerful degrees of freedom of external beam radiotherapy, at least with regard to its ability to shape the delivered dose distributions to conform with target volume. In the future we will therefore most likely see an increased use of nonuniform dose delivery for the optimization of external beam radiation therapy by the new methods discussed here.

Acknowledgements. The continuing fruitful discussions and collaborations with Bengt Lind, Patric Källman, Svante Söderström, Anders Gustafsson, Anders Eklöf, Annakarin Ågren, Roger Svensson, Pirjo Aaltonen, and Anders Ahnesjö are gratefully acknowledged.

References

Aaltonen P (1992) An inventory of dose specification in the Nordic centres and a suggestion to a standardized procedure. Radiation dose in radiotherapy from prescription to delivery. IAEA, Leuven, Belgium, 16–20 September 1991

Abel NH (1823) Magazin for Naturvidenskaberne 1. See also: Abel NH (1881) Oeuvres completes 1. Grøndal and Son, Christiania, pp 2, 97

Ågren A-K, Brahme A, Turesson I (1990) Optimization of uncomplicated control for head neck tumors. Int J Radiat Oncol Biol Phys 19: 1077–1085

Ahnesjö A (1984) Application of transform algorithms for calculation of absorbed dose in photon beams. Proc. 8th int. conf. on the use of computers in radiation therapy. IEEE Computer Society, San Diego, Calif., pp 227–230

Ahnesjö A, Trepp A (1991) Acquisition of the effective lateral energy fluence distribution for photon beam dose calculations by convolution models. Phys Med Biol 36: 973–985

Ahnesjö A, Andreo P, Brahme A (1987) Calculation and application of point spread functions for treatment planning with high energy photon beams. Acta Oncol 26: 49–56

Andrew JR (1985) Benefit, risk and optimization by ROC analysis in cancer radiotherapy. Int J Radiat Oncol Biol Phys 11: 1557–1562

Bahr GK, Kereiakes JG, Horwitz H, Finney R, Galvin J, Goode K (1968) The method of linear programming applied to radiation treatment planning. Radiology 91: 686–693

Barth NH (1990) An inverse problem in radiation therapy. Int J Radiat Oncol Biol Phys 18: 425–431

Birkhoff GD (1940) On drawings composed of uniform straight lines. J de Mathematiques pures et appliquées 19: 221–236

Blattman H (1979) Therapy planning and dosimetry for the pion accelerator at the Swiss Institute of Nuclear Research (SIN). Rad Environm Biophys 16: 205–209

Bochner S (1948) Vorlesungen über Fouriersche Integrale. Chelsea, New York

Bollman R, Schmidt K-P, Tabbert E (1981) Verbesserung der Bestrahlungsplanung in der Hochvolttherapie durch mathematische Optimierung. Radiobiol Radiother 22: 594–601

Boltzmann L (1927) Vorlesungen über Gas-theorie. Barth, Leipzig

Bortfeld T, Bürkelbach J, Boesecke R, Schlegel W (1990) Methods of image reconstruction from projections applied to conformation radiotherapy. Phys Med Biol 35: 1423–1434

Boyer A, Antonuk L, Fenster A, Van Herk M, Meertens H, Munco P, Reinstein L, Wong J (1992) A review of electronic portal imaging devices (EPIDs). Med Phys 19: 1–16

Brahme A (1981) Correction of a measured distribution for the finite extension of the detector. Strahlentherapie 157: 258–259

Brahme A (1982) Physical and biologic aspects on the optimum choice of radiation modality. Acta Radiol Oncol 21: 469–479

Brahme A (1984) Dosimetric precision requirements in radiation therapy. Acta Radiol Oncol 23: 379–391

Brahme A (1985) Developments of external beam treatment units and the role of high energy electrons and photons. Teaching lecture ECCO 3, Stockholm, p 46

Brahme A (1987) Design principles and clinical possibilities with a new generation of radiation therapy equipment. Acta Oncol 26: 403–412

Brahme A (1988a) Optimization of stationary and moving beam radiation therapy techniques. Radiother Oncol 12: 129–140

Brahme A (1988b) Clinical possibilities with a new generation of radiation therapy equipment. Proc. int. symp. on dosimetry in radiotherapy. IAEA SM 298/67: 321–334

Brahme A (1992a) Biological and physical dose optimization in radiation therapy. In: Fortner JG, Rhoads JE (eds) Accomplishments in cancer research 1991. General Motors Cancer Research Foundation, pp 265–298

Brahme A (1992b) Which parameters of the dose distribution are best related to the radiation response of tumors and normal tissues? Radiation dose in radiotherapy from prescription to delivery. IAEA, Leuven, Belgium 16–20 September 1991

Brahme A (1993) Optimization of radiation therapy and the development of multileaf collimation. Int J Radiat Oncol Biol Phys 25: 373–375

Brahme A, Ågren A-K (1987) Optimal dose distribution for eradication of heterogeneous tumours. Acta Oncol 26: 377–385

Brahme A, Roos J-E, Lax I (1982) Solution of an integral equation encountered in rotation therapy. Phys Med Biol 27: 1221–1229

Brahme A, Lind B, Näfstadius P (1987) Radiotherapeutic computed tomography with scanned photon beams. Int J Radiat Oncol Biol Phys 13: 95–101

Brahme A, Källman P, Lind B (1989) Optimization of proton and heavy ion therapy using an adaptive inversion algorithm. Radiother Oncol 15: 189–197

Brahme A, Lind B, Källman P (1990) Inverse radiation therapy planning as a tool for 3D dose optimization. Phys Med 6: 53–63

Brahme A, Källman P, Lind B (1991) Optimization of the probability of achieving complication free tumor control using a 3D pencil beam scanning technique for protons and heavy ions. In: (eds) Itano A, Kanai T, Proceedings from the NIRS int. workshop on heavy charged particle therapy and related subjects. Chiba, Japan, pp 124–142

Brix F, Christiansen R, Hancken C, Quirin A (1988) The field integrated dose modification (FIDM): three typical clinical applications of a new irradiation technique. Radiother Oncol 12: 199–207

Buck B, Macaulay VA (1991) Maximum entropy in action, a collection of expository essays. Clarendon , Oxford

Burgers JMV, Awwad HK, Van der Laarse R (1985) Relationship between local cure and dose-time-volume factors in interstitial implants. Int J Radiat Oncol Biol Phys 11: 715–723

Bush RS (1982) The complete oncologist: Franz Buschke lecture. Int J Radiat Oncol Biol Phys 8: 1019–1027

Campbell A, Converse PE, Rodgers WL (1976) The quality of life perceptions, evaluation, satisfactions. Russel Sage Foundation, New York

Censor Y, Altschuler MD, Powlis WD (1988a) A computational solution of the inverse problem in radiation therapy treatment planning. Appl Math Comput 25: 57–87

Censor Y, Altschuler MD, Powlis WD (1988b) On the use of Cimmino's simultaneous projections method for computing a solution of the inverse problem in radiation therapy treatment planning. Inv Pro 4: 607–623

Chui CS, Mohan R (1988) Extraction of pencil beam kernels by the deconvolution method. Med Phys 15: 138–144

Cohen L (1960) The statistical prognosis in radiation therapy. Am J Roentgenol 84: 741–753

Cohen L (1987) Optimization of dose-time factors for a tumor and multiple associated normal tissues. Int J Radiat Oncol Biol Phys 13: 251–258

Convery DJ, Rosenbloom ME (1992) The generation of intensity-modulated fields for conformal radiotherapy by dynamic collimation. Phys Med Biol 37: 1359–1374

Cooper REM (1978) A gradient method of optimizing external-beam radiotherapy treatment plans. Radiology 128: 235–243

Cormack AM (1987) A problem in rotation therapy with x-rays. Int J Radiat Oncol Biol Phys 13: 623–630

Cormack AM, Cormack RA (1987) A problem in rotation therapy with x-ray: dose distributions with an axis of symmetry. Int J Radiat Oncol Biol Phys 13: 1921–1925

Cormack AM, Quinto ET (1989) On a problem in radio-therapy: questions of non-negativity. Int J Imag Syst Technol 1: 120–124

Davis AR (1982) On the maximum likelihood regularization of Fredholm convolution equations of the first kind. In: (eds) Baker CTH, Miller GS Treatment of integral equations by numerical methods. Academic Press, New York, pp 95–105

Deans RS (1983) The radon transform and some of its applications. John Wiley, New York

Djordjevich A, Bonham DJ, Hussein EMA, Andrew JW, Hale ME (1990) Optimal design of radiation compensators. Med Phys 17: 397–404

Dunn WL (1981) Inverse Monte Carlo analysis. J Comput Phys 41: 154–166

Ebert U (1977) Computation of optimal radiation treatment plans. J Comput Appl Math 3: 99–104

Fisher JJ (1969) Theoretical considerations in the optimization of dose distribution in radiation therapy. Br J Radiol 42: 925–930

Fowler JF (1981) Nuclear particles in cancer treatment. Adam Hilger, Bristol (Medical Physics Handbooks 8)

Franke DS (1980) Die Anwendung der mathematischen Optimierung in der Strahlentherapie zur Findung optimaler Bestrahlungstechniken. Radiobiol Radiother 21: 668–676

Goitein M, Niemierko A (1988) Biologically based models for scoring treatment plans. In: Zink S (ed) Proc. Joint US-Scandinavian symp. on future directions computer aided radiation therapy, San Antonio, Texas, National Cancer Institute

Gremmel H, Hebbinghaus D, Wendhausen H (1978) An optimization criterion for dose distributions, minimising the radiation effect in healthy tissue. In: Rosenow W (ed) Proc. 6th international Conference on the use of computers in radiation therapy, Myet-Druck, Dransfeld, Göttingen, pp 199–209

Gustafsson A, Lind BK Brahme A (1993) A general pencil beam algorithm for optimization of radiation therapy. Med Phys 21: 343–356

Hodes L (1974) Semiautomatic optimization of external beam radiation treatment planning. Radiology 110: 191–196

Holmes T, Mackie RT, Simpkin D, Reckwerdt P (1991) A unified approach to the optimization of brachytherapy and external beam dosimetry. Int J Radiat Oncol Biol Phys 20: 859–873

Hope CS, Orr HS (1965) Computer optimization of 4 MeV treatment planning. Phys Med Biol 10: 365–373

Hope CS, Laurie J, Orr JS, Halnan JS (1967) Optimization of x-ray treatment planning by computer judgement. Phys Med Biol 12: 531–542

ICRU Report 33 (1980) Radiation quantities and units. International Commission on Radiation Units and Measurements, Bethesda.

Ishigaki T, Itoh Y, Horikawa Y, Kobayashi H, Obata Y, Sakuma S (1990) Computer-assisted conformation radiotherapy. AMPI Med Phys Bull 15: 185–211

Jaynes ET (1957) Information theory and statistical mechanics. Phys Rev 106: 620–630

Karnofsky DA, Abelmann WH, Craver LL, Burchenal JH (1948) The use of the nitrogen mustards in the palliative treatment of carcinoma. Cancer 1: 634–658

Källman P (1992) Optimization of radiation therapy planning using physical and biological objective function. Thesis, Stockholm University, Sweden

Källman P, Lind BK, Brahme A (1992a) An algorithm for maximizing the probability of complication free tumor control in radiation therapy. Phys Med Biol 37: 871–890

Källman P, Ågren A, Brahme A (1992b) Tumor and normal tissue responses to fractional non uniform dose delivery. Int J Radiat Biol 62: 249–262

Källman P, Lind B, Eklöf A, Brahme A (1988) Shaping of arbitrary dose distributions by dynamic multileaf collimation. Phys Med Biol 33: 1291–1300

Kirkpatrick S, Gelatt CD, Vecci MP (1983) Optimisation by simulated annealing. Science 220: 671–680

Lane DP (1992) Guardian of the genome. Nature 358: 15–16

Langer M, Leong J (1987) Optimization of beam weights under dose-volume restrictions. Int J Radiat Oncol Biol Phys 13: 1255–1260

Langer M, Kijewski P, Brown R, Ha C (1991) The effect on minimum tumor dose of restricting target-dose inhomogeneity in optimized three-dimensional treatment of lung cancer. Radiother Oncol 21: 245–256

Larsson B, Lidén K, Sarby B (1974) Irradiation of small structures through intact skull. Acta Radiol TPB 13: 513–534

Lax I, Brahme A (1982) Rotation therapy using a novel high-gradient filter. Radiology 145: 473–478

Leemann C, Alonso J, Grunder H, Hoyer E, Kalnins G, Rondeau D, Staples J, Voelker F (1977) A 3-dimensional beam scanning system for particle radiation therapy. IEEE Trans Nucl Sci NS-24: 1052–1054

Legras J, Legras B, Lambert J-P (1982) Software for linear and non-linear optimization in external radiotherapy. Comput Programs Biomed 15: 233–242

Leibel SA, Ling CC, Kutcher GJ, Mohan R, Cordon-Cordo C, Fuks Z (1991) The biological basis for conformal three-dimensional radiation therapy. Int J Radiat Oncol Biol Phys 21: 805–811

Lind BK (1990) Properties of an algorithm for solving the inverse problem in radiation therapy. Inverse Problems 6: 415–426

Lind BK (1991) Radiation therapy planning and optimization studied as inverse problems. Thesis, Stockholm University Sweden

Lind B, Brahme A (1985) Generation of desired dose distributions with scanned elementary beams by deconvolution methods. Proc. VII ICMP Espoo, Finland, p 953

Lind B, Brahme A (1987) Optimization of radiation therapy dose distributions with scanned photon beams. In: (eds) Bruinvis IAD, Van der Giessen PH, Van Kleffens HJ Proc. 9th int. conf. on the use of computers in radiation therapy. Elsevier, Amsterdam, pp 235–239

Lind BK, Brahme A (1992) Photon field quantities and units for kernel based radiation therapy planning and treatment optimization. Phys Med Biol 37: 891–909

Lind BK, Källman P, Sundelin B, Brahme A (1993) Optimal radiation beam profiles considering uncertainties in beam patient alignment. Acta Oncol 32: 331–342

Luenberger DG (1973) Introduction to linear and nonlinear programming. Addison-Wesley, Menlo Park

Mantel J, Perry H, Weinkam JJ (1977) Automatic variation of field size and dose rate in rotation therapy. Int J Radiat Oncol Biol Phys 2: 697–704

McDonald SC, Rubin P (1977) Optimization of external beam radiation therapy. Int J Radiat Oncol Biol Phys 2: 307–317

McShan DL, Fraass BA (1987) Integration of multi-modality imaging for use in radiation therapy treatment planning. In: Lemke HU, Rhodes ML, Jaffe CC, Felix R (eds) Computer assisted radiology, Springer, Berlin Heidelberg New York, pp 300–304

Metropolis N, Rosenbluth A, Rosenbluth M, Teller H, Teller E (1953) Equation of state calculations by fast computing machines. J Chem Phys 21: 1087–1092

Miller GF (1974) Fredholm equations of the first kind. In: Delves LM, Walsh J (ed) Numerical solutions of integral equations. Oxford University press, London, pp 175–188

Miller DW (1985) Optimization of attenuator shapes for multiple field radiation treatment. In: Paliwal BR, Herbert DE, Orton CG (eds) Optimization of cancer radiotherapy. Am Assoc Physic Symp Proc No 5. American Institute of Physics, New York

Mohan R, Chui C, Lidofsky L (1986) Differential pencil beam dose computation model for photons. Med Phys 13: 64–72

Mohan R, Mageras GS, Baldwin B, Brewster LJ, Kutcher GJ, Leibel S, Burman CM, Ling CC, Fuks Z (1992) Clinically relevant optimization of 3D conformal treatments. Med Phys 19: 933–944

Moore DH, Mendelsohn ML (1972) Optimal treatment levels in cancer therapy. Cancer 30: 95–106

Morrill SM, Lane RG, Rosen II (1990) Constrained simulated annealing for optimized radiation therapy treatment planning. Comput Meth Progr Biomed 33: 135–144

NACP (Nordic Association of Clinical Physicists) (1994) Specification of dose delivery in radiation therapy. (submitted)

Natterer F (1986) The mathematics of computerized tomography. John Wiley, Stuttgart

Niemierko A (1993) Random search algorithm (RONSC) for optimization of radiation therapy with both physical and biological end points and constraints. Int J Radiat Oncol Biol Phys 23: 89–108

Norman A, Kagan AR, Chan SL (1988) The importance of genetics for the optimization of radiation therapy. Am J Clin Oncol 11: 84–88

Ortega JM, Rheinboldt WC (1970) Iterative solution of nonlinear equations in several variables. Academic Press, New York

Popp FA, Bothe B, Goedecke R (1975) Prinzipien zur Optimierung der Bestrahlungsplanung. Strahlentherapie 150: 389

Powlis WD, Altschuler MD, Censor Y, Buhle EL (1989) Semi-automated radiotherapy treatment planning with a mathematical model to satisfy treatment goals. Int J Radiat Oncol Biol Phys 16: 271–276

Raju MR (1980) Heavy particle radiotherapy. Academic Press, New York

Redpath AT, Vickery B, Wright DH (1976) A new technique for radiotherapy planning using quadratic programming. Phys Med Biol 21: 781–791

Rosen II, Lane RG, Morrill S, Belli JA (1991) Treatment plan optimization using linear programming. Phys Med Biol 18: 141–152

Schafer RW, Mersereau RM, Richards MA (1981) Constrained interative restoration algorithms. Proc IEEE 69: 432–450

Schultheiss TE, Orton CG, Peck RA (1983) Models in radiotherapy Volume effects. Med Phys 10: 410–415

Schwartz JL, Rotmensch J, Giovanazzi BS, Cohen MB, Weichselbaum RR (1988) Faster repair of DNA double-strand breaks in radioresistant human tumor cells. Int J Radiat Oncol Biol Phys 15: 907–912

Shannon CE (1948) A mathematical theory of communication. Bell System Tech J 27: 379–423, 623–659

Sonderman D, Abrahamson PG (1985) Radiotherapy design using mathematical programming models. Oper Res 33: 705–725

Söderström S, Brahme A (1993a) Selection of suitable beam orientations in radiation therapy using entropy and Fourier transform measures. Phys Med Biol 37: 911–924

Söderström S, Brahme A (1993b) Optimization of multiple field techniques using radiobiological objective functions. Med Phys 20: 1201–1210

Söderström S, Gustafsson A, Brahme A (1993) The clinical value of different treatment objectives and degrees of freedom in radiation therapy optimization. Radiother Oncol 29: 148–163

Starkschall G (1984) A constrained least-squares optimization method for external beam radiation therapy treatment planning. Med Phys 11: 659–665

Starkschall G, Eifel PJ (1992) An interactive beam-weight optimization tool for three-dimensional radiotherapy treatment planning. Med Phys 19: 155–163

Sterling TD, Perry H, Weinkam JJ (1965) Automation of radiation treatment planning V. Calculation and visualisation of the total treatment volume. Br J Radiol 38: 906–913

Svensson R, Källman P, Brahme A (1994) An analytical solution for the dynamic control of multileaf collimators. Phys Med Biol 39: 37–61

Swindell W, Simpson RG, Oleson JR Chen CT, Grubbs EA (1983) Computed tomography with a linear accelerator with radiotherapy applications. Med Phys 10: 416–420

Takahashi S (1965) Conformation radiotherapy, rotation techniques as applied to radiography and radiotherapy. Acta Radiol Suppl 242

Thames HD, Hendry JH (1987) Fractionation in radiotherapy. Taylor & Francis, London

Timme TL, Moses RE (1988) Review: diseases with DNA damage-processing defects. Am J Med Sci 295: 40–48

Ulsø N, Brahme A (1988) Computer aided irradiation technique optimization. In: Zink S (ed) Proc. Joint US-Scandinavian symp. on future directions of computer aided radiation therapy, San Antonio, Texas, National Cancer Institute

Van der Laarse R, Strackee J (1976) Pseudo optimization of radiotherapy treatment planning. Br J Radiol 49: 450–457

Van Santvoort J, Huizenga H (1991) The use of elementary electron beams and inverse planning techniques. Paper presented at a Workshop on Developments in Dose Planning and Treatment Optimization, The Karolinska Institute, Stockholm

Webb S (1989) Optimization of conformal radiotherapy dose distributions by simulated annealing. Phys Med Biol 34: 1349–1369

Webb S (1990) Non-standard CT scanners: their role in radiotherapy. Int J Radiat Oncol Biol Phys 19: 1589–1607

Webb S (1993) The physics of three-dimensional radiation therapy. I.O.P. Publishing, Bristol

Weeks KJ, Sontag MR (1991) 3-D dose-volume compensation using nonlinear least squares regression technique. Med Phys 18: 474–480

Wolbarst AB (1984) Optimization of radiation therapy II: the critical-voxel model. Int J Rad Oncol Biol Phys 10: 741–745

12 Recent Developments in Basic Brachytherapy Physics

Jeffrey F. Williamson

12.1 Introduction

Brachytherapy physics research has experienced a renaissance of creative and innovative developments over the last decade which have only begun to influence clinical practice. The purpose of this chapter is to review the major innovations in single-source brachytherapy dosimetry introduced during the last 10 years. Among these developments are:

1. Development of new low-energy isotopes for brachytherapy, including ^{103}Pd, ^{241}Am, ^{143}Sm, and ^{169}Yb, with photon energies in the range of 23–100 keV
2. Introduction of new physical configurations of conventional isotopes (^{137}Cs and ^{192}Ir) in response to the proliferation of high- and low-dose-rate remote afterloading devices
3. Validation of brachytherapy dose-measurement techniques and acceptance of directly measured dose distributions for clinical treatment planning
4. Validation of Monte Carlo photon-transport simulation as a clinical dosimetry tool
5. Development of new dose-calculation formalisms and source-strength specification quantities to facilitate clinical dose computation using directly measured or Monte Carlo-generated single-source dose distributions
6. Intensive investigation of dose perturbations arising from applicator shielding, inter-source and-applicator attenuation, and variations in tissue density and composition.

These developments constitute a significant departure from the conventional approach to brachytherapy dosimetry. Traditionally, brachytherapy treatment techniques, dose prescriptions, and knowledge of normal-tissue and tumor dose-response relationships have evolved empirically, guided by observed control and complication rates in large groups of patients treated in a uniform fashion over many years. This evolutionary dynamic placed relatively little emphasis on physically accurate dose computation and "physics-based" optimization, especially in gynecologic brachytherapy. Prior to 1980, treatment planning and dosimetric evaluation of brachytherapy had changed relatively little since the introduction of afterloading techniques and new reactor-produced radium substitutes such as ^{137}Cs and ^{192}Ir in the 1950s and early 1960s. With the exception of ^{125}I interstitial sources, the armamentarium of brachytherapy sources and techniques was relatively stable. Dose distributions were calculated

Jeffrey F. Williamson, PhD, Mallinckrodt Institute of Radiology, Radiation Oncology Center, Physics Section, 510 S. Kingshighway, St. Louis, MO 63110, USA

by superposition, which approximates dose at a point by the sum of contributions from each source. Each source contribution was estimated from a single-source dose matrix, usually derived from simple semiempirical models such as the Sievert integral (SIEVERT 1921; YOUNG and BATHO 1964). These models assumed unbounded water medium and accounted only for oblique filtration of primary photons by the source capsule. Heterogeneities were ignored, including tissue-composition and density variations, applicator attenuation and shielding effects, air–tissue interfaces, and inter-source and -applicator shielding effects. Direct measurement of absorbed dose near brachytherapy sources, using thermoluminescent dosimeters, radiographic film, solid-state detectors, or small ion chambers, was relatively rare even in the research laboratory and had little impact on clinical practice.

In contrast to the traditional "minimalist" philosophy of brachytherapy physics, interest in the current developments, listed above, is motivated by the premise that physical and radiobiologic optimization of implant therapy can improve clinical outcome. For example, a major rationale for investigating new low-energy brachytherapy sources, as well as new applications of ^{125}I in temporary implantation (LING et al. 1988), is the ease with which sensitive tissues can be shielded from low-energy photons. In addition, some of these sources offer possibilities of improving clinical outcome through manipulation of dose-rate effects (LING 1992). The increasing use of high-dose-rate (HDR) remote afterloading is also driving development of more accurate dosimetry. Because HDR brachytherapy sacrifices the inherent biologic sparing of dose-limiting normal tissues characteristic of low-dose-rate (LDR) brachytherapy (BRENNER and HALL 1991), excessive normal tissue complications can only be avoided by delivering lower doses to critical structures than are required for LDR implants having an equivalent tumoricidal effect. This has focused attention on the problem of optimizing the design of shielded applicators. Another emerging research area, image-based 3D treatment planning (LING et al. 1987; SCHOEPPEL et al. 1993) seeks to minimize the very large random errors inherent in conventional treatment evaluation, by basing dose specification on a quantitative geometric model of patient anatomy. The goal is to extract technique-independent normal tissue and tumor control dose-response information to facilitate future optimization of this therapy. Accurate, prospective, and reasonably rapid estimation of dose-rate distribu-

tions, often in the presence of tissue and applicator heterogeneities, is essential to each of these developments.

12.2 New Sources and Isotopes for Brachytherapy

Table 12.1 contrasts the properties of several novel isotopes for brachytherapy with conventionally used sources. Two basic rationales for utilizing these isotopes have been discussed in the literature. For isotopes with mean photon energies in the 60–150 keV range, the prospect of using thin high-atomic-number foils for customized shielding of dose-limiting tissues is the central motivation. As Fig. 12.1 shows, ^{169}Yb and ^{241}Am produce depth doses in water qualitatively similar to those of conventional radium-substitute isotopes, which closely approximate inverse-square law falloff. In contrast, ultra-low-energy sources such as ^{103}Pd and ^{125}I have depth-dose characteristics significantly less penetrating than that predicted by inverse-square law. The photons emitted by ^{169}Yb and ^{241}Am sources, averaging 93 and 60 keV, respectively, interact with water largely by approximately elastic Compton interactions, giving rise to multiply scattered photons which compensates for attenuation of primary photons. However, unlike higher-energy radium-substitute photons, these low-energy photons interact photoelectrically in lead and other shielding materials, offering the prospect of effectively shielding critical structures with thin foils on the order of 0.5 mm thick. Unlike ^{137}Cs and ^{192}Ir, which require 2–5 mm of tungsten shielding to produce a twofold dose reduction, these new isotopes make possible customization of shielding to improve clinical outcome of individual patients (NATH and GRAY 1987).

The rationale for introducing ^{103}Pd as an interstitial source is to enlarge the domain of permanent implantation. The conventional low-energy source, ^{125}I, has a half-life of 59.6 days, resulting in delivery of 100–200 Gy to the tumor periphery in 6 months at

Table 12.1. Properties of low-energy implant isotopes

Isotope	Mean energy	Half-life	TVL Pb
^{137}Cs	662 keV	30.0 years	18.5 mm
^{192}Ir	360 keV	74.2 days	7.1 mm
^{125}I	28 keV	59.0 days	0.025 mm
^{103}Pd	22 keV	17.0 days	0.013 mm
^{241}Am	60 keV	432.2 years	0.41 mm
^{145}Sm	43 keV	340.0 days	0.20 mm
^{169}Yb	93 keV	32.0 days	1.6 mm

Fig. 12.1. Plot of radial dose function as a function of distance for various conventional and investigational brachytherapy sources. The radial dose function is proportional to the product of dose rate and the square of the distance and is normalized to unity at a distance of 1 cm from the source. The data for ^{103}Pd, ^{145}Sm, and ^{169}Yb are derived from the author's unpublished Monte Carlo calculations

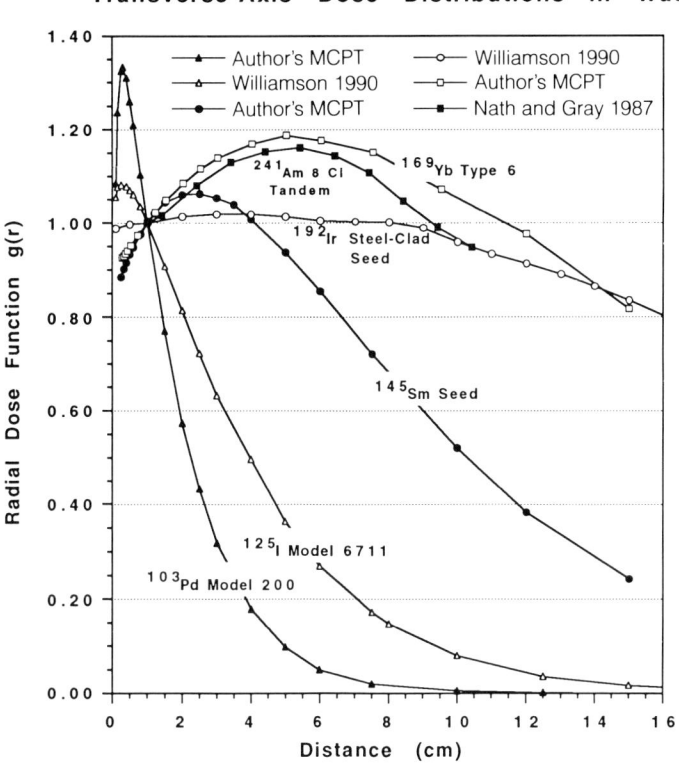

Fig. 12.1. Plot of radial dose function as a function of distance for various conventional and investigational brachytherapy sources. The radial dose function is proportional to the product of dose rate and the square of the distance and is normalized to unity at a distance of 1 cm from the source. The data for ^{103}Pd, ^{145}Sm, and ^{169}Yb are derived from the author's unpublished Monte Carlo calculations

ultra-low initial dose rates of 5–10 cGy/h. At such low dose rates, cell proliferation may compete with radiation-induced cell lethality and leave too large a surviving fraction of malignant cells at the end of treatment to guarantee control for tumor histologies containing rapidly proliferating cells, i.e., cells with small potential doubling times, T_{pot} (MEIGOONI et al. 1990; LING 1992). ^{103}Pd, which emits characteristic x-rays of 21 keV, has all of the radiation protection advantages of ^{125}I along with a significantly shorter half-life of 17 days. With this source, an implant can deliver 105 Gy in approximately 8 weeks at initial peripheral dose rates of 20 cGy/h. Using the linear-quadratic theory, LING (1992) has shown that ^{103}Pd

implants yield both a significantly higher biologically effective dose and smaller surviving fraction than ^{125}I for tumors with a T_{pot} less than 5 days.

12.2.1 Palladium-103 Seeds for Permanent Implantation

Palladium-103 sources were developed by Dr. John Russell in conjunction with Theragenics Corporation in the early 1980s and are now commercially available with strengths up to 2 mCi/seed (maximum air-kerma strength, S_K, of 2.6 cGy·cm²·h⁻¹). The external dimensions of the model 200 seed (Fig. 12.2)

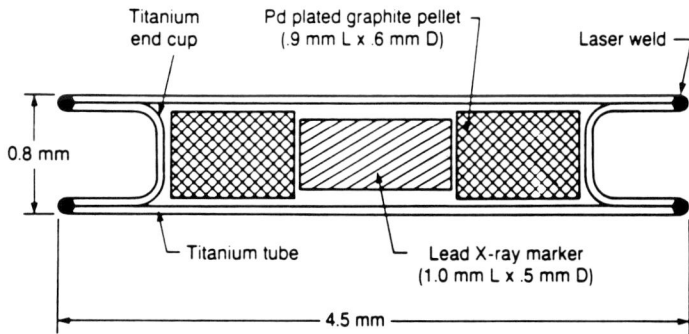

Fig. 12.2. Schematic diagram of the ^{103}Pd model 200 interstitial source (from MEIGOONI et al. 1990)

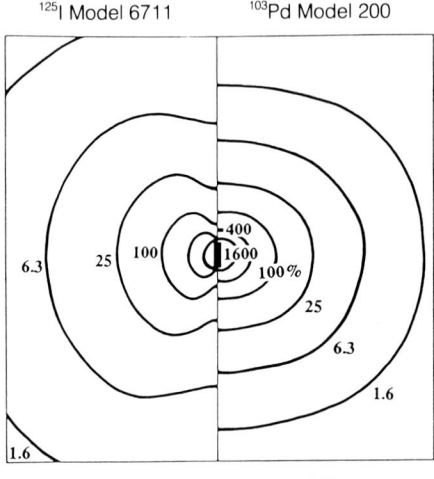

Doses expressed as % of Dose
at 1 cm on transverse axis

Fig. 12.3. Comparison of isodose curves arising from the model 6711 ^{125}I seed and the model 200 [103]Pd seed. Both data sets are normalized to 100% at a distance of 1 cm form the source on the transverse axis. The [103]Pd data and ^{125}I data are derived from Meigooni et al. (1990) and Williamson and Quinterro (1988) respectively

though its dose distribution falls off faster (Fig. 12.1, 12.3) than that of ^{125}I, [103]Pd volume implants have only slightly more heterogeneous dose distributions than their ^{125}I counterparts (Nath et al. 1992a). [103]Pd interstitial seed isodose curves (Fig. 12.3) are slightly more anisotropic than those of [125]I.

Two complete dosimetric studies exist in the published literature (Meigooni et al. 1990; Chiu-Tsao and Anderson 1991) giving specific dose constants for water, two-dimensional angular profiles, and transverse-axis dose distributions in forms suitable for clinical treatment planning. Both studies used calibrated LiF TLD dosimeters to measure the dose-rate distribution in solid-water medium, as described more fully in Sect. 12.3.1.1. Chiu-Tsao and Anderson performed measurements over the distance range of 0.2–4 cm while Meigooni et al.'s measurements encompassed the distance range 0.5–7.5 cm. Although the two data sets (Fig. 12.4) are in good agreement beyond 1 cm, along the transverse axis they disagree by 19% and 8% at 0.5 cm and 1.0 cm, respectively. Based on averaging the two data sets, and correcting the result by 4% for the nonliquid water equivalence of solid water (see Sect. 12.3.4.1 below), a dose-rate constant, Λ_0, of 0.74 cGy·h^{-1} per unit air-kerma strength can be recommended for treatment planning.

are identical to its ^{125}I permanent-implant counterpart (model 6711) and it can be implanted using conventional seed inserters and Mick guns. Even

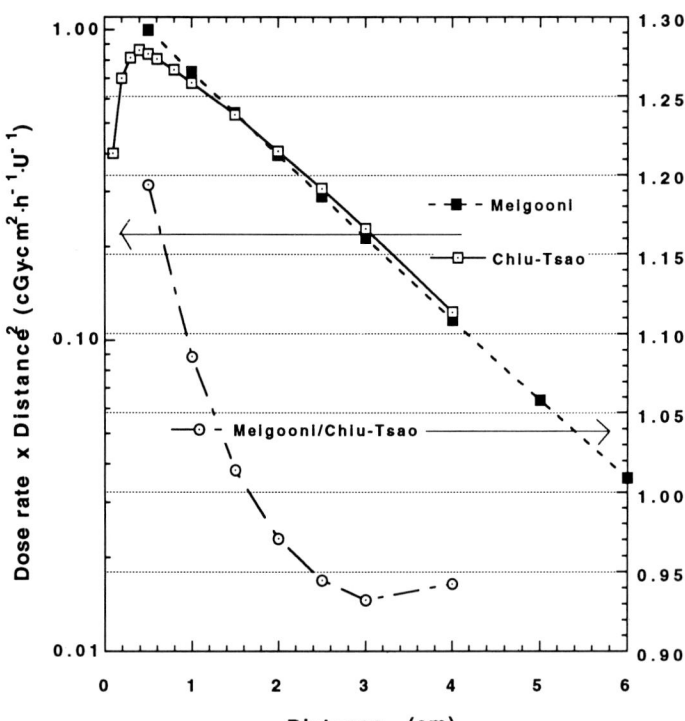

Fig. 12.4. Comparison of [103]Pd model 200 transverse-axis dose rates as measured by Meigooni et al. (1990) and Chiu-Tsao and Anderson (1991). The product of dose rate and the square of the distance is plotted on the left axis. The right axis shows the ratio of the two measured data sets

12.2.2 Americium-241 Sources
for Intracavitary Therapy

Radioactive [241]Am sources have been developed for intracavitary treatment of gynecologic malignancies by Dr. Nath and his colleagues at Yale University and are the most extensively investigated of the new radioisotopes for brachytherapy. Review of this experience clearly illustrates the potential clinical advantages of low-energy sources as well as the challenging dosimetric, engineering, and radio-biologic problems that must be solved to make these sources clinically practical. [241]Am decays by emission of alpha particles with a half-life of 432.2 years, emitting photons with an energy of 59.537 keV (NATH and GRAY 1987). Its exposure rate constant $(\Gamma_\delta)_x$ and specific activity are 0.122 R·h^{-1}·cm^2·mCi^{-1} and 3.432 Ci·g^{-1}, respectively. The [241]Am radial dose distribution (Fig. 12.1) is very similar (within 15%) to that of ^{192}Ir but has a much smaller half-value layer in lead, about 0.125 mm. The clinical advantages of this source include increased personnel protection, ability to shield critical structures with thin (0.5-mm) lead foils, and a long half-life which eliminates cost of source replacement.

Cylindrical sources with external diameters of 0.8–1.0 cm and lengths of 1.6–8.8 cm (Fig. 12.5) have been developed (NATH et al. 1987b). They contain a mixture of [241]AmO$_2$ and aluminum powders doubly encapsulated in 1 mm of titanium. Contained activities range from 2 Ci for the smallest source to 8 Ci for the tandem-like source. Because of the low energy of the emitted photons and the high atomic number of [241]Am, self-absorption is extremely high. This dictates a source design with a large surface-to-volume ratio, resulting in bulky sources with minimum diameters of about 1 cm. The 2-, 5-, and 8-Ci sources have air-kerma strengths of 44, 142, and 197 cGy·cm^2·h^{-1}, respectively, which implies that approximately 80% of the emitted photons are absorbed within the sources. The 2- and 5-Ci sources are placed parallel to one another in molded rubber vaginal plaques (Fig. 12.6), giving rise to 5-5, 5-4-5, and 5-4-4-5 Ci loadings (NATH et al. 1988). This applicator system was designed to duplicate the Morris system dose distribution (a variant of the Stockholm system) and gives point A and B dose rates of 60 and 23 cGy/h. Isodose curves for a typical implant are shown in Fig. 12.7.

The Yale group has extensively studied the dosimetric characteristics of [241]Am (NATH and GRAY 1987; NATH et al. 1987b, 1990a, 1992b), mostly using calibrated LiF TLD-100 dosimeters in solid-water

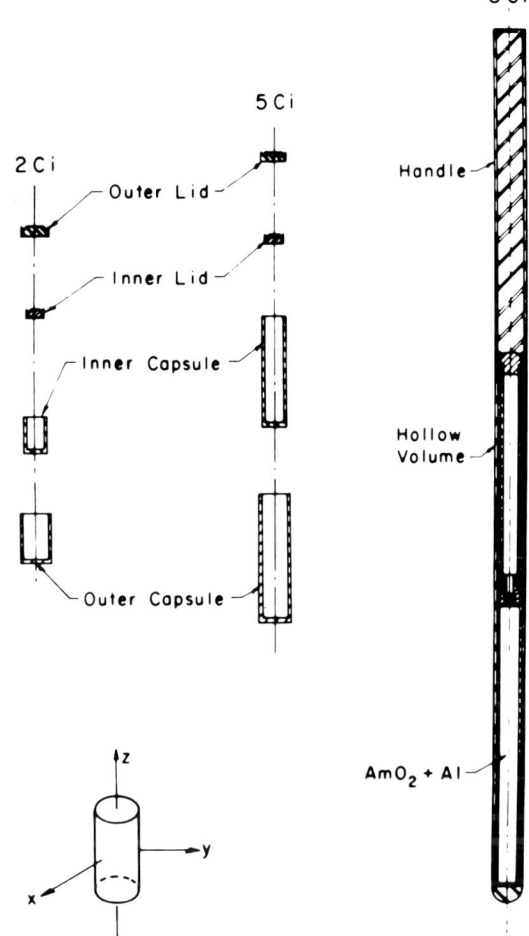

Fig. 12.5. Schematic drawings showing the design of the 2-, 5-, and 8-Ci cylindrical [241]Am sources (from NATH et al. 1988)

phantoms and 0.3 cm^3 ion-chambers in liquid water to measure dose-rate distributions. The Yale dose-computation model (see Sect 12.4.2) consists of a three-dimensional generalization of the Sievert integration model with elaborate empirical corrections necessitated by the anisotropic distribution of scattered radiation about single sources (NATH and GRAY 1987). Multiple-source [241]Am plaques are even more computationally challenging since the bulky sources give rise to large source-to-source shielding effects, reducing dose rates by as much as 40% relative to simple superposition calculations. These dose perturbations are due to both line-of-sight primary- and scattered-photon attenuation as well as global reduction of the multiply scattered photon-dose component (NATH et al. 1990a). Lead-foil shielding gives rise to dose-reduction factors that are highly dependent on shield diameter and can reduce dose rates in the unshielded region by as much as 20% (MUENCH

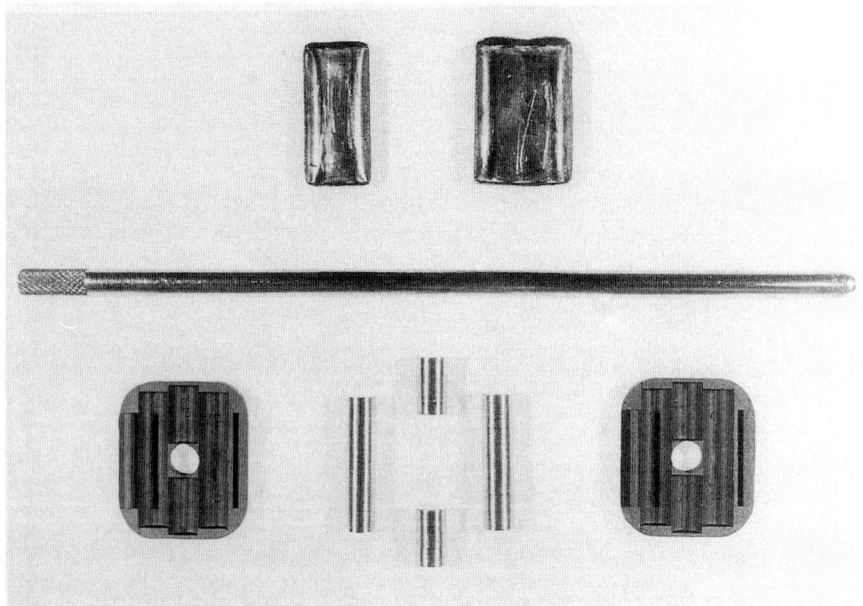

Fig. 12.6. Photograph of the 5-4-5 Ci [241] Am vaginal applicator, which includes two 5-Ci and two 2-Ci sources arranged in the illustrated pattern. Also shown is the 8-Ci tandem source and lead foils designed to shield one side of a two- or three- segment vaginal plaque. (From NATH at el. 1988)

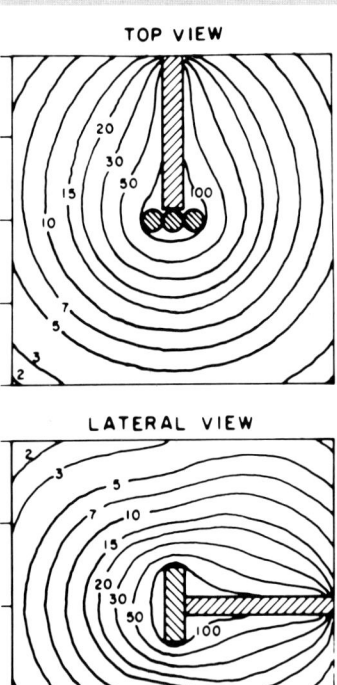

Fig. 12.7. Calculated isodose-rate curves (cGy/h) produced by the 5-4-5 Ci plaque in combination with the 8-Ci tandem. The *hash marks* on the side of the diagrams indicate 5-cm intervals. (From NATH et al. 1988)

complex than their radium-substitute counterparts, making both dose measurement and dose-calculation algorithm design correspondingly more difficult.

Figures 12.8 and 12.9 illustrate two strategies for protecting normal tissues made possible by low-energy sources such as [241]Am (NATH et al. 1988). A 0.5-mm-thick lead foil (Fig. 12.9) reduces dose rate by a factor of 17 at 2 cm distance, creating a nearly unidirectional applicator. Figure 12.8 shows that introduction of 560 mg/g $BaSO_4$ or 50% Hypaque solutions into the rectum can reduce dose to the posterior rectal wall by a factor of 6.

12.2.3 Samarium-145 and Ytterbium-169 Sources for Intracavitary Therapy

Samarium-145 interstitial seeds, encapsulated in titanium of the same dimensions as currently available ^{125}I seeds, have been developed by FAIRCHILD et al. (1987) at the Brookhaven National Laboratory. This isotope decays by electron capture producing a cascade of characteristic x-rays (38–45 keV) and one weak gamma ray of 61 keV, resulting in an average energy, 43 keV, slightly higher than that of [125]I. [145]Sm has a half-

and NATH 1992). Dose-calculation algorithms that accurately predict such shielding effects are yet to be developed. The Yale experience clearly demonstrates that low-energy dose distributions are much more

Fig. 12.8. Measured dose rates in water (cGy/h) arising from a 5-4-5 Ci vaginal applicator with and without test tubes containing BaSO$_4$ and Hypaque solution. Both the test tube and the vaginal plaque were placed in a thank and surrounded by water. (From NATH et al. (1988)

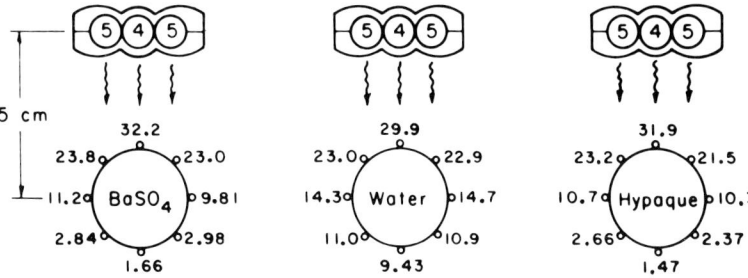

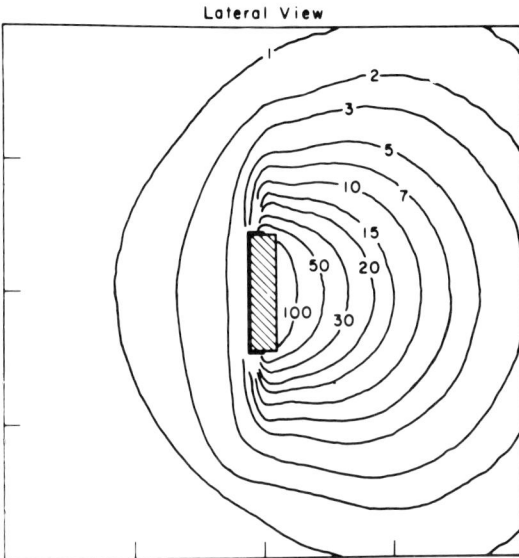

Lateral View

Fig. 12.9. Isodose-rate curves (cGy/h) produced by a shielded 5-5 Ci ^{241}Am applicator. The shield on the left side of the applicator consists of a 0.5-mm-thick lead foil. The *hash marks* indicate 5-cm intervals. (From NATH et al. 1988)

life of 340 days, an air-kerma rate constant of 0.775 cGy·h^{-1}·cm^2·mCi^{-1}, and has been produced with strengths of up to 3 mCi, although activation to strengths of 40 mCi is, in principle, achievable (FAIRCHILD et al. 1987). ^{145}Sm sources have depth-dose characteristics (Fig. 12.1) intermediate between those of ^{125}I and radium-substitute isotopes. The potential clinical advantages of ^{145}Sm include (a) improved logistics and cost-effectiveness as a temporary interstitial implant isotope compared to ^{125}I because of its longer half-life, (b) improved dosimetric characteristics (less anisotropy, fewer seeds required in large implants) compared to ^{125}I because of higher energy, and (c) enhanced radiation protection and local shielding capability compared to radium-substitute isotopes.

Samarium-145 decays to ^{145}Pm (promethium-145), creating a technical problem so far not addressed by the published literature. ^{145}Pm has a much longer half-life (17.7 years) than ^{145}Sm and a photon yield (37–62 keV) per disintegration about one-half

that of ^{145}Sm (FAIRCHILD et al. 1987). Thus radioactive equilibrium is never established during the useful life of the source: 6 and 24 months after activation, the activity of ^{145}Pm is 3.4% and 76%, respectively, of the contained activity of ^{145}Sm. In addition to anomalous decay, the dose-rate constant and relative dose distribution will vary with the age of the source. The extent to which this unusual decay scheme creates a clinical problem is unknown.

An important biologic rationale for ^{145}Sm and other intermediate-energy isotopes is radiosensitization of tumor cells that have incorporated halogenated thymidine analogs, such as IUdR and BUdR, into their DNA (FAIRCHILD et al. 1982; NATH et al. 1990c). These halogenated thymidine analogs are preferentially incorporated into actively proliferating cells (i.e., tumor rather than normal tissue cells) and are known to be potent radiosensitizers in themselves. These investigators hypothesize that by using low-energy brachytherapy sources emitting photons just above the K-absorption edge of iodine (33.2 keV), additional radiosensitization may result from Auger electron cascades arising from photoelectric absorption of 40- to -60-keV photons by the incorporated iodine atoms. Because of the limited range of the Auger electrons, this additional cytotoxic enhancement is highly localized to those cells with significant IUdR or BUdR uptake. NATH et al. (1990c) have shown that when Chinese hamster lung cells are incubated in a growth medium containing 10^{-4} M IUdR followed by ^{241}Am irradiation, cell kill is enhanced by a factor of 3.04 relative to cells in normal growth medium. In comparison, irradiation by ^{226}Ra and and ^{125}I sources enhances radiosensitivity by factors of 1.89 and 2.48, respectively. Theoretically, even greater radiosensitization should by achievable by ^{145}Sm, since its 43-keV photons are better matched to the K-edge of iodine.

Ytterbium-169 has a number of characteristics (MASON et al. 1994; PIERMATTEI et al. 1992; PERERA et al. 1994) that warrant its continued investigation as a brachytherapy isotope. It decays by electron capture, producing x-rays and gamma rays ranging from 49.8 keV to 307.7 keV (average = 93 keV) with a half-life of 32 days. In addition to enhanced radiation protection and local shielding capabilities [half-value layer (HVL) = 0.2 mm Pb], its nonradio-active precursor, [168]Yb, has an extremely large neutron-capture cross-section, making activity concentrations as large as 10 Ci/mm^3 possible. Unlike [241]Am or ^{125}I, miniaturized [169]Yb interstitial and intracavitary sources can be easily constructed for LDR brachy-therapy. Potentially, high-intensity [169]Yb sources could be used to deliver HDR intraoperative brachytherapy treatments in lightly shielded operating rooms, thereby broadening the clinical indications of brachytherapy. Permanent implantation with this isotope may not be practical since the 308-keV photon line may pose a significant radiation-safety hazard.

Several prototypes (types 4–6) of a titanium-encapsulated [169]Yb interstitial seed have been manu-factured with activities up to 200 mCi (Fig. 12.10; see also Fig. 12.18). Although these various prototypes differ in internal construction, they all have external dimensions (0.8 mm diameter by 4.2 mm long) similar to the currently available ^{125}I seed and are designed to compete with LDR ^{192}Ir seeds in tempo-rary interstitial implants.

Several sets of [169]Yb dosimetric data have been published. Mason et al.'s (1992) Monte Carlo-based investigation of the now-obsolete prototype models 4 and 5 interstitial seeds provides much useful infor-

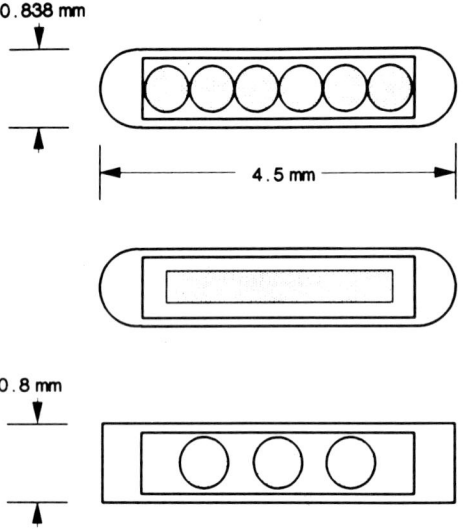

Fig. 12.10. Cross-sectional views of the (*top*) type 4 and (*middle*) type 5 [169]Yb prototype interstitial seeds. The seeds are encapsulated in a titanium tube (0.076 mm wall thickness) and closely match the external dimensions of the currently available ^{125}I seeds (*bottom*). The type 4 core consists of solid Yb$_2$O$_3$ spheres (3.1 g/cc) and the type 5 core of an aluminum-Yb$_2$O$_3$ mixture with a density of 3.1 g/cc. (From Mason et al. 1992)

mation, including HVLs, the air-kerma constant (1.577 cGy·h^{-1}·cm^2·mCi^{-1}), and 2D dose-rate tables and anisotropy factors for the two seed types. These theoretical studies predicted a dose-rate constant value, Λ_0, of 1.225 cGy·h^{-1} per unit air-kerma stren-gth for water medium. Piermattei et al. (1992) used calibrated TLDs to measure the transverse-axis dose-rate distribution about a type 6 seed (similar to type 4, Fig. 12.10a) in a muscle-equivalent phantom. By using an approximate theoretical correction for self-absorption and filtration, they inferred air-kerma strength from the vendor's contained activity assay. Using this value to normalize their measured dose rates, they reported a dose-rate constant in water of 1.52 cGy·h^{-1} per unit air-kerma strength, a value 28% higher than that of Mason. In addition to Monte Carlo calculations about the type 6 seed, Perera et al. (1993) inferred the transverse-axis dose-rate distribution in water from calibrated p-type silicon diode readings made in water medium. They also measured the air-kerma strength with a 100 cm^3 spherical ion chamber equipped with NIST-traceable x-ray and ^{137}Cs gamma-ray beam air-kerma calibration factors. Taking Λ_0 to be the ratio of mea-sured dose rate at 1 cm in water to measured air-kerma strength yielded a dose-rate constant of 1.22 cGy·h^{-1} per unit air-kerma strength, a value that agreed closely with both their own and Mason's Monte Carlo calculations. When Perera normalized his measured dose rate at 1 cm to the vendor's specified activity, the resultant Λ_0 value of 1.92 cGy·h^{-1}· mCi^{-1} closely agreed with Piermattei's value of 1.86 cGy·h^{-1}·mCi^{-1}. These data suggest that the vendor's experimental implementation of their activity stand-ard does not agree with their definition, resolving the apparent 28% discrepancy between theory and measurement. Until a validated air-kerma strength standard is implemented, Perera and Williamson recommend using the value 1.96 cGy·h^{-1} per vendor mCi as an interim value of Λ_0.

Perera et al. (1993) also measured heterogeneity correction factors for cylindrical shields of lead, steel, titanium, and aluminum shielding materials. Figure 12.11 shows that a 0.4-mm- and 1.1-mm-thick lead shields reduce dose rates by 20%–80%, indicating that 0.4–0.6 mm of lead confers the same level protection immediately downstream of the shield as 3–5 mm of tungsten with conventional ^{137}Cs sources. This work demonstrates that customized shielding of [169]Yb-bearing intracavitary applicators is feasible. Note that the heterogeneity correction factors (HCFs) depend as much or more on shield diameter and distance as they do on thickness.

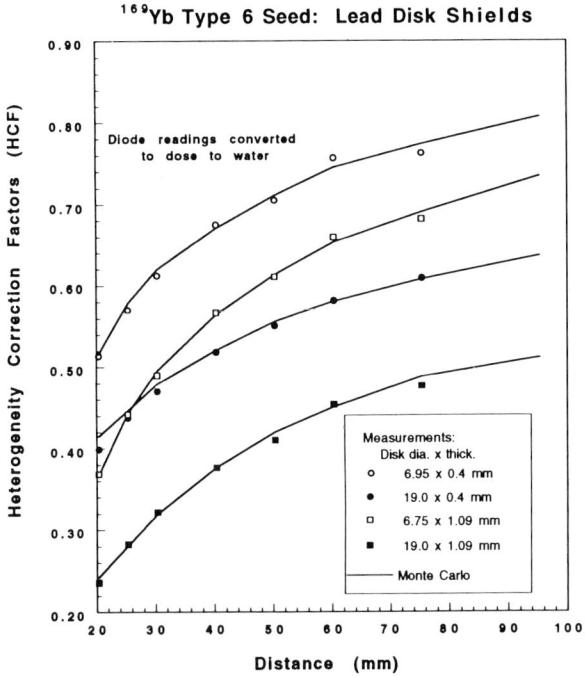

¹⁶⁹Yb Type 6 Seed: Lead Disk Shields

Fig. 12.11. Heterogeneity correction factors (HCFs: dose with heterogeneity/dose in homogeneous water) for ¹⁶⁹Yb type 6 seeds in the presence of cylindrical lead shields placed 15 mm from the source on its transverse bisector. The graph compares Monte Carlo HCFs (*solid lines*) to experimental diode measurements of HCFs (symbols) which have been corrected for energy response by Monte Carlo simulation. See Sect. 12.4.1 for more details. (From PERERA et al. 1993)

12.2.4 New Applications
for Conventional Brachytherapy Isotopes

New applications have been developed over the last 10 years for the long-familiar isotopes ¹²⁵I, ¹³⁷Cs, and ¹⁹²Ir. Several groups (GENEST et al. 1985; GOFFINET et al. 1987; LING et al. 1988) have proposed using currently available ¹²⁵I seeds to replace ¹⁹²Ir seeds in temporary interstitial implants in all body sites, including head and neck and breast. Prior to these investigations, ¹²⁵I sources were used as temporary implant sources in relatively specialized procedures such episcleral plaque treatment of intraocular melanomas and stereotaxically guided brain implants. A practical advantage of this approach is improvement of dose distributions through variable spacing and differential loading of ribbons (LING et al. 1988). This is an example of transforming adversity into an advantage, since seeds must be custom loaded for every patient anyway as they are not available in ribbon form. Other advantages include substantially reduced radiation exposures to

nursing and physics personnel, the possibility of locally shielding dose-limiting normal structures with 200 μm-thick lead foils, and more rapid fall-off of dose outside the implanted volume. The disadvantages include (a) tissue-composition-dependent dose distributions, (b) custom-loading of every ribbon with its attendant investment in specialized equipment and personnel time, and (c) eightfold higher source cost compared to conventional ¹⁹²Ir ribbons. One can partially compensate for (c) by using the seeds in several procedures. In support of this clinical application, Medi-Physics has introduced the model 6712 seed (AHMAD et al. 1992) which has a smaller external diameter (0.5 mm vs 0.8 mm for the 6711 seed), allowing ribbons with the same diameter as nylon-encapsulated ¹⁹²Ir to be fabricated. In reducing personnel exposures and increasing technical flexibility, ¹²⁵I temporary implantation competes with single stepping-source remote afterloaders designed for LDR brachytherapy.

From the point of view of basic dosimetry, curative-intent ¹²⁵I temporary implantation has stimulated investigation of tissue heterogeneity effects. Based upon one-dimensional Monte Carlo studies of the ¹²⁵I radial dose distribution in various tissue media, DALE (1983) observed that, relative to water, the specific dose constant fell 33% and 45% when an ¹²⁵I seed was implanted in adipose tissue and body fat, respectively. This was subsequently confirmed experimentally by HUANG et al. (1990), using TLD dosimetry. These investigators found that when model 6702 seeds were implanted in breast phantom medium, the dose–rate constant fell by 24% and the radial dose distribution was significantly more penetrating than in solid-water medium. These data can be extrapolated to heterogeneous geometries: a 25%–40% dose reduction is expected at a water–adipose interface while a 4.3-fold dose increase is expected at water–bone interfaces. Clearly, tissue heterogeneities can profoundly alter the dose distribution in temporary implants, rendering conventional treatment planning, which assumes homogeneous liquid–water medium, an unreliable indicator of the dose distribution administered to the patient.

The clinical implications of tissue heterogeneities were further investigated by LING and YORKE (1989) and YORKE et al. (1991), who studied the ¹²⁵I transition-zone dose distributions near water–adipose and bone–soft tissue interfaces, respectively. Simple mathematical models (SPIERS 1949), confirmed by biologic dosimeters consisting of mammalian cell monolayers, showed the existence of a narrow

10-μm-thick transition zone due to the short range of the secondary electrons liberated by photon interactions in the various media. For the tissue–adipose interface (LING and YORKE 1989), three-quarters of the 40% dose transition occurred within 5 μm of the interface, indicating that only a single layer of water-equivalent cells on the interface boundary or an isolated water-equivalent cell embedded in adipose tissue would receive an underdose of 30%. Based on these data, they postulated a therapeutic benefit for ^{125}I seed breast implants: tumor cells, assumed to be water equivalent, would receive the full dose predicted by water-medium dosimetry while normal cells, consisting of adipose, would be spared by 40%. The bone–soft tissue interface study (YORKE et al. 1991) came to similar conclusions: the four- to fivefold increase in dose, arising from increased photoelectric absorption in the mineral matrix of the bone, affects only a 10-μm-thick layer of living cells located on the bone–soft tissue interface. Since bone necrosis is due to cell depopulation within soft-tissue cavities containing blood vessels, osteocytes, and undifferentiated cells for which the average diameter is 50 μm, the bone-mineral dose may greatly overstate the risk of bone necrosis. Another factor that mitigates ^{125}I irradiation of bone is the much more rapid falloff of dose in bone compared to soft tissue: 20% per mm versus 5% per mm in water. Compared to ^{192}Ir, the rapid attenuation of ^{125}I photons by bone mineral spares soft-tissue cavities located more than 3 mm beneath the bone surface. Combining these two effects, Yorke estimates that the volume of soft tissue within bone that receives a moderately high dose from an implant adjacent to the bone surface is one-sixth that of an identical ^{192}Ir implant. This suggests that use of ^{125}I to implant floor of the mouth and tongue lesions may reduce the significant incidence of mandible necrosis which accompanies curative ^{192}Ir brachytherapy of these sites. These elegant studies suggest that the complex behavior of tissue heterogeneities can be manipulated to therapeutic advantage. However, manipulation of these factors requires (a) use of imaging studies to document the composition and geometry of tissue heterogeneities in common implant sites and (b) more sophisticated dose calculations that accurately characterize the effects of bounded heterogeneities on the equilibrium dose distribution. Relatively little investigation of these effects (MEIGOONI and NATH 1992b is an exception) has been reported to date.

Over the last 10 years, a large variety of ^{137}Cs source designs have been introduced, mostly to satisfy the mechanical requirements of the many LDR remote-afterloading devices available (see WILLIAMSON 1992 for a review). Spherical sources, 6-cm-long flexible sources with differentially loaded segments, and ^{137}Cs seeds (WILLIAMSON and SEMINOFF 1987), encapsulated in stainless steel carriers as an alternative to ^{192}Ir ribbons, are now available. Unfortunately, published 2D dose-rate tables exist only for conventional intracavitary tubes. However, both experimental (DIFFEY and KLEVENHAGEN 1975; METCALFE 1988) and Monte Carlo (WILLIAMSON and SEMINOFF 1987; WILLIAMSON 1988a) studies demonstrate that the Sievert filtered line-source model quite accurately predicts ^{137}Cs single-source dose distributions. The practical problem reduces to correctly approximating the source geometry, within the limits imposed by one's treatment planning system, and using appropriate effective attenuation coefficients to describe the filtration effect. Appropriate data for both high-density extruded gold-wire sources and low-density ceramic or glass sources are given by WILLIAMSON (1988b).

A variety of high-activity (1–10 Ci) ^{192}Ir sources are now available for HDR remotely afterloaded brachytherapy treatments. Despite the fact that more than 500 HDR devices are in active use throughout the world, the author knows of only two peer-reviewed publications (CERRA and RODGERS 1990; MELI et al. 1988) experimentally documenting the 2D dose distribution about such sources. The latter used a diode detector and a small ion chamber to measure angular anisotropy profiles and the transverse-axis dose distribution, respectively, for a Gamma-Med IIi source consisting of a 5.5-mm-long by 0.5-mm-wide ^{192}Ir pellet encapsulated in an 8.5 mm × 1.1 mm stainless-steel capsule. They showed that the classic MEISBERGER et al. (1968) polynomials accurately characterized the water-to-air exposure-rate ratio on the transverse axis. The angular profiles demonstrated a moderate level of dose anisotropy, resulting in longitudinal-axis dose rates, relative to the corresponding transverse-axis values, ranging from 60% at 3 cm to 80% at 10 cm. The ion-chamber measurements of MELI et al. (1988) confirmed the applicability of relative LDR ^{192}Ir seed data transverse-axis dose distributions to HDR sources. In view of large position-dependent energy-response artifacts of silicon diode detectors (as large as 75%: see Sect. 12.4.1), the quantitative accuracy of the Cerra-Rodgers profile data is suspect. Secondly, HDR vendors are constantly revising their source designs so as to make them shorter or thinner. For example, the Nucletron pulse dose-rate (PDR) source core is only 1.2 mm long by 0.6 mm in diameter. Thus Cerra-Rodgers data do not describe the 2D dose distribution of the current generation of sources. There is no evidence that the Sievert line-source model accurately describes ^{192}Ir relative dose distributions, rendering this avenue of extrapolation to new source designs suspect. Given the nearly total absence of HDR source-specific dosimetry data, probably the best interim clinical solution is to calibrate each HDR source in terms of air-kerma strength and use conventional LDR data to approximate its transverse-axis absolute dose-rate distribution. Finally, almost nothing has been published documenting the dose distributions about the many shielded colpostats and rectal/vaginal cylinders available for HDR brachytherapy. Although reasonably accurate 1D computational models exist for shielded ^{137}Cs applicators (see Sect. 12.4.2.1), none of these has been validated for ^{192}Ir dosimetry. *Because HDR brachytherapy places increased emphasis on physical optimization to compensate for unfavorable radiobiology, dosimetric investigation in this area is urgently needed.*

12.2.5 Conclusions and Future Prospects for Low-Energy Sources

Figure 12.12 illustrates several photon-energy-dependent trade-offs involved in selecting radio-nuclides for various brachytherapy applications. In the 200-keV to 2-meV range, all isotopes are clearly radium-equivalent. The dose-rate constant (represented in Fig. 12.12a by dose in medium/water kerma in free space) is a constant that is independent of both photon energy and composition of the surrounding medium. In addition (Fig. 12.12b), inverse-square law holds within a few percent out to 5 cm. Thus 200 keV is the lower limit of energy where radium-substitute dosimetry remains applicable. Although the dose-rate distribution is independent of photon energy, the tenth-value layer (Fig. 12.12c) of lead falls from 60 mm at 2 MeV to about 1 mm at 200 keV. In the 60- to 200-keV energy range, energy transfer to biologic media is dominated by approximately elastic Compton interactions. Consequently, build-up of scattered photons overcompensates for primary photon attenuation and the dose-rate constant becomes both energy- and medium-dependent. From the point of view of dose-algorithm design, this energy range is the most challenging since multiple-scattering effects make shielding correction factors and source-to-source shielding effects highly dependent upon the implant and applicator geom-

etry. From the point of view of dose measurement, this energy range is difficult since distance-dependent energy-response artifacts are potentially large. In the ultra-low energy range (<40 keV), photoelectric effect in tissue-equivalent media becomes significant. The dose-rate constant becomes profoundly dependent upon tissue composition and photon energy, and the transverse dose-rate distribution falls off significantly faster than predicted by inverse-square law (Fig. 12.12b). Sources with a mean photon energy of less than 50 ke V are probably inappropriate for gynecologic intracavitary brachytherapy (MEIGOONI and NATH 1992a). Tissue heterogeneities can significantly perturb the dose distribution, adding a new dimension of complexity to dose-calculation algorithm design. However, accurate dose measurement in this energy range is aided by the relatively position-independent photon spectrum, eliminating the problem of distance-dependent energy response artifacts even for medium-atomic number detector such as silicon diode.

Both clinicians and physicists must be aware that the biology of radiation response is highly dependent upon photon energy, rendering tumor- and normal-tissue dose-response data derived from radium-substitute clinical experience suspect in the low-energy brachytherapy source domain. For example, NATH et al. (1990d) have shown for Chinese hamster cells that the relative biologic effectiveness (RBE) of [241]Am and ^{125}I relative to [226]Ra ranges from 1.20 to 1.57 and from 1.30 to 1.46, respectively, depending upon dose rate. In vivo and in vitro biologic studies, followed by phase I/II human clinical studies based upon meticulous dosimetry, are therefore required to develop a base of clinical experience that can serve as guide for designing future clinical trials using these new and promising sources.

12.3 Advances in Brachytherapy Dosimetry

The existing experience with low-energy sources, especially [241]Am, clearly demonstrates that the traditional minimalist approach to brachytherapy dosimetry, i.e., simple Sievert-like models which ignore applicator and tissue heterogeneities, is completely inadequate in the low-energy domain. The dominance of photoelectric effect and nearly elastic scattering of photons gives rise to dose distributions that are highly dependent on both the geometry and the composition of the measurement medium. Interest in these new sources, along with increasing clinical concern over the status of ^{125}I dosimetry, has

Dose Rate at 1 cm from a Point Source

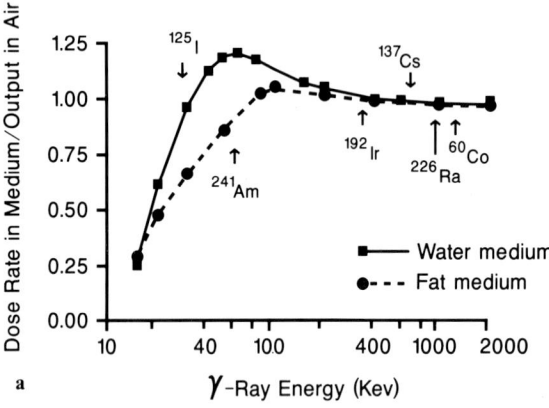

a

Dose Rate at 5 cm from a Point Source in Water

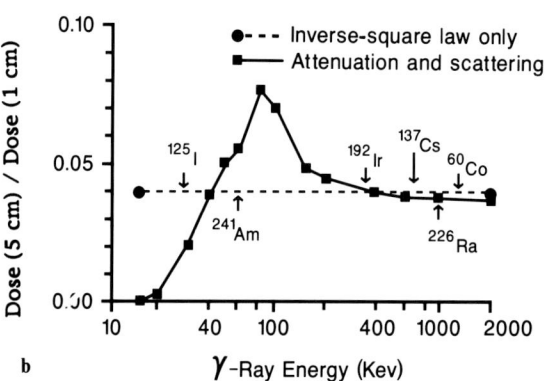

b

Thickness of Lead Required for 10% Transmission

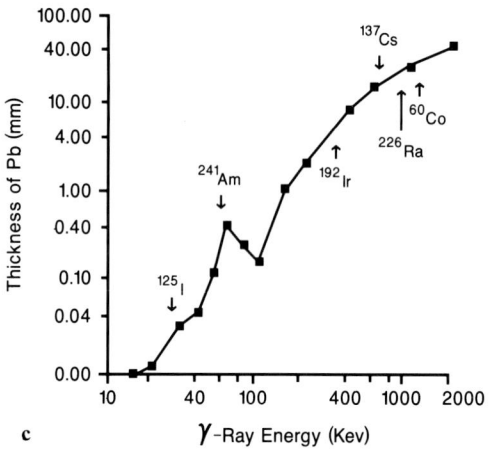

c

prompted a renewal of interest in both experimental and computational brachytherapy dosimetry.

12.3.1 Experimental Brachytherapy Dosimetry

Until about 1980, direct measurement of dose around brachytherapy sources and applicators in support of clinical treatment planning and quality assurance was relatively uncommon even within the research setting, let alone the clinical environment. Current clinical quality assurance practice is limited to experimental confirmation of the strength and geometry of purchased sources (WILLIAMSON 1991b). Historically, this is due not only to the difficulties and labor intensity of such measurements, but also to a consensus view that objective and reproducible dose measurement was impossible and that even simplistic theoretical models were more reliable. Indeed, brachytherapy dose measurement does place severe demands on dose detectors since the dose distributions are characterized by large dose gradients, a large range of dose rates, and relatively low photon energies. A suitable detector must have a wide dynamic range, flat energy response, small size, and high sensitivity. All currently used detectors, including organic detectors such as radiochromic film and plastic scintillator, are subject to artifacts: volume averaging, self-attenuation, anisotropy, and energy response. Traditionally used dose detectors include small ion chambers (SAYLOR and DILLARD 1976; MEERTENS and VAN DER LAARSE 1985), diode detectors (MOHAN et al. 1985; METCALFE 1988; LING et al. 1983), TLD dosimeters (see below), and silver-halide radiographic film. Emerging detector technologies include GAF-radiochromic film (MCLAUGHLIN et al. 1991; MUENCH and NATH 1992) and plastic scintillator (PERERA et al. 1992), consisting of organic dyes dispersed in a solid plastic matrix.

Detector response, relative to dose in water, is plotted as a function of energy for source-detector distances of 1 and 10 cm in Fig. 12.13 for several common detectors. It is clear, even for plastic detectors (Z_{eff}= 5.9), that the measured reading/unit dose depends significantly on photon energy. A more

Fig. 12.12. a Graph illustrating the dependence of dose rate in water or fat medium per unit output in air as a function of photon energy for monoenergetic isotropic point sources. Output in air is quantified in terms of water kerma in free space at 1 cm distance. (From unpublished Monte Carlo calculations by the present author.) **b** Graph showing dose at 5 cm in water form a point source as a fraction of dose at 1 cm as a function of photon energy for monoenergetic point sources. (From unpublished Monte Carlo Calculations of the present author.) **c** Graph plotting thickness of lead in mm required to reduce photon particle fluence by a factor of 10 from monoenergetic point sources of various energies

serious problem is the variation of detector response with source-to-detector distance, due to softening of the photon spectrum secondary to buildup of lower-energy scattered photons and attenuation of primary photons. This can lead to dose-conversion factors which vary with location of the measurement point in the phantom and cannot be estimated without a complete characterization of the photon spectrum throughout the measurement geometry. Because of the relatively high atomic number of silicon (Z_{eff} = 14), diode response/unit dose in water varies by a factor of 2 over the distance range 1–10 cm for a 100-keV source and by 1.6 for [192]Ir. Even for ^{137}Cs, 10%–15% energy-response artifacts have been demonstrated. Although some investigators have developed distance-dependent correction factors (MOHAN et al. 1985; LING and SPIRO 1984; WEEKS and DENNETT 1990), these methods cannot be used to estimate dose behind a thick shield since the latter perturbs the scatter-to-primary ratio as much as moving 10 cm away from the source (WILLIAMSON et al. 1993a). Despite the limited utility of diode detectors above 40 keV, diode response appears to be position-independent (CHIU-TSAO et al. 1990; WILLIAMSON et al. 1993b) around ultra-low energy sources, such as ^{125}I and [103]Pd, and is widely used as a relative dose detector in this domain. Several published Monte Carlo calculations, as well as limited experimental data, indicate that the ^{125}I photon spectrum changes very little with respect to distance in medium. However, the highly anisotropic angular response of widely used Rikner p-type diodes has raised concern (CHIU-TSAO et al. 1990) that these detectors might progressively underrespond to dose as distance increases, since the fraction of dose due to backscattered photons increases with the scatter-to-primary ratio, which in turn rapidly increases with distance. Radiographic film is probably the most widely used relative detector in brachytherapy, despite having a very large atomic number (Z_{eff} = 53). Although no published data exist documenting its response in medium around low-energy sources, silver-halide film may give rise to very large energy-response artifacts (MUENCH et al. 1991) and should be avoided for quantitative work. Even purely organic detectors, such as plastic scintillator, have a surprisingly large 21% variation of response with distance at 100 keV (PERERA et al. 1992).

Another major factor governing detector utility in brachytherapy is sensitivity, which is summarized in Table 12.2 for several detectors. The need for high sensitivity and small size effectively eliminates ion chambers as an option for LDR sources. In addition,

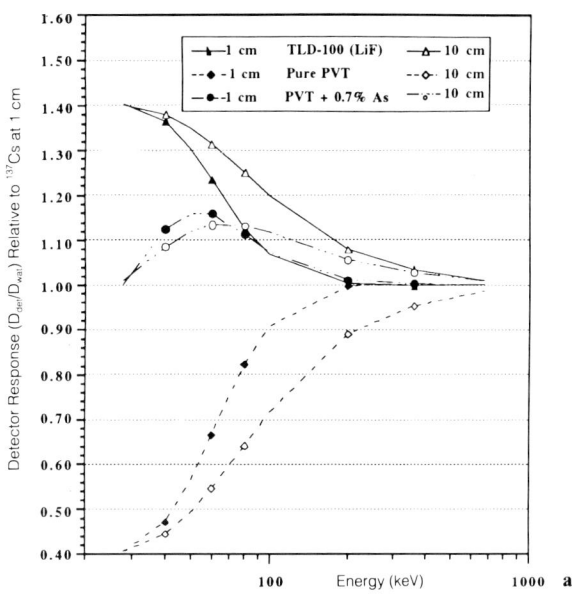

Energy Response: TLD and Plastic Scintillator
1 cm and 10 cm Distances from Point Sources

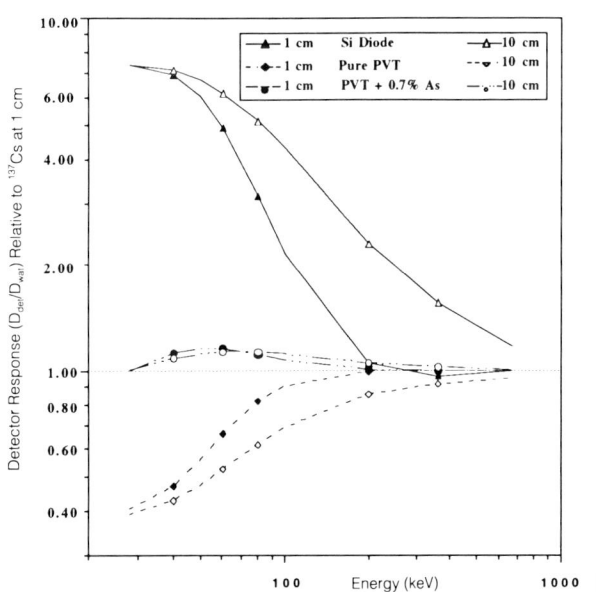

Energy Response: Diode and Plastic Scintillator
1 cm and 10 cm Distances from Point Sources

Fig. 12.13. a Theoretical detector response (dose to detector/dose to water) of pure PVT plastic scintillator, arsenic-doped plastic scintillator and TLD-100 (LiF) detectors at 1 cm and 10 cm distances in water from point sources of various energies. **b** Same as in a except TLD-100 detector is replaced by a silicon diode detector. The results of each detector type are normalized to unity at 1 cm from the ^{137}Cs source. (From PERERA et al. 1992)

Table 12.2 Absolute response of 1 mm³ Detector to 0.1 cGy absorbed dose. Data from PERERA et al. (1992)

Detector	Energy dissipated/ observed quantum (eV)	No. quanta emitted	Typical quantum effciency × geometric collection efficiency	Practical signal relative to ion chamber
TLD-100	8400	1.8×10^6	0.20 × 0.08	0.120
Silicon diode	3.6	4.0×10^9	1.00 × 1.00	17.000
Ion chamber (air)	33.8	2.4×10^5	1.00 × 1.00	1.00
Plastic scintillator	100	6.2×10^7	0.20 × 0.05	2.60

the problem of correcting for displacement of medium by the detector in the presence of large dose gradients remains unsolved for a scattering medium. Even though TLD-100 has very poor intrinsic sensitivity, its practical response per unit volume approaches that of ion chamber, largely because TLD is solid. By irradiating TLDs long enough, integrated signals with good signal-to-noise ratio characteristics have been obtained over the distance range of 1-7.5 cm near LDR sources. In contrast to ion chamber, such long 6- to 48-h irradiations are practical since many TLD detectors can be simultaneously irradiated in a single experiment. Silicon diode detectors are four orders of magnitude more sensitive than ion chambers and have been used in conjunction with scanning water phantoms to map out complex dose distributions, but must be used cautiously because of the potential for large energy-response artifacts.

Two emerging planar detector technologies deserve mention: plastic scintillator (PERERA et al. 1992) and radiochromic film (MUENCH et al. 1991). Both detectors are available in the form of thin sheets and offer the prospect of simultaneous measurement of absorbed dose within an entire plane at an en–hanced spatial resolution, which is currently limited to about 1 mm for conventional single-element diode and TLD detectors. Plastic scintillator consists of a solid polystyrene or polyvinyltoluene base, along with one or more wave-shifting dyes. It converts about 3% of the energy absorbed from ionizing radiation to visible light through a complex multi-step process (PERERA et al. 1992). Recently, the author's group (PERERA et al. 1992) has shown that optical scintillation, measured using a CCD digital camera, is linear with respect to absorbed dose arising from ^{137}Cs and ^{125}I brachytherapy sources using a 3 × 3 × 1 mm plastic scintillator detector. Plastic scintillator has a number of attractive properties, including a practical efficiency nearly three times that of ion chamber and 10–20 times that of TLD. In its pure form (Z_{eff}= 5.7), its energy response is not quite as good as TLD, showing as much as 60% underresponse with respect to energy

and as much as 19% variation with respect to distance at 100 keV. However, plastic scintillator containing small amounts of high-atomic number impurities can be manufactured, resulting in an almost perfectly flat energy response. Radiochromic film (MCLAUGHLIN et al. 1991) consists of a 6- to 23-µm- thick colorless leuco dye bonded to a 100 µm-thick mylar base. In response to ionizing radiation, the dye gives rise to an ultrafine-grain image (1200 line pairs/mm) which can be reproducibly quantified using 600- to 680-nm light transmission densito-metry. Unlike silver-halide film, image formation requires no processing. MUENCH et al. (1991) have proposed using this detector for brachytherapy dosimetry. Using low-energy x-ray beams, they found that radiochromic film response, relative to dose to water in free space, varied by only 30% in the 28- to 1710-keV energy range, compared to an 11-fold variation of response for silver-halide film. Currently available radiochromic requires a minimum exposure of 800 cGy, which severely limits its use in LDR brachytherapy dosimetry. However, by increasing concentration of the dye and the thickness of the sensitive layer, it may be possible to increase sensitivity by an additional factor of 10.

12.3.1.1 Validation of TLD Dosimetry in Brachytherapy

A major accomplishment of the last decade is validation and acceptance of TLD dosimetry as an accurate and comprehensive source of directly mea-sured single-source dose distributions for clinical treatment planning. This development is largely, but not exclusively, due to the efforts of the recently completed Interstitial Brachytherapy Dosimetry Contract supported by the National Cancer Institute (ANDERSON et al. 1990). The associated collaborative working group (ICWG) consisted of independent teams of researchers based at three institutions (Yale, Memorial Sloan-Kettering, and University of California at San Francisco). The ICWG made available three independently measured sets (WEAVER et al. 1989; NATH et al. 1990b; CHIU-TSAO et al. 1990) of dose distributions for ^{125}I and ^{192}Ir interstitial sources for comparison to validate their TLD measurement methodology. The achievements of other groups also deserve praise, most notably the TLD measurement of LUXTON et al. (1990) and PIERMATTEI et al. (1988) and the relative diode measurements of LING et al. (1983, 1985) and SCHELL et al. (1987). The TLD methodology developed by these investigators has been extended to provide complete 2D dose distributions about ^{125}I, ^{192}Ir, and ^{103}Pd brachytherapy sources (NATH et al. 1993; CHIU-TSAO et al. 1990). Currently, TLD dosimetry is regarded as the most reliable and best-validated experimental approach in brachytherapy.

Although instrumentation and TLD annealing practices vary from investigator to investigator, potential experimental artifacts such as volume averaging, angular anisotropy, energy response, positioning inaccuracies, background corrections, and interchip attenuation and scattering effects must be carefully controlled and correction factors derived from ancillary experiments or Monte Carlo simulations. Most experimentalists use LiF TLD-100 in the form of powder or extruded chips. Relative calibration factors should be derived for each dosimeter by exposing the TLD array to a known dose of megavoltage x-rays in a uniform broad beam under conditions of charged-particle equilibrium. To extract estimates of dose rate (cGy/h) per unit strength near a brachytherapy source, rather than dose relative to a reference point, an accurate TLD reading (TL)/unit absorbed dose calibration must be established. This is achieved by exposing TLDs to a known dose in free space in an x-ray beam which has a spectrum that matches that of the brachytherapy source of interest. Such reference beams must be calibrated with an ion chamber having an NIST-traceable air-kerma calibration factor. Response of TLD-100 relative to response to 4-MV x-rays is shown in Fig. 12.13a as a function of equivalent energy. Note that below 30 keV and above 200 keV, TLD response is independent of energy. For ^{125}I dosimetry, a multiple-seed array can be used as a calibration beam, provided a calibrated ion chamber is used to measure the dose rate at the TLD location. Using these techniques, MEIGOONI et al. (1988b), LUXTON et al. (1990), and WEAVER (1984) have found that the response of TLD-100 to ^{125}I x-rays to be 1.41, 1.44, and 1.39, respectively, relative to 4-MV x-rays or ^{60}Co gamma rays.

Positioning artifacts are controlled by using a plastic phantom with precisely machined detector slots whose location relative to the source can be accurately measured. For ^{192}Ir dosimetry, absorbed dose is independent of phantom composition (MELI et al. 1988), allowing use of lucite or polystyrene plastic as a water substitute. However, at ^{125}I energies, the dose distribution is extremely sensitive to small deviations in phantom composition from water, with the result that doses measured in polystyrene at 5 cm distance deviate from those measured in water by as much as factor of 3 (MEIGOONI et al. 1988b). The ICWG settled on solid-water plastic phantom material as a measurement medium, which contains 2.3% calcium by weight to match the effective atomic number of liquid water. However, subsequent Monte Carlo investigations by the author (WILLIAMSON 1991a) demonstrated that this material was not water equivalent, giving rise to dose underestimates ranging from 4% at 1 cm to 25% at 10 cm in the ^{125}I energy range. At higher photon energies, distant-dependent energy response artifacts must be accounted for. Usually, Monte Carlo simulation is used to calculate the ratio of LiF kerma to water kerma as a function of distance from a point source. For ^{192}Ir, this leads to overresponse corrections of 5%–10% at 10 cm distance. By comparing TLD and ion chamber response as a function of distance from an HDR ^{192}Ir source, MEIGOONI et al. (1988a) confirmed these predictions. For ^{125}I dosimetry, interchip shielding effects, which occur when one TLD detector is shielded by others, as large as 6% have been measured (MEIGOONI et al. 1988b). These correction factors vary with distance and phantom material as well as detector arrangement. finally, at distances of less than 1 cm, averaging of absorbed dose over the extended detector volume becomes significant. Most investigators have simply averaged an inverse-square law factor over the chip volume, which leads to correction factors of 3%–6% in the 2–5 mm distance range (WEAVER et al. 1989; MEIGOONI et al. 1988b).

The ICWG investigators claim an accuracy of 3%–6% for transverse-axis TLD dosimetry for LDR ^{125}I and ^{192}Ir seeds. Review of the data published by the three participating institutions reveals agreement in absolute dose rate (not relative) ranging from 2% to 5% over the 1–5 cm distance range, supporting this claim. However, at shorter and longer distances, discrepancies of 10%–30% appear, indicating that at large distances, signal-to-noise ratio problems limit TLD precision, and at distances of less than 10 mm, positioning or volume-averaging artifacts limit measurement accuracy. Measured brachytherapy dose distributions will be compared to conventionally used data and Monte Carlo calculations in Sect. 12.3.4.

12.3.2 Monte Carlo Simulation: a New Clinical Dosimetry Tool

Another exciting development in brachytherapy dosimetry is the emergence of Monte Carlo photon-transport simulation as a reliable and accurate source of brachytherapy dosimetry data. This approach is made feasible by the availability of more accurate photon cross-section libraries and 3D geometric modeling techniques allowing the effects of internal source structure, dose measurement geometry, and source calibration geometry to be modeled (WILLIAMSON 1989). Under certain conditions, Monte Carlo simulation can be used to calculate actual dose rates in medium per unit source strength, as well as relative dose distributions.

Monte Carlo simulation is a specific numerical solution to a general problem, namely the Boltzmann transport equation. A central result of transport theory is that this equation, given the distribution of ionizing radiation sources and absorbing media in a system along with a description of the collisional dynamics underlying the transport, scattering, and absorption of ionizing radiation in matter, completely characterizes the resultant dose distribution. Unfortunately, this integrodifferential equation is too complex to be solved accurately by analytic or even deterministic numerical methods in any but the simplest of 1D and 2D geometries. Thus, Monte Carlo simulation, which solves the transport equation by random sampling, is currently the only practical theoretical method of calculating absorbed dose in the presence of geometrically complex boundary conditions (MILLER and LEWIS 1984; JENKINS et al. 1988). Using probability distributions derived from total and differential cross-sections, a small (10^5–10^7) subset of photon or electron histories is randomly constructed by following each photon from birth through successive scattering events and, eventually, to absorption or escape from the system, using random sampling to decide its fate at each decision point. The result is equivalent to randomly selecting a small number of photon histories from the set of all those possible. A statistical estimate of absorbed dose rate at a point is obtained by calculating the dose contribution from each simulated collision (a process known as "estimation") and taking the average over all contributions (WILLIAMSON 1987). Because particle histories can be accurately and efficiently constructed even in the presence of complex, 3D boundary conditions, "exact" but statistically uncertain solutions, derived from first principles with little

reliance on approximations, are possible for a wide range of clinically relevant brachytherapy problems.

In principle, it is straightforward to include transport of secondary electrons in the simulation, which is clearly necessary to solve problems of clinical interest in megavoltage photon domain. However, this reduces the efficiency of the calculation by two orders of magnitude, limiting Monte Carlo simulation to relatively simple 1D and 2D geometries given the computing resources generally available to medical physicists. In contrast, Monte Carlo photon transport (MCPT) problems of great complexity can be readily solved on small minicomputers or workstations. MCPT approximates absorbed dose in medium by collision kerma, an assumption which is valid only when secondary charged-particle equilibrium (CPE) obtains. Although the influence of secondary electron transport has not been studied in brachytherapy, significant CPE-failure artifacts are expected only near media interfaces and very close to (<5 mm) higher-energy sources since brachytherapy photon energies are relatively low (ROESCH 1958). This is fortunate, since realistic modeling of brachytherapy source, applicator, and detector geometries has proven to be essential for accurate results.

Monte Carlo simulations and other 1D solutions of the transport equation have been used since the 1960s to calculate radial dose distributions arising from isotropic point sources in medium. Two widely cited references, the point-source buildup factors of BERGER (1964) and the tissue-attenuation and scatter- buildup factors of MEISBERGER et al. (1968), are based upon, in part or in whole, such calculations. However, MCPT has been applied to geometrically complex problems in brachytherapy only relatively recently. One of the earliest 3D studies (WILLIAMSON 1983) used Monte Carlo simulation to assess the accuracy of the Sievert model for platinum-encapsulated ^{226}Ra and ^{192}Ir sources suspended in free space. This study showed large discrepancies (5%-200%) between Sievert and Monte Carlo calculations for low-energy (<500 keV) sources encapsulated in high-atomic-number media. For clinical ^{226}Ra and ^{192}Ir sources, with strength specified in terms of output on the transverse axis, the Sievert model was found to be accurate within 2% except near the longitudinal axis, where discrepancies of 3%–23% were observed. The present author continued to utilize MCPT to calculate dose distributions in water medium about ^{137}Cs interstitial seeds (WILLIAMSON and SEMINOFF 1987), 2D dose distributions about ^{137}Cs intracavitary sources (WILLIAMSON 1988a), dose-rate constants (WILLIAMson 1988b), and 2D dose distributions (WILLIAMSON and QUINTERRO 1988) about ^{125}I interstitial seeds, 3D dose distributions about shielded gynecologic colpostats (WILLIAMSON 1990), and, more recently, dose perturbations arising from high-density metal heterogeneities near ^{125}I, ^{169}Yb, ^{192}Ir, and ^{137}Cs clinical sources (WILLIAMSON et al. 1993b). This series of papers was the first to report theoretically calculated dose rates per unit source strength (cGy· h^{-1}· U^{-1}), in contrast to most investigators who had limited MCPT to calculation of relative dose distributions. Significant contributions have been made by many other investigators. BURNS and RAESIDE were the first to use MCPT to calculate relative 2D dose distributions around ^{125}I seeds (1987) and to calculate interseed perturbation corrections for multiple-seed implants (1989). CHIU-TSAO (1988) used the multiple-group cross-section code MORSE to calculate 2D dose distributions for ^{125}I seeds (1990) and dose perturbations arising from gold-backed ophthalmic plaques loaded with ^{125}I seeds (1988). MCPT calculations of absolute dose rates have also been reported by MARCHESE et al. (1990) for ^{125}I seeds and by MASON et al. (1992) for ^{169}Yb seeds. Recent work, which shows agreement of 2%–5% between MCPT calculations and experimental measurements, will be reviewed in Sect. 12.3.4.1.

12.3.2.1 Technical Aspects of MCPT Simulation in Brachytherapy

Accurate and reliable MCPT-based dosimetry requires painstaking attention to many technical details including: choice of physical model of photon scattering, choice of cross-section library, accuracy and reliability of the underlying geometric model, and appropriate choice of estimators and other variance-reduction techniques. Each of these issues will be reviewed in the following sections.

Physics of Photon Scattering and Choice of Cross-Section Library

Figure 12.14 illustrates the photon-scattering mechanisms of importance in the brachytherapy energy range. In general, an MCPT code requires total cross-sections for each photon collision process, tabulated as a function of incident photon energy and medium. These data are used to randomly choose the interaction mechanism at each simulated collision and to randomly choose the distances between successive collisions. Differential cross-sections are needed to randomly select the outcome of each collision, in terms of scattering angle and scattered photon energy. Finally, mass-energy absorption coefficients are needed to convert energy fluence into estimates of absorbed dose in the CPE approximation. Many public-domain codes (e.g., ITS and MORSE) neglect the influence of electron binding on photon scattering, approximating coherent scattering by uncollided primary photons and incoherent scattering by the free-electron Klein-Nishina differential cross-section. The author's work (WILLIAMSON et al. 1984) suggests that the free-electron approximation is of questionable accuracy below energies of 100 keV. The binding of target atomic electrons to positively charged nuclei significantly modifies the angular distribution of scattered

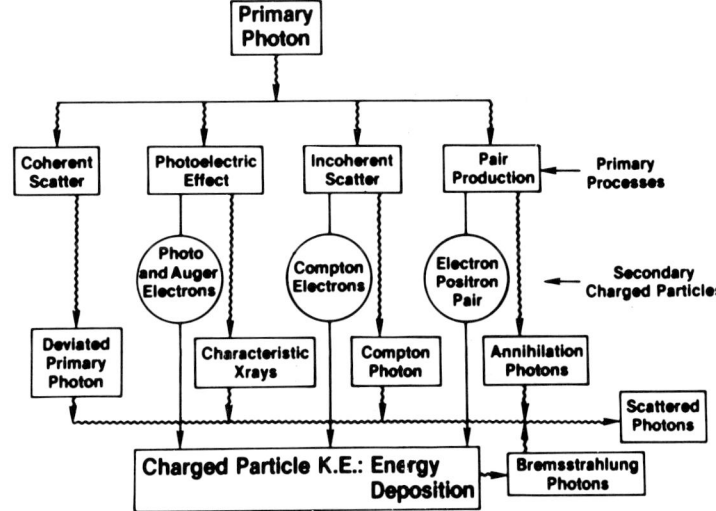

Fig. 12.14. Principal mechanisms of photon scattering and energy deposition in the 2-keV to 10-meV energy range, illustrating the various photon scattering processes and secondary photon production processes that a Monte Carlo photon transport code must simulate. (From Williamson 1988c)

photons, particularly in high-atomic-number media. Although orbital-electron binding effects have less influence on energy deposition, the free-electron approximation was found to overestimate energy fluence transmitted through thick barriers by 2%–15% (Fig. 12.15). Below 100 keV, the author recommends that coherent scattering be explicitly simulated and that the Compton free-electron distribution be corrected for orbital-electron binding effects. Such corrections are easily made by correcting the Thompson and Klein-Nishina differential cross-sections by atomic form factors (AFFs) and incoherent scattering factors (ISFs) respectively. Experimentally validated tabulations of these data are available (Hubbell et al. 1975; Hubbell and Øverbø 1979). In the special but important case of liquid-water medium, the liquid-water AFF data of Morin (1982) are recommended for calculation of coherent-scattering total cross-sections and angular scattering distributions. Coherent scattering, in contrast to other mechanisms, is strongly influenced by molecular bonds and intermolecular forces. These extra-atomic forces increase the value of the coherent-scattering cross-section by 20%–40% in the 10- to 40-keV energy region compared to the predictions of the mixture-rule model, which accounts only for atomic binding forces. MCNP is currently the only public-domain MCPT code available that simulates both coherent and incoherent scattering.

An important secondary scattering mechanism in brachytherapy is production of characteristic x-rays following photoelectric collisions in medium- and high-atomic-number materials. For example, approximately 25% of the photon fluence emitted by the model 6711 ^{125}I seed is due to emission of characteristic x-rays within the silver wire upon which the radioactive material is adsorbed (Ling et al. 1983). Cygler et al. (1990) have shown that a silver plaque behind a ^{125}I seed increases dose rate by as much as 30% due to backscattered characteristic x-rays. The significant contribution of low-energy (4.5-keV) characteristic x-rays emitted by the titanium encapsulation of ^{125}I seeds to the free-air exposure rate has been experimentally observed by Kubo (1985) and demonstrated by Williamson (1988b) to significantly influence NIST calibration standard for these sources. The influence of L-edge characteristic x-rays emitted by gold ophthalmic plaques containing iodine seeds has been studied by several investigators and is believed to alter the dose distribution by 3%–6%. Fundamental data needed to quantitate characteristic x-ray emission, and practical methods of sampling the K- and L-edge x-ray spectrum, have been described by Williamson (1988b). Among the public domain codes, only MCNP simulates both L- and K-characteristic x-ray production. Accurate treatment of this phenomenon by MCPT is essential whenever the isotope under study emits photons with energies near the K-edge binding energy of any shielding materials present.

Accurate MCPT-aided brachytherapy dosimetry would not be possible without the evolution in accurate photon cross-section libraries made over the last 20 years. The success of these efforts owes much to John Hubbell of the National Institute of Standards and Technology (NIST), who has systematically compiled available cross-section measurements, critically evaluated published cross-section libraries, and made available comprehensive libraries with improved accuracy. The current libraries, DLC-99 (Roussin et al. 1983) and its more recent successor, DLC-103 (Trubey et al. 1989), are available on easy-to-access magnetic tape format from the Radiation Shielding Information Center. These libraries are based on theoretical calculations validated by extensive comparison with the NIST experimental cross-section data base, including many recent "direct" measurements of photoionization (Saloman and Hubbell 1986, 1987; Saloman et al. 1988). The photoelectric effect cross-sections are the theoretical values of Scofield (1973), based upon a relativistic Hartree-Slater model renormal-

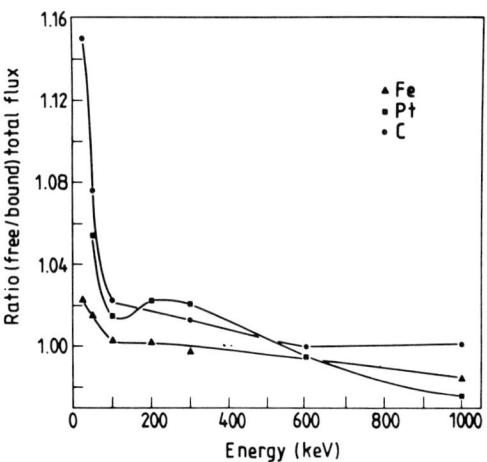

Fig. 12.15. Monte Carlo estimates of total photon energy flux transmitted through two mean free-path thick absorbers consisting of carbon, iron, and platinum. The free-electron Monte Carlo result, relative to the bound electron estimate, is shown as a function of incident photon energy. (From Williamson et al. 1984)

ized to the Hartree-Fock model for low-atomic-number elements (McMasters et al. 1969). DLC-99 and DLC-103 coherent and incoherent scattering cross sections are derived from Hubbell's relativistic atomic form factors (1979) and nonrelativistic incoherent scattering factors (1975), respectively. Our MCPT calculations utilize Hubbell's mass-energy absorption coefficients (1982) and the transition probabilities, K- and L-shell vacancy probabilities, and characteristic x-ray yields tabulated by Plechaty et al. (1978). The potential MCPT user is warned that low-energy dose distributions are sensitive to the choice of cross-section library. For example, ^{125}I radial dose functions, based upon the older DLC-7F library (Roussin 1978; Plechaty et al. 1978), differ from those calculated using the newer DLC-99 library by 2%, 5%, and 10% at 2 cm, 5 cm, and 10 cm distances, respectively (Williamson 1991a). Due solely to adopting a new cross-section library, the present author has revised his calculated dose-rate constants, for models 6711 and 6702 seeds downward by 3.4% (Williamson 1991a).

Geometric Modeling

Accurate MCPT dose calculation requires a flexible and general system of geometric modeling. Brachytherapy sources and applicators have complex internal designs often leading to highly anisotropic dose distributions. In addition, dose distributions around sources emitting photons with energies greater than 40 keV are sensitive to the shape and size of the surrounding scattering medium. Calculation of absolute dose rates may require simulation of the experimental geometry used to standardize air-kerma strength for the source. finally, comparison of MCPT predictions with measured results may require simulation of detector response. Simulation of detectors such as commercially available silicon diodes poses a difficult modeling challenge, since these detectors are both small and are endowed with intricate internal structures. The approach of choice is surface- or volume-based combinatorial modeling (Li and Williamson 1992). This approach, sometimes called complex combinatorial geometry or CCG, defines complex spatial regions as set-theoretic intersections, unions, and differences of elementary volumes such as cuboids, ellipsoids, elliptic cylinders, cones, and half-planes. The modeling code must allow the composition, location, and dimensions of each geometric region to be independently specified, must allow complex structures to be

nested inside of one another, and must support point classification, line-segment classification, and ray tracing. Most of the public-domain codes support CCG modeling packages. Many of these packages are difficult to use and give rise to artifacts and numerical instabilities, particularly when complex regions share common boundaries or when the ratio of the largest to smallest linear dimension within the system is very large. As an alternative, the author and his colleagues (Li and Williamson 1992) have developed a CCG modeling package which is based upon topologically rigorous definitions of boundary, interior, and exterior of a complex surface, which eliminates most of these artifacts.

Figures 12.16 and 12.17 illustrate the models used by the author to simulate dose distributions around

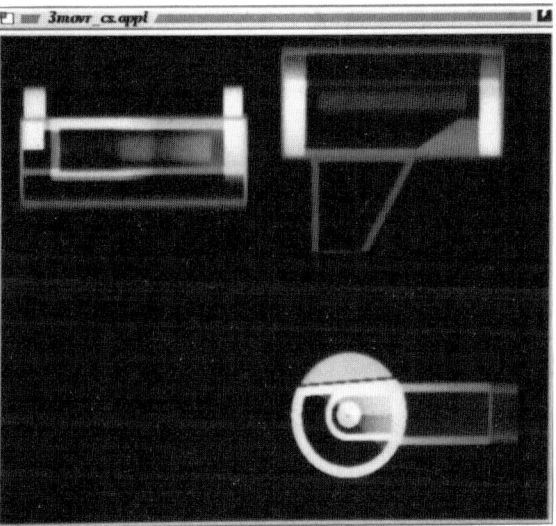

Fig. 12.16. Simulated orthogonal radiographs of a 3M fletcher-Suit-Delclos colpostat. These gray-scale images were derived by applying ray casting to a mathematical model of the colpostat using the volume-based modeling library of Li and Williamson (1992). The model consists of six different media and approximately 45 elementary geometric shapes

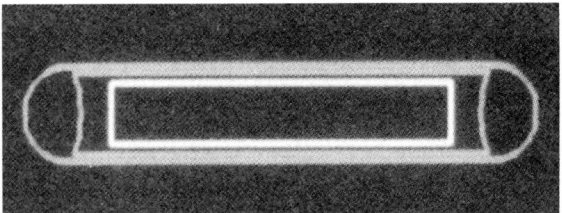

Fig. 12.17. Intersection of model 6711 ^{125}I interstitial seed with a plane containing the seed axis derived from the volume-based combinatorial geometry modeling package described by Li and Williamson (1992). Elliptical end welds are modeled by representing the titanium capsule as the union of a right cylinder and two spheres and the air cavity as the intersection of a right cylinder with the compliments of two spheres

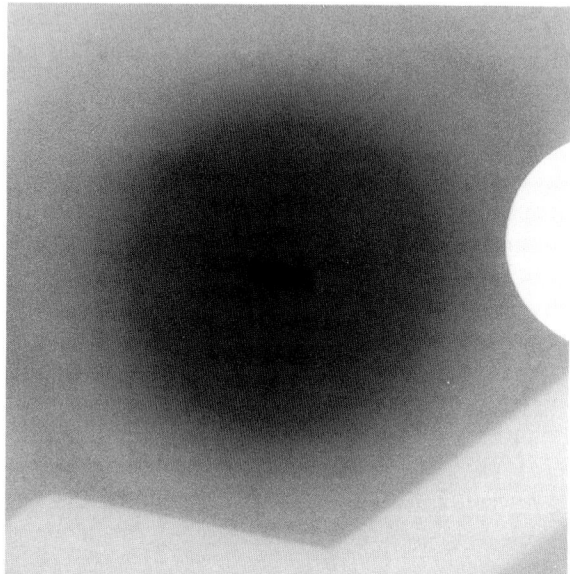

Fig. 12.18. Radiographic images obtained by the author's group of the type 6 ^{169}Yb interstitial source manufactured by Amersham. Top: a pin-hole autoradiograph, exposed using the technique proposed by ALBERTI et al. (1993), which illustrates the geometric arrangement of the active source components. Bottom: a contact transmission radiograph which illustrates the mechanical structure of the source

the model 6711 iodine seed and the fletcher-Suit-Delclos colpostat distributed by the 3M company. Accurate identification of the geometric structure of low-energy interstitial seeds is not always straightforward. Imaging techniques, such as pinhole autoradiography (Fig. 12.18) and contact transmission microradiography, should be used to verify the vendor's mechanical drawings whenever the dose distribution is expected to be sensitive to design uncertainties.

Choice of Estimator
and Other Variance Reduction Techniques

An estimator is a computational device for extracting a statistical estimate of the quantity of interest from a simulated photon trajectory. Most MCPT applications utilize so-called analog estimators to estimate absorbed dose-at-a-point. This approach consists in tallying energy transfers from those photon collisions that occur in an imaginary detector volume centered about the point of interest. Then, dose at the detector center can be approximated by the ratio of energy transferred to the detector volume to its mass. This approach must be used cautiously in brachytherapy. If dose gradients are large over the detector dimensions, spatial resolution will be lost due to volume averaging. If the detector volume is constrained by nearby media interfaces or the need for high spatial resolution, efficiency may be unacceptably low since only a small fraction of the simulated collisions will occur within the detector. A simple modification of the analog estimator, the tracklength estimator, greatly increases simulation efficiency by analytically estimating the energy transferred to the detector volume by all photon flight paths whose projections intersect the detector (WILLIAMSON 1987). The next-flight estimator, described below, increases efficiency, relative to the desired level of spatial resolution, even more, especially for low-energy photons. The simplest device for reducing variance and improving spatial resolution is to deterministically calculate the primary-photon collision kerma component rather than to stochastically score randomly constructed primary-photon trajectories. This calculation consists of straightforward 3D numerical integration of the operator, $\rho\,(\vec{r})\cdot e^{-\Sigma\mu\cdot l}\cdot(\mu_{en}/\rho)\cdot|\vec{r}|^{-2}$, over the distribution of radioactivity, $\rho(\vec{r})$ in terms of mCi/cm^3. By limiting Monte Carlo simulation to scoring of scattered-photon trajectories, the variance of the total-dose estimates near sources is reduced by at least a factor of 2.

Calculation of "absolute" absorbed dose rate in medium per unit air-kerma strength poses a unique estimation problem. MCPT must be used to calculate both dose rate in medium and air-kerma strength, per photon emitted by the active source. Then the air-kerma strength/emitted photon estimate can be used to normalize the condensed medium dose-rate estimate in terms of air-kerma strength. Simulation of air-kerma strength requires calculation of air-kerma rate on the transverse axis of the source in a vacuum or gaseous medium. Neither the track-length nor analog estimators work well under these circumstances because of the low density of simulated collisions. The author (1987) has solved this problem by using a more sophisticated analytic technique known as bounded next-flight estimation. This point-dose estimator forces every simulated collision to contribute to the quantity of interest by analytically calculating its expected contribution to the point detector, accounting for inverse-square law, attenuation of scattered photons through the intervening media, and probability that the scattered photon emerging from the collision intersects the detector point. Unfortunately the $1/r^2$ singularity in the estimator destroys the statistical stability of the dose estimate, since a single collision occurring close to the detector may make a huge but random dose contribution. This problem can be solved by replacing the $e^{-\mu r}/r^2$ term by its average value within a small critical sphere surrounding the point. This modified next-flight estimator solves the problem of accurately calculation of kerma at a point in a dilute medium. In addition, next flight works well in condensed media as long as the point of interest remains 1–2 mm from any media interface and the critical averaging sphere encloses a single medium. We are currently validating a generalization of this estimator, once-more collided flux estimation, which eliminates these restrictions. This will allow MCPT to calculate collision kerma at media interfaces and within small, geometrically complex detectors such as silicon diodes and TLD chips.

Numerous other variance-reduction techniques have been developed and implemented which can greatly reduce the CPU time required to achieve the desired level of statistical precision for complex problems. Common strategies (JENKINS et al. 1988; MILLER and LEWIS 1984) employed in brachytherapy include Russian roulette and splitting, forced scattering, and biased angle sampling. Each of these techniques uses biased sampling distributions in conjunction with correction factors designed to eliminate any bias in the sample mean of the quantity of interest. These sampling distributions and photon-weight correction factors are cleverly designed so as to force the computer to focus CPU effort on simulated random events that contribute significantly to the quantity of interest.

12.3.3. A New Dose-Calculation Formalism for Brachytherapy

Historically, single-source dose distributions for clinical brachytherapy treatment planning were calculated using semianalytical models such as the Sievert integral (SIEVERT 1921; YOUNG and BATHO 1964).

These models estimate the 2D dose distribution around an extended source by integrating the isotropic point-source distribution over the assumed spatial distribution of radioactive material within the source. Input data consist only of effective attenuation coefficients for the encapsulating materials, source dimensions, and scatter buildup and attenuation data, usually obtained from photon transport calculations based upon an unfiltered point source in unbounded water medium. For interstitial seed dosimetry, the theoretical isotropic point-source distribution was usually used directly, often in conjunction with an average distance-independent anisotropy correction factor. Such simple models allow even a small treatment planning computer to rapidly create a 2D dose matrix sufficiently dense to permit accurate calculation dose rate at any point during clinical treatment planning by linear interpolation. WILLIAMSON (1988a) has identified the assumptions underlying this model and has demonstrated that the model is in excellent agreement with MCPT calculations for ^{137}Cs sources.

The new approach to treatment planning is based upon 2D dose distributions obtained by direct measurement or Monte Carlo simulation. The 2D data matrices are necessarily sparse, both because of the difficulty in obtaining accurate dose-rate data, the limited data entry capabilities of the user, and space limitations in scientific journals. Such developments have prompted development of a new dose calculation formalism which takes as its input these measured or theoretically calculated dose rates, supports interpolation between sparsely distributed data points, and allows accurate reconstruction of the entire 2D dose distribution. This formalism was originally developed by the NCI-supported Brachytherapy Contract Group (ICWG) (ANDERSON et al. 1990) and has been refined by the AAPM Task Group 43 (NATH 1994), which recommends its use in clinical treatment planning.

12.3.3.1 General Formalism for the Two-Dimensional Case

We restrict consideration to cylindrically symmetric sources, such as that illustrated by Fig. 12.19. The dose distribution is two-dimensional and can be described in terms of a polar co-ordinate system, (r,θ), with its origin at the source center where r is the distance to the point of interest and θ is the angle with respect to the long axis of the source. Then dose rate,

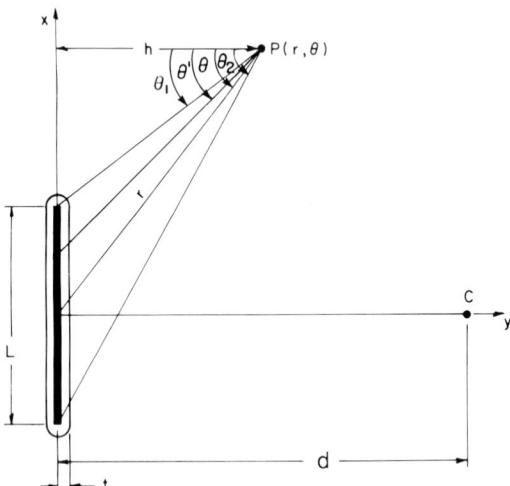

Fig. 12.19. Schematic drawing illustrating the geometry of the ICWG formalism for a cylindrically symmetric filtered line-source. Point *P* indicates the point of interest in the medium at which dose rate is to be calculated and point c designates the point of output measurement in free space where $d<L$. (From WILLIAMSON and NATH 1991)

$D(r,\theta)$, at (r,θ) can be written as:

$$\dot{D}(r,\theta) = S_k \cdot \Lambda \cdot \frac{G(r,\theta)}{G(r_0,\theta_0)} \cdot F(r,\theta) \cdot g(r), \qquad (12.1)$$

where

S_k = source strength in terms of air-kerma strength in units of U (see below)

$G(r,\theta)$ = geometry function in units of cm^{-2}

Λ = is the dose-rate constant for the source and surrounding medium and has units of cGy·h^{-1}·U^{-1}

(r_0,θ_0) = the reference point near the source where the dose distribution is normalized to unity. (r_0,θ_0) is assumed to lie on the transverse bisector of the source at a distance of 1 cm from its center, i.e., $r_0 = 1$ cm and $\theta_0 = \pi/2$

$F(r,\theta)$ = the dimensionless dose anisotropy function which takes the value unity for θ_0 at all r

$g(r)$ = the dimensionless radial dose function and takes the value unity at $r = r_0$

Each of these functions will now be described.

Air-Kerma Strength

Air-kerma strength is defined (NATH et al. 1987c) as the product of air-kerma rate in free space, $K(d)$, measured along the transverse bisector of the source, and the square of the measurement distance, d,

Output Specification

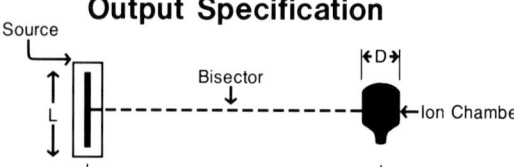

Conditions

1. Large distance d: L << d, D << d
2. Free in space
   - measured in air
   - corrected for air attenuation
   - corrected for scattering from air, walls, etc.

Fig. 12.20. Schematic drawing showing the experimental setup used to define brachytherapy source-strength output standards which is the basis of such quantities as apparent activity, equivalent mass of radium, reference exposure rate, reference air-kerma rate, and air-kerma strength. The measurement distance *d* need not be the same as used in the definitions of reference exposure rate or reference air-kerma rate. (From WILLIAMSON and NATH 1991)

$$S_k = K(d) \cdot d^2. \qquad (12.2)$$

The location of the point of output measurement and the filtered source is illustrated by Fig. 12.20. The distance *d* must be large enough that both source and detector may be treated as mathematical points. Air-kerma rate standardization measurements are performed in air using air attenuation and scattering corrections where needed. If kerma, time, and distance are assigned units of μGy, h, and m respectively, S_k will have units of μGy·m^{-2}·h^{-1} as recommended by the TG-32 report (NATH et al. 1987c). These units are commonly denoted by the symbol U, that is,

$$1\ U = 1 \text{ unit of air-kerma strength}$$
$$= 1\ \mu Gy \cdot m^2 \cdot h^{-1} \qquad (12.3)$$
$$= 1\ \mu cGy \cdot cm^2 \cdot h^{-1}.$$

The relationship of air-kerma strength to NIST brachytherapy calibration standards and practices, and conversion factors needed to renormalize calibration certificates using alternative source-strength formalisms (e.g., mgRaEq, mCi$_{app}$, etc.) have been discussed by WILLIAMSON and NATH (1991).

Dose-Rate Constant

The dose-rate constant is defined as

$$\Lambda = \frac{\dot{D}(r_0,\theta_0)}{S_k}. \qquad (12.4)$$

This constant depends on the medium surrounding the source and includes the effects of source geometry, spatial distribution of radioactivity, encapsulation, self-filtration in the source, and attenuation and scattering of photons in the surrounding medium. It also depends on the standardization measurements to which the air-kerma strength calibration of the source is traceable. Task Group 43 (NATH et al. 1994) recommends that liquid water be used as the medium for specification of absorbed dose in clinical brachytherapy.

Geometry Function

$G(r,\theta)$ describes the dose falloff about an extended source due solely to the effects of inverse-square law at (r,θ). It depends on the spatial distribution of radioactivity within the source but ignores the effects of attenuation and scattering in the source or surrounding medium, which are included in other terms described below. In the general three-dimensional case:

$$G(\mathbf{r}) = \frac{\int \dfrac{\rho(\mathbf{r}')}{|\mathbf{r}'-\mathbf{r}|^2} \cdot dV'}{\int \rho(\mathbf{r}') \cdot dV'}, \tag{12.5}$$

where $\rho(\mathbf{r})$ represents the density of radioactivity at the point $\mathbf{r} = (x, y, z)$ within the source in units of Bq/cm^3. It is used to suppress variation in anisotropy profiles and transverse-axis dose distributions due to inverse-square law variations with respect to distance and angle. Especially near the source, this greatly improves the accuracy of 2D interpolation and permits use of sparse data matrices. When the distribution of radioactivity can be approximated by a cylindrically symmetric line source or a point source $G(r,\theta)$ reduces to

$$G(r,\theta) = \begin{cases} r^{-2} & \text{Point source} \\[2mm] \dfrac{\Delta\theta}{L \cdot r \cdot \sin\theta} & \text{Line source of active length } L \end{cases}$$

where $\Delta\theta = \theta_2 - \theta_1$ is the angle subtended by the active source with respect to the point (r,θ) (see Fig. 12.19).

Dose Anisotropy Distribution

The anisotropy function is defined as

$$F(r,\theta) = \frac{\dot{D}(r,\theta) \cdot G(r,\theta_0)}{\dot{D}(r,\theta_0) \cdot G(r,\theta)}. \tag{12.6}$$

This two-dimensional function gives the angular variation of dose about the source at each distance due to self-filtration, oblique filtration of primary photons through the encapsulating material, and photon attenuation and scattering in the surrounding medium. The role of the geometry distribution in Eq. 12.6 is to suppress the influence of inverse-square law on the angular dose distribution arising from the extended spatial distribution of radioactivity within the source.

The value of suppressing inverse-square law induced dose gradients by means of the geometry function is illustrated by Fig. 12.21, which shows $F(r,\theta)$ profiles for $r = 0.25$ and 1.0 cm derived from WILLIAMSON and QUINTERO's (1988) Monte Carlo study of the model 6702 ^{125}I seed. If the line-source rather than the point-source $G(r,\theta)$ function is applied to a 0.25-cm angular dose profile, the large point of inflection at 25° disappears, producing a profile very similar to the 1-cm profile. This approach significantly improves the accuracy of extrapolation from angular profiles measured at distances greater than 10 mm to smaller distances as well as interpolation between measured profiles.

Because the distribution, $\rho(\mathbf{r})$ is uncertain for many sources such as ^{125}I and because the choice of $G(r,\theta)$ influences only the accuracy of interpolation, Task Group 43 recommends approximating $G(r,\theta)$ by the line-source formula.

Radial Dose Function

The radial dose function, $g(r)$ is defined as

$$g,(r) = \frac{\dot{D}(r,\theta_0) \cdot G(r_0,\theta_0)}{\dot{D}(r_0,\theta_0) \cdot G(r,\theta_0)}, \tag{12.7}$$

where θ_0 defines the reference axis (normally the transverse-bisector, $\pi/2$) and r_0 (normally 1 cm) defines the reference distance where $g(r)$ is normalized to unity. This function defines the falloff of dose along the transverse axis due to attenuation and scattering in the medium and is influenced by filtration effects. The function $g(r)$ is similar to a normalized transverse-axis tissue-attenuation factor or absorbed dose-to-kerma in free space ratio. In contrast to these other ratios, the geometry distribution used to suppress inverse-square law effects has been generalized to allow for the nonpoint spatial distribution of radioactivity.

Each of the quantities used to calculate absorbed dose rate is measured or calculated for the specific

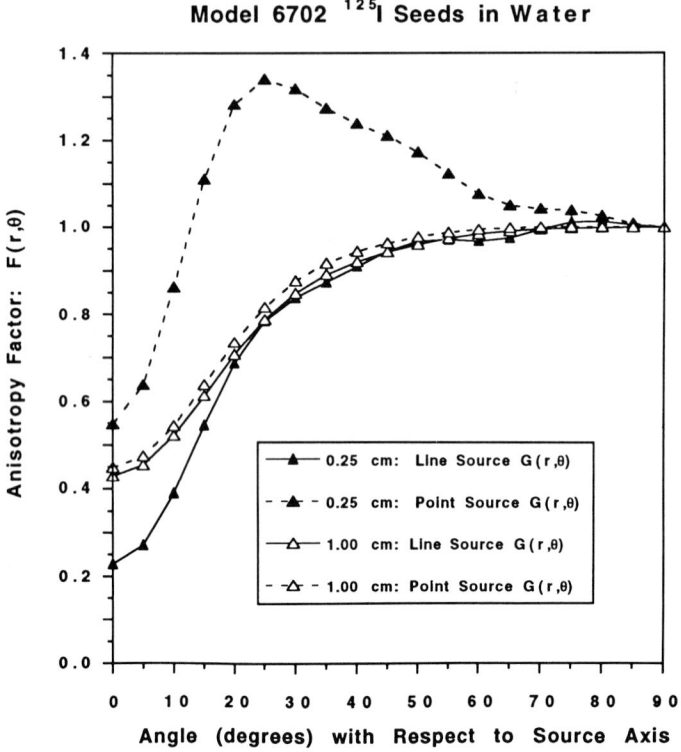

Fig. 12.21. Polar anisotropy function, $F(r,\theta)$, data for the model 6702 ^{125}I seed in water calculated by WILLIAMSON and QUINTERRO (1988) using Monte Carlo simulation. Note that for $r = 0.25$ cm, the function F has a completely different shape than at $r = 1$ cm distance when a point-source geometry distribution is used to suppress inverse-square law variations. When a line source $G(r,\theta)$ is used (3.6 mm active length), the resultant anisotropy profiles (solid line) are very similar at both distances, facilitating more accurate interpolation

source in question and therefore depends on source construction and geometry in addition to the primary photon spectrum and medium. In contrast, many of the input data to the older semianalytical models, including exposure-rate constants and buildup factors, are fundamental properties of the radionuclide.

12.3.3.2 One-Dimensional Isotropic Source Approximation

Most commercial computer-assisted treatment planning systems utilize the 1D isotropic source distribution to compute doses around interstitial sources. In this approximation, dose depends only on distance from the source. If the seeds are randomly oriented, or the degree of anisotropy is limited, the dose contribution to tissue from each seed can be well approximated by the average radial dose as estimated by integrating about the anisotropic source:

$$\dot{D}(r) = \frac{1}{4\pi} \int_{4\pi} \dot{D}(r,\theta) \, d\Omega \qquad (12.8)$$

where $d\Omega = 2\pi \sin\theta \cdot d\theta$ for a cylindrically symmetric dose distribution. Substituting Eq. 12.1 into Eq. 12.8, we obtain

$$\dot{D}(r) = S_k \cdot \Lambda \cdot \frac{G(r,\theta_0)}{G(r_0,\theta_0)} \cdot g(r) \cdot \phi_{an}(r), \qquad (12.9)$$

where the average anisotropy factor, $\phi_{an}(r)$, is defined as

$$\phi_{an}(r) = \frac{\int_0^\pi \dot{D}(r,\theta) \cdot \sin\theta \cdot d\theta}{2 \cdot \dot{D}(r,\theta_0)}. \qquad (12.10)$$

The factor $\phi_{an}(r)$ is the ratio of the dose at distance r, averaged with respect to solid angle, to dose on the transverse axis at the same distance. For ^{125}I sources, the distance dependence of the anisotropy factor is negligible, so that it may be approximated by a constant, $\bar{\phi}_{an}$, called the anisotropy constant. At distances greater than twice the active length of the source, the absorbed dose-rate equation (12.9) simplifies to:

$$\dot{D}(r) = \frac{S_k \cdot \Lambda}{r^2} \cdot g(r) \cdot \bar{\phi}_{an}. \qquad (12.11)$$

12.3.3.3 Consequences of Source-Strength Specification

A major advantage of the dose-calculation formalism, outlined above, is that it is constructed around a physically well defined and readily measurable source-strength quantity: air-kerma rate in free space on the transverse bisector of the source. Air-kerma strength standards are maintained by NIST for LDR ^{192}Ir seeds, for models 6702 and 6711 ^{125}I seeds, and for all currently available LDR ^{137}Cs sources. Prior to the acceptance of air-kerma strength as a source-specification standard, radium-substitute radionuclide strength was quantified in terms of equivalent mass of radium (units: mgRaEq) and ^{125}I was quantified in terms of effective activity (units: mCi$_{app}$). Although these descriptions were conceptually confusing, at least they were derived from well-defined physical quantities: the exposure standards maintained by NIST. However, NIST-traceable standards are not available for ^{198}Au or ^{103}Pd seeds, HDR ^{192}Ir sources, or any of the brachytherapy sources using developmental radionuclides. With regard to HDR ^{192}Ir sources, each end user is expected to measure air-kerma strength in a free-air geometry using an ion chamber, correcting for room scatter, air attenuation, and gradient effects. As described by GOETSCH et al. (1991), an ^{192}Ir calibration factor for the ion chamber is derived by interpolating between NIST-traceable hard orthovoltage and ^{137}Cs external-beam air-kerma calibration factors. For physically meaningful results, the ion chamber must be equipped with a sufficiently thick buildup cap, to ensure charged-particle equilibrium for ^{192}Ir gamma rays, must be in place both during HDR source calibration and during calibration of chamber against NIST air-kerma standards. Reentrant chambers, with AAPM- accredited ^{192}Ir calibration factors derived by intercomparison with this secondary interpolative in-air calibration procedure, are becoming available.

Unfortunately, no accepted primary or secondary interpolative standards exist for ^{198}Au, ^{103}Pd, ^{169}Yb, or other new isotopes. Vendors standardize the strength of these sources using a variety of instrumentation and generally report calibration values in terms of activity. Usually, end users accept the vendor's activity calibration for clinical dose calculation without verifying it against an independent standard. Some vendors use a calibrated ion chamber as the basis of their in-house standard, while others use an NaI(Tl) scintillation probe or a lithium-drifted germanium spectrometer to measure the emission rate of one or more spectral lines, using published spectra to convert this count rate to apparent activity. NIST-traceable radioactivity standards, with photon energies close to that of the targeted emission peak, are used in various ways to calibrate the detector or to define its collection efficiency. Such source-strength reports cannot be assumed to be well-defined physical quantities with known relationships to air-kerma strength, reference exposure rate, contained activity, or dose rate in medium. In particular, there is no assurance that air-kerma strength is given by

$$S_K = A_{app} \cdot (\Gamma_\delta)_x \cdot \left(\frac{W}{e}\right), \qquad (12.12)$$

a relationship that is true by definition for apparent activity specifications derived from NIST brachytherapy standards.

The accuracy of clinical dose calculations based upon *measured* dose-rate distributions is independent of how accurately the vendor has implemented their calibration standard. Dosimetric accuracy requires that (a) the source vendor reproducibly transfers its in-house but possibly arbitrary calibration standard to each clinical source, (b) the measured dose rates are normalized to the same standard, and (c) the end clinical user utilizes source-strength values consistent with the vendor's standard as the basis of dose calculation. Often published dose-measurements are normalized (see, for example, MEIGOONI et al. 1990; CHUI-TSAO and ANDERSON 1991) to air-kerma strength which has been calculated from the vendor's apparent activity assay using some assumed value of $(\Gamma_\delta)_x$. Obviously, the clinical user must utilize the same conversion factor to avoid dosimetric error.

In the absence of a well-defined and accurately measured source standard, no *theoretical method* of dose calculation can be trusted to yield accurate absolute dose rates in medium. For example, the dose distribution around an isotropic point source emitting photons with energies above 200 keV is accurately predicted by the following simple analytic model:

$$\dot{D}(r) = S_K \cdot \left(\overline{\mu_{en}/\rho}\right)_{air}^{med} \cdot B(\mu r) \cdot e^{-\mu r} \cdot r^{-2}, \qquad (12.13)$$

where B is the scatter buildup factor as defined by MEISBERGER et al. (1968). However, this theoretical relationship between dose rate in medium and S_K obtains only if the S_K value used for dose calculation accurately represents the true air-kerma strength of

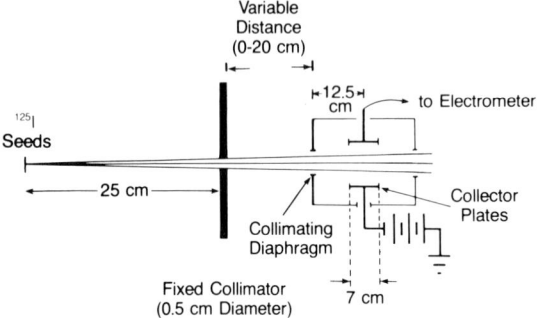

Fig. 12.22. Schematic drawing of the Ritz low-energy free-air chamber (FAC), showing the air-attenuation measurement geometry. For the measurement of air-kerma rate, the additional fixed collimator is not used. (From WILLIAMSON 1988b)

the implanted source. Although Monte Carlo simulation can be used to accurately calculate air-kerma rate in free space per contained mCi on the transverse axis of a geometrically complex source, the calculated dose rates in medium will be inaccurate if the experimental implementation of the air-kerma strength standard applied to the actual source is inaccurate.

The ^{125}I air-kerma strength standard implemented by NIST is an example of a calibration standard which challenges all but the most sophisticated and thorough approaches to theoretical modeling. The NIST ^{125}I standard (LOFTUS 1984) uses the low-energy free-air chamber (FAC) illustrated by Fig. 12.22 (RITZ 1960). The FAC is a collimated detector which potentially upsets the compensation between primary-photon attenuation and scattering in air. In addition, it has no condensed-medium wall, allowing low-energy contaminant x-rays to contribute to the measured ionization. KUBO (1985) has experimentally demonstrated that K-shell x-rays (4.5 keV) arising from photoelectric absorption of primary ^{125}I photons by the titanium capsule constitute 5%–20% of the total transverse-axis exposure rate in air over the distance range of 1–27 cm. However, these low-energy x-rays have a penetration of less than 0.1 mm in a unit-density medium and therefore do not contribute to dose rate in medium at distances of therapeutic interest. Inclusion of these contaminant x-rays in the FAC measurements inflates the air-kerma strength standard, leaves dose rates in water unchanged, and lowers the dose-rate constant, Λ.

The present author (WILLIAMSON 1988b) performed an extensive Monte Carlo study of the NIST calibration geometry with the goal of identifying the dosimetric consequences of both FAC collimation and the low-energy characteristic x-rays. The under-

lying geometric model included internal ^{125}I seed structure and the FAC collimator and carefully duplicated the measurement distances, multiple-seed arrays, and air-attenuation measurement methods employed by Loftus at NIST. Figure 12.23 shows the final results. According to our MCPT calculations, the FAC collimation has no significant effect and the Ti characteristic x-rays inflate ionization readings only by 1%–2% at the source-to-collimator distances (25 and 50 cm) used by Loftus for his final measurements. Unfortunately, the rapid falloff of contaminant x-rays with respect to distance in air gave rise to an anomalously large air-attenuation correction, which Loftus applied to his final exposure-rate measurements. As indicated by the broken lines in Fig. 12.23, this extrapolation led to a 6%–8% overestimate of the free-air-kerma rate due to the penetrating component of the ^{125}I spectrum. This finding was supported by close agreement between the air-attenuation correction measured by Loftus and that calculated by MCPT and by the 1% agreement between MCPT and Kubo's measured distribution of Ti characteristic x-rays in air. In general, any theoretical calculation of absolute dose rates in medium about a brachytherapy source must include corrections for any systematic experimental artifacts inherent in the basic air-kerma strength standard or in the transfer of calibration from the standard to clinical sources. Fortunately, in the case of ^{125}I, detailed descriptions of the NIST calibration were available, as well as ancillary experimental data to validate the MCPT findings.

12.3.4 Recent Interstitial Seed Dose Measurements and Monte Carlo Simulations

New dosimetry data sets of both experimental and theoretical origin are now available for the most commonly used interstitial implant sources. In this section, data related to ^{125}I and ^{192}Ir sources will be reviewed. Our goals are twofold: to compare the recent data with dose tables conventionally used for treatment planning and to compare dose-rate distributions calculated by MCPT with their empirically measured counterparts.

12.3.4.1 Data for ^{125}I Interstitial Implant Dosimetry

Table 12.3 summarizes the published dose-rate constant data for ^{125}I interstitial sources. The data used conventionally are represented by those of

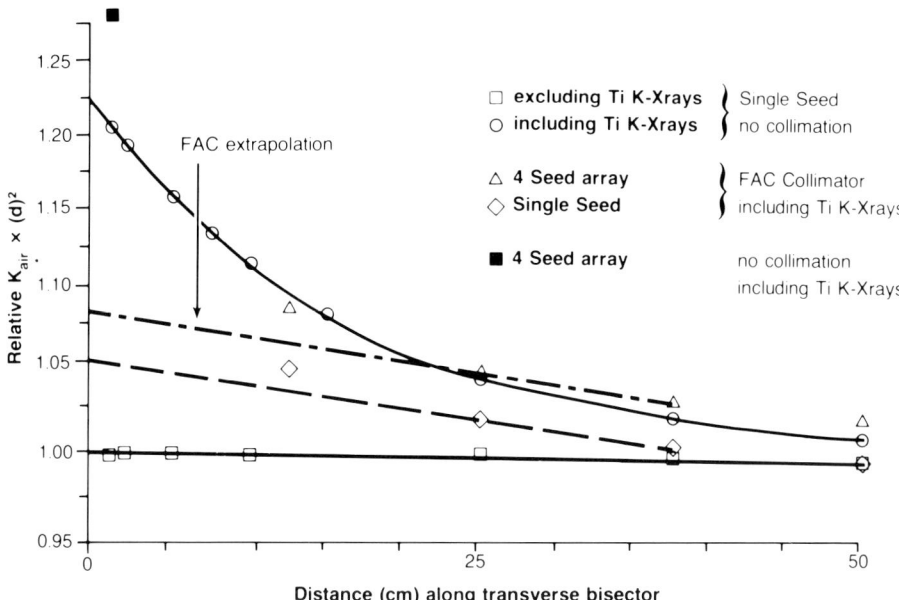

Fig. 12.23. Product of air-kerma rate in dry air and square of the distance along the transverse bisector of a model 6711 seed as estimated by Monte Carlo simulation for various calibration geometries. The *broken lines* illustrate calculation of S_k by extrapolation from simulated NIST measurements. (From WILLIAMSON 1988b)

LING et al. (1983) and KRISHNASWAMY (1978). The latter used the theoretically calculated point-source buildup data of BERGER (1964) to obtain a dose-rate constant of 1.043 cGy·h^{-1}·U^{-1} for the now obsolete model 6701 seed. Ling derived the dose-rate constant from the radium-substitute approximation $\Lambda = (\overline{\mu_{en}/\rho})_{air} = f_{wat} \cdot (W/e)^{-1}$, where f_{wat} is the Roentgen-

to-cGy conversion factor, and obtained the same result: 1.035 cGy·h^{-1}·U^{-1}. Both assumed that the NIST calibration standard accurately quantifies air-kerma strength and that seed structure does not influence the D_{wat}/S_K ratio. Λ was assumed to be the same for the 6711 and 6702 seed models despite the fact that the photon spectrum of the 6711 seed was known (LING et al. 1983) to differ from that of the 6702 seed due to the presence of silver characteristic x-rays. In contrast, the NCI-sponsored brachy-therapy contract participants (CHIU-TSAO et al. 1990; NATH et al. 1990b; WEAVER et al. 1989; ANDERSON et al. 1990; hereafter denoted by "ICWG"), using TLD

Table 12.3. Specific dose rate constant in water for ^{125}I seeds

Author	Method	Measurement Medium	Λ_o(cG.cm^2·h^{-1}·U^{-1})	
			Model 6702	Model 6711
Williamson (1988b)	Monte Carlo DLC-7F	Atomic water	0.962	0.909
Williamson (1991a)	Monte Carlo DLC-99	Liquid water	0.932	0.877
		Solid water	0.899	0.841
		Lucite	–	0.945^a
NCI Contract Group				
NATH et al.(1990b)	Measurement	Solid water	0.903	0.853
WEAVER et al. (1989)	Measurement	Solid water	0.926	0.832
CHIU-TSAO et al. (1990)	Measurement	Solid water	0.932	0.853
LUXTON et al. (1990)	Measurement	Lucite	–	0.932
PIERMATTEI et al. (1988)	Measurement	Water	–	0.933
LING et al. (1983)	Monte Carlo	Water	1.035	1.035
KRISHNASWAMY (1978)	Berger Buildup data	Water	1.043 (6701)	–
HILARIS (1975)	Measurement	MIX-D	1.38 (6701)	–

^{a} Including Luxton's estimated water-to-lucite fluence ratio (0.954)

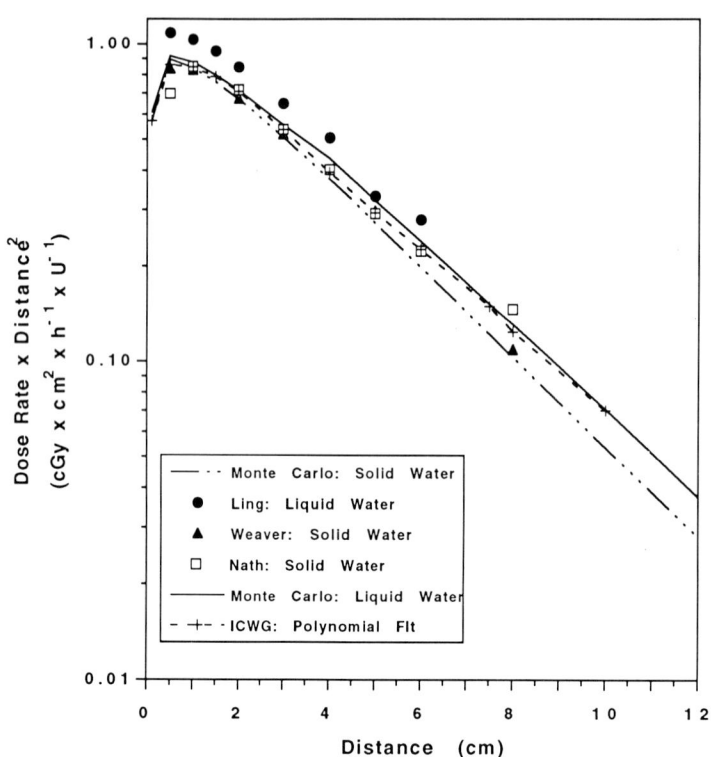

Transverse Axis Dose Distribution
Model 6711 ^{125}I Seed

Fig. 12.24. Comparison of measured and calculated (Monte Carlo) absolute dose rates per unit air kerma strength for the model 6711 seed. Monte Carlo estimates (from WILLIAMSON 1991a) of dose rate to water are plotted for both liquid- and solid-water measurement medium. The measured data are taken from the following references: LING et al. (1983), WEAVER et al.(1989). NATH et al. (1990b), and ANDERSON et al. (1990). (From WILLIAMSON 1991a)

dosimetry in solid-water medium, found Λ-values of 0.846 and 0.920 cGy·h^{-1}·U^{-1} for the models 6711 and 6702 seeds respectively. This shows that the conventional data overestimate dose rate in medium by 13% and 23%, respectively, for the two seed models. In addition, the two seed models, for the same strength lead to dose rates at 1 cm that differ by 8.8%. The measurements of PIERMATTEI et al. (1988) and LUXTON et al. (1990) support the conclusion that conventional ^{125}I dosimetry overestimates dose by the same magnitude. WILLIAMSON'S (1988b, 1991a) Monte Carlo Λ-values, assuming solid-water measurement medium (liquid-water kerma in solid-water medium), are in remarkably close agreement with the ICWG measurements (+1.3% and –0.6% for models 6702 and 6711 respectively). Figures 12.24 and 12.25 demonstrate the excellent agreement between measured transverse-axis dose rates (dose rate × distance^2/unit U) and the author's MCPT calculations (WILLIAMSON 1991a) for the models 6711 and 6702 seeds. It is obvious that the conventional data (represented by Ling's 6711 data) significantly overestimate absorbed dose at all distances.

The ICWG investigators have argued that solid water (MEIGOONI et al. 1988b) is an accurate liquid-

water substitute and that their measured dose-rate distributions accurately characterize the dose distribution in this reference medium. However, WILLIAMSON'S (1991a) Monte Carlo data (Fig. 12.26) show that solid water leads to dose underestimates relative to liquid water of 4%, 16%, 24%, and 38% at distances of 1, 5, 10, and 20 cm respectively. Solid water has an atomic composition of H:C:N:O:Ca = 100: 70: 2: 15: 1 compared to H:O = 100 : 50 for liquid water where the calcium (2.3% by weight) is intended to compensate for the replacement of oxygen by carbon in the resin base of the plastic phantom. At ^{125}I energies (25 keV), solid-water total, photoeffect, and scattering cross-sections are 4% larger, 13% larger, and 3% smaller than those of liquid water.

Figures 12.27 and 12.28 illustrate the ratio of absorbed dose rate, as measured by the ICWG, to that calculated by MCPT, for both solid water and liquid water as a function of transverse-axis distance. When the ICWG data are compared to MCPT assuming solid-water measurement medium, agreement between theory and measurement is generally within 1%–5% in the 1–5 cm distance range. When measurement is compared to liquid-water MCPT calculations, discrepancies of 5%–10% result. Note

Fig. 12.25. Comparison of measured and calculated dose rates per unit air kerma strength for the model 6702 seed. Monte Carlo estimates of dose rate to water are plotted for both liquid- and solid-water measurement media. The measured data are taken from the following references: WEAVER et al. (1989), (NATH et al. (1990b), and SCHELL et al. (1987). The relative dose measurements of Schell have been normalized to the Λ_0 value for liquid-water medium (0.932 cGy·cm²·h⁻¹·U⁻¹) recommended by WILLIAMSON (1991a). (From WILLIAMSON 1991a)

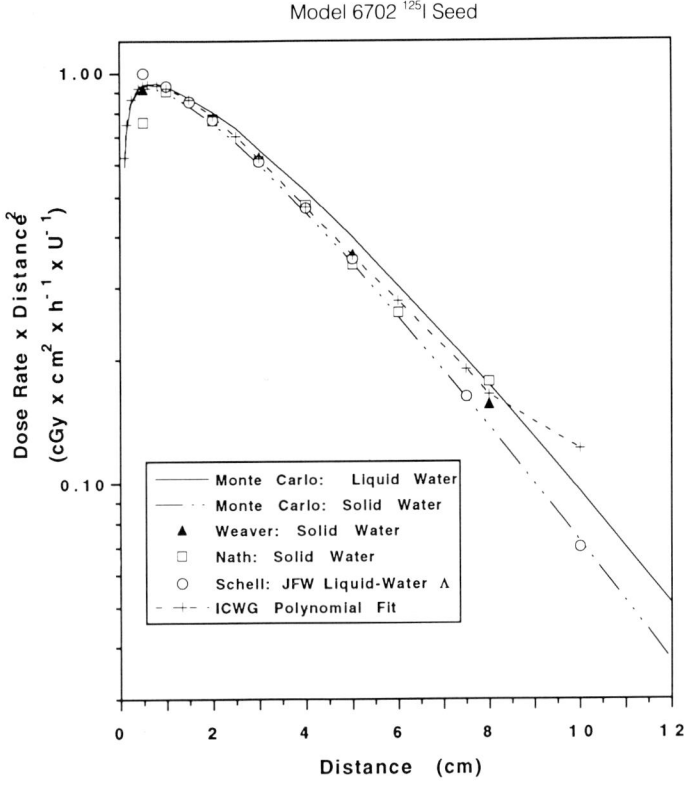

Fig. 12.26. Comparison of transverse-axis dose distributions for model 6711 and 6702 ¹²⁵I seeds in solid and liquid-water medium as calculated by Monte Carlo simulation. The graph shows the ratio of dose rate to water in water medium, to dose rate to water in solid-water medium, both per unit air-kerma strength as a function of distance. (From WILLIAMSON 1991a)

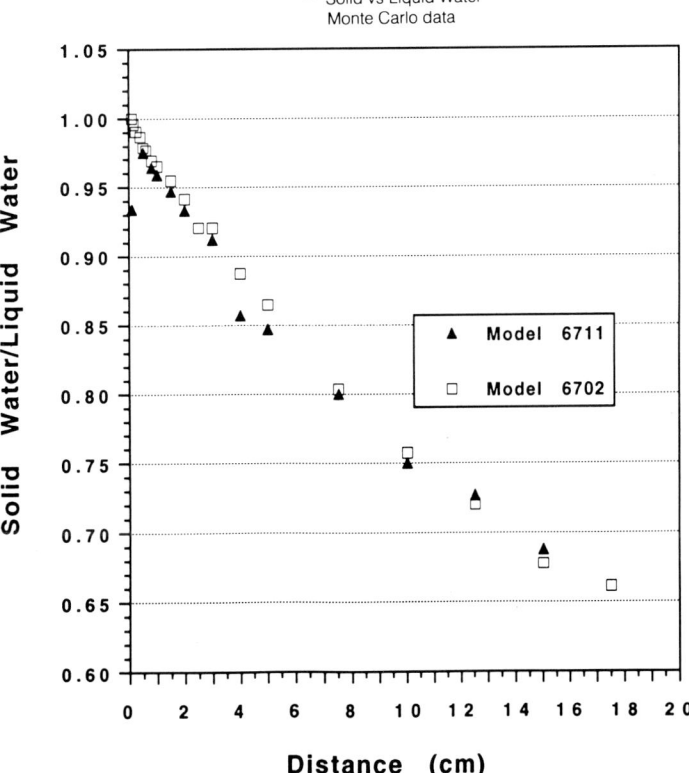

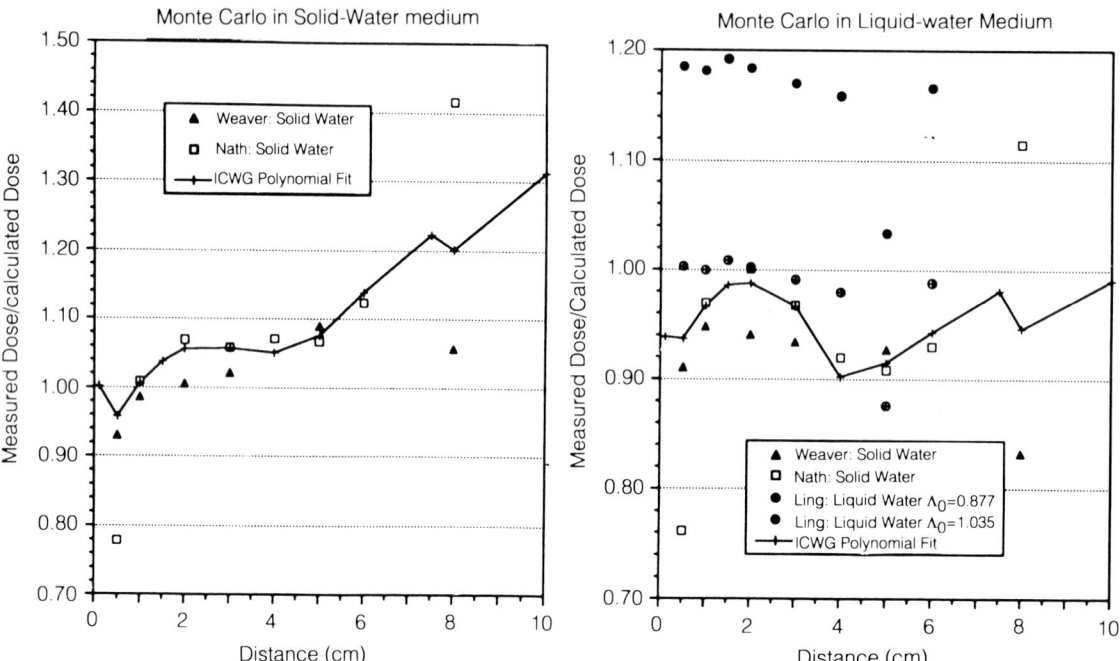

Fig. 12.27. Measured versus calculated transverse-axis dose rates for the model 6711 ^{125}I seed. The *left panel* presents measured dose rates relative to Monte Carlo calculations assuming solid-water medium while the right panel shows measured dose relative to calculated dose assuming liquid-water measurement medium. The data of Ling are presented both to his recommended Λ_0 value and to the Monte Carlo generated value (0.877 cGy·cm^2·h^{-1}·U^{-1}) recommended by WILLIAMSON (1991a). (From WILLIAMSON 1991a)

that the relative dose data of Ling, measured in liquid water with a silicon diode, agree with MCPT within 2%. For both larger and smaller distances outside the 1–5 cm range, agreement between theory and measurement is much poorer. However, the poor agreement (as large as 35%) between different experimental measurements at small and large distances suggests that positioning artifacts and signal-to-ratio limitations preclude precise TLD dose measurement outside the 1–5 cm range.

Angular anisotropy factors, $F(r,\theta)$, measured by TLD dosimetry, are also available in the literature. LING et al. (1985) and SCHELL et al. (1987) measured dose profiles around the models 6711 and 6702 seeds, respectively, in water using a silicon-diode detector. Unfortunately, their papers contain data in tabular form only in the 0–30° angular range. More complete data sets, based upon interpolated TLD, diode, or TLD-MCPT hybrid methods, have been published (AHMAD et al. 1992; ANDERSON et al. 1990; NATH et al. 1993; CHIU-TSAO et al. 1990; WILLIAMSON and QUINTERRO 1988). A sample of CHIU-TSAO's work (CHIU-TSAO et al. 1990), comparing TLD measurements to MORSE-code Monte-Carlo calcula-

tions, is shown in Fig. 12.29. Excellent agreement between MCPT and measurement was found for the model 6702 seed. However, significant disparities between calculated and measured F-values were found for the model 6711 seed, a finding that Chiu-Tsao attributes to an inadequate geometric model of the silver-wire active core. For use in implementing the isotropic point-source model (Eq. 12.11), the ICWG (ANDERSON et al. 1990) recommended anisotropy factors, ϕ_{an}, of 0.937 and 0.961 for models 6711 and 6702 ^{125}I seeds, respectively, which were derived from a more complete but unpublished description of the Ling-Schell 2D diode measurements. Based upon recent TLD measurements, NATH et al. (1993) obtained very similar values. Although CHIU-TSAO et al. (1990) did not publish ϕ_{an} values, her data indicate that ^{125}I dose distributions are slightly less anisotropic than the diode data of Ling and Schell. This suggests that Chiu-Tsao's data might yield somewhat larger ϕ_{an} values than those of Nath. Note that all recent measurements yield anisotropy corrections significantly closer to unity than the conventional value of 0.87 (LING et al. 1983), which was derived from in-air measurements.

Measured Data Vs Monte Carlo
Model 6702 ^{125}I seed

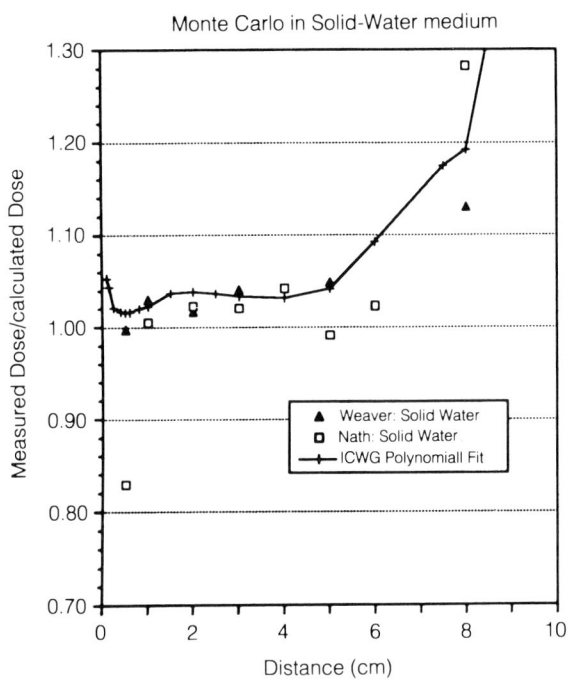

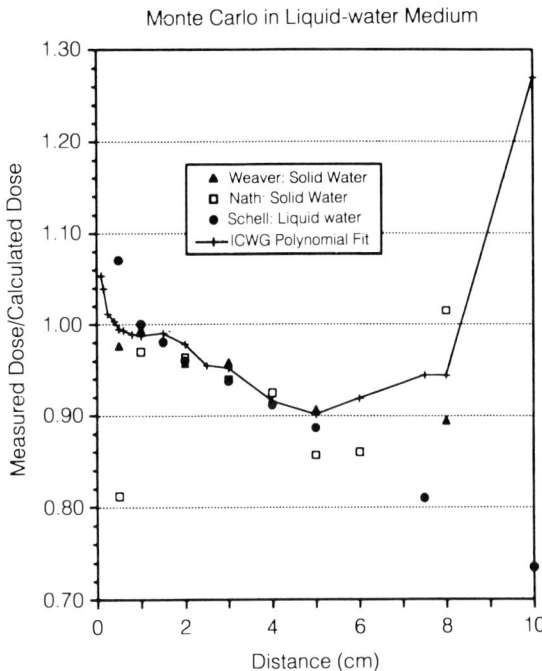

Fig. 12.28. The ratio of measured transverse-axis dose rate to dose rate calculated by Monte Carlo simulation as a function of distance for the model 6702 ^{125}I seed. The *left panel* shows measured dose rate relative to Monte Carlo calculations based on solid-water medium while the *right panel* shows measured dose rate relative to calculated dose rate based on liquid-water medium.(From WILLIAMSON 1991b)

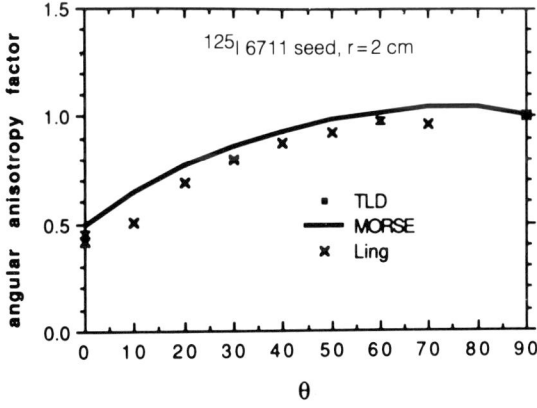

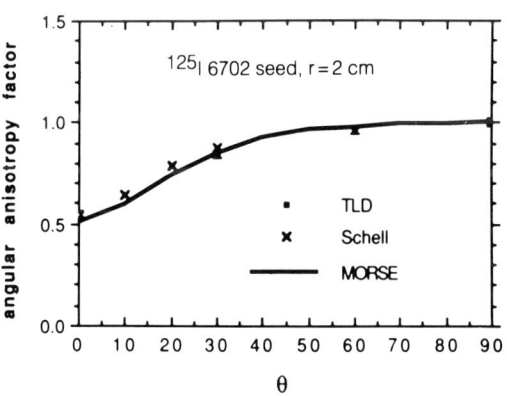

Fig. 12.29. The angular anisotropy function at 2 cm distance from the center of a model 6711 ^{125}I seed (left panel) and 2 cm from the center of a model 6702 ^{125}I seed (*right panel*). The solid-water measurements performed by CHIU-TSAO et al. (1990) using TLD dosimetry in a solid-water phantom (*squares*) are compared to Monte Carlo calculations using the MORSE Monte Carlo code (*solid line*) and to the relative diode measurements (*crosses*) of LING et al. (1985) and SCHELL et al. (1987). (From CHIU-TSAO et al. 1990)

In conclusion, recent dosimetry ^{125}I seed measurements and Monte Carlo calculations are in very close agreement along the transverse axis, with respect to

both relative and absolute dose rates. It is clear that conventionally used dosimetry data overestimate absorbed dose rates in water by 13%–20% and stand in urgent need of revision. To obtain good agreement between theory and measurement for this radio-nuclide requires accurate theoretical simulation of the NIST calibration geometry, an accurate geometric model of seed internal structure, and accurate modeling of the measurement medium. Due to the non-water-equivalence of solid water, the ICWG data are expected to underestimate absorbed dose in liquid water by 4%–15% in the 1–5 cm distance range. About one-third the discrepancy between the recent and the conventional dosimetric data can be attributed to inclusion of low-energy contaminant x-rays in the ^{125}I air-kerma strength standard maintained by NIST. The remaining discrepancy may be due to the effect of seed internal construction and encapsulation on the ^{125}I spectrum. The difference between model 6711 and 6702 dose-rate constants is due to the lower-energy 22- and 25-keV silver characteristic x-rays found in the model 6711 seed free-space spectrum. As Fig. 12.12a shows, Λ varies rapidly with photon energy in the ^{125}I-energy range. The status of 2D dose distributions is not as clear. Three complete measured 2D data sets have been published for the model 6711 seed and two for the model 6702 seed, but they have not been critically compared. The limitations and advantages of MCPT in predicting 2D anisotropy functions have not been thoroughly studied.

12.3.4.2 Data for [192]Ir Interstitial Implant Dosimetry

Recently measured and calculated dose-rate constants for the stainless-steel clad ^{192}Ir seed, commonly used for LDR interstitial implants, are tabulated in Table 12.4. Excellent agreement (±1%) among different experimental measurements (WEAVER et al. 1989; NATH et al. 1990b), Monte Carlo calculations (WILLIAMSON 1991a), and the traditionally used value of MEISBERGER et al. (1968) is evident. The Monte Carlo results indicate that solid-water and liquid-water measurement media yield nearly identical results, and that the geometry of the measure-ment phantom does not significantly influence dose rates near the seed.

Figure 12.30 compares MCPT dose-rate esti-mates to those measured by the ICWG investigators (WILLIAMSON 1991a). When the Monte Carlo calculations assume unbounded phantom, good agreement out to 5 cm is observed, but 5%–15% discrepancies appear at larger distances. This differ-ence can be explained by the finite, bounded phantoms used by the experimentalists. NATH et al. (1990b) used a $30 \times 25 \times 20$ cm^3 phantom enclosed in a lead vault. When the measurement-phantom geometry is included in the Monte Carlo simulation, agreement between measured and calculated dose rates is within ±2%. Although the differences between bounded and unbounded media are not clinically significant, this experience indicates that the accurate geometric modeling of all relevant experimental parameters is essential to meaningful comparison of theoretical and measured dose rates. All of the recent transverse-axis dose-rate distribu-tions are in close agreement (±2%) with the widely used Meisberger data. To supplement the 1D transverse-axis data, NATH et al. (1993) have recently measured and tabulated the anisotropy function, $F(r,\theta)$, for this interstitial source.

12.3.4.3 Comparison of Monte Carlo and Diode Transverse-axis Dosimetry

To more rigorously define the accuracy limits of Monte Carlo simulation for ^{125}I, ^{137}Cs, [169]Yb, and ^{192}Ir sources, a series of precision, transverse-axis diode measurements was designed by the author and his colleagues (WILLIAMSON et al. 1993b). A Scandi-tronix diode detector was chosen because of its high

Table 12.4. Specific dose rate constant for ^{192}Ir seeds stainless-steel clad seeds: water medium

Author	Method	Measurement Medium	$\Lambda_0 (cGy \cdot h^{-1} \cdot U^{-1})$
Williamson (1991a)	Monte Carlo DLC-99	Liquid water: unbounded	1.110 ± 0.2%
		Solid water: unbounded	1.121 ± 0.3%
		Solid water: Nath et al. (1990b) phantom	1.119 ± 0.2%
NCI Contract Group			
Nath et al. (1990b)	Measurement	Solid water	1.12 ± 2.7%
Weaver et al. (1989)	Measurement	Solid water	1.111 ± 1.5 %
Chiu-Tsao et al. (1990)	Measurement	Solid water	1.10
Meisberger et al. (1968)	Measurement/ transport theory	Water	1.118

Monte Carlo Calculations vs Measurement: ^{192}Ir

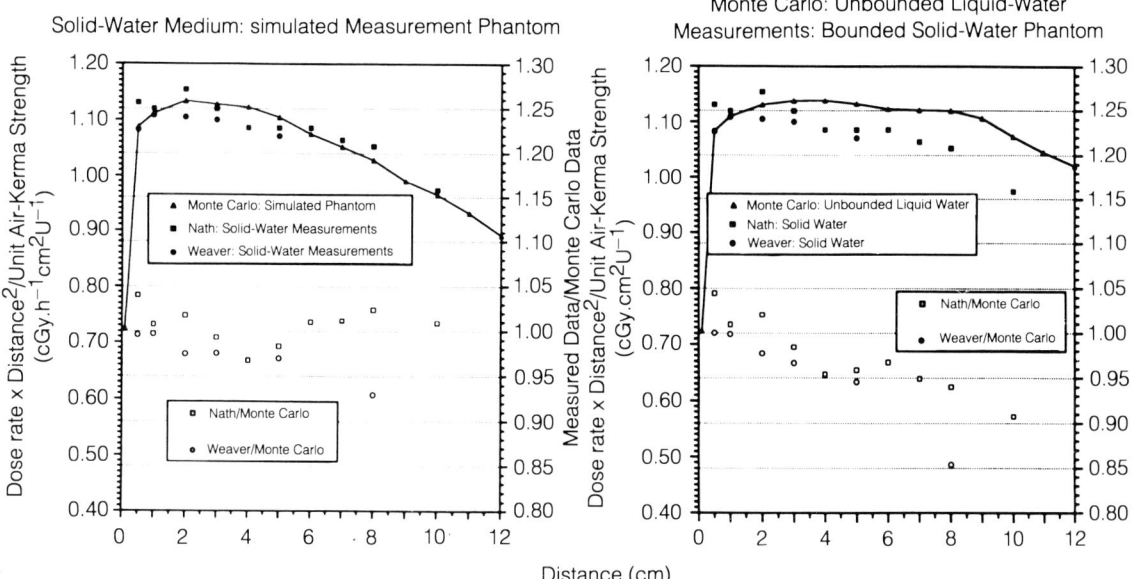

Fig. 12.30. Comparison of calculated and measured transverse-axis dose rates for the steel-clad ^{192}Ir seed. The *right axis of each panel* shows the ratio of measured dose rate to dose rate calculated by Monte Carlo simulation, both per unit air-kerma strength. The *right panel* compares measurements with Monte Carlo calculations assuming unbounded liquid water medium while the *left panel* compares measurements to calculations modeling the bounded rectangular solid-water phantom used by the experimentalists to obtain their data. (From WILLIAMSON 1991a)

sensitivity and excellent reproducibility and linearity of response. This detector was used in conjunction with a standard electrometer equipped with a timer-driven integrator allowing measurements with a precision of 1%–3% to be obtained at distances as large as 10–15 cm. Source internal structure was carefully defined using pin-hole autoradiography and transmission contact microradiography. A water phantom with a 2D micrometer-driven positioning jig (see Fig. 12.34) was designed which allowed source–detector distance to be measured with an accuracy of 20 μm. Measurements were obtained over the distance range of 0.23–15 cm. Measured and MCPT-calculated doses to water cannot be compared "cleanly" since conversion of diode readings to dose in water requires location-dependent energy-response and volume-averaging correction factors. Since MCPT is the only practical and available tool for defining such corrections, direct verification of MCPT *with respect to dose rate in medium* is necessarily circular. To avoid this methodological difficulty, the *uncorrected* diode readings themselves were compared to *simulated* detector readings, i.e., absolute rate of specific energy absorption in the active detector element as predicted by MCPT. The underlying geometric model included source internal structure, calibration geometry, water phantom geometry, and internal structure of the detector.

Figure 12.31 compares measured diode readings per unit air-kerma strength, to MCPT-simulated readings, normalized to the measurements at 1 cm distance. Excellent agreement (±3%) between theory and measurements within was observed for source-to-detector distances ranging from 2.85 mm to 95.35 mm. Figure 12.32 shows the ratio of absolute energy transferred to the diode detector active volume per unit mass, as calculated by MCPT from the measured air-kerma strength of each source (S_k), to the measured diode reading as a function of source-to-detector distance. This ratio is constant within ±3%, showing that MCPT is able to predict the variation of diode response not only with respect to distance but also with respect to photon spectrum. This demonstrates that MCPT is able to "linearize out" diode energy-response, volume-averaging, and self-attenuation artifacts, relative to dose in water, that span nearly an order of magnitude. Finally, Fig. 12.33 shows the ratio of dose to the detector volume to dose to water at the detector center, as calculated by MCPT, as a function of distance. For ^{137}Cs, ^{192}Ir, and ^{169}Yb, diode readings deviate from

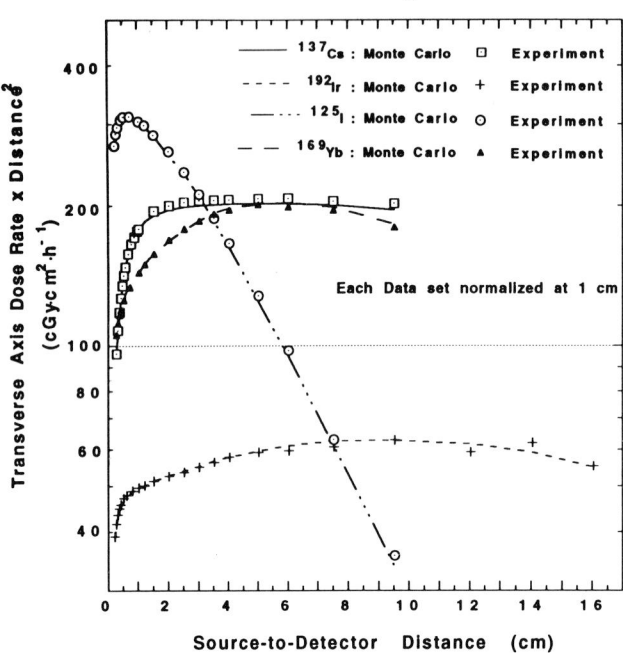

Fig. 12.31. Comparison of Monte Carlo predicated diode response (*lines*) and diode measurements (*symbols*) along the transverse axes of ^{137}Cs, ^{192}Ir, ^{125}I, and [169]Yb sources in homogeneous water. The data are plotted as the product of dose rate as estimated by Monte Carlo from the measured source strength and the square of the distance. For each isotope, the measured and calculated data are normalized to one another at 1 cm distance. [Adapted from WILLIAMSON et al. (1993b) and PERERA et al. (1994)]

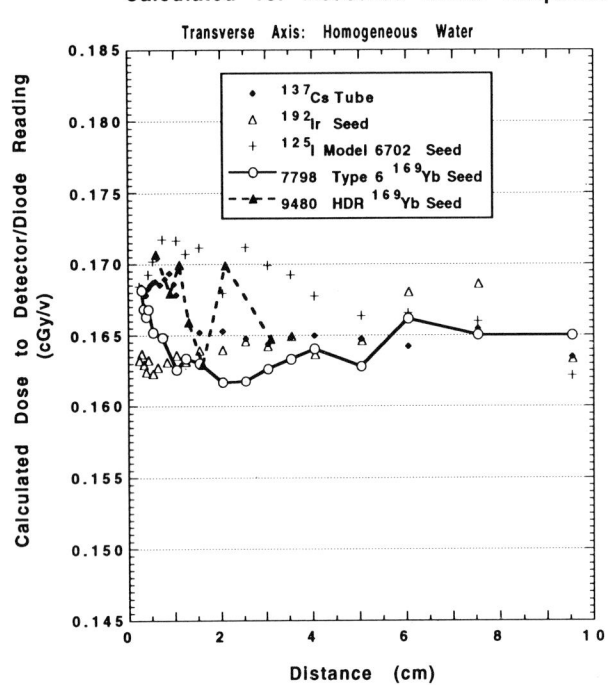

Fig. 12.32. Calculated dose to the active detector volume/measured diode reading as a function of distance along the transverse axes of ^{137}Cs, ^{192}Ir, ^{125}I, and [169]Yb sources in homogeneous water. Dose to the active collection volume of the diode was calculated by Monte Carlo simulation using the measured source strength, decayed to the time of the experiment, as the only directly measured input parameter. [Adapted from WILLIAMSON et al. (1993b) and PERERA et al. (1994)]

Fig. 12.33. Graphical representation of the ratio, simulated diode response per unit dose in water at the measurement point, as predicted by Monte Carlo simulation. The data are plotted as a function of distance along the transverse axes of ^{137}Cs, ^{192}Ir, ^{125}I, and ^{169}Yb sources in homogeneous water. Relative reading/dose is plotted on the *left side* for ^{137}Cs and ^{192}Ir and on the *right side* of the graph for the model 6702 ^{125}I and type 6 ^{169}Yb interstitial seeds. The calculations included the effects of detector attenuation and averaging dose cover its active volume at distance less than 2.2 cm, explaining the sharp fall in relative detector response per unit dose at very small treatment distances. [Adapted from WILLIAMSON et al. (1993b) and PERERA et al. (1994)]

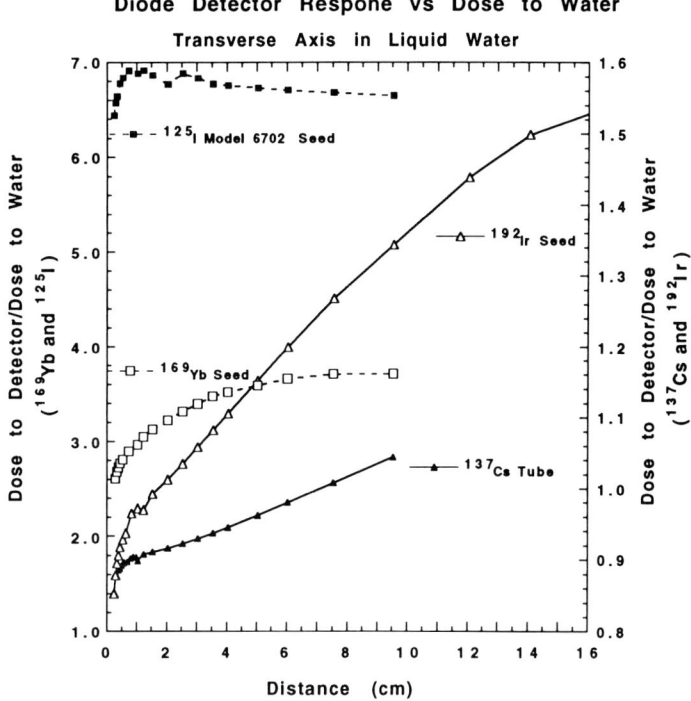

proportionality to dose in water by 14%, 75%, and 40% respectively. Only for ^{125}I is diode response, relative to dose in water, constant throughout the medium.

12.3.4.4 Summary: Empirical and Theoretical Dosimetry in Brachytherapy Dosimetry

Recent dose measurements performed by several independent investigators for interstitial brachytherapy are in remarkably close quantitative agreement with Monte Carlo dose-rate estimates. Reevaluation of basic dosimetry data for ^{125}Ir is clearly indicated, and is the subject of Task Group 43 of the American Association of Physicists in Medicine, which expects to publish its recommendations in 1994. In contrast, recent dose measurements and MCPT calculations about ^{192}I interstitial sources are in close agreement with historically used data.

The excellent agreement between Monte Carlo calculations and TLD and diode measurements enhances confidence both in the reliability and utility of Monte Carlo simulation as a clinical dose-computation tool and in the credibility of direct dose measurements to reproducibly characterize brachytherapy dose distributions. Comparison of these data also clarifies the limitations of both approaches. TLD, due to its relatively flat energy response, results

in accurate and reproducible dose rates in the 1–5 cm distance range. However, agreement between different investigators is poor outside of this range, which may be attributable to positioning errors at short distances and poor signal-to-noise ratio at large distances. At intermediate energies (40–200 keV), MCPT and some experimental evidence indicates that TLD response/dose may vary by as much 10%–15% with distance, indicating the need for sophisticated energy-response corrections, especially in heterogeneous geometries. Silicon diode detectors, due to high sensitivity and a mechanical design that permit precise positioning in liquid media, yield reproducible readings over a wider range of distances (0.2–8 cm). However, outside of the ultra-low-energy photon range, distance-dependent energy-response artifacts limits diode use to verification of theoretical calculations.

Monte Carlo simulation overcomes many of the limitations of purely empirical dosimetry approaches. Errors due to detector displacement and volume-averaging artifacts near sources and signal-to-noise ratio problems far from sources can be limited to a few percent through selection of appropriate estimators and variance-reduction techniques. Monte Carlo simulations are not subject to positioning or energy-response artifacts. However, in contrast to experimental methodologies, meaningful simulation requires precise knowledge of the geometric configuration of sources and applicators. In addition, the influence of relatively small uncertainties (±2%–3%) in low-energy photon cross-sections is amplified at large distances. To calculate absolute dose rates knowing only the source strength, the method of source-strength standardization must be well understood. Any deviation of stated air-kerma strength from its formal definition will introduce errors into the calculated dose rates. The available evidence indicates that when all of these conditions are met, Monte Carlo dose-rate estimates have an accuracy on the order of a few percent.

The optimal approach to brachytherapy dosimetry is probably a combination of empirical and Monte Carlo dose-estimation techniques. Because its accuracy depends on many types of input data, Monte Carlo simulation should not be used as the sole source of clinical dosimetry data. Among the documented "surprises" that a purely theoretical method may fail to anticipate are the contamination of the NIST ^{125}I calibration standard by low-energy photons, described above, and contamination of a prototype [103]Pd seed by high-energy photons due to neutron activation of trace elements (MEIGOONI et al. 1990). Monte Carlo (or any other theoretical) dose calculations about any source or applicator involving previously unverified geometry, photon spectrum, or calibration standards should be experimentally verified. At minimum, the measured benchmark data should include transverse-axis measurements and at least one angular dose profile. A calibrated detector, which allows measurement of absolute dose rates (not just relative doses), is necessary. As the author's own work shows, the detector need not respond linearly to dose in medium: for the purpose of verification, it is sufficient to compare measured and simulated detector responses per unit air-kerma strength. Having verified accuracy of MCPT for the given source type, the 2D dose-rate distribution in water may be calculated.

Experimentally verified MCPT simulation is a powerful clinical dosimetry tool. In addition to providing artifact-free single-source dose-rate distributions, MCPT calculations can be performed prospectively without construction of prototype sources. Thus Monte Carlo simulation can serve as a design tool, helping to identify the mechanical specifications of sources and applicators that give rise to desired dosimetric characteristics. Monte Carlo can also be used to calculate basic dosimetry data which, in practice or in principle, are inaccessible to measurements. Examples include

scatter-to-primary ratios and dose-spread arrays needed as input data to clinical dose-computation algorithms. Characterization of poorly understood phenomena, such as dosimetric influence of applicator shielding and tissue heterogeneities, can be greatly accelerated. finally, an emerging application of MCPT is characterization of dosimeter artifacts and optimization of dose detector design.

12.4 Heterogeneity Corrections in Brachytherapy Dosimetry

In brachytherapy dosimetry, in contrast to external beam, comparatively little work has been done either to empirically characterize dose in heterogeneous geometries or to develop clinically useful heterogeneity-correction algorithms. Commercially available treatment planning programs almost universally ignore heterogeneities such as air–tissue interfaces, local tungsten shielding in vaginal applicators, and tissue-composition variations in ^{125}I seed implants despite the fact that such heterogeneities may perturb dose by as much as 50% (DALE 1983; WILLIAMSON 1990). Fortunately, this topic has received increasing attention over the last 10 years. In this section, the published literature dealing with brachytherapy heterogeneity effects will be briefly reviewed and a number of new approaches to dose calculation in heterogeneous geometries discussed.

The most extensively studied heterogeneity effect is the influence of shielded gynecologic colpostats on ^{137}Cs dose distributions. Both experimental studies (LING and SPIRO 1984; MEERTENS and VAN DER LAARSE 1985; MOHAN et al. 1985; SAYLOR and DILLARD 1976; WEEKS and DENNETT 1990) and Monte Carlo calculations (WILLIAMSON 1990) show that applicator shielding reduces doses by as much as 50% for a single applicator. For typical clinical combinations of applicators, dose computation by superposition, which accounts only for source filtration, overestimates doses at bladder and rectal reference points by 12%–25% (WILLIAMSON 1990; LING and SPIRO 1984). One-dimensional algorithms (VAN DER LAARSE and MEERTENS 1984; WEEKS and DENNETT 1990) have been developed which apply effective pathlength attenuation corrections to those primary photons passing through the high-density applicators or shields. These approaches, reviewed in detail below, ignore shield size and location, which significantly influence the dose distribution, especially for low- and medium-energy sources. An alternative approach, described by MOHAN et al. (1985), uses a 3D relative dose matrix measured about a single applicator, using a silicon diode detector, directly in treatment planning. However, all of these methods have serious limitations. Purely empirical approaches cannot be used to optimize shielding or applicator design without constructing a prototype at each iteration of the design cycle. The 1D computational approaches require extensive comparison with Monte Carlo or measured data before they can yield accurate dose estimates for a given source–heterogeneity combination.

In interstitial implant dosimetry, few publications treat heterogeneity phenomena. MEISBERGER et al. (1968) has shown that implantation of ^{192}Ir seeds near an air–water interface results in dose underestimates of 7%. For ^{137}Cs sources in the presence of 2-cm-thick aluminum, air, and bone slabs, PRASAD et al. (1983) has reported dose correction factors, based upon film dosimetry, ranging from 3% to 8%, suggesting that tissue heterogeneities are probably not significant for higher-energy brachytherapy sources. Both shielded ^{241}Am intracavitary applications (MUENCH and NATH 1992) and shielded ^{125}I episcleral plaques, giving rise to a 8% dose reduction (WEAVER 1986) in the unshielded region, have been described. For ^{241}Am, NATH and GRAY (1987) have shown that lead shielding correction factors are highly dependent upon lateral shielding dimensions, indicating that heterogeneity corrections in this energy range depend significantly on geometric boundary conditions.

Low-energy (≤ 40 keV) sources pose an additional challenge since absorbed dose is highly dependent on the atomic composition of the tissue. DALE (1983), using 1D Monte Carlo calculations, has shown that the specific dose rate constant for ^{125}I in adipose tissue in 40% smaller than in water. Experimentally, HUANG et al. (1990) have demonstrated that the ^{125}I dose-rate constant for breast phantom material is only 76% that of liquid water. For ^{125}I breast implants, such sparing of normal adipose tissue may confer therapeutic benefit (LING and YORKE 1989). In the first published study of bounded tissue heterogeneities near low-energy sources. MEIGOONI and NATH (1992b) showed that a 2.1-cm cylindrical annulus of polystyrene perturbed doses by as much as 125%, 53%, and 10% for ^{103}Pd, ^{125}I, and ^{241}Am sources respectively.

12.4.1 Measured and Calculated Single-Source Heterogeneity Correction Factors

To illustrate the complexity of single-source heterogeneity corrections, and the potential role MCPT calculations can play in characterizing these phenomena, the author will review the measurements and calculations recently published by his group (WILLIAMSON et al. 1993b). Our intent was (a) to systematically study heterogeneity correction factors (HCFs) as a function of thickness, lateral dimensions, composition, and location relative to the point of measurement and (b) to obtain a set of precision benchmarks for assessing the accuracy of MCPT in predicting dose distributions in the presence of bounded heterogeneities. Using a Scanditronix silicon diode detector (electron field type), as described earlier, HCFs in water were measured downstream of cylindrical heterogeneities of lead, steel, titanium, silver, aluminum, and air positioned on the transverse axes of ^{125}I, ^{137}Cs, ^{169}Yb, and ^{192}Ir sources as illustrated by Fig. 12.34. Cylindrical heterogeneities were positioned with their axes aligned with the transverse source bisector and their centers positioned 15 mm from the source center. Manual diode readings were obtained at 5–15 points along the transverse source axis spanning the source center-to-detector face distances of 1.8–16 cm. Generally, for each source–material combination, measurements were made for four different geometries: small- and large-diameter (6.3-mm and 19-mm) cylinders of two thicknesses (15%–25% or 35%–60% transmission) each. To quantify the effects of heterogeneities, the measured HCF was defined as:

$$\text{HCF}_{\text{meas}} = \frac{\text{diode reading/s with heterogeneity}}{\text{diode reading/s in homogeneous water}}$$

at the same point in space.   (12.14)

Each of the measurements was simulated using MCPT calculations, which took into account the 3D geometric structure of the source, measurement phantom, heterogeneity, and detector. For each measured HCF_{meas}, the corresponding theoretical ratio, HCF_{det}, defined as the ratio of simulated detector readings, was calculated. The corresponding HCF_{wat}, defined as the ratio of MCPT dose rates in water, was also calculated. Figures 12.35 and 12.36 compare a sample of the nearly 1200 manually measured HCFs to their MCPT-simulated counterparts. Agreement was excellent, rarely exceeding 5%,

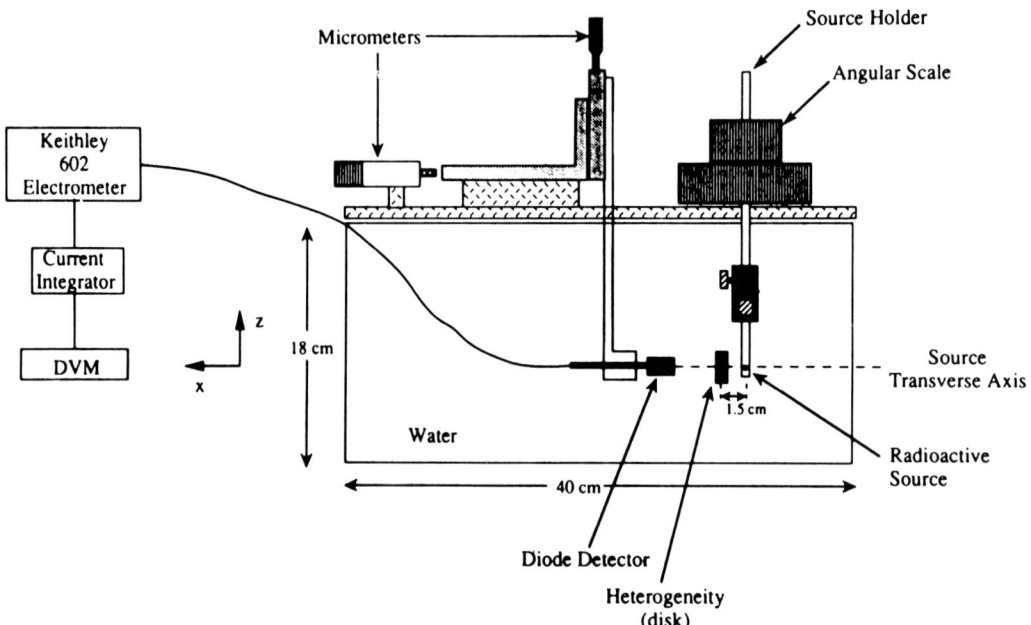

Fig. 12.34. Illustration of the micrometer-driven source positioning jig and associated water phantom used to measure silicon diode response as function of distance along the transverse axis of a brachytherapy source. The source holder is mounted on a micrometer-driven turntable allowing angular dose profiles to be measured. Micrometers allow positioning of the detector holder along the axis parallel to the source bisector and along its axis of symmetry. In addition, the source holder has fine adjustment screws allowing the source to be moved in the x-y plane to ensure that the longitudinal source axis intersects and is bisected by the axis of turntable rotation. Source alignment is observed through a magnifying periscope to reduce personnel exposure. Not illustrated is the lead-brick vault surrounding the apparatus. (From WILLIAMSON et al. 1993b)

and generally was in the 1%–3% range. Several conclusions can be drawn:

1. Heterogeneity effects can be quite large. A 1-cm-thick air void near a ^{125}I seed increases dose downstream by 21%–37%. The magnitude of the HCF is often significantly smaller (by 50%–90%) than predicted by attenuation of the primary photons, emphasizing the importance of scattered radiation to these phenomena.
2. Even for relatively high-energy sources such as ^{192}Ir, measured· HCFs vary significantly, with heterogeneity in diameter (up to 50% variation) and measurement distance (up to a factor of 3). For ^{125}I, HCF variations with distance and diameter exceed the importance of thickness as a variable, with the larger thin disk conferring more protection than the small thick disk. In general, HCF behavior is complex, involving an interplay between primary photons transmitted by the

barrier, scattered photons diffusing around the barrier, and scattered photons originating in the barrier. For clinical source arrangements that approximate a single-source geometry, dose calculation in the presence of bounded high-density heterogeneities is intrinsically a three-dimensional problem that cannot be successfully attacked by simple 1D algorithms.

3. Comparison of measured HCFs with those calculated by MCPT reveals excellent agreement, on the order of 1%–3% for distances up to 7.5 cm, at which point the precision of the experimental readings begins to deteriorate. The mean of percentage deviations of theory from measurement is +1.1%, –0.6%, and –1.1% for ^{125}I, ^{137}Cs, and ^{192}Ir sources respectively. Our experience suggests that Monte Carlo simulation is a powerful, convenient, and accurate tool for investigating this long-neglected area.
4. Comparison of the Monte Carlo ratios, HCF_{meas} and HCF_{wat}, demonstrates that ratios of silicon diode readings can significantly over- or underestimate the "true" HCF, defined as the ratio of doses in water. For ^{192}Ir and ^{169}Yb, errors as large as –35% and +30% result. Even for ^{137}Cs, lead-cylinder HCFs are overestimated by 5%–15%. This is disturbing since most recent measurements about shielded ^{137}Cs-bearing colpostats have utilized Scanditronix diodes (WEEKS and DENNETT 1990; MEERTENS and VAN DER LAARSE 1985). Only for ^{125}I does diode appear to be a suitable detector for characterizing heterogeneity corrections.

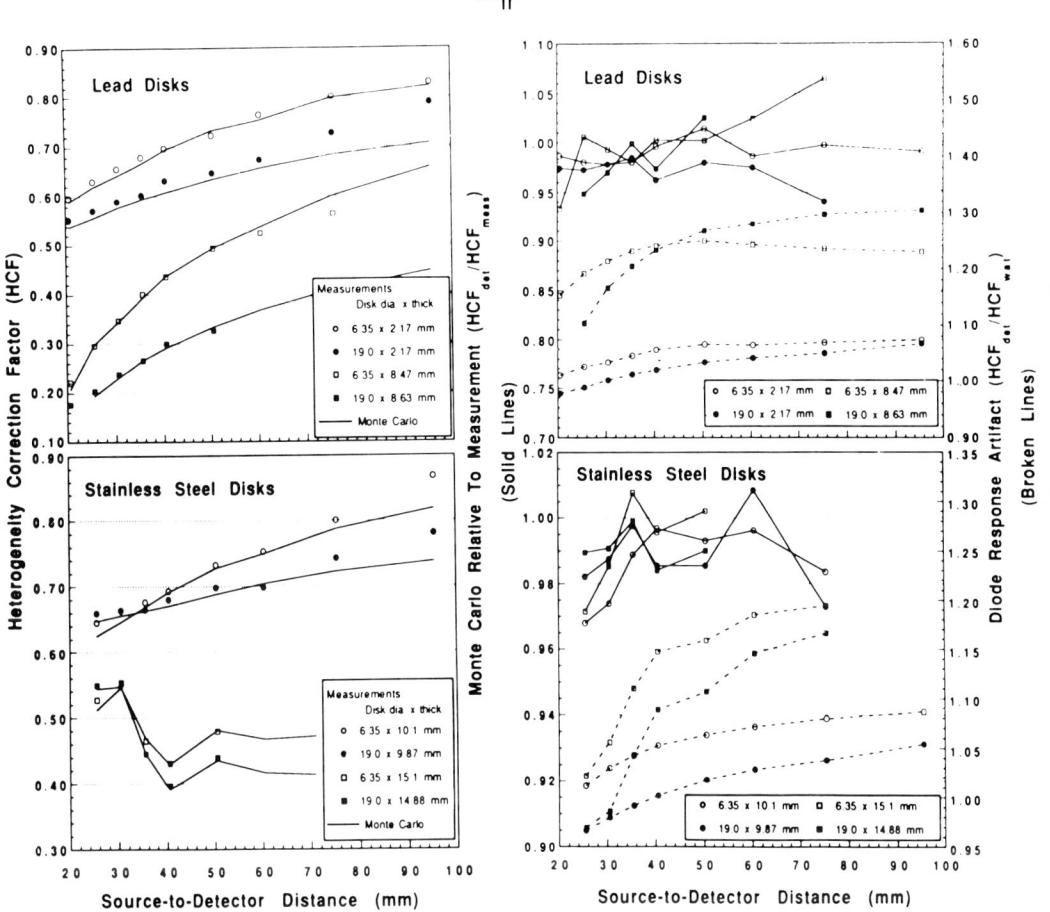

Fig. 12.35. Comparison of HCFs, as measured with a Scanditronix silicon diode detector to Monte Carlo calculations for a steel-clad ^{192}Ir interstitial seed in the presence of disk-shaped lead and steel heterogeneities. For Figs. 12.35 and 12.36, each horizontal panel, consisting of two graphs, describes HCFs for a single combination of source and shielding material. The *left panels* plot HCF_{meas}, consisting of ratios of measured diode readings (*symbols*), along with its Monte Carlo counterpart, HCF_{det} (ratio of simulated diode responses), represented as *solid lines*. Each *right panel* shows the deviation of calculated HCF det from measured HCF_{meas} (*solid lines* using left-hand scale). Also plotted on each right-hand graph (*broken lines*, right-hand scale) is the error in the simulated diode response ratio, HCF_{det}, relative to the "true" HCF_{wat}, calculated by Monte Carlo as the ratio of point doses to water. The *solid* and *open symbols* denote large- and small-diameter heterogeneities while *circles* and *squares* denote thin and thick heterogeneities respectively. (From WILLIAMSON et al. 1993b)

Many important questions remain to be studied, including (a) the effects of bounded tissue-composition heterogeneities on low-energy seed dosimetry and (b) the conditions under which the complex behavior of single-source heterogeneities is ameliorated by the presence of multiple sources in clinical implants.

12.4.2 *"Practical" Dose-Calculation Algorithms for Heterogeneous Geometries*

Accurate but practical dose-calculation algorithms are needed both to facilitate applicator design for low-energy isotopes and to extract meaningful normal-tissue and tumor-control dose response data from image-based evaluation of patients treated with intracavitary therapy. In addition, accurate dosimetric treatment of applicator shielding and tissue composition variations is a prerequisite to incorporating physical optimization into clinical brachytherapy. A useful dose calculation algorithm must be fast enough to allow evaluation of absorbed dose throughout large 3D dose matrices with reasonable turnaround time, must be accurate, and must be sufficiently general to handle the range of source arrangements, applicator geometries, and tissue heterogeneities likely to be encountered in the targeted clinical applications. Ideally, such an algorithm should be applicable to both high- and low-energy radionuclides, should require only a limited base of well-defined physical data, and

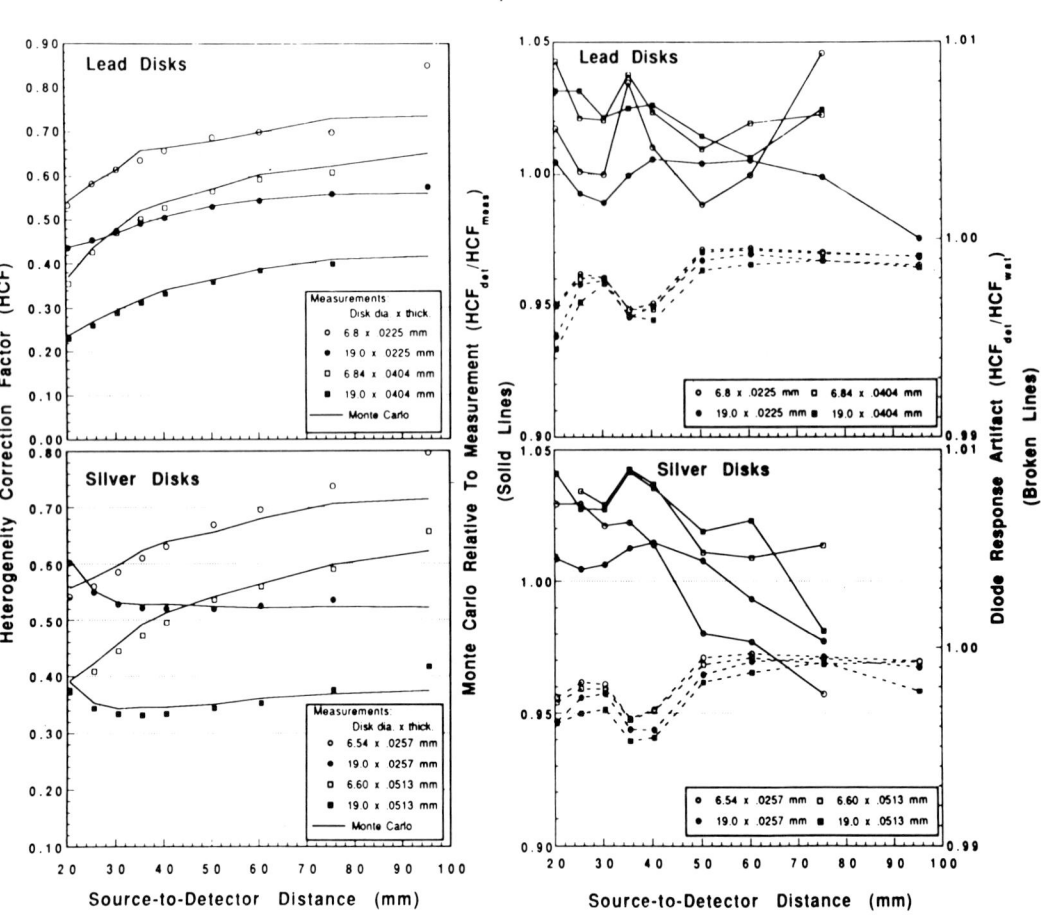

Fig. 12.36. Same in Fig. 12.35 for a ^{125}I model 6702 seed in the presence of lead-and sliver-foil shields (From WILLIAMSON et al. 1993b)

should be fully three-dimensional. Monte Carlo simulation, although highly general and accurate, is too CPU-intensive to support the volume of dose calculations required by clinical treatment planning given the computing resources currently available for this task. This has stimulated a revival in brachytherapy algorithm development with the goal of striking a more practical compromise between accuracy and speed.

12.4.2.1 One-Dimensional Primary-Photon Pathlength Algorithms

One-dimensional (1D) heterogeneity corrections depend only on to thickness, t_h, of heterogeneity traversed by primary photons. These models are fundamentally generalizations of the Sievert integral, which is widely used to calculate dose distributions about sources that can be approximated by radioactivity uniformly distributed along a line segment which is encapsulated in a cylindrical filter, usually stainless steel or platinum. In the case of an isotropic point source, the essential feature of the 1D primary-photon pathlength approach can be expressed simply:

$$\dot{D}_{\text{het}}(\mathbf{r}) = \dot{D}_{\text{hom}}(\mathbf{r}) \cdot e^{-\mu_h \cdot t_h}, \qquad (12.15)$$

where $\dot{D}_{\text{hom}}$ and $\dot{D}_{\text{het}}(\mathbf{r})$ denote the dose rate at the point of interest in homogeneous medium and in the presence of the heterogeneity, respectively. The thickness of heterogeneous material traversed by primary photons traveling from the source to the point of interest, $\mathbf{r}$, is denoted by t_h, and μ_h is the effective linear attenuation coefficient of the heterogeneous medium. Since the exponential factor corrects both the primary and scattered-photon dose contributions, μ_h is an empirical parameter that can deviate significantly from either the linear attenuation or the energy absorption coefficient. 1D pathlength algorithms are very fast since ray tracing need only be applied between each source element and each point of interest. The 1D pathlength algorithm and its empirical multidimensional generalizations are the only clinically usable computational models for heterogeneous geometries currently available. So far, development and validation of this approach has been limited to high-density, high-atomic-number heterogeneities, such as internal shielding found in gynecologic colpostats.

A general form of the 1D pathlength model, applicable to geometrically complex shielded source geometries, is illustrated by Fig. 12.37. The problem geometry consists of a brachytherapy source with an active core source of radius s, encapsulated in a metal filter of radial thickness t, in the presence of a high-density metal internal shield. The active source is decomposed into N_i small elements each of volume ΔV_i and located at $\mathbf{r}_1$. Then, just as in the Sievert integral, inverse-square law, tissue-attenuation, and tissue-buildup corrections, and effective attenuation corrections, are separately applied to each source element, ΔV_i. The N_i dose-rate contributions are then summed, giving an estimate of the dose rate at the point of interest, $\mathbf{r}$. The dose rate is given by

$$\dot{D}_{\mathrm{het}}(\mathbf{r}) = S_{\mathrm{K}} \cdot (\mu_{\mathrm{en}}/\rho)_{\mathrm{air}}^{\mathrm{wat}} \cdot \frac{\displaystyle\sum_{i=1}^{N_S} \left[\Delta V_i \cdot B(\lambda_{i3}) \cdot (\mathbf{r} - \mathbf{r}_i')^{-2} \cdot \exp\left(-\sum_{j=1}^{4} \mu_j \cdot \lambda_{ij} \right) \right]}{|\mathbf{r}_c|^2 \cdot \displaystyle\sum_{i=1}^{N_s} \left[\Delta V_i \cdot (\mathbf{r}_c - \mathbf{r}_i')^{-2} \cdot \exp(-\mu_1 \cdot s - \mu_2 \cdot t) \right]}, \qquad (12.16)$$

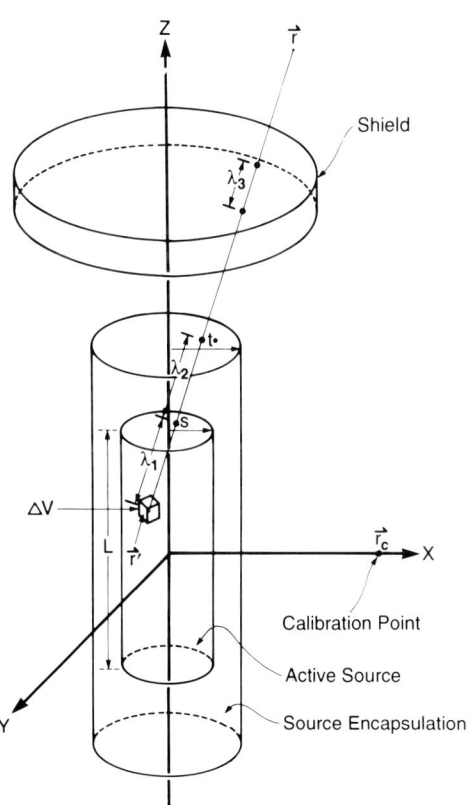

Fig. 12.37. Diagram illustrating the geometry of the 1D primary-photon pathlength dose-calculation model for calculating absorbed dose about geometrically complex sources and shielded applicators. The diagram illustrates an intracavitary source consisting of uniformly distributed radioactivity in a cylindrical region of volume V, length L, and radius s which is concentrically and symmetrically placed in a larger cylinder of radius t which denotes the source capsule. Also pictured is a disk-shaped shield located above the source. dV' denotes a differential volume element located at r'. (Adapted from WILLIAMSON 1988a)

where the indices $j = 1,...,4$ denote the media composing the active source core, the source encapsulation, the surrounding water, and the internal shielding, respectively. The variables $\lambda_{i1},.....\lambda_{i4}$ denote the distances traversed by primary photons through each of these four media as they travel from the source element ΔV_i at $\mathbf{r}_i$ to the point of interest, $\mathbf{r}$. The other symbols in Eq. 12.16 are defined as follows:

S_{K} = strength of the source in terms of air-kerma strength in units of $\mu\mathrm{Gy} \cdot \mathrm{m}^2 \cdot \mathrm{h}^{-1}$ where 1 $\mu\mathrm{Gy} \cdot \mathrm{m}^2 \cdot \mathrm{h}^{-1} = 1$ $\mathrm{cGy} \cdot \mathrm{m}^2 \cdot \mathrm{h}^{-1} = 1$ U. The point on the transverse source axis (usually 1 m from the source) where the air-kerma rate in free space needed to define S_{K}, is denoted by $\mathbf{r}_c$.

$(\mu_{\mathrm{en}}/\rho)_{\mathrm{air}}^{\mathrm{wat}}$ = the mean ratio of mass-energy absorption coefficients, averaged over the photon spectrum in free space with respect to air kerma, for water to that of air.

$B(d)$ = total dose in water/primary dose in water at distance d from an isotropic point source, i.e., the buildup factor for water.

μ_j = the effective attenuation coefficient for active source, filter, and shielding material for $j = 1, 2,$ and 4 respectively. For the special case of water ($j = 3$), μ_j denotes the narrow-beam attenuation coefficient, as required by the definition of buildup factor.

A 3D ray-tracing subroutine library is required to calculate the sequence of pathlengths, $\lambda_{i1},......\lambda_{ij}$ for each pair of points $(\mathbf{r}_i, \mathbf{r})$ from a model of the source-shielding geometry. Mathematically, λ_{ij} for a given $\mathbf{r}$ can be represented by:

$$\lambda_{ij} = \int_0^{|\mathbf{r}_i - \mathbf{r}|} \pi_j \left(\mathbf{r}_i + \frac{\mathbf{r} - \mathbf{r}_i}{|\mathbf{r} - \mathbf{r}_i|} \cdot s \right) \cdot ds, \qquad (12.17)$$

where $\pi_j(\mathbf{r})$ = if $\mathbf{r}$ lies in medium j and $\pi_j(\mathbf{r})=0$ if it does not.

$\pi_j(\mathbf{r})$ is the point classification function, which simply classifies an arbitrary point, $\mathbf{r}$, as to whether it is inside or outside the union of regions composed of medium j. The sums in the numerator and denominator of Eq. 12.16 are proportional to the dose rate in water and air-kerma strength, respectively, *per unit activity contained in the source.* The denominator is required to renormalize the numerator (dose rate in water) to the quantity air-kerma strength, which is the output produced by the *filtered* source. Omitting this correction will result in a "double" oblique filtration correction. It is obvious that the above model could easily be generalized to more than four media and to extended sources of arbitrary shape.

The 1D primary pathlength approach was introduced independently by VAN DER LAARSE and MEERTENS (1984, 1985) and WEEKS and DENNETT (1990) for ^{137}Cs sources positioned in vaginal colpostats containing tungsten-alloy shielding, although not in the general form just described. Weeks used a small ion chamber to measure the dose downstream of a sample of each material in the problem (tungsten, steel, styrofoam, lucite, and nylon) with cross-sectional dimensions of 1×5 cm and known thicknesses. By normalizing each measurement to the reading measured in homogeneous water at the same point, he was able to calculate the effective "water replacement" attenuation coefficient, μ_j, for each material in this representative geometry. Weeks verified his model by comparing its predictions to extensive diode measurements made around a plastic colpostat containing tungsten-alloy shields, which were converted to absolute dose rates by normalizing homogeneous-medium readings to corresponding calculated dose rates. He found excellent agreement (3%) between the measured and calculated results. The model of Meertens and Van der Laarse is very similar except that the effective attenuation coefficient, is treated as a parameter of best fit. These investigators used a small ion chamber to map the relative dose distribution in several planes arising from a Selectron shielded colpostat containing a sequence of four spherical ^{137}Cs sources. Since the distance between each measurement plane and applicator center could not be specified exactly, they varied both the distance and μ_h until the optimal fit between measured and calculated isotransmission lines was achieved. They claimed an accuracy of 4% away from the shield edges and 10% or 3 mm near the boundaries of the region shadowed by the shields. Interestingly, Meertens and Van der Laarse's value of μ_h (0.10 mm^{-1}) for tungsten was reasonably close to that of Weeks and Dennett (0.12 mm^{-1}). In comparison, the linear attenuation and energy absorption coefficients take the values 0.170 and 0.091 mm^{-1}, respectively, assuming that the tungsten alloy has a density of 17 g/cm^3. Figure 12.38 shows the isotransmission lines (dose with shields/dose without shields) calculated by Meertens and Van der Laarse for a typical clinical loading consisting of two colpostats and an intrauterine tandem.

Although one could argue that the dose measurement techniques used to validate the 1D primary-photon pathlength models are no more accurate than 10%, the model probably has an accuracy better than 5% for multiple-applicator combinations typically used in clinical practice. Under these conditions, dose sparing due to internal shields is limited to approximately 25%. This level of accuracy is quite remarkable in view of the fact that the 1D pathlength model completely neglects the influence of shield location and cross-sectional area on the resultant HCF. For a ^{137}Cs point source, Fig. 12.39 shows the effect of tungsten shield diameter on the dose downstream of the heterogeneity. This Figure shows that up to distances of 5 cm the variation in HCF with respect to distance is 17% and with respect to shield diameter is about 20%. Thus for points of interest less than 2 cm downstream of the shield (the region of most clinical relevance), the 3D scattering effects neglected by the 1D pathlength model probably introduce errors on the order of ±5%, especially when large fractions of the dose are delivered by unshielded sources. Although one should be cautious in applying this model to unidirectional applicators, e.g., segmentally shielded vaginal cylinders, this simple and reasonably accurate approach can be recommended to those interested in more accurate dose calculation around intracavitary implants utilizing ^{137}Cs.

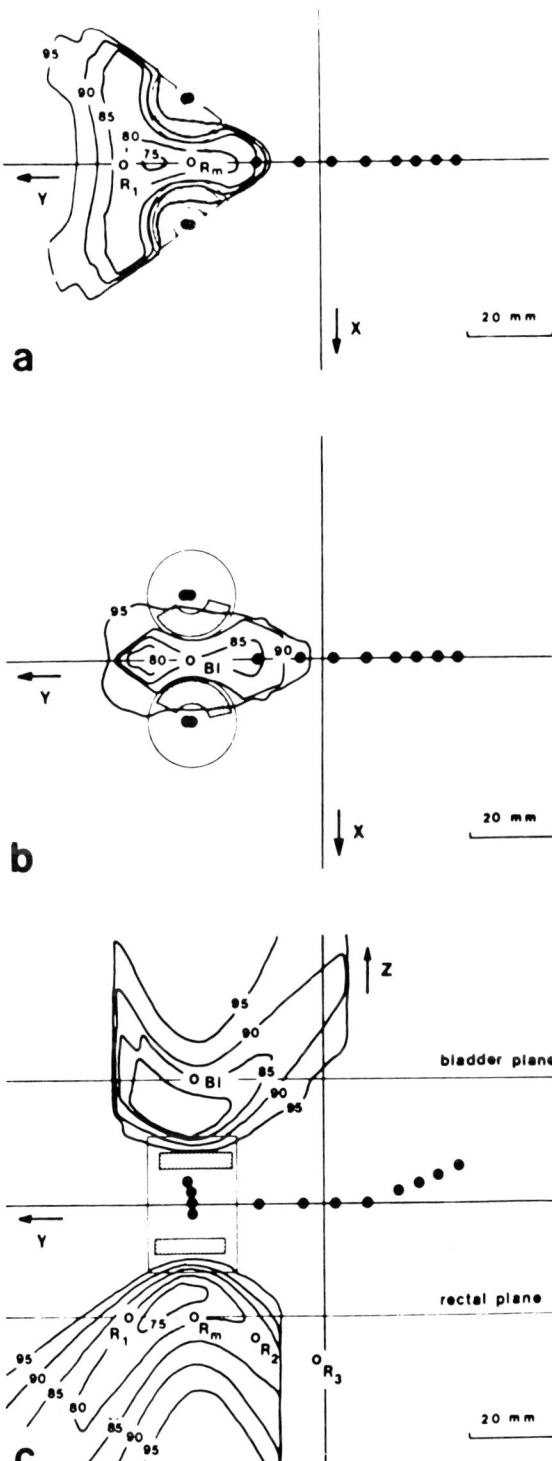

a

20 mm

b

20 mm

bladder plane

rectal plane

c

20 mm

Generalizations of 1D Primary-Photon Pathlength Algorithms

Figures 12.35 and 12.36 indicate that for photon energies at and below that of ^{192}Ir, shielding correction factors become more dependent on shield dimensions and locations. For unidirectional ^{192}Ir applicators, variations as large as 45% with respect to shield diameter and distance from the shield penumbra can be expected. It is unlikely that the 1D pathlength approach is applicable to lower energy sources such as ^{192}Ir and [169]Yb, which have larger scatter-to-primary ratios, and photon scattering interactions that more closely approximate elastic collisions. This has stimulated development of more sophisticated and accurate algorithms. In an effort both to retain the efficiency characteristic of primary-photon ray tracing and to improve upon the accuracy of the basic 1D pathlength model, some investigators have turned to empirical modeling of the scatter-dose distribution. These approaches (a) explicitly separate the primary and scattered-photon dose components and (b) apply empirical corrections to scatter-dose component which have a relatively complex dependency on the 3D geometry of interest point, source, and heterogeneity. Thus, the complex 3D behavior of the often-dominant scatter-dose distribution is partially reintroduced into the 1D pathlength model.

The simplest example of an empirical 3D scatter-correction model is the scatter-separation method developed by the author (WILLIAMSON 1990). This method evolved from the observation that, despite factor-of-two variations in primary dose around shielded vaginal applicators containing ^{137}Cs and [226]Ra sources, the scatter-dose component is approximately isotropically distributed (±20%). This suggested that scatter dose can be treated as a distance-dependent but angle-independent term, obviating the need to calculate scatter dose at each point by Monte Carlo. Most of the residual anisotropy was eliminated by observing that the perturbation of the scatter dose by the applicator at each point was linearly related to the additional primary-photon attenuation contributed by the applicator, expressed in mean-free paths (MFPs) (Fig. 12.40).

Fig. 12.38a–c. Isotransmission curves produced by a typical clinical loading of a pair of shielded fletcher-Suit colpostats (four spherical ^{137}Cs sources each) and an intrauterine tandem (eight sources) manufactured by Nucletron corporation for the Selectron remote-afterloading device. The data are calculated by the 1D path-length model developed by MEERTENS and VAN DER LAARSE (1985). The internal shields consists of 3.5-mm-thick tungsten. Isotransmission curves are shown for **a** the rectal plane (coronal plane 25 mm posterior to the tandem), **b** the bladder plane (coronal plane 27 mm anterior to the tandem), and **c** a sagittal plane containing the tandem. (From MEERTENS (and VAN DER LAARSE 1985)

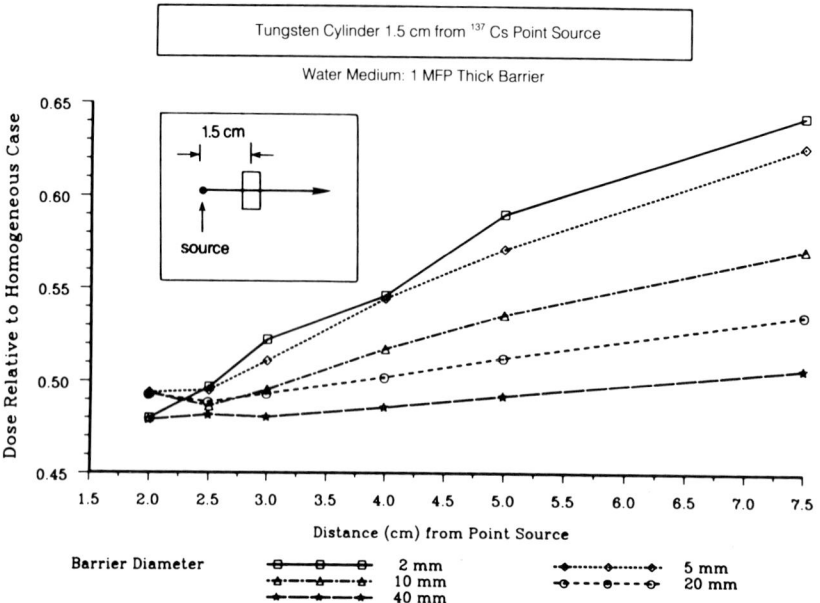

Fig. 12.39. Transmission ratios along the axis of a one mean-free path (5.4-mm) cylindrical tungsten shield located 1.5 cm from a ^{137}Cs point source in water. The data were calculated by Monte Carlo simulation for shields ranging from 2 to 40 mm in diameter. The *left scale* denotes dose in water with the shield in place relative to dose in water at the same point in homogeneous medium. (From WILLIAMSON 1990)

Basic data for the model were derived from Monte Carlo simulations: scattered-photon dose rates at 30 polar angles with respect to the longitudinal source axis at distances of 1.2, 1.5, 2.0, 3.0, and 5.0 cm in the transverse plane of each source–applicator combination. These simulations were based upon realistic geometric models of the source internal structure, colpostat body, source restraining mechanism, and bladder and rectal shields, involving as many as 55 different geometric shapes and seven media. In addition, scatter dose rates per unit air-kerma strength for the filtered source alone, $\dot{D}_{s,s}(\mathbf{r})$ were calculated at the same points. By applying curve fitting to this limited base of data, the ratio of scatter with applicator to scatter from source alone, $\alpha(r,\mathrm{MFP})=\dot{D}_{s,a}(\mathbf{r})/\dot{D}_{s,s}(\mathbf{r})$, was reduced to a coarse 2D table depending only on distance from the applicator center and applicator attenuation. Then the total dose rate, $\dot{D}_a(\mathbf{r})$ at any point $\mathbf{r}$ in the presence of a shielded applicator is given by:

$$\dot{D}_a(\mathbf{r}) = S_k \cdot [\dot{D}_{p,a}(\mathbf{r}) + \dot{D}_{s,s}(|\mathbf{r}|,\cos\theta) \cdot \alpha(|\mathbf{r}|, \mathrm{MFP})], \tag{12.18}$$

where $D_{p,a}(\bar{r})$ and $D_{p,s}(\bar{r})$ denote the primary dose rate per U with the shielded applicator and for the

filtered source alone. The primary-photon attenuation due to the applicator, MFP (**r**) is given by

$$\mathrm{MFP}(\mathbf{r}) = -\ln\left[\frac{\dot{D}_{p,a}(\mathbf{r})}{\dot{D}_{p,s}(|\mathbf{r}|,\cos\theta)}\right]. \tag{12.19}$$

Thus calculation of scatter dose at any point around an applicator requires exact calculation of the primary dose, three 2D table lookups from precalculated arrays: scatter dose, $\dot{D}_{s,s}(r,\cos\theta)$ and primary dose, $\dot{D}_{p,s}(r,\cos\theta)$, from the source alone and the scatter perturbation correction, $\alpha(r, \mathrm{MFP})$. For three widely used applicator-source combinations (3M applicator with ^{137}Cs and rectangular-handled fletcher- Suit applicator with ^{137}Cs and [226]Ra), dose-calculation accuracy and computational efficiency, both relative to Monte Carlo simulation, were found to be 3% and 15 000, respectively. Although computational complexity is similar to that of the simple primary ray-path correction models, scatter separation explicitly corrects for distance dependence of heterogeneity corrections and anisotropy of the scattered- and primary-photon dose profiles arising from the source alone. Our model ignores the variation in HCF with shiled area at a fixed distance.

A different approach to empirical scatter correction has been developed by NATH and Co-workers (NATH and GRAY 1987; NATH et al. 1990a) in an effort to model the dose distribution around vaginal plaques containing multiple [241]Am sources. Nath's basic model is very similar to the generic 1D primary ray-tracing model described by Eq. 12.16, except

Fig. 12.40. Scatter-dose ratio, α (scatter dose with colpostat/ scatter dose from encapsulated source only), plotted as a function of average primary photon attenuation expressed in units of mean-free path, contributed by the applicator. The *symbols* denote the Monte Carlo data for rectangular handled fletcher-Suit colpostats or 3M fletcher-suit-Delclos colpostats and the *lines* indicate the piecewise linear fit to the data. (From WILLIAMSON 1990)

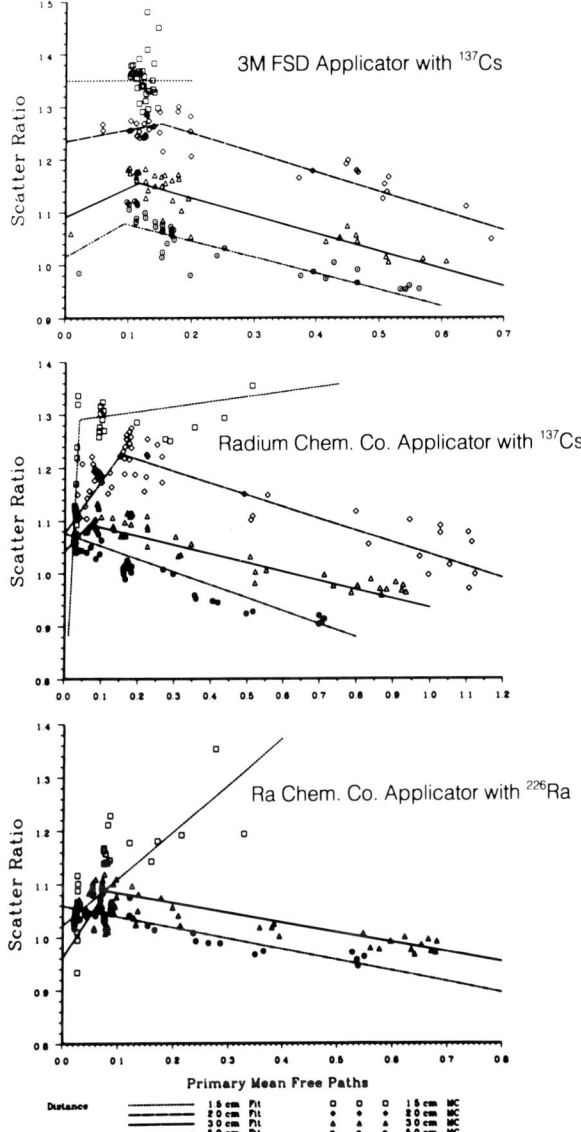

that his geometric model includes as many as five 1-cm-diameter cylindrical sources (Fig. 12.41). However, the buildup factor, B, for water medium is replaced by an anisotropic buildup factor, B_{anis}:

$$B_{anis} = 1+(B_{iso} - 1)\cdot a(\theta)^{n}, \qquad (12.20)$$

where B_{iso} is the isotropic point-source buildup factor and θ is the angle between the longitudinal axis of the source currently being calculated and the point of interest, $\mathbf{r}$. The function $a(\theta)$ corrects for the anisotropic distribution of scatter around each single source and has the form $a+(1-a)\cdot\cos\theta$ for a cylindrical source where a is an empirical value

varying from 0.75 to 0.85. The integer n denotes the number of sources in the plaque and corrects for global reduction of the multiply scattered photon dose-rate component due to the presence of multiple, bulky extended sources. The generally good agreement between the model predictions and dose rates measured by thermoluminescent dosimetry is illustrated in Fig. 12.42 for a 5-4-5 Ci vaginal applicator. The RMS average deviation of the model predictions from the measurements varied from 7% to 13% although relative errors at individual points as large as 35% were reported. That the average errors are somewhat larger than those of the 1D

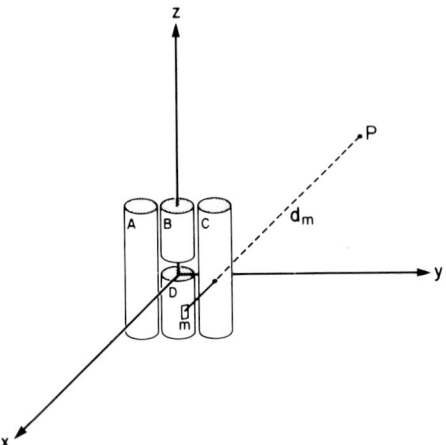

Fig. 12.41. Schematic diagram, showing the 5-4-5 Ci ^{241}Am vaginal plaque, illustrating the geometry used by Nath's generalized 1D pathlength model for calculation of dose rate near multiple, bulky source arrays. The angle, θ, used to correct for anisotropic distribution of scatter dose, is that between the z axis and the line segment PE. Integration over surface-area elements, E, is used to obtain line-of-sight primary- and scattered-photon source-to-source shielding corrections. (From NATH et al. 1990a)

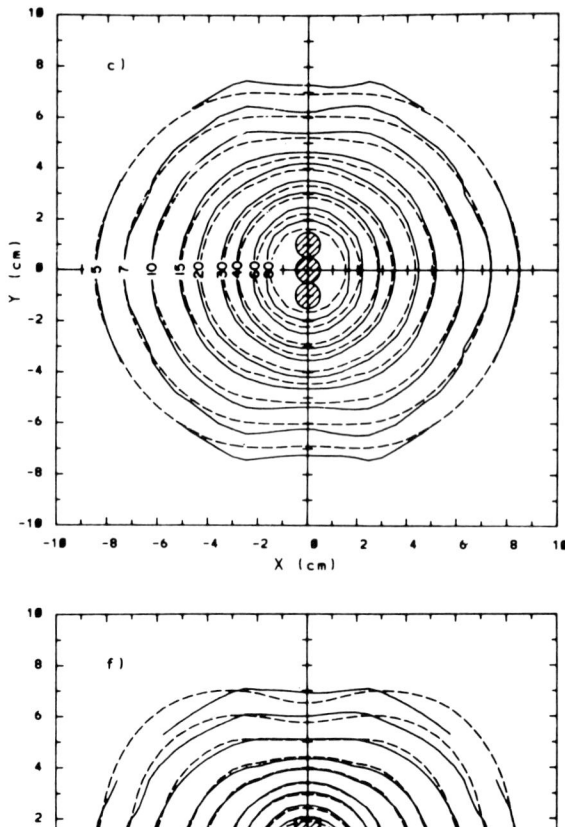

Fig. 12.42. Measured (*solid lines*) and calculated (*broken lines*) isodose-rate curves (cGy/h) produced by a 5- 4-5 Ci suit ^{241}Am vaginal applicator. The *top panel* shows the dose distribution in a plane through the center and perpendicular to the source axes and the *bottom panel* represents the dose distribution in a plane perpendicular to the plaque through the center of the central source. The dose calculations, based on Nath's model, include corrections for both line-of-sight corrections and corrections for global reduction in scatter dose. (From NATH et al. 1990a)

primary-photon pathlength model is a reflection of how much more difficult algorithm design is for 60-keV photons than for ^{137}Cs gamma rays. Of interest is Nath's conclusion (NATH et al. 1990b) that multiple-source arrays give rise to nearly factor-of-two source-to-source or "line-of-sight" shielding corrections and additional "indirect" dose correction factors averaging 19% due to reduction of multiply scattered photon fluence. Use of partial- or full-transmission lead-foil shields further complicates the problem, giving rise to a dose distribution that is highly dependent on cross-sectional shield shape as well as thickness (MUENCH and NATH 1992). A major lesson of the Yale experience is that simple, Sievert-like models, based on the universally used dose-superposition principle, do not model dose distributions with adequate accuracy in this energy range.

Despite the advantage of computational simplicity, empirical generalizations of the 1D pathlength model have several shortcomings. First, dose-calculation accuracy for low energy (<400 keV) appears to be limited to 10% for typical multiple-source arrays characteristic of clinical practice. In addition, these models require extensive validation against measured or Monte Carlo dose rates in order to define the empirical scatter-dose corrections. This laborious process must be repeated for each applicator-source combination. For multiple source arrays of new low-energy sources such as ^{241}Am, each type of shielding geometry requires empirical

validation. Clearly, to fully exploit the potential of customized shielding of low-energy source implants, a more general and accurate dose-calculation algorithm is necessary. An important potential application of the 1D algorithms is the computational dosimetry of ^{192}Ir high-dose-rate intracavitary therapy, for which no validated dose-calculation algorithm has been reported to date.

12.4.3 Emerging Developments in Dose Calculation: Explicitly 3D Algorithms

Our review of the brachytherapy dosimetry literature suggests that in many clinical applications of low-energy sources, intersource shielding, local shielding, and possibly tissue heterogeneities give rise to large dose perturbations that have a complex dependence upon the geometric distribution of sources and applicators, and upon the lateral dimensions and location of shielding media. An algorithm which is efficient, is fully three-dimensional, is derived from principles of radiation transport, and supports accurate prospective dose calculation over a wide range of heterogeneous geometries would significantly enhance our capability to optimize dose distributions in brachytherapy. Unfortunately, no such algorithm exists in clinically usable form at the time of writing. In fact, the only published contributions in the area of explicitly 3D dose calculation are those of the present author. His group has developed two 3D algorithms: 3D scatter convolution (WILLIAMSON et al. 1991) and scatter subtraction (WILLIAMSON et al. 1993a). Our current implementations of these approaches are limited to point sources in the presence of geometrically simple heterogeneities and demonstrate, at best, proof of underlying principles. However, these studies demonstrate that (a) a high level of accuracy is potentially achievable without resorting to Monte Carlo simulation, (b) the tradeoff between accuracy and numerical complexity is a steep one, and (c) there at least two promising pathways for future work in this area.

12.4.3.1 3D Convolution Algorithm for Brachytherapy Dosimetry

The convolution method was first proposed for megavoltage photon-beam dose calculations (BOYER and MOK 1985; MACKIE et al. 1985; MOHAN et al. 1986) in heterogeneous media and was successfully adapted to brachytherapy dose calculation by the author's group (WILLIAMSON et al. 1991). The algorithm requires a limited base of well-defined physical data, is fully three-dimensional, and predicts absolute dose rates as well as relative correction factors. It takes into account not only the thickness of heterogeneities, but also their composi-

tion, size, and location relative to the source and point of interest. Our prototype brachytherapy convolution is accurate within 3% for both low-(^{125}I) and high-energy (^{137}Cs) sources in the presence of shields, air voids, and air-tissue boundaries. The basic principle of the convolution method is illustrated by Fig. 12.43, which shows a point source near an irregularly shaped heterogeneity.

The 3D space surrounding the source is partitioned into small scattering voxels of volume dV'. The primary energy-fluence rate, per unit air-kerma strength, $\Psi(\mathbf{r}')$ calculated at the center $\mathbf{r}'$ of each voxel, including the attenuating effects of heterogeneities. The scatter-dose contribution to the point of interest, $\mathbf{r}'$, from each voxel, $\mathbf{r}'$, is estimated from the scatter-dose kernel, $K(\mathbf{r}-\mathbf{r}')$. The process of adding together these differential scattered-photon dose contributions amounts to convolving energy fluence density, $\Psi(\mathbf{r}')\cdot\mu(\mathbf{r}')$ against the scatter-dose kernel $K(\mathbf{r}-\mathbf{r}')$. Thus, the total dose rate, $\dot{D}(\mathbf{r})$, at point $(\mathbf{r})$ is

$$\dot{D}(\mathbf{r}) = S_K \cdot [\dot{D}_p(\mathbf{r}) + \int_v \Psi(\mathbf{r}') \cdot \mu(\mathbf{r}') \\ \cdot K(|\mathbf{r}-\mathbf{r}|, \theta') \cdot dV'].\tag{12.21}$$

The basic data required by the algorithm are the "dose-spread array" or scatter-dose kernel, $K(t,\theta)$, which gives the distribution of scatter dose arising from monodirectional primary photons going into first collision at $t=0$. These data were precalculated by Monte Carlo simulation, illustrating utility of this method in calculating basic treatment planning data that are inaccessible to direct measurement. Isodose curves of ^{125}I and ^{137}Cs dose spread arrays are shown in Fig. 12.44. Our prototype convolution code used 3D numerical integration to evaluate the integral

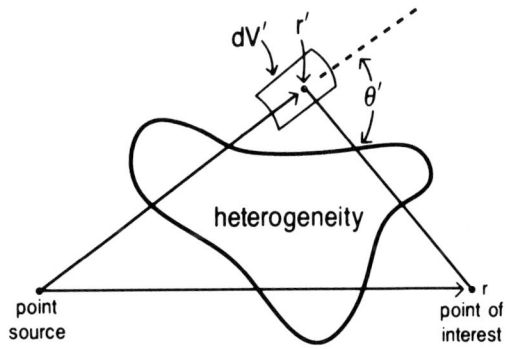

Fig. 12.43. Geometry of the dose-convolution method. dV' denotes a typical scatter-dose voxel, defined as a differential volume-centered about the primary photon collision point $\mathbf{r}$ in a spherical coordinate system. (From WILLIAMSON et al. 1991)

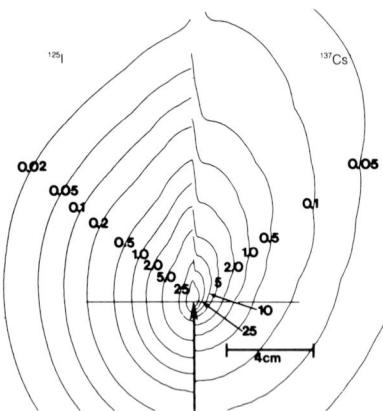

Fig. 12.44. Isodose representation of dose-spread arrays or scatter-dose kernels calculated for unbounded water medium by Monte Carlo simulation. The *curves to the left and right of the bold arrow* represent the dose-spread arrays for the ^{125}I and ^{137}Cs photon spectra, respectively. The *bold arrow* represents the primary-photon and collision site. The display is normalized to a value of 100 at a point of 1 cm downstream of the primary-photon collision site. (From WILLIAMSON et al. 1991)

and required 100 000 to 200 000 integrand evaluations and careful attention to integrand singularities $r' = 0$ and $r' = r$ to obtain numerically stable and accurate results.

Our results show that correcting only the primary-photon energy fluence rate for heterogeneities upstream of the scatter voxel at r is inadequate: scaling corrections to the scattering kernel, $K(t,\theta)$, to account for the perturbing effect of heterogeneities intersecting the scattered-photon path between r' and r were necessary as well. This gave rise to the following scaling correction:

$$K_{\text{het}}(t',\theta')$$

$$= K_{\text{wat}}(\rho',\theta') \cdot \left(\frac{\rho'}{t'}\right)^2 \cdot \left(1 - \frac{\mu_{\text{en}}(r')}{\mu(r')}\right)_{\text{wat}}^{\text{het}} \quad (12.22)$$

where $t = |\mathbf{r}-\mathbf{r}'|$, $\cos \theta' = (\mathbf{r}-\mathbf{r}').\mathbf{r}'/t$, and K_{wat} is the scattering kernel calculated for homogeneous water medium. ρ' represents the effective pathlength relative to water over which the scattered photons are attenuated and is given by:

$$\rho' = \frac{1}{\mu'_{\text{wat}}} \cdot \int_0^{t'} \mu' \left[\mathbf{r} + l \cdot \frac{(\mathbf{r}-\mathbf{r}')}{|\mathbf{r}-\mathbf{r}'|} \right] \cdot dl. \quad (12.23)$$

$\mu(\mathbf{r})$ and μ'_{wat} denote the linear attenuation coefficients in the heterogeneous and homogeneous geometries respectively. The last bracketed term in Eq. 12.22 is the ratio of scattered photon energy emitted per primary photon collision in the heterogeneity relative to that in water. This correction, important for low-energy sources such as ^{125}I, varies from unity only when the scatter voxel at r' falls inside a heterogeneity of composition different than that of water. Figure 12.45 shows that the model accurately predicts the HCF arising from a 1 MFP-thick titanium disk near an ^{125}I point source, including the 30% variation in dose with disk diameter.

The form of the kernel-scaling correction, Eq. 12.22, has important implications for practical implementation of the algorithm. Our recommended correction makes the scattering kernel spatially variant, i.e., K depends not only on $|\mathbf{r} - \mathbf{r}'|$ but on $\mathbf{r}$ and $\mathbf{r}'$ individually. First, evaluation of the integral (actually a superposition integral) requires ray tracing between every scattering voxel r' and every point of interest, r. Such algorithms have been classified as "scatter ray-trace methods" by WONG and PURDY (1990). Secondly, spatial variance implies that 3D fast Fourier transform (FFT) techniques (BOYER and MOK 1986b), which require a spatially invariant kernel, cannot be used to accelerate numerical evaluation of the integral. Although convolution calculations are very accurate, our

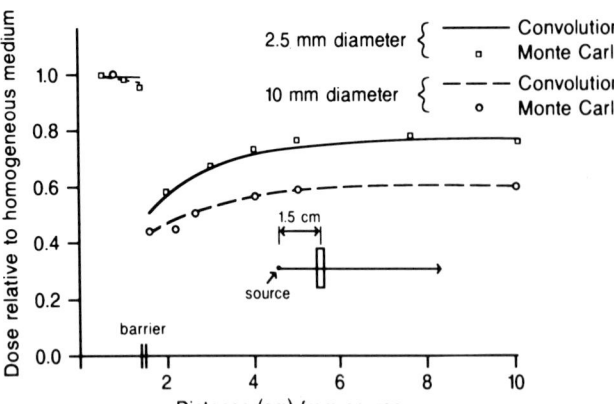

Fig. 12.45. Dose-correction factors (dose in heterogeneous geometry/dose in unbounded homogeneous water) as a function of distance from a ^{125}I point source in the presence of a flat disk (0.4 mm thick by 2.5 or 10 mm diameter) of titanium aligned perpendicular to the dose-profile axis and located at 1.5 cm downstream from the source. The *lines* indicate the results of the convolution calculation and the *symbols* the corresponding Monte Carlo estimates. (From WILLIAMSON et al. 1991)

implementation (not optimized for speed) was only 20–50 times faster than Monte Carlo simulation, which is clearly too slow for treatment planning. We found that the first-order Taylor expansion of K_{het}, successfully used by BOYER and MOK (1986a) in external-beam dosimetry to elimi–nate explicit scatter-voxel ray tracing and to make the convolution integral spatially invariant, fails in the presence of high-density brachytherapy shields. Thus methods such as adaptive multidimen–sional integration and exploitation of parallel processing, rather than FFT methods, will have to be used to accelerate the calculations. These approaches, along with adaptive recursively defined 3D calculation grids, e.g., octrees (YAU and SHIVHARI 1983), offer the potential of reducing the computational burden by as much as two orders of magni–tude. Common to adaptive approaches is selective concentration of grid points and integrand evaluations to those spatial regions in which dose gradients are large.

12.4.3.2 The Scatter-Subtraction Method of Brachytherapy Dose Calculation

To achieve a better compromise between computational efficiency and physical accuracy, the author's group (WILLIAMSON et al. 1993a) has developed a variant of the 3D scatter-integration approach, which we call the "scatter-subtraction" method. The underlying principle, scatter substraction, is widely utilized in external beam dosimetry to estimate the scatter dose under small blocks positioned in extended photon-beam fields: scatter dose contributed by an extended field of cross-sectional area F to a point under a small block of area B at depth d is proportional to

$SMR(d, F)$–$SMR(d, B)$. LULU and BJARNGARD (1982) first applied scatter subtraction to 2D bounded heterogeneities in ^{60}Co beams, reducing these problems to simpler 1D problems that could be solved with conventional heterogeneity corrections, such as the Batho method. Our prototype scatter-substraction computer code for brachytherapy exploits this principle to reduce the problem of calculating the dose behind a 2D bounded hetero-geneity to two simpler 1D problems. Thus scatter subtraction reduces the dimensionality of the scatter-convolution integral by one, in principle allowing 3D heterogeneity problems to be solved using 2D numerical integration.

The scatter-subtraction algorithm describes the perturbing effect of bounded heterogeneities on the scattered-photon dose distribution in terms of a fundamental dosimetric ratio: the collimated point source scatter-to-primary ratio or $SPR(r,\theta)$. Mathematically, $SPR(r,\theta)$ is defined (Fig. 12.46) as

$$SPR(r,\theta) = \frac{\text{Scatter dose in water}}{\text{primary dose in water}} \qquad (12.24)$$

at distance r from a collimated point source.

Isotropic emission of primary photons is understood to be restricted or collimated to a cone of half-angle θ relative to the line connecting the source and calculation point. These data are precalculated by Monte Carlo simulation for homogeneous liquid-water medium and stored as a 2D look-up table for use in subsequent dose calculations. Figure 12.47 graphically illustrates $SPR(r,\theta)$ data sets for the ^{125}I photon spectrum. The collimated SPR is functionally similar to the SMR of external beam dosimetry except that the angle $2.\theta$ serves as a measure of "brachytherapy field size."

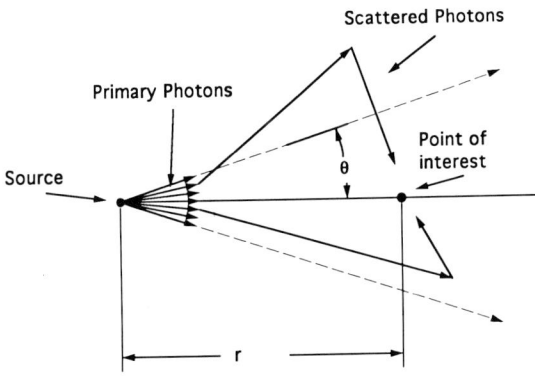

Fig. 12.46. Drawing illustrating the concept of collimated scatter-to-primary ratio, $SPR(r,\theta)$. Primary photon emission from an isotropic point source, embedded in homogeneous liquid- water medium, is theoretically restricted to a cone of half-angle θ, with respect to the source-to-calculation point axis by the Monte Carlo code subroutine which samples the primary-photon trajectories. Both first- and multiply scattered photon dose contributions are then calculated. For $\theta = \pi$, $SPR(r,\theta)$ is equivalent to the isotropic point-source build-up factor. (From WILLIAMSON et al. 1993a)

Figure 12.48 illustrates the simple 2D benchmark problem our prototype code is designed to solve: find the dose rate at distance r from a point source positioned on the axis of a cylindrical water-equivalent heterogeneity. This 2D problem is reduced to two more tractable 1D cylindrically symmetric problems by partitioning the point source into two disjoint collimated sources. These conical scattering regions are bounded by the cone of half-angle θ which subtends the heterogeneity at its center when the cone apex is positioned at the source point. The smaller cone defines a brachytherapy "minibeam" in the presence of a slab heterogeneity, while the complementary-cone primary photons interact only with homogeneous medium, contributing scatter dose to the point of interest by diffusion of multiply scattered photons around and through the barrier. The minibeam problem can be solved by a 1D correction functionally similar to those of external beam dosimetry. Applying the scatter subtraction principle, the dose rate per unit air-kerma strength, $D_i(r)$ in the inhomogeneous geometry, can be written as

$$D_i(r) = D_{p,i}(r) \cdot \left[1 + \text{SPR} \ (r,\theta) \cdot C_1 \right]$$

$$+ D_{p,h}(r) \cdot \left[\text{SPR} \ (r,\pi) - \text{SPR}(r,\theta) \right] \cdot C_2,$$

$$(12.25)$$

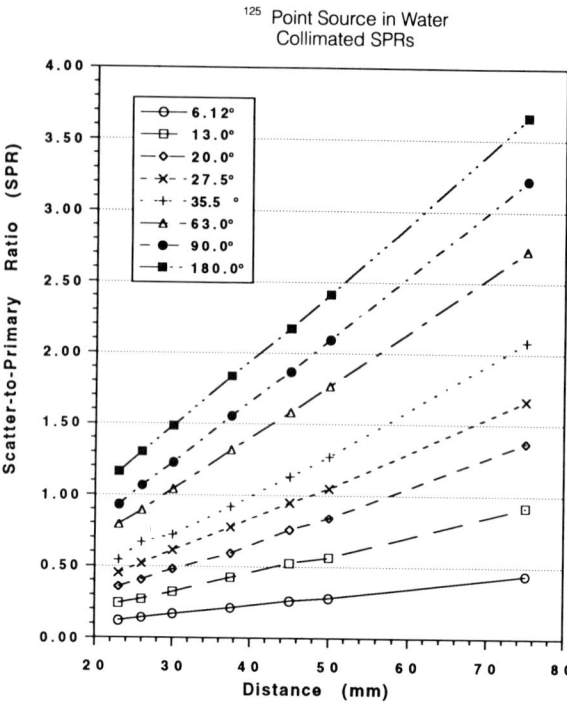

Fig. 12.47. Graphical representation of SPR(r,θ) plotted as a function of distance along the cone axis for various angles θ, for primary-photon spectra consisting of ^{125}I photons. All data were calculated by Monte Carlo simulation. (From WILLIAMSON et al. 1993a)

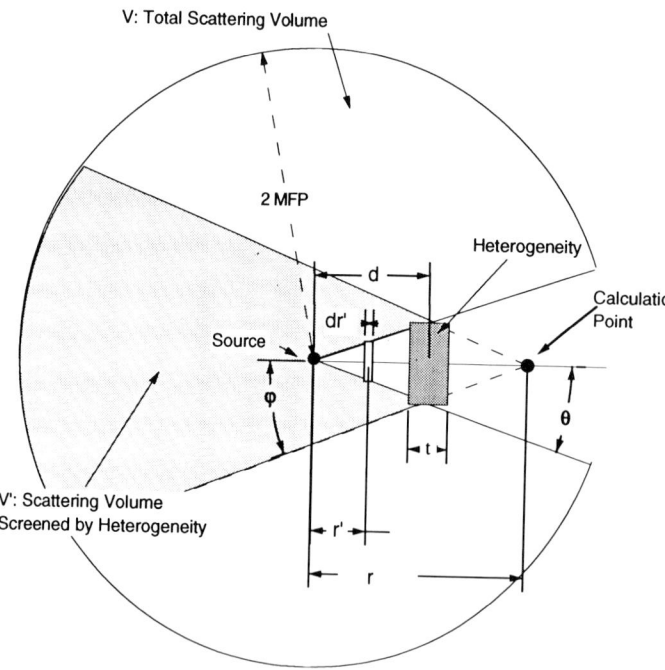

Fig. 12.48. Geometry of the scatter-subtraction model, as applied to a two-dimensional cylindrically symmetric heterogeneous geometry. The minibeam boundaries are formed by the cone with its apex located at the photon source, its axis passing through the calculation point and solid angle which subtends the disk-shaped heterogeneity at its center. (From WILLIAMSON et al. 1993a)

where $D_{p,i}(r)$ and $D_{p,h}(r)$ denote the primary dose rates for the inhomogeneous and homogeneous geometries respectively. The first term represents the primary and scatter dose arising at r from primary photons initially confined to the minibeam. The second term represents scatter dose that diffuses around and through the heterogeneity due to primary photons colliding in the homogeneous medium outside of the minibeam. C_1 is the heterogeneity correction factor for the minibeam slab problem and C_2 is a barrier attenuation correction to the scatter contribution originating outside the minibeam.

The factor C_1 corrects the homogeneous minibeam SPR for the presence of the slab heterogeneity accounting for (a) attenuation of primary photons by the barrier which influences the magnitude of

scattered-photon dose accumulated dowstream of the barrier, (b) perturbation of scatter radiation accumulating downstream of the barrier due to changes in scatter production within the barrier, and (c) perturbation of the scattered radiation component generated upstream of the barrier due to traversing the barrier. Evaluation of this parameter is more complex in brachytherapy than in external beam therapy because the minibeam cannot be approximated by a nondivergent beam. The scatter-subtraction method uses a relatively simple SPR-rescaling approximation that we call "1D scatter integration." Despite its simplicity, the approach is sufficiently accurate because both numerator and denominator of C_1 use the same simplistic assumptions resulting in significant cancellation of errors:

$$C_1 = \frac{\left[\int_0^\infty \bar{\mu}(r') \cdot \exp\left(-\int_0^r \mu(l) \cdot dl\right) \cdot G(r' \cdot \tan\theta, |r - r'|) \cdot T\left(\int_{r'}^r \mu(l) \cdot dl\right) \cdot dr'\right] \bigg/ \dot{D}_{p,1}(r)}{\left[\int_0^\infty \bar{\mu}(r') \cdot e^{-\mu \cdot r} \cdot G(r' \cdot \tan\theta, |r - r'|) \cdot T(|r' - r|) \cdot dr'\right] \bigg/ \dot{D}_{p,h}(r)}. \qquad (12.26)$$

The numerator and denominator of Eq. 12.26 are proportional, respectively, to the heterogeneous and homogeneous SPRs in the minibeam geometry, both evaluated in the 1D scatter-integration approximation. Each integral term is approximately proportional to scatter dose, which is approximated by the sum of infinitesimal disk-source contributions. This approximation is derived by partitioning the minibeam into thin disk-shaped scattering sources (Fig. 12.48), and (a) correcting each scattering element for number of once-scattered photons liberated due to primary collisions, (b) correcting the first-order scatter fluence for the inverse-square law, and (c) correcting for attenuation of once-scattered photons and subsequent buildup of multiply scattered photons. The function $G(B,r)$ corrects the once-scattered photon fluence at r for the fact that each disk is a geometrically extended source of scattered photons:

$$G(B,r) = \frac{1}{4\pi B^2} \cdot \ln\left[\frac{B^2 + r^2}{r^2}\right], \qquad (12.27)$$

where r is distance from a disk-source of radius B to the point of interest. The function $T(r)$ accounts for the attenuation of once-scattered photons emitted by the disk and the consequent buildup of multiply scattered photons over the distance $r–r'$:

$$T(r) = e^{-\mu r} \cdot [1 + \text{SPR}(r, \pi)],$$

$$= \frac{\text{Total dose in water}}{\text{dose in free space}} \text{ at distace } r \text{ from a}$$

$$\text{point source} \qquad (12.27)$$

and is approximated by the primary-photon SPR data. The terms of the form $\int_{r'}^r \mu(l) \cdot dl$ described 1D ray tracing along the minibeam central ray and account for the attenuating effect of the heterogeneity on the primary- and scattered-photon dose components.

Since the non-unit density of the heterogeneity modifies the transmission of first- and multiply scattered leakage photons that traverse the heterogeneity, simple subtraction of SPRs does not adequately characterize this scatter-dose component. This phenomenon gives rise to the scatter-substraction correction, C_2, in Eq. 12.25. A simple geometric correction factor was found to model this phenomenon with an adequate degree of accuracy. This correction assumes that the fraction of leakage scatter which is screened by the heterogeneity (volume V' in Fig. 12.48) is equal to that fraction of the total scattering volume (volume V in Fig. 12.48) that is shadowed by the barrier with respect to the calculation point. Photons originating in the unscreened region are assumed to be unperturbed while those shadowed by the barrier are assumed to be exponentially attenuated so that C_2 becomes:

$$C^2 = \left(1 - \frac{V'}{V'}\right) + \left(\frac{V'}{V'}\right) \cdot e^{-\bar{\mu} \cdot (\rho_{\text{het}} - 1) \cdot t} \qquad (12.28)$$

where ρ_{het} and t denote the mass density and axial thickness of the heterogeneous region respectively. The volumes V and V' are given by simple analytic formulae.

Scatter-subtraction predictions were compared to extensive Monte Carlo calculations for ^{125}I, ^{192}Ir, and 100-keV point sources near disk-shaped water-equivalent barriers ranging in diameter from 3.6 mm to 24 mm, and in primary-photon transmission from +36% to –70%. Figure 12.49 illustrates these comparisons for ^{125}I and ^{192}Ir sources near 36% transmission disks. For all problems evaluated, the RMS error in our absolute dose-rate predictions ranges from 0.6% to 6.6% with a maximum error of 7% in the worst case. Relative to Monte Carlo simulation, scatter subtraction was 500–1000 times faster. Extension of this promising approach to 3D geometries, to high-atomic-number shields of irregular shape, and to tissue composition heterogeneities is underway.

12.5 Conclusions

We have reviewed a number of new and promising developments in basic brachytherapy dosimetry that have emerged during the last decade. These include development of new low-energy sources that offer greatly enhanced potential for radiobiologic and physical optimization of implant therapy, and new uses for conventional sources, such as ^{192}Ir in high-dose-rate brachytherapy. These opportunities for physical optimization of brachytherapy have reawakened interest in basic experimental and theoretical brachytherapy dosimetry. Through the efforts of a relatively small number of investigators, both dose measurement and Monte Carlo simulation techniques have been perfected and validated, resulting in accurate and reliable clinical treatment planning data for low-energy sources such as ^{125}I and [103]Pd and basic single-source dosimetry data for

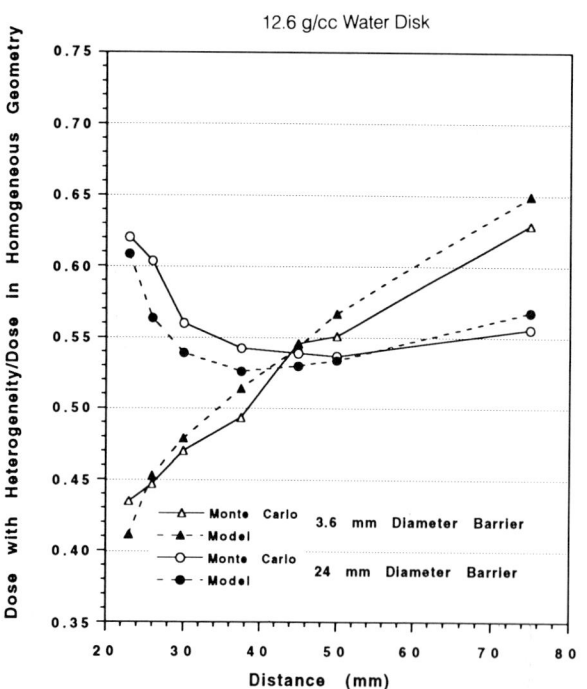

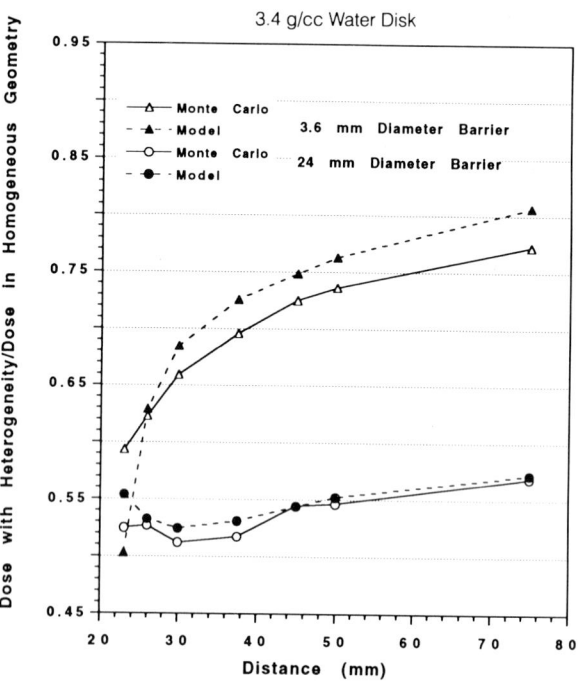

Fig. 12.49. Comparison of 1D scatter-subtraction predictions (*closed symbols*) with corresponding Monte Carlo calculations (*open symbols*) for water-equivalent disk-shaped density heterogeneities located 15 mm from isotropic point sources. The *left panel* shows a 12.6 g/cc water disk near a ^{192}Ir point source.

The *right panel* shows a 3.4 g/cc disk-shaped water heterogeneity in the presence of an ^{125}I point source. Both graphs compare the model predictions to Monte Carlo calculations for a small 3.6-mm-diameter as well as large 24-mm-diameter barriers. (From WILLIAMSON et al. 1993a)

developmental brachytherapy sources. An area of current investigation is development of algorithms capable of supporting accurate, fast, and prospective calculation of absorbed dose in the presence of tissue heterogeneities, applicator shielding, and inter-applicator and -source shielding effects. Success of these efforts is essential to large-scale clinical exploitation of the opportunities offered by new isotopes, high-dose-rate remote afterloading, and image-based treatment planning for improving clinical outcome. While much has been achieved and a number of promising new directions identified, this practical problem remains unsolved.

References

Ahmad M, Fontenla DP, Chiu-Tsao S-T, Chui CS, Reiff JE, Anderson LL (1992) Diode dosimetry of models 6711 and 6712 ^{125}I seeds in a water phantom. Med Phys 19: 391–399

Alberti W, Divoux S, Pothmann B, Tabor P, Hermann K-P, Harder D (1993) Autoradiography for iodine-125 seeds. Int J Radiat Oncol Biol Phys 25: 881–884

Anderson LL, Nath R, Weaver KA, et al. (Interstitial Collaborative Working Group) (1990) Interstitial brachytherapy, physical, biological, and clinical considerations. Raven, New York

Berger MJ (1964) Energy deposition in water by photons from point isotropic sources. MIRD Pamphlet No. 2, J Nucl Med

Boyer AL, Mok EC (1985) A photon dose distribution model employing convolution calculations. Med Phys 12: 169–177

Boyer AL, Mok EC (1986a) Calculation of photon dose distribution in an inhomogeneous medium using convolutions. Med Phys 13: 503–509

Boyer AL, Mok EC (1986b) Brachytherapy seed dose distribution calculation employing the fast Fourier transform. Med Phys 13: 525–529

Brenner DJ, Hall EJ (1991) Fractionated high dose-rate versus low dose-rate regimens for intracavitary brachytherapy of the cervix. Br J Radiol 64: 133–141

Burns GS, Raeside DE (1983) Monte Carlo estimates of specific absorbed fractions for an I-125 point source in water. Med Phys 10: 197–198

Burns GS, Raeside DE (1987) Monte Carlo simulation of the dose distribution around ^{125}I seeds. Med Phys 14: 420–424

Burns GS, Raeside DE (1989) The accuracy of single-seed dose superposition for I-125 implants. Med Phys 16: 627–631

Cerra F, Rodgers JE (1990) Dose distribution anisotropy of the Gamma Med IIi brachytherapy source. Endocurie Hypertherm Oncol 6: 71–80

Chiu-Tsao S-T, Anderson LL (1991) Thermoluminescent dosimetry for ^{103}Pd seeds (model 200) in solid water phantom. Med Phys 18: 449–452

Chiu-Tsao S-T, O'Brien K, Sanna R, et al. (1986) Monte Carlo dosimetry for ^{125}I and ^{60}Co in eye plaque therapy. Med Phys 13: 678–682

Chiu-Tsao S-T, Anderson LL, Stabile I (1988) TLD dosimetry for I-125 eye plaque (abstract). 1988 World Congress on Medical Physics and Biomedical Engineering, San Antonia, TX Chiu-Tsao S-T, Anderson LL, O'Brien K, Sanna R (1990) Dose rate determination for ^{125}I seeds. Med Phys 17: 815–825

Cygler J, Szanto J, Soubra M, Rogers DWO (1990) Effects of

fold and silver backings on the dose rate around an ^{125}I seed. Med Phys 17: 172–178

Dale RG (1983) Some theoretical derivations relating to the tissue dosimetry of brachytherapy nuclides with particular reference to iodine-125. Med Phys 10: 176–183

Diffey BL Klevenhagen SC (1975) An experimental and calculated dose distribution in water around CDC K-type caesium-137 sources. Phys Med Biol 20: 446–454

Fairchild RG, Grill AB, Ettinger KV (1982) Radiation enhancement with iodinated deoxyuridine. Radiology 17: 407–415

Fairchild RG, Kalef-Erza J, Packer S et al. (1987) Samarium-145: a new brachytherapy source. Phys Med Biol, 32: 847–858

Genest P, Hilaris BS, Nori D, et al. (1985) Iodine-125 as a substitute for Ir-192 in temporary interstitial implants. Endocurie Hypertherm Oncol 1: 223–228

Goetsch SJ, Attix FH, Pearson DW, Thomadsen BR (1991) Calibration of ^{192}Ir high-dose-rate afterloading systems. Med Phys 18: 462-467

Goffinet DR, Ling CC, Mariscal M, Phillips TL (1987) 125 Iodine removable breast implants. Preliminary report. Endocurie Hypertherm Oncol 3: 121–125

Hilaris BS, Holt GJ, St German J (1975) The use of iodine-125 for interstitial implants. US Deprtment of Health, Education and Welfare, DHEW/FO 476–8022, Rockville, MD

Huang DYC, Schell MC, Weaver KA, Ling CC (1990) Dose distribution of ^{125}I sources in different tissues. Med Phys 17: 826–832

Hubbell JH (1982) Photon mass attenuation and energy absorption coefficients from 1 keV to 20 MeV. Int J Appl Radiat-Isot 33: 1269–1290

Hubbell JH, Øverbø I (1979) Relativistic atomic form factors and photon coherent scattering cross sections. J Phys Chem Ref Data 8: 69–105

Hubbell JH, Veigele WJ, Briggs EA, Brown RT, Cramer DT, Howerton RJ (1975) Atomic form factors, incoherent scattering functions, and photon scattering cross sections. J Phys Chem Ref Data 4: 471–538

Jenkins TM, Nelson WR, Rindi A (1988) Monte Carlo transport of electrons and photons. Plenum, New York

Krishnaswamy V (1978) Dose distribution around an ^{125}I seed source in tissue. Radiology 126: 489–491

Kubo H (1985) Exposure contribution from Ti K x rays produced in the titanium capsule of the clinical 1-125 seed. Med Phys 12: 215–220

Li Z, Williamson JF (1992) Volume-based geometric modeling for radiation transport calculations. Med Phys 19: 667–678

Ling CC (1992) Permanent implants using Au-198, Pd-103, and I- 125: radiobiological considerations based on the linear quadratic model. Int J Radiat Oncol Biol Phys 23: 81–87

Ling CC, Spiro IJ (1984) Measurement of dose distribution around fletcher-Suit-Delcos colpostats using a Therados radiation field analyzer (RFA-3). Med Phys 11: 326–330

Ling CC, Yorke ED (1989) Interface dosimetry for ^{125}I seeds. Med Phys 16: 376–381

Ling CC, Yorke ED, Spiro IJ, Kubiatowicz D, Bennett D (1983) Physical dosimetry of ^{125}I seeds of a new design for interstitial implant. Int J Radiat Oncol Biol Phys 9: 1747–1752

Ling CC, Schell MC, Yorke ED (1985) Two-dimensional dose distribution of ^{125}I seeds. Med Phys 12: 652–655

Ling CC, Yorke ED, Schell MC, Goffinet D, Phillips TL (1986) Physical advantages of using iodine-125 in temporary implants of the breast. Endocurie Hypertherm Oncol 2: 216–217

Ling CC, Schell MC, Working KR, Jentzsch K, Harisiadis L, Carabell S, Rogers CC (1987) CT-assisted assessment of bladder and rectum dose in gynecological implants. Int J Radiat Oncol Biol Phys 13: 1577–1582

Ling CC, Huang DY, Barnett C, et al. (1988) Improved dose distribution with customized 1–125 source loading in temporary interstitial implants. Int J Radiat Oncol Biol Phys 15: 769–774

Loftus TP (1984) Exposure standardization of iodine-125 seeds used for brachytherapy. J Res Natl Bur Stand 89: 295–303

Lulu BA, Bjarngard BE (1982) Batho's correction factor combined with scatter summation. Med Phys 9: 372–377

Luxton G, Astrahan MA, Findley DO, Petrovich Z (1990) Measurement of dose rate from exposure-calibrated ^{125}I seeds. Int J Radiat Oncol Phys 18: 1199–1207

Mackie TR, Scrimger JW, Battista JJ (1985) A convolution method for calculating dose for 15-MV x rays. Med Phys 12: 188–196

Marchese MJ, Goldhagen PE, Zaider M, Brenner DJ, Hall EJ (1990) The relative biological effectiveness of photon radiation from encapsulated iodine-125, assessed in cells of human origin: I. Normal diploid fibroblasts. Int J Radiat Oncol Biol Phys 18: 1407–1413

Mason DL, Battista JJ, Barnett J, Porter AT (1992) Ytterbium-169: calculated physical properties of a new radiation source for brachytherapy. Med Phys 19: 695–704

McLaughlin WL, Yun-Dong C, Soares CG, Miller A, Van Dyk G, Lewis DF (1991) Sensitometry of the response of a new radiochromic film dosimeter to gamma radiation and electron beams. Nucl Instrum Methods Phys Res A302: 165–176

McMasters WH, Kerr N, Mallett JH, Hubbell JH (1969) Compilation of X-ray cross sections. Lawrence Livermore Laboratory, Livermore, CA, UCRL-50174, Sec. II, Rev. 1

Meertens H, van der Laarse R (1985) Screens in ovoids of a Selectron cervix applicator. Radiother Oncol 3: 69–80

Meigooni AS, Nath R (1992a) A comparison of radial dose functions for ^{103}Pd, ^{125}I, ^{145}Sm, ^{241}Am, ^{169}Yb, ^{192}Ir, and ^{137}Cs brachytherapy sources. Int J Radiat Oncol Biol Phys 22: 1125–1130

Meigooni AS, Nath R (1992b) Tissue inhomogeneity correction for brachytherapy sources in a heterogeneous phantom with cylindrical symmetry. Med Phys 19: 401–407

Meigooni AS, Meli JA, Nath R (1988a) Influence of the variation of energy spectra with depth in the dosimetry of ^{192}Ir using LiF TLD. Phys Med Biol 33: 1159–1170

Meigooni AS, Meli JA, Nath R (1988b) A comparison of solid phantoms with water for dosimetry of ^{125}I brachytherapy sources. Med Phys 15: 695–701

Meigooni AS, Sabnis S, Nath R (1990) Dosimetry of palladium-103 brachytherapy sources for permanent implants. Endocurie Hypertherm Oncol 6: 107–117

Meigooni AS, Meli JA, Nath R (1992) Interseed effects on dose for ^{125}I brachytherapy implants. Med Phys 19: 385–390

Meisberger LL, Keller RJ, Shalek RJ (1968) The effective attenuation in water of the gamma rays of gold 198, iridium 192, cesium 137, radium 226, and cobalt 60. Radiology 90: 953–957

Meli JA, Meigooni AS, Nath R (1988) On the choice of phantom material for the dosimetry of ^{192}Ir sources. Int J Radiat Oncol Biol Phys 14: 587–594

Metcalfe PE (1988) Experimental verification of cesium brachytherapy line source emission using a semiconductor detector. Med Phys 15: 702–706

Miller WF, Lewis EE (1984) Computational methods of neutron transport, John Wiley, New York

Mohan R, Ding IY, Martel MK, Anderson LL, Nori D (1985) Measurements of radiation dose distributions for shielded cervical applicators. Int J Radiat Oncol Biol Phys 11: 861–868

Mohan R, Chui C, Lidofsky L (1986) Differential pencil beam dose computation model for photons. Med Phys 13: 64–73

Morin LRM (1982) Molecular from factors and photon coherent scattering cross sections of water. J Phys Chem Ref Data 11: 1091–1098

Mortin J, Yabuki H, Porter EA, Rockwell S, Nath R (1989) Relative biological effectiveness of ^{241}Am relative to ^{192}Ir for continuous low-dose-rate irradiation of BA 1112 rat sarcomas. Radiat Res 119: 478–488

Muench PJ, Meigooni AS, Nath R, McLaughlin WL (1991) Photon energy dependence of the sensitivity of radiochromic film and comparison with silver halide film and LiF TLDs used for brachytherapy dosimetry. Med Phys 18: 769–775

Muench PJ, Nath R (1992) Dose distributions produced by shielded applicators using ^{241}Am for intracavitary irradiation of tumors in the vagina. Med Phys 19: 1299–1306

Nath R, Gray L (1987) Dosimetry studies on prototype ^{241}Am sources for brachytherapy. Int J Radiat Oncol Biol Phys 13: 897–905

Nath R, Bongiorni P, Rossi PI, Rockwell S (1987a) Enhancement of IUdR radiosensitization by low energy photons. Int J Radiat Oncol Biol Phys 13: 1071–1079

Nath R, Gray L, Park CH (1987b) Dose distributions around cylindrical ^{241}Am sources for a clinical intracavitary applicator. Med Phys 14: 809–817

Nath R, Anderson L, Jones D, Ling C, Loevinger R, Williamson JF, Hanson W (1987c) Specification of brachytherapy source strength. A report by Task Group 32 of the American Association of Physicists in Medicine. American Institute of Physics, New York

Nath R, Peschel RE, Park CH, fischer JJ (1988) Development of an ^{241}Am applicator for intracavitary irradiation of gynecologic cancers. Int J Radiat Oncol Biol Phys 14: 969–978

Nath R, Park CH, King CR, Muench P (1990a) A dose computation model for ^{241}Am vaginal applicators including the source- to-source shielding effects. Med Phys 17: 833–842

Nath R, Meigooni AS, Meli JA (1990b) Dosimetry on the transverse axes of ^{125}I and ^{192}Ir interstitial brachytherapy sources. Med Phys 17: 1032–1040

Nath R, Bongiorni P, Rossi PI, Rockwell S (1990c) enhanced IUdR radiosensitization by ^{241}Am photons relative to ^{226}Ra and ^{125}I photons at 0.72 Gy/hr. Int J Radiat Oncol Phys 18: 1377–1385

Nath R, Bongiorni P, Rockwell S (1990d) The relative biological effectiveness of iodine-125 and americium-241 photons relative to radium-226 photons for continuous low dose rate irradiations at dose rates of 0.17 to 0.73 Gy/hr. Endocurie Hypertherm Oncol 6: 81–91

Nath R, Bongiorni P, Rossi PI, Rockwell S (1990e) Iododeoxyuridine radiosensitization by low- and high-energy photons for brachytherapy dose rates. Radiat Res 124: 249–258

Nath R, Meigooni AS, Melillo A (1992a) Some treatment planning considerations for ^{103}Pd and ^{125}I permanent interstitial implants. Int J Radiat Oncol Biol Phys 22: 1131–1138

Nath R, Rockwell S, King CR, Bongiorni P, Kelley M, Carter D (1992b) Development of a shielded ^{241}Am applicator for continuous low dose rate irradiation of rat rectum. Int J Radiat Oncol Biol Phys 23: 175–181

Nath R, Anderson L, Luxton G, Weaver K, Williamson JF, Meigooni AS (1994) Dosimetry of interstitial brachytherapy sources: Recommendations of the AAPM Radiation Therapy Committee Task Group 43 Med Phys (in press)

Nath R, Meigooni AS, Muench P, Melillo A (1993) Anisotropy functions for ^{103}Pd, ^{125}I, and ^{192}Ir interstitial brachytherapy sources. Med Phys 20: 1465–1473

Perera H, Williamson JF, Monthofer SP, Binns WR, Klammen JC, Fuller GA, Wong JW (1992) Rapid two-dimensional dose measurement in brachytherapy using plastic scintillator sheet: linearity, signal-to-noise ratio and energy response characteristics. Int J Radiat Oncol Biol Phys 23: 1059–1069

Perera H, Williamson JF, Li Z, Mishra V, Meigooni A (1994) Rapid two-dimensional dose measurement in brachytherapy using plastic scintillator sheet: linearity, signal-to-nois ratio and energy response characteristics of a new ytterbium-169 seed: an experimentally-validated Monte Carlo investigation. Int J Radiat Oncol Biol Phys 28: 953–971

Peschel RE, Dowling S, Nath R, et al. (1988) An intracavitary vaginal applicator using americium-241. Endocurie Hyper–therm Oncol 4: 91–96

Piermattei A, Arcovito G, Bassi FA (1988) Experimental dosimetry of ^{125}I new seeds (model 6711) for brachytherapy treatments. Physica Medica 1: 59–70

Piermattei A, Arcovito G, Azario L, Rossi G, Soriani A, Montemaggi P (1992) Experimental dosimetry of ^{169}Yb seeds prototype 6 for brachytherapy treatment. Physica Medica 8: 163–169

Plechaty EF, Cullen DE, Howerton RJ (1978) Tables and graphs of photo-interaction cross sections from 0.1 keV to 100 MeV derived from the LLL evaluated-nuclear-data library, report no. UCRL-50400 (Lawrence Livermore Laboratory, Livermore, CA), vol 6, Rev 2

Prasad SC, Bassano DA, Kubsada SS (1983) Buildup factors and dose around a ^{137}Cs source in the presence of inhomogeneities. Med Phys 10: 705–708

Pratt RH, Ron A, Tseng HK (1973) Atomic photoelectric effect above 10 keV. Rev Modern Phys 45: 273–324

Ritz VH (1960) Standard free-air chamber for the measurement of low energy x-rays (20 to 100 kilovolts-constant-potential). J Res Natl Bur Stand 64C: 49–53

Roesch WC (1958) Dose for nonelectronic equilibrium conditions. Radiat Res 9: 399–410

Roussin RW (1978) Documentation for DLC7/HPICE Data Package, Oak Ridge National Lab, RSIC Data Library Collection, Radiation Shielding Information Center

Roussin RW, Knight JR, Hubbell JH, Howerton RJ (1983) Description of the DCL-99/Hugo package of photon interactions. Oak Ridge National Laboratory, RSIC Data Library Collection, Radiation Shielding Information Center, December, Report ORN/RSIC-46

Saloman EB, Hubbell JH (1986) X-ray attenuation coefficients (total cross sections): Comparison of the experimental data base with the recommended values of Henke and theoretical values of Scofield for energies between 0.1 keV–100 keV. National Bureau of Standards, Report no. NBSIR 86–3431, Washington D.C.

Saloman EB, Hubbell JH (1987) Critical analysis of soft X-ray cross section data. Nucl Instr Meth Phys Res A255: 38–42

Saloman EB, Hubbell JH, Scofield JH (1988) X-ray attenuation cross sections for energies 100 keV to 100 keV and elements Z=1 to Z=92. Atomic Data and Nuclear Data Tables 38:1–197

Samuels M, Peschel RE, Papadoulos D et al. (1991) A feasibility study of intracavitary americium-241 for recurrent pelvic malignancies. Endocurie Hypertherm Oncol 7: 131–137

Saylor WL, Dillard M (1976) Dosimetry of ^{137}Cs sources with the fletcher-Suit gynecological applicator. Med Phys 3: 117–119

Schell MC, Ling CC, Gromadzki ZC, Working KR (1987) Dose distributions of model 6702 ^{125}I seeds in water. Int J Radiat Oncol Biol Phys 13: 795–799

Schoeppel SL, Ellis JH, La Vigne ML, Martel MK, McShan DL, Fraass BA, Roberts JA (1993) 3-D treatment planning of intracavitary gynecologic implants: analysis of ten cases and implications for dose specification. Int J Radiat Onc Biol Phys 28: 277–283

Scofield JH (1973) Theoretical photoionization cross sections from 1 to 1500 keV. Lawrence Livermore Laboratory, Livermore, CA, UCRL–5132

Sievert RM (1921) Die Intensitätsverteilung der primaren-Strählung in der Nähe medizinischer Radiumpräparate. Acta Radiol 1: 89–128

Spiers FW (1949) The influence of absorption and electron range on dosage in irradiated bone. Br J Radiol 22: 251–533

Trubey DK, Berger MJ, Hubbell JH (1989) Photon cross sections for ENDF/B-VI. Presented to: Advances in Nuclear Computation and Radiation Shielding and American Nuclear Society Topical Meeting, Santa Fe, New Mexico

van der Laarse R, Meertens H (1984) An algorithm for ovoid shielding of a cervix applicator. In: Cunningham JR, Ragan D, Van Dyk D. (eds) The proceedings of the 8th international conference on the use of computers in radiation therapy, Toronto, Canada. IEEE Computer Society, Los Angeles, pp 365–369

Weaver KA (1984) Response of LiF powder to ^{125}I photons. Med Phys 11: 850–854

Weaver KA (1986) The dosimetry of ^{125}I seed eye plaques. Med Phys 13: 78–83

Weaver KA, Smith V, Huang D, Barnett C, Schell MC, Ling C (1989) Dose parameters of ^{125}I and ^{192}Ir seed sources. Med Phys 16: 636–643

Weeks KJ, Dennett JC (1990) Dose calculation and measurements for a CT-compatible version of the fletcher applicator. Int J Radiat Oncol Biol Phys 18: 1191–1198

Williamson JF (1983) Monte Carlo evaluation of the Sievert integral for brachytherapy dosimetry. Phys Med Biol 28: 1021–1032.

Williamson JF (1987) Monte Carlo evaluation of kerma at a point for photon transport problems. Med Phys 14: 567–576

Williamson JF (1988a) Monte Carlo and analytic calculation of absorbed dose near ^{137}Cs intracavitary sources. Int J Radiat Oncl Biol Phys 15: 227–237

Williamson JF (1988b) Monte Carlo evaluation of specific dose constants in water for ^{125}I seeds. Med Phys 15: 686–694

Williamson JF (1988c) Monte Carlo simulation of photon transport phenomena. In: Morin RL (ed) Monte Carlo simulation in the radiological sciences. CRC Press, Boca Raton, Fl, pp 53–102

Williamson JF (1989) Radiation transport calculation in treatment planning. Comput Med Imaging Graph 13: 251–268

Williamson JF (1990) Dose calculations about shielded gynecological colpostats. Int J Radiat Oncol Biol Phys 19: 167–178

Williamson JF (1991a) Comparison of measured and calculated dose rates in water near I-125 and Ir-192 seeds. Med Phys 18: 776–783

Williamson JF (1991b) Practical quality assurance for low dose-rate brachytherapy. In: Starkshall G, Horton J (eds) Proceedings of an American Collage of Medical Physics Symposium. Medical Physics Publishing Company, Madison, WI, pp 139–182

Williamson JF (1992) Dosimetry, treatment planning and quality assurance in gynecological intracavitary therapy. In: Purdy JA (ed) Advances in radiation oncology physics. Medical Physics Monograph 19, American Institute of Physics, New York, pp 258–288

Williamson JF, Nath R (1991) Clinical implementation of AAPM Task Group 32 recommendations on brachytherapy source strength specification. Med Phys 18: 439–448

Williamson JF, Quinterro F(1988) Theoretical evaluation of dose distributions in water about models 6711 and 6702 ^{125}I seeds. Med Phys 15: 891–897

Williamson JF, Seminoff T (1987) Template-guided interstitial implants: Cs-137 reusable sources as a substitute for Ir-192. Radiology 165: 265–269

Williamson JF, Deibel FC, Morin RL (1984) The significance of electron binding corrections in Monte Carlo photon transport calculations. Phys Med Biol 29: 1063–1073

Williamson JF, Baker R, Li Z (1991) A convolution algorithm for brachytherapy dose computations in heterogeneous geometries. Med Phys 18: 1256–1265

Williamson JF, Li Z, Wong JW (1993a) One-dimensional scatter- subtraction method for brachytherapy dose calculation near bounded heterogeneities. Med Phys 20: 233–244

Williamson JF, Perera H, Li Z, Lutz WR (1993b) Comparison of calculated and measured heterogeneity correction factors for ^{125}I, ^{137}Cs and ^{192}Ir brachytherapy sources near localized heterogeneities. Med Phys 20: 209–222

Wong JW, Purdy JA (1990) On methods of inhomogeneity corrections for photon transport. Med Phys 17: 807–814

Yau MM, Shirhari SN (1983) A hierarchical data structure for multidimensional digital images. Communications of the ACM Journal 26: 504–515

Yorke ED, Huang YCD, Schell MC, Wong R, Ling CC (1991) Clinical implications of I-125 dosimetry of bone and bone-soft tissue interfaces. Interfaces. Int J Radiat Oncol Biol Phys 21: 1613–1620

Young MEJ, Batho HF (1964) Dose tables for linear radium sources calculated by an electronic computer. Br J Radiol 37: 38–44

13 Stereotactic Brachytherapy Physics

CONTENTS

13.1 Introduction

The word "stereotactic" (or "stereotaxic") is compounded from the Greek words "stereos," meaning "solid" or "three-dimensional," and "taxis," meaning "arrangement" or "positioning." Stereotactic brachytherapy refers to the accurate placement of radioactive sources with the aid of a special mechanical frame. Such frames allow a user to specify target points as precise three-dimensional (3-D) sets of coordinates, and to guide a probe along a variety of trajectories to hit a target with great accuracy. A high-resolution 3-D imaging technique, such as computed tomography (CT) or magnetic resonance imaging (MRI), is usually required for identifying the targets.

In principle, stereotactic techniques could be applied to any part of the body. In practice, however, the body's soft tissue makes rigid, stable fixation of a stereotactic frame very difficult for most anatomic sites. This difficulty does not exist for the head, where only a relatively thin layer of skin overlies the rigid cranium. In addition, the accuracy required for probing intracranial sites is greater than for most other parts of the body. Because of these two factors, stereotactic techniques have so far only been applied to sites in the head.

Therapy via stereotactic placement of radioactive sources was introduced by Mundinger and co-workers in 1954, with ^{192}Ir first being used in 1966 and ^{125}I in 1979 (MUNDINGER and HOEFER 1974; MUNDINGER et al. 1980; MUNDINGER and WEIGEL 1984). In 1979 workers at the University of California, San Francisco (UCSF), began using stereotaxy to treat brain tumors with high-activity ^{125}I sources in temporary implants (GUTIN et al. 1981, 1984; LEIBEL et al. 1989). The general procedure is roughly the same for all brachytherapy applications: First the frame is attached firmly to the patient. Next, images (usually contrast-enhanced CT) of the head and frame are obtained. The target volume (tumor plus margin) is chosen. Implantation targets are defined on the images, and the target coordinates are converted into the frame reference system (if required). After source preparation, the guidance system of the frame is used to place the sources, usually in roughly straight catheters, in the planned locations. Finally, since even stereotaxy may involve significant placement errors, some method is normally used to determine the final, true source positions. The technical aspects of stereotactic brachytherapy are reviewed in detail in the following sections.

13.2 Stereotactic Frames

Several types of stereotactic frame have been used successfully for brachytherapy. Though a detailed

KEITH A. WEAVER, PhD, University of California, Department of Radiation Oncology, Long Hospital, Room L-75, Parnassus Avenue, San Francisco, CA 94143-0226, USA

description of each type has been published, an overview of the most common types is provided in this section.

13.2.1 Brown–Roberts–Wells System

The stereotactic system most frequently used in the United States is the Brown–Roberts–Wells (BRW)

Fig. 13.1. The BRW ring and arc positioning assembly with stylet

Fig. 13.2. The BRW ring and arc assembly mounted on the phantom base

system (supplied by Radionics, Inc.) (BROWN 1979; BROWN et al. 1980; HEILBRUN et al. 1983). It is composed of four main functional sections (Figs. 13.1, 13.2). The base ring is the section that attaches to the patient by four pins screwed into the skull. It is usually left in place for the entire procedure, though a system supplied by the manufacturer for accurate reattachment has been used successfully (FINDLAY et al. 1985; COFFEY and FRIEDMAN 1987). A framework of nine graphite rods (Fig. 13.3) can be attached to the base ring during CT scans. The positions of the rod cross-section images on transverse CT planes define a mathematical transformation between the CT coordinate system and the BRW coordinate system (KELLY et al. 1984; WEAVER et al. 1990). For MRI imaging, the carbon fiber localizer can be replaced by a localizer having tubes filled with oil.

When the patient is ready for the implant, the graphite-rod frame is replaced by a ring-and-arc assembly. This component has four angular degrees of freedom: azimuthal rotation of the arc around the base ring (angle α), pivot of the arc diameter away from the ring diameter (angle β), translation of the implantation guide along the curve of the arc (angle γ), and pivot of the guide about the attachment point on the arc (angle δ) (Fig. 13.4). In addition, a stylet can be advanced by a measured amount through the guide. For each target point, a trajectory of approach can be specified via either a second (entry) point or spatial angles, e.g., azimuth and declination. As discussed below, computer programs provide computation of frame angles corresponding to any

Fig. 13.3. Framework of nine graphite rods used to compute the CT-BRW coordinate transformation

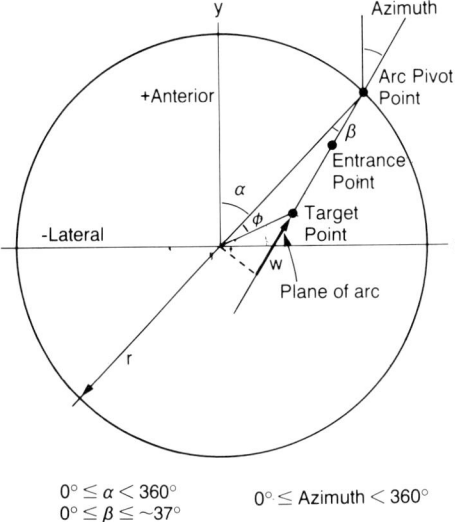

$0° \leq \alpha < 360°$
$0° \leq \beta \leq \sim 37°$

$0° \leq$ Azimuth $< 360°$

a   **Ring-view from above**

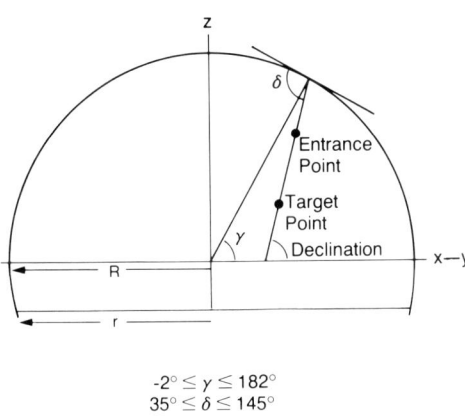

$-2° \leq \gamma \leq 182°$
$35° \leq \delta \leq 145°$

$-45° \leq$ Declination $\leq 90°$

b   **Arc-side view**

Fig. 13.4a,b. Schematic showing the angles α, β, γ, and δ that define a trajectory in the BRW system. **a** View from above. **b** Side view

desired approach trajectory, consistent with the mechanical limits of the frame.

A fourth component of the BRW system, the phantom base (Fig. 13.2), allows the frame angles and stylet depth to be verified: A pointer on the phantom is set to the BRW Cartesian coordinates of a target point. The ring-and-arc assembly, set to the computed frame angles, is transferred to the phantom base, and the stylet is advanced to meet the target. Any errors in the depth setting can be corrected, and errors in the angle settings can be investigated.

13.2.2 Cosman–Roberts–Wells System

A system compatible with the BRW system is the Cosman–Roberts–Wells (CRW) system (supplied by Radionics, Inc.). The CRW frame (Fig. 13.5) fits on the BRW base ring, but is a target-centered design. Instead of the BRW ring-and-arc assembly, the CRW frame features an arc assembly that translates linearly in either the anterior-posterior (y) direction or the lateral (x) direction (depending on how the assembly is mounted), and in the superior-inferior (z) direction. The arc assembly consists of a u-shaped arc support that pivots on circular gimbals about an axis through the target point, and an arc that can be moved linearly across the support to place its center of curvature at the target point. This last translation is normal to the other two, for full 3-D positioning. An implantation guide slides along the curve of the arc. Since it cannot pivot about its attachment point, it always points at the target defined by the three linear translations.

The CRW frame can be used with the graphite-rod localizer shown in Fig. 13.3. Alternatively the frame can be rigidly attached to a scanner (CT or MRI) table, and aligned precisely with the scanner coordinate system. A localizer frame, with vertical rods on the x- and y-axes and diagonal rods that indicate z position, can be attached to the base ring. The target coordinates are then read directly from a transverse image without the need for further computation.

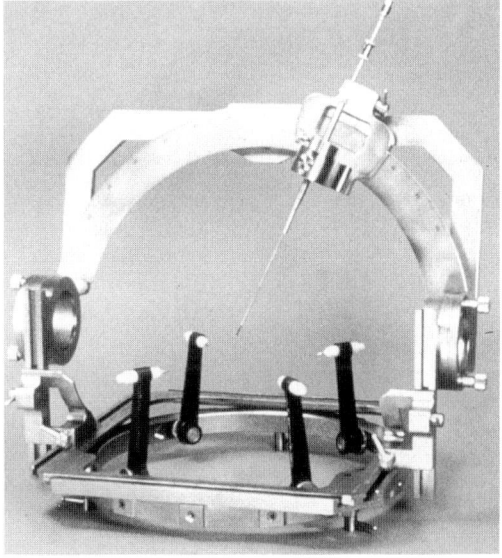

Fig. 13.5. The CRW stereotactic frame

13.2.3 Leksell Frame

Another frame frequently used for stereotactic brachytherapy (supplied by Electa Instruments, Inc.) has been described by Leksell and others (LEKSELL and JERNBERG 1980; LEKSELL et al. 1985). This frame is target centered, like the CRW frame. It must be fixed to the scanner table and aligned with its coordinate system. The frame consists of four vertical posts joined at their tops by laterally extending bars, and at their bottoms by bars along the anterior-posterior (A-P) direction (Fig. 13.6). Rigid attachment to the skull is via three or more carbon fiber pins. An arc assembly moves on the frame along graduated rails parallel to the A-P (y-coordinate) frame bars. The rails themselves can be moved in the superior-inferior (z-coordinate) direction. Gimbal rings mount to the rails and in turn support laterally extending bars. The arc slides laterally (x-coordinate) along these bars. The insertion guide slides along the arc and always points at the arc's center of curvature, which defines the target.

Because the frame is fixed to a scanner table and aligned to the scanner coordinate axes, no computation of special frame coordinates is required. Instead, an assembly of fiducial plates can be attached to the frame during scans. The plates have vertical rods adjusted to lie on the scanner's lateral (x) axis. Additionally, diagonal rods produce images

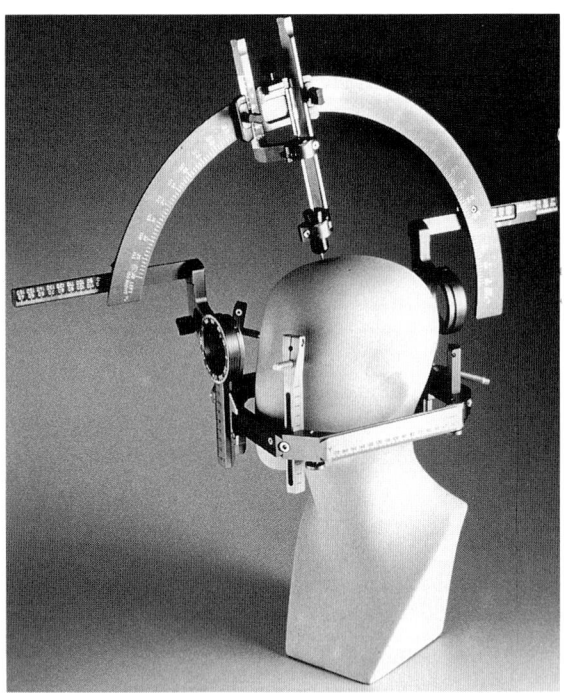

Fig. 13.6. The Leksell stereotactic frame

that indicate the z-coordinate of an image point when the transverse scanner image is superposed onto a calibrated scale.

13.2.4 Riechert–Mundinger Frame

A frame commonly used in Europe has been described by RIECHERT and MUNDINGER (1955). This frame has subsequently undergone revision for compatibility with CT scans (STURM et al. 1983). In this system a ring is affixed to the patient's head with clamps that support graphite pins. To this ring is attached a coronal arc; the point of fixation can be adjusted anteroposteriorly over a space of about 5 cm. The arc can pivot about its points of attachment. The implantation guide slides along the arc and in addition can pivot about its point of attachment. A computer program converts frame positions and angles to and from scanner coordinates.

13.2.5 Other Frames

Other frames have been used for stereotactic implants (e.g., MORAN et al. 1982; ABRATH et al. 1986). These generally have been constructed and used only at single institutions. Most have somewhat less versatility than the other systems described, but otherwise function in similar ways.

13.3 Type of Implant

13.3.1 Source Type

Several types of sources have been used in stereotactic implants. The most common are ^{125}I and ^{192}Ir seeds. In addition, ^{198}Au and ^{252}Cf have been used to treat brain tumors. ^{103}Pd sources have been proposed for permanent implants. The characteristics of all these source types have been described extensively in the literature, but a summary will be given here.

13.3.1.1 Iodine-125 Sources

Iodine-125 is a radionuclide with a 59.6-day half-life that decays by electron capture to the first excited state of ^{125}Te. Characteristic x-rays of energies 27.4 keV and 31.4 keV are produced, along with some 35.5-keV gamma rays (7% of decays).

Iodine-125 seed sources are available commercially (supplied by MediPhysics, Inc.) in two models and a range of strengths. Dimensions and encapsulation of both models are identical: The seeds are 4.5-mm-long cylinders with 0.8-mm diameters. The active material is enclosed in a 0.05-mm-thick Ti tube sealed by end welds. The model 6702 seed holds three to five resin spheres onto which ^{125}I is absorbed; this model is available in strengths ranging from 6.3 to 51 U (5–40 mCi) (U, a unit of air kerma strength, equals cGy cm² h⁻¹). The model 6711 seed contains a 3-mm-long, 0.5-mm diameter Ag wire with the radionuclide in the form of AgI on the surface. This results in the production of some fluorescent Ag x-rays of energies 22.0 and 24.9 keV. Model 6711 seeds are available in strengths up to 6.3 U (5 mCi), though strengths below 0.9 U (0.7 mCi) are most common. Dosimetric properties of both seed types have been described in detail (WEAVER et al. 1989; Interstitial Collaborative Working Group 1990; WILLIAMSON 1991). Both seed models are available as loose seeds. In addition, the model 6711 is available in sterile strands of absorbable suture. Each strand holds ten seeds spaced 1 cm apart.

The cost of ^{125}I seeds is substantial. One way to lessen the financial impact is to reuse seeds, especially the model 6702. However, seeds should be leak checked prior to reuse. The procedure recommended by the manufacturer is to soak the recovered seeds for 16 h or more in a solution of $Na_2S_2O_3$ (known as "hypo" in photography) and detergent. The hypo dissolves the iodine, while the detergent acts as a wetting agent and an aid to penetrating cracks. The soak solution can be counted with a scintillator to verify absence of more than 5 nCi of activity.

13.3.1.2 Iridium-192 Sources

Iridium-192 has a 74.2-day half-life and decays to both [192]Pt and [192]Os. Many photon energies are produced, with the most probable ones being around 310 and 470 keV.

Iridium-192 seeds are available in the United States from several commercial suppliers. The cylindrical seeds have a core of Ir–Pt alloy surrounded by a cladding of either Pt or stainless steel. (The seed design has little influence on the dose properties.) Typical dimensions are 3 mm long by 0.5 mm in diameter. Commercial vendors usually supply seeds inside nylon tubing 0.8 mm in outside diameter; normal spacing is 1 cm, though custom spacing can be ordered. The nylon strands can hold up to 20 seeds. Seed strengths up to 7.2 U (1.8 mCi = 1 mg Ra eq) are normally available, with higher strengths available on special order.

13.3.1.3 Gold-198 Sources

Gold-198 has a 2.70-day half-life, and beta decays to excited states of [198]Hg. Deexcitation produces gamma rays with average energy of about 0.42 MeV. The seeds are 3 mm long by 0.5 mm in diameter and have no cladding or encapsulation. Strengths up to 51 U (25 mCi) are available. Due to its short half-life, [198]Au is used only in permanent implants. The strength must be specified at a particular time as well as a date.

13.3.1.4 Palladium-103 Sources

Palladium-103 has a 17.0-day half-life and decays via electron capture to excited states of [103]Rh. Photon emission is largely fluorescent x-rays with average energy of 21 keV. Two gamma rays are produced with low probability (each 0.001 per decay) and energies of 40 and 360 keV. This latter photon group has high enough energy to produce a measurable background when a typical shielding thickness of a few mm Pb is used. The sources are similar to ^{125}I seeds in size and encapsulation, though the construction details are slightly different. Internally, the seeds feature Pd plated on two graphite pellets with a Pb marker in between. Though no reports of [103]Pd therapy have yet appeared, it has been proposed for permanent stereotactic implantation. Potential advantages include shorter half-life and higher initial dose rate than ^{125}I, but more convenient logistics than [198]Au.

13.3.2 Source Calibration

To avoid problems caused by possible vendor errors, the strength of ordered sources should be checked by the user, who should not merely rely on the value from the supplier (Interstitial Collaborative Working Group 1990). For implants with many (more than ten) sources of the same strength, a sample of that batch may be checked. Typically ten sources or 10% of the sources, whichever is larger, should be measured. If the measured strength agrees with the stated value to within 5%, the supplier's value can usually be used with confidence. Greater disagreement should trigger additional inquiries. For implants with few high-strength sources, each source should be individually checked.

The preferred device for source checks is the well ionization chamber. Several models are available commercially. Calibration factors are usually assigned by measuring a source whose strength has been certified by the National Institute of Standards and Technology or by an Accredited Brachytherapy Calibration Laboratory. Since encapsulation can affect chamber readings, the source used for calibration must be the same design as those to be measured. Also important for accurate measurement is constancy of source position. This can be assured through use of a positioning jig. Additional information on the use of well ionization chambers is available in the literature (e.g., Interstitial Collaborative Working Group 1990).

13.3.3 Temporary Versus Permanent Implants

Initially stereotactic brachytherapy of brain tumors was by permanent implantation of low-activity ^{125}I, ^{192}Ir, or [198]Au seeds (GUTIN et al. 1981; MUNDINGER and WEIGEL 1984). Total doses were usually 100–120 Gy. Table 13.1 lists the initial dose rates necessary to deliver 100 Gy by permanent implant of four radionuclides. Advantages of this technique included low cost for the sources and no need for catheter placement prior to implantation. The main disadvantage for fast-growing gliomas was the low initial dose rate for ^{125}I and ^{192}Ir, which were the most frequently used radionuclides. Other disadvantages included possible migration of sources with time, and radiation hazards to members of the patient's family.

In 1979 workers at UCSF began using high-activity ^{125}I sources in temporary implants. Minimum tumor dose rates were about 40–50 cGy/h, and minimum tumor doses were 60–80 Gy. The main advantage of this technique is that the dose is delivered in only a few days, which leads to higher probability for control of rapidly growing tumors. Since the sources are in removable catheters, their positions are stable. After therapy, all sources are removed, and the patient presents no radiation hazard. Disadvantages include the greater cost of the high-activity sources, and the greater complexity of the implant procedure.

13.3.4 Templates Versus Individual Catheter Placement

Two approaches to implant design are in common use: implantation with the aid of a template, and implantation of independently targeted catheters. The first approach usually involves insertion of more catheters and more seeds than the second. Also, with the template approach, sources tend to be uniformly distributed throughout the target volume, while the other technique tends to employ more centrally located seeds.

Several designs of template guides have been described in the literature (ANDERSON 1985; FINDLAY et al. 1985; ABRATH et al. 1986). As a typical example, Fig. 13.7 shows a guide used with the BRW frame. In this case the guide holes are spaced 1 cm apart in a hexagonal pattern. The catheters inserted with this guide are constrained to be parallel.

A template-guided insertion will generally result in a conventional volume implant. The advantages of this technique include simplicity of trajectory calculation (same angles for all catheters), use of inexpensive low-activity seeds, and perhaps a more uniform dose through the target volume if the interior seeds are weaker. Automatic optimization techniques are easier to apply to regular seed patterns (see Sect. 13.5.4). The peripheral dose tends to fall off more rapidly than for a more central insertion with stronger seeds; however, this may or may not be an advantage. More sparing of normal brain will result,

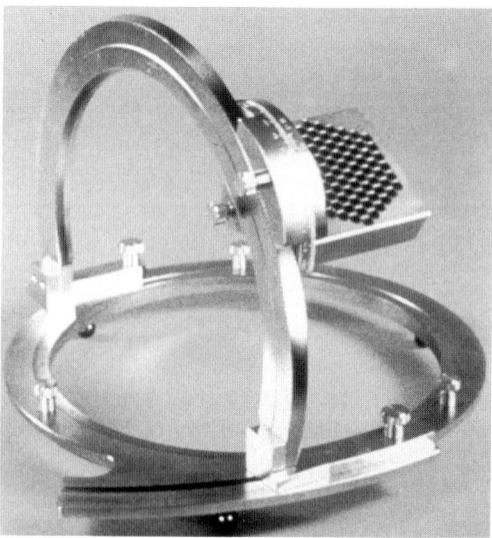

Fig. 13.7. An implantation template guide attached to the BRW frame

Table 13.1. Initial dose rates needed to deliver 100 Gy by permanent implant

	[125]I	[192]Ir	[198]Au	[103]Pd
Half-life (days)	59.6	74.2	2.70	17.0
Initial dose rate (cGy/h)	4.85	3.89	107	17.0

but the chance of underdosing occult marginal disease will increase. A major disadvantage of this technique is that insertion of multiple catheters tends to produce greater patient trauma. Another disadvantage is that misplacement of a peripheral catheter can more easily result in a marginal "cold spot" than for a more central insertion. Finally, the source preparation and surgical times are usually increased when multiple catheters are used.

The relative advantages and disadvantages of insertions with individually targeted catheters are roughly the opposites of those for template-guided insertions. Major advantages include the ability to match the catheter placement to the target volume (see Fig. 13.9), less surgical trauma for the patient, less sensitivity of peripheral isodoses to small source displacements (if sources are centrally located), and shorter source preparation and surgery times. The time required to plan an optimum implant of independent catheters tends to be longer than for template techniques, though experience leads to significant improvements in planning speed. Use of high-activity seeds increases expense and may not be cost effective for institutions where stereotactic implants are uncommon. Finally, implants with relatively few strong sources located centrally have decreased dose uniformity. However, while more uniform dose in a target volume seems theoretically desirable, clinical superiority has not been demonstrated for brain implants.

A technique that combines some aspects of both approaches discussed above has been described (ANDERSON et al. 1993). In this case catheter trajectories are related to a common geometric point. If the point lies inside the patient, the catheters are converging, while an external point produces diverging catheters. A point at infinity yields parallel trajectories. This approach removes some of the limitations of the strict parallelism required by most template guides, but retains the orderly seed array required for efficient use of optimization routines.

13.4 Technical Aspects of Implantation

13.4.1 Catheter Design

Sources for temporary implants are placed in stereotactically positioned catheters. For implants using sources in preformed strands (e.g., ^{192}Ir ribbons), these strands are usually inserted directly into the catheters and are fixed in place. Single (usually ^{125}I) seeds are inserted into a second, smaller catheter, often with appropriate spacers. The smaller cath-

eters are then guided into the larger, prepositioned catheters.

Iridium-192 ribbons are usually ordered with the required seed spacing (usually 1 cm). For catheters assembled from free seeds, spacers must be used to achieve the proper, planned array. The nylon leader from ^{192}Ir strands is useful for making spacers, in that it is readily available, the right diameter, and easily cut. This material is also useful for making retainers that hold the seed array in place in the catheter. A short (4–5 mm) piece cut on a sharp diagonal and dipped in, for example, silicone rubber adhesive can be pushed between the outer seed and the catheter wall to act as a retaining wedge.

A commonly used catheter system is made by Radionics Corp. This coaxial system features a larger (2.2-mm outer diameter) catheter of silicone rubber, and a smaller (1.4-mm outer diameter) catheter of Teflon. A silicone fixation collar that slides over the larger catheter can be either sutured to the scalp or glued to the skull to provide source stability. The inner catheter can be glued to the outer; however, fixation with surgical clips allows postimplant adjustments in source positions.

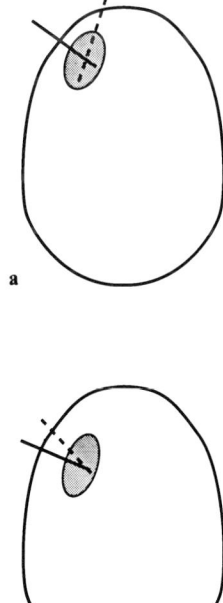

Fig. 13.8a,b. Possible catheter trajectories for implanting a tumor. **a** The *dashed line* enters the skull at an undesirably acute angle. The *solid line* represents a better approach. **b** The *dashed line* does not match a symmetry axis of the tumor. Achieving a conforming dose distribution would be more difficult with such an approach. The *solid line* shows a better trajectory

13.4.2 Implantation Guidelines

Those who design implants will likely produce better implants, from both surgical and dosimetric perspectives, if certain geometric guidelines are observed. For most implants, both template-guided and independently targeted, fewer surgical problems will be encountered if the catheter trajectories are nearly normal to the skull (Fig. 13.8a). Oblique angles of approach are more likely to cause significant deviations from plan in the achieved catheter positions. However, one should also give consideration to the target volume geometry. A conforming dose distribution is easier to achieve with catheters that are roughly parallel to any major symmetry axis of the tumor (Fig. 13.8b). If a choice is possible, a shorter catheter path will usually be more accurate than a longer one. A volume implant will usually have seeds extending close to the target volume periphery. If relatively fewer seeds are used to treat a larger (> 4 cm) target volume, the peripheral seeds are likely to lie at least 5 mm inside the target volume boundary. For additional information on implant planning, see Sects. 13.5.3 and 13.6.

13.4.3 Stereotactic Accuracy

Not even stereotactic implantation can ensure total accuracy in achieving planned catheter placements. A study of 50 implants with 278 seeds at UCSF revealed a mean deviation between planned and achieved positions of 3.8 mm, with a standard deviation of 3.3 mm (WEAVER et al. 1990). Resolution of misses into directions parallel and normal to the catheter revealed that the magnitude of misses was about equal in these two orthogonal directions. However, the misses in the parallel direction were more likely to be too shallow than too deep. Reasons for the misses are not totally clear, but may include deflections due to mineralizations or gliotic tissue, pressure from the bony edges of skull apertures, or necessary "play" in the implantation guides. The tendency for catheters to be a few millimeters too shallow may indicate a rebound effect of unknown cause (possibly edema). Errors due to wrong stereotactic frame settings are unlikely, since each set of frame angles was checked on a target phantom to verify accuracy.

13.5 Computer Applications

13.5.1 Imaging and Coordinate Transformations

Unlike most brachytherapy procedures, stereotactic brachytherapy planning is usually done on CT images of the target volume. Magnetic resonance images are also useful and can at times reveal tumor more readily than contrast-enhanced CT scans. However, the only means of using MRI so far reported has been through viewing films, though planning programs that read both CT and MRI are under development. For planning on MRI images, freedom from spatial distortion would have to be assured.

For many of the frames described in Sect. 13.2, the frame and scanner coordinates are adjusted to be readily convertible without extensive computation. However, the popular BRW frame is not attached to the scanner and requires the coordinate transformation to be calculated. A computer for this purpose is supplied by the frame manufacturer. Alternatively, the conversion may be coded into general planning programs. The mathematics of the transformation have been described in detail (WEAVER et al. 1990).

13.5.2 Modeling of Source Data

The spatial dose variations around interstitial brachytherapy sources have been reported by numerous authors (LING et al. 1985, SCHELL et al. 1987; THOMASON and HIGGINS 1989; WILLIAMSON and QUINTERO 1988; WILLIAMSON 1991). While the largest variation is due to geometric (e.g., inverse-square) effects, attenuation and scatter are also important for ^{125}I and other low-energy sources. The implant planner must verify that accurate data are used for the dose computations, for either commercial or independently developed systems.

Both point and line source models are used successfully. The line source model yields a more accurate dose computation, but requires that source orientation be specified. This can be done easily by recording the coordinates of the two seed end points. The point source model has the advantage of simplicity in both data storage and source localization. A point source model works best for sources with little spatial dose heterogeneity (such as ^{192}Ir seeds) or for many-seed implants. Reference dose rates in the literature are usually specified at points in a central plane normal to the source's long axis. Point data for sources with considerable spatial heterogeneity, such as ^{125}I, should be adjusted by the dose distribution averaged over solid angle.

Source data can be specified either in a lookup table (1-D for point source, 2-D for line source) or by an analytic expression. The latter generally provides better computation speed than the former. Data in a lookup table are usually stored with geometric variation removed, for more accurate interpolation. For analytic expressions, care must be taken that the expression is valid throughout the entire volume (distance and angle) of interest.

13.5.3 Treatment Planning

The goal of treatment planning is the design of a seed array that will produce an acceptable dose distribution. Many approaches to solving this problem have been described; in this section the philosophy and techniques used at UCSF for 12 years will be given as an example. Catheters are independently targeted, rather than implanted with a template.

Prior to starting the plan, a therapist or physicist marks the target volume by outlining the contrast-enhanced tumor image in each CT slice. (For targets with no tumor visible, such as margins of resections, a therapist selects the target volume based on MRI or other information.) The planner (physicist or dosimetrist) views the marked volume from a sufficient number of perspectives to get a clear idea of the target size, shape, orientation, and depth. The neurosurgeon may indicate limitations on possible angles of approach: for example, catheters near midline or in certain temporal areas are restricted.

The geometric considerations described in Sect. 13.4.2 usually suggest a preferred angle of approach. A helpful first step for large tumors is to define a catheter with the chosen trajectory and with its target point in the center of the tumor volume. The USCF planning program presents the option of reconstructing a CT plane perpendicular to the tip of a catheter. This view represents the cross-section of the tumor from the direction of approach of the catheters. Usually the size and shape of the tumor section suggest the number and placement of catheters: smaller (< 3 cm) tumors may require two or three catheters, while four to six catheters will be required to treat larger tumors. Regular (e.g., spheroidal) volumes lend themselves to symmetric catheter placement, but irregular volumes often require asymmetric geometries. While placing more catheters will make conformal isodoses easier to achieve, the desire to lessen surgical trauma leads to using the fewest catheters that will produce a good plan.

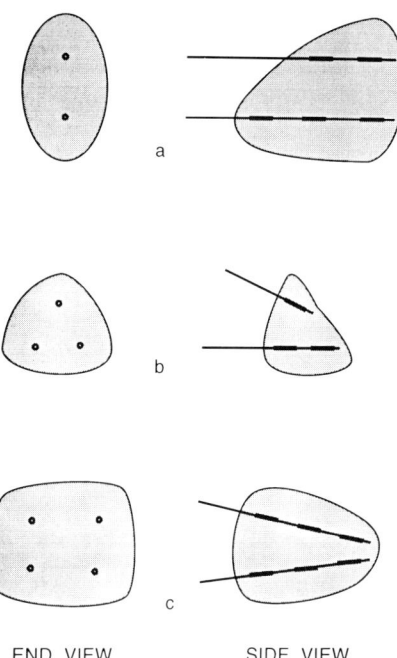

END VIEW   SIDE VIEW

Fig. 13.9a–c. Possible catheter arrays for the indicated tumor shapes. **a** Small tumor with greater superior-inferior size than anterior-posterior, and increased inferolateral extent. **b** Small flattened tumor with superolateral extension. **c** Large tumor that tapers medially

Figure 13.9 shows how catheters might be placed in several tumor shapes.

Catheters with their target points selected in the mid-tumor viewing plane need to be advanced along their lengths to traverse the tumor. In many cases the catheter tips are placed at or beyond the tumor periphery, and tip spacers are used to locate the seeds inside the tumor. Use of tip spacers allows positional adjustments to be made (see Sect. 13.5.5) and facilitates recovery of used sources. Once the catheter tips are positioned, the trajectory angles can be adjusted to match the tumor geometry. Figure 13.9 shows the utility of nonparallel approaches.

When the catheter array is determined, seeds are added. In general larger tumors will require stronger seeds than smaller ones. For nearly parallel catheters implanted in regular volumes, the loading will usually be symmetric, with somewhat weaker seeds in the tumor center to minimize hot spots. Converging catheters will require weaker tip seeds, while diverging catheters need stronger seeds in the tips. Irregular volumes will usually require special seed placements, and will in general entail more normal brain in the treatment volume than will regularly shaped tumors.

When a tentative seed array is designed, isodoses are calculated and compared to the marked target volume. In an iterative process, catheter positions and seed positions and strengths are adjusted to best match the desired isodose rate surface (usually 40–50 cGy/h) to the target volume. The entire planning process usually takes less than an hour for an experienced operator.

13.5.4 Automatic Optimization of Plans

As mentioned above, manual planning is not difficult for an experienced worker. However, those with less experience may take several hours to produce an acceptable plan. At centers where stereotactic implants are less common, an optimum level of expertise may take a long while to achieve. To address these problems, several computerized implant optimization routines have been described (ANDERSON 1985; BAUER-KIRPES et al. 1988; ROSENOW and WOJCICKA 1991). These algorithms have some general features in common.

A typical approach to computerized implant optimization is to define points, often on the target volume periphery, where some value of dose rate is desired. For a given source array the sum of the variances (deviations squared) between the desired and achieved rates is computed. By automatically varying the source numbers, positions, and strengths according to some orderly, iterative scheme, the variance sum is minimized. Additional modifying factors may include doses to points outside the target volume and near the tumor center (both to be kept below specified values).

With unlimited time, catheters, source strengths, etc., an arbitrarily good isodose fit could be achieved. However, practical constraints limit the possible variations. Optimization routines usually require either parallel catheters with some regular spacing or catheters referenced to some common point, as described in Sect. 13.3.4. Seed shifts are allowed only along catheter trajectories. In some cases adjacent catheters may be combined into one. Seed strengths are restricted to those available in inventory. Even with such limitations an optimization routine will usually produce a good plan with less than an hour's computing time, depending on the speed of the host computer.

13.5.5 Verification of Source Positions

As described above, even stereotactically guided implantation cannot be counted on to achieve perfect positioning accuracy. Since the target misses can be substantial, the isodoses evaluated for tumor coverage must be based on the true, rather than the planned, source positions. Some method is needed not only for determining the positions but also for transferring them, and the resulting isodoses, onto CT images for evaluation.

13.5.5.1 Orthogonal Films

The simplest source localization method consists of taking a set of orthogonal films. This may be done on, for example, a therapy simulator, or in the operating room if equipment to assure orthogonality is available. ^{192}Ir and model 6711 ^{125}I seeds are readily visible on radiographs. Model 6702 ^{125}I seeds are harder to see, though the end welds will usually be apparent for most implants lying superior to the bony orbital structures. As a location aid, small slivers of lead can be inserted into hollow pieces of spacer material. Any brachytherapy planning program can reconstruct 3-D coordinates from orthogonal films. The problem remains of accurately transferring the seed positions to CT images. This could be done by identifying three non-colinear points (e.g., markers fixed to the patient's skin) visible both on the orthogonal radiographs and on the preimplant CT. The two sets of point coordinates (CT space and "source space") would define a transformation that could be used to map the source coordinates into CT space.

13.5.5.2 Postimplant CT Images

An alternative to the procedure outlined above is to take a postimplant CT scan on which sources are visible. Theoretically the seed positions and orientations could all be determined from such a scan, and doses subsequently calculated. In practice, finite CT slice thickness and artifacts due to the high atomic number of the seeds can lead to significant position uncertainties for some of the seeds. An easier and more practical approach is to use orthogonal films to calculate doses in planes corresponding to the transverse CT slices. The correct orientation can be determined by comparing the films and CT scout view. By printing the isodoses on transparencies at the correct magnification, the isodose plots can be superposed onto CT images and registered by matching visible seed positions.

13.5.5.3 Fiducial Box

A seed localization technique that does not require a postimplant scan was described by SIDDON and

BARTH (1987). This technique uses a four-sided plastic box or frame that attaches to the stereotactic base ring. The box has a rectangular pattern of four lead markers in each of its faces. After seeds are implanted the guide assembly is replaced by the fiducial box, and paraorthogonal films are taken. Strict orthogonality is not required; each film need only show the eight fiducial marker images and all of the sources. The rectangular pattern of markers will in general appear as two trapezoids on each film. From the relative size and shape of the trapezoids and from projective geometry, the positions of the sources can be calculated by computer in frame coordinates. The transformation matrix calculated previously then transfers the seed coordinates to CT space for display and dose calculation.

13.5.6 Dose Calculations

13.5.6.1 Isodoses

Once the true seed positions are determined, isodoses can be calculated. The minimum dose rate that encompasses the target volume then fixes the duration of the implant. Isodoses should be evaluated in a sufficient number of orientations (multiple transverse, coronal, sagittal, and/or selected oblique) to provide a complete picture of the marginal tumor dose. In cases where seed strengths or positions are changed during an implant (e.g., to improve the dose distribution), a total dose calculation, based on the actual durations for each seed configuration, is needed. An example of isodose rates superposed on a CT image is given in Fig. 13.10.

13.5.6.2 Dose–Volume Histograms

In addition to isodoses, a calculation of dose–volume data can be helpful in evaluating tumor coverage. Table 13.2 shows a sample of doses versus

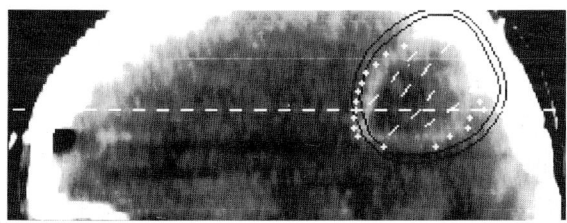

Fig. 13.10. Isodose-rate contours superposed onto a sagittal CT image with marked target volume. The lines shown are the 30 and 40 cGy/h lines

Table 13.2. Dose–volume data for isodose contours shown in Fig. 13.10

Dose rate (cGy/h)	Total volume inside isodose contour (cm³)	Target volume inside isodose contour (cm³)	Volume outside target (cm³)
10	204.4	36.3	168.1
20	102.2	36.2	66.0
30	66.4	35.5	30.9
35	56.6	35.0	21.6
40	49.0	33.8	15.1
45	42.8	32.7	10.2
50	38.0	31.2	6.9
60	31.1	27.3	3.8
80	21.2	20.0	1.2

volumes for the implant shown in Fig. 13.10. This example shows not only the total volume inside each listed isodose-rate surface, but also the amount of marked target and the volume of normal brain (i.e., region outside the marked target). The practice at UCSF is to strive for more than 95% of the marked target volume inside the chosen isodose surface, with a minimum of normal brain. This percentage reflects the fact that tumor markings are inherently somewhat imprecise and subjective, especially for irregular volumes with little CT contrast.

13.6 Brachytherapy and Hyperthermia

The medical aspects of combined stereotactic brachytherapy and hyperthermia have been discussed in a recent paper by SNEED et al. (1991). The catheters placed for seed implantation are also ideal vehicles for insertion of microwave antennas and thermometry probes. Because the spacing requirements for heating are rather stringent, at UCSF the details of implant planning are somewhat different than for cases without heat.

Generally, cases to be heated require more catheters than would otherwise be used. The catheters are usually not more than 1.5 cm apart. Most will be placed about 0.5 cm inside the target periphery, though a central catheter is also commonly inserted. Total catheter number is generally four to eight, depending on tumor size. Perhaps not all inserted catheters will be loaded with seeds, since some are for thermometry. In cases where more catheters are employed, the seed strengths are usually lower than would otherwise be used.

The practice at UCSF is to implant dummy sources on the day of surgery if hyperthermia is planned. Pieces of small wire solder are cut to seed length and are inserted in small catheters in exactly the planned seed array. Localization films are made with the inserted dummies, which are clearly visible on the films. Any adjustments in seed positions or strengths that may be required due to positioning errors are easily made when the active catheters are prepared. Usually implantation of active seeds occurs the morning after surgery, following the tumor heating.

A review of dose–volume information suggests that similar dose coverage is achieved in both heat and nonheat cases. This may be because the somewhat less advantageous catheter placement required for hyperthermia is compensated for by the opportunity to adjust seed positions and strengths.

References

Abrath FG, Henderson SD, Simpson JR, Moran CJ, Marchosky JA (1986) Dosimetry of CT-guided volumetric Ir-192 brain implants. Int J Radiat Oncol Biol Phys 12: 359–363

Anderson L (1985) Physical optimization of afterloading techniques. Strahlentherapie 161: 264–269

Anderson L, Harrington P, Osian A, Arbit E, Leibel S, Malkin M (1993) A versatile method for planning stereotactic brain implants. Med Phys 20: 1459–1464

Bauer-Kirpes B, Sturm V, Schlegel W, Lorenz WJ (1988) Computerized optimization of ^{125}I implants in brain tumors. Int J Radiat Oncol Biol Phys 14: 1013–1023

Brown RA (1979) Stereotactic headframe for use with CT body scanners. Invest Radiol 14: 300–304

Brown RA, Roberts TS, Osborn AG (1980) Stereotactic frame and computer software for CT-directed neurosurgical localization. Invest Radiol 15: 308–312

Coffey RJ, Friedman WA (1987) Interstitial brachytherapy of malignant brain tumors using computed tomography-guided stereotaxis and available imaging software: technical report. Neurosurgery 20: 4–7

Findlay PA, Wright DC, Rosenow U, Harrington FS, Miller RW (1985) ^{125}I interstitial brachytherapy for primary malignant brain tumors: technical aspects of treatment planning and implantation methods. Int J Radiat Oncol Biol Phys 11: 2021–2026

Gutin PH, Phillips TL, Hosobuchi Y, et al. (1981) Permanent and removable implants for brachytherapy of brain tumors. Int J Radiat Oncol Biol Phys 7: 1371–1381

Gutin PH, Phillips TL, Wara WM, et al. (1984) Brachytherapy of recurrent malignant brain tumors with removable high-activity iodine-125 sources. J Neurosurg 60: 61–68

Heilbrun M, Roberts T, Apuzzo M, Wells T, Sabshin J (1983) Preliminary experience with Brown-Roberts-Wells (BRW) computerized tomography stereotaxic guidance system. J Neurosurg 59: 217–222

Interstitial Collaborative Working Group (1990) Interstitial brachytherapy: physical, biological and clinical considerations. Raven, New York

Kelly P, Kall B, Goerss S (1984) Transposition of volumetric information derived from computed tomography scanning into stereotactic space. Surg Neurol 21: 465–471

Leibel SA, Gutin PH, Wara WM, et al. (1989) Survival and quality of life after interstitial implantation of removable high-activity iodine-125 sources for the treatment of patients with recurrent malignant gliomas Int J Radiat Oncol Biol Phys 17: 1129–1139

Leksell L, Jernberg B (1980) Stereotaxis and tomography: a technical note. Acta Neurochir (wien) 52: 1–7

Leksell L, Leksell D, Schwebel J (1985) Stereotaxis and nuclear magnetic resonance. J Neurol Neurosurg Psychiatry 48: 14–18

Ling C, Schell M, Yorke E, Palos B, Kubiatowicz D (1985) Two-dimensional dose distribution of ^{125}I seeds. Med Phys 12: 652–655

Maruyama Y, Chin HW, Young AB, Wang PC, Tibbs P, Beach JL, Goldstein S (1984) Implantation of brain tumors with Cf-252. Radiology 152: 177–181

Moran C, Naidich T, Marchosky JA, Barbier J (1982) A simple stabilization device for intracranial aspiration procedures guided by computed tomography. Radiology 144: 183–184

Mundinger F, Hoefer T (1974) Protracted long-term irradiation in inoperable midbrain tumors by stereotactic Curie therapy using iridium-192. Acta Neurochir Suppl (wien) 21: 93–100

Mundinger F, Weigel K (1984) Long-term results of stereotactic interstitial curietherapy. Acta Neurochir Suppl (wien) 33: 367–371

Mundinger F, Ostertag CB, Birg W, Weigel K (1980) Stereotactic treatment of brain lesions: biopsy, interstitial radiotherapy (iridium-192 and iodine-125) and drainage procedures. Appl Neurophysiol 43: 198–204

Riechert T, Mundinger F (1955) Beschreibung und Anwendung eines Zielgerätes für stereotaktisches Hirnoperationen (II. Modell). Acta Neurochir Suppl (wien) 3: 308–337

Rosenow U, Wojcicka J (1991) Clinical implementation of stereotaxic brain implant optimization. Med Phys 18: 266–272

Schell M, Ling C, Gromadzki Z, Working K (1987) Dose distributions of model 6702 I-125 seeds in water. Int J Radiat Oncol Biol Phys 13: 795–799

Siddon R, Barth N (1987) Stereotaxic localization of intracranial targets. Int J Radiat Oncol Biol Phys 13: 1241–1246

Sneed P, Stauffer P, Gutin P, et al. (1991) Interstitial irradiation and hyperthermia for the treatment of recurrent malignant brain tumors. Neurosurgery 28: 206

Sturm V, Pastyr O, Schlegel W, et al. (1983) Stereotactic computer tomography with a modified Riechert-Mundinger device as the basis for integrated stereotactic neuroradiological investigations. Acta Neurochir (wien) 68: 11–17

Thomason C, Higgins P (1989) Radial dose distribution of ^{192}Ir and ^{125}I seed sources. Med Phys 16: 254–257

Weaver KA, Smith V, Huang D, Barnett C, Schell MC, Ling CC (1989) Dose parameters of ^{125}I and ^{192}Ir seed sources. Med Phys 16: 636–643

Weaver KA, Smith V, Lewis JD, et al. (1990) A CT-based computerized treatment planning system for I-125 stereotactic brain implants. Int J Radiat Oncol Biol Phys 18: 445–454

Williamson J (1991) Comparison of measured and calculated dose rates in water near I-125 and Ir-192 seeds. Med Phys 18: 776–786

Williamson J, Quintero F (1988) Theoretical evaluation of dose distributions in water about models 6711 and 6702 ^{125}I seeds. Med Phys 15: 891–897

14 Hyperthermia Therapy Physics

PETER FESSENDEN and JEFFREY W. HAND

CONTENTS

14.1 Introduction

Hyperthermia involves the use of heat as a cancer treatment modality. It is occasionally used alone, but most often is employed in conjunction with ionizing radiation, and more recently, with chemotherapy

PETER FESSENDEN, PhD, Stanford University School of Medicine, Department of Radiation Oncology, Room S-044, 300 Pasteur Drive, Stanford, CA 94305, USA
JEFFREY W. HAND, PhD, Department of Medical Physics, Royal Postgraduate Medical School, Hammersmith Hospital, Du Cane Road, London W12 ONN, UK

(DAHL and MELLA 1990). The goal of hyperthermia is to raise the temperature of the targeted region to therapeutic levels (approximately 42°–45°C) while keeping adjacent normal tissues at subtherapeutic temperatures.

Hyperthermia, in relation to radiotherapy, is a very inefficient process, energy-wise. For example, often courses of ionizing radiation are very successful when a total of about 50 Gy of absorbed dose is localized to the tumor region. This represents an energy deposition of 50 j/kg of tissue, which, even if delivered instantaneously to a well-insulated region of tissue with no blood flow, would raise the temperature by only 0.01°C. The dosimetry for radiotherapy is well developed, and even in the most complex situations, accuracy of 10% or better is usually attainable. In contrast, a heat dose for hyperthermia is not defined, nor possibly is even definable, although for a long time laboratory evidence has indicated that there exists at least a doubling of effect per °C within the therapeutic temperature range (HAHN 1982). Since the propagation of energy from a source into, its absorption by, and its retention within tissue are critically dependent on the nature and local thermodynamics (particularly blood flow and conduction) of the tissue, hyperthermia is very challenging and will likely remain much less precise than radiotherapy.

Hyperthermia holds much promise, but it has yet to become widely accepted. There is a very strong radiobiologic and clinical rationale for the use of hyperthermia (OVERGAARD 1993), but technical problems associated with its delivery have proven to be substantial. The complicated nature of energy absorption in something as thermodynamically complex as the living human body, coupled with the usually detrimental processes of heat transfer, provide a challenge to the task of raising the temperature of a target (cancerous region) preferentially with respect to normal tissues. The situation is made worse by our ability at the present time to monitor the temperature within only a small percentage of the volume of tissue (targeted and otherwise) being heated. Even with the state of the art whereby multiple catheters are inserted surgically or transcutaneously to allow continuous mapping of temperatures along their lengths, less than 1% of the affected tissues is sampled. Since the ideal of a uniform temperature distribution is rarely even approximated (gradients of greater than 2°–3°C/cm are typically found to be prevalent within the target region), the lack of adequate temperature monitoring hinders assessment of heating ability, forcing a generally conservative treatment philosophy as well as preventing optimal analysis of results.

We are still very much on the learning curve with respect to hyperthermia, and are a long way from having perfected techniques and accomplishing predictable heating in patients. It is imperative that the underlying physics be well understood by the team delivering hyperthermia and developing equipment, so that limitations are accounted for, progress can be made, and necessary safety and cautions can be exercised.

Techniques utilizing energy deposition directly into tissue from adjacent or distant electromagnetic (EM) and ultrasound (US) sources have been developed, as have conduction techniques utilizing energy transfer from hot sources. Both external and invasive means are used. There have been some very encouraging developments in local control of power deposition, both spatially and temporally, but the progress is certainly limited by the fundamental nature of energy propagation and transfer, as well as by our current suboptimal temperature monitoring ability. Progress is being made, however, and some phase III randomized trials are in progress (VAN DER ZEE 1988; VAN DER ZEE and VAN RHOON 1993; VERNON et al. 1990) and others planned. Increased theoretical understanding of power deposition coupled with advances in computer technology have led to computerized hyperthermia treatment simulation that is an important step towards prospective treatment planning.

14.2 External Electromagnetic Methods

External methods involve the use of extracorporeal sources and these have accounted for most of the hyperthermia treatments delivered to patients to the present time. However, in certain circumstances, the sources of energy can be brought physically into the body by insertion into body cavities or by surgical incision; these methods are discussed in Sect. 14.4.

14.2.1 Interaction with Tissue

The frequencies encountered in practice in external EM hyperthermia fall roughly in the 10–1000 MHz range. At these frequencies, photon energies are at least five orders of magnitude too small to be ionizing, and the radiations are part of a group referred to generically as nonionizing. Energy transfer from the EM radiations (or fields) to the tissue results from kinetic energy imparted to atoms and molecules directly or indirectly from the electric (E) or magnetic (B) components, respectively, of the EM field. It has been suggested that there are important nonthermal hyperthermic effects, basically deriving from molecular excitation, but there is not yet sound theoretical or experimental demonstration of such in the frequency region where clinical hyperthermia is practical (NCRP 1986).

An EM field exerts a force directly on a charge q by virtue of the electric field E. This force is in the same direction as E. The magnetic part of the EM field, fully designated by the vector field B[1], exerts a force on q only if q has a velocity v component perpendicular to B. This magnetic force serves only to change the direction of q, without imparting any energy (as the electric field does). Thus for static fields, the force on q is the Lorentz force $\bar{F} = q\bar{E} + q\bar{v} x \bar{B}$ (where x represents the vector curl operation). However, if the magnetic field is changing in time, the first of Maxwell's equations, $\bar{\nabla}_x \bar{E} = -\partial \bar{B}/\partial t$ (Faraday's law), shows that there is an induced electromotive force set up which is parallel to the charge motion, and therefore an energy exchange can occur. A well-known example of this is the acceleration of electrons in a betatron. The situation is relatively simple for free space. For clinical hyperthermia, however, the important interactions take place in tissue. The tissue and other local environment affects not only the reaction of the charge to external EM fields, but also the EM field itself. Accurate calculations of hyperthermia EM devices must concern themselves with the entire applicator/tissue/local environment. These situations are extremely challenging and much of the information about EM applicator characterization and performance is empirical.

---

[1] B is really the magnetic induction, or magnetic flux density, associated with a magnetic field, but for our purposes we will term B simply the magnetic field. (The magnetic field strength, H, is equal to B/μ, where μ is the permeability.)

The target tissue (or the phantom) material may be considered a collection of charges, some free, contributing mainly to conduction currents and Joule heating via Ohm's law, and some bound, essentially comprising a dielectric. What is important is the value of the internal electric field at the point of interest. The applied external EM field serves as a guide only, the resultant internal electric field being the superposition of the external electric field (suitably attenuated) plus the electric field induced by changing magnetic fields, all somewhat modified by the fields existing due to the very charges contained within the tissue.

Maxwell's equations can be combined to form the wave equation $\nabla^2 E - \omega^2 \mu \varepsilon = 0$ governing the characteristics of E and H in the presence of a dielectric (Sect. 14.6.1). The plane wave solution for E has the sinusoidally varying time dependence $e^{j\omega t}$, where ω is the angular frequency, and a multiplying spatial part $e^{-j\omega\sqrt{\mu\varepsilon}\,z}$ where ε is the permittivity, which loosely relates the effect that the tissue (i.e., the dielectric) behavior has on the free space E field, and z is the direction of propagation. The quantity $\varepsilon = \varepsilon_o \varepsilon_r$, where ε_o is the free space permittivity and ε_r is the complex relative permittivity usually designated as $\varepsilon_r = \varepsilon_r' - j\varepsilon_r''$. The imaginary component $\varepsilon_r'' = \sigma/\omega\varepsilon_o$ and is responsible for the energy loss of the EM wave (σ = electrical conductivity).

The generalization is often made that the conduction currents due to free charges as well as the instantaneous currents associated with the bound charges of the dielectric (via vibrational, rotational, and translational molecular disturbances) are characterized by a single conductivity. Then, considering that the integral over a small volume of $\mathbf{E}\cdot\mathbf{J}\, dV$, where $\mathbf{J} = \sigma\mathbf{E}$ is the current density, is the time rate of change of the dissipated energy W, and that the specific absorption rate (SAR) is defined as the time rate of change of the absorbed energy per unit mass,

or $\dfrac{d}{dt}\left(\dfrac{dW}{\rho dV}\right)$ where ρ is the local mass density, we find SAR $= \sigma E^2/\rho$. E, here, is the instantaneous value of the internal field at the point of interest, and if it can be measured along with the electrical conductivity σ, leads to a direct determination of the absorbed energy per unit volume per unit time. SAR is the most important dosimetric quantity in hyperthermia. If it is known along with the appropriate local thermal conductivity, temperature rise (at least initially) can be predicted accurately (Sect. 14.6). SAR $= \sigma E^2/2\rho$ if E is interpreted as the electric field amplitude.

14.2.2 Localization and Attenuation

Electromagnetic hyperthermia does not provide for much localization, particularly at the lowest, most penetrating, frequencies. Even accounting for the $1/\sqrt{\varepsilon}$ reduction of free space wavelength due to the presence of tissue, the wavelengths vary from about 4 cm to 200 cm for the 1000 MHz to 10 MHz "practical" range. If one were able to take full advantage of interference effects, spatial resolutions of about one-half of these values, at best, would result.

Another spatial aspect of EM hyperthermia relates to penetration, which, at these frequencies, is mainly determined by energy absorption and characterized by an attenuation coefficient. Scatter as deriving from reirradiation does not exist because of intermolecular interactions leading to relaxation phenomena destroying any tendency for nonthermal emission. Geometric divergence from finite-sized sources and apertures affects penetration markedly, as can interference effects. As stated above, the electric field can be written as:

$$E = E_o e^{j\omega t - j\omega\sqrt{\mu\varepsilon_o\varepsilon_r}\,z}. \qquad (14.1)$$

The amplitude attenuation coefficient $\omega\sqrt{\mu\varepsilon_o\varepsilon_r}$ clearly increases with frequency. Also since ε_r contains a term proportional to the electrical conductivity, the attenuation is going to be influenced strongly by σ. There is a slight frequency dependence of ε and σ over the range 100–1000 MHz due to rotation of large polar molecules in the presence of the E field and this results in a modest increase in attenuation. (Above 1000 MHz, there is a rapid increase in σ due to the polar properties of the water molecules in tissue and this results in a rapid increase in attenuation at these higher frequencies.) This is intuitively expected, since a higher conductivity will lead to larger local currents and larger energy losses.

Calculations that assume plane wave propagation have been made using average values of conductivity for high-water-content (muscle) and low-water-content (fat, bone) tissues (JOHNSON and GUY 1972). These are summarized in Table 14.1. The selection of frequencies here is dictated partly by the historically "allowed" frequencies for which much industrial development was done, and which became emphasized naturally in medical work. Since 2450 MHz has such a shallow penetration, the practical hyperthermia range really extends from 27.12 to 915 MHz. Table 14.1 illustrates that the lower conductivity and magnitude of the relative permittivity

Table 14.1. Attenuation of electromagnetic waves (Johnson and Guy 1972)

Frequency (MHz)	λ (air) (cm)	λ (muscle) (cm)	σ (muscle) (s/m)	ε (muscle)	Penetration depth (cm)	Air/muscle reflection	Muscle/fat reflection
10	3000	110	0.62	160	21.6	0.96	
27.1	1106	68.1	0.60	113	14.3	0.92	0.65
433	69.3	8.8	1.18	53	3.6	0.80	0.56
915	32.8	4.5	1.28	51	3.0	0.77	0.52
2450	12.2	1.8	2.17	47	1.7	0.75	0.50

Frequency (MHz)	λ (fat) (cm)	σ (fat) (s/m)	ε (fat)	Penetration depth (cm)	Air/fat reflection	Fat/muscle reflection
27.1	241	0.01–0.04	20	159	0.66	0.65
433	28.8	0.04–0.12	5.6	26.2	0.43	0.56
915	13.7	0.06–1.5	5.6	17.7	0.42	0.52
2450	5.2	0.10–0.21	5.5	11.2	0.40	0.50

Muscle and fat are representative of high and low water content tissue, respectively

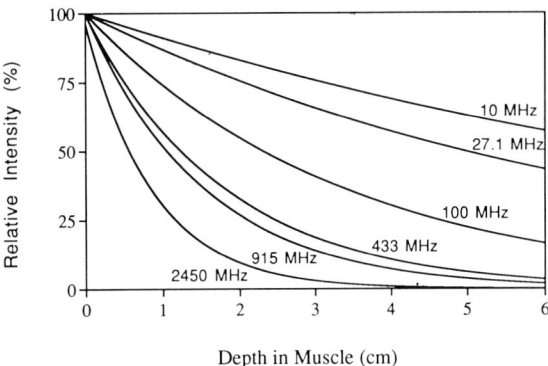

Fig. 14.1. Electromagnetic plane wave attenuation in muscle-like material. (After Johnson and Guy 1972)

(sometimes called the dielectric constant) for low-water-content fat and bone results in larger wavelengths and increased penetration compared to muscle. The term attenuation depth refers to where the electric field amplitude has been reduced to e^{-1}, or more importantly, where the relative intensity of the field has been reduced to e^{-2}, or about 13%. More clinically relevant would be the depth at which the relative intensity is reduced to 50%, which for these ideal plane waves is ln $2/2 = 0.346$ times the penetration depth. Inspection of Fig. 14.1 or Table 14.1 shows that the 50% depths for muscle are only about 1.0 and 1.2 cm for 915 and 433 MHz, respectively, emphasizing that radiating EM energy in this part of the EM spectrum is likely suitable for superficial hyperthermia only.

Figure 14.1 illustrates a sobering aspect, showing the relative intensity as a function of depth in muscle-like material for plane waves, the most ideal situation. As the aperture lateral dimensions decrease to less than a wavelength, a diffraction-limited situa-

tion results, whereby there is more and more spreading (beam divergence) and this geometric effect diminishes the penetration. In fact, as the aperture size continues to decrease, it is this, rather than the frequency, that dominates the penetration. It is quickly seen that a guiding, unfortunate, principle is that increased penetration can usually come only at the expense of increased applicator size (Hand 1990a).

Interference effects can influence penetration. For example, a closely packed array of coherent sources operated in proper phase can increase the penetration along a specified direction (not necessarily the central axis). One does not get something for nothing, however, and the result is an increased penetration along a narrow path only, at the expense of less energy deposited below the array applicator elsewhere where there is destructive, rather than constructive, interference (Magin and Peterson 1989). More subtle are the interference effects that can occur as the result of reflection from interfaces between regions of significantly different dielectric properties. As Table 14.1 shows, this is the case for muscle/air, muscle/fat, etc. At a fat/muscle interface, for example, we see the classically expected nearly 180° phase change, which leads to the reflecting wave canceling the transmitted wave near the interface (cold spot) and an interference maximum (hot spot) in the fat $\lambda/4$ from the interface. Straightforward calculations and accompanying experimental measurements have been done for idealized plane waves and parallel layered phantoms (Johnson and Guy 1972). Although such ideal conditions do not strictly apply to clinical hyperthermia, they are an important reminder that we are dealing with waves and certainly some interference phenomena will occur, very dependent on local tissue composition and

Table 14.2. Near field considerations

Frequency (MHz)	Approximate muscle wavelength (cm)	Typical maximum aperture lateral dimension L (cm)	$2L^2/\lambda$ (cm)	Typical aperture-to-tumor distance (cm)
27.1	68	10–30	2.9–26	3–15
433	9	5–15	5.5–50	1–5
915	4.5	3–10	4–44	1–5

geometry. This has been demonstrated clinically (CHOU et al. 1990, 1991).

A wave effect that strictly also influences penetration is the dramatic change in SAR that occurs at interfaces irrespective of interference (DURNEY 1990). Application of Maxwell's equation to continuity at an interface between two dielectrically different tissues raises a warning. The parallel component of the electric field is continuous across the boundary, leading to an enhanced heating (i.e., $1/2\ \sigma E^2$) in the muscle boundary relative to the fat due to the higher conductivity in muscle than fat. For any perpendicular component of the electric field, however, it is the electric displacement $D = \varepsilon E$ that must be continuous. Conserving D across the boundary leads to the relative fat heating $= (\sigma_f/\sigma_m)(\varepsilon_m^2/\varepsilon_f^2)$. Even though $\sigma_f < \sigma_m$, the square of the ratio of muscle to fat dielectric constants dominates. Thus, if there is a strong perpendicular component to the electric field, there may be much enhanced power deposition in the fat close to a fat–muscle boundary. Since often the fat has reduced blood flow and vascular cooling, enhanced temperature rises and burns sometime result.

14.2.3 Radiative Electromagnetic

A major portion of the hyperthermia applicators in preclinical and clinical use are of the radiative EM type, which often are viewed simply as apertures from which a plane wave emanates with the E and H fields perpendicular to the direction of propagation z. This idealization is rarely true, however, since one almost always is dealing with apertures that are comparable in size to (most often smaller than) the wavelength and the volume of tissue to be heated is usually located close to the applicator and is thus exposed to its near field. A geometric analysis of the Fresnel diffraction pattern expected from a finite aperture indicates that the transition from the near field to the "well-behaved" far field, the region loosely characterized as applicable to plane wave analysis, is $2L^2/\lambda$, where L is the maximum aperture dimension and λ is the wavelength in tissue. The near field is further subdivided into a radiating region, where there is the typical diffraction pattern with oscillatory character in all directions, and the very near field region within one or two wavelengths of the aperture surface. This latter very near field is reactive, or quasistatic, where there can be a strong perpendicular component to the E field, and where much of the energy is exchanged back and forth between the field and the applicator, rather than irradiated (NCRP 1981). Thus irradiation of tissue with the very near field is to be avoided since in addition to enhanced intensity fall off due to geometric effects, there is the risk of severe overheating at fat/muscle interfaces.

A fully satisfactory analytical description of the behavior of EM field interactions in situations as complex as an aperture source coupled via some medium to dielectrically inhomogeneous living tissues in the near field is not possible. Thus numerical approximations are required to predict the resulting fields and extensive experimental characterization is mandatory. Table 14.2 emphasizes the problem.

14.2.3.1 Superficial

When the dimensions of an applicator are comparable with the wavelength, radiation efficiency is increased. A class of applicators for which this is true is the so-called waveguide type (SANDHU et al. 1978), illustrated in Fig. 14.2. Here, the applicator is based on hollow waveguides with various cross-sections (for example rectangular, circular, or ridged rectangular) and with a closed end. Typically a single feed antenna is positioned within the guide to excite a desired mode. A common design is based on waveguides of rectangular cross-section whose dimensions are chosen to support a TE mode. In this case the propagating wave within the guide is characterized by the fact that the electric field is restricted to the transverse plane (parallel to the plane of the aperture) and there is a magnetic field in the direction of propagation (along the guide). By a suitable choice of dimensions, the guide can be made to support a single mode and hence produce a

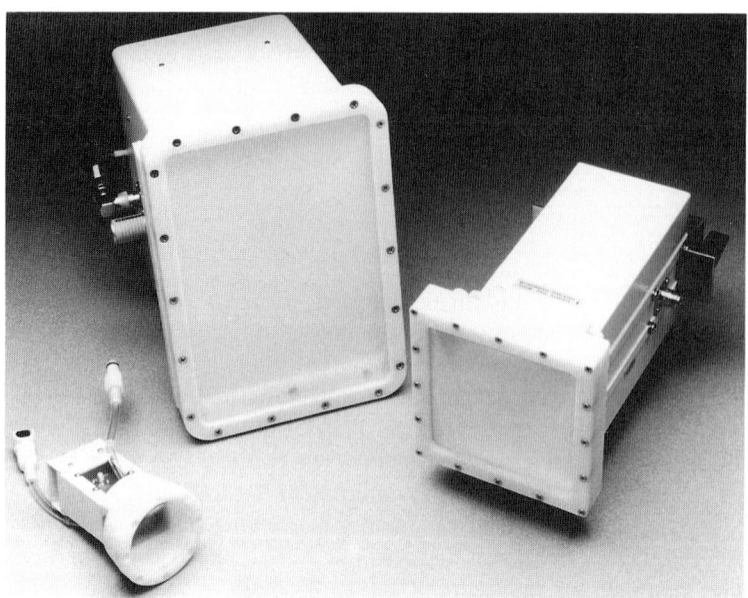

Fig. 14.2. Examples of waveguide applicators. (Courtesy of BSD Medical Corporation)

predictable field distribution. Other modes, which are likely to be present particularly close to the feed antenna, are attenuated rapidly along the guide.

The waveguide applicator dimensions are related to the wavelength within the guide, and if the guide is air filled, the dimensions are particularly large (Table 14.2). The practice of dielectric loading is common, whereby all or a portion of the cavity interior is filled ("loaded") with high dielectric material. Solid ceramic with $\varepsilon = 85$, resulting in a reduction in some dimensions by almost an order of magnitude, has been used, although it is more common to use a more workable material with ε on the order of 10 (HAND and HIND 1986). Considerations of impedance matching sometimes influence the nature of the dielectric filling material.

In theory, the variation in electric field in the aperture plane of many rectangular waveguide applicators shows a cosine-like behavior along the larger dimension (and so the SAR can be expected to fall off as $\cos^2$ from the central axis). In practice, however, the field distribution can be perturbed to some extent by the presence of the bolus and inhomogeneous tissues. Often the clinically useful area beneath the aperture is defined as that in which the SAR is at least 50% of that on the central axis. Applying this criterion to the typical field distribution produced by most rectangular waveguide applicators results in clinically useful areas only 60% or less of the aperture size. Some interesting work over the past few years (LEYBOVICH et al. 1991) in the use of multiple antenna feeds demonstrates progress

in extending the useful field area somewhat, and indicates potential for partial control of the SAR pattern locally during a treatment.

Occasionally, more than one surface MW waveguide applicator is employed for a specific treatment. The Rotterdam group (VAN RHOON et al. 1993) has used up to five 433-MHz, 10 cm × 10 cm aperture applicators simultaneously in order to treat larger areas. Two waveguide applicators have been used in angled geometry to the intact breast, as shown in Fig. 14.3 (LAGENDIJK et al. 1988).

A class of applicators that has found some use is based on horn antennas. These appear to produce a transverse field radiating out from the aperture as well as a particularly strong quasistatic field between the edges of the horn behaving like the curved fringing field associated with the edges of a parallel plate capacitor. Conical shaped horns operating at 915 MHz show some promise, particularly when operated coherently in groups orthogonally or parallel-opposed (GROSS et al. 1990). Pyramidal horns are available commercially (see Sect. 14.5.1.1.2) that operate in the range near 90–300 MHz. Because of the nature of the potentially large perpendicular component, these latter devices must be used with a very thick coupling bolus (5–10 cm) to prevent intense local hot spots near the surface. These applicators are capable of delivering hyperthermia to somewhat larger and deeper regions than waveguide or microstrip designs. There is no control over the SAR distribution, and the unusual nature of the electric field patterns demands particularly

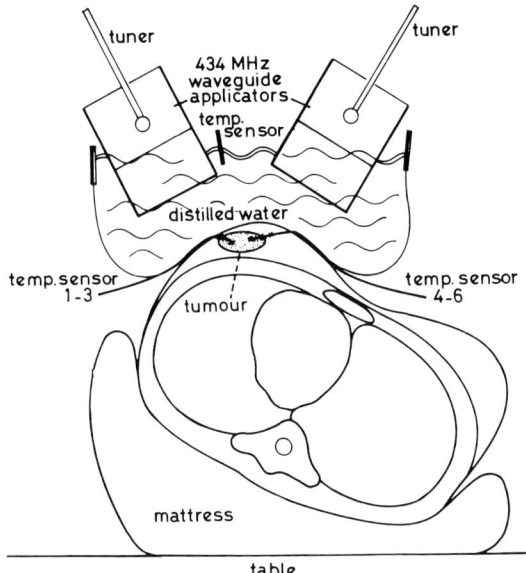

Fig. 14.3. Setup using two 433-MHz waveguide applicators to treat inoperable breast tumors. (Courtesy of Jan Lagendijk; by permission, Taylor & Francis)

exhaustive characterization by the user using simulated clinical situations with both high-water-content and low-water-content materials.

It cannot be emphasized enough that experimental characterization is paramount. Some revealing work related to the cautious use of waveguide applicators (in some ways the most familiar and best understood applicators) has been done by the Loma Linda group in California. Their work with commercial 915-MHz waveguide applicators shows the possibility of overheating in the subcutaneous fat that is very dependent on fat layer thickness (CHOU et al. 1990). It appears that the standing wave phenomena predicted by a straightforward analysis of propagation and reflection in successive layers (see Sect. 14.2.1) occurs. Also, the important influence on SAR patterns of limb curvature and relative direction of the *E* polarization has been documented carefully (CHOU et al. 1991).

In order to create equipment that is more practical for clinical purposes, Johnson and others developed low-profile applicators that use microstrip MW antennas as the radiating element (for example JOHNSON et al. 1984). The microstrip antenna is basically a several-mm-thick layer of a solid dielectric substrate (usually with relative permittivity <10) sandwiched between a flat conducting ground plane and a printed circuit conducting antenna pattern, or patch, fed by the extended center conductor of a small coaxial cable inserted through the sandwich

from the ground plane side and electrically connected to the antenna. With proper dimensions and coupling to the load (tissue, phantom, etc.), there is rather broad resonance whereby most of the energy radiated backward from the printed circuit is reflected to combine with the forward radiating energy. The early designs generally had rectangular patches and although important for the overall development aspect of low-profile hyperthermia applicators, did not make their way into the clinic in any significant manner. One drawback was that in many of these devices, the clinically effective area was considerably smaller than the device itself.

A group of researchers from Varian made a systematic study of microstrip designs and concluded that the single-arm Archimedean spiral was particularly promising (TANABE et al. 1983). The Stanford group then developed these ideas into some very practical applicators (reviewed by FESSENDEN et al. 1993). Figure 14.4 shows a cross-section of a single Archimedean antenna. The spiral antennas provide stable operation with very low reflected power over several hundred MHz when several cm thickness of deionized water is used as a coupling bolus. This thickness of coupling bolus also helps to minimize the potential for the very near field to cause fat overheating. The first significant development

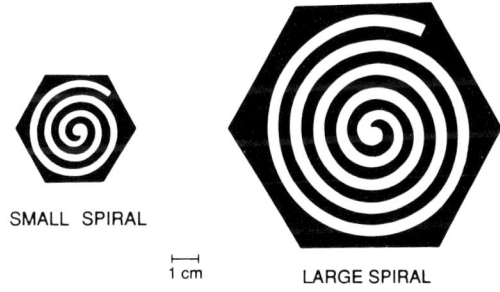

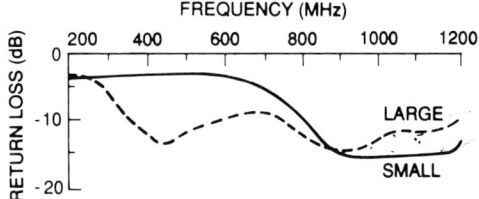

Fig. 14.4. Archimedean spiral microstrip antennas and accompanying return loss plots showing reflected power as a function of frequency (muscle-like phantom, 1.5-cm-thick deionized water coupling bolus). The substrate with a dielectric constant of 2.33 separating the copper spiral from the copper back plane is 1.6 mm and 3.2 mm thick for the small ("915-MHz") and large ("433-MHz") antennas, respectively. (By permission, Pergamon Press)

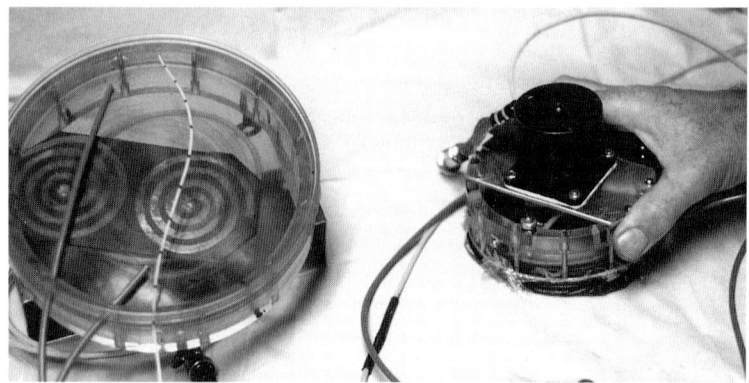

Fig. 14.5. Photograph of the scanning double spiral (*left*) and scanning single spiral (*right*) applicators collectively used in over 500 patient treatments for surface hyperthermia at Stanford University. A computer-controlled stepping motor, visible on the right, provides the circular scanning (SAMULSKI et al. 1990)

Fig. 14.6. The 5 × 5 element 915-MHz antenna array applicator (described in LEE et al. 1992). The independent mounting of each antenna support at the base of the applicator allows conformal flexing in all directions

resulted in a family of scanning spiral applicators (SAMULSKI et al. 1990a). These circularly scan one or two 8.5-cm 433-MHz antennas over planar regions, giving areas with 70% SAR coverage of up to 33 and 170 cm², respectively (Fig. 14.5). During treatment, the antenna power levels are independently controlled as they consecutively scan through eight 45° arc segments electronically defined. A family of static applicators has also been developed using the smaller 915-MHz 3.5-cm-diameter spiral antennas as basic elements closely spaced in rectangular conformal arrays (LEE et al. 1992). These arrays (Fig. 14.6) extend 70% SAR coverage to up to 340 cm² (5 × 5 array) with excellent uniformity, as well as allowing improved spatial definition of power control, ease of matching adjacent fields, and conformation to contoured surfaces. Independent power control to each antenna is achieved via computer-controlled electromechanical switches involving limited numbers of noncoherent MW sources. These versatile scanning and array spiral applicators, operated noncoherently, allow larger areas to be treated with

significantly improved control of the local SAR. The control is in response to temperature feedback from several temperature sensors periodically stepped through catheters implanted invasively approximately 1.0 cm deep in tissue, as well as from arrays of surface temperature probes, one to two positioned within each 45° octant for the scanning, or one under each antenna for the conformal array, applicator and on several scars or potential hot spots. Manual temperature assessment and subsequent control of the power distribution is cumbersome and time consuming, but computer automation based on recursive algorithms guided by the effect of a given individual antenna power change on all surrounding regions holds promise for improving the resulting clinical temperature distributions further (ZHOU and FESSENDEN 1993). In a sense, these devices may already have too much control. Even for surface tumors, the areas can easily exceed several hundred cm² and volumes exceed 500 cm³. The typical 20–32 temperature probes only sample a few mm³ of tissue each, which is certainly insufficient in the presence of temperature distributions that often spatially change very rapidly (see Sect. 14.5).

The spiral antenna is characterized by a Gaussian-type transverse SAR distribution (SAMULSKI et al. 1990a) which limits the sharpness, or spatial resolution, of the local control for both the scanning and the array devices. A smaller antenna, and/or one that had a sharper delta-function-like square edge to its SAR profile, would be desirable. Unfortunately, as revealed in systematic, mainly empirical, studies (TANABE et al. 1983), a sharp edge is obtained only at the expense of significantly increased perpendicular E components near straight edges and particularly at corners. This relates to Maxwell's equation $\bar{\nabla} \cdot \varepsilon_o \bar{E} = \rho_c$, where ρ_c is the charge density, which implies that electric field lines begin and end on charges (DURNEY 1990). Whenever there is a sharp discontinuity in the charge density, for example at a point, around a line source, or close to the edge of an electrode, the electric field can be expected to be high as the field lines concentrate, giving rise to a potential hot spot.

As discussed previously, some improvement in spatial resolution at depth for surface applicators is possible by operating arrays coherently with proper control of relative phases (HAND 1990a). Some degree of focusing is indeed possible, but due to the large wavelengths and relatively high attenuation involved, this cannot be done anywhere near as well as it can with ultrasound, where for a surface array or scanning arrangement SARs at depth can exceed that at the surface. A curved surface is the most promising site to possibly take advantage of the limited EM phase focusing, and some work using both microstrip and segmented waveguide applicator designs has been done (NIKAWA et al. 1993).

14.2.3.2 Deep Regional

Several radiative EM designs have evolved which operate in the lower frequency range near and below 100 MHz. As seen from Table 14.2, this allows increased penetration. The large wavelength, however, results in two negative features: first, large apertures and bulky devices are necessary, and second, the localization of SAR is so poor that the heating is considered regional rather than local. The resultant temperature distributions are extremely dependent on tissue compositions, geometry, and physiology (particularly blood flow) in a manner considerably more restrictive than for surface EM hyperthermia. It is virtually impossible to avoid

heating large volumes of normal tissue, and here, more than most elsewhere in clinical hyperthermia, the phrase "dump and pray" applies. Even with some recent advances in true 3D simulation of EM hyperthermia (SULLIVAN 1992), the approach remains very conservative with very incomplete documentation of the actually achieved temperature distributions.

The most widely used approach to regional hyperthermia is to use an applicator or an array of applicators which produces an inwardly radiating electromagnetic field at a frequency in the range 70–100 MHz with the E-field vector directed parallel to the patient's longitudinal axis (TURNER 1984; DE LEEUW et al. 1990; VAN DIJK et al. 1990). The present version of the BSD Medical commercial system is the Sigma 60, an annular ring of eight large "bow-tie" electric dipole antennas, each 40 cm long flared to 4 cm wide at their tips, rigidly fixed to a 60-cm-diameter plastic cylindrical shell (Fig. 14.7). An idealization to the principle of operation is simply that there exist eight converging plane waves each having the form $E_o e^{-j\omega t} e^{-j\omega r \sqrt{\mu \varepsilon_0 \varepsilon_r}}$, radiating towards the center along eight equally spaced radii. For equal amplitudes, phases and equivalent path-lengths, the standing wave at the center would have an amplitude of $8 E_o$, giving 64 times the SAR from a single wave. Whereas a single wave passing through muscle material for 15 cm would be attenuated by a factor of about 20, this idealized analysis predicts an actual enhancement of the SAR at the center relative to the surface by a factor of 3. An effect of this magnitude has never been seen, nor is it really expected, realizing that a simple plane wave analysis is not applicable (Sect. 14.2.2 and Table 14.2). Some enhancement at depth has been observed in certain situations (e.g., SAMULSKI et al. 1987a; SULLIVAN et al. 1993). The Sigma 60 device provides for independent amplitude, and to a certain extent phase, control, and SAR pattern steering is possible. The degree to which this can be controlled and best used to advantage is still evolving. It is still very much hampered, however, by seriously incomplete thermometry. Computer modeling and treatment planning help to a certain extent, but results are far from the desired level of necessary accuracy.

A device developed by the Amsterdam group operates also on the principle of simultaneously converging radiating EM waves (VAN DIJK et al. 1990). This has four independent EM sources which are apertures of large waveguides, and allows some phase steering similar to the Sigma 60 device. An interesting device called the TEM applicator has been developed by the Utrecht group (DE LEEUW

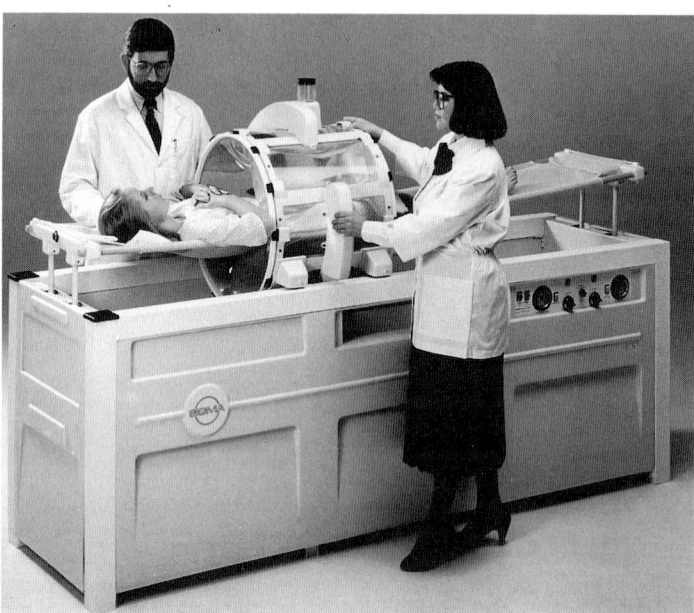

Fig. 14.7. The "Sigma 60" eight-antenna annular regional hyperthermia device. Each quadrant (pair of adjacent antennas) can be controlled independently. (Courtesy of BSD Medical Corporation)

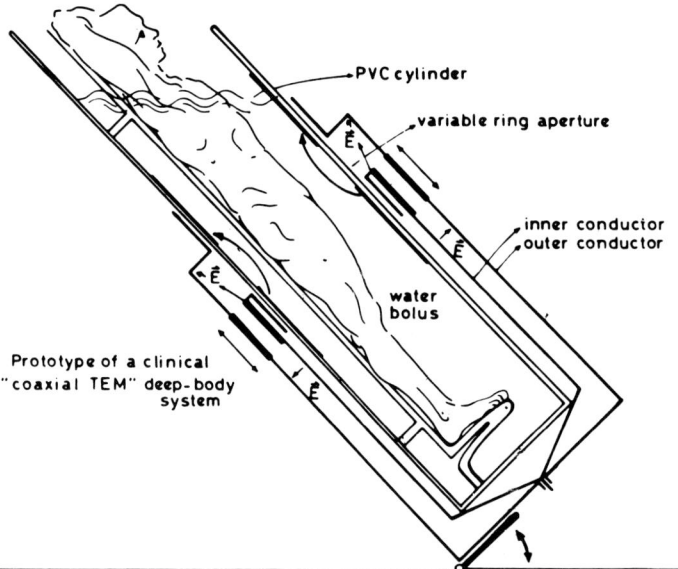

Fig. 14.8. Prototype of the Coaxial TEM regional deep-body hyperthermia applicator. A very similar version is involved in clinical trials. (Courtesy of Astrid De Leeuw; by permission, Taylor & Francis)

1993). This provides a continuous distributed annular source of converging radiating EM via the annulus aperture created by spacing two large 50-cm-diameter cylindrical electrodes end to end with a 30-cm gap. Since the geometry of this device supports a TEM mode (in which both electric and magnetic fields have components in the plane transverse to the direction of propagation) it is similar in nature to a coaxial line and is therefore inherently broad band. For deep-body hyperthermia, however, the applicator is used at 70 MHz. Figure 14.8 is a schematic of the device, illustrating the gap that launches the converging TEM wave, as well as the concept of submerging the patient in the water-filled cylinder for ideal coupling to the EM source. Phase steering is not possible, but pattern steering can be done by shifting the patient relative to the geometric center. The treatment technique involves inserting several arrays of thermocouple temperature sensors in the patient. Preliminary short low-power runs are performed which, via the related rates of temperature rise monitored by the thermocouples, yield a good idea of the SAR pattern actually achieved, guiding beneficial patient translations. (If constant specific heat is assumed and conductive and convective heat losses are ignored, then the temperature rise is a direct measure of the absorbed power and of the relative E^2 (see Sect. 14.6.3).

14.2.4 Capacitive

An alternating current exists within the region between the plates of a capacitor when there is an applied oscillating voltage across the plates. The capacitor is a quasistatic device normally operated at the ISM frequency of 27.12 MHz ($\lambda = 68$ cm in muscle, 241 cm in fat, Table 14.1) or lower. There is no radiation, the theory and construction are relatively simple, and I^2R joule heating is the main source of energy dissipation (HEINZL et al. 1990). For hyperthermia, plates of the capacitor are normally placed parallel-opposed across tissue. The capacitor applicator is well characterized by electric field lines starting on one plate and terminating on the other. Even accounting for fringing effects and inhomogeneities, this creates a situation where the E field is predominately perpendicular to tissue interfaces, resulting in the potential for overheating of the low-water-content side of the interface. This is a particularly serious problem for the subcutaneous fat regions within a few cm of the surface.

A commercial device, the Thermotron 8, has been developed for clinical use. Although data are sparse, there are reports of its successful use in Japan. It may be that the generally smaller body cross-section and low amount of fat typical in Japanese patients play an important role. The plates of this device are coupled to the patient with very cold circulating water (5°–20°C). This maneuver helps, but is limited, particularly for the United States population, due to the poor thermal conductivity of tissue in general, and fat in particular, which prevents surface cooling from being effective much beyond 1 cm.

Some degree of localization is possible by reducing the size of the plate on the body surface nearest the tumor region, thus forcing the E field lines to be more concentrated and resultant current density larger on that side. A novel improvement on this technique is the use of three capacitor plates, where essentially one plate of a conventional capacitive applicator is split into two. By varying the size and placement of the three plates, as well as adjusting the relative voltages across any given two plates, some partially controlled changes in the current density distribution are possible, helping to overcome hot-spot problems during clinical use (NUSSBAUM et al. 1986).

The inhomogeneities and accompanying large variation in conductivities and geometry throughout a human body are very influential in determining the E field pattern. Restricted regions, caused simply by reduced-area cross-sections or by the presence of low conductivity bone and particularly air, result in regions of field concentration and potential hot spots (with accompanying cold spots in some cases in regions of sparse field lines). The hot spots could be particularly severe if there is also the $E_\perp$ phenomenon going on nearby. This emphasizes caution, and points out the desire for rather precise local control, which is not possible.

14.2.5 Inductive

Inductive-type applicators rely on the principle of magnetic induction, whereby any changing magnetic field creates an electromotive force on charges such that induced currents are set up. These currents create magnetic fields which tend to oppose the effect of the original magnetic field change. As mentioned in Sect. 14.2.1, the important coupling between a changing magnetic field and the induced electric field is seen explicitly in the first of Maxwell's equations.

Early attempts to employ concentric coil induction applicators whereby one or more tightly wound coils encircled a portion of the body were very marginally successful (STORM et al. 1981). Faraday's law of magnetic induction, derivable from Maxwell's equations, states that:

$$\int_L \mathbf{E} \cdot \mathbf{dl} = -d\,\Phi_B/dt, \qquad (14.2)$$

where the left side is the line integral around a path L of the induced electric field along that path, and the right side is the negative time rate of change of magnetic flux through the area enclosed by the path L. Applying this to an idealization of a human torso cross-section as a circle shows that the value of E is dependent on the πr^2 cross-sectional area. Since E decreases as the square of the radius, and SAR depends on E^2, the induced SAR will decrease very rapidly (essentially as r^4) from the surface inward. In fact, this predicts identically zero SAR at the center (STROHBEHN 1982). Again, caution must be exercised in the full interpretation of such a simple analysis of a quasistatic device coupled to an electrically and geometrically complex region such as the human body. In fact, evidence suggests that the ever-present

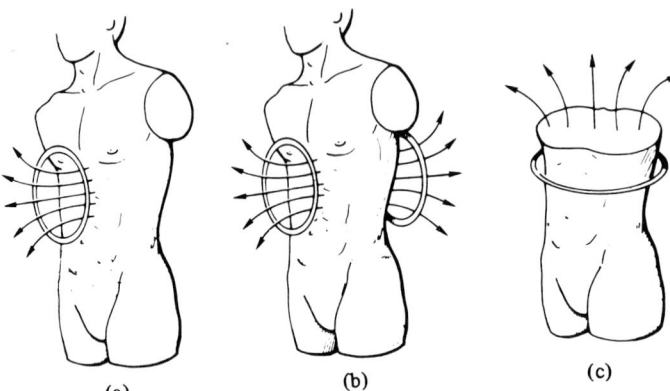

(a)   (b)   (c)

Fig. 14.9. Three configurations for inductive applicators: **a** single surface coil, **b** coaxial coil pair, **c** concentric body-encircling coil. The *arrows* represent magnetic field lines. The high SAR regions will mimic annular rings near the surface (**a,c**) or a cylindrical shell through the body (**b**). (Courtesy of James Oleson; by permission, the Institute of Electrical and Electronics Engineers, Inc.)

inhomogeneities result in islands of induced current regions where there are magnetic and electric discontinuities. This is loosely analogous to eddy currents. The dominating r^4 behavior is so important, however, that this encircling concentric coil technique is not effective.

There are, on the other hand, developments that make use of coaxial coils, particularly when there are two opposed with tissue placed on the axis between the coils as shown in Fig. 14.9 (CORRY and BARLOGIE 1982). The result can be viewed as a somewhat restricted high SAR region resulting from the intersection of an imaginary cylindrical annulus, with a radius equal to that of the RF coils, and intervening tissue. This kind of device has been used sparingly.

Another class of magnetic applicators are those in which the loop of current carrying conductors does not encircle the patient but is still in a plane perpendicular to the skin surface. In this case there is a conducting surface parallel to the skin and the conductor carrying the return current is located behind this surface, away from the patient. This type of applicator can be considered in terms of a magnetic dipole moment parallel to the skin (MORITA and BACH ANDERSEN 1982) and examples include distributed current devices (BACH ANDERSEN et al. 1984; JOHNSON 1986) and a twin dipole applicator (FRANCONI et al. 1986).

The type of applicator described by JOHNSON (1986) led to the development of the compact current sheet applicator (CSA) (JOHNSON et al. 1987; Gopal et al. 1992). The applicator consists of a folded piece of highly conducting material (aluminium) with a U-shaped cross-section when viewed on the side. Figure 14.10 is a schematic of the CSA. The overall capacitance and inductance is adjusted by proper spacing and dielectric filling so that the device resonates over a range of frequencies providing large currents. These devices can be operated in arrays in

a similar fashion to the spiral microstrip conformal array discussed in Sect. 14.2.3.1. The compact CSAs are each approximately 6×8 cm^2 and are less conformal than the spiral microstrip arrays with 3.5×3.5 cm^2 elements. Although the SAR pattern cannot be optimized as well as for the spiral arrays that have 50% SAR areas greater than the physical antenna element footprint (allowing very uniform SAR over large areas in the noncoherent array configuration), the CSAs can be operated usefully in either the coherent or the noncoherent mode, providing a versatile amount of control over the SAR pattern. Unlike the spiral arrays that have almost insurmountable difficulties associated with coherent operation due to their circular polarization, the relatively linear polarization of the CSA allows for multiple element arrays with overall well-behaved smooth interference patterns. This allows some focusing for increased penetration centrally.

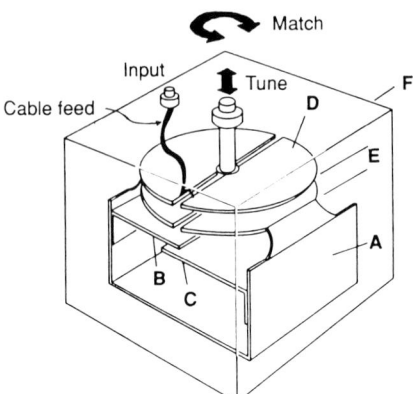

Fig. 14.10. Schematic of current sheet applicator (CSA) showing the "U"-shaped high conductivity piece *A* with added capacitor plates *B* and *C*. Power is coupled through the split disc *D*, and the resonant frequency is adjusted by changing the distance *E* and the dielectric filling. A screening box *F* encloses the complete structure (HAND 1990a)

14.3 External Ultrasound Methods

14.3.1 Ultrasound Interaction with Tissues

To a first-order approximation, an ultrasound wave propagating in a nonattenuating medium satisfies the longitudinal wave equation:

$$\frac{\partial^2 u}{\partial \chi^2} = \frac{1}{c^2}\frac{\partial^2 u}{\partial t^2}, \qquad (14.3)$$

where c, the wave velocity, is $\sqrt{K/\rho}$; K and ρ are the bulk modulus and the density, respectively, of the medium. Particles in the medium undergo a simple harmonic displacement $u(x,t) = u_A e^{j(\omega t - kx)}$ of amplitude u_A. The particle velocity $v(x,t) = u_A j\omega e^{j(\omega t - kx)} = v_A e^{j(\omega t - kx)}$ and the instantaneous pressure $p(x,t) = \rho c v = \rho c u_A j\omega e^{j(\omega t - kx)} = p_A e^{j(\omega t - kx)}$ also satisfy this form of equation. $k = 2\pi/\lambda$ is the wave number and $\omega = 2\pi f = 2\pi c/\lambda$ is the angular frequency. The product ρc is known as the characteristic acoustic impedance of the medium, Z. Differences in Z determine the behavior of ultrasound at tissue interfaces, as discussed later.

In reality, ultrasound is attenuated as it propagates through tissues. The amplitude of a plane wave in an homogeneous attenuating medium decreases exponentially with distance and so $p_A(x + \Delta x) = p_A(x)e^{-\mu\Delta x}$. The amplitude attenuation coefficient μ (Neper m^{-1}) consists of contributions from both absorption (μ_a) and scattering (μ_s); absorption is dominant for propagation through tissues (i.e., $\mu \approx \mu_a$). The acoustic intensity I, the rate of energy flow through unit area normal to the direction in which the wave is propagating, is given by the product of the wave velocity and the energy density, namely $c\rho v_A^2/2$ W m^{-2}. In an attenuating medium, the intensity of a plane wave decreases over a distance Δx according to $I = I_o e^{-2\mu\Delta x}$. The absorbed power density (APD) (in W m^{-3}) is $-dI/dx = 2\mu_a I \cong 2\mu I$. In general, the time-averaged absorbed power density $\langle \text{APD} \rangle_{time}$ is $\mu_a p_a^2/\rho c$. The specific absorption rate (SAR) is related to the APD by SAR = APD/ρ.

The velocity of sound in most soft tissues is around 1500 m s^{-1} although it is slightly less in fat ($\approx$ 1480 m s^{-1}) and significantly less in lung ($\approx$ 600–1000 m s^{-1} and strongly dependent upon the degree of inflation). Since the compressibility (the inverse of the bulk modulus K) is low for bone, the velocity in bone is considerably higher (1800–3700 m s^{-1} for longitudinal waves and $\approx$ 1940 m s^{-1} for transverse, or shear, waves). Except for inflated lung, tissues exhibit very little dispersion up to a frequency of 15 MHz. For most soft tissues, the temperature

coefficient for velocity lies in the range 0.04%–0.08%/°C^{-1}. However, because of the high lipid content of adipose tissue, the velocity in fat decreases with increasing temperature ($\approx$ –0.2 – –0.5%/°C^{-1}) (NASONI and BOWEN 1989). The temperature coefficients for both longitudinal and shear waves in bone are also negative. The characteristic acoustic impedance of most soft tissues is around 1.6×10^6 kg m^{-2}s^{-1} although it is slightly lower in fat. There is therefore little reflective loss during propagation from one soft tissue to another. In contrast, the values of Z for bone and lung differ considerably from those of soft tissues. When ultrasound is incident on an interface between tissues of differing Z both reflection and transmission occur. If θ_i, θ_r, θ_t are the angles of incidence, reflection, and transmission, respectively, then $\theta_r = \theta_i$ and

$$\frac{\sin \theta_t}{\sin \theta_i} = \frac{c_2}{c_1} \text{ (Snell's law)},$$

where c_1, c_2 are the velocities in the two tissues. If $c_2 \geq c_1$, total reflection will occur when $\theta_i \geq \sin^{-1}(c_1/c_2)$. When the boundary is large compared with the wavelength, plane and perpendicular to the plane of propagation, the intensity reflection and transmission coefficients are:

$$\frac{I_r}{I_i} = \frac{(Z_2 \cos\theta_i - Z_1 \cos\theta_t)^2}{(Z_2 \cos\theta_i + Z_1 \cos\theta_t)^2} = \Gamma$$

$$\frac{I_T}{I_i} = \frac{\cos\theta_i}{\cos\theta_t}(1 - \Gamma). \quad (14.4)$$

The high reflections which occur at interfaces between soft tissue and bone or soft tissue and gas are the cause of the major limitation in using ultrasound to induce hyperthermia. Thus ultrasound cannot be used in the thoracic cavity and there are difficulties in treating tumors in the abdomen and those immediately overlying bones. This is particularly troublesome, leading to patient pain, when nonfocused ultrasound fields are used and may be due to the fact that significant volumes of normal tissue are heated. When bone is present, the possibility of converting some of the energy present in longitudinal waves to transverse waves arises at the soft tissue–bone interface and, since shear waves are very rapidly attenuated, leads to excessive heating and pain in the periosteum.

The amplitude attenuation coefficient in most soft tissues is generally around 10 Np m^{-1}MHz^{-1} although it is slightly lower in fat. The unit Np(neper)m^{-1} is such that the amplitude of the wave decreases by a factor $1/e$ (to approximately 36.8% of

Table 14.3. Acoustic properties of mammalian tissues at 1 MHz and 37°C (except a = 23°C, b = 40°C, c = temperature not reported). In the case of brain, d = gray matter, e = white matter. After data collated by HYNYNEN (1990)

Tissue	Velocity (m s^{-1})	Density (kg m^{-3})	Acoustic impedance ($\times 10^6$ kg m^{-2} s^{-1})	Attenuation (Np m^{-1})	Absorption (Np m^{-1})
Bone	1500–3700	1380–1810	3.75–7.38	150–350	
Brain	1516–1575	1030	1.56–1.62	4–29	1.2^d–6.4^e
Fat	1400–1490	921	1.29–1.37	5–9	
Kidney	1564–1640	1040	1.62–1.71	3–10	3.3
Liver	1540–1640	1060	1.70–1.74	3.2–18	2.3–3.2
Lung	470–658	400	0.188–0.263	430–480	7
Muscle	1508–1630	1070–1270	1.61–2.07	4.4–15^b	2–11
Skin	1498^a	1200	1.80	14–66^c	

the initial amplitude) over unit distance. The intensity decreases by a factor $1/e^2$ (to approximately 13.5%) over the same distance. Attenuation in terms of intensity or energy is often stated in units of db m^{-1}. For comparison, 1 Np m^{-1} is equivalent to 8.686 db m^{-1}. Attenuation is very high in lung (430–480 Np m^{-1} at 1 MHz) and bone (150–350 Np m^{-1} and $\approx 10^4$–$5\cdot 10^5$ Np m^{-1} for longitudinal and shear waves, respectively, at 1 MHz). The absorption coefficient in soft tissues is around 10 Np m^{-1} MHz^{-1}. The dependence of attenuation and absorption on frequency f (MHz) may be expressed as $\mu = \mu_o f^n$ where μ_o is the amplitude attenuation coefficient per MHz and n is between 1 and 1.5, dependent upon tissue type (for tissues other than bone for which $n \approx 2$) (Goss et al. 1979; LYONS and PARKER 1988). In general, attenuation is temperature dependent but in many tissues there is apparently little change at frequencies around 1 MHz over the temperature range relevant to hyperthermia.

A summary of acoustic properties of tissues is given in Table 14.3.

14.3.2 Localization and Penetration

A simple estimate of a suitable frequency for ultrasound-induced superficial hyperthermia can be made from the expressions for APD for plane waves and the frequency dependence of absorption referred to in Sect. 14.3.1. From those expressions it follows that:

$$\frac{\text{APD}}{I_o} = 2\mu_a f^n e^{-2\mu f^n} x. \qquad (14.5)$$

In choosing a suitable frequency, one must take into account the trade-off between increasing the local APD (favoring a higher frequency) and reducing the attenuation which occurs in intervening tissue

between the transducer and the target (favoring a lower frequency). If we assume that $\mu_a \approx \mu$ and $n = 1$, and consider the change in APD at depth x as the frequency is varied, we see that there is a stationary point in this relationship at the frequency $1/(2\mu x)$. It follows that a suitable frequency for plane wave heating in muscle at depths of 2, 5, 10, and 15 cm is around 2.5, 1, 0.5, and 0.3 MHz, respectively, since μ_a is about 10 Np m^{-1}. At these frequencies the wavelength in soft tissues is 0.6–5 mm. Thus ultrasound offers good penetration with waves of short wavelength (amplitude attenuation per wavelength is roughly 1%) which contrasts greatly with the case of electromagnetic waves.

The wavelength at these frequencies is considerably smaller than the dimensions of applicators (several cms or larger) envisaged for noninvasive hyperthermia treatment. Such applicators may therefore be expected to produce well-collimated beams; this is discussed in the next section. Another consequence of the small wavelength is that beams may be focused to small volumes. This gives rise to the possibility of selective heating at depth if the gain in intensity due to focusing is greater than the loss due to attenuation in passing through the tissue. This topic is taken up in Sect. 14.3.4.

14.3.3 Planar Transducers

Some materials, with an anisotropic lattice structure, exhibit the property that when pressure is applied, a voltage, proportional to the applied pressure, is produced across a crystal. These materials are known as piezoelectric. Likewise, application of a voltage across a crystal of these materials results in a mechanical deformation. Lead zirconate titanate is such a material and is the basis of the most widely used family of materials, known as PZT, employed as ultrasound transducers on account of their power handling and efficiency in converting RF power to ultrasound.

A common ultrasound transducer is the plane disc which acts as a piston source. The sound field produced may be divided into two parts – one close to the transducer where a complex field resulting from interference phenomena is clearly present (the near field or Fresnel diffraction region) and the other, more distant from the transducer, in which the field profiles are relatively smooth with maxima on the central axis (far field or Fraunhofer diffraction region). The complexity of the sound field also increases as the ratio disk radius/wavelength (a/λ) increases. The pressure amplitude exhibits maxima and minima along the central axis of the transducer and, in the absence of attenuation, the positions of the last of these are approximately a^2/λ and $a^2/2\lambda$, respectively. The transition between the near and far fields occurs at a distance of approximately 0.75 a^2/λ and the width of the beam profile is at a minimum, about $0.3\,a$, at this range. Detailed discussions of fields from piston sources may be found in ZEMANEK (1970) and LOCKWOOD and WILLETTE (1973). NYBORG and STEELE (1985) and SWINDELL

(1986) discuss means of introducing attenuation into these fields which results in some smoothing of the near-field structure and a shift of the on-axis maxima towards the transducer. An example of the intensity distribution from such a source is shown in Fig. 14.11.

The simplest type of ultrasound hyperthermia applicator consists of a plane circular, air-backed transducer mounted in a housing containing a column of temperature-controlled and degassed water. By arranging an air–transducer interface at the back of the transducer, almost complete reflection of the energy within the transducer is achieved, improving the transducer's efficiency. On the other hand, energy is transmitted from the transducer at its front surface on account of the smaller difference in impedances between the transducer and water (a situation which can be further improved by the addition of an impedance matching layer). It is necessary to degas the water (by boiling or exposing the water to vacuum) to avoid the creation of air bubbles due to cavitation; these lead to scattering of the ultrasound beam and can even coat the surface of the transducer. The water column should be sufficiently long to

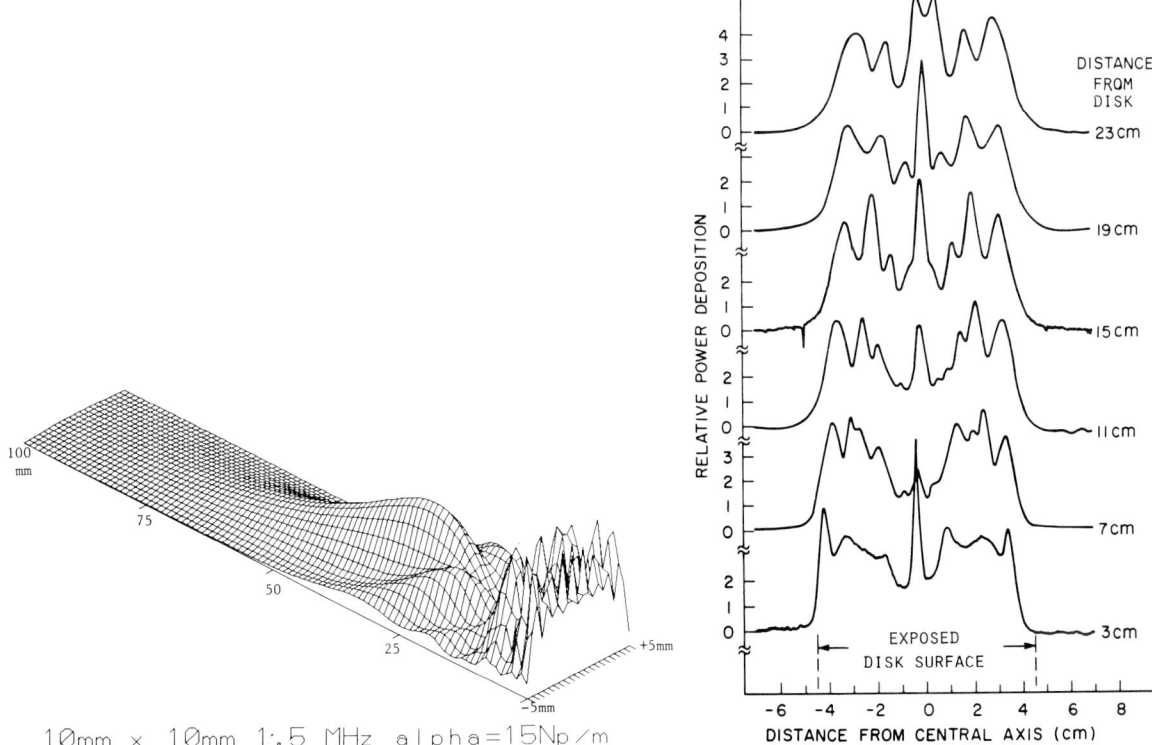

Fig. 14.11. Ultrasound intensity distributions. On the *right* are experimental scans for a 1.1-MHz 10-cm diameter circular transducer in water (courtesy of Eric Lee: by permission, Karger). The *left figure* is the theoretical distribution for a square 10 mm × 10 mm transducer operating at 1.5 MHz ($\lambda = 1$ mm) directed into muscle-like material with an attenuation coefficient of 15 Np/m

contain the intense peaks of the near field and is usually closed by a thin latex membrane. The applicator is placed in direct contact with the tissue and ultrasound coupling is aided through the use of a thin layer of gel or degassed water. Even with the water column, tissues are usually exposed to part of the near field and so the driving frequency is often modulated to reduce hot spots; thermal conduction and blood flow cause additional smoothing of the resultant temperature distribution. As is somewhat the case for superficial microwave applicators, the depth to which effective heating can be achieved is determined by selection of ultrasound frequency and power and temperature of the water bolus. Transducers used for this type of applicator are usually 3–12 cm in diameter, driven at a frequency in the range 0.5–3 MHz and at power densities of 0.3–2 W cm^{-2} (CORRY et al. 1984; MARCHAL et al. 1982). They are useful for treating tumors to depths of 3–4 cm. Pain is often a limiting factor, particularly when the tumor overlies bone. The construction of such applicators has been described by POUNDS and BRITT (1984) and HUNT (1990).

An early demonstration of the potential of ultrasound to heat deeply located tumors was the work of FESSENDEN et al. (1984a), who used an applicator in which six transducers, 7 cm in diameter and driven at their resonant frequencies close to 0.35 MHz, were mounted on a spherical surface. The beams converged to a "focus" of variable width enabling selective heating of perfused volumes of approximately 200 cm^3 at depths of about 10 cm to be achieved. Many of the patients treated in this study experienced pain which was attributed to the significant ultrasound intensity beyond the target volumes.

Another simple source is the plane, rectangular transducer. As described by OCHELTREE and FRIZZELL (1989), the field from a rectangular source of dimensions $a \times b$ is dependent on a/b in addition to a/λ and b/λ, and it lacks the on-axis nulls and many of the lateral variations seen in the near field of a circular source.

Plane, rectangular transducers have been used in multielement applicators designed for treatment of superficial or intermediate depth fields with good spatial control of ultrasound power deposition (DICKINSON 1984; BENKESER et al. 1989). One commercial system uses a transducer with a radiating area of 15 cm × 15 cm which is subdivided into 16 segments, each 3.5 cm × 3.5 cm and driven separately by one of 16 amplifiers. This allows some geometric field shaping and adjustment of the absorbed power distribution during treatment. In addition, the trans-

ducers may be driven at either their fundamental frequency of 1 MHz or at the harmonic 3 MHz and so some control over the profile of power deposition with increasing depth is possible. In their experience with this applicator, SAMULSKI et al. (1990b) found that better treatment temperatures could be achieved in tumors located in the groin and body trunk than in other sites. However, treatment setup was difficult and time consuming and complaints of pain during treatment were common.

14.3.4 Focusing

Ultrasound may be focused from single sources using curved radiators, lenses, or reflectors or from multiple sources by electronic control of phase. For a spherical shell transducer of diameter d and radius of curvature R (Fig. 14.12), the intensity I_{gf} at the geometrical focus is $I_s(\pi d^2/8R\lambda)^2$ where I_s is the average intensity over the transducer surface. The acoustic pressure profile $p(R,\theta)$ near the focal plane is described approximately by $p_{gf} 2J_1(ka\sin\theta)/ka\sin\theta$ where p_{gf} is the pressure at the geometrical focus and J_1 is the first-order Bessel function. The ratio R/d (the F-number of the transducer) and the frequency are parameters which determine focusing properties. For strongly focusing transducers, the focal volume can be considered as an elongated ellipsoid with an aspect ratio of approximately 6.5 d/R (SWINDELL 1986). By taking these factors into account and assuming that the attenuation coefficient for soft tissue is around 10 Np m^{-1} MHz^{-1}, it can be shown that a useful frequency for focused transducers is around 1 MHz for many applications in hyperthermia (HYNYNEN et al. 1981).

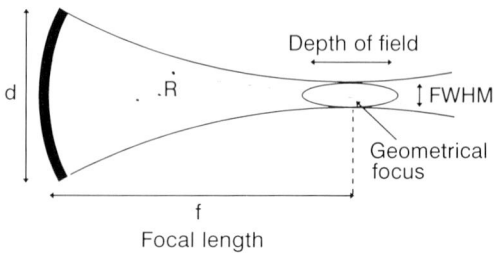

Fig. 14.12. Spherical shell transducer showing tranducer diameter d, radius of curvature R and focal length f. The prolate spheroidal focal volume (defined by the surface within which the intensity is greater than 50% of the peak value) is characterized by the depth of field and the full width at half maximum (FWHM)

14.3.5 Mechanically Scanned Arrays

Since the focal volume of a focused ultrasound transducer is small compared with tumor volumes, a method of delivering ultrasound energy over the whole target is required. One technique for heating a larger volume is to use several focused transducers orientated such that their focal volumes are close but offset from each other. A recent system of this type employs a large aperture array (≈ 50 cm in diameter) of 30 transducers with weakly focusing polystyrene lenses arranged in four concentric rings on a spherical surface (SEPPI et al. 1985; NUSSBAUM et al. 1991). By adjusting the positions of the transducers, a focal ring of variable diameter (0–11 cm) can be formed at a distance of approximately 32 cm from the array. The complete array can be translated in three orthogonal directions and rotated about its central axis. Oscillatory rotation over $\pi/2$ radians is also possible. The transducers may be driven from 0.5 to 1 MHz and the power applied to each ring of transducers may be set independently. The array is immersed in a water bath, on top of which the patient is positioned. Preclinical experience with this system suggests that selective deep heating of tissue volumes of lateral dimensions of 4–8 cm at depths of at least 11 cm in soft tissue may be possible.

An alternative approach to treating clinically relevant volumes is to scan the small focal volume around a predetermined trajectory within the target volume. The intensity gain due to focusing arises because the energy in the beam passes through a smaller area at the focus than at the transducer surface; the ratio of these areas, and hence the intensity gain, may be changed when a transducer is mechanically scanned. To maintain intensity gain, the transducer should be moved over a large window while continuously being directed at the target volume, for example, as in the system described by LELE (1983).

In other systems, a noncoherent array of focused transducers positioned such that their focal volumes coalesce is mounted on a gantry which may be translated along three orthogonal axes, rotated, and tilted. The common focal volume may be positioned within the target volume using an integral ultrasound imaging system or other diagnostic images and its trajectory is determined by scanning around one or more octagonal paths in the horizontal plane (Fig. 14.13). Details of individual systems are given by HYNYNEN et al. (1987, 1990) and HAND et al. (1992). These systems have shown that bulky tumors in the breast and other superficial sites can be heated quite well, as can some tumors in the pelvic region

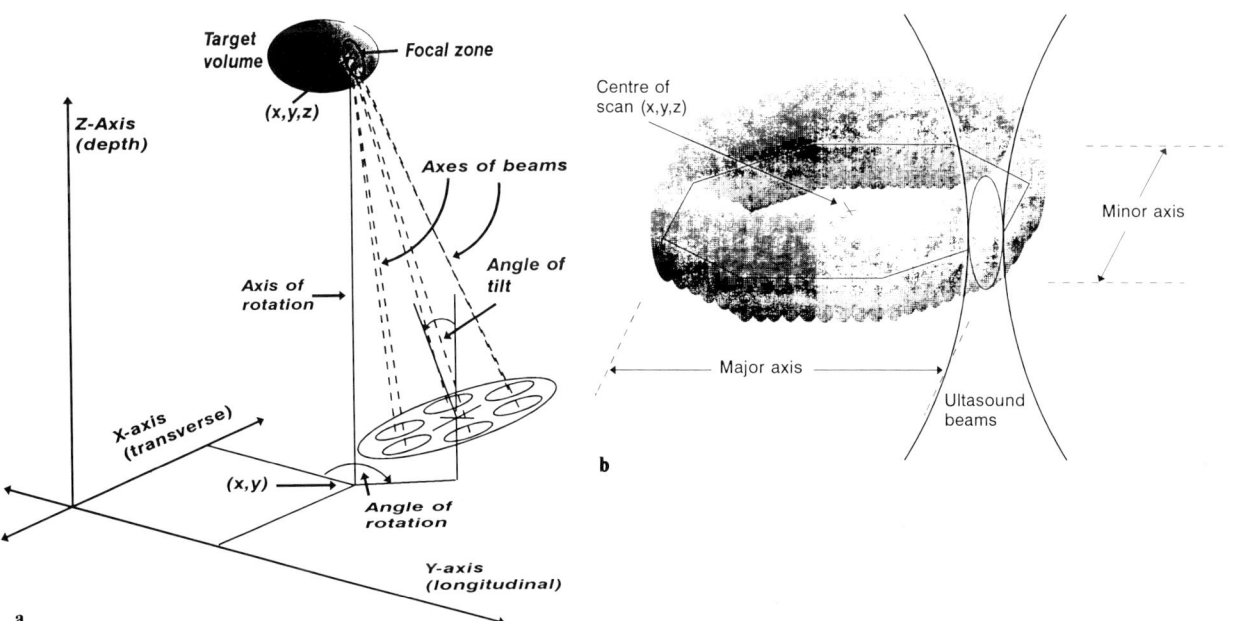

Fig. 14.13a,b. A scanned focused ultrasound system (Sonotherm 6500). **a** shows the position of the target volume relative to the machine's coordinate system. The transducer array may be translated in the x, y, and z directions, rotated through angle ϕ about an axis parallel to z and passing through the position (x,y,z), and tilted through angle α about (x,y,z,ϕ). Scanning is achieved by programming (x, y) to follow one or more octagonal paths. In this way a toroidal power deposition pattern, as shown in **b**, is achieved in an x–y plane at the target volume

when an adequate acoustic window is available. Recently, by adding higher frequency transducers ($\approx 4\,\text{MHz}$), the technique has been extended successfully to the treatment of chest wall tumors of large area (ANHALT et al. 1992).

14.3.6 Ultrasound Phased Arrays

An ultrasound phased array consists of small transducers which are coherent and individually driven. The distribution of phases and amplitudes applied to the elements can be chosen to produce either a focused beam which may be moved through the treatment volume or to synthesize directly a more diffuse heating pattern.

The first of these methods, referred to as "spot scanning," is the electronic analogue of mechanical scanning described in Sect. 14.3.5. Such electronic scanning may remove the need for mechanical movement and so simplify procedures for coupling between the ultrasound source and the patient. It can also take greater advantage of the limited windows through which the ultrasound can be transmitted to the tumor volume. Although phased arrays offer faster scanning than mechanical systems, a disadvantage of spot scanning remains in that high-intensity ultrasound is required as a consequence of attempting to heat clinically relevant volumes with relatively small foci. In theory, non-linear effects associated with moderate intensity ultrasound might be beneficial in heating tumors, particularly with respect to limiting power deposition distal to the tumor. However, it appears that the benefit occurs only within certain intensity/frequency combinations, and this is not completely understood (SWINDELL 1986).

In the second method, SAR distributions consisting of multiple foci are synthesized directly. Since the heating pattern is more diffuse than in spot scanning, the need for high-intensity fields is avoided, enabling even the potential benefits of nonlinear effects to be investigated. Furthermore, complex heating patterns can be produced by switching periodically between two or more predetermined multifoci patterns. The possible advantages of this method are obtained at the cost of increased complexity in the associated electronics. Problems such as the production of unwanted secondary foci and inefficient excitation of the array elements must also be addressed (EBBINI and CAIN 1991a).

IBBINI and CAIN (1989) investigated focal patterns associated with a concentric ring array. Using a field conjugation method to determine the amplitudes

and phases of the velocities at the surfaces of the rings, they demonstrated that, in addition to single spot focusing on the central axis, both single and multiple concentric focal rings of variable width can be produced. However, the usefulness of these patterns is compromised by unwanted secondary foci produced before and beyond the focal plane. IBBINI and CAIN (1990) suggested that single and multiple focusing on axis combined with mechanical scanning of the applicator might reduce this problem and result in useful heating patterns for small, intermediate to deeply located tumors.

A further limitation of the concentric ring applicator is that only annular patterns may be synthesized because of its circular symmetry. This restriction is removed if true two-dimensional arrays are considered. An example of a true two-dimensional array is the sector-vortex array (UMEMURA and CAIN 1989, 1992), which consists of a transducer disc partitioned into I tracks each divided into N sectors of equal size (Fig. 14.14). The array is geometrically focused and the phase of the driving signals applied to the sectors rotates M times per rotation along each track of N sectors. The signal applied to the nth sector on the ith track is:

$$A_i(\theta_n) = A_o e^{j[M(\theta_n + \beta(\theta_n)) + \gamma_i - \omega t]}$$

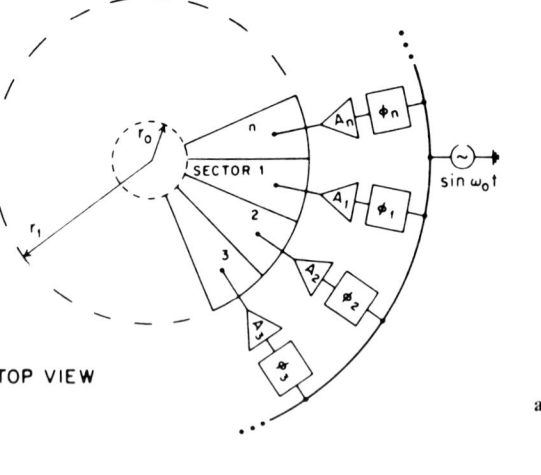

Fig. 14.14a,b. Schematic of sector-vortex 2D ultrasound array capable of synthesizing a variety of SAR distributions at depth directly. **a** Top view, **b** side view. (Courtesy of Charles Cain; by permission, The Institute of Electrical and Electronics Engineers, Inc)

where $n = 1...N$; $i = 1...$ I; $\theta_n = 2\pi n/N$ is the polar angle of the nth sector; $-N/2 \leq M \leq N/2$; γ_i is the constant phase difference between tracks; and ω is the angular frequency. $\beta(\theta)$ is a modulation function which determines the shape of the field patterns; when $\beta(\theta) \neq 0$ patterns without cylindrical symmetry can be produced. For $\gamma = 0$, a single focal spot on the axis of the array is formed when $M = 0$. When $M \neq 0$, annular focal patterns which increase in diameter with increasing M can be produced without creating unwanted secondary foci. When $\gamma \neq 0$ multiple focal spots ($M = 0$) or annular rings ($M > 0$) are formed, providing control of the acoustic power distribution in depth. Simulations suggest that at least $4M$ sectors per track are needed to produce smooth temperature distributions. A prototype device consisting of two tracks with 32 sectors per track, driven at 0.5 MHz and operated with $M = 8$ is under investigation (UMEMURA and CAIN 1992).

Other two-dimensional applicators which have been investigated include $N \times N$ square element arrays with plane, cylindrical, or spherical geometry (IBBINI and CAIN 1989; IBBINI et al. 1990, EBBINI and CAIN 1991a; McGOUGH et al. 1992). When choosing the size of the elements of a phased-array, thought must be given to the intensity of secondary (side or grating) lobes which will be produced. If possible, elements should be smaller than $\lambda/2$ (e.g., 1.5 mm at 0.5 MHz) but this implies a very large number of elements since the overall dimensions of the array will be on the order of 10 cm for clinical applications. IBBINI and CAIN (1989) used elements of 1.5 λ in a 20×20 element array and obtained grating lobe intensities less than 13 db below the focal intensity; in the case of 2-λ elements the lobe intensities increased to -6 db with respect to the focal intensity.

Other phased arrays, designed with reduced complexity in mind, include a series of linear phased arrays in which electronic steering is achieved in one plane while adjustment of the heating pattern in the orthogonal plane is achieved by switching to neighboring arrays (OCHELTREE et al. 1984) or, in the case of tapered arrays, by adjusting the driving frequency to elements with different thickness (BENKESER et al. 1987). The intensity gain achievable with one-dimensional arrays can be enhanced through geometrical focusing as in the cylindrical array considered by EBBINI and CAIN (1991b).

The success of ultrasound hyperthermia will depend greatly upon progress in treatment planning, including the ability to determine optimum acoustic windows on a patient-specific basis. This may well include advances made in incorporating US imaging into US hyperthermia devices. In the case of mechanically scanned systems, optimization of scanned paths and scanning speed and matching of frequency to treatment site should lead to improved treatments. Clinical evaluation of phased-array systems is imminent and assessment of their performance is eagerly awaited.

14.4 Interstitial and Intracavitary Methods

14.4.1 Radiofrequency Electrodes

In radiofrequency (RF) methods for interstitial hyperthermia, currents are forced to flow between pairs of electrodes and give rise to resistive heating within the intervening tissue. The choice of frequency for these techniques must be such that conduction currents dominate over displacement currents. This imposes an upper limit of a few tens of MHz in high-water-content tissue. On the other hand, direct electrical stimulation of nerve and muscle fibers must be avoided. This implies a lower limit of a few hundred kHz. Many systems operate at around 500 kHz, although recent systems take advantage of the capacitive coupling which is possible at 27 MHz.

The electric field distribution within the implant may be found by solving Laplace's equation with boundary conditions set by the voltages applied to the electrodes. The resulting current density is high at the electrodes and falls off rapidly with distance. Theoretical calculations and practical experience both suggest that interelectrode spacing should be limited to about 15 mm. It is important that parallelism of electrodes is achieved during implantation since regions in which electrodes converge are likely to be heated excessively. The distribution of absorbed power is also modified by spatial variations in the dielectric properties of the tissues.

Rigid metallic needles insulated from normal tissues at entry and exit points only were used in early studies of the technique. However, flexible electrodes were subsequently developed to overcome a number of practical problems encountered with these early types. One method uses electrodes consisting of a central metallic section, the length of which is chosen to match the dimensions of the tumor to be heated, with a flexible plastic section attached at each end (COSSET 1990). Electrical contact to the central section is made via a metallic probe inserted through the lumen of one of the plastic sections. In another approach, the electrodes consist of hollow plastic tubes covered with wire braid which in turn is covered with a thin layer of vinyl (KAPP and PRIONAS 1992). The electrodes are customized for particular applications by removing appropriate lengths of the vinyl insulation.

Several methods of applying power to arrays of interstitial (RF) electrodes have been investigated. These include (a) using several generators, each connected to a predetermined electrode pair, (b) switching the power from a single generator to predetermined pairs in such a way that the sequencing, dwell time, and power for each pair can be used to control the temperature at the electrodes, (c) switching the power from a single generator to any two electrodes within the array. The latter method (Fig. 14.15) offers the best control over the temperature distribution but requires that the temperature is measured at each electrode (GOFFINET et al. 1990). Prionas has developed a preclinical RF electrode with four electrically independent segments allowing a fair measure of axial control of the SAR distribution (PRIONAS et al. 1993).

Fig. 14.15. Schematic of a versatile RF interstitial electrode power control system. Pulsed 0.5 MHz power is applied to the coldest pair of electrodes for a dwell time calculated on line to maintain a specified target temperature (Courtesy of Stavros Prionas; by permission, Pergamon Press)

In a variation of the traditional RF technique which relies upon direct (galvanic) contact between electrodes and tissues, Visser and other developed a system which operated at 27 MHz (VISSER et al. 1989). This system takes (Fig. 14.16) advantage of the capacitive coupling between an electrode within an insulating catheter and the surrounding tissue, avoiding direct electrical contact with the internal electrode. Electronically, this arrangement can be considered in terms of a circuit in which the capacity of the catheter C_{cath} is connected in series with the parallel arrangement of a resistance R_{tiss} and a capacity C_{tiss} representing the tissue. Since the impedance associated with C_{cath} is large, the current flowing through the tissue between electrodes is less dependent on variations in interelectrode spacing; the capacitively coupled electrode behaves as a current source. This is in contrast with the methods using lower frequency which depend upon galvanic contact so that the current density at an electrode depends upon the impedance presented to the electrode in the presence of neighboring electrodes (VISSER et al. 1993). The active length of this type of electrode can be selected by making the diameter of the electrode within the catheter large in the region of interest, thus causing a local increase in coupling between electrode and tissue. Each electrode is associated with an external ground plane electrode

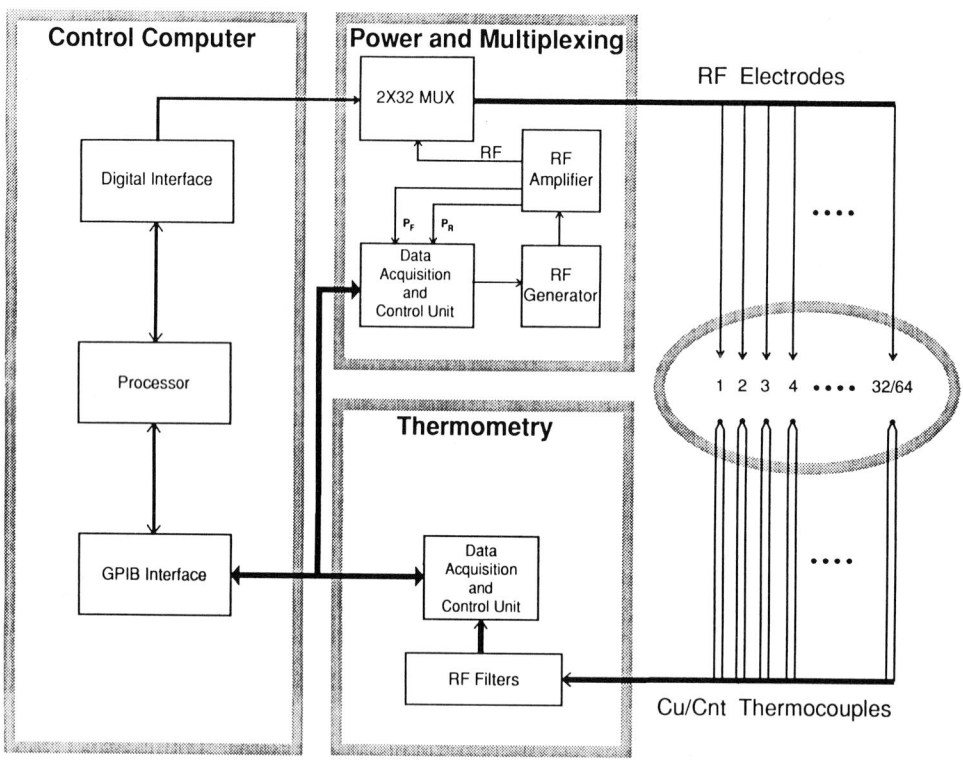

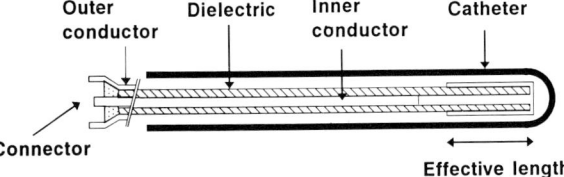

Fig. 14.16. Schematic diagram of an RF capacitively coupled interstitial electrode. The effective heating is limited to the section of conductor with enlarged diameter (over which coupling to the tissue is enhanced). The device is used with an externally located ground plane

and is driven by an RF amplifier via an impedance matching network. Since direct electrical contact is not necessary, this method is readily combined with brachytherapy procedures using standard after-loading catheters and, like the flexible wire braid electrodes (GOFFINET et al. 1990; PRIONAS et al. 1994), extends the use of RF interstitial hyperthermia to treatments in the head and neck region where the use of curved catheters with small radii of curvature may be encountered. Several applicator configurations have been utilized (DEURLOO et al. 1991).

Capacitive electrodes may be used singly, each with respect to a grounded electrode on the skin, or in balanced pairs, in which case current passes from one electrode to the other. A third possibility is to have a pair of electrodes (or multiple pairs) within the same catheter. The last configuration has been used in a system in which two capacitively coupled electrodes are driven with signals having 180° phase difference; current flows between the pair of small electrodes resulting in local heating around that region of the catheter. Arrays of catheters can be driven by coherent signals and control over the SAR distribution can be achieved by varying the phase relationships between signals at the electrode pairs (LAGENDIJK 1994).

For intracavitary applications, an electrode usually in the form of a coil or wire is placed within a housing which is often water-cooled, and this is inserted into the body cavity. A much larger area return electrode is placed externally on the patient's skin. Frequencies in the range of 8–27 MHz are used such that capacitive coupling is possible. The applicator may be considered as a rod-like conductor to which electric field lines converge (in a manner similar to the case of an external capacitive electrode described in Sect. 14.2.4). This leads to a region of relatively high current density around the applicator and hence to local heating. The use of the technique for tumors of the esophagus has been described (SUGIMACHI and MATSUDA 1990).

14.4.2 Microwave Antennas

A common form of microwave interstitial applicator is a dipole antenna made from flexible miniature coaxial cable with an extension of the inner conductor at the distal end (Fig. 14.17). Such antennas may be used within plastic catheters. Since the length of the antenna is comparable with the wavelength, energy is radiated into the surrounding tissue. However, fields decay rapidly with distance from the antenna due to geometric factors (the antenna is essentially a line source) in addition to attenuation in the tissue. Nevertheless, this type of interstitial device can achieve higher levels of absorbed energy away from the antenna relative to that close to it than can radiofrequency devices. The performance of an interstitial dipole antenna is dependent upon its dimensions [in particular its radius and the distances from the junction, where the extension of the inner conductor begins, to the tip (h_A) and back to the tissue–air interface (h_B), the radius, wall thickness, and dielectric properties of the catheter in which it is inserted, and the dielectric properties of the surrounding tissues (Fig. 14.17). The behavior of this type of antenna has been described theoretically (KING et al. 1983; CASEY and BANSAL 1986; ISKANDER and TUMEH 1989; ZHANG et al. 1988). There are benefits (e.g., antenna impedance close to 50 ohms) if h_A and h_B are approximately one-quarter of the wavelength (TREMBLY 1985; JONES et al. 1988, 1989). In practice, the insertion depth (the distance from the

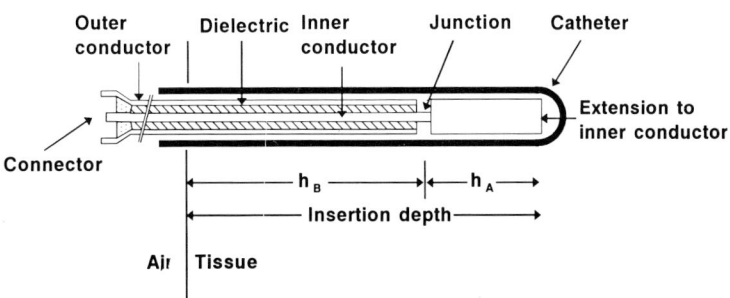

Fig. 14.17. Schematic diagram of a microwave dipole interstitial applicator within a catheter showing the characteristic dimensions h_A, h_B and insertion depth

insertion point to the tip of the antenna) is such that h_B can differ considerably from h_A, so that the dipole antenna becomes asymmetrical. Asymmetrical dipole antennas have been studied (ZHANG et al. 1988; JONES et al. 1988, 1989). A practical difficulty in using conventional dipole antennas is that SAR distributions are dependent upon insertion depth (KING et al. 1981; JONES et al. 1988) but designs to minimize this, for example by incorporating a quarter wavelength sleeve, or "choke," on the feeding line proximal to the junction, have been reported (HÜRTER et al. 1991; RYAN et al. 1990).

The SAR distribution around an implanted antenna depends upon catheter size and wall thickness (JAMES et al. 1989; RYAN et al. 1991). Temperatures close to the catheter may be reduced by cooling the antenna/catheter. TREMBLY et al. (1991) suggest that it is better to use air as a cooling agent rather than water since the high value of relative permittivity for water results in an antenna/cooling medium/catheter/tissue combination for which fields decay with increasing radial distance more rapidly than when the cooling medium has low permittivity.

A disadvantage of conventional dipole antennas is that power deposition decreases towards the tip of the antenna, leading to inadequate heating in that area and the need to implant the antenna beyond the target volume. Modified dipole antennas which provide increased coupling to the tissue in the region of the tip have been developed (TURNER 1986; ROOS and HUGANDER 1988; TUMEH and ISKANDER 1989).

In view of these drawbacks of conventional dipole antennas, other antenna designs such as multinode antennas (LEE et al. 1986), sleeved antennas (LIN and WANG 1987), and several types of helical coil antennas (WU et al. 1987; SATOH et al. 1988) have been proposed. The most popular alternatives appear to be helical coil antennas in which a helical winding is positioned over the distal section of a semirigid coaxial cable along which the outer conductor has been removed (Fig. 14.18). Performance is dependent upon the operating frequency, pitch of the helix, and the manner in which the helix is connected to the coaxial cable. These antennas can produce a marked shift of the heated volume toward the tip and a distribution which is more cylindrical than those

Fig. 14.18A–D. Schematic diagrams of alternative designs for microwave interstitial applicators with lengths in mm indicated (not to scale). **A** Multisection antenna based on 0.95-mm-diameter coaxial cable in which sections (i) and (v) are collars of 1.07 mm diameter, (ii) and (iii) are sections of the coaxial cable with and without the outer conductor, respectively, and (iv) is a metallic tube 0.86 mm in diameter (TURNER 1986). **B** Helical coil antenna designed for operation at 915 MHz (SATOH et al. 1988). The coaxial cable is 0.95 mm in diameter and the coil, wound from 0.32-mm-diameter nichrome wire, has 35 turns with 1 mm pitch and is less than 1.2 mm in outer diameter. **C** Dipole antenna modified by the addition of quarter wave-length choke sections over both proximal and distal sections (RYAN et al. 1990). **D** Three-node antenna (LEE et al. 1986) in which sections of the outer conductor of 0.95 mm diameter coaxial cable are removed, leaving the dielectric and inner conductor

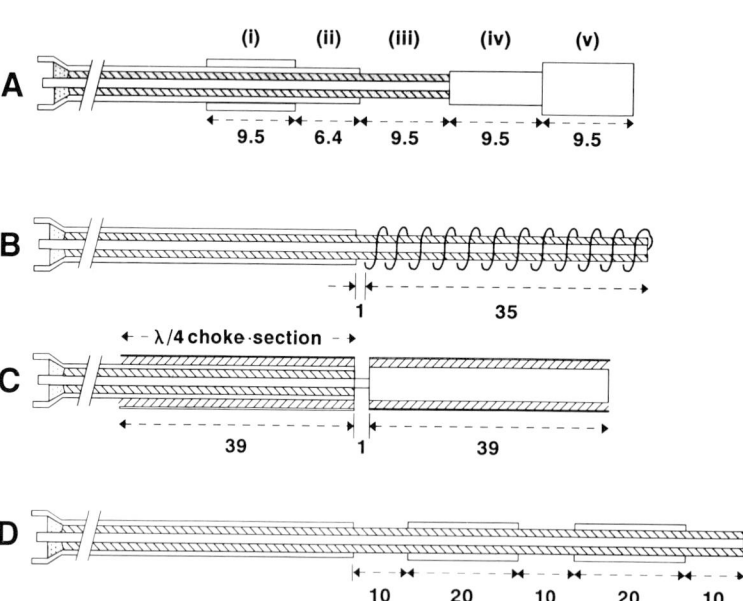

associated with conventional dipole antennas and fairly insensitive to insertion depth (Satoh et al. 1988; Sathiaseelan et al. 1991; Ryan 1991).

Since heating patterns associated with single interstitial microwave antennas have a radial penetration of no more than about 10 mm, an array of implanted antennas is usually required in clinical applications. The antennas in such arrays may be driven either incoherently or coherently; in the latter case the SAR at certain locations can be enhanced locally by constructive interference. Suitable phase and amplitude modulation of coherent arrays of dipole antennas can result in either a relatively uniform SAR across a significant area of the array or a distribution in which the SAR in peripheral regions is greater than that in the central region of the array (Trembly et al. 1986; Furse and Iskander 1989; Zhang et al. 1990a,b, 1991a). Although theoretical studies show that control over SAR in the junction-plane of an array is feasible, factors such as small differences between individual antennas and feed lines, antennas being inserted to different depths, and dielectric heterogeneities within the tissues complicate their use in practice.

The dependence of the SAR distribution on insertion depth for arrays of dipole antennas has been studied experimentally and the observed behavior may be explained theoretically in terms of basic transmission line models (Chan et al. 1989; Denman et al. 1988; Zhang et al. 1991b; Mechling et al. 1992). In practice, the clinician performing the surgical implantation sometimes has difficulty in achieving the perfect geometry most physical models assume. Clibbon et al. (1993) have examined the consequences of a nonideal configuration of antennas. Their work suggests that a correction to the phase distribution amongst antennas should overcome distortions in the SAR distribution when the antennas are implanted to differing depths. In the case of helical coil antennas, the SAR in the central region of transverse planes relative to that near antennas is less than in the case of dipole antennas (Sathiaseelan et al. 1991; Ryan 1991).

Intracavitary devices are often designed around a microwave antenna similar to those used for interstitial purposes but enclosed within an insulating sleeve which normally incorporates water cooling. Dipoles, sleeved antennas, and helical coils have been used (Visser et al. 1992). In some devices the antenna is located eccentrically within this sleeve so that heating occurs predominantly within a local region close to the applicator. In other designs, a reflector is mounted within the insulation to achieve

the same result. Intracavitary and intraluminal applicators are usually designed for operation at 915 MHz. Intracavitary applicators have also been designed for use with brachytherapy (Astrahan et al. 1989).

14.4.3 Ultrasound

The use of interstitial ultrasound transducers is discussed by Hynynen (1992). Simulations suggest that adequate penetration could be achieved by driving small cylindrical transducers of outer diameter approximately 1 mm at 6–10 MHz. Preliminary experiments in which such transducers were inserted into standard plastic brachytherapy catheters and coupled to tissue by degassed water showed that this was a feasible method to heat tissue in which perfusion was comparable to that in many tumors. Among possible advantages are heating to the tip of the transducer, independence of the ultrasound field from insertion depth, and sufficient radial penetration to allow a relatively large intercatheter spacing of up to 2 cm.

As mentioned in Sects. 14.4.1 and 14.4.2, electromagnetic intracavitary devices produce heating patterns which cannot readily be adjusted in the longitudinal direction and which generally are characterized by very limited radial penetration. Recently, ultrasound devices have been developed which overcome these important limitations. In one type of device, a linear array of cylindrical transducers, each approximately 10λ long, 10–15 mm in diameter, and driven sequentially from a single power amplifier, is housed within a cylindrical membrane through which cooled, degassed water is circulated. The depth at which the maximum temperature in the tissue occurs can be up to 15 mm from the surface of such applicators operating at 1.07 and 1.6 MHz. When the applicator is used transrectally, therapeutic temperatures at depths of 2–3 cm can be achieved (Diederich and Hynynen 1990). If the application does not require a heating pattern with cylindrical geometry, a more directional pattern may be achieved by using plane transducers within the applicator. A radial penetration of 5 cm is claimed for this type of device operating at 1 MHz (Chitnalah et al. 1991). Applicators using phased linear arrays of cylindrical elements are also under development. In the case of devices designed for operation at 0.5 MHz, it is possible to use elements which are small enough to avoid grating lobe problems (see Sect. 14.3.6) yet which in practically realizable numbers can form an array of appropriate length for many applications. Such an array is expected to selectively heat regions between 2 and 5 cm from its surface (DIEDERICH and HYNYNEN 1991). HAND et al. (1993) considered a linear array of 25 planar transducers, each 1 wavelength wide and 15 mm high and operating at 0.5 MHz, which produced a focal region SAR greater than 50% of the peak value approximately 4 mm wide and 15 mm high with a depth of focus of 20 mm at a range of up to 6 cm. The fact that small focal regions may be produced offers the possibility of treatment by high temperature, short duration hyperthermia, and thermal ablation (HUNT et al. 1991; DORR and HYNYNEN 1992) in addition to conventional hyperthermia.

14.4.4 Hot Source Techniques

Hot source techniques differ from the electromagnetic and ultrasound ones discussed in Sects. 14.4.1–14.4.3 in that there is no (or, in the case of ferromagnetic seeds, very little) direct energy deposition from the implanted devices in the intervening tissues. Thus heterogeneities in the dielectric or ultrasound properties of the tissues do not directly affect the temperature distribution. Factors which are important in these methods are the geometry of the implant, the temperatures of the implanted devices, tissue perfusion, and tissue conductivity.

14.4.4.1 Hot Water Tubes

Arguably the simplest of these techniques is one in which hot water is passed through an implanted array of steel tubes or plastic catheters. By adjusting the flow rate (preferably high enough to ensure turbulent flow) and the temperature of the water (typically $46°$–$48°C$), sufficient power is available from an array of tubes spaced by approximately 10–15 mm to heat tissues having moderate perfusion (HAND et al. 1992a).

The essential components of a practical system are a thermostatically regulated reservoir of water, manifolds to supply water to the implanted tubes, and pumps to circulate the water. Designs incorporating unidirectional flow through open-ended tubes or counter-flow within closed catheters have been reported (SCHREIER et al. 1990). The facts that the maximum temperature within the tissue cannot exceed that of the water, that some degree of regulation of the temperature distribution along the tube occurs since the local heat transfer is dependent upon

the difference in temperatures of the water and the local tissue, and that no additional problems are posed when using conventional thermocouple or thermistor thermometry, are some of the advantages of hot water systems. On the other hand, like all hot source techniques, performance is susceptible to changes in blood perfusion. To minimize this, tubes should have the largest possible diameter and be implanted with the minimum spacing, subject to practical restrictions. In practice, tubes 1.6 mm in diameter have been used with unidirectional flow and 1.8 mm in diameter in the case of the counter-flow method; spacing between tubes is typically 8–14 mm (BUDIHNA et al. 1992; HANDL-ZELLER and HANDL 1992).

14.4.4.2 Electrically Heated Implants

In this technique, catheters approximately 2 mm in diameter and containing an electrical resistance element heated by DC or low-frequency current are implanted into the tissue at a spacing of approximately 15 mm (DEFORD et al. 1991). The lengths of the heating elements may be matched to the dimensions of the tumor. The temperature within catheters is monitored and this information is used in controlling the power applied to the heating elements. Since the heaters produce uniform power per unit length, there is no control of the temperature distribution along an individual catheter when heat losses along it are nonuniform, as is usually the case. This problem could be resolved by using segmented heaters in each catheter with individual power control to each segment.

14.4.4.3 Ferromagnetic Seeds

In this technique, implanted ferromagnetic material in the form of thin cylindrical seeds is subjected to an alternating (typically at 100–200 kHz) magnetic field. These seeds are heated predominantly as a result of induced eddy currents. ATKINSON et al. (1984) discuss how the heating power per unit length of long seeds depends upon the relative orientation of the seed with respect to the magnetic field. It is given by $\pi H_o^2 a \sqrt{f(x) \omega \mu / 2\sigma}$ when they are parallel and $8 \pi H_o^2 a \sqrt{g(x, \mu) \omega \mu / 2\sigma}$ when perpendicular. In these expressions a, σ, μ are the radius, electrical conductivity, and permeability of the implanted seed, $x = \sqrt{\omega \sigma \mu}$, and ω and H_o are the angular frequency and strength of the magnetic field. $f(x)$ is a function of $ber(x)$, $bei(x)$ (i.e., Kelvin functions containing a Bessel function of the first kind) and their derivatives while $g(x,\mu)$ is also dependent upon the relative permeability of the implanted material. In practice the implanted seeds should be approximately parallel with the magnetic field. If we assume a cylindrical tissue load within a solenoidal coil, then the local absorbed power density in tissues due directly to the alternating magnetic field is proportional to $\sigma_t(\mu_t H_o fr)^2$ where σ_t, μ_t are the conductivity and permeability of the tissue and r is the radial distance from the central axis to the point in question. BREZOVICH (1988) suggests that the product $H_o f$ should not exceed 4.3×10^8 amp $m^{-1} s^{-1}$ to avoid direct heating of the tissue.

The fact that the spontaneous magnetization of ferromagnetic material vanishes above a temperature known as the Curie temperature (T_c) enables ferromagnetic seeds to be used in one of two ways. If the seed is made of a material for which T_c is much higher than temperatures normally encountered in hyperthermia, its "heating power" depends upon ($H_o f$) and is essentially independent of its temperature. The temperature attained by this type of seed is dependent upon the local heat transfer mechanisms in the tissue and the power delivered to the seed. Such seeds are referred to as constant power seeds; stainless steel seeds are an example of this type. On the other hand, seeds may be made from materials with T_c close to the desired operating temperature. In these cases the dependence of heating power P on temperature T may be expressed in the form $P = (1 + A e^{BT})^{-1}$ where A and B are constants associated with a particular type of implant (HAIDER et al. 1987). Alloys of nickel-silicon, nickel-copper, and nickel-palladium have been investigated for this purpose (CHEN et al. 1988; BREZOVICH et al. 1984; KOBAYASHI et al. 1991). In the case of these so-called constant temperature seeds, the "heating power" is strongly dependent upon temperature in the region of T_c and thermal regulation occurs if sufficient power is supplied. For most seeds investigated to date, the transition from ferromagnetic to nonferromagnetic states takes place over a range of about 10°C but new techniques with Ni-Pd are very encuraging (Fig. 14.19).

Clinical investigations of ferromagnetic seed techniques show that the seed spacing should not be greater than about 1–1.2 cm. Temperature distributions may be improved through judicious choice of seed materials within the implant aided by treatment planning. Although to date ferromagnetic seeds have been used within plastic catheters for

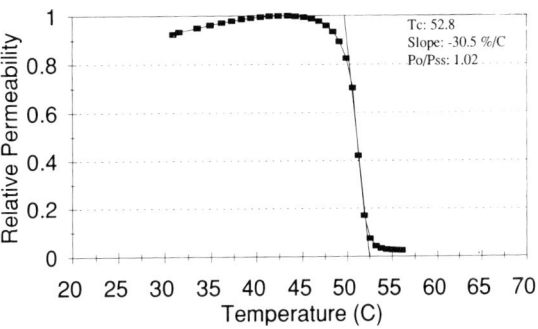

Fig. 14.19. Relative permeability (nearly proportional to power absorption) as a function of temperature for thermally regulating ferromagnetic seeds. The Curie point, where the magnetic permeability value falls to that for free space, is determined by extrapolation of the linear part of the curve to zero power. (Courtesy of Thomas Cetas)

temporary implants, the techniques offer the possibility of using permanent implants when problems of seed migration and biocompatibility have been satisfactorily addressed.

14.4.5 Summary of Interstitial and Intracavitary Methods

1. *RF techniques*

LCF electrodes
Advantages:
– Relatively simple technique
– Commercially available systems
– Longitudinal control of SAR using segmented electrodes

Disadvantages:
– Flexible electrodes must be custom made
– Parallel implants highly desirable
– Setup can be tedious

Capacitively coupled electrodes
Advantages:
– Compatible with plastic brachytherapy catheters
– Less sensitive to deviations from perfect geometry
– Single catheter configuration possible
– Multiple electrode in a single catheter possible
– SAR control using coherent arrays of multiple electrode catheters

Disadvantages:
– No commercial systems at present
– Setup can be tedious

2. *Microwave techniques*
Advantages:
– Commercially available systems
– Compatible with plastic brachytherapy catheters
– Offer greatest spacing between implants
– Most robust performance in regions of high blood flow
– SAR control by phase and amplitude adjustments

Disadvantages:
– Longitudunal control of SAR difficult
– SAR distribution depends upon insertion depth
– Optical thermometry required

3. *Hot source techniques*
General features
Advantages:
– Heating patterns dependent only on thermal properties of tissue, not directly on dielectric properties
– Unrestricted choice for geometry and length of implant
– Tolerant of imperfect geometry

Disadvantages:
– In general, closer spacing between sources than RF or microwave methods
– Least robust performance in regions of high blood flow

Hot water tubes
Advantages:
– Commercially available
– Simple thermometry
– Maximum temperature known
– Essentially constant temperature sources

Disadvantages:
– Setup tedious

Ferromagnetic seeds
Advantages:
– Thermally self-regulating
– Potential for long-term implants

Disadvantages:
– Treatment planning critical
– Difficult to change temperature of implants during treatment

Electrically heated implants
Advantages:
– Commercially available
– Simple thermometry

Disadvantages:
– Difficult to control SAR along implants

14.5 Thermometry

Excellent thermometry is just as important to clinical hyperthermia as is the development of good, practical heating applicators. We still have much to learn about the temperature level and duration of heating relative to cure and complications, but important progress has been made (OLESON et al. 1993; LEOPOLD et al. 1993). Knowledge of temperatures obtained and retrospective analysis is crucial to this progress. On-line thermometry and patient feedback are the basis for safety and flexible power control.

14.5.1 Invasive Thermometry

Preclinical research and development and clinical experience indicate the requirements listed in Table 14.4. There are trade-offs, of course. For example, if one is willing to recalibrate before each laboratory experiment or clinical treatment, a drift of 0.1 °C/h would be tolerable. Assuming the sensor temperature equilibrates at a rate proportional to temperature difference ΔT, ΔT will be equal to $\Delta T_o (1 - e^{t/\tau})$. ΔT and ΔT_o are the measured and actual temperature changes in the local surrounding tissue respectively, and t is time. With a time constant $\tau = 3$ s, the sensor reading for most situations will be within 0.1°C of the final reading in about 10 s, an upper limit if extensive thermal mapping is performed.

With few exceptions, invasive thermometry, unfortunately, is the rule. The 1 mm maximum size goal is more than satisfactory for most laboratory work, and is very desirable from the viewpoint of patient comfort.

14.5.1.1 Electromagnetic

The fact that the temperature probe's sensor and leads are not tissue equivalent can cause interference with the temperature measurement. Metallic parts can lead to a high likelihood of currents induced by

Table 14.4. Hyperthermia thermometry requirements

Parameters	Performance goal
Accuracy	≤ 0.2°C overall
Precision	≤ 0.1°C
Drift	≤ 0.1°C between calibrations
Response time	≤ 3 s thermal time constant τ
Sensor size	≤ 1 mm
Calibration frequency	≥ 24 h
Durability	≥ 6 month lifetime under clinical use
Artifacts	≤ 0.1°C (well characterized in any case)

electromagnetic interference (EMI), and may even result in a perturbation of the heating source EM fields.

14.5.1.1.1 Thermocouple There is no way of avoiding metallic parts with a thermocouple temperature sensor. When two metals with dissimilar conduction electron Fermi levels (i.e., work functions) are brought into good electrical contact, a voltage difference of generally a few millivolts (mV) numerically equal to the difference in work functions develops at the metal–metal junction. Temperature affects the energy of the conduction electrons, and thus the Fermi levels, work function difference, and thermocouple junction (thermoelectric) voltage. In a closed circuit of the two dissimilar metals (fine wires) there must of course be a second junction. This second junction is kept in a constant temperature environment (usually formed by a temperature-regulated electronic heater), and the net voltage across the first junction (the temperature "probe") is then accurately related to the first junction's temperature. Since this is related only to the work functions, a universal calibration applies to all thermocouple probes for a given set of two metals (SAMULSKI and FESSENDEN 1990). The most popular thermocouple probe in the United States is a copper–constantan wire pair, while in Europe it is often a manganin–constantan wire pair.

Small wire gauges with diameters of 25 µm or less may be used, so even multisensor thermocouple junction probes, which require a bundle of wires, cause no problem with respect to size (CARNOCHAN et al. 1986; WATERMAN and HOH 1994). In fact, referring to Table 14.4, the only problems with these probes are the artifacts related mainly to the unavoidable metal present. The high thermal conductivity causes thermal smearing (discussed later). In addition, there is significant perturbation due to EMI and, to a lesser degree, disturbance of the very EM field which is causing the tissue heating.

In typical use, the power applied to the applicator must be turned off for a short time on the order of a second to allow measuring the thermocouple voltage unperturbed by EMI. The power off time also allows for (some) decay of the artifactually high temperature caused by I^2R heating of the junction due to EMI-induced currents in the thermocouple circuit. The assumption is usually made that the artifactual heating decays very rapidly so that a T measurement after about 1 s of power off is the correct tissue temperature. The accuracy of this procedure, however, is not so clear. Alternatively, the technique of extending the "power off" time to several seconds and extrapolating successive measurements back to the true "power on" tissue T is sometimes employed, a reasonable thing to do realizing that the cooling of the local tissue can be significant, depending on the local blood flow (DE LEEUW et al. 1993). The problems are worse for the higher range of EM frequencies where EMI is very difficult to minimize in the exposed thermocouple leads. For frequencies below several tens of MHz, however, particularly in the 500-kHz range at which RF interstitial hyperthermia is often performed, the lower capacitive coupling of the external fields to the thermocouple probe and leads is less of a problem and can be electronically filtered out to a great extent. Even here, however, measurement errors of 0.1°–0.2°C are common. Uncertainties easily approach 0.3°–0.5°C at higher frequencies, even with the mandatory "power off" techniques.

The prevalent technology, low expense, and relatively minor attention needed to calibration quality assurance resulted in thermocouples gaining wide acceptance for hyperthermia early on. This is less the case now, due to better recognition of the thermal smearing and EM-field-related artifacts. For systems at low frequency (KAPP et al. 1988) and/or with very large numbers of temperature sensors (DE LEEUW et al. 1993), however, they still find use.

14.5.1.1.2 Thermistor At first, thermistors did not find much use in hyperthermia applications. A thermistor is a type of electrical resistance thermometer whereby the temperature dependence of the resistivity of a small amount of material (the "probe") is used to advantage. The probe material for resistance thermometry could be a metal, but use of a semiconductor is preferably due to sensitivity and EMI. A typical material is a metal oxide with a temperature coefficient of resistivity equal to about 4%/°C. Two leads are necessary to supply a small current through the semiconductor. In addition, however, two additional leads are necessary to allow measurement of the voltage drop across the probe which, by virtue of the constant current source, allows the determination of the resistance via $V = IR$. Even with the use of a semiconductor material instead of metal for the probe, the four metallic leads of a conventional-type thermistor circuit is a serious problem with respect to EMI and possible EM field disturbance.

A development (BOWMAN 1976) whereby high-resistance carbon-impregnated Teflon leads were used in place of the four metallic wires, was very important. The carbon-impregnated Teflon approximates the electrical properties of soft tissue, eliminating the severe impedance mismatch and potential for large EMI-induced currents that metallic wires embedded in tissue present. BSD Medical Corporation (2188 West 2200 South, Salt Lake City, UT 84119, USA) has developed these probes commercially. With reference to the performance goals of Table 14.4, the only problems are size (due to the necessary four carbon-impregnated Teflon leads, a single sensor probe requires a 16-gauge, typically 1.4 mm outside diameter, catheter) and cost (presently well over $1000 per replacement probe). These probes have proven particularly rugged and accurate in clinical use.

14.5.1.1.3 Fiberoptic In response to the EMI problems, much work has gone into developing nonperturbing probes based on fiberoptic technology. Two approaches have been rather successful. One is based on the temperature dependence of the energy gap between the valence and conduction bands of the semiconductor gallium arsenide (GaAs). This first successful nonperturbing fiberoptic probe was originally developed by CHRISTENSEN (1977), and brought into commercial use by Clinitherm Corporation (Dallas, Texas, USA; Clinitherm products are now supported by the TEX-L Company, 2709 Avenue E. East, Suite 307, Arlington, TX 76011). Here, infrared light from an LED is directed onto a small prism of GaAs. The light that is not absorbed is reflected back through the same optical fiber and detected. The light absorption is dependent on the degree of overlap between the infrared light energy band and the energy gap ΔE via $e^{\alpha(h\upsilon - \Delta E)/kT}$ where α is a material sensitive constant, h is Planck's constant, υ the infrared light frequency, k is Boltzmann's constant, and T is the absolute temperature in $^\circ K$. The infrared light is chosen such that its central $h\upsilon$ and ΔE are close in value, resulting in a sensitive dependence of absorption (and thus on the light reflected) on temperature. Precision, accuracy, and drift have been problems with GaAs probes for some users, but the later versions appear to be more reliable. The probe size is not a problem, but although overall systems with up to 12 individual optical fibers (each with their own sensor) are available, multiple sensors within a single bundle have not been developed. Replacement probes are almost as expensive as the Bowman thermistor probes.

A different optical approach originally developed by SAMULSKI et al. (1982) and made available commercially by Luxtron (2775 Northwestern Parkway, Santa Clara, CA 95051, USA) relies on the photoluminesence of certain materials. In the first design, the aspect that was emphasized was simply the temperature dependence of the amount of photoluminesence. Light of one frequency is directed through an optical fiber to a small amount of a rare earth substance. Some light is reflected back, and this is in addition to the different frequency light that is emitted by the phosphor after absorption of the original light (i.e., phosphorescence). The ratio of light at the two frequencies is then a measure of the temperature. This technique is appealing because the important aspect is the ratio of the two returning frequencies, and thus does not depend on a well-controlled source intensity (as the GaAs system does). A rather serious problem relates to the change in the optical fiber's transmission upon bending which, unfortunately, is different for the two returning frequencies. Thus if the fiber were bent, the ratio of returning light intensities would be affected, and introduces temperature errors of several tenths of a degree or more under normal clinical use.

A later version of the Luxtron optical probe interrogates the decay time of the phosphoresence, rather than the light intensity ratio (WICKERSHEIM 1986). The decay time dependence on temperature is sufficiently sensitive for certain materials such that only minimal amounts of material are necessary. These later versions of the Luxtron probe, therefore, are free of the bend artifact error and also do not require a well-controlled source intensity. They can be packaged in tight small fibers allowing linear arrays of four sensors in a rugged unit that can be placed in a single implanted catheter. These have proven to meet all the criteria of Table 14.4, except possibly cost. (Present replacement cost of four-sensor brobes is about $600). Rather complex instrumentation is required for this and other fiberoptic thermometry systems (SAMULSKI and FESSENDEN 1990).

14.5.1.2 Thermometry in Ultrasound (US) Fields

If care is used to minimize EM leakage from the US power sources, there is virtually no concern about EMI or other EM effects associated with US thermometry, and therefore metallic leads are not a problem. For this reason, thermocouple probes are often used. The metallic leads are small enough that interference (reflection or scattering) by them of the US field is not a problem, even when multisensor probes consisting of linear arrays of thermocouple junctions are used. A straightforward application of applying Huygen's principle to the shadowing by a probe leads to very minimal effect if the probe diameter is $\le \sqrt{\lambda}/4$ (HYNYNEN and EDWARDS 1989).

A problem with thermocouple thermometry for US occurs with the use of plastic-type catheters for mapping the thermocouple junction(s) within, and even with the Teflon or parylene coating commonly used to insulate the leads and junctions from tissue. Such catheter and coating materials contain complex long molecular chains that absorb US energy more readily than tissue, thus giving rise to artifactually increased heating. The best material appears to be polyurethane (WATERMAN and LEEPER 1990), but in order to avoid errors of a few tenths of a °C or greater, only minimal amounts of the insulating material should be used. The problem is avoided if the thermocouples are mapped within implanted stainless steel catheters. Thin-walled tubing should be used to minimize thermal conduction smearing. In addition, if arrays of the metal catheters or tubing are used, interference with the US field could be a problem, as discussed above, especially with tubing having outside dimensions of 1.0 mm or greater.

It is important to have a snug fit of the probe mapped within the catheter or tubing in order to maintain good thermal contact among all three of the probe, catheter, and tissue. The stiffer Luxtron fiberoptic probes appear to offer a consistently better fit. Although the plastic fiber and coating of the Luxtron probes precludes their use as bare probes in an US field, employing them with a snug fit within small thin-walled stainless steel tubing is recommended.

14.5.1.3 Summary of Invasive Thermometry

There is no single thermometry system that will satisfy all the requirements of Table 14.4 for all situations. Metal should be avoided for EM hyperthermia, especially for frequencies in the few MHz

and above range. Other than the cost aspect, it appears that fiberoptic probes are ideal for EM. If multisensor probes are not required for the particular applications, then the Bowman type with high-resistance carbon-impregnated Teflon leads is suitable, particularly with respect to reliability and stability. For US thermometry, it is best to avoid direct exposure of the probe to the US field. Mapping is a highly efficient way to collect a large amount of temperature data from a single invasive catheter, but care should be used to maintain good thermal contact and dwell times should be sufficient to establish thermal equilibrium before moving the probe to the next measurement position within the catheter. Mapping should never be attempted without a catheter, as increased local tissue damage and likely artifactual high temperature readings due to blood clotting, etc., will result.

Artifacts of many types can occur. These should be anticipated and studied for each specific probe type/hyperthermia modality. Thermal smearing cannot be avoided when metallic leads, catheters (needles), or tubing are used. This has been well studied (SAMULSKI et al. 1985; CARNOCHAN et al. 1986; WATERMAN and HOH 1994), and its awareness can lead to prudent system designs. Copper has a thermal conductivity almost 100 times that of constantan or manganin, so eliminating copper will help the thermal conduction problem immensely. As pointed out by DICKINSON (1985), however, this makes the artifact associated with multijunction thermocouple arrays of the type where a common wire is used more of a potential problem. Unless extreme care is experienced in construction (LAGENDIJK and DE LEEUW 1993), the multijunction thermocouples are best made by packaging together junctions each with two independent leads. The common artifacts are listed below in Table 14.5. As a rule, the desired overall accuracy of 0.2°C can often be met for EM hyperthermia, but is not so easy or common for US thermometry.

Table 14.5. Hyperthermia thermometry artifacts

Artifact	Typical range of errors (°C)	Where encountered
EMI	0.1–1.0	Metal present in EM field
Preferential energy absorption	0.1–0.4	Plastic in US field
Thermal smearing	0.2–1.0	Metal catheters or exposed leads near large T gradients
Thermal shielding	0.1–0.5	Poor thermal contact near large T gradients
Bend artifact	0.1–0.4	Small radius bend of fiberoptic when light intensity is important
Junction offset	0.1–2.0	Multijunction thermocouple
Viscous	0.1–0.3	Loose fit in higher frequency US fields (FESSENDEN et al. 1984b)

14.5.2 Noninvasive Techniques

Even employing advanced 2D mapping techniques (TARCZY-HORNOCH et al. 1992), invasive thermometry will always be severely limiting with respect to the amount of information necessary for optimal safety, control, and analysis. Much work has gone into noninvasive temperature monitoring, inspired by the dream that someday one might be able to visualize accurately the 3D distribution completely noninvasively. Infrared, computerized axial tomography (CT), US, MW radiometry/tomography, applied potential tomography (APT), and magnetic resonance imaging (MRI) are some of the areas of investigation. The latter three are somewhat encouraging.

14.5.2.1 Microwave Thermometry

Blackbody radiation is emitted by any object at a temperature greater than absolute zero. The integral amount is proportional to T^4, and the relative intensity peaks in the infrared (IR) region. This is suitable for IR vision of surface objects, but due to the very rapid absorption of IR by tissue, tells us little or nothing about tissue deeper than about 1 mm. Microwave frequencies are significantly more penetrating, and the sensing of MW radiation emanating from hot tissue has been exploited in hyperthermia research. A system in use for some time in France uses a MW radiometer at the ISM frequency of 2450 MHz to control the power applied to MW applicators. Since only one frequency is detected, there is no way to tell whether the MW radiation is from a warm region close to the surface or a hot region deeper, and this imposes severe limitations on the technique (CHIVE 1990).

In order to explore MW radiometry more fully, recognition must be made of the fact that the attenuation of MW is greater the higher the frequency (PRIONAS and HAHN 1985). Mizushina, in particular, has done significant research in this area and has developed a succession of multiband radiometers. These can quantify the amount of radiation transmitted through the tissue surface in up to five different MW bands between 1.2 and 3.6 GHz. Then, making assumptions such as there exist successive layers at different depths, each at an unknown but specific temperature, simultaneous equations can be solved giving the temperature at a number of depths equal to the number of MW bands. A fundamental drawback is the poor spatial resolution relating to

the MW wavelength in tissue, resulting in the temperatures obtained being averages over tissue dimensions of several cm. (Recall from Table 14.1, for example, that the muscle tissue wavelength of 915 MHz is 4.5 cm.) Another serious problem is the dependence of the detected MW intensity on the nature of the surface. Also, the radiometer response is extremely sensitive to the temperature of the radiometer, cables, etc. (MIZUSHINA et al. 1989, 1992). Use of a thin temperature-controlled water bolus and a method of controlling the radiometer receiver temperature can help.

A different approach is MW tomography. Here, several MW radiometers, each tuned to a different frequency, are positioned around the object and simultaneously detect MW radiation emitted in different directions. Analogous to CT scanning reconstruction, the information is used to reconstruct the spatial distribution of sources. In this instance, however, the sources are regions of tissue at various temperatures. Much must be known about the attenuation of various MW frequencies by different kinds of tissues. Using reasonable assumptions based on the limited information available, Bardati and others have shown some success in static head phantoms (BARDATI et al. 1987). The MW tomography approach is further complicated by coherent effects, since many biologic structures have dimensions on the order of a few mm, which is close to the tissue wavelengths of the band of MW radiation (near 2.45 GHz) that the radiometers are sensitive to.

14.5.2.2 Applied Potential Tomography

This tomographic approach, sometimes called electrical impedance tomography (EIT), relies on the fact that the surface distributions of voltage and current are dependent on the intervening resistivity and dielectric distributions, which are in turn temperature dependent. In practice, APT consists of placing an orderly circular array of electrodes on the surface around a flat or cylindrical region of interest. Interelectrode voltages are measured while currents are injected between adjacent electrodes, leading to sets of complex impedance measurements which are deconvoluted, usually by techniques analogous to those used for CT, to give a crude 2D map of tissue resistivities. These resistivities are tracked over the heating time, and via known or assumed resistivity dependencies on temperature (1%–4%/°C), a map of temperature changes results (BOLOMEY and HAWLEY 1990).

Promising homogeneous phantom experiments indicate spatial resolution equal to approximately 10% of the diameter of the electrode array radius and temperature resolution of a few tenths °C (CONWAY 1987; GRIFFITHS and AHMED 1987). These values were significantly deteriorated when the experiments were extended to human volunteers. Improvements in data collection by using both a larger number of electrodes (16 were used in the work cited) and more sophisticated software and instrumentation will certainly help. The reconstruction algorithms are another key feature. The challenge is more difficult than CT scanning, due to the complexities of tissue resistivity temperature dependence and the nature of the APT electrode current and voltage fields that are dependent on the details of inhomogeneous objects.

The equipment is relatively inexpensive, since low frequencies (few hundred kHz or less) or employed and conventional electrodes such as those used for EEG studies are suitable. As the understanding of tissue properties improves, and better algorithms develop (BERNTSEN et al. 1991), the APT technique should yield practical results for at least partial control (along with limited invasive thermometry) of hyperthermia treatments. More work is also necessary to establish compatibility with conventional hyperthermia techniques, but this at present does not seem to be a big problem.

14.5.2.3 Magnetic Resonance Imaging

For some time, it has been realized that the MRI basic relaxation time constants T1 (spin-lattice) and T2 (spin-spin) are temperature dependent. Research has gone in to exploiting this, but there have been serious problems, including difficulties in isolating the temperature dependence from other influences on T1 and T2.

LeBihan has postulated and shown that the effects of molecular diffusion on MRI image intensity reduction due to brownian motion are temperature dependent, and hold much promise for a 3D noninvasive temperature measuring technique (LEBIHAN et al. 1991). The diffusion of a water-like substance is governed by $e^{-E/kT}$, where E is the activation energy, k is Bolzmann's constant, and T is the absolute temperature. The low activation energy of approximately 0.2 eV leads to a linear relationship between temperature and the diffusion coefficient D in the hyperthermia temperature region of 2.0%–2.7%/°C. Determination of D on a pixel-by-pixel basis is accomplished by noting the MRI image intensity reduction (exponentially dependent on D)

for different values of the MRI magnetic field gradient (SAMULSKI et al. 1992).

A concern with any MRI temperature measurement technique is the electromagnetic and mechanical compatibility between the MRI imaging system and the hyperthermia apparatus. Delannoy and co-workers demonstrated technique feasibility and promising results by sequential time switching of the MRI RF and hyperthermia RF fields (DELANNOY et al. 1990). For these studies they used a 25-cm diameter version of the BSD Medical mini annular phased array (MAPA) modified, in particular, to remove all ferromagnetic material. This was operated at 168 MHz and placed (with a 12-cm-diameter homogeneous phantom) in a 21-MHz, 0.5-T MRI unit. Appropriate π RF filters resonant at 21 or 168 MHz were used, but sparingly to preserve satisfactory signal to noise, so that time switching between imaging and heating was necessary. The results of a 7-min scan at quasi-steady state gave 1°C accuracy and resolution for temperature increases from a baseline of 15°C to a 20°–32°C range. The spatial resolution of the selected regions of interest was 1 cm wide by 11 cm long.

Recently SAMULSKI et al. (1992) have used improved RF filtering and isolation techniques to allow simultaneous heating and conventional MRI multislice imaging of a phantom at 130 MHz and 64 MHz (RF part), respectively. Results with about 4 min imaging time led to 0.5°C and 1 cm³ resolutions. It is believed that using fast echo planar MRI imaging instead of the conventional spin-echo technique, along with other enhancements achievable with current technology, will soon allow reaching the goal of only 1 min scan time needed to obtain 0.5°C, 1 cm³ resolution (SAMULSKI et al. 1994). This is exciting, particularly in view of the fact that the resultant, almost on-line, MRI diffusion-produced T images can have the conventional MRI anatomic images superimposed. Although measurements in phantoms are very promising, there are still some major difficulties relating to blood perfusion and tissue anisotropy to be overcome when MRI is applied in vivo (YOUNG et al. 1994).

14.5.3 Conclusion

Invasive thermometry equipment and techniques are satisfactory at present, particularly for EM hyperthermia *to the extent that invasiveness is permitted.* This is the problem, of course, and hyperthermia researchers and practitioners will remain severely handicapped for some time. The noninvasive progress is very encouraging, but early clinical utilization will most certainly require limited use of simultaneous invasive thermometry to establish baselines and/or for "calibrating" the noninvasive systems on line. The advances being made in multi-probe temperature feedback for automatic power control (HARTOV et al. 1993; ZHOU and FESSENDEN 1993) and numerical modeling of temperature distributions (CLEGG and ROEMER 1989; SAMULSKI et al. 1994) will go hand in hand with the semi-noninvasive techniques. This increases the complexity of hyperthermia treatments, but is a necessary and worthwhile step for advancement.

14.6 Modeling and Treatment Planning

The ultimate goal of planning a clinical hyperthermia treatment is to predict accurately the complete spatial and time dependence of the temperature. This first requires determining the power deposition or SAR pattern, and secondly, using knowledge of the SAR throughout space and time, coupled with the thermal properties and physiologic nature of the patient, predicting the developing and steady state temperature field. The sensitivity of these two steps to the detailed nature of living tissue is immensely greater than the case for radiation treatment planning. Ideally, we need accurate patient-specific anatomic and physiologic models, and complete understanding of the EM or US interaction between the power sources and patient, as well as the thermal interaction with the patient and between the patient and at least the immediately surrounding environment. This is a tall order, presently incompletely fulfilled.

One-dimensional and 2D modeling, sometimes employing realistic patient anatomy, has received much effort and even led to qualitative agreement with clinical data (STROHBEHN et al. 1986). Some 3D modeling with unrealistic homogeneous or simple layered strata, concentric cylinders, or spheres, has also been undertaken. These studies have given some guidelines as to expectations and been somewhat useful for comparing devices, but it is now generally recognized that nothing short of full 3D patient-specific calculations will be useful for prospective hyperthermia treatment planning.

14.6.1 Electromagnetic Modeling

As discussed in Sect. 14.2.1, the quantity of interest is SAR = $\sigma E^2/\rho$, so one must determine the electric field E at all points in the patient. This amounts to solving two of Maxwell's equations:

$$\nabla \times \mathbf{E} = -\mu \partial \mathbf{H}/\partial t \qquad \text{and} \qquad (14.6)$$

$$\nabla \times \mathbf{H} = \varepsilon \partial \mathbf{E}/\partial t + \sigma \mathbf{E} \quad \text{for} \qquad (14.7)$$

our system. $\mathbf{E}$, $\mathbf{H}$, and ε are complex quantities with real and imaginary components. Since useful prospective patient-specific treatment planning is the goal, our system consists of an anatomically correct patient model usually fully in the near field (Table 14.2), as well as the applicator itself. This situation does not allow solution in closed form, and numerical techniques are needed.

Much attention has been paid to EM modeling over the last 10 years, encompassing the areas of (1) comparative, (2) prospective, (3) concurrent, and (4) retrospective dosimetry (ROEMER and CETAS 1984). Comparative, concurrent, and retrospective dosimetry deal with comparing the potential heating ability of hyperthermia applicators under various hypothetical clinical conditions, enlisting modeling to aid in applicator control during treatment, and analyzing clinical (or animal) heating after the fact, respectively. For example, a hybrid combination of the finite element method (FEM) and the boundary element method (BEM) has been used to compare in 2D the optimization of fixed and moveable EM arrays for deep regional heating (YUAN et al. 1990). A straightforward application of basic EM theory assuming known aperture fields from planar arrays of noninteracting small sources has been used to predict advantages to be gained from coherent operation (HAND et al. 1986). These are just two examples out of many where EM modeling has been applied with varying degrees of sophistication to hyperthermia. Much has been learned, but unfortunately, progress in prospective dosimetry (i.e., realistic patient-specific treatment planning) has been slow.

Very encouraging work has been accomplished by several research groups applying the finite difference time domain (FDTD) numerical modeling technique to the full 3D EM regional hyperthermia problem (LAU et al. 1986; SULLIVAN 1990; CHEN and GANDHI 1992; PIKET-MAY et al. 1992). Sullivan and colleagues have developed a system whereby the simulated SAR distributions are used routinely in the prospective patient-specific treatment planning process for the BSD sigma 60 four-quadrant applicator (SULLIVAN et al. 1993). In the FDTD technique, the time-dependent forms of the two curl Maxwell equations 14.6 and 14.7 are solved by representing the spatial and time derivatives involved with centered difference approximations resulting from Taylor's series expansion and first-order truncation of the E and H. The two vector equations 14.6 and 14.7 thus simplify to six linear scalar equations. For example the x component of Eq. 14.6 is:

$$\frac{\partial E_y}{\partial z} - \frac{\partial E_z}{\partial y} = -\mu_o \frac{\partial H_x}{\partial t}$$

and this becomes:

$$\frac{E_z(j) - E_z(j+1)}{\Delta y} + \frac{E_y(k+1) - E_y(k)}{\Delta z}$$

$$= -\mu_o \frac{H_x(i+1/2) - H_x(i-1/2)}{\Delta t}. \tag{14.8}$$

Note μ is replaced by the constant permeability of free space μ_o, recognizing that the human body is nonmagnetic. The 3D space is divided into an orderly rectilinear grid with increments Δx, Δy, and Δz. The indices locate a particular grid point. For example: $i-1/2, j, k+1$ represents $(i-1/2)\Delta x, j\Delta y, (k+1)\Delta z$. The six-field components $E_x, E_y, E_z, H_x, H_y, H_z$ are specified near every spatial point on this uniform grid, forming a basic unit, the so-called Yee cell, in the calculation space. The simulation begins by "turning on" a localized sinusoidal EM excitation and "time stepping" (i.e., propagating) the incident and scattered EM wave throughout the entire space. The interlaced E and H components are evaluated alternately as the time stepping proceeds. Each evaluation amounts to solving a large number of simple scaler equations. This is carried on for at least several full wave periods until convergence is obtained. The size of the spatial increments is strongly correlated with the computer memory requirements, while the size of the time step Δt strongly influences the computation time and, through the relation of Δt to the frequency and wavelength, the accuracy. Typically the spatial and time increments are 1 cm and 1/20 of a period, respectively, requiring a few minutes of calculation time on a CRAY XMP or YMP supercomputer to simulate a multiapplicator treatment of a realistic model of the human torso.

Much of the FDTD success to date relates to the development of efficient methods to construct the realistic patient-specific anatomic model used as input for the computer SAR simulations (JAMES and SULLIVAN 1992). Typically 60 CT slices at 1-cm intervals were obtained from the patient to be treated, and using an approximation relating the CT

numbers to one of four tissue or air classes, frequency-dependent values of ε and σ are assigned to each pixel. These are in turn averaged to yield the ε and σ values for each FDTD unit Yee cell. Further improvements in efficiency, including a very nice adaptation of frequency-dependent, or so-called $(FD)^2TD$ method (SULLIVAN 1992), allow the entire patient-specific SAR treatment planning to be done in approximately 1 day. The CRAY YMP supercomputer is used for the $(FD)^2TD$ calculation for each applicator segment. A local Silicon Graphics IRIS 4D 240/GTX workstation is used for the anatomic model creation and the determination of somewhat optimized selection of applicator frequency, amplitudes, and relative phases.

The FDTD technique is one of the most computational efficient numerical modeling techniques (PAULSEN 1990). It allows for particularly efficient storage strategies, and therefore only modest memory requirement if not too fine a computational grid is used. The time stepping often requires calculation at 100 or more sequential times, and this detracts somewhat from computational efficiency via extending the time necessary. It also has a drawback in that the necessity of a uniform rectilinear grid makes conformation to irregular external and internal boundaries (i.e., tissue interfaces and possibly the interior of applicators) difficult, resulting in errors derived from the "stair step" approximation to smooth contours (CANGELLARIS and WRIGHT 1991).

The finite element method (FEM) is attractive from at least the point of view of the calculational grid that it uses (STROHBEHN et al. 1986). Unlike the FDTD method, FEM breaks the calculational space into small regions (i.e., finite elements) which usually are triangular or four-sided in 2D, and in 3D either simply two of these adjacent structures connected or simple shapes like tetrahedrons. There is no need to have all the small regions, or cells, alike, so that curved and jagged boundaries can be matched almost exactly. (A nonuniform grid is advantageous for the subsequent thermal calculations also.) A particular form of Maxwell's equations are solved for each small region separately, a relatively simple computational requirement. The more difficult part is assuring that the solution at the boundaries of a given cell is compatible with the solutions at all the surrounding neighboring boundaries. The following boundary consistency conditions are imposed, and calculations are done iteratively until agreement at all cell boundaries is obtained:

$$\mathbf{n} \cdot (\varepsilon_{r1}\mathbf{E}_1 - \varepsilon_{r2}\mathbf{E}_2) = 0 \tag{14.9}$$

$$\mathbf{n} \times (\mathbf{E}_1 - \mathbf{E}_2) = 0 \qquad (14.10)$$

$$\mathbf{n} \times (\mathbf{H}_1 - \mathbf{H}_2) = 0 \qquad (14.11)$$

$$\mathbf{n} \cdot (\mu_1 \mathbf{H}_1 - \mu_2 \mathbf{H}_2) = 0, \qquad (14.12)$$

where $\mathbf{n}$ is the unit vector normal to an interface. These derive from the last of Maxwell's equations (not Eqs. 14.6 or 14.7) and we see that Eq. 14.9 tells us, as discussed earlier, that the ratio of the perpendicular electric field components across a boundary is inversely proportional to the ratio of the complex relative permittivities.

The FEM problem solution can be worked out in the so-called frequency domain. If the very reasonable assertion is made that any steady-state solution will be sinusoidal, the time dependence of $\mathbf{E}$ and $\mathbf{H}$ is expressed by $e^{j\omega t}$. Substituting this in Eq. 14.6 and 14.7 and then dividing through by $e^{j\omega t}$ yields:

$$\nabla \times \mathbf{E} = j\omega\mu\mathbf{H} \qquad \text{and} \qquad (14.13)$$

$$\nabla \times \mathbf{H} = -j\omega\varepsilon\mathbf{E} + \sigma\mathbf{E}. \qquad (14.14)$$

This is the frequency domain expression of Maxwell's equations, where the vectors (now technically phasors) are functions of space and frequency, but not time. Recognizing that σ may be written as $-jj\sigma$ ω/ω, Eq. 14.14 simplifies to:

$$\nabla \times \mathbf{H} = -j\omega\varepsilon_o\varepsilon_r, \qquad (14.15)$$

where ε_r is the complex relative permittivity $\varepsilon/\varepsilon_o + j\sigma/$ $\varepsilon_o\omega = \varepsilon' - j\varepsilon''$. Further, Eqs. 14.13 and 14.14 can be combined to reduce to an equation in either $\mathbf{E}$ or $\mathbf{H}$ only. For example, taking the curl of Eq. 14.13 and substituting $\nabla \times \mathbf{H}$ from Eq. 14.15 gives:

$$\nabla \times \nabla \times \mathbf{E} - \omega^2\mu\varepsilon_r = 0. \qquad (14.16)$$

We see the FEM here has changed a very difficult problem of dealing with the partial differential Eqs. 14.6 and 14.7 throughout the problem space to one of a collection of the much simpler Eq. 14.16.

Although the FEM was quickly and somewhat successfully brought to bear for solving complex hyperthermia problems in 2D (STROHBEHN et al. 1986), it has to date lagged behind the success of FDTD for 3D. Although the computational time is not too severe, the computer storage and memory requirements are still serious, relating partly to the matrix inversions required to solve the large sets of relatively simple equations like Eq. 14.16 (PAULSEN 1990). The advantageous aspect of nonuniform grids allowing good matching to boundaries is not yet practical due to the necessity of still undeveloped automatic 3D mesh (grid) generators. Of a more fundamental nature is that, unlike the FDTD

method for example, the FEM needs to have the EM source distribution specified at all the patient (plus bolus) external boundaries. However, most real clinical situations result in a significant amount of unbounded surface. A promising area of numerical modeling research is now dealing with hybrid methods, whereby the FEM is used for the patient, while the exterior boundary regions are solved by the boundary element method (BEM) (PAULSEN 1990; PAULSEN et al. 1988). The BEM is an integral equation method capable of handling boundaries at infinity.

Several other numerical modeling methodologies, including particularly the domain integral equation (DIE) (often referred to as the method of moments) technique, have received a lot of attention historically for solving EM problems. Much research and development will be needed to find the best compromise between practicality (efficient use of time and computer resources) and accuracy (PAULSEN 1990; WUST et al. 1993). Even though much can be gained from computer simulation on its own, experimental verification of modeling predictions is absolutely essential. An excellent start in the FDTD arena has been made (SULLIVAN et al. 1992), but overall there has been an unfortunate and somewhat surprising lack of useful tests of 3D numerical modeling appearing in the literature. A versatile standard phantom which could be reproduced or possibly shared among institutions is an important consideration (LAGENDIJK et al. 1992).

Less sophistication than the above discussion on EM modeling implies may be sufficient in certain regions where low frequencies are involved. This would be for situations where large wavelengths and slowly varying (in time) E and H fields are involved. Currents in this case are essentially resistive, leading to purely I^2R conductive power deposition, and the well-known Laplace's equation $\nabla^2 E = 0$ may apply. This is likely the case for frequencies below 10 MHz, and certain problems in hyperthermia dealing with induction, capacitive, or interstitial radiofrequency heating are greatly simplified, even in 3D (Doss 1982; STROHBEHN and ROEMER 1984).

14.6.2 Ultrasound Modeling

In analogy to the situation in EM theory where E and H are the origin of the forces on charges, the pressure p is the origin of the forces on particles and molecules subject to a sound wave, which is the propagation of alternating regions of contraction and rarefaction moving through matter. The local displacement for frequencies and intensities useful in hyperthermia is only on the order of 100 A° (SWINDELL 1986). Thus even though the particle vibration is damped by viscous forces leading to ultrasound energy absorption, the basic premise that the molecular elastic restoring forces are linearly proportional to the displacement is usually valid (i.e., Hooke's law). Thus Newton's second law of motion that force equals mass times acceleration leads to the scalar wave equation:

$$\nabla^2 p + k^2 p = 0, \qquad (14.17)$$

where p is the instantaneous pressure (Sect. 14.3) and k is a complex wave number with real part $2\pi/\lambda$. In general there is energy propagation in all directions, but Eq. 14.17 is a set of three scalar equations, not a vector equation like Eq. 14.16 for the EM case where the three components are strongly coupled. In some situations only the longitudinal aspects of motion in only one direction are of importance, so that p would be considered to be in only one direction. If we call x the direction of propagation, and recognize the sinusoidal nature of the sound wave $u(x,t) = u\hat{A}e^{j(\omega t - kx)}$ we have the longitudinal work equation:

$$\frac{\partial^2 u}{\partial x^2} = \frac{1}{c^2}\frac{\partial^2 u}{\partial t^2}, \qquad (14.18)$$

appearing at the beginning of Sect. 14.3.

In hyperthermia one deals with finite sized sources (piezoelectric discs) and works in or close to the near field, so diffraction effects are usually encountered. Huygen's principle is evoked to determine the resultant acoustic pressure at any given point in space resulting from Huygen wavelets emanating from each infinitesimal element of surface area dS from the transducer face. This process can be represented mathematically by using the velocity potential

$$\psi = \frac{1}{2\pi}\int_s \frac{v_s}{r}e^{-jkr}ds, \qquad (14.19)$$

when r is the distance from dS to the given point in space and v_s is the normal component of the particle velocity adjacent to the region dS on the transducer face of area S (STRUTT 1945). The pressure at the field point of interest is:

$$p = \rho\frac{d\Psi}{dt} = j\omega\Psi, \qquad (14.20)$$

from which the SAR may be computed using the development in Sect. 14.3.1. by SAR $= \mu_a p_a^2/\rho^2 c$.

For the real case of an attenuating medium, there must be a negative exponential multiplying factor in the integrand of Eq. 14.19 to account for attenuation between dS and the field point. This will be different for every set of a particular surface element dS and field point, but is greatly simplified if the medium is homogeneous.

The solution of Eq. 14.19 requires numerical methods (ZEMANEK 1971), whereby the surface S is usually divided into a small 2D grid. Since the US wavelengths are on the order of mm, there are appreciable phase differences among pathlengths representing the distance from a given field point to various elements on the transducer surface. Thus the surface elements must be small, resulting in very computationally demanding calculations, even assuming uniform intensity and constant phase at the transducer surface. Various investigators have used different approximations to solve the problem for individual transducer shapes, and this has resulted in valuable transducer characterization and design considerations (SWINDELL 1986; HUNT 1990).

If the hyperthermia target area is not near the surface, it is advantageous to have the transducer design produce a fairly strong (Sect. 14.3.4) focus. This will result in a cigar, or elongated ellipsoid, shaped region giving a FWHM of a few λ with a depth (cigar length) of approximately 7 λ (HUNT 1990). Thus a useful system will need arrays of many individual transducers that could effectively carry out mechanical scanning by selectively aiming and powering individual transducers, or actually mechanically scanning one or a group of transducers (Sect. 14.3.5). This is necessary not only to provide a clinically useful volume with sufficient power deposition, but also to work within the constraints of the anatomic "windows" presented by the bone, lung, air-filled bowel and critical structures of the particular patient anatomy. Consideration must be given to the scanning speed to assure the focus (or foci) has returned to any given spot before the temperature has decayed significantly. This is usually indicated to be 10 s or less (HYNYNEN et. al. 1986).

Scanning can also be done by electronic means if the US source(s) is compartmentalized into many small transducers that are operated independently, but coherently, to use to advantage the phase relations among the different elements. (Since the US pressure is a scalar, this maneuver is not as fundamentally challenging with respect to optimizing constructive and destructive interference patterns as is the case for working with large coherent arrays of radiative MW sources, because of the vector nature of the latter.) As discussed earlier in this chapter

(Sect. 14.3.6), the sector-vortex array, consisting of several annular rings each with many elements (sectors), is an advanced phased array with the design philosophy to provide arbitrary US intensity or SAR patterns by direct synthesis (CAIN and UMEMURA 1986). In the sector-vortex simulation, ideally the sector-vortex elements are spaced apart by only a distance equal to $\lambda/4$ or less, which for 1 MHz would be much less than 1 mm! The small size assures that little error results from assuming each element is a point, rather than a distributed, source. This simplifies calculations, but the task of handling hundreds of mm-size elements and optimizing the sequential phase and amplitude configuration requires a CRAY XMP-48 supercomputer. Simulations and limited experimental verification have shown that direct synthesis is a very promising, powerful technique. The research by Cain and associates is concentrating on efficiencies of miniature transducer fabrication techniques and calculation algorithms. It has recently extended to large cylindrical and spherical sector arrays, and to powerful inverse methods whereby all the driving phases and amplitudes can be derived from the specifications of a complex desired 3D US intensity pattern, and eventually even from the desired temperature field (McGOUGH et al. 1992).

Ultrasound power deposition simulations are certainly playing an important role in applicator systems design, even though the sector-vortex concept, for example, is tested conceptionally assuming insonation of homogeneous tissue with a constant 1.0 dB/cm/MHz for α (neglecting scatter). Semi-quantitative treatment planning is accomplished at several clinical sites (HYNYNEN et al. 1992) by choosing an optimal frequency related to accessible paths as indicated on CT scans, assuming approximate representative attenuation coefficients, transducer availability, and calculated time-averaged US intensities superimposed on the CT scans. The CT scans help immensely with respect to warning where potential problems, such as insonation at bone and air interfaces, may arise, and evaluating possible transducer positions and orientations.

Fully quantitative patient-specific SAR treatment planning will have to deal with a number of problems. As implied by the ranges of values given in Table 14.3, there is much uncertainty in the acoustic properties of tissue. Lack of ability to determine the specific attenuation and absorption constants, for example, prevents accurate calculation of beam intensity and also the subsequent energy absorption at each point. Even though present CT scan data can determine mass density fairly well, the lack of good US velocity information prevents accurate accounting of reflection and refraction. These effects can be very severe (Sect. 14.3.1). Reflection at a fat–muscle interface can be 100% for incident angles greater than about 70°, and at even smaller angles for a muscle–bone interface. At interfaces of soft tissue with air-like regions, reflection is usually complete. The reflected energy can combine with incident energy (coherent or otherwise) to give greatly enhanced SAR and temperature that create clinical problems (HYNYNEN 1990b; PRIONAS et al. 1990).

In addition to at least partially blocking the US beam, bone also can be a source of pain. Even if the bone is some distance "downstream" of the tumor region where the US intensity is presumably much reduced, its significantly enhanced attenuation and absorption coefficients (Table 14.3) can result in painful temperature elevation at the bone surface where the nerve cells are located (HYNYNEN and DE YOUNG 1988; WILLIAMS 1983). The absorption in bone is complicated by the fact that, depending on the angle of incidence, mode conversion can be important, and in some instances, dominant (DUNN and FRIZZELL 1982). Bone has sufficient structural rigidity that it can support shear waves, and indeed if a longitudinal US wave is incident upon a muscle–bone interface at an angle greater than 20°, the transmitted wave will be mostly shear with an absorption coefficient enhanced by an additional factor of almost 2. (This nonlinear effect is separate from the high-intensity nonlinear energy absorption alluded to in Sect. 14.3.6.) At sufficiently high intensity accompanying strong focusing, higher order frequency modes are mixed in to the fundamental frequency such that a shock wave may result. This gives enhanced nonlinear energy absorption with much more abrupt fall off distally, a desired situation (SWINDELL 1986). Lack of complete understanding of the phenomena, coupled with our poor knowledge of specific acoustic tissue properties, however, cautions avoidance of shock waves at this time.

Testing in phantoms should play a role in verifying SAR modeling and treatment-planning strategies. Even though an array of US tissue substitutes are available (HYNYNEN et al. 1992), useful phantom development lags behind the EM situation. In reality, much of the verification of US treatment planning development has so far relied on patient temperature data and expressed comfort level (pain).

The problems with advancing to true patient-specific treatment planning are indeed severe for both EM and US. They are not insurmountable, and as one explores and follows the progress, it becomes clearer that EM and US hyperthermia complement each other (PAULSEN 1990).

14.6.3 Thermal Modeling

Thermal modeling will build on the SAR simulations, using the results from the (ideally 3D) EM or US energy absorption calculations as just one very important part of the input. Although it is not yet possible, patient-specific thermal modeling also needs accurate specification of the specific heat, thermal conductivity, and blood flow throughout the target and surrounding regions. These properties are temperature dependent, and thus time dependent. The applied SAR, as well as metabolic heat generation, provides the energy input to the target and system. Energy removal is also of prime importance, and this is caused by conduction, convection via blood flow, and even radiation. (Under circumstances whereby the patient surface is in contact with warmer objects, conduction, of course, and even radiation can be involved in adding energy to the patient or phantom.)

The human body has over 10^6 m of blood vessels, and these vary in size from the roughly 1-cm diameter aorta and vena cava down to the fractional mm diameter arterioles and venules, and the even smaller capillaries (LAGENDIJK 1990). The velocities of the blood in the largest to the smallest vessels cover a range of 50 cm/s to less then 1 mm/s, and these can change significantly during a hyperthermia treatment due to the body's thermoregulatory response to heat. It is clear that blood flow is of utmost

importance and is a serious hindrance to accurate prospective hyperthermia treatment planning. It is the extra parameter of energy removal, as well as the dynamic aspect, that prevents even a rough analogy to the task of patient-specific 3D radiotherapy treatment planning.

The time rate of change of the temperature at a point P within a patient undergoing hyperthermia is determined by the many sources of energy exchange (Fig.14.20). Multiplying the rate of temperature change by the mass density ρ (kg/m^3) and the specific heat c (J/kg°C) expresses the quantity as the time rate of energy change per unit volume:

$$\frac{\rho c \partial T}{\partial t} = \bar{\nabla} \cdot k \bar{\nabla} T - \rho c \bar{\upsilon} \cdot \bar{\nabla} T + \rho SAR + Q_{met} + Q_{rad},$$

$$(14.21)$$

where k is the thermal conductivity (watt/m°C), $\bar{\upsilon}$ is the local blood velocity vector, SAR is the rate of absorbed energy per unit mass from EM or US applicators, Q_{met} is the rate of metabolic energy generated per unit volume, and Q_{rad} is related to the net rate of absorbed energy due to radiation exchange with the surrounding environment. Q_{rad} is important in a discussion of surface cooling by air contact, the physics of which can be discussed by considering the T^4 black body radiation and departures thereto. Except for a discussion of the intentional radiant heating in some types of whole-body hyperthermia (ROBINS et al. 1985), it is of little significance for the thermal modeling of local or regional hyperthermia and is usually neglected (LAGENDIJK 1990). The metabolic generation of heat in an average person is close to 100 W, depending on

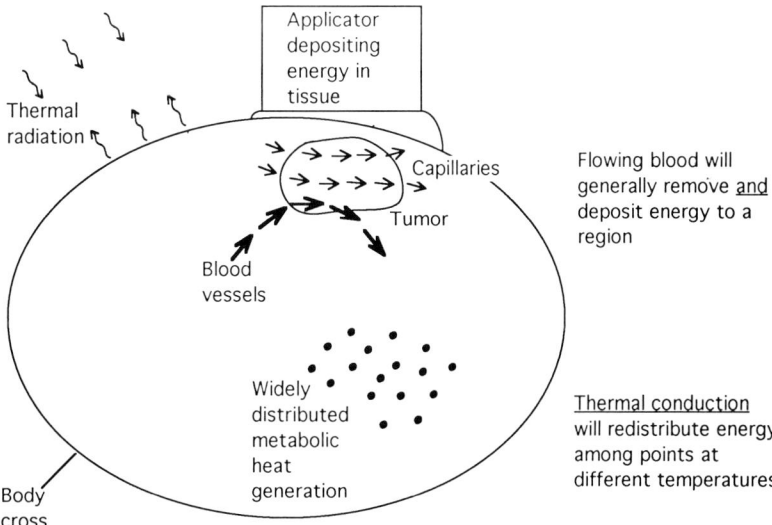

Fig. 14.20. Simplified schematic of energy exchange affecting the time rate of change of temperature at a point within tissue (see text)

temperature. It increases about 10% for each °C rise, and varies from almost zero in subcutaneous fat to 50 times the average for strenuously working muscle (Sekins and Emery 1982). Again, except for the internal energy changes associated with a well-insulated person undergoing whole-body hyperthermia, the Q_{met}, averaging less that 1 W/kg, is usually neglected.

The first term on the right-hand side of Eq. 14.21 represents thermal conduction. This is the 3D version of Fourier's law which shows that thermal conduction is proportional to the spatial gradient $\bar{\nabla}T$. (In a rectangular slab of area A, thickness Δx, and ΔT across the slab, the time rate of energy flow perpendicular to the area in the direction of decreasing temperature is proportional to $-A\Delta T/\Delta x$. The constant of proportionality is k, the thermal conductivity in x direction. Looking at an infinitesimal volume with heat transfer in at x and out at $x + \Delta x$, and expanding the kdT/dx outflow in a Taylor's series, leads to the net rate of heat transfer equal to kd^2T/dx^2 (Roemer 1990).)

The very important term $-\bar{v} \cdot \bar{\nabla}T$ accounts for the energy transport due to blood flow. A positive (negative) value to $\bar{v}$ represents a net removal (addition) of energy to the point of interest. As a dramatic example of the importance of blood perfusion, it has been shown that the power necessary to keep the steady state in a human breast tumor is ten times that needed for steady state in a static breast phantom (Lagendijk et al. 1988). Ideal modeling would account for this term accurately, including the blood flow distribution everywhere within the 10^6 m of human vessel network, and tracking this accurately as the perfusion changes within the tumor and normal tissue during the entire hyperthermia treatment time.

For practical reasons due to limitations of knowledge of true blood flow distributions, most thermal modeling is presently done via approximating the blood velocity term in Eq. 14.21 by $wc_b(T - T_b)$ (originally due to Pennes 1948). Eq. 14.21 then reduces to:

$$\rho c \frac{\partial T}{\partial t} = \bar{\nabla} \cdot k\bar{\nabla}T - wc_b(T - T_b) + \rho\text{SAR}, \quad (14.22)$$

where w is the volumetric blood perfusion rate in (kg blood/m^3 tissue/s) = (6 ml blood/100 g tissue/ min), c_b is the specific heat of blood, and T_b is the normal blood temperature. T_b is almost universally considered to be the "arterial blood" temperature of 37°C, even though Pennes and, recently, others (Kapp et al. 1993) have shown by measurement that

normal temperatures, particularly near the skin surface, can vary by 1° or 2° (usually lower than 37°C). The interpretation of the blood flow term as presented in Eq. 14.22 is that blood of temperature $T_b = 37$°C flows into the region, instantly comes to equilibrium at temperature T, and flows out of the region. This is sometimes called the Pennes bioheat equation heat sink term, referring to the removal of energy from the region instantly and completely. This is a tremendous simplification, in that both the warming up of blood as it passes through a heated region and the subsequent cooling down as it passes through unheated tissue is actually gradual. It can be characterized by a thermal length equal to the distance over which the blood must travel before the differences between the initial blood temperature and that of a constant temperature environment is reduced by the factor $1/e$. This varies from approximately 10^{-5} cm for the capillaries, to 7 cm for secondary arteries and veins, to about 300 cm for the many large vessels, but may be significantly reduced where arteries and veins are found in tightly coupled counterflow pairs (Lagendijk 1990).

In spite of the limitations of Eq. 14.22, it serves as the basis for much thermal dosimetry. For example, if a measurement is done quickly upon power turn on before ∇T or $T - T_b$ is significant, Eq. 14.22 becomes simply $c\partial T/\partial t = $ SAR. SAR can usually be measured accurately in this manner in the first minute of power-on for power levels typical of clinical treatments. Similarly, assuming a region with minimal temperature gradient, Eq. 14.22 reduces to:

$$\rho c\partial T/\partial t = -wc_b(T - T_b) \quad (14.23)$$

at power-off. Equation 14.23 is easily solved to give a simple decaying exponential such that measurements of the "thermal washout" temperatures during the first minute or so of cool down after power-off yields the value of w (for example, Samulski et al. 1987b). Due to the very simple approximation $wc_b(T - T_b)$ for the blood velocity vector expression $\rho c\bar{v} \cdot \bar{\nabla}T$ in the more accurate energy balance Eq. 14.21, such a measured w must be interpreted with great caution. Rather than interpret and use w as a measure of real blood flow, it is better interpreted as a thermal clearance parameter indicating for the particular system (patient) temperature distributions existing at the time, simply how fast local tissue cooling occurs. This measurement is then directly interpretable as what SAR might be needed, and in fact the "thermal washout" procedure can lead to determination of SAR (Roemer et al. 1985).

Equation 14.22 must be solved by numerical methods to give a complete temperature distribution in 3D, or even in 2D. When the finite difference (FD) approach is used, as alluded to above in the discussion of EM modeling (Sect. 14.6.1), the differentials in Eq. 14.22 are expressed as differences, small enough to allow linearity in the immediate vicinity of each element or grid point. It is usually a reasonable assumption to assume k to be isotropic, so that $\bar{\nabla} \cdot k \, \bar{\nabla} T$ becomes simply $k \nabla^2 T$, a set of three scaler equations. After doing both a forward and backward Taylor's series expansion of T and combining, the x part of the second derivative $\nabla^2 T$ becomes in the central FD approximation:

$$\frac{T(x-dx) - 2T(x) + T(x+dx)}{dx^2}. \qquad (14.24)$$

Thus Eq. 14.22 is easily linearized, and the full set of such equations throughout the calculation space results in a set of simultaneous linear equations to be solved. As before, the space is divided up into usually a uniform grid with cubic (3D) or square (2D) elements, having temperatures specified and solved at the corners (nodes or grid points). The boundary conditions are conveniently applied by specifying known temperatures at certain peripheral grid points. Otherwise there are methods of specifying specific zero or finite heat flux values at peripheral nodes if this is the form of boundary condition (CHATO 1990).

The other numerical method most often used is the finite element method (FEM). This has been applied most successfully to the steady state, and often explicitly referencing the unknown temperatures to a background temperature assumed to be T_b (STROHBEHN et al. 1986). Eq. 14.22 then simplifies to:

$$0 = \bar{\nabla} \cdot k \bar{\nabla} T - wc_b T + \rho \text{SAR}. \qquad (14.25)$$

An element is one of the individual small areas (2D) or volumes (3D) formed by connecting straight lines between neighboring grid points. Each grid point has a weighting function ϕ_i, and then T at an arbitrary position within the element is (for example in 2D cartesian coordinates):

$$T(x,y) = \sum_{i=1}^{N} T_i \phi_i, \qquad (14.26)$$

where $i = 1$ to N designates nonzero values only at that specific set of grid points defining the element within which $T(x,y)$ resides. Combining Eq. 14.26 and Eq. 14.25 for the entire space (for the thermal modeling, this usually is simply the patient or phantom, but possibly the cooling bolus, etc., if

temperatures at the patient surface are not fixed) results in a set of linear equations of a number equal to the number of grid points. The solution is arrived at iteratively, requiring that the T values agree on each boundary common to all contiguous elements. Note that via Eq. 14.26, T will then be known at all points within each element in addition to at grid points and element boundaries.

The FD technique is well suited to transient analysis because of its inherent aspect of stepping through time. The FEM, on the other hand, can easily accommodate grids with variable element size, and thus for both the SAR and thermal simulations can more accurately represent realistic shapes. With respect to the thermal simulations, unfortunately both FD and FEM suffer because of the inability to use the proper thermal parameters. Tissue density ρ is known with sufficient precision from CT scans, and thermal conductivity k and specific heat c are known within reasonable ranges (for example CHATO 1990). The biggest problem in thermal modeling, however, is not being able to input the local, time varying, blood flow. Several groups have used FD or FEM numerical modeling with generic literature values of blood flow to analyse, usually in 2D, potential heating abilities of various classes of applicators (PAULSEN et al. 1985), to assess the effects of various distributions of blood flow (ROEMER 1990), and to estimate complete temperature fields from limited measurements (CLEGG and ROEMER 1989; RINE et al. 1992). Such studies are invaluable with respect to characterizing many aspects of clinical hyperthermia, and certainly advance the modeling field.

Lagendijk and his colleagues make it very clear that eventual clinically useful prospective treatment planning awaits increased ability to handle the blood flow problem, and that more than the traditional bioheat transfer equation Eq. 14.21, is needed (LAGENDIJK 1990). Those researchers present a modified theory whereby small and medium vessels can be handled relatively well via an enhanced conductivity term like $\bar{\nabla} \cdot k \, \bar{\nabla} T$, but with a $k_{eff} > k$ which varies locally and includes conduction plus a collective blood flow component. In addition, it will be mandatory to account for large arteries and veins (i.e., "thermally significant" vessels) on an individual basis. The computer requirements are still significant, and likely the next generation of computers will be necessary before patient-specific prospective planning incorporating thermal modeling can be done. Singling out of only the largest vessels for individual consideration, along with using average blood perfusion values provided by advanced MRI or positron emission tomography (PET) techniques, however, seems like a compromise that may work. Several other research groups have put forth different means of handling blood flow in an accurate bioheat equation, but the Lagendijk group has made the most impressive progress to date (e.g., CREZEE et al. 1993).

14.7 Quality Assurance

Quality assurance (QA) procedures which ensure correct and safe functioning of equipment and define good practice for administering therapy are an essential part of any clinical investigation into the efficacy of hyperthermic treatment. Since, at present, one is unable to predict accurately the temperature distributions achieved during treatment, QA guidelines are currently based on the results of idealized comparative studies and practical experience. Characteristics of applicators such as field size and heating depth are often described in terms of absorbed power measurements in phantoms. While this provides a convenient method of monitoring the performance and correct functioning of a device, and for making comparisons between devices, caution is needed in extrapolating these values to assessment of clinical performance.

It is also very important to remember that hyperthermia equipment must satisfy stringent electrical safety criteria as laid down by appropriate national and international agencies (e.g., the International Electrotechnical Commission's regulation 601–1).

14.7.1 QA for Heating Devices

The advantages offered by standard measurements for QA have led to definitions of effective field size and heating depth in terms of SAR measurements in simple phantoms (DUNSCOMBE et al. 1989; HAND et al. 1989; IBBOTT et al. 1989; SHRIVASTAVA et al. 1989). These parameters are defined in terms of SAR contours (e.g., 50% of SAR_{peak}) usually measured at or with respect to a depth of 10 mm from the surface of a plane, homogeneous muscle-like phantom (to avoid artifacts which might occur in measurements made closer to the surface). The choice of SAR contour is somewhat arbitrary and some have argued for a more conservative estimate of the useful field based on 70%–80% of SAR_{peak}. Interstitial applicators may be characterized in an analogous

manner in terms of effective heating length and width.

The minimum requirement when selecting an applicator for use in a particular treatment is that effective treatment volume (in practice, usually expressed in terms of effective field size and heating depth) should encompass the tumor volume. However, it is often good practice to ensure that the effective field size provides a margin of approximately 5 cm beyond the lateral dimensions of the tumor.

The effective heating depth for most electromagnetic applicators used for localized, superficial treatments in approximately 10 mm, although radiative applicators with large apertures operating at relatively low frequencies and large capacitive electrodes may achieve a modest increase over this value. Effective heating depth for ultrasound applicators for superficial treatment are more dependent upon operating frequency.

Clinical experience has shown that many electromagnetic applicators can induce temperatures in the therapeutic range at depths of about 2 cm. In the case of stationary ultrasound transducers operating at 1 MHz, the typical useful depth is about 3 cm. These values are usually increased when there is an underlying fat layer present. Since the presence and thickness of a layer of fat and a bolus can have a considerable influence on the SAR distribution, it is good practice to supplement the measurements in simple phantoms alluded to above with those in phantoms simulating fat layers of different thickness (for example, CHOU et al. 1990, 1991 for EM; CHAN et al. 1974 for US). In addition, realistic phantoms which simulate treatment setup and important anatomic features should be used whenever possible; this is particularly important in the case of radio-frequency capacitive electrodes where SAR depends strongly upon the separation and orientation of the electrodes and the variation in electrical conductivity between tissue types.

Guidelines describing implantation for interstitial hyperthermia should not deviate from those established for brachytherapy except possibly for the need of additional catheters for thermometry. Clinical experience suggests that intersource spacing of more than 15 mm can result in inadequate heating in well-perfused tissue.

Quality Assurance guidelines for deep-body hyperthermia are currently under discussion. Standard elliptical phantoms for testing radiative applicators for deep-body hyperthermia have been developed (e.g., ALLEN et al. 1988). In Europe a larger phantom, filled with saline solution (3 g NaCl per liter) and containing an elliptical matrix of light-emitting diodes spaced at 2 cm (SCHNEIDER and VAN DIJK 1991), has been recommended for rapid qualitative assessment of the fields produced by regional heating devices. Small nonperturbing E-field probes are recommended for rapid quantitative measurement of SAR distributions in these phantoms.

A summary of phantom materials which simulate electromagnetic properties of tissues has been given by HAND (1990b). These materials, which include gels, liquids, and solids, simulate not only typical properties of tissues with high or low water content but also those of specific tissues such as brain and lung. Frequently used phantoms for local hyperthermia include those described by CHOU et al. (1984) and BINI et al. (1984). SAR is usually determined from measurements of local rates of increse in temperature in gel or solid phantoms using arrays of temperature sensors, infra-red thermography, or liquid crystal sheet thermography. As discussed in Sect. 14.6.3, the measurement time is an important consideration. Alternatively, SAR may be determined through measurements of the electric field using a probe which is scanned through a liquid phantom. It is highly recommended that calorimetric measurements also be performed to measure the efficiency of applicators and to provide a rapid check that devices are functioning correctly.

The measurement of acoustic fields from ultrasound applicators is important since these may be influenced by many factors, some of which, such as mounting and clamping the transducer, are difficult to account for in computations. Measurements of the power output of the transducer, the spatial peak intensity, and the intensity distribution characterize ultrasound devices. Degassed water may be used as a general phantom material (speed of sound 1480 ms^{-1}, absorption coefficient at f MHz, and 20°C = 0.0025 f^2 Np m^{-1}) although measured field profiles must be adjusted mathematically to account for attenuation which is present in tissue. Alternatively, SAR distributions may be determined in phantoms which have similar absorption to tissues (EDMONDS et al. 1985; BASSEN et al. 1985). A list of US tissue mimicking materials has been compiled by HAND (1990b).

The overall hyperthermia system has many components: applicators, power sources, control and monitoring computers, water or other cooling systems, and thermometry. Routine procedures for checking individual component and overall system performance must be developed (IBBOTT et al. 1989).

14.7.2 Thermometry QA Considerations

Temperatures within heated fields are currently monitored by invasive multisensor probes. Mechanical scanning of single sensor probes along catheter tracks is highly recommended. Guidelines which discuss general QA requirements for thermometry such as thermometer performance and calibration, preferred locations for sensors in tumors and normal tissues, and documentation of this information have been discussed by HAND et al. (1989), SHRIVASTAVA et al. (1989), DEWHIRST et al. (1990), and SAMULSKI and FESSENDEN (1990). Specific requirements for interstitial and deep-body hyperthermia are discussed by EMAMI et al. (1991) and SAPOZINK et al. (1991), respectively. As many temperature sensors as possible, compatible with patient tolerance and clinical safety, should be used in every hyperthermic treatment. Potential artifacts due to interactions between the heating fields and the probes, the thermal conductivity of the probe, and the response time of the sensor within its sheath and/or catheter must be evaluated (Sect. 14.5). If the artifact in measuring temperature is less than or equal to 1°C, then temperature data should be corrected so that the error is reduced to less than 0.2°C. Data containing artifacts greater than 1°C should be considered suspect and should not be used in any subsequent analysis. Although microwave radiometry as a noninvasive method for thermometry during superficial hyerthermia is undergoing clinical evaluation, it should not, at present, be considered an alternative to invasive thermometry as a source of data for dosimetry. However, investigation of its role in controlling hyperthermia treatments is strongly encouraged.

In the case of small superficial tumors (dimensions less than or equal to 5 cm), it is recommended that intratumoral temperatures be measured at 5-mm intervals along the catheter tracks. In the case of larger tumors, a somewhat larger distance between measurements, about 10 mm, is often acceptable in view of practical considerations (e.g., number of sensors per thermometry probe and the excessive time required for a large number of thermal mappings). Interstitial temperature measurements should be extended into normal tissue beyond the tumor/normal tissue boundary. The minimum number of sensors compatible with good practice is site specific (e.g., HAND et al. 1989), but in all cases measurements in the peripheral regions of tumors are essential. If an array applicator or a mechanically scanned applicator is used the surface should be divided into appropriate spatial elements and at least one temperature sensor must be located in each elemental region.

Relative positions of tumor, applicator, and thermometry probes should be recorded and the depths to which sensors are inserted should be estimated. CT, MR, or ultrasound imaging techniques should be used whenever possible to ascertain probe positions. The goal should be to attain this information with a resolution of 2 mm.

In the case of interstitial treatments, temperatures should be measured at locations representative of maximum and minimum temperatures. Thus, in addition to monitoring within or close to electrodes, antennas, or other sources, preferred locations are midway between two sources, centrally between groups of sources (e.g., four microwave antennas) and at the periphery (deepest and lateral extremes) of the tumor (EMAMI et al. 1991). Temperatures should also be monitored at tissue interfaces and in regions susceptible to overheating.

The placement of invasive thermometry probes for deep-body hyperthermia is often limited by anatomic constraints. The use of CT or MRI guidance is often a necessity for accuracy safety. The absolute minimum requirement for thermometry during a deep-body hyperthermia treatment is that the temperatures at the deep margin of the tumor and at the tumor/normal tissue boundary should be monitored via one implanted catheter. Additional monitoring of temperatures via body cavities should be used whenever appropriate.

14.8 Summary

In this chapter the various methods which are available for inducing and monitoring hyperthermia treatments have been reviewed. Here, we summarize the fundamental limitations of these methods, identify areas where recent progress has occurred, and point to those areas where future progress is likely.

The percieved technical difficulties encountered in hyperthermia have been summarized succinctly by HALL (1982), who commented that "although biology is clearly on our side [in hyperthermia], the picture is considerably different for the physics of localized hyperthermia, the basic principles of which appear to be against us." This conclusion was based on the facts that noninvasive electromagnetic methods are unable to achieve good localization and penetration simultaneously (Sect. 14.2.2) and that the presence of gas and bone are problematic for

ultrasound methods. In addition, it can be added that a routine means of monitoring temperatures noninvasively during hyperthermia treatments is not yet available.

At first sight these limitations might suggest that there has been little chance of improving techniques over the past decade. However, much of the material discussed in this chapter suggests that this has not been the case. First, consider superficial hyperthermia, where it has been demonstrated that clinical treatments are feasible despite the restrictions imposed by the spatial resolution and penetration possible with electromagnetic devices. Here, progress has been made as evidenced by the move away from the use of large single waveguide applicators to conformable arrays of small applicators which offer treatment of large fields such as those involving the chest wall with a spatial control of the temperature distribution on the scale of a few cm. Scanned focused ultrasound methods are now also available to treat large superficial tumors. However, in order to take full advantage of these new approaches, further work is necessary in developing methods for monitoring and controlling such treatments. Some promising progress involving automatic computer control based on multitemperature probe feedback was cited in this chapter. Microwave radiometry and MRI appear to offer some promise in this respect.

The current generation of electromagnetic applicators designed for deep-body hyperthermia have the capability of modifying SAR distributions. However, judged in terms of the volumes of tumors in which the temperature/time relationships reach levels and durations apparently necessary when hyperthermia is used as an adjunct to radiotherapy, the progress with these devices has been less encouraging. There still is, however, a lack of consensus in quantifying the temperature/time relationships that are actually required, but some optimistic phase I/II clinical studies continue to be reported (OLESON et al. 1993). The temperature/time relationships required when hyperthermia is used in conjunction with some chemotherapeutic agents may be less stringent. There has been rapid progress in computing three-dimensional electric field distributions within patient-specific models and the development of treatment-planning procedures based on these models. Such studies improve our understanding of these complex treatments and in time should help us to use existing heating systems more effectively as well as to identify those aspects where improvement may lead to better equipment. There has also been progress in thermal modelling and some models which take account of the vascular structure at a 1-mm scale, albeit in relatively small volumes ($10 \times 10 \times 10$ cm^3), are becoming available. However, further work is required in acquiring more detailed description of the vascular structure and in developing models which encompass both discrete and continuum descriptions of vessels.

There has been a recent upsurge in interest in ultrasound techniques. The use of single transducers for superficial treatments reported a decade ago has given way to techniques which offer greater control of treatment and in many cases extend the clinical use of this modality to deep-seated tumors. There has been progress in the use of scanned focused ultrasound and devices based on recent theoretical and experimental work on electronically steered arrays are now becoming available for clinical evaluation. Such methods should enable the theoretical advantages of inducing hyperthermia by ultrasound to be realised in practice.

There has been steady progress in all methods of interstitial hyperthermia. For example, new antenna designs have addressed problems such as heating near the tips of antennas, and an understanding of the behavior of antennas as a function of insertion depth is now fairly well understood. Radiofrequency electrode systems which offer control over SAR and which are more tolerant to imperfect implant geometry have been developed, while ferromagnetic seed systems are now in clinical use and new seed materials continue to be developed. Other hot source techniques such as hot water tubes have been developed to the level of clinical use.

The lesson that an extensive quality assurance program is an important component of clinical hyperthermia has been learned over the past decade. The principles for assessing and checking applicators have been identified and procedures using standard phantoms have been defined. Invasive thermometry and associated sources of artifacts are now well understood.

There is little doubt that hyperthermia treatment will be more acceptable to clinicians if thermometry can be performed noninvasively. Section 14.5.2 discussed several techniques which are being investigated for this purpose. There has been considerable progress in microwave radiometry; the only method to have been used clinically to date. For example, systems can now accommodate a water bolus between the radiometer antenna and the skin, thus improving compatibility between radiometry and heating systems. Although the presence of a water bolus increases signal loss, there are also advantages

such as stabilizing the temperature of the receiving antenna and creating a well-defined thermal boundary condition. Of great practical importance is the fact that radiometer performance has been assessed under realistic conditions, such as assuming incomplete information regarding the dielectric inhomogeneity of the tissues (the normal situation in clinical hyperthermia). For example, an accuracy of 1.5°–2°C is expected for the ill-defined problem encountered at a site such as the chest wall. MRI techniques are unique in that they offer good temperature ($<1^\circ$C) and spatial (a few mm) resolutions at depth and further investigations in vivo are warranted in this area to optimize procedures and to determine artifacts. Compatibility between MRI and several heating systems (RF interstitial electrodes, RF capacitive electrodes, and radiative applicators) has already been established. In other methods, dielectric and acoustic characterization of tissues, improved reconstruction algorithms, and problems related to compatibility between heating and measurement systems are areas where further research is required. Finally, since most noninvasive methods are unable to meet all specifications required for thermometry per se, the case for qualitative or semi-quantitative monitoring of hyperthermia treatment through changes in image data should be considered.

Clearly, routine application of hyperthermia to most tumors is still a technical challenge. However, considerable progress has been made over the past decade; further progress leading to improved clinical performance is expected in several areas in the near future. In other areas such as patient-specific thermal treatment planning, where a general impact on clinical practice lies further ahead, progress will be dependent upon further advances in imaging and computing technologies. The facts that optimistic phase I/II clinical trials of hyperthermia combined with radiotherapy or chemotherapy continue to be reported and some phase III trials are nearing completion continue to be incentives to find solutions to these technical challenges in clinical hyperthermia.

Acknowledgement. The authors thank Jan Newland for manuscript preparation and coordination. Juanita Clack's early secretarial contributions are appreciated, and one of us (PF) is indebted to Alfred Smith for his encouragement.

References

Allen S, Kantor G, Bassen H, Rugerra P (1988) CDRH RF phantom for hyperthermia systems evaluation. Int J Hyperthermia 4: 17–24

Anhalt D, Hynynen K, Roemer RB, Nethonson SM, Stea B, Cassady JR (1982) Scanned ultrasound hyperthermia for treating superficial disease. In: Gerner EW (ed) Hyperthermic oncology 1992, vol 1. Arizona Board of Regents, Tucson, p 34

Astrahan MA, Sapozink MD, Luxton G, Kampp TD, Petrovich Z (1989) A technique for combining microwave hyperthermia with intraluminal brachytherapy of the oesophagus. Int J Hyperthermia 5: 37–51

Atkinson WJ, Brezovich IA, Chakraborty DP (1984) Usable frequencies in hyperthermia with thermal seeds. IEEE Trans Biomed Eng 31: 70–75

Bach Andersen J, Baun A, Harmark K, Heinzl L, Raskmark P, Overgaard J (1984) A hyperthermia system using a new type of inductive applicator. IEEE Trans Biomed Eng 31: 21–27

Bardati F, Bertero M, Mongiardo M, Solimini D (1987) Singular system analysis of the inversion of microwave radiometric data: applications to biological temperature retrieval. Inverse Problems 3: 347–370

Bassen H, Allen S, Herman B, Kantor G, Robinson R (1985) Quality assurance of RF and ultrasound cancer hyperthermia systems. Proc 7th Annual Conference of the IEEE Engineering in Medicine and Biology Society. IEEE, New York, pp 346–351

Benkeser PJ, Frizzell LA, Ocheltree KB, Cain CA (1987) A tapered phased array ultrasound transducer for hyperthermia treatment. IEEE Trans Ultrason Ferroelec Freq Contr 34: 446–453

Benkeser PJ, Frizzell LA, Goss SA, Cain CA (1989) Analysis of a multielement ultrasound hyperthermia applicator. IEEE Trans Ultrason Ferroelec Freq Contr 36: 319–325

Berntsen S, Bach Andersen J, Gross E (1991) A general formulation of applied potential tomography. Radio Sci 26: 535–540

Bini M, Ignesti A, Millanta L, Olmi R, Rubino N, Vanni R (1984) The polyacrylamide as a phantom material for electromagnetic hyperthermia studies. IEEE Trans Biomed Eng 31: 317–322

Bolomey JC, Hawley MS (1990) Noninvasive control of hyperthermia. In: Gautherie M (ed) Methods of hyperthermia control. Springer, Berlin Heidelberg New York, pp 35–111 (Clinical thermology, subseries thermotherapy)

Bowman RR (1976) A probe for measuring temperature in radio-frequency heated material. IEEE Trans Microwave Theory Tech MTT-24: 43–45

Brezovich IA (1988) Low frequency hyperthermia: capacitive and ferromagnetic seed methods. In: Paliwal B, Hetzel FW, Dewhirst MW (eds) Biological, physical and clinical aspects of hyperthermia. American Institute of Physics, New York, pp 82–110

Brezovich IA, Atkinson WJ, Chakraborty DP (1984) Temperature distributions in tumor models heated by self-regulating nickel-copper alloy thermoseeds. Med Phys 11: 145–152

Budihna M, Lesnicar H, Handl-Zeller L, Schreier K (1992) Animal experiments with interstitial water hyperthermia. In: Handl-Zeller (ed) Interstitial hyperthermia. Springer, Vienna, pp 155–163

Cain CA, Umemura SI (1986) Concentric-ring and sector vortex phased-array applicators for ultrasound hyperthermia. IEEE Trans Microwave Theory Tech MTT-34: 542–551

Cangellaris AC, Wright DB (1991) Analysis of the numerical error caused by the stair-stepped approximation of a conducting boundary in FDTD simulations of electromagnetic phenomena. IEEE Trans Antennas Propagat 39: 1518–1525

Carnochan P, Dickinson RJ, Joiner MC (1986) The practical use of thermocouples for temperature measurement in clinical hyperthermia. Int J Hyperthermia 2: 1–19

Casey JP, Bansal R (1986) The near field of an insulated dipole in a dissipative dielectric medium. IEEE Trans Microwave Theory Tech 34: 459–463

Chan AK, Sigelmann RA, Guy AW (1974) Calculations of therapeutic heat generated by ultrasound in fat-muscle-bone layers. IEEE Trans Biomed Eng 21: 280–284

Chan KW, Chou CK, McDougall JA, Luk KH, Vora NL, Forell BW (1989) Changes in heating patterns of interstitial microwave antenna arrays at different insertion depths. Int J Hyperthermia 5: 499–507

Chato JC (1990) Fundamentals of bioheat transfer. In: Gautherie M (ed) Thermal dosimetry and treatment planning. Springer, Berlin Heidelberg New York, pp 1–56 (Clinical thermology, subseries thermotherapy)

Chen JS, Poirier DR, Damento MA, Demer LJ, Biencaniello F, Cetas TC (1988) Development of Ni-4wt%Si thermoseeds for hyperthermia cancer treatment. J Biomater Res 22: 303–319

Chen JY, Gandhi OP (1992) Numerical simulation of annular-phased arrays of dipoles for hyperthermia of deep-seated tumors. IEEE Trans Biomed Eng 39: 209–216

Chitnalah A, Marchal C, Prieur G (1991) Sonde ultrasonore plane pour hyperthermie intracavitaire. Innovation et Technologie en Biologie et Medecine 12: 114–125

Chive M (1990) Use of microwave radiometry for hyperthermia monitoring radiometry and as a basis for thermal dosimetry. In: Gautherie M (ed) Methods of hyperthermia control. Springer, Berlin Heidelberg New York, pp 113–128 (Clinical thermology, subseries thermotherapy)

Chou CK, Chen GW, Guy AW, Luk KH (1984) Formulas for preparing phantom muscle tissue at various radiofrequencies. Bioelectromagnetics 5: 435–441

Chou CK, McDougall JA, Chan KW, Luk KH (1990) Effects of fat thickness on heating patterns of the microwave applicator MA-151 at 631 and 915 MHz. Int J Radiat Oncol Biol Physics 19: 1067–1070

Chou CK, McDougall JA, Chan KW, Luk KH (1991) Heating patterns of microwave applicators in inhomogeneous arm and thigh phantoms. Med Phys 18: 1164–1170

Christensen DA (1977) A new non-perturbing temperature probe using semiconductor band edge shift. J Bioeng 1: 541–545

Clegg ST, Roemer RB (1989) Towards the estimation of three-dimensional temperature fields from noisy temperature measurements during hyperthermia. Int J Hyperthermia 5: 467–484

Clibbon KL, McCowen A, Hand JW (1993) SAR distributions in interstitial microwave antennas with a single dipole displacement. IEEE Trans Biomed Eng 40: 925–932

Conway J (1987) Electrical impedance tomography for thermal monitoring of hyperthermia treatment: an assessment using in vitro and in vivo measurements. Clin Phys Physiol 8 (Suppl A): 147–153

Corry PM, Barlogie B (1982) Clinical application of high frequency methods for local hyperthermia. In: Nussbaum GH (ed) Physical aspects of hyperthermia. American Institute of Physics, New York, pp 307–328

Corry PM, Jabboury K, Armour EP, Kong JS (1984) Human cancer treatment with ultrasound. IEEE Trans Sonics Ultrasonics 31: 444–455

Cosset JM (1990) Interstitial hyperthermia. In: Gautherie M (ed), Interstitial, endocavitary and perfusional hyperthermia, Springer, Berlin Heidelberg New York, pp 1–41

Crezee J, Mooibroek J, Lagendijk JJW (1993) Thermal model verification in interstitial hyperthermia. In: Seegenschmiedt MH, Sauer R (eds) Interstitial and intracavitary thermoradiotherapy. Springer, Heidelberg Berlin New York, pp 147–153

Dahl O, Mella O (1990) Hyperthermia and chemotherapeutic agents. In: Field SB, Hand JW (eds) An introduction to the practical aspects of clinical hyperthermia. Taylor & Francis, London, pp 108–142

DeFord JA, Babbs CF, Patel UH, Bleyer MW, Marchosky JA, Moran CJ (1991) Effective interstitial hyperthermia. Int J Hyperthermia 7: 441–453

De Leeuw AAC, Crezee J, Lagendijk JJW (1993) Temperature and SAR measurements in deep-body hyperthermia with thermocouple thermometry. Int J Hyperthermia 9: 685–697

De Leeuw AAC, Lagendijk JJW, Van den Berg PM (1990) SAR distribution of the 'coaxial TEM' system with variable aperture width: measurements and model computations. Int J Hyperthermia 6: 445–451

De Leeuw AAC, Crezee J, Lagendijk JJW (1993) Temperature and SAR measurements in deep-body hyperthermia with thermocouple thermometry. Int J Hyperthermia 9: 685–697

Delannoy J, LeBihan D, Hoult DI, Levin RL (1990) Hyperthermia system combined with a MRI unit. Med Phys 17: 855–860

Denman DL, Foster AE, Cooper Lewis G, et al. (1988) The distribution of power and heat produced by interstitial microwave antenna arrays. II. The role of antenna spacing and insertion depth. Int J Radiat Oncol Biol Phys 14: 537–545

Deurloo IKK, Visser AG, Morawska M, van Geel CAJF, van Rhoon CG, Levendag PC (1991) Applications of a capacitive-coupling interstitial hyperthermia system at 27 MHz: study of different applicator configurations. Phys Med Biol 36: 119–132

Dewhirst MW, Philips TL, Samulski TV, et al. (1990) RTOG quality assurance guidelines for clinical trials using hyperthermia. Int J Radiat Oncol Biol Phys 18: 1249–1259

Dickinson RJ (1984) A non-rigid mosaic applicator for local ultrasound hyperthermia. In: Overgaard J (ed) Hyperthermic oncology 1984, vol 1. Taylor and Francis, London, pp 671–674

Dickinson RJ (1985) Thermal conduction errors of manganin-constantan thermocouple arrays. Phys Med Biol 30: 445–453

Diederich CJ, Hynynen K (1990) The development of intra-cavitary ultrasonic applicators for hyperthermia: a design and experimental study. Med Phys 17: 626–634

Diederich CJ, Hynynen K (1991) The feasibility of using electrically focused ultrasound arrays to induce deep hyperthermia via body cavities. IEEE Trans Ultrason Ferroelec Freq Contr 38: 207–219

Dorr LN, Hynynen K (1992) The effects of tissue heterogeneities and large blood vessels on the thermal exposure induced by short high-power ultrasound pulses. Int J Hyperthermia 8: 45–60

Doss JD (1982) Calculation of electric fields in conductive media. Med Phys 9: 566–573

Dunn F, Frizzell LA (1982) Bioeffects of ultrasound. In: Lehman JF (ed) Theraputic heat and cold, 3rd edn. Williams & Wilkins, Baltimore, pp 388–390

Dunscombe PB, Cetas TC, Connor WG, et al. (1989) Hyperthermia treatment planning (AAPM Report No 27). American Institute of Physics, New York

Durney CH (1990) Electromagnetic field propagation and interaction with tissue. In: Field SB, Hand JW (eds) An introduction to the practical aspects of clinical hyperthermia. Taylor & Francis, London, pp 242–274

Ebbini ES, Cain CA (1991a) Optimization of the intensity gain of multiple-focus phased-array heating patterns. Int J Hyperthermia 7: 953–973

Ebbini ES, Cain CA (1991b) Experimental evaluation of a prototype cylindrical section ultrasound hyperthermia phased-array applicator. IEEE Trans Ultrason Ferroelec Freq Contr 38: 510–520

Edmonds PD, Ross WC, Lee ER, Fessenden P (1985) Spatial distributions of heating by ultrasound transducers in clinical use indicated in a tissue-equivalent phantom. Proc 1985 IEEE Ultrasonics Symposium, New York, IEEE, pp 908–912

Emami B, Stauffer P, Dewhirst MW, et al. (1991) RTOG quality assurance guidelines for interstitial hyperthermia. Int J Radiat Oncol Biol Phys 20: 1117–1124

Fessenden P, Lee ER, Anderson TL, Strohbehn JW, Meyer JL, Samulski TV, Marmor JB (1984a) Experience with a multi-transducer ultrasound system for localized hyperthermia of deep tissues. IEEE Trans Biomed Eng 31: 126–135

Fessenden P, Lee ER, Samulski TV (1984b) Direct temperature measurement. Cancer Res (Suppl) 44: 4799s–4804s

Fessenden P, Kapp DS, Lee ER, Samulski TV (1988) Clinical microwave applicator design. In: Paliwal BR, Hetzel FW, Dewhirst MW (eds) Biological, physical and clinical aspects of hyperthermia. AAPM Medical Physics Monograph 16. American Institute of Physics, New York, pp 123–131

Fessenden P, Lee ER, Kapp DS, et al. (1993) Review of the Stanford experience developing non-focusing scanning and array surface microwave (MW) applicators. In: Gerner EW, Cetas TC (eds) Hyperthermic oncology 1992, vol 2. Arizona Board of Regents, Tucson, pp 183–186

Franconi C, Tiberio CA, Raganella L, Begnozzi L (1986) Low frequency RF twin dipole applicator for intermediate depth hyperthermia. IEEE Trans Microwave Theory Tech 34: 612–619

Furse CM, Iskander MF (1989) Three-dimensional electromagnetic power deposition in tumors using interstitial antenna arrays. IEEE Trans Biomed Eng 36: 977–986

Goffinet DR, Prionas SD, Kapp DS, et al. (1990) Interstitial ^{192}Ir flexible catheter radiofrequency hyperthermia treatments of head and neck and recurrent pelvic carcinomas. Int J Radiat Oncol Biol Phys 18: 199–210

Gopal MK, Hand JW, Lumori MLD, Alkhairi S, Paulsen KD, Cetas TC (1992) Current sheet applicator arrays for superficial hyperthermia of chestwall lesions. Int J Hyperthermia 8: 227–240

Goss SA, Frizzell LA, Dunn F (1979) Ultrasonic absorption and attenuation in mammalian tissues. Ultrasound Med Biol 5: 181–186

Griffiths H, Ahmed A (1987) Applied potential tomography for non-invasive temperature mapping in hyperthermia. Clin Phys Physiol 8 (Suppl. A): 147–153

Gross EJ, Cetas TC, Stauffer PR, Liu RL, Lumori MLD (1990) Experimental assessment of phased-array heating of neck tumors. Int J Hyperthermia 6: 453–474

Hahn GM (1982) Hyperthermia and cancer. Plenum, New York

Haider SA, Chen ZP, Cetas TC, Roemer RB (1987) Interstitial ferromagnetic implant heating: practical guidelines for use. Proceedings 9th Annual Conference of IEEE Engineering in Medicine and Biology Society (vol 3). IEEE, New York, pp 1626–1628

Hall EJ (1982) Hyperthermia: an overview. In: Dethlefsen LA, Dewey WC (eds) Third international symposium: cancer therapy by hyperthermia, drugs and radiation (NCI Monograph 61), National Cancer Institute, Bethesda, pp xv–xvi

Hand JW (1990a) Biophysics and technology of electromagnetic hyperthermia. In: Gautherie M (ed) Methods of external hyperthermic heating. Springer, Berlin Heidelberg New York, pp 1–59 (Clinical thermology, subseries thermotherapy)

Hand JW (1990b) Quality assurance in hyperthermia. In: Field SB, Hand JW (eds) An introduction to the practical aspects of clinical hyperthermia. Taylor and Francis, London, pp 513–532

Hand JW, Hind AJ (1986) A review of microwave and RF applicators for localised hyperthermia. In: Hand JW, James JR (eds) Physical techniques in clinical hyperthermia. Research Studies Press, Letchworth, Hertfordshire, England, pp 98–148

Hand JW, Cheetham JL, Hind AJ (1986) Absorbed power distributions from coherent microwave arrays for localized hyperthermia. IEEE Trans Microwave Theory Tech 34: 484–489

Hand JW, Lagendijk JJW, Andersen JB, Bolomey JC (1989) Quality assurance guidelines for ESHO protocols. Int J Hyperthermia 5: 421–428

Hand JW, Trembly BS, Prior MV (1992a) Physics of interstitial hyperthermia: radiofrequency and hot water tube techniques. In: Urano M, Douple E (eds) Hyperthermia and oncology, vol 3. Interstitial hyperthermia: physics, biology and clinical aspects. VSP, Utrecht, pp 99–134

Hand JW, Vernon CC, Prior MV (1992b) Early experience of a commercial scanned focused ultrasound hyperthermia system. Int J Hyperthermia 8: 587–607

Hand JW, Ebbinni E, O'Keeffe D, Israel D, Mohammadtaghi S (1994) An ultrasound phased array for use in intracavitary applicators for thermotherapy of prostatic diseases. Proc IEEE Ultrasonics 1993 Symposium, vol 2, IEEE, NY, pp 1225–1228

Handl-Zeller L, Handl O (1992) Simultaneous application of combined interstitial high-or low-dose rate irradiation with hot water hyperthermia. In: Handl-Zeller L (ed) Interstitial hyperthermia. Springer, Vienna, pp 165–170

Hartov A, Colacchio TA, Strohbehn JW, Ryan TP, Hoopes PJ (1993) Performance of an adaptive MIMO controller for a multiple-element ultrasound hyperthermia system. Int J Hyperthermia 9: 563–579

Heinzl L, Hornsleth SN, Raskmark P, Andersen JB (1990) Electromagnetic applicators. In: Field SB, Hand JW (eds) An introduction to the practical aspects of clinical hyperthermia. Taylor & Francis, London, pp 275–304

Hunt JW (1990) Principles of ultrasound used for generating localized hyperthermia. In: Field SB, Hand JW (eds) An introduction to the practical aspects of clinical hyperthermia. Taylor & Francis, London, pp 371–422

Hunt JW, Lalonde R, Ginsberg H, Urchuk S, Worthington A (1991) Rapid heating: critical theoretical assessment of thermal gradients found in hyperthermia treatments. Int J Hyperthermia 7: 703–718

Hürter W, Reinbold F, Lorenz WJ (1991) A dipole antenna for interstitial microwave hyperthermia. IEEE Trans Microwave Theory Tech 39: 1048–1054

Hynynen K (1990a) Biophysics and technology of ultrasound hyperthermia. In: Gautherie M (ed) Methods of external hyperthermic heating. Springer, Berlin Heidelberg New York, pp 61–115 (Clinical thermology, subseries thermotherapy)

Hynynen K (1990b) Hot spots created at skin-air interfaces during ultrasound hyperthermia. Int J Hyperthermia 6: 1005–1012

Hynynen K (1992) The feasibility of interstitial ultrasound hyperthermia. Med Phys 19: 979–987

Hynynen K, DeYoung D (1988) Temperature elevation at muscle-bone interface during scanned, focused ultrasound hyperthermia. Int J Hyperthermia 4: 267–279

Hynynen K, Edwards DK (1989) Temperature measurements during ultrasound hyperthermia. Med Phys 16: 618–626

Hynynen K, Watmough DJ, Mallard JR (1981) Design of ultrasonic transducers for local hyperthermia. Ultrasound Med Biol 7: 397–402

Hynynen K, Roemer R, Moros E, Johnson C, Anhalt D (1986) The effect of scanning speed on temperature and equivalent thermal exposure distributions during ultrasound hyperthermia in vivo. IEEE Trans Microwave Theory Tech 34: 552–559

Hynynen K, Roemer RB, Anhalt D, Johnson C, Xu ZK, Swindell W, Cetas TC (1987) A scanned focused, multiple transducer ultrasonic system for localised hyperthermia treatments. Int J Hyperthermia 3: 21–35

Hynynen K, Shimm D, Anhalt D, Stea B, Sykes H, Cassady JR, Roemer RB (1990) Temperature distributions during clinical scanned, focused ultrasound hyperthermia treatments. Int J Hyperthermia 6: 891–908

Hynynen K, Frederiksen F, Gautherie M, et al. (1992) Ultrasound hyperthermia (Tor Vergata Medical Physics Monograph Series, vol 2). Postgraduate School of Medical Physics, II University of Rome, Rome, pp 24–25

Ibbini MS, Cain CA (1989) A field conjugation method of direct synthesis of hyperthermia phased-array heating patterns. IEEE Trans Ultrason Ferroelec Freq Contr 36: 3–9

Ibbini MS, Cain CA (1990) The concentric-ring array for ultrasound hyperthermia: combined mechanical and electrical scanning. Int J Hyperthermia 6: 401–419

Ibbini MS, Ebbini ES, Cain CA (1990) N x N square-element ultrasound phased array applicator: simulated temperature distributions associated with directly synthesized heating patterns. IEEE Trans Ultrason Ferroelec Freq Contr 37: 491–500

Ibbott GS, Brezovich I, Fessenden P, et al. (1989) Performance evaluation of hyperthermia equipment (AAPM Report No 26). American Institute of Physics, New York

Iskander MF, Tumeh AM (1989) Design optimization of interstitial antennas. IEEE Trans Biomed Eng 36: 238–246

James BJ, Sullivan DM (1992) Direct use of CT scans for hyperthermia treatment planning. IEEE Trans Biomed Eng 39: 845–851

James BJ, Strohbehn JW, Mechling JA, Trembly BS (1989) The effect of insertion depth on the theoretical SAR patterns of 915 MHz dipole antenna arrays for hyperthermia. Int J Hyperthermia 5: 733–747

Johnson CC, Guy AW (1972) Nonionizing electromagnetic wave effects in biological materials and systems. Proc IEEE 60: 692–717

Johnson RH (1986) New type of compact electromagnetic applicator for hyperthermia in the treatment of cancer. Proc IEE 22: 591–593

Johnson RH, James JR, Hand JW, Hopewell JW, Dunlop PRC, Dickinson RJ (1984) New low-profile applicators for local heating of tissues. IEEE Trans Biomed Eng 31: 28–37

Johnson RH, Preece AW, Hand JW, James JR (1987) A new type of lightweight low-frequency electromagnetic hyperthermia applicator. IEEE Trans microwave Theory Tech 35: 1317–1321

Jones KM, Mechling JA, Trembly BS, Strohbehn JW (1988) SAR distributions for 915 MHz interstitial microwave antennas used in hyperthermia for cancer therapy. IEEE Trans Biomed Eng 35: 851–857

Jones KM, Mechling JA, Strohbehn JW, Trembly BS (1989) Theoretical and experimental SAR distributions for interstitial dipole arrays used in hyperthermia. IEEE Trans Microwave Theory Tech 37: 1200–1209

Kapp DS, Prionas SD (1992) Experience with radiofrequency-local current field interstitial hyperthermia: biological rationale, equipment development and clinical results. In: Handl-Zeller L (ed) Interstitial hyperthermia. Springer, Vienna, pp 95–119

Kapp DS, Fressenden P, Samulski TV, et al. (1988) Stanford University institutional report. Phase I evaluation of equipment for hyperthermia treatment of cancer. Int J Hyperthermia 4: 75–115

Kapp DS, Peters Brown AN, Cox W, Cox RS (1993) Temperature differentials between treatment and pretreatment temperatures correlate with local control following radiotherapy and hyperthermia. Int J Radiat Oncol Biol Phys 27: 331–344

King RWP, Shen LC, Wu TT (1981) Embedded insulated antennas for communication and heating. Electromagnetics 1: 115–117

King RWP, Trembly BS, Strohbehn JW (1983) Electromagnetic field of an insulated antenna in a conducting or dielectric medium. IEEE Trans Microwave Theory Tech 31: 574–583

Kobayashi T, Kida Y, Tanaka T, Hattori K, Matsui M, Amemiya Y (1991) Interstitial hyperthermia for brain tumors using ferromagnetic implants with low Curie temperature. J Neuro-oncol 4: 153–163

Lagendijk JJW (1990) Thermal models: principles and implementation. In: Field SB, Hand JW (eds) An introduction to the practical aspects of clinical hyperthermia, Taylor & Frances, London, pp 478–512

Lagendijk JJW, Visser AG, Kaatee RSJP et al. (1994) The 27 MHz multi-electrode current source interstitial hyperthermia method. Activity, International Selectron Brachytherapy Journal 8(3)

Lagendijk JJW, De Leeuw AAC (1993) Technical note. Temperature errors using multi-sensor thermocouple probes with a common constantan wire. Int J Hyperthermia 9: 763–764

Lagendijk JJW, Hofman P, Schippr J (1988) Perfusion analysis in advanced breast carcinoma during hyperthermia. Int J Hyperthermia 4: 479–495

Lagendijk JJW, van den Berg PM, Hand JW, et al. (1992) Task Group Report 4: treatment planning and modelling in hyperthermia. (Tor Vergata Medical Physics Monograph Series) Postgraduate School of Medical Physics, II University of Rome, Rome

Lau RWM, Sheppard RJ, Howard G, Bleehen NM (1986) The modelling of biological systems in three dimensions using the time domain finite-difference method. II. The application and experimental evaluation of the method in hyperthermia applicator design. Phys Med Biol 31: 1257–1266

LeBihan D, Turner R, Moonen CTW, Pekar J (1991) Imaging of diffusion and microcirculation with gradient sensitization: design, strategy and significance. J Magn Reson Imaging 1: 7–28

Lee DJ, O'Neill MJ, Lam KS, Rostock R, Lam WC (1986) A new design of microwave interstitial applicator for hyperthermia with improved treatment volume. Int J Radiat Oncol Biol Phys 12: 2003–2008

Lee ER, Wilsey TR, Tarczy-Hornoch P, Kapp DS, Fessenden P, Lohrbach A, Prionas SD (1992) Body conformable 915 MHz microstrip array applicators for large surface area hyperthermia. IEEE Trans Biomed Eng 39: 470–483

Lele PP (1983) Physical aspects and clinical studies with ultrasonic hyperthermia. In: Storm FK (ed) Hyperthermia in cancer therapy. G.K. Hall, Boston, pp 333–367

Leopold KA, Dewhirst MW, Samulski TV, et al. (1993) Cumulative minutes with T90 greater than TEMP index is predictive of response of superficial malignancies to hyperthermia and radiation. Int J Radiat Oncol Biol Phys 25: 841–847

Leybovich LB, Emami B, Myerson RJ, Straube WL, Sathiaseelan V (1991) Dual-antenna applicators for hyperthermia of tumors at intermediate depth. Int J Hyperthermia 7: 455–464

Lin JC, Wang YJ (1987) Interstitial microwave antennas for thermal therapy. Int J Hyperthermia 3: 37–47

Lockwood JC, Willette JG (1973) High-speed method for computing the exact solution for the pressure variations in the nearfield of a baffled piston. J Acoust Soc Am 53: 735–741

Lyons M, Parker KJ (1988) Absorption and attenuation in soft tissues. II. Experimental results. IEEE Trans Ultrason Ferroelec Freq Contr 35: 511–521

Magin RL, Peterson AF (1989) Noninvasive microwave phased arrays for local hyperthermia: a review. Int J Hyperthermia 5: 429–450

Marchal C, Bey P, Metz R, Gaulard ML, Robert J (1982) Treatment of superficial human cancerous nodules by local ultrasound hyperthermia. Br J Cancer 45 (Suppl V): 243–245

McGough RJ, Ebbini ES, Cain CA (1992) Direct computation of ultrasound phased-array driving signals from a specified temperature distribution for hyperthermia. IEEE Trans Biomed Eng 39: 825–835

Mechling JA, Strohbehn JW, Ryan TP (1992) Three-dimensional theoretical temperature distributions produced by 915 MHz dipole antenna arrays with varying insertion depths in muscle tissue. Int J Radiat Oncol Biol Phys 22: 131–138

Mizushina S, Hamamura Y, Sugiura T (1989) A method of solution for a class of inverse problems involving measurement errors and its application to medical microwave radiometry. IEEE MTT-S International Symposium Digest 171–174

Mizushina S, Shimizu T, Sugiura T (1992) Non-invasive thermometry with multi-frequency microwave radiometry. Front Med Biol Eng 4: 129–133

Morita N, Bach Andersen J (1982) Near field absorption in a circular cylinder from electric and magnetic line sources. Bioelectromagnetics 3: 253–274

Nasoni RL, Bowen T (1989) Ultrasonic speed as a parameter for non-invasive thermometry. In: Mizushina S (ed) Non-invasive temperature measurement. Gordon and Breach, New York, pp 95–107

NCRP (1981) NCRP Report No. 67. Radiofrequency electromagnetic fields: properties, quantities and units, biophysical interaction and measurements. National Council on Radiation Protection and Measurements, Washington, D.C., pp 25–44

NCRP (1986) NCRP Report No. 86. Biological effects and exposure criteria for radiofrequency electromagnetic fields. National Council on Radiation Protection and Measurements, Bethesda, Maryland, pp 38–39

Nikawa Y, Kikuchi M, Kaneko R, Matsuda T (1993) Design and evaluation of a lens applicator for a 430 MHz heating system. In: Gerner EW, Cetas TC (eds) Hyperthermic oncology 1992, vol 2. Arizona Board of Regents, Tucson, pp 199–202

Nussbaum GH, Side J, Rouhanizadeh N, et al. (1986) Manipulation of central axis heating patterns with a prototype, three-electrode capacitive device for deep-tumor hyperthermia. IEEE Trans Microwave Theory Tech 34: 620–625

Nussbaum GH, Straube WL, Drag MD, et al. (1991) Potential for localized, adjustable deep heating in soft-tissue environments with a 30-beam ultrasonic hyperthermia system. Int J Hyperthermia 7: 279–299

Nyborg WL, Steele RB (1985) Nearfield of a piston source of ultrasound in an absorbing medium. J Acoust Soc Am 78: 1882–1891

Ocheltree KB, Frizzell LA (1989) Sound field calculation for rectangular sources. IEEE Trans Ultrason Ferroelec Freq Contr 36: 242–248

Ocheltree KB, Benkeser PJ, Frizzell LA, Cain CA (1984) An ultrasonic phased array applicator for hyperthermia. IEEE Trans Sonics Ultrasonics 31: 526–531

Oleson JR, Samulski TV, Leopold KA, Clegg ST, Dewhirst MW, Dodge RK, George SL (1993) Sensitivity of hyperthermia trial outcomes to temperature and time: implications for thermal goals of treatment. Int J Radiat Oncol Biol Phys 25: 289–297

Overgaard J (1993) The future of hyperthermic oncology. In: Gerner EW, Cetas TC (eds) Hyperthermic oncology 1992, vol 2. Arizona Board of Regents, Tucson, pp 87–92

Paulsen KD (1990) Calculation of power deposition patterns in hyperthermia. In: Gautherie M (ed) Thermal dosimetry and treatment planning. Springer, Berlin Heidelberg New York, pp 57–117 (Clinical thermology, subseries thermotherapy)

Paulsen KD, Lynch DR, Stronbehn JW (1988) Three dimensional finite boundary and hybrid element solutions of the Maxwell equations for lossy dielectric media. IEEE Trans Microwave Theory Tech 36: 682–693

Paulsen KD, Strohbehn JW, Lynch DR (1985) Comparative theoretical performance for two types of hyperthermia systems. Int J Radiat Oncol Biol Phys 11: 1659–1671

Pennes HH (1948) Analysis of tissue and arterial blood temperatures in the human resting forearm. J Appl Physiol 1: 93–122

Picket-May MJ, Taflove A, Lin WC, Katz DS, Sathiaseelan V, Mital BB (1992) Initial results for automated computational modeling of patient-specific electromagnetic hyperthermia. IEEE Trans Biomed Eng 39: 226–236

Pounds DW, Britt RH (1984) Single ultrasonic crystal techniques for generating uniform temperature distributions in homogeneously perfused tissues. IEEE Trans Sonics Ultrasonics 31: 482–496

Prionas SD, Hahn GM (1985) Noninvasive thermometry using multiple frequency-band radiometry: a feasibility study. Bioelectromagnetics 6: 391–404

Prionas SD, Kapp DS, Sokol JL, Fessenden P (1990) Absorption of ultrasound (US) near tissue to air interfaces. Abstracts of papers for the tenth annual meeting of the North American Hyperthermia Group, New Orleans, Louisiana, p 18

Prionas SD, Kapp DS, Goffient DR, Bagshaw MA, Ben-Yosef R, Sokol JL, Fessenden P (1993) Interstitial radiofrequency-induced hyperthermia. In: Gerner EW, Cetas TC (eds) Hyperthermic oncology 1992, vol 2. Arizona Board of Regents, Tucson, pp 249–253

Prionas SD, Kapp DS, Goffinet DR, Ben-Yosef R, Fessenden P and Bagshaw MA (1994) Thermometry of interstitial hyperthermia given as an adjuvant to brachytherapy for the treatment of carcinoma of the prostate. Int J Radiat Oncol Biol Phys 28(1): 151–162

Rine GP, Dewhirst MW, Cobb ED, Clegg ST, Coleman EN, Samulski TV, Wallen CA (1992) Feasibility of estimating the temperature distribution in a tumor heated by a waveguide applicator. Int J Radiat Oncol Biol Phys 23: 1009–1019

Robins HI, Dennis WH, Neville AJ, et al. (1985) A non-toxic system for 41.8°C whole body hyperthermia: results of a phase I study using a radiant heat device. Cancer Res 45: 3937–3944

Roemer RB (1990) Thermal dosimetry. In: Gautherie M (ed) Thermal dosimetry and treatment planning. Springer, Berlin Heidelberg New York, pp 119–214 (Clinical thermology, subseries thermotherapy)

Roemer RB, Cetas TC (1984) Applications of bioheat transfer simulations in hyperthermia. Cancer Res 44: 4788S–4798S

Roemer RB, Fletcher AM, Cetas TC (1985) Obtaining local SAR and blood perfusion data from temperature measurements: steady-state and transient techniques compared. Int J Radiat Oncol Biol Phys 11: 1539–1550

Roos D, Hugander A (1988) Microwave interstitial applicators with improved longitudinal heating patterns. Int J Hyperthermia 4: 609–615

Ryan TP (1991) Comparison of six microwave antennas for hyperthermia treatment of cancer: SAR results for single antennas and arrays. Int J Radiat Oncol Biol Phys 21: 403–413

Ryan TP, Mechling JA, Strohbehn JW (1990) Absorbed power deposition for various insertion depths for 915 MHz interstitial dipole antenna arrays: experiment versus theory. Int J Radiat Oncol Biol Phys 19: 377–387

Ryan TP, Hoopes PJ, Taylor JH, Strohbehn JW, Roberts DW, Douple EB, Coughlin CT (1991) Experimental brain hyperthermia: techniques for heat delivery and thermometry. Int J Radiat Oncol Biol Phys 20: 739–750

Samulski TV, Clegg ST, Das S, MacFall J, Prescott DM (1994) Application of new technology in clinical hyperthermia. Int J Hyperthermia 10: 389–394

Samulski TV, Fessenden P (1990) Thermometry in therapeutic hyperthermia. In: Gautherie M (ed) Methods of hyperthermia control. Springer, Berlin Heidelberg New York, pp 1–34 (Clinical thermology, subseries thermotherapy)

Samulski TV, Chopping PT, Haas B (1982) Photoluminescent thermometry based on europium-activated calcium sulfide. Phys Med Biol 27: 107–114

Samulski TV, Lyons BE, Britt RH (1985) Temperature measurements in high thermal gradients. II. Analysis of conduction effects. Int J Radiat Oncol Biol Phys 11: 963–971

Samulski TV, Kapp DS, Fessenden P, Lohrbach A (1987a) Heating deep seated eccentrically located tumors with an annular phased array system: a comparative clinical study using two annular array operating configurations. Int J Radiat Oncol Biol Phys 13: 83–94

Samulski TV, Fessenden P, Valdagni R, Kapp DS (1987b) Correlations of thermal washout rate, steady-state temperatures and tissue type in deep seated recurrent or metastatic tumors. Int J Radiat Oncol Biol Phys 13: 907–916

Samulski TV, Fessenden P, Lee ER, Kapp DS, Tanabe E, McEuen A (1990a) Spiral microstrip hyperthermia applicators: technical design and clinical performance. Int J Radiat Oncol Biol Phys 18: 233–242

Samulski TV, Grant WJ, Oleson JR, Leopold KA, Dewhirst MW, Vallario P, Blivin J (1990b) Clinical experience with a multi-element ultrasonic hyperthermia system: analysis of treatment temperatures. Int J Hyperthermia 6: 909–922

Samulski TV, MacFall J, Zhang Y, Grant W, Charles C (1992) Non-invasive thermometry using magnetic resonance diffusion imaging: potential for application in hyperthermic oncology. Int J Hyperthermia 8: 819–829

Sandhu TS, Kowal HS, Johnson RJR (1978) The development of microwave hyperthermia applicators. Int J Radiat Oncol Biol Phys 4: 515–519

Sapozink MD, Corry PM, Kapp DS, et al. (1991) Quality assurance guidelines for clinical trials using hyperthermia for deep seated malignancy. Int J Radiat Oncol Biol Phys 20: 1109–1115

Sathiaseelan V, Leybovich L, Emami MS, Stauffer P, Straube W (1991) Characteristics of improved microwave interstitial antennas for local hyperthermia. Int J Radiat Oncol Biol Phys 20: 531–539

Satoh T, Stauffer PR, Fike JR (1988) Thermal distribution studies of helical coil microwave antennas for interstitial hyperthermia. Int J Radiat Oncol Biol Phys 15: 1209–1218

Schneider C, van Dijk JDP (1991) Visualization with a matrix of LEDs of interference effects from a radiative four applicator hyperthermia system. Int J Hyperthermia 7: 355–366

Schreier K, Budhina M, Lesnicar H, et al. (1990) Preliminary studies of interstitial hyperthermia using hot water. Int J Hyperthermia 6: 431–434

Sekins KM, Emery AF (1982) Thermal science for physical medicine. In: Lehmann JF (ed) Therapeutic heat and cold, 3rd edn. Williams & Wilkins, Baltimore, pp 70–132

Seppi E, Shapiro E, Zitelli L, Henderson S, Wehlau A, Wu G, Dittmer C (1985) A large aperture ultrasonic array system for hyperthermia treatment of a deep seated tumors. In: Proceedings IEEE Ultrasonics Symposium, IEEE, New York, pp 942–948

Shrivastava P, Luk K, Oleson JR, et al. (1989) Hyperthermia quality assurance guidelines. Int J Radiat Oncol Biol Phys 16: 571–587

Storm FK, Harrison WH, Elliott RS, Kaiser LR, Silberman AW, Morton DL (1981) Clinical radiofrequency hyperthermia by magnetic loop induction. J Microwave Power 16: 179–184

Strohbehn JW (1982) Theoretical temperature distributions for solenoidal-type hyperthermia systems. Med Phys 9: 673–682

Strohbehn JW, Roemer RB (1984) A survey of computer simulations of hyperthermia treatments. IEEE Trans Biomed Eng 31: 136–149

Strohbehn JW, Paulsen KD, Lynch DR (1986) Use of the finite element method in computerized thermal dosimetry. In: Hand, JW, James JR (eds) Physical techniques in clinical hyperthermia. Research Studies Press, Letchworth, Hertfordshire, England, pp 383–451

Strutt JW (1945) Theory of sound, vol 2. Dover, New York

Sugimachi K, Matsuda H (1990) Experimental and clinical studies of hyperthermia for carcinomas of the esophagus. In: Gautherie M (ed) Interstitial, endocavitary and perfusional hyperthermia. Springer, Berlin Heidelberg New York, pp 59–76

Sullivan D (1990) Three-dimensional computer simulation in deep regional hyperthermia using the finite-difference time-domain method. IEEE Trans Microwave Theory Tech 38: 204–211

Sullivan DM (1992) A frequency-dependent FDTD method for biological applications. IEEE Trans Microwave Theory Tech 40: 532–539

Sullivan DM, Buechler D, Gibbs FA (1992) Comparison of measured and simulated data in an annular phased array using an inhomogeneous phantom. IEEE Trans Microwave Theory Tech 40: 600–604

Sullivan DM, Ben-Yosef R, Kapp DS (1993) Standford 3-D hyperthermia treatment planning system. Int J Hyperthermia 9: 627–643

Swindell W (1986) Ultrasonic hyperthermia. In: Hand JW, James JR (eds) Physical techniques in clinical hyperthermia. Research Studies Press, Letchworth, Hertfordshire, England, pp 288–326

Tanabe E, McEuen A, Norris CS, Fessenden P, Samulski TV (1983) A multi-element microstrip antenna for local hyperthermia. IEEE MTT-S International Microwave Symposium Digest, pp 183–185

Tarczy-Hornoch P, Lee ER, Sokol JL, Prionas SD, Lohrbach AW, Kapp DS (1992) Automated mechanical thermometry probe mapping systems for hyperthermia. Int J Hyperthermia 8: 543–554

Trembly BS (1985) The effects of driving frequency and antenna length on power deposition within a microwave antenna array used for hyperthermia. IEEE Trans Microwave Theory Tech 32: 152–157

Trembly BS, Wilson AH, Sullivan MJ, Stein AD, Wong TZ, Strohbehn JW (1986) Control of the SAR pattern within an interstitial mocrowave array through variation of antenna driving phase. IEEE Trans Microwave Theory Tech 34: 568–571

Trembly BS, Douple EB, Hoopes PJ (1991) The effect of air cooling on the radial temperature distribution of a single microwave hyperthermia antenna in vivo. Int J Hyperthermia 7: 343–354

Tumeh AM, Iskander MF (1989) Performance comparison of available interstitial antennas for microwave hyperthermia. IEEE Trans Microwave Theory Tech 37: 1126–1133

Turner PF (1984) Regional hyperthermia with an annular phased array. IEEE Trans Biomed Eng 31: 106–114

Turner PF (1986) Interstitial equal-phased arrays for EM hyperthermia. IEEE Trans Microwave Theory Tech 34: 572–578

Umemura S, Cain CA (1989) The sector-vortex phased array: acoustic field synthesis for hyperthermia. IEEE Trans Ultrason Ferroelec Freq Contr 36: 249–257

Umemura S, Cain CA (1992) Acoustical evaluation of a prototype sector-vortex phased-array applicator. IEEE Trans Ultrason Ferroelec Freq Contr 39: 32–38

van der Zee J (1988) ESHO 5–88: protocol for phase III trial involving reirradiation of recurrent breast cancer with or without hyperthermia. Dr. Daniel den Hoed Cancer Center, Rotterdam

van der Zee J, van Rhoon GC (1993) Eindverslag Ontwikkelingsgeneeskunde project OG 89–23: de waarde van hyperthermie bij toevoeging aan radiotherapie bij de behandeling van inoperabele en stralingsresistente tumoren. A report to the Dutch Ministry of Health Care

van Dijk JDP, Schneider C, van Os R, Blank LE, Gonzalez DG (1990) Results of deep body hyperthermia with large waveguide radiators. Adv Exp Med Biol 267: 315–319

van Rhoon GC, Rietveld PJM, Broekmeyer-Reurink MP, Verloop-van't Hof EM, van der Ploeg SK, van der Zee J (1993) A 433 MHz waveguide applicator system with an improved effective field size for hyperthermia treatment of superficial tumors on the chest wall. In: Gerner EW, Cetas TC (eds) Hyperthermic oncology 1992, vol 2. Arizona Board of Regents, Tucson, pp 187–190

Vernon CC, Hand JW, Field SB, Machin D (1990) A study of the use of hyperthermia in the treatment of breast and head and neck tumours: a MRC multi-centre phase III trial. Trial closed for accrual Dec 1993. Results to be published in 1994

Visser AG, Deurloo IKK, Levendag PC, Ruifrok ACC, Cornet B, van Rhoon GC (1989) An interstitial hyperthermia system at 27 MHz. Int J Hyperthermia 5: 265–276

Visser AG, Chive M, Hand JW, et al. (1992) Interstitial and intracavitary hyperthermia: a task group report of the European Society for Hyperthermic Oncology, Postgraduate School of Medical Physics, University of Rome. Tor Vergata, Rome

Visser AG, Kaatee RSJP, Levendag PC (1993) Radiofrequency techniques for interstitial hyperthermia. In: Seegenschmiedt MH, Sauer R (eds) Interstitial and intracavitary thermoradiotherapy. Springer, Berlin Heidelberg New York, pp 35–41

Waterman FM, Hoh LLS (1994) A rcommended revision in the RTOG thermometry guidelines for hyperthermia administered by ultrasound. Int J Hyperthermia

Waterman FM, Leeper JB (1990) Temperature artifacts produced by thermocouples used in conjunction with 1 and 3 MHz ultrasound. Int J Hyperthermia 6: 383–399

Wickersheim KA (1986) A new fiberoptic thermometry system for use in medical hyperthermia. SPIE Proceedings, vol 713

Williams RA (1983) Ultrasound: biological effects and potential hazards. Academic, London, pp 107–110

Wu A, Watson ML, Sternick ES, Bielawa RJ, Carr KL (1987) Performance characteristics of a helical coil microwave interstitial antenna for local hyperthermia. Med Phys 14: 235–237

Wust P, Fahling H, Nadobny J, Felix R, Seebass M (1993) Potential of radiofrequency hyperthermia: planning, optimization, technological development. In: Gerner EW, Cetas TC (eds) Hyperthermia oncology 1992 vol 2. Arizona Board of Regents, pp 65–72

Young IR, Hand JW, Oatridge A, Prior M, Forse G (1994) Further observations of the measurement of tissue T1 to monitor temperature in vivo by MR. Magn Reson Med 31: 342–345

Yuan X, Strohbehn W, Lynch DR, Johnsen M (1990) Theoretical investigation of a phased-array hyperthermia system with moveable aperatures. Int J Hyperthermia 6: 227–240

Zemanek J (1971) Beam behaviour within the nearfield of a vibrating piston. J Acoust Soc Am 49: 181–191

Zhang Y, Dubal NV, Takemoto-Hambleton R, Joines WT (1988) The determination of the electromagnetic field and SAR pattern of an interstitial applicator in a dissipative medium. IEEE Trans Microwave Theory Tech 36: 1438–1443

Zhang Y, Joines WT, Oleson JR (1990a) The calculated and measured temperature distribution of a phased interstitial antenna array. IEEE Trans Microwave Theory Tech 38: 69–77

Zhang Y, Joines WT, Oleson JR (1990b) Microwave hyperthermia induced by a phased interstitial antenna array. IEEE Trans Microwave Theory Tech 38: 217–221

Zhang Y, Joines WT, Oleson JR (1991a) Heating patterns generated by phase modulation of a hexagonal array of interstitial antennas. IEEE Trans Biomed Eng 38: 92–97

Zhang Y, Joines WT, Oleson JR (1991b) Prediction of heating patterns of a microwave interstitial array at various insertion depths. Int J Hyperthermia 7: 197–207

Zhou LJ, Fessenden P (1993) Automation of temperature control for large-array microwave surface applicators. Int J Hyperthermia 9: 479–490

15 Dosimetry of Radiolabeled Antibodies

BARRY W. WESSELS and DONALD J. BUCHSBAUM

CONTENTS

15.1 Introduction

Even before the advent of monoclonal antibodies, clinicians and scientists had been intrigued by the possibility of selectively targeting tumors and metastatic disease with a variety of agents including hormones, transmitters, drugs, metabolites, and polyclonal antibodies (WAGNER 1968). Naturally, if it was found that any of these agents were not sufficiently cytotoxic to the tumor, a radionuclide label was added to promote selective tumor lethality

BARRY W. WESSELS, PhD, George Washington University Medical Center, Department of Radiation Oncology, 901 Twenty Third St. NW, Washington, DC 20037, USA
DONALD J. BUCHSBAUM, PhD, University of Alabama-Birmingham, Department of Radiation Oncology, 619 S. 19th Street, Birmingham, AL 35233, USA

and "action at a distance" tumor cell kill. This addition of the radionuclide to the biologically targeted therapy also made it possible to externally monitor the time-dependent distribution of the agent in animals and humans.

KOHLER and MILSTEIN's discovery (1975) of the hybridoma technology for producing highly specific, tumor-associated monoclonal antibodies renewed the interest in biologically targeted imaging and therapy. Over the last 15 years there has been both academic and industrial interest in producing effective agents for use in the clinic. Correspondingly, quantitation of the efficacy of these new radiolabeled monoclonal antibody agents has also become necessary.

Along this line of reasoning, one might ask, "What role does radiation dosimetry play in the development of radiolabeled antibodies for clinical use?" To physicists, this question seems to be somewhat rhetorical and obvious. However, to many of the potential users of radiolabeled antibodies for radioimmunodiagnosis (RAID) or radioimmunotherapy (RIT), including clinical oncologists, nuclear medicine physicians, and those who find their academic roots in immunology or genetic engineering, quantification of the effects of radiolabeled antibody products dose not necessarily mean the measurement of absorbed dose. In fact to many investigators, the measurement of product viability is thought to correlate with in vitro stability assays and/or the prescription of the radiopharmaceutical in units of activity per patient. Specifications of units of absorbed dose for these products is not only foreign to most investigators, but does not permit an obvious interpretation or prediction of response. The compounding difficulty of heterogeneity of antigen expression and antibody uptake in tumors, which occurs in the diagnosis or treatment of patients, lends increased credence to those who would specify a "dose" in the simplest units possible (see Sect. 15.3.5).

This chapter outlines the historical development of the dosimetry methods used to evaluate RAID/

RIT through a comprehensive review of the literature. In addition, it is the goal of this summary to enable the reader to acquire a perspective on the fundamental principles of internal dosimetry and radiobiology in relation to exponentially decaying, low-dose-rate radiation associated with radio-labeled antibody delivery. Citations have also been included which provide references to specific methodologies, along with an overview of the pertinent clinical results to date.

15.2 Factors Influencing the Accumulation of Absorbed Dose in Tumors and Normal Tissue

15.2.1 Antigen Expression

Many investigators have described the multifactorial parameters that affect the pharmacology of monoclonal antibody targeted therapy (COLCHER et al. 1983; SCHLOM et al. 1990). In order to localize. the antibody will usually bind to a tumor-associated antigenic site located on the cell surface. GALLAGHER (1983) has described the requirements for an appropriate receptor antigen as: (a) nonshedding, (b) nonmodulating, (c) present on cell variants, (d) high density of expression, (e) tumor-specific or -associated and (f) having expression independent of cell cycle. The properties of shedding and antigenic modulation are usually associated with a decrease in target localization by an in situ change of the antigen's characteristics after antibody coupling or by interference with cell surface localization from circulating antigen. The antigen should be preferentially expressed in high density on abnormal cells as compared to normal tissue cells. Those antigens which are considered tumor-specific should not be expressed by normal tissues. Thus, antibodies reactive with tumor-specific antigens would lack "cross-reactivity" with normal tissues.

15.2.2 Antibody Properties and Uptake Criteria

It has been observed that most nonspecific proteins have some small degree of localization in well-vascularized tumors or organs compared to the extravascular compartments (JAIN 1990). Hence, it is important to quantitate the degree of specific localization since the absorbed dose target (tumor) to nontarget (remainder of body) ratio shares a direct relationship with radiolabeled antibody uptake. GALLAGHER (1983) defines a localization index (L.I.) for animal models as:

$$\text{L.I.} = \frac{\dfrac{\text{Tumor }^{125}\text{I activity/weight}}{\text{Tumor }^{131}\text{I activity/weight}}}{\dfrac{\text{Blood }^{125}\text{I activity/weight}}{\text{Blood }^{131}\text{I activity/weight}}}.$$

In this example, ^{125}I was used to label the specific monoclonal antibody and ^{131}I was labeled to a nonspecific antibody. Studies have shown that smaller tumors have higher localization indices (MOSHAKIS et al. 1981; WILLIAMS et al. 1988). This is thought to be due to many factors such as increased vascularity in small tumors, necrosis and regions inaccessible to blood flow for large tumors, differences in extracellular fluid space, lymphatic drainage, and increased intratumoral pressures for large tumors (SANDS 1990; JAIN 1990).

Requirements for the carrier antibody include:

1. Homogeneity-only tumor-specific active light chains
2. High selectivity with little cross-reactivity with normal tissues
3. High antigen affinity
4. Pharmacokinetics matched to radionuclide physical properties of radioconjugate
5. High purity (no retroviruses)
6. High stability under biologic and chemical challenge
7. Surface groups available for conjugation
8. Nonimmunogenic to host

Recent studies have also shown that the mass amount of antibody administered to animals or humans has a substantial impact on distribution heterogeneity in tumors as well as antibody size (KLEIN et al. 1989; BADGER et al. 1990; JAIN 1990). In general, as more antibody is delivered to the tumor, decreases in heterogeneity of deposition result, up to tumor saturation. This is very dependent on degree of tumor necrosis, size, antigen expression, and vascularity.

15.2.3 Tumor Properties

Radiolabeled monoclonal antibodies, which have molecular weights ranging from 50 kDa for Fab fragments, 120 kDa for F(ab')$_2$ fragments, and 180 kDa for intact IgG to 900 kDa for IgM, are very large relative to classic drugs and hormones, which have sizes of several hundred to thousands of daltons (SANDS 1990). Most intact immunoglobulins clear much more slowly than their smaller fragment counterparts and remain in the vascular compartment with half-lives of 12 h to several days. Although tumors have been shown to be more permeable to antibodies than normal organs (JAIN 1990), these immunoconjugates must pass through the vascular compartment into the interstitial space before localization to tumor cell surface antigens is possible. The specific process of transport through the vascular epithelial barrier includes (but is not limited to): (a) size of the antibody, (b) affinity, (c) stereochemical configuration, (d) charge, (e) number and size of the vascular pores, (f) rate of diffusion, (g) rate of convection, and (h) intratumoral pressure (JAIN 1990; SANDS 1990; WEINSTEIN and VAN OSDOL 1992b). WILLIAMS et al. (1988) showed that there was a linear relation between uptake of antibody and size for smaller tumors that were presumably well perfused. However, as tumors grow larger and outstrip their blood supply, specific uptake of antibody may plateau and then decrease. This phenomenon has been thought to be observed in patients with liver metastases undergoing either RAID (NABI and DOERR 1992) or RIT (LEICHNER et al. 1989).

15.2.4 Radionuclide Properties and Selection

The selection of optimum radionuclides for both RAID and RIT has been an important area for both theoretical and experimental studies for over a decade. Using HOSAIN and HOSAIN's (1978) list of radionuclides for biologically targeted therapy, many investigators have compiled selective lists of radionuclides which match the multivariable criteria necessary for optimum imaging or therapy. WESSELS and ROGUS (1984) cited a list with some 25 variables, under the major subheadings of (a) radionuclide physical properties, (b) radionuclide labeling chemistry, (c) radionuclide/antibody biodistribution and biologic half-life, (d) target (tumor) to nontarget time-dependent ratio, (e) immunologic purity of antibody, and (f) characteristics of the imaging system with respect to radiolabel and marketability and convenience of the final product. Among these six general categories, the most important criterion for optimum radionuclide selection was found to be the matching of the physical half-life of the radionuclide with the initial peak in the target to nontarget ratio of the antibody carrier. WESSELS and ROGUS went on to show by using both human and animal data after searching the chart of radionuclides from $A = 50$–250, that ^{186}Re, ^{90}Y, ^{67}Cu, and ^{131}I are among the best radionuclides for intact antibody RIT. A rule of thumb given for selecting the

optimum radionuclide half-life is that this physical half-life should be 1.5–3.0 times longer than the time it takes for the target to nontumor biodistribution ratio to maximize or reach a steady state. In a recent publication by YORKE et al. (1991), it was shown that this criterion is correct over a broad range of half-lives, and that its application is consistent with successful radiolabeled antibody therapy results. It was also shown that an approximate criterion for selecting imaging radionuclides versus RIT radionuclides may be found in examining the nonpenetrating (NP) to penetrating (P) ratio of particulate emissions. Optimum imaging agents were shown to have NP/P ratios of less than 0.5 whereas optimal therapy radionuclides had ratios well above 1.0 (WESSELS and ROGUS 1984).

JUNGERMAN et al. (1984) surveyed the chart of nuclides for possible RIT radionuclides, but limited their search to nuclides with half-lives from 1 to 3 days. They were concerned that a longer half-life would lead to increased normal tissue radiation doses in patients due to shedding of the antibodies from the cell surface of tumors before completion of the radioactive decay. It has been demonstrated in the literature that after initial excretion, antibody/radionuclide pairs usually clear more slowly in the tumor as compared to normal tissues (YORKE et al. 1991; GOLDENBERG et al. 1990). Other requirements cited were (a) stability of radioactive daughter product to avoid an increase in normal tissue dose, (b) minimization of gamma rays and high-energy x-rays for therapeutic agents to reduce normal tissue dose, and (c) production availability by a cyclotron or accelerator.

In 1986, HUMM broadened the discussion of radionuclide selection by proposing a classification system to match radionuclides with the heterogeneous architecture and specific radiobiologic characteristics of the tumor target. This classification based on particulate emission type and range includes: (a) radionuclides which decay by electron capture and/or internal conversion (mean range $< 10 \, \mu m$), (b) alpha emitters (mean range $= 100 \, \mu m$), (c) low-range beta emitters (mean range $< 200 \, \mu m$), (d) medium-range beta emitters [(200 μm < mean range < 1 mm), and (e) high-range beta emitters] (mean range > 1 mm). An analysis of the effects that "cold" regions without radionuclide uptake had on radiobiologic response versus particle range showed that, depending on the size of the heterogeneity, some radionuclides were better than others. However, "cold" regions always dominated potential tumor regrowth and in turn tumor response. This

Table 15.1. Radionuclides for RIT

Radionuclides	Mean[a] Range (mm)	$T_{1/2p}$(h)	NP/P[b]	Δ_i[c] (rad·g)/ (μCi·h)	Imaging Gamma Energy (keV)	Antibody Match[d]
Y-90	2.76	64.0	–	1.99	–	F(ab')$_2$
Re-188	2.43	17.0	13.6	1.66	155	Fab
P-32	1.85	342.2	–	1.48	–	Mab(IgM)
Re-186	0.92	90.6	16.7	0.73	137	Mab
Pd-109	0.91	13.4	37.3	0.93	88	Fab
Au-198	0.82	64.7	0.81	0.70	412	F(ab')$_2$
Sm-153	0.53	46.7	4.33	0.57	103	F(ab')$_2$
I-131	0.40	193.0	0.50	0.41	364	Mab
Sc-47	0.30	80.3	1.51	0.35	159	Mab
Lu-177	0.28	161.7	4.19	0.31	208	Mab
Cu-67	0.27	61.9	1.35	0.33	185	F(ab')$_2$
Au-199	0.16	75.3	1.62	0.30	158	Mab
At-211	0.06[e]	7.2	64.1	5.32	–	Fab
Bi-212	0.06[e]	1.0	26.3	5.79	–	Fab

[a] Mean Range: X_{60} is the distance in which 60% of decay energy is deposited (Loevinger 1956 – Chapter 16-eq. 10 and 41). These values were calculated from principal beta emissions with 6% abundance or greater and are in reasonable agreement with Berger's Table 10, MIRD Pamphlet No. 7, 1971

[b] Equilibrium dose constant ratio of nonpenetrating to penetrating radiation

[c] Average equilibrium dose constant for NP radiation

[d] Antibody Match: According to the results reported by Wessels and Rogus (1984) and Yorke et al. (1991), the antibody class to be coupled with each radionuclide is optimal if the $T_{1/2\,phys}$ of the radionuclide is 1.5–3.0 times greater than the $T_{1/2\,biol}$ for antibody tumor uptake normal tissue clearance

[e] Maximum alpha particle range

All values are obtained from MIRD : Radionuclide Data and Decay Schemes, 1989
except: those
for Lu-177 which are from ICRP 38 (p 756) provided by Dr. J.B. Stubbs
Maximum energies for mean range calculation are taken from Table of Isotopes Sixth Edition 1968 by Lederer et al. J Wiley & Sons Inc

Table 15.2. Imaging radionuclides

Radionuclide	$T_{1/2p}$ (h)	NP/P ratio[a]	Imaging gamma (keV)/abundance (%)	Antibody match[b]
^{99m}Tc	6.0	0.13	140/89	Fab
^{111}In	67.9	0.09	245/94	F(ab')$_2$
^{131}I	193.0	0.50	364/81	Mab
^{186}Re	90.6	16.7	137/9	Mab
^{123}I	13.2	0.16	159/83	Fab

[a] Average equilibrium dose constant for NP radiation

[b] Antibody match: According to the results reported by Wessels and Rogus (1984) and Yorke et al. (1991), the antibody class to be coupled with each radionuclide is optimal if the $T_{1/2\,phys}$ of the radionuclide is 1.5–3.0 times greater than the $T_{1/2\,biol}$ for antibody tumor uptake and normal tissue clearance

was later theoretically confirmed in studies by Yorke et al. (1991) which showed that all "cold" regions in the tumor were detrimental with respect to tumor curability.

More recently several excellent review papers have examined radionuclide selection for RIT from the viewpoints of chemistry (Mausner and Srivastava 1993; Volkert et al. 1991; Fritzberg et al. 1988) and radiobiology (Langmuir et al. 1993; Fowler 1990). An abridged compilation from these references and new derived mean range values are shown in Tables 15.1 and 15.2. There are several common misconceptions regarding the primary determinants of RAID or RIT optimization. First, the absolute uptake in the tumor is not the primary criterion for therapeutic or diagnostic efficacy, as is commonly thought. Rather, this is only the case if limitations are imposed by tumor saturation, substantial tumor necrosis, and/or problems with low specific activity of the radionuclide/antibody combination. The *most important criterion* for optimum therapy or successful diagnosis using radiolabeled antibodies is maximization of the target (tumor) to nontarget (remainder of the body) ratio. The uptake in nontarget normal organs (e.g., bone marrow, kidneys, lung, and liver) is the primary limitation for therapy. Hence, the maximum target to nontarget localization ratio is most important for both diagnosis and therapy.

The next common misconception involves the specification of the localization ratio or target to nontarget ratio. Imaging agents may be compared by the maximum peak in the target to nontarget ratio, usually resulting in ratios between 5 and 20 (YORKE et al. 1991). Therapeutic efficacy, however, must be quantified or assessed by using a time-averaged target to nontarget ratio which for antibodies and fragments usually ranges between 3 and 8 and rarely exceeds 10. There is a common tendency to use the maximum target/nontarget ratio as an efficacy parameter for therapy. This usually results in an overestimation of the radiolabeled antibody's cytotoxic capability.

15.2.5 Chemical Stability of Radioimmunoconjugate

Methods aimed at direct and indirect labeling of an antibody with radionuclides are well established and have been the subject of several reviews (FRITZBERG et al. 1988; MOI et al. 1990; BHARGAVA and ACHARYA 1989). A major limitation of both methods is that the radioimmunoconjugate breaks apart as a result of in vivo metabolism after administration. These effects have been well documented for ^{131}I (YOKOYAMA et al. 1989; BADGER et al. 1990; FERENS et al. 1984), ^{90}Y (ROSELLI et al. 1989; KOZAK et al. 1989; HNATOWICH et al. 1985), ^{111}In (COLE et al. 1987), ^{99}Tc and ^{186}Re (FRITZBERG et al. 1988; PAIK et al. 1986), and ^{212}Bi HUNEKE et al.1992). The rate at which these agents catabolize has been the subject of intense investigation over the past 15 years. Reduction of free circulating radionuclide levels and/or radionuclide/chelate conjugates is critical to reduce unwanted normal tissue toxicity. These problems have been observed to be most severe for ^{90}Y localized in the bone, ^{111}In in the liver, ^{131}I in the thyroid, and ^{186}Re in the kidneys.

15.2.6 Methods to Improve Antibody Targeting

There are several approaches aimed at improving the localization of radioimmunoconjugates to the target (tumor), reducing or clearing the antibody from normal tissues, and potentiating damage to tumors or protecting the normal organs. These techniques are potentially important for the enhancement of the time-dependent target/nontarget (T/NT) ratio, which is the critical indicator of success of RIT. GOODWIN et al. (1985) originally described accelerated chemical clearance of unwanted free chelate with ethylene diamine tetra-acetic acid (EDTA) administered after the antibody conjugate localized in the target. Similarly, methods using bifunctional antibodies (GOODWIN et al. 1988; STICKNEY et al. 1991) or streptavidin–biotin (HNATOWICH et al. 1985) systems are methods used to localize radionuclide in a two-step process. These approaches first utilize the localization of an unlabeled bifunctional antibody to an appropriate antigen in the tumor, while non-bound antibody is cleared from the blood and extravascular spaces. Subsequently, the radionuclide is administered attached to a relatively small hapten which will selectively bind to the free arm of tumor-bound antigen–antibody complex in the tumor. The introduction of a second clearing antibody (BEGENT et al. 1987) and extracorporeal immunoabsorption (STRAND et al. 1993) are other methods used to reduce the normal tissue toxicity by clearing the blood pool of unbound radioactivity. Biologic response modifiers such as cytokines used to upregulate tumor antigen expression (GREINER et al. 1987), interleukin-2 and vasodilators to increase vascular flow to tumors (GOLDENBERG et al. 1990; HENNIGAN et al. 1991), growth factors (interleukin-1 and GM-CSF) for radioprotection and repair of bone marrow (BLUMENTHAL et al. 1992), and antibodies conjugated with radioprotectors or radiosensitizers (LANGMUIR 1991; BUCHSBAUM et al. 1993a) are all viable means of reducing toxic effects on normal tissues or sensitizing tumor tissues.

15.3 Developing a Standard of Practice for Antibody Dosimetry

The Nuclear Medicine Task Group II, "Dosimetry of Radiolabeled Antibodies," of the American Association of Physicists in Medicine has been engaged in surveying the literature associated with antibody dosimetry as well as suggesting practical ways to bring these methods to the laboratory and clinic (WESSELS et al. 1990; SIEGEL et al. 1990; BUCHSBAUM et al. 1993b). The following discussion has been abstracted from these review articles and others (LANGMUIR 1991) to provide the reader with an overview of the literature and current techniques.

15.3.1 Microscopic Scale Dosimetry

Microdosimetry is the estimation of radiation energy deposition within microscopic volumes encompassing the cell or even the DNA of the cell ($<10 \mu$m). The size of the target volume introduces stochastic effects. Alpha particle emitters and internalized Auger electron emitters may be useful in RIT because of their high linear energy transfer (LET) and relative biological effectiveness (RBE). The methodology to calculate tumor dosimetry for these short-range particles must consider energy deposition at the cellular and subcellular level. HUMM and co-workers (1993) have summarized the microdosimetric approaches that have been used in RIT. Such a microdosimetric approach has been adopted by a number of investigators (FISHER 1986; HUMM 1987; HUMM and CHARLTON 1989; HUMM and CHIN 1990). Two methods are used for the calculation of microdosimetric spectra in RIT, the Fourier convolution technique and Monte Carlo simulation.

These microdosimetric techniques have become increasingly important when examining the heterogeneous patterns of uptake for antibodies radiolabeled with alpha emitters (HUNEKE 1992; ZALUTSKY et al. 1989; HARRISON and ROYLE 1987; MACKLIS et al. 1992). Since many alpha emitters have a maximum range of approximately 60 μm, the heterogeneous uptake patterns for radiolabeled antibodies are liable to be more deleterious to short-range particles such as alpha emitters (many of which have a maximum range of 60 μm) and the action at a distance principle between cells that have bound radiolabeled antibodies is minimal. This effect, of course, is very dependent on the size of the heterogeneity. If the size of the heterogeneity pattern is smaller than the range of the alpha particle and there is a more or less even distribution of the radiolabeled antibody on the tumor cells, then a high absorbed dose pattern with minimal cool spots would be observed. Similarly for Auger emitters that internalize into the cell (BRADY et al. 1990; WOO et al. 1990), the action at a distance principle is not available at all and is not advantageous to this treatment modality. This type of localized emission will yield very precise boundaries in terms of normal tissue toxicity. However, the mechanism of action is more like that of an internalized toxin or chemotherapy agent, where one must attach the complex to the cell for internalization in order for the same cell to be destroyed. Heterogeneity of dose also appears to be very important in calculating or estimating the relationship between absorbed dose

and toxicity in normal tissues where there are complex interfaces, such as the trabecular bone in bone marrow and critical structures in organs such as the kidney and lung (FISHER 1986; ECKERMAN 1985; SIEGEL et al. 1990b; MATTHEWS et al. 1991; BREITZ et al. 1992).

15.3.2 Multicellular Level Dosimetry

The distance scale used to describe "multicellular" or "small scale" dosimetry ranges from 10 to 10^4 μm. This size scale encompasses a substantial portion of the heterogeneity patterns encountered with radiolabeled antibody administration and has important radiobiologic significance, ultimately leading to the success or failure of treatment.

Radionuclide activity variations within tumors can be measured with quantitative autoradiography. However, quantitative autoradiography alone cannot provide total dose measurements, because of the temporal change in radiolabeled monoclonal antibody uptake, penetration, and clearance. GRIFFITH and co-workers (1988) have used small thermoluminescent dosimeter(s) (TLD) implanted into tumor xenografts to directly measure the total absorbed radiation dose from radiolabeled monoclonal antibodies. Micro-TLD have been removed from serial tumor slices and their light output measured, and the results compared with dosimetry calculations using autoradiography to measure the distribution of radioactivity in tumor sections. ROBERSON et al. (1992a) have used autoradiography and a brachytherapy treatment planning system to obtain three-dimensional isodose rate distributions and dose rate–volume histograms in tumor xenografts injected with ^{131}I-labeled monoclonal antibody. Point source calculation methods (BERGER 1971; KWOK et al. 1985; GRIFFITH et al. 1988) become an important tool in examining the relative efficacy of using [131]I (average range, 0.4 mm) or [90]Y (average range, 2.8 mm) as the tumor burden becomes larger. For micrometastases (<1 mm in diameter), the initial dose rate of ^{131}I will be higher than that of ^{90}Y for equal initial concentrations even though ^{90}Y yields five times more NP radiation energy per decay. The lack of "cross-fire" or "build-up" of radiation dose in the case of ^{90}Y is responsible for this effect (LANGMUIR and SUTHERLAND 1988). "Cross-fire" may be defined as irradiation of distant targets by cells that have bound radiolabeled antibodies. If there are not many radiolabeled antibodies attached in a local region, and the range of the particulate radiation emitted from the radionuclide is small, the situation of lack of cross-fire or the generation of a cold spot is likely to occur. For example, the use of chemotherapy or internalization of toxins in a cell, provides no cross-fire obtained since every cell must incorporate the toxin or chemotherapy agent for it to be destroyed.

As the tumor size increases, this effect is diminished in proportion to the added cellular mass since equilibrium is established for ^{90}Y at a tumor diameter of 1 cm or more (HOWELL et al. 1989). Further investigation is required to relate the dose-rate distributions to time averaged dose distributions, cell kill, and therapeutic efficacy (YORKE et al. 1991). YORKE et al. (1993) and LEICHNER and KWOK (1993) have reviewed these calculation methods and experimental procedures in detail.

15.3.3 Macroscopic Dosimetry and Imaging Methods

The approach developed by the Medical Internal Radiation Dose (MIRD) Committee of the Society of Nuclear Medicine for the estimation of average absorbed dose from internally deposited radionuclides has been applied to RIT in animals and humans (WATSON et al. 1993). The approach includes the collection of serial blood, total-body, organ, and tumor time-activity data. Integration of these data yields source volume cumulated activities. It is assumed that there is a uniform distribution of cumulated activities of radiolabeled antibody within each source region and a uniform deposition of energy within each target region. In the MIRD approach, the average absorbed dose to a given target region from a particular source region is estimated for a specific anatomic model (standard man), and the total target volume absorbed dose is the summation of the source volume absorbed dose contributions from the sources within the target tissue and sources in adjacent tissues (WATSON et al. 1993).

The term "MIRD formalism," commonly used when referring to dose calculations associated with RIT, should be specified with more precision. These calculations can be grouped into three categories: (a) infinite media/equilibrium dose constant calculations (DILLMAN et al. 1975), (b) human "S factor" calculations (SNYDER et al. 1975), and (c) point source calculations (BERGER 1971). The first two methods are used for macroscopic calculations (>1 cm) and the third for microscopic calculations; the

third method is applicable to the problem of dose heterogeneity frequently encountered in RIT. The term "MIRD formalism" is usually associated with the macroscopic treatment, but is something of a misnomer since all MIRD pamphlets taken together give both macroscopic and microscopic calculation methods. This macroscopic treatment yields average dose estimates which may not properly describe the dose spatial heterogeneity produced by alpha particle and Auger electron emitters. With this class of radionuclides, a microdosimetric approach will be required, as described previously.

LEICHNER et al. (1981b) and WESSELS and ROGUS (1984) have modified the macroscopic treatment of the MIRD formalism to include methods to calculate absorbed dose to tumors receiving radiolabeled antibodies. LEICHNER et al. (1990) have utilized tumor volume calculations from sequences of computerized tomography (CT) or single-photon emission computerized tomography (SPECT) scan slices, quantitation of radiolabeled antibody in a patient's tumor, determination of kinetic parameters of radiolabeled antibody for tumor, and calculation of dose rates and dose for tumor based on the macroscopic MIRD method. In these studies, the beta particle radiation was treated as nonpenetrating because the tumor and organ volumes were large compared with the range of ^{131}I or ^{90}Y beta particles.

Several investigators have attempted to take into account the distribution of radiolabeled monoclonal antibodies in a tumor when calculating absorbed dose (GRIFFITH et al. 1988; HUMM 1986; JUNGERMAN et al. 1984; KWOK et al. 1985; LANGMUIR and SUTHERLAND 1988; PRESTWICH et al. 1989; SGOUROS et al. 1990). In modeling of absorbed dose distributions, analytical, numerical, and Monte Carlo methods have been used to investigate the effects of uniform and nonuniform activity distributions. LEICHNER and KWOK (1993) have written a review summarizing the calculation techniques that have been used for beta particle tumor dosimetry in RIT.

In the past, planar scintillation camera imaging was used to estimate the biodistribution of radiolabeled antibodies. Absorbed dose may be calculated from time-dependent activity curves obtained from the conjugate view technique (THOMAS et al. 1976; HAMMOND et al. 1984). Using this technique, the activity in a defined region of interest (ROI) is measured, corrected for background and calibrated with a known standard. This method remains a "work horse" technique to measure activity uptake in tumor and normal tissue and yields reasonably accurate data (± 20%) for organs of 3 cm or greater when corrected for partial volume effects of overlaying background (EARY et al. 1989; GREEN et al. 1990).

More recently, SPECT, positron emission tomography (PET), CT, and magnetic resonance imaging (MRI) have been used to estimate radionuclide concentrations and tumor volume through image registration (GREEN et al. 1990; LEICHNER et al. 1990; ERDI Y. et al. 1993; ERDI A. et al. 1993). LEICHNER et al. (1993) have reviewed these imaging techniques and treatment planning procedures that have been used in clinical RIT. The topics addressed in this review include tumor and normal organ volume computations from CT and MRI data, quantitation of the activity of radiolabeled antibodies from planar gamma camera imaging, quantitative SPECT and PET, and correlative image analysis and treatment planning for RIT.

15.3.4 Normal Tissue Dosimetry

The subject of normal tissue dosimetry has become increasingly important in understanding the efficacy of treatment with radiolabeled antibodies, since it is the radiation of the normal tissues (bone marrow, kidneys, lung, gut, and liver) which is most likely to limit the amount of administered radiolabeled antibody. Thus, as in the case of external beam radiation, the maximum tolerated dose of any of these organs will limit the amount of absorbed dose that the tumor target may receive. In most nuclear

medicine reports (e.g., LOEVINGER, MIRD Primer 1991), homogeneous dose distribution patterns are nominally reported for normal organ systems. With respect to the dosimetry of radiolabeled antibodies, specialized reports dealing with bone marrow irradiation and toxicity (SIEGEL et al. 1990b) and kidney and lung doses (PRESS et al. 1990; MEREDITH et al. 1993; BREITZ et al. 1992) as well as liver doses (ORDER 1982; LARSON et al. 1985) have been included in the overall dosimetry information appropriate for patient studies. Table 15.3 shows the absorbed dose delivered to the tumor, which must be considered in most cases to be the maximum dose received by the tumor when a reasonable normal tissue toxicity is taken into account. The articles listed in Table 15.3 contain many citations of normal tissue absorbed dose measurements. Biologically, it is interesting to note that many of the reports coming from the University of Washington, the Virginia Mason Clinic, and the NeoRx Corporation (PRESS et al. 1990; BREITZ et al. 1992) find that the absorbed dose values computed from imaging data exceed the published values of normal organ dose tolerances as understood for external beam therapy (LANGMUIR 1991). For the most part, this is probably due to a decreased toxicity per gray for low-dose-rate radiation (WESSELS 1990; FOWLER 1990; LANGMUIR et al. 1993). The dosimetry of normal organs, the heterogeneity of dose to those organs, and the underlying radiobiology for exponentially decaying low-dose-rate radiation remain under active investigation.

Table 15.3. Tumor doses and response rates reported from clinical trials of RIT (adapted from MEREDITH et al. 1993)

Antibody	Tumor	Radionuclide; mCi/infusion	Tumor (cGy/mCi)	Dose range total (Gy)	Response rate[b]	Reference
MB-1 combined	Non-Hodgkin's lymphoma	[131]I; 232–608	2.1–8.8	8.5–42.6	5/6 CR, 1/6 PR	PRESS et al. (1989)
MB-1	Non-Hodgkin's lymphoma	[131]I; 25–161	0.55–5.7	0.3–4.6	6/9	KAMINSKI et al. (1992)
LYM-1	Non-Hodgkin's lymphoma	[131]I; 20–60		12–120[a]	2/18 CR 8/18 PR 3/18 stable	DeNARDO et al. (1988)
LL-2	Non-Hodgkin's lymphoma	[131]I	2.7–9.3	0.23–2.63	80%	GOLDENBERG et al. (1991)
OKB7	Non-Hodgkin's lymphoma	[131]I; 2	3.0–7.9		Tracer dose only	SCHEINBERG et al. (1990)
Anti-idiotype	B-cell lymphoma	[90]Y; 370			1/1 PR	PARKER et al. (1990)
T-101	Cutaneous T-cell lymphoma	[131]I; 100–150		0.40–51	2/5 PR 3/5 MR	ROSEN et al. (1987)

(Contd.)

Table 15.3. (*Contd.*)

Antibody	Tumor	Radionuclide; mCi/infusion	Tumor (cGy/mCi)	Dose range total (Gy)	Response rate[b]	Reference
Anti-ferritin	Hodgkin's disease	[131]I 30, day 1 20, day 5 ^{90}Y; ≤ 30		1 pt. 27	1/37 CR 14/37 PR 4/8 CR	LENHARD et al. (1985) VRIESENDORP et al. (1989) ORDER et al. (1988)
Anti-ferritin	Hepatoma	[90]Y; 8–37		2.7–11.5 9–21.5	3/9 PR	LEICHNER et al. (1981a; 1988) ORDER et al. (1988)
	Hepatoma	[90]Y	11.5–20.4	9–21.5		LEICHNER et al. (1988)
	Hepatoma	[131]I; 30, day 1 20, day 5		10–12 per cycle; 105 patients	7% CR 41% PR	ORDER et al. (1985)
Ch B72.3	Metastatic colon	[131]I; 40–67	2.3–5.9	0.91–2.7	4/12 stable at 6 weeks	MEREDITH et al. (1992)
Anti-CEA	Metastatic colon	[131]I; 38–152	0.5–5.1	0.25–2.6	1/16 PR 1/16 symptom relief	BEGENT et al. (1989)
CO17-1A	Gastro-intestinal	[125]I; 7.1–128.1 total, 3–25/ infusion			1/53 PR	MARKOE et al. (1990)
Specific MoAB	Colorectal and gastric	[131]I; mean dose 10		40.2 mean for abdominal metastasis; 24.6 mean for liver metastasis	2/14 PR 5/14 stable	RIVA et al. (1989)
Anti-CEA	Liver tumors	[131]I	0.7–217.8			SIEGEL et al. (1990)
EGFRI H17E2	Gliomas	[131]I; 40–140		≤ 12.5	6/10 clinical improvement 2/10 radiographic regression ≥ 25%	KALAFONOS et al. (1989)
BC-2	Gliomas	[131]I; mean 10.9	Mean 2200, range 340–7000		2/7 stable 3/7 PR	RIVA et al. (1991)
Ch L6	Metastatic breast	[131]I; 20/m² to 70/m²	5.4–21.6		3/5 PR	DENARDO et al. (1991b)
Ch L6	Advanced breast	[131]I; 20/m² to 70/m²	5.4–18.9		3/4 PR	DENARDO et al. (1991a)
A6H	Renal cell carcinoma	[131]I; 50	20–26	0.55–13	1 tumor regression in 4 patients	VESSELLA et al. (1991)
Anti-CEA	Cholangio-carcinoma	[131]I; 20 day 0 10 day 5			37 patients 7% CR 26% PR	STILLWAGON et al. (1987)
Anti-P97	Melanoma	[131]I		38–85	1/3 PR	CARRASQUILLO et al. (1984)
UJ13A	Neuro-blastoma	[131]I; 35–55	37.5 ± 23		1/5 CR	LASHFORD et al. (1987)
NR CO-02 F(ab')₂	Metastatic colon, gastric, lung CEA+	[186]Re; 25–350	0.43–18.6	6–20 for PR patient	1/31 PR 1/10 PR	BREITZ et al. (1992, 1993) SCHROFF et al. (1990)
NR-LU-10	Metastatic, lung renal, ovarian	[186]Re; 44–259	0.35–17.7	0.6–1.6	No objective response	BREITZ et al. (1992, 1993)
HMFG1	Advanced ovarian	[90]Y; 15–25	–	–	16/35 CR	STEWART et al. (1990)

[a] Cumulative dose estimate from multiple courses of 30–60 mCi/injection to a total of 120 Gy
[b] CR, complete tumor regression; PR, >50% tumor regression; MR, tumor regression of <50%

15.3.5 Standardization of Terminology and Units

Following the lead from the Oak Ridge Associated Universities and the MIRD Committee, standardization of many of the definitions and units can be found in Oak Ridge publications under the auspices of the DOE (Symposium Series) and the MIRD pamphlets 1–13 (Society of Nuclear Medicine). Nevertheless, the multidisciplinary nature of radio-labeled antibody therapy draws upon the conceptual knowledge from the nuclear medicine, radiation oncology, medical oncology, and immunology subspecialties. This can cause unforeseen difficulties when scientists from these different communities describe what data they acquire and measure in their own terms. As previously mentioned in the Introduction, the concept of "dose" has different meanings to an immunologist and nuclear medicine specialist and the radiation oncologist. To the immunologist, "dose" means the amount of mass associated with injection of antibody; to the nuclear medicine specialist, it connotes the activity administered; and to the radiation oncologist, it connotes the amount of absorbed dose that the irradiated tissue has received. The suggestion has been made that an additional modifier needs to be added to the term "dose"; namely, "mass dose," "activity dose," and "absorbed dose" should be used for describing mass, activity, and radiation dose, respectively.

In addition, the term percent injected dose per gram (%ID/g) has been widely used to indicate the amount of uptake of radiolabeled antibodies in an organ or tumor. However, this is in no way to be confused with the amount of radiation absorbed dose that a tissue may receive. The %ID/g is a unit which does not take into account the amount of activity that has been injected into the host. The maximum activity that can be injected is determined largely by the maximum absorbed radiation dose that normal tissues can receive such that the host will survive the administration of the radiolabeled antibody. This amount differs greatly among different animal species as well as humans. It also depends on the mass of antibody delivered, the molecular size, and the antibody type. Hence, the %ID/g is of no use when comparing absolute amounts in terms of an interspecies comparison or a comparison between antibody classes. The concentration of radioactivity in μCi/g in the target and nontarget tissues seems to be the most reliable predictor of therapeutic index. These and other problems still remain to be clarified and work is continuing in several task groups and regulatory entities to increase the standardization of these terms and methods.

15.4 Correlation of Dosimetry and Radiobiology in Predicting Response

15.4.1 Low Dose Rate and the "4 R's"

KNOX et al. (1992) and LANGMUIR et al. (1993) have reviewed the comparative dosimetry and radiobiology of low-dose-rate external beam radiation

Table 15.4. Radiobiologic responses of several tumor and monoclonal antibody combinations (adapted from LANGMUIR et al. 1993)

Reference	External beam			Tumor response		
	Type	Shoulder	VDT[a]	Antibody	Radionuclide	Max. dose rate
NEACY et al. (1986) WESSELS (1990a, 1991)	Colorectal xeno-graft LS174T	+	Moderate	B72.3	^{131}I	15
WESSELS et al. (1989)	Renal cell cancer xenografts TK39, TK82	++	Slow	A6H	^{131}I	50
KNOX et al. (1990)	B cell lymphoma syngeneic 38C13	–	Rapid	4C8	^{131}I	25
BUCHSBAUM et al. (1990)	Colorectal xeno-graft LS174T	+	Moderate–slow	17–1A	^{131}I	15
WILLIAMS et al. (1991)	Grade IV glioma xenograft u-251	+++	Moderate	P96.5 QCI	^{90}Y	50
BURAS et al. (1991)	Colorectal xeno-grafts LS174T	+				12–18
	WiDr	++	Moderate–slow	ZCE 025	^{90}Y	9–15

[a] VDT, volume doubling time; moderate VDT = 3–4 days; slow VDT > 4 days; rapid VDT < 3 days

Table 15.5. Comparison of external beam irradiation schemata and relative tumor responses observed following dose equivalent RIT and external beam irradiation (adapted from LANGMUIR et al. 1993)

Reference	External beam			Tumor response		
	SF	MF	WB/ local	Method	Relative efficacy factor[a]	
					SF	MF
NEACY et al. (1986) WESSELS (1990a, 1991)	4 MV		Local	VDT	1.6[b]	
WESSELS et al. (1989)	10 MV	250 cGy × 4 in 2 weeks	Local	VDT	1.5–1.7[c]	2.5[c]
KNOX et al. (1990)		250 kV x-ray 10 fx/2 wks	WB	Cumulative % per day; Tumor size (VT) decrease 0–12 days		3.5 (1.19)[d]
BUCHSBAUM et al. (1990)	^{60}Co		Local	SDT growth delay	0.32[e]	
WILLIAMS et al. (1991)		^{137}Cs:1–10 fx; 1.5–10 Gy per fx		Growth delay to $V/V_0 = 1$[f]		0.33[c]
BURAS et al. (1991)	^{60}Co		Local	Growth delay[g] to 2 g	0.5 WiDr[c] 1.0 LS174T	

SF, single fraction; MF, multiple fractions; WB, whole body; VDT, volume doubling time; VT, volume transformation

$= \sqrt{\text{length} \times \text{width}} \times \sqrt{\text{height}}$ in mm; SDT, size (product of two tumor dimensions) doubling time

[a] Relative efficacy of RIT compared with external beam irradiation

[b] Ratio of VDT RIT/external beam at equivalent doses

[c] Ratio of dose external beam/dose RIT required to give the same VDT

[d] Ratio of the slope defined as the averaged cumulative percentage of tumor size decrease/Gy for RIT/external beam; ( ) = corrected for specific uptake of ^{131}I-MAB by tumor by dividing the relative efficacy factor by 1.7 = averaged cumulative concentration ratio of ^{131}I-MAB tumor/^{131}I-MAB whole body

[e] Ratio of the dose external beam/dose RIT required to give the same size doubling time

[f] $V/V_0 = 1$: tumor volume/initial tumor volume at the time of treatment = 1; delay in regrowth to original volume

[g] Tumor size was calculated using the formula for volume of an ellipsoid rotated about its major axis; the longest and shortest dimensions of the tumor were measured

and RIT. It was concluded that tumors most likely to respond to RIT would be those that are inherently radiosensitive, those with a poor capacity to repair radiation damage or with long repair times, those that are susceptible to blockade in sensitive phases of the cell cycle, and those that reoxygenate rapidly. KNOX et al. (1992) and LANGMUIR et al. (1993) have reviewed five studies comparing the therapeutic efficacy of RIT and external beam radiation. RIT was less effective, equally effective, or more effective than external beam irradiation in inhibiting tumor growth in different model systems (Tables 15.4, 15.5). Preliminary conclusions are that tumors characterized by a large shoulder (greater capacity to repair), a low alpha/beta ratio associated with the linear-quadratic (L-Q) model of cell survival, and a short doubling time had a significant dose-rate effect for RIT irradiation compared to external beam irradiation. A comparison of alpha and beta particle emitters for RIT indicates an advantage for beta particle emitters if the L-Q alpha/beta ratio for tumors is greater than that for the critical organ of toxicity, as is usually the case. However, better therapeutic ratio is predicted for alpha particle radiation when bone marrow (high L-Q alpha/beta ratio) is considered as the critical organ (LANGMUIR et al. 1993).

15.4.2 Summary of Animal RIT

BUCHSBAUM et al. (1993b) summarized the results that have been obtained in experimental RIT with tumor and tumor cell spheroid models and a variety of murine syngeneic tumors and human tumor xenografts including colon cancer, leukemias and lymphomas, hepatoma, renal cell cancer, neuroblastoma, glioma, mammary carcinoma, small cell lung carcinoma, ovarian cancer, bladder cancer, and cervical carcinoma (Table 15.6). The approaches

Table 15.6. Animal RIT results (adapted from Buchsbaum et al. 1993)

Tumor	MoAB	Radionuclide	Administered activity (MBq)	Radiation dose to tumor (mGy)	Tumor dose/activity (mGy/MBq)	Reference
GB-39 colon (hamster)	Polyclonal anti-CEA	^{131}I	37	13 250 (MIRD)	358[b]	Goldenberg et al. (1981)
GW-39 colon (hamster)	NP-4	^{131}I	18.5	11 960 (MIRD)	646	Sharkey et al. (1987)
COLO 205 colon	250-30.6	^{131}I	37	7000 (MIRD)	189	Zalcberg et al. (1984)
LS174T	17-1A	^{131}I	11.1	19 060 (MIRD)	1717	Buchsbaum et al. (1990)
LS174T	17-1A	^{131}I	11.1	5000 ext. beam	450	Buchsbaum et al. (1990)
LS174T	17-1A	^{131}I	11.1, 11.1, 11.1	9200 ext. beam	829	Buchsbaum et al. (1990)
LS174T	B72.3	^{131}I	7.4	8100 (TLD) 8240 (MIRD)	1095	Griffith et al. (1988)
LS174T	17-1A	^{131}I	11.1	19 200 (MIRD); 0-150 mGy/h day 1; 15-100 mGy/h day 2 3D autoradiography	1728	Roberson et al. (1992a,b)
T380 colon	CEA MoAbs	^{131}I	18.5, 18.5	94 200 (MIRD)	2546	Buchegger et al. (1990)
T380 colon	CEA MoAbs F(ab')$_2$	^{131}I	92.5, 92.5, 92.5	91 700 (MIRD)	330	Buchegger et al. (1990)
SW948 colon	17-1A	^{90}Y	7.4	33 400-41 600 (MIRD)	4514-5622	Washburn et al. (1991)
GW-39 colon (mice)	NP-2	^{90}Y	1.85	16 030	8665	Sharkey et al. (1988)
LS174T colon	ZCE025	^{90}Y	4.44	34 000 (MIRD)	7658	Buras et al. (1990)
LS174T colon	ZCE025	^{90}Y	4.44	17 500 (MIRD size correction)	3941	Buras et al. (1993)
LS174T colon	17-1A	^{90}Y	9.25	17 900 (MIRD)	1935	Buchsbaum et al. (1993c)
Murine erythroleukemia	Polyclonal IgG	^{131}I	5.92	187 000 (MIRD)	3041	Redwood et al. (1984)
Murine erythroleukemia	103A	^{90}Y	1-1.85	ND	ND	Anderson-Berg et al. (1987)
EL4 leukemia	Thy 1.2	^{212}Bi	5.55-8.51	ND	ND	Macklis et al. (1988, 1989)
SL2 lymphoma (Thy 1.2 mice)	Thy 1.2	^{131}I	18.5	16 000 (MIRD)	865	Badger et al. (1985)
SL2 lymphoma (Thy 1.1 mice)	Thy 1.1	^{131}I	55.5	16 000 (MIRD)	288	Badger et al. (1986)
Raji lymphoma	Lym-1	^{131}I	24.27	3920-17 400 TLD	162-717	Griffith et al. (1988)
EL4 lymphoma	Lyl	^{90}Y	5.18	1000-2000 ext. beam	193-386	Schmidberger et al. (1991)

Cell/tissue	Antibody	Isotope	Activity	Dose		Reference
H-4-II-E hepatoma	Polyclonal antiferritin	^{131}I	18.5	4500 (MIRD)	243	Rostock et al. (1983)
HepG2 hepatoma	QCI054	^{90}Y	7.4–11.1	75 000–124 000 TLD	10 135–11171	Klein et al. (1989)
TK-82 renal cell	A6H	^{131}I	3.7	50 000 (MIRD imaging)	6757	Vessella et al. (1987)
TK-177G renal cell	A6H	^{131}I		(MIRD)	10 260	Chiou et al. (1988)
TK-82 renal cell	A6H	^{131}I		(MIRD)	15 930	Chiou et al. (1988)
TK-82 renal cell	A6H	^{131}I	5.55	7000–24 000 TLD	1261–4324	Vessella et al. (1988)
TK-82 renal cell	A6H	^{131}I	3.7	3410 TLD	922	Wessels et al. (1989)
			7.4	3830 TLD	518	
			14.8	8860 TLD	599	
			22.2	10 340 TLD	466	
#1 neuroblastoma	3F8	^{131}I	18.5–37	42 000 (MIRD)	1135–2270	Cheung et al. (1986)
D-54 MG glioma (s.c.)	81C6	^{131}I	37	97 190 (MIRD)	2627	Lee et al. (1988b)
D-54 MG glioma, intracerebral	81C6	^{131}I	46.25	15 850 (MIRD)	343	Lee et al. (1988a)
D-54 MG glioma, intracerebral	Mel-14 F(ab')$_2$	^{131}I	55.5, 55.5	9150 (MIRD)	82	Colapinto et al. (1990)
U-251 glioma	P96.5	^{90}Y	3.7	37 700 TLD	10189	Williams et al. (1990)
U-87MG glioma	425 (Fab)$_2$	^{131}I	11.1	190 (MIRD)	17	Bender et al. (1992)
U-87MG glioma	425 (Fab)$_2$	^{131}I	11.1	1590 (MIRD)	143	Bender et al. (1992)
Human mammary carcinoma	BW 495/36	^{131}I	7.4, 7.4	77 000 (MIRD)	5203	Senekowitsch et al. (1989)
TNSC-1 small cell lung cancer	TFS-4	^{131}I	11.1–18.5	10 380 (MIRD)	561–935	Yoneda et al. (1988)
SHT-1 small cell lung cancer	NR-LU-10	^{186}Re	7.88, 10.32 8.07, 2.22 10.36, 1.67	20 120 (MIRD) 26 710 (MIRD)	1105 1197	Beaumier et al. (1991)
ME-180 cervical carcinoma	TNT-1	^{131}I	11.1	55 880 (MIRD)	1678	Chen et al. (1989)
NIH:OVCAR – ovarian carcinoma	139H2	^{131}I	18.5	13 000 (MIRD)	703	Molthoff et al. (1992)
Bl-17 bladder cancer	BLCA-38	^{153}Sm	37	19 090 (MIRD)	516	Lightfoot et al. (1991)

used to estimate tumor dosimetry in the experimental animal studies include the MIRD formalism, thermoluminescent dosimetry (TLD), autoradiography, and comparison to external beam irradiation. The results indicate that monoclonal antibodies radiolabeled with a variety of radionuclides (^{131}I, ^{90}Y, ^{186}Re, ^{67}Cu, ^{153}Sm, ^{177}Lu, and ^{212}Bi) have been effective in inhibiting tumor growth and producing cures. The effective tumor dose per quantity of injected radioactivity has shown a large range (0.1–40 cGy/μCi) in these experimental studies. The effectiveness of RIT depends on a number of factors relating to the antibody, antigen expression, and physiologic factors, as described above. A goal is an understanding of how heterogeneity in radiolabeled monoclonal antibody deposition in tumors and spheroids using TLD and autoradiography affects absorbed dose distribution and the radiobiologic consequences.

15.4.3 Comparison of Radiobiology of Animal Tumor and Normal Tissue Responses to RIT and External Beam Radiation

As described in the previous section, KNOX et al. (1992) and LANGMUIR et al. (1993) reviewed studies comparing the therapeutic efficacy of RIT and external beam radiation in five animal models. The significance of these experiments is that they provide a framework for the comparison of RIT to external beam radiation therapy in humans, as discussed below. WESSELS (1990) discussed the limits of applicability of animal modeling to the clinical setting in terms of intrinsic radiosensitivity, tumor volume effects, tumor bed effects, and host defense mechanisms. It was noted that the D_0 (dose producing 37% survival on the linear portion of the cell survival curve) for mammalian cells ranges from 70 to 150 cGy for low LET radiation, indicating that the intrinsic radiosensitivity of different mammalian cells is quite similar. ROFSTAD (1985) compared the usefulness of animal tumor xenograft experiments using external beam radiation with clinical external beam therapy studies. In general, the human tumor intrinsic radiosensitivity is preserved in the animal model and is the most important determinant of tumor growth delay and tumor control dose (WESSELS 1990). Tumor growth delay or tumor control dose results showed a correlation with human trials according to type of neoplasm. However, ROFSTAD (1985) has cautioned against direct extrapolation of animal results with external beam radiation to the clinical setting based on differences

in tumor volume-doubling times, tumor bed effects, and host defense mechanisms. Additional differences between animal models and the human cancer problem with regard to RIT include volume differences, antigen cross-reactivity with normal tissues, the existence of established metastases, tumor cell diversity, and pharmacokinetics of the administered antibody and its immunogenicity.

Wessels and co-workers proposed that radiobiologic characterization of a tumor be performed in animals to help predict RIT clinical efficacy, by correlating animal model data of external beam radiation therapy and RIT for the same tumor model and measuring the absorbed dose in each (WESSELS et al. 1989; WESSELS 1990). A similar comparative approach would be used to extrapolate from animal RIT data to clinical RIT trials as has been used to compare animal and clinical external beam radiation therapy data (KNOX et al. 1992; FOWLER 1990; WESSELS 1990; ORTON and COHEN 1988; DALE 1985). By deriving a ratio of radiobiologic response between RIT and external beam radiation therapy in animals for the same tumor cell line, a predictive response ratio would be obtained when examining the potential efficacy of different radiolabeled antibodies for clinical trials. In addition, YORKE et al. (1991) described a procedure for predicting the relative efficacy of different radionuclide antibody combinations in man based on their relative tumor/normal tissue ratios in animal models and the known ratio in man for one of the radionuclides.

15.4.4 Summary of Human RIT

A number of clinical RIT studies have included tumor and normal tissue dosimetry estimates. Reviews have been written on the approaches and results that have been obtained in lymphoma (SIEGEL et al. 1993), solid tumors (MEREDITH et al. 1993), and intraperitoneal therapy (ROESKE et al. 1993).

Radiation dosimetry in B-cell lymphoma patients has been done using the MIRD approach. Organ and tumor radionuclide activity measurements have usually been performed with conjugate view planar scintillation camera imaging (SIEGEL et al. 1993). Organ and tumor volumes have been obtained by CT, SPECT or the published values of the MIRD committee. Response rates in patients with B-cell lymphoma of greater than 90% with many complete remissions have been reported by several groups of investigators using either low (5–10 mCi) or high

(232–608 mCi) doses of ^{131}I-labeled monoclonal antibodies against B-cell tumor-associated antigens, as summarized by SIEGEL et al. (1993). Tumor doses have ranged from 1.9 to 20 cGy/mCi of ^{131}I administered (Table 15.3). There has been little correlation of tumor response with estimated tumor absorbed dose. Estimated absorbed doses delivered to normal organs have ranged from 0.7 to 8.1 cGy/mCi administered. Toxicity has been limited by bone marrow suppression with the larger amounts of administered ^{131}I-labeled antibody. When used in combination with bone marrow transplant, second organs of toxicity must be carefully monitored (lung, kidney, gut).

For solid tumors, the MIRD approach, planar imaging, and tumor volumetrics have been performed in a similar manner as in lymphoma studies (MEREDITH et al. 1993). There have been wide variations in estimated tumor doses in different studies (Table 15.3), and no definite dose-response relationship has been observed. The spatial resolution limits of planar and SPECT imaging devices prevent detection of the nonuniformity of radiolabeled monoclonal antibody uptake and thus permit only the estimation of average dose to tumor. As summarized by MEREDITH et al. (1993), solid tumor dosimetry estimates generally assume homogeneous activity distribution in source organs. It is predicted that as computation tools become available for incorporating inhomogeneous cellular level activity distributions, the currently used "average dose" will likely be replaced with a statistical distribution based on the number of viable cells in the tumor volume. Estimates of tumor control doses could then be based on a linear extension of absorbed dose coupled with a threshold dose for cell sterilization (MEREDITH et al. 1993).

Regional administration of radiolabeled monoclonal antibodies has been used in the peritoneum, the cerebrospinal fluid, and the pleural/pericardial cavity, and within cystic brain tumors. As summarized by ROESKE et al. (1993), intraperitoneal and intracavitary RIT differ from other RIT approaches in that high activity and dose gradients exist near the solution–tumor interface. Dose to tumor and normal tissue at the interface is a function of depth and is due to: (a) the activity concentration as a function of time within the compartment, (b) the spatial distribution of radiolabeled antibody as a function of depth and time as the antibody binds to and permeates tumor and normal tissues, and (c) the physical characteristics of the radionuclide in relation to the depth of antibody penetration. ROESKE

et al. (1993) have discussed the biologic and physical aspects of intraperitoneally administered radiolabeled antibodies and the state of experimental and calculational studies for this site, and have made recommendations concerning the types of measurements and calculations which are required for accurate dosimetry.

15.5 Future Directions of Radiolabeled Antibody Dosimetry

It is clear from a cursory examination of the rather dismal results of clinical RIT (Table 15.3), and from the more encouraging results for animal RIT (Table 15.4–15.6), that several major problems with regard to low tumor/nontarget ratios and tumor uptake heterogeneity remain to be resolved. This "gray area" or "no man's land" between the encouraging animal results and generally discouraging solid-tumor clinical results provides a marvelous opportunity for the radiation physicist and dosimetrist. If all was well with the performance of the radiolabeled antibodies in the laboratory and the clinic, the clinicians would not be so obliged to rely on dosimetry as an indicator of success or failure of this modality (as is the case with ^{131}I treatment of thyroid carcinoma). If this modality did not work in either setting, there would be little reason to continue. Hence the immunologist, chemist, and physicist find themselves in a perpetual cycle of optimization of these treatment regimens in order to bring some limited success to the clinic. Similar arguments can be made for radiosensitizers, fast neutron radiation therapy, hyperthermia, and perhaps gene therapy for the treatment of cancer. Nevertheless, improvements in humanized and "designer" molecules for radiolabeled antibodied therapy, two-step and three-step delivery approaches, along with bone marrow transplantation, extracorporeal immunoabsorption, and normal tissue regeneration mechanisms or protectors should provide the radiation physicist and radiation biologist with an enormous number of new treatment enhancement schemes to keep them busy for the next decade. Let us hope that all this activity eventually leads to an improved final product for the treatment of our patients.

Besides these rather exciting challenges, there is a renewed emphasis in many of our national organizations on the compilation of "how to" manuals regarding existing procedures. To date, many of the joint efforts with regard to radiolabeled antibody dosimetry have been extensive literature searches and fact-finding missions (April 1993 supplement to *Medical Physics*, "Radiolabeled Antibody Tumor Dosimetry" and "Bone Marrow Dosimetry and Toxicity for Radioimmunotherapy", SIEGEL et al. 1990b). Clinical protocols for imaging and prospective therapy trials will necessitate the increased use of standardized dosimetry methods and treatment-planning computer programs.

In summary, as more imaging products using radiolabeled antibodies become approved by the regulatory agencies, a constant driving force to produce better localization ratios for antibodies will be a permanent source of impetus for those interested in RIT. This should result in the increased opportunity to bring new and improved therapy products into clinical use with their ultimate success highly dependent on the target/nontarget time-dependent localization ratio and careful selection of smaller, radiosensitive tumor targets, radiochemical chelate stability, and minimal antibody uptake heterogeneity.

Acknowledgements. We would like to thank the Elaine Snyder Research Award Trust at George Washington University and NIH grant #CA 44173 for their sponsorship in part of this work. Thanks are also due to Ms. Lydia Lucuesta, Ms. Alev Erdi, Mr. Yusuf Erdi, and Ms. Hannah Wolken for their proofreading and preparation of the manuscript.

References

Anderson-Berg WT, Squire RA, Strand M (1984) Specific radioimmunotherapy using ^{90}Y-labeled monoclonal antibody in erythroleukemic mice. Cancer Res 47: 1905–1912

Badger CC, Krohn KA, Peterson AV, Shulman H, Bernstein ID (1985) Experimental radiotherapy of murine lymphoma with ^{131}I-labeled anti-Thy 1.1 monoclonal antibody. Cancer Res 45: 1536–1544

Badger CC, Krohn KA, Shulman H, Flournoy N, Bernstein ID (1986) Experimental radioimmunotherapy of murine lym-phoma with ^{131}I-labeled anti-T-cell antibodies. Cancer Res 46: 6223–6228

Badger CC, Wilbur DS, Hadley SW, Fritzberg AR, Bernstein ID (1990) Biodistribution of *p*-iodobenzoyl (PIP) labeled antibodies in a murine lymphoma model. Int J Radiat Appl Instrum [B] 17: 381–387

Beaumier PL, Venkatesan P, Vanderheyden J-L et al. (1991) ^{186}Re radioimmunotherapy of small cell lung carcinoma xenografts in nude mice. Cancer Res 51: 676–681

Begent RH, Bagshawe KD, Pedley RB et al. (1987) Use of second antibody in radioimmunotherapy. NCI Monogr 3: 59–61

Begent RHJ, Ledermann JA, Green AJ et al. (1989) Antibody distribution and dosimetry in patients receiving radiolabeled antibody therapy for colorectal cancer. Br J Cancer 60: 406–412

Bender H, Takahashi H, Adachi K et al. (1992) "Immunotherapy of human glioma xenografts with unlabeled, ^{131}I-, or ^{125}I-labeled monoclonal antibody 425 to epidermal growth factor receptor. Cancer Res 52: 121–126

Berger MJ (1971) Distribution of absorbed dose around point sources of electrons and beta particles in water and other media. MIRD Pamphlet No. 7. J Nucl Med (Suppl) 12: 1–23

Bhargava KK, Acharya SA (1989) Labeling of monoclonal antibodies with radionuclides. Semin Nucl Med 19: 187–201

Blumenthal RD, Sharkey RM, Goldenberg DM (1992) Dose escalation of radioantibody in a mouse model with the use of recombinant human interleukin-1 and granulocyte-macrophage colony-stimulating factor intervention to reduce myelo-suppression. J Natl Cancer Inst 84: 399–406

Brady LW, Markoe AM, Woo DV, Rackover MA, Koprowski MA, Steplewski Z, Pyester RG (1990) Iodine 125 labeled anti-epidermal growth factor receptor-425 in the treatment of malignant astrocytomas. A pilot study. J Neurosurg Sci 34: 243–249

Breitz H, Weiden P, Vanderheyden JL et al. (1992) Clinical experience with Re-186-labeled monoclonal antibodies for radioimmuniotherapy: results of phase I trials. J Nucl Med 33: 1099–1112

Breitz HB, Fisher DR, Weiden PL et al. (1993) Dosimetry of rhenium-186-labeled monoclonal antibodies: Methods, prediction from technatium-99m-labeled antibodies and results of Phase I trials. J Nucl Med 34: 908–917

Buchegger F, Pelegrin A, Delaloye B, Bischof-Delaloye A, Mach J-P (1990) Iodine-131-labeled MAb F(ab')$_2$ fragments are more efficient and less toxic than intact anti-CEA antibodies in radioimmunotherapy of large human colon carcinoma grafted in nude mice. J Nucl Med 31: 1035–1044

Buchsbaum DJ, Ten Haken RK, Heidorn DB, Terry VH, Guilbault DM, Stelewski Z, Lichter AS (1990) A comparison of I^{131}-labeled monoclonal antibody 17-1A treatment to external beam irradiation on the growth of LS174T human colon carcinoma xenografts. Int J Radiat Oncol Biol Phys 18: 1033–1041

Buchsbaum DJ, Khazaeli MB, Davis MA, Lawrence TS (1993a) Sensitization of radiolabeled monoclonal antibody therapy using bromodeoxyuridine. Cancer 73: 999–1005

Buchsbaum DJ, Langmuir VK, Wessels BW (1993b) Experimental radioimmunotherapy. Med Phys 20: 551–567

Buchsbaum DJ, Lawrence TS, Roberson PL, Heidorn DB, Ten Haken RK, Steplewski Z (1993c) Comparison of ^{131}I- and ^{90}Y-labeled monoclonal antibody 17-1A for treatment of human colon cancer xenografts. Int J Radiat Oncol Biol Phys 25: 629–638

Buras RR, Beatty BG, Williams LE, Wanek PM, Harris JB, Hill R, Beatty JD (1990) Radioimmunotherapy of human colon cancer in nude mice. Arch Surg 125: 660–664

Buras RR, Wong JYC, Kuhn JA, Beatty BG, Williams LE, Wanek PM, Beatty JD (1993) Comparison of radioimmunotherapy and external beam radiotherapy in colon cancer xenografts. Int J Radiat Oncol Biol Phys 25: 473–479

Carrasquillo JA, Krohn KA, Beaumier P et al. (1984) Diagnosis of and therapy for solid tumors with radiolabeled antibodies and immune fragments. Cancer Treat Rep 68: 317–328

Chen F-M, Taylor DK, Epstein AL (1989) Tumor necrosis treatment of ME-180 human cervical carcinoma model with ^{131}I-labeled TNT-1 monoclonal antibody. Cancer Res 49: 4578–4585

Cheung N-K, Landmeier B, Neely J et al. (1986) Complete tumor ablation with iodine-131-radiolabeled disialoganglioside GD2-specific monoclonal antibody against human neuroblastoma xenografted in nude mice. J Natl. Cancer Inst 77: 739–745

Chiou RK, Vessella RL, Limas C, Shafer RB, Elson MK, Arfman EW, Lange PH (1988) Monoclonal antibody-targeted radiotherapy of renal cell carcinoma using a nude mouse model. Cancer 61: 1766–1775

Colapinto EV, Zalutsky MR, Archer GE, Noska MA, Friedman HS, Carrel S, Bigner DD (1990) Radioimmunotherapy of intracerebral human glioma xenografts with ^{131}I-labeled F(ab')$_2$ fragments of monoclonal antibody Mel-14. Cancer Res 50: 1822–1827

Colcher D, Zalutsky M, Kaplan W, Kufe D, Austin F, Scholm J (1983) Radiolocalization of human mammary tumors in athymic mice by a monoclonal antibody. Cancer Res 43: 736–742

Cole WC, De Nardo SJ, Meares CF et al. (1987) Comparative serum stability of radiochelates for antibody radiopharmaceuticals. J Nucl Med 28: 83–90

Dale RG (1985) The application of the linear quadratic dose-effect equation to fractionated and protracted radiotherapy. Br J Radiol 58: 515–528

DeNardo SJ, DeNardo GL, O'Grady LF et al. (1988) Pilot studies of radioimmunotherapy of B cell lymphoma and leukemia using I-^{131} LYM-1 monoclonal antibody. Antib, Immunoconj Radiopharm 1: 17–33

DeNardo SJ, Warhoe KA, O'Grady LF, DeNardo GL, Hellstrom I, Hellstrom KE, Mills SL (1991a) "Radioimmunotherapy with I-131 chimeric L-6 in advanced breast cancer," in Breast Epithelial Antigens, edited by R.L. Ceriani. Plenum, New York

DeNardo SJ, Warhoe KA, O'Grady LF et al. (1991b) Response to ^{131}I chimeric MoAb L-6 radioimmunotherapy in patients with advanced metastatic breast cancer. J Nucl Med 32: 922

Dillman LT, Von der Lage FC (1975) Radionuclide decay schemes and nuclear paramters for use in radiation-dose estimation. MIRD Pamphlet No 10. Society of Nuclear Medicine, New York

Eary JF, Appelbaum FR, Durack L, Brown P (1989) Preliminary validation of the opposing view method for quantitative gamma camera imaging. Med Phys 16: 382–387

Eckerman KF (1985) Aspects of the dosimetry of radionuclides within the skeleton with particular emphasis on the active marrow. Proceedings of the Fourth International Symposium on Radiopharmaceutical Dosimetry. US Depart-ment of Energy, DE86010102 (CONF 851113), 514–534, Oak Ridge, Tenn

Erdi AK, Wessels BW, DeJager R et al. (1994) Tumor activity confirmation and isodose curve display for patients receiving iodine-131-16.88 human monoclonal antibody. Cancer 73: 932–44

Erdi YE, Wessels BW, DeJager R et al. (1994). A new fiducial alignment system to overlay abdominal computed tomography or magnetic resanance anatomical images with radiolabeled antibody single-photon emission computed tomographic scans. Cancer 73: 923–931

Ferens JM, Krohn KA, Beaumier PL et al. (1984) High-level iodination of monoclonal antibody fragments for radiotherapy. J Nucl Med 25: 367–370

Fisher DR (1986) The microdosimetry of monoclonal antibodies labeled with alpha particles. In Schlafke-Stelson AT, Watson EE (eds) Fourth International Radiopharmaceutical Dosimetry Symposium. Oak Ridge Associated Universities, Oak Ridge, Tenn. pp 446–457

Fowler JF (1990) Radiobiological aspects of low dose rates in radioimmunotherapy. Int J Radiat Oncol Biol Phys 18: 1261–1269

Fritzberg AR, Berninger RW, Hadley SW, Wester DW (1988) Approaches to radiolabeling of antibodies for diagnosis and therapy of cancer. Pharm Res 4: 325–334

Gallagher BM (1983) Monoclonal antibodies: the design of appropriate carrier and evaluation systems. Animal Models in Radiotracer Design 3: 61–105

Goldenberg DM, Gaffar SA, Bennett SJ, Beach JL (1981) Experimental radioimmunotherapy of a xenografted human colonic tumor (GW-39) producing carcinoembryonic antigen. Cancer Res 41: 4354–4360

Goldenberg DM, Sharkey RM, Blumenthal RD, Goldenberg H, Murthy S, Hansen HJ, Pinsky CM (1990) Problems and prospects for radioimmunodetection and radioimmunotherapy of cancer. Antibody Immunoconj Radiopharm 3: 151–167

Goldenberg DM, Horowitz JA, Sharkey RM et al. (1991) Targeting, dosimetry, and radioimmunotherapy of B-cell lymphomas with ^{131}I-labeled LL2 (EPB-2) monoclonal antibody J Clin Oncol 9: 548–564

Goodwin DA, Smith SI, Meares CF, David GS, McTigue M, Finston RA (1985) Chelate chase of radiopharmaceuticals reversibly bound to monoclonal antibodies improves dosimetry. In: Schlafke-Stelson AT, Watson EE (eds) Fourth International Radiopharmaceutical Dosimetry Symposium. Oak Ridge Associated Universities, Oak Ridge, Tenn., pp 477–492

Goodwin DA, Meares CF, McCall MJ, McTigue M, Chaovapong W (1988) Pre-targeted immunoscintigraphy of murine tumors with I-111-labeled bifunctional haptens. J Nucl Med 29: 226–234

Green AJ, Dewhurst SE, Begent RH, Bagshawe KD, Riggs SJ (1990) Accurate quantification of 131-I distribution by gamma camera imaging. Eur J Nucl Med 16: 362–365

Greiner JW, Guadagni F, Noguchi P, Pestka S, Colcher D, Fisher PB, Schlom J (1987) Recombinant interferon enhances monoclonal antibody-targeting of carcinoma lesions in vivo. Science 235: 895–898

Griffith MH, Yorke ED, Wessels BW, DeNardo GL, Neacy WP (1988) Direct dose confirmation of quantitative autoradiography with micro-TLD measurements for radioimmunotherapy. J Nucl Med 29: 1795–1809

Hammond MD, Moldofsky PJ, Beardsley MR, Mulhern CB Jr (1984) External imaging techniques for quantitation of distribution of I-131 F(ab') fragments of monoclonal antibody in humans. Med Phys 11: 778–783

Harrison A, Royle L (1987) Efficacy of astatine-211 labeled monoclonal antibody in treatment of murine T-cell lymphoma. NCI Monogr 3: 157–158

Hennigan TW, Begent RHJ, Allen-Mersh TG (1991) Histamine leukotriene C4 and interleukin-2 increase antibody uptake into a human carcinoma xenograft model. Br J Cancer 64: 872–874

Hnatowich DJ, Virzi F, Doherty PW (1985) DTPA-coupled antibodies labeled with yttrium-90. J Nucl Med 26: 503–509

Hosain F, Hosain P (1978) In: Spencer R (ed) Therapy in nuclear medicine. Grune and Stratton, New York, pp 33–34

Howell RW, Rao DV, Sastry KSR (1989) Macroscopic dosimetry for radioimmunotherapy: nonuniform activity distributions in solid tumors. Med Phys 16: 66–74

Humm JL (1986) Dosimetric aspects of radiolabeled antibodies for tumor therapy. J Nucl Med 27: 1490–1497

Humm JL (1987) A microdosimetric model of astatine-211 labeled antibodies for radioimmunotherapy. Int J Radiat Oncol Biol Phys 13: 1767–1773

Humm JL, Charlton DE (1989) A new calculational method to assess the therapeutic potential of Auger electron emission. Int J Radiat Oncol Biol Phys 17: 352–360

Humm JL, Chin LM (1990) Cellular dosimetry. In: Adelstein SJ, Kassis AI, Burt RW (eds) Dosimetry of administered radionuclides. American College of Nuclear Physicians, Washington, DC, pp 306–330

Humm JL, Roeske JC, Fisher DR, Chen GTY (1993) Microdosimetric concepts in radioimmunotherapy. Med Phys 20: 535–542

Huneke RB, Pippin CG, Squire RA, Brechiel MW, Gansow OA, Strand M (1992) Effective alpha-particle-mediated radioimmunotherapy of murine leukemia. Cancer Res 52: 5818–5820

Jain RK (1990) Tumor physiology and antibody delivery. Front Radiat Ther Oncol 24: 32–46

Jungerman JA, Yu K-HP, Zanelli CI (1984) Radiation absorbed dose estimates at the cellular level for some electron-emitting radionuclides for radioimmunotherapy. Int J Appl Radiat Isot 35: 883–888

Kalofonos HP, Pawlikowska TR, Hemingway A et al. (1989) Antibody guided diagnosis and therapy of brain gliomas using radio-labeled monoclonal antibodies against epidermal growth factor receptor and placental alkaline phosphatase. J Nucl Med 30: 1636–1645

Kaminski MS, Fig LM, Zasadny KR et al. (1992) Imaging, dosimetry, and radioimmunotherapy with iodine-131-labeled anti-CD37 (MB-1) antibody in B-cell lymphoma. J Clin Oncol 10: 1696–1711

Klein JL, Nguyen TH, Laroque P et al. (1989) Yttrium-90 and iodine-131 radioimmunoglobulin therapy of an experimental human hepatoma. Cancer Res 49: 6383–6389

Knox SJ, Goris ML, Wessels BW (1992) Overview of comparative studies comparing radioimmunotherapy with dose equivalent external beam irradiation. Radiother Oncol 23: 111–117

Kohler G, Milstein C (1975) Continuous culture of fused cells secreting antibody of predefined specificity. Nature 256: 495–497

Kozak RW, Raubitschek A, Mirzadeh S, Brechbiel M, Junghaus R, Gansow O, Waldmann T (1989) Nature of the bifunctional chelating agent used for radioimmunotherapy with yttrium-90 monoclonal antibodies: critical factors in determining in vivo survival and organ toxicity. Cancer Res 49: 2639–2644

Kwok CS, Prestwich WV, Wilson BC (1985) Calculation of radiation doses for non-uniformly distributed beta and gamma radionuclides in soft tissue. Med Phys 12: 405–412

Langmuir VK (1991) Radioimmunotherapy: clinical results and dosimetric considerations. Nucl Med Biol 19: 213–225

Langmuir VK, Sutherland RM (1988) Dosimetry models for radioimmunotherapy. Med Phys 15: 867–873

Langmuir VK, Fowler JF, Knox SJ, Wessels BW, Sutherland RM, Wong JYC (1993) Radiobiology of radiolabeled antibody therapy as applied to tumor dosimetry. Med Phys 20: 601–610

Larson SM, Carrasquillo JA, McGuffin RW et al. (1985) Use of I-131 labeled, murine Fab against a high molecular weight antigen of human melanoma: preliminary experience. Radiology 155: 487–492

Lashford L, Jones D, Pritchard J, Breatnach F, Kemshead JT (1987) Therapeutic application of radiolabeled monoclonal antibody UJ13A in children with disseminated neuroblas-toma. NCI Monogr 3: 53

Lee Y, Bullard DE, Humphrey PA (1988a) Treatment of intracranial human glioma xenografts with ^{131}I-labeled anti-tenascin monoclonal antibody 81C6, Cancer Res 48: 2904–2910

Lee Y-S, Bullard DE, Zalutsky MR (1988b) Therapeutic efficacy of antiglioma mesenchymal extracellular matrix ^{131}I-radiolabeled murine monoclonal antibody in a human glioma xenograft model, Cancer Res 48: 559–566

Lederer CH, Hollander JM, Perlman I (1968) Table of isotopes, 6th edn. John Wiley & Sons Inc., New York

Leichner PK, Kwok CS (1993) Tumor dosimetry in radioimmunotherapy: methods of calculation for beta particles. Med Phys 20: 529–534

Leichner PK, Klein JL, Sieglman SS, Ettinger DS, Order SE (1981a) Dosimetry of I-labeled antiferritin hepatoma. Cancer Treat Rep 67: 647–658

Leichner PK, Klein JL, Garrison JB, Jenkins RE, Nickoloff EL, Ettinger DS, Order SE (1981b) Dosimetry of ^{131}I-labeled anti-ferritin in hepatoma: A model for radioimmunol-globulin dosimetry. Int J Radiat Oncol Biol Phys 7: 323–333

Leichner PK, Yang NC, Frenkel TL, Loudenslager DM, Hawkins WG, Klein JL, Order SE (1988) Dosimetry and treatment planning for ^{90}Y-labeled antiferritin in hepatoma. Int J Radiat Oncol Biol Phys 14: 1033–1042

Leichner PK, Stillwagon GB, Order SE (1989) 194 Hepatocellular cancers treated by radiation and chemotherapy combinations: toxicity and response: a Radiation Therapy Oncology Group study. Int J Radiat Oncol Biol Phys 17: 1223–1229

Leichner PK, Hawkins WG, Yang NC (1990) Quantitative SPECT in radioimmunotherapy. Antibody Immunoconj Radiopharm 4: 25

Leichner PK, Koral KF, Jaszczak RJ, Green AJ, Chen GTY, Roeske JC (1993) An overview of imaging techniques and physical aspects of treatment planning in radioimmunotherapy. Med Phys 20: 569–578

Lenhard RE, Order SE, Spunberg JJ, Asbell SO, Leibel SA (1985) Isotopic immunoglobulin: A new systemic therapy for advanced Hodgkin's disease. J Clin Oncol 3: 1296–1300

Lightfoot DV, Walker KK, Boniface GR, Hetherington EL, Izard ME, Russell PJ (1991) Dosimetric and therapeutic studies in nude mice xenograft models with ^{153}samarium-labelled monoclonal antibody, BLCA-38 Antib. Immunoconj Radiopharm 4: 319–330

Loevinger R, Japha E, Brownell G (1956) Discrete radioisotope sources. Radiation dosimetry. Academic Press, New York

Loevinger R, Budinger TF, Watson EE (1991) MIRD primer for absorbed dose calculations. Society of Nuclear Medicine

Macklis RM, Kinsey BM, Kassis AI et al. (1988) Radioimmunotherapy with alpha-particle-emitting immunoconjugates. Science 240: 1024–1026

Macklis RM, Kaplan WD, Ferrara JLM, Atcher RW, Hines JJ, Burakoff SJ, Coleman CN (1989) Residents essay award: Alpha particle radio-immunotherapy: Animal models and clinical prospects. Int J Radiat Oncol Biol Phys 16: 1377–1387

Macklis RM, Lin JY, Beresford B, Atcher RW, Hines JJ, Humm JL (1992) Cellular kinetics, dosimetry, and radiobiology of alpha-particle radioimmunotherapy: induction of apoptosis. Radiat Res 130: 220–226

Markoe AM, Brady LW, Woo D et al. (1990) Treatment of gastrointestinal cancer using monoclonal antibodies. In: Frontiers of Radiation Therapy Oncology, edited by JM Vaeth and JL Meyer. Karger, Base Vol. 24: pp 214–224

Matthews DC, Appelbaum FR, Eary JF et al. (1991) Radiolabeled anti-CD45 monoclonal antibodies target lympho-hematopoietic tissue in the macaque. Blood 78: 1864–1874

Mausner LF, Srivastava SC (1993) Selection of radionuclides for radioimmunotherapy. Med Phys 20: 503–510

Meredith RF, Khazaeli MB, Plott G et al. (1992) Phase I trial of ^{131}I-chimeric B72.3 in metastatic colorectal cancer. J Nucl Med 33: 23–29

Meredith RF, Johnson TK, Plott G et al. (1993) Dosimetry of solid tumors. Med Phys 20: 583–592

Moi MK, De Nardo SJ, Meares CF (1990) Stable bifunctional chelates of metals used in radiotherapy. Cancer Res (Suppl) 50: 789s–793s

Molthoff CFM, Pinedo HM, Schluper HMM, Boven E (1992) Influence of dose and schedule on the therapeutic efficacy of ^{131}I-labelled monoclonal antibody 139H2 in a human ovar-ian cancer xenograft model. Int J Cancer 50: 474–480

Moshakis V, McIlhinney RA, Raghavab D, Neville AM (1981) Localization of human tumor xenografts after IV adminis-tration of radiolabeled monoclonal antibodies. Br J Cancer 44: 91–99

Nabi HA, Doerr RJ (1992) Radiolabeled monoclonal antibody imaging (immunoscintigraphy) of colorectal cancers; current status and future perspective. Am J Surg 163: 448–456

Neacy WP, Wessels BW, Bradley EW, Kovandi S, Justice T, Danskin S, Sands H (1986) Comparison of radioimmuno-therapy (RIT) and 4 MV external beam radiotherapy of human tumor xenografts in athymic mice. J Nucl Med 27: 902–903

Order SE (1982) Monoclonal antibodies: potential role in radiation therapy and oncology. Int J Radiat Oncol Biol Phys 8: 1193–1201

Order SE, Stillwagon GB, Klein JL et al. (1985) Iodine-131-antiferritin, a new treatment modality in hepatoma: a Radiation Therapy Oncology Group Study. J Clin Oncol 3: 1573–1582

Order SE, Vriesendorp HM, Klein JL, Leichner PK (1988) A phase I study of ^{90}yttrium antiferritin: dose escalation and tumor dose. Antib Immunoconj Radiopharm 1: 163–168

Orton CG, Cohen LA (1988) A unified approach to dose-effect relationships in radiotherapy. I. Modified TDF and linear quadratic equations. Int J Radiat Oncol Biol Phys 14: 549–556

Paik CH, Eckelman WC, Reba RC (1986) Transchelation of Tc-99m from low affinity sites to high affinity sites of antibodies. Int J Radiat Appl Instrum [B] 13: 359–362

Parker BA, Vassos AB, Halper SE et al. (1990) Radioimmuno-therapy of human B-cell lymphoma with ^{90}Y-conjugated anti-idiotype monoclonal antibody. Cancer Res Suppl 50: 1022–1028

Press OW, Eary JF, Badger CC et al. (1989) Treatment of refractory non-Hodgkin's lymphoma with radiolabeled MB-1 (anti-CD37) antibody. J Clin Oncol 7: 1027–1038

Press OW, Eary JF, Badger CC et al. (1990) High-dose radio-immunotherapy of B cell lymphomas. Front Radiat Ther Oncol 24: 204–213

Prestwich WV, Nunes J, Kwok CS (1989) Beta dose point kernels for radionuclides of potential use in radioimmuno-therapy. J Nucl Med 30: 1036–1046 and 1739–1740

Redwood WR, Tom TD, Strand M (1984) Specificity, efficacy, and toxicity of radioimmunotherapy in erythroleukemic mice. Cancer Res 44: 5681–5687

Riva P, Moscatelli G, Lazzari S et al. (1989) Systemic and locoregional administration of labelled MOABS for gastrointestinal cancer radioimmunotherapy: Biokinetics, pharmacokinetics and dosimetry aspects. J Nucl Med 30: 779

Riva P, Arista A, Mariani G et al. (1991) Improved tumor targeting by direct intralesional injection of radiolabeled monoclonal antibody: A phase I study in brain glioma. J Nucl Med 32: 922

Roberson PL, Buchsbaum DJ, Heidorn DB, Ten Haken R (1992a) Three-dimensional tumor dosimetry for radio-immunotherapy using serial autoradiography. Int J Radiat Oncol Biol Phys 24: 329–334

Roberson PL, Buchsbaum DJ, Heidorn DB, Ten Haken RK (1992b) Variations in 3-D dose distributions for ^{131}Iodine-labeled monoclonal antibody. Antib Immunoconj Radio-pharm 5: 397–402

Roeske JC, Chen GTY, Brill AB (1993) Dosimetry of intra-peritoneally administered radiolabeled antibodies. Med Phys 20: 593–600

Rofstad R (1985) Human tumor xenografts in radiothera-peutic research. Radiother Oncol 3: 35–46

Rosen ST, Zimmer AM, Goldman-Leikin R et al. (1987) Radioimmunodetection and radioimmunotherapy of cuta-neous T cell lymphomas using an 131I-labeled monoclonal antibody: An Illinois Cancer Council study. J Clin Oncol 5: 562–573

Roselli M, Schlom J, Gansow OA, Raubitschek A (1989) Comparative biodistributions of yttrium- and indium-labeled monoclonal antibody B72.3 in athymic mice bearing human colon carcinoma xenografts. J Nucl Med 30: 672–682

Rostock RA, Klein JL, Leichner P, Kopher KA, Order SE (1983) Selective tumor localization in experimental hepa-toma by radiolabeled antiferritin antibody. Int. J Radiat Oncol Biol Phys 9: 1345–1350

Sands H (1990) Experimental studies radioimmunodetection of cancer: an overview. Cancer Res (Suppl) 50: 809s–813s

Scheinberg DA, Straus DJ, Yeh SD et al. (1990) A phase I toxicity, pharmacology, and dosimetry trial of monoclonal antibody OKB7 in patients with non-Hodgkin's lym-phoma: Effects of tumor burden and antigen expression. J Clin Onc 8: 792–803

Schlom J, Hand PH, Greiner JW et al. (1990) Innovations that influence the pharmacology of monoclonal antibody guid-ed tumor targeting. Cancer Res (Suppl) 50: 820s–827s

Senekowitsch R, Reidel G, Mollenstadt S, Kriegel H, Pabst H-W (1989) Curative radioimmunotherapy of human mam-mary carcinoma xenografts with iodine-131-labeled monoclonal antibodies. J Nucl Med 30: 531–537

Schroff RW, Weiden PL, Appelbaum J et al. (1990) Rhenium-186 labeled antibody in patients with cancer: Report of a pilot Phase I study. Antib Immunoconj Radiopharm 3: 99–110

Schmidberger H, Buchsbaum DJ, Blazar BR, Everson P, Vallera DA (1991) Radiotherapy in mice with yttrium-90-labeled anti-Lyl monoclonal antibody: Therapy of the T cell lymphoma EL4. Cancer Res 51: 1883–1890

Sgouros G, Barest G, Thekkumthaia J, Chui C, Mohan R, Bigler R, Zanzonico P (1990) Treatment planning for internal radionuclide therapy: three-dimensional dosimetry for nonuniformly distributed radionuclides. J Nucl Med 31: 1884–1891

Sharkey RM, Pykett MJ, Siegel JA, Alger EA, Primus FJ, Goldenberg DM (1987) Radioimmunotherapy of the GW-39 human colonic tumor xenograft with ^{131}I-labeled murine monoclonal antibody to carcinoembryonic antigen. Cancer Res 47: 5672–5677

Sharkey RM, Kaltovich FA, Shih LB, Fand I, Govelitz G, Goldenberg DM (1988) Radioimmunotherapy of human colonic cancer xenografts with ^{90}Y-labeled monoclonal antibodies to carcinoembryonic antigen. Cancer Res 48: 3270–3275

Siegel JA, Pawlyk DA, Lee RE, Sasso NL, Horowitz JA, Sharkey RM, Goldenberg DM (1990a) Tumor, red marrow, and organ dosimetry for ^{131}I-labeled anti-carcinoembryonic antigen monoclonal antibody. Cancer Res Suppl 50: 1039–1042

Siegel JA, Wessels BW, Watson EE et al. (1990b) Bone marrow

dosimetry and toxicity for radioimmunotherapy. Antibody Immunoconj Radiopharm 3: 213–233

Siegel JA, Goldenberg DM, Badger CC (1993) Radio-immunotherapy dose estimation in patients with B-cell lymphoma. Med Phys 20: 579–582

Snyder WS, Ford MR, Warner GG, Watson SB (1975) "S", absorbed dose per unit cumulated activity for selected radionuclides and organs. MIRD Pamphlet No. 11. Society of Nuclear Medicine, New York

Stewart, JSW, Hird V, Snook D et al. (1990) Intraperitoneal yttrium-90-labeled monoclonal antibody in ovarian cancer. J Clin Oncol 8: 1941–1950

Stickney DR, Anderson LD, Slater JB, Ahlem CN, Kirk GA, Schweighardt SA, Frincke JM (1991) Bifunctional antibody: a binary radiopharmaceutical delivery system for imaging colorectal carcinoma. Cancer Res 51: 6650–6655

Stillwagon GB, Order SE, Klein JL et al. (1987) "Multi-modality treatment of primary nonresectable intrahepatic cholangiocarcinoma with ^{131}I anti-CEA-a Radiation Therapy Oncology Group study". Int J Radiat Oncol Biol Phys 13: 687–695

Strand SE, Zanzonico P, Johnson T (1993) Pharmacokinetic modeling. Med Phys 20: 515–528

Thomas SR, Maxon R, Kereiakes G (1976) In vivo quantitation of lesion activity using external counting methods. Med Phys 3: 325–335

Vessella RL, Alvarez V, Chiou R-K et al. (1987) Radio-immunoscintigraphy and radioimmunotherapy of renal cell carcinoma xenografts. Natl Cancer Inst Monogr 3: 159–167

Vessella RL, Lange PH, Palme II DF, Chiou RK, Elson MK, Wessels BW (1988) Radioiodinated monoclonal antibodies in the imaging and treatment of human renal cell carcinoma xenografts in nude mice. In: Targeted Diagnosis and Therapy, edited by J.D. Rodwell (Marcel Dekker, New York, NY, 1988), Vol. I, pp 245–282

Vesselia RL (1991) Radioimmunoconjugates in renal cell carcinoma. In: Immunotherapy of Renal Cell Carcinoma, edited by F.M.J. Debruyne, R.M. Bukowski, J.E. Pontes, and P.H.M. de Mulder Springer Verlag, Heidelberg, pp 38–46

Volkert WA, Goeckeler WF, Ehrhardt GJ, Ketring AR (1991) Therapeutic radionuclides: production and decay property considerations. J Nucl Med 32: 174–185

Wagner H (1968) Principles of nuclear medicine. W.B. Saunders, London, UK

Washburn LC, Sun TTH, Lee Y-CC et al. (1991) Comparison of five bifunctional chelate techniques for ^{90}Y-labeled monoclonal antibody CO17-1A. Nucl Med Biol 18: 313–321

Watson EE, Stabin MG, Siegel JA (1993) MIRD formulation. Med Phys 20: 511–514

Weinstein JN, Van Osdol W (1992a) The macroscopic and microscopic pharmacology of monoclonal antibodies. Int J Immunopharmacol 14: 457–463

Weinstein JN, Van Osdol W (1992b) Early intervention in cancer using monoclonal antibodies and other biological ligands: micropharmacology and the "binding site barrier". Cancer Res 52 (Suppl): 2747s–2751s

Wessels BW (1990) Current status of animal radioimmunotherapy. Cancer Res (Suppl) 50: 970s–973s

Wessels BW, Rogus RD (1984) Radionuclide selection and model absorbed dose calculations for radiolabeled tumor associated antibodies. Med Phys 11: 638–645

Wessels BW, Vessella RL, Palme DF, Berkopec JM, Smith GK, Bradley EW (1989) Radiobiological comparison of exter-nal beam irradiation and radioimmunotherapy in renal cell carcinoma xenografts. Int J Radiat Oncol Biol Phys 17: 1257–1263

Wessels BW, Yorke ED, Bradley EW (1990) Dosimetry of heterogeneous uptake of radiolabeled antibody for radio-immunotherapy. Front Radiat Ther Oncol 24: 104–108

Wessels BW, Neacy WP, Yorke ED, Sands H (1991) External beam and radioimmunotherapy dosimetry comparison of colorectal xenografts. US Department of Energy, (DE-AC05-760R00033), 65–76, Oak Ridge, Tenn

Williams JR, Dillehay LE (1989) Radiological characteristics of radiolabeled immunoglobulin therapy. Proceedings of the 1988 ACNP/SNM Joint Symposium on the Biology of Radionuclide Therapy. American College of Nuclear Physicians, Washington, DC, pp 262–269

Williams LE, Duda RB, Profitt RT et al. (1988) Tumor uptake as a function of tumor mass: a mathematical model. J Nucl Med 29: 103–109

Williams JA, Wessels BW, Wharam MD, Order SE, Wanek PM, Poggenburg JK, Klein JL (1990) Targeting of human glioma xenografts in vivo utilizing radiolabeled antibodies. Int J Radiat Oncol Biol Phys 18: 1367–1375

Woo DV, Li DR, Brady LW, Emrich J, Mattis J, Steplewski Z (1990) Auger electron damage induced by radioiodinated iodine-125 monoclonal antibodies. Front Radiat Ther Oncol 24: 47–63

Yokoyama K, Carrasquillo JA, Chang AE et al. (1989) Differences in biodistribution of indium-111 and iodine-131-labeled B72.3 monoclonal antibodies in patients with colorectal cancer. J Nucl Med 30: 320–327

Yoneda S, Fujisawa M, Watanabe J, Okabe T, Takaku F, Homma T, Yoshida K (1988) Radioimmunotherapy of transplanted small cell lung cancer with ^{131}I-labelled monoclonal antibody. Br J Cancer 58: 292–295

Yorke ED, Beaumier PL, Wessels BW, Fritzberg A, Morgan C (1991) Optimal antibody-radionuclide combinations for clinical radioimmunotherapy: a predictive model based on mouse pharmacokinetics. Nucl Med Biol 18: 827–835

Yorke ED, Williams LE, Demidecki AJ, Heidorn DB, Roberson PL, Wessels BW (1993) Multicellular dosimetry for beta-emitting radionuclides: autoradiography, thermoluminescent dosimetry and three-dimensional dose calculations. Med Phys 20: 543–550

Zalcberg JR, Thompson CH, Lichtenstein M, McKenzie FC (1989) Tumor immunotherapy in the mouse with the use of ^{131}I-labeled monoclonal antibodies. J Natl Cancer Inst 72: 697–704

Zalutsky MR, Garg PK, Friedman HS, Bigner DD (1989) Labeling monoclonal antibodies and F (ab') fragments with the alpha-particle-emitting nuclide astatine-211: preser-vation of immunoreactivity and in vivo localizing capacity. Proc Natl Acad Sci USA 86: 7149–7153

16 Radiation Dose-Response Models

Timothy E. Schultheiss

CONTENTS

16.1 Introduction

In studying radiation responses, there are primarily two motivations. One motivation is to learn the basic mechanisms involved in the response of living tissue to radiation. The other, but not altogether different motivation, is to learn what response is likely in a human when a patient is irradiated for therapeutic (or diagnostic) purposes. Clearly these objectives are not mutually exclusive. However, both the investigators and the methods they employ are likely to be different for the two arenas of investigation. The basic research is more likely to study the mechanisms of effects and less likely to be concerned with the quantification and prediction of effects. The more clinically oriented research is likely to work at the cellular level and above, whereas the basic research generally works at the cellular level and below. In clinical research, investigators are often interested in modeling the responses to radiation so that these responses may be predicted. There are primarily two types of response to high-dose radiation that are

modeled mathematically. These are cell survival and organ response. Although lethality (of organisms) has been the subject of investigation, such lethality is often a surrogate for the response of a particular organ.

This chapter will be directed toward an exposition of the *human* radiation biology – especially those responses that are modeled mathematically. Radiation physicists are often involved in this modeling process or in the evaluation of radiobiological data. It is important that the physicist or whoever is responsible for designing and evaluating experimental or clinical response data has an understanding of the stated and unstated assumptions in the models used and that these assumptions are tested appropriately. The following is a list of common assumptions that are generally made in radiobiological model building:

1. Cell survival after radiation is stochastic and obeys binomial or poisson statistics.
2. Organ response is determined by the death or survival of target cells.
3. All target cells respond identically.
4. All subjects respond identically.
5. Isoeffect relationships are independent of the level of response.
6. Equal effects are elicited from equal dose fractions when sufficiently separated in time.

These assumptions are not valid for all responses, and the usefulness of response models incorporating these assumptions may or may not depend heavily on their validity. It should be emphasized that this is a list of common assumptions; other more specific assumptions may be made for models that are used to describe a specific situation. The most commonly used models relevant to human radiobiology will be described. We will start by examining models of cell survival and then examine the models of organ response.

An inescapable aspect of dose-response modeling is the statistical analysis of the dose-response data. Physicists may be somewhat unaccustomed to

Timothy E. Schultheiss, PhD, Radiation Oncology, Fox Chase Cancer Center, 7701 Burholme Avenue, Philadelphia, PA 19111, USA

statistical analysis because their experiments can usually be designed to exclude nearly all sources of variation in the measured parameter except that which is due to the applied stimulus. In biological systems, not only are the mechanisms less well understood from first principles, but many more mechanisms are at work in a given experiment than can be modeled or controlled. Thus, experimental results will vary from one experiment to the next, from one laboratory to the next, and from one subject to the next. To elicit information on the mechanism at work in a given experiment requires both thoughtful experimental design and appropriate statistical analysis. Measured or inferred biological parameter values obtained without properly designed, executed, and analyzed experiments are uninformative to scientists and dangerous when applied to patients. For these reasons, considerable attention is devoted to the statistical analysis of dose-response experiments in this chapter. However, a complete exposition of the statistics of dose response is well beyond the scope of this chapter, and the reader should consult the references in the text for greater detail.

16.1.1 Bioassay

Radiation dose-response modeling is not substantially different from dose-response modeling involving other toxic agents. The fundamental aspects of dose-response modeling are the administration of a known level of stimulus to a population of subjects whose response (the dependent variable) is generally binary, i.e., dichotomous. "Events" may be death of the subject, death of a cell, control of the tumor, or failure of an organ. In dose-response analysis, the endpoint is reached for an individual when that individual's tolerance has been exceeded. However, it is not possible to predict what a specific individual's tolerance may be. The objective is to determine the distribution of tolerances for the entire population, and if possible, determine that distribution for similar populations that may be exposed to the stimulus in the future.

Dose-response analysis for radiation oncology differs from conventional applications in several respects. The dose has several components – number of fractions, dose per fraction, interfraction interval, and overall treatment time. The response is usually delayed, thereby resulting in censoring of patients as a result of intercurrent deaths or loss to follow-up. Local control can be difficult to verify. For endpoints related to control of disease, patients often must be

stratified according to initial stage and grade of tumor and factors other than dose that may be related to tumor response. In experimental studies, some of these difficulties may be obviated by an appropriate experimental design. In clinical data analysis, one also faces the probable lack of stratification according to dose. That is, the dose to the tissue that is under study may be different for each individual in the study. All of these difficulties can be overcome through the proper statistical analytic techniques, but too often ad hoc short cuts are employed instead (HERBERT, 1993).

This chapter will cover several of the dose responses investigated in radiation oncology and the appropriate statistical techniques in experimental design and analysis.

16.2 Cell Survival

Although it may seem to be a trivial statement, radiation kill cells. Of course, in clinical radiation oncology what must always be remembered is that some of every type of cell in the radiation field will be killed. In studies of normal tissue responses to radiation, sometimes the emphasis is placed so heavily on identifying a specific target cell that investigators ignore the fact that all cell types within an organ will be damaged to some extent and therefore will contribute to the radiation effect. Thus, the statement that radiation kills cells applies to cells whose response is not being modeled as well as to cells which are the target of the modeling.

16.2.1 Cell Survival Models

There are many mathematical models that describe the dependence of the probability of cell survival, S, on dose, d, and fraction number, N. (Lowercase d will be used to represent single doses and D will be used to denote total fractionated doses.) If cells were killed as a result of a single, dose-related event occurring at a single critical site in the cell, then the cell survival would be exponential (NIAS 1990). However, there is a shoulder to the cell survival curve for mammalian cells exposed to low LET radiation. The mechanistic models of cell survival derive the shape of this curve based on assumptions about the nature of the radiation-induced lesions produced in the cell that are responsible for cell death. Most cell survival curves are linear at the higher doses (lower survival), and this fact is used to produce an empirical model with two shape parameters. These parameters are

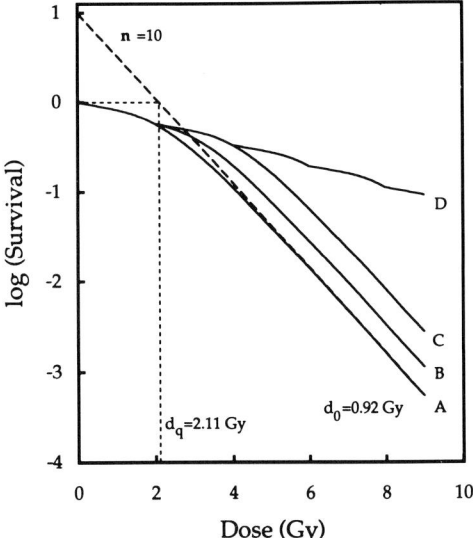

Fig. 16.1. Cell survival curves. *Curve A* is a two-component multitarget cell survival curve with $n = 10$, $d_1 = 5.5$ Gy, and $d_2 = 1.1$ Gy. This effectively yields the values of d_0 and d_q that are shown. *Curve B* shows the cell survival after a second fraction given before repair is complete. *Curve C* shows the cell survival after a second fraction given after complete repair. *Curve D* shows cell survival from multiple fractions of 2 Gy

shown in Fig. 16.1. d_0 is the inverse of the slope of the linear portion of the curve (the mean lethal dose for exponential cell kill) and n is the y-intercept, usually called the extrapolation number. A third parameter, the quasi-threshold dose (d_q), is related to these parameters by the equation $d_q = d_0 \cdot \ln(n)$. Typical values of d_0 are in the range of 1–2 Gy, and n is generally larger than 2.

The shoulder on the cell survival curve has been attributed to the repair of lesions that are not lethal (sublethal damage or SLD) but could combine with other similar events into a lethal lesion if additional radiation were given. Repair of sublethal damage takes place over a period of hours. After a sufficient interval following a first dose of radiation, the survival probability of cells surviving that first dose attains the same dose dependence as unirradiated cells. In Fig. 16.1, curve C shows the survival of cells that have had time for complete repair of SLD after an initial dose of 2 Gy. The portion of the curve C starting at 2 Gy is identical in shape (only displaced) to the curve starting at 0 Gy. If a second dose of radiation is given before the repair is complete, then the cell survival will be intermediate between curve A and curve C. This is depicted in curve B. If an erroneous assumption is made that SLD is complete, then the cell survival will be less than anticipated, and

for normal tissues, the damage will be greater than anticipated. This problem is addressed by the model of incomplete repair discussed below.

Of course the relevant value of S is that which is obtained for tumors irradiated in vivo. However, in vivo measurements of the surviving fraction are difficult to make and therefore subject to many criticisms.

Among the simplest mechanistic models of radiation cell kill is the simple multitarget model. For this model the surviving fraction after a dose d is given by:

$$S = 1 - \left[1 - \exp\left(-\frac{d}{d_0} \right) \right]^n, \qquad (16.1)$$

where d_0 is the mean lethal dose and n is the extrapolation number or the number of targets that must be inactivated for the cell to die. However, this model gives an initial slope of zero, which is usually not in accord with experimental data obtained in vitro. A post hoc addition to the model is suggested to correct for the lack of a nonzero initial slope. In this case, the form of S given in Eq. 16.1 is multiplied by $\exp(-d/d_1)$ so that we have:

$$S = \exp\left(-\frac{d}{d_1} \right) \cdot \left| 1 - \left[1 - \exp\left(-\frac{d}{d_2} \right) \right]^n \right|, \qquad (16.2)$$

where d_1 is determined from the initial slope and d_2 is determined from (but not equal to) the final slope. The factor $\exp(-d/d_1)$ accounts for cells killed by single hits, giving rise to the name two-component multitarget model.

Currently, the most popular cell survival model is the linear-quadratic (LQ) model, which can be derived from several different sets of assumptions (CHADWICK and LEENHOUTS 1973; DOUGLAS et al. 1979; KELLERER and ROSSI 1971; NEARY 1965). In this model, the probability of cell survival is given by:

$$S = \exp(-\alpha d - \beta d^2), \qquad (16.3)$$

where α and β can be viewed as simply the coefficients of the linear and quadratic terms of an unspecified function or the coefficient of terms of single-hit (e.g., inducing a double strand break) and two-hit (e.g., single strand breaks combining into a double strand break) events. This model is qualitatively different from many other cell survival models in that it is continuously concave downward rather than being linear at high doses. As a result, some have suggested that the LQ model is valid only over a limited dose range (DOUGLAS et al. 1979; HERBERT 1993; NIAS 1990; WITHERS et al. 1983). At least part of the LQ model's widespread use is a result of its mathematical simplicity – it is a linear model. (Note that the term linear model is used to describe linearity in the unknown parameters, and consequently, the LQ model is something of a misnomer if viewed from a statistical perspective). Because of its linearity, the LQ model can easily be generalized to describe fractionated radiation. Under the assumptions mentioned earlier (primarily that equal effects are elicited from equal dose fractions sufficiently separated in time), the survival probability after N equal fractions of dose d is simply S^N, and N enters are terms in the exponent as a linear coefficient. Considerable use is made of this feature, as will be discussed below.

Another model for cell survival is the LPL model (CURTIS 1986). In this model, long-lived repairable lesions are created in proportion to dose. These repairable lesions (B lesions) may combine over time to produce irreparable lesions (C lesions) in proportion to the square of the number of B lesions, or they may be repaired. Irreparable lesions may also be produced in proportion to dose from the rapid combination of two repairable lesions formed in close proximity. Two differential equations describing the time rate of change of the numbers of repairable and irreparable lesions can be written along with their boundary values. These are:

$$\frac{dN_B}{dt} = -\varepsilon_{B \to A} N_B - \varepsilon_{B \to C} N_B^2 \tag{16.4}$$

and

$$\frac{dN_C}{dt} = \varepsilon_{B \to C} N_B^2 \tag{16.5}$$

where ε's are rate constants for various processes. [This notation primarily follows THAMES and HENDRY (1987)]. B$\to$A represents the repair process, and B$\to$C represents the process of combination of B lesions into an irreparable lesions. The boundary values are:

$$N_B(0) = \eta_{A \to B} d; \; N_C(0) = \eta_{A \to C} d, \tag{16.6}$$

where $\eta_{A \to B}$ and $\eta_{A \to C}$ are the constants of proportionality for initial productions of B and C lesions, respectively. Assuming that both repairable (but unrepaired) and irreparable lesions are lethal, the cell survival is given by:

$$S = \exp(-N_B - N_C), \tag{16.7}$$

which at time, Δt, after the administration of dose d can be shown to be:

$$S = \exp(-(\eta_{A \to C} + \eta_{A \to B})d) \cdot$$
$$[1 + (\eta_{A \to B} d/\varepsilon)(1 - \exp(-\varepsilon_{B \to A} \Delta t))]^\varepsilon, \tag{16.8}$$

where $\varepsilon = \varepsilon_{B \to A}/\varepsilon_{B \to C}$. The B lesions represent potentially lethal damage (PLD) rather than SLD, and the

shoulder arises as a result of fixation of this damage rather than repair of SLD.

This is a four-parameter model that includes the additional variate, Δt. This Δt can be used as the interfraction interval for fractionated treatment. However, the formula for S for fractionated treatment must be determined from Eqs. 16.4 and 16.5 with the boundary conditions appropriately modified for each new dose fraction.

This model is a good illustration of the fact that even simple mechanistic models can become complex. Furthermore, with increasing aggressiveness in producing more realistic models, more parameters will be necessary. Cell survival functions can be reasonably well described with two-parameter models, so cell survival data are not likely to be capable of distinguishing between models with much statistical power. More direct tests must be made to evaluate mechanisms of cellular responses to radiation.

Because the actual lesions responsible for radiation cell death are not known with certainty, the concept of "unrepaired dose" has been used as a surrogate for the residual damage. THAMES (1985) used this concept to express the effective dose remaining at a time Δt. Following OLIVER (1964), Thames expresses the unrepaired dose at time Δt as $\theta \cdot d$, where d is the initial dose that was given and θ is the fraction that remains at time Δt with:

$$\theta = e^{-\mu \Delta t}, \tag{16.9}$$

where μ is the repair rate constant. The probability of cell survival, $S_N(d, \theta)$, following N doses of d which are separated by a time Δt is:

$$S_N(d, \theta) = \frac{\prod_{j=1}^{N} S\left(\sum_{l=0}^{j-1} \theta^l d\right)}{\prod_{j=1}^{N-1} S\left(\sum_{l=1}^{j} \theta^l d\right)}$$

$$= S(d) \cdot \frac{S(d + \theta d)}{S(\theta d)} \cdot \frac{S(d + \theta d + \theta^2 d)}{S(\theta d + \theta^2 d)}$$

$$\tag{16.10}$$

If $S(d) = \exp(-\alpha d - \beta d^2)$, then:

$$S_N(d, \theta) = \exp(-\alpha d N - \beta d^2 N h_N(\theta)), \tag{16.11}$$

where

$$h_N(\theta) = \frac{2}{N} \frac{\theta}{1-\theta}\left[N - \frac{1-\theta^N}{1-\theta}\right]. \tag{16.12}$$

Equation 16.10 is the incomplete repair model for cell survival. It can be generalized for continuous, low-dose-rate irradiation, bilinear repair kinetics, and nonconstant intervals between fractions (as in BID treatments).

Finally, a different class of mechanistic models has been proposed, that is not based solely on the production and accumulation of lesions within the cell, but rather on repair kinetics of potentially lethal lesions. Traditionally, the shoulder on the cell survival curve has been explained in terms of the repair of SLD. If a second dose of radiation immediately follows an initial dose, then the cell survival is found to be a result of the sum of the two doses. The longer the second dose is delayed, the more apparent the shoulder on the second survival curve becomes, as in Fig. 16.1. Alternatively, the shoulder of the cell survival curve has been hypothesized to be the result of dose-dependent repair of SLD that saturates at higher doses. That is, at doses where the survival is exhibiting straight line exponential behavior, the repair process is overwhelmed by the increasing dose, but at lower doses, some repair is achieved, giving rise to the shoulder. The exact mathematical form of the cell survival curve depends on the assumed nature of the lesions and the repair mechanism.

Several factors affect the probability of cell survival. PLD may be repaired in cells that are not actively proliferating during and following irradiation. This increases cell survival. The probability of cell survival also depends on the stage of the cell cycle at the time of irradiation. The phases of the cell cycle are: M, the phase during which the cells undergo mitosis; S, the phase when cells synthesize DNA prior to mitosis; G_1, a "gap" following M and preceding S; and G_2, a "gap" following S and preceding M. The biochemical processes that occur during G_1 and G_2 are not well understood. Finally, cells may not always proceed directly from G_1 to S, but may go into G_0, a phase during which the cells become nonproliferating.

Cells in mitosis are most sensitive to radiation. Cells at the G_1-S transition are next in sensitivity. Cells in mid-S are least sensitive. The relative sensitivities of G_1 and G_2 depend on many factors, among which are the relative lengths of these phases. Cells that are not cycling are usually considered to be in G_0, a relatively resistant phase of the cell cycle. The local environment also has an effect on cell survival. The most widely studied environmental factor is vascularization or oxygenation. Cells which are viable but poorly oxygenated have a higher probability of surviving a given dose than well-oxygenated cells. Because factors such as hemoglobin levels, age,

nutrition, and anemia have been implicated in tumor and normal tissue responses, it is assumed that they may directly or indirectly affect cell survival. The degree of differentiation of tumor cells (that is, how much the tumor cells resemble morphologically those normal cells from which the malignant transformation occurred) also affects their intrinsic radiation sensitivity, although perhaps this is an indirect effect of other factors.

16.2.2 Cell Survival Statistics

Because of the many factors that affect cell survival, it is not reasonable to use a mechanistic model of cell survival relevant to synchronized cells irradiated in vitro to model the clinical response of tumors (or normal tissues) without taking some account of the factors that may alter the in vitro response. When tumor cells are irradiated in vivo, they exist in different phases of the cell cycle, they have different oxygen levels, and different degrees of differentiation within the same tumor. Thus, the cell survival that is actually achieved is an average over all of the factors that affect the sensitivity. Let $S(\boldsymbol{\beta}|d, \boldsymbol{\gamma})$ be the cell survival model for synchronized cells of a specific type where $\boldsymbol{\beta}$ is a vector of parameters to be estimated, d is the dose, and $\boldsymbol{\gamma}$ is a vector of nuisance parameters that may alter the response of the cells to dose d. After irradiating a mixture of cells to dose d, the cell survival will be:

$$\langle S \rangle = \int S(\boldsymbol{\beta}|d, \gamma) \cdot f(\gamma) d\gamma, \qquad (16.13)$$

where $f(\boldsymbol{\gamma})$ is the multivariate distribution of $\boldsymbol{\gamma}$ over all cells. Thus, it is the expected value of S, $\langle S \rangle$, that is observed and not S itself. The dose dependence of $\langle S \rangle$ is not likely to be the same as the dose dependence of S (SCHULTHEISS et al. 1987). That is $S(d)$ and $\langle S(d) \rangle$ will not have the same functional form. As a consequence, a model that may work in vitro may not have the same mathematical behavior in vivo, even if the fundamental mechanisms regulating cell survival are the same. Furthermore, it is unwise to fit cell survival models to in vitro data unless the cells are synchronized and have the same radiation sensitivity. Otherwise any values of model parameters will be biased and their variances will be inflated (SCHULTHEISS et al. 1987).

The LQ model provides a good example for the fitting of cell survival models to data. The LQ model is a log-linear poisson model. It has the form:

$$m_i = \exp(\boldsymbol{x}_i^T \cdot \boldsymbol{\beta}), \qquad (16.14)$$

where m_i is the measured response (number of cells surviving), $\boldsymbol{x}_i$ is the vector of dose variates, and $\boldsymbol{\beta}$ is the vector of coefficients. For the LQ model:

$$x_i = (1, d_i, d_i^2).   (16.15)$$

Note that in Eq. 16.15, the inclusion of '1' in the dose vector is necessary for there to be a constant term, β_0, in the expression for $\log(m_i)$. This constant term is the estimate of the zero dose log cell count. Investigators often make the mistake of deleting this term, and then performing a regression through the origin of the expression for $\log(S)$. This technique is equivalent to giving infinite weight to the control values of the cell survival, implying absolute certainty in the colony counts for the unirradiated samples. Ignoring the constant term will bias the estimates of the other terms and inflate their variances.

A regression of $\log(m)$ on x provides parameter estimates of α and β of the LQ model (β_1 and β_2 in the regression model). Of course, cell count, rather than cell survival, is being modeled. Care must be taken to ensure that each point is properly weighted (using a weighting from the poisson, not the normal distribution), and some consideration should be given to the problem of the effects of the correlation between d and d^2. The output of the regression will provide point as well as interval estimates of β, goodness-of-fit statistics, and residual values. If the ratio of parameters is desired (such as α/β in the LQ model), the confidence limits of this ratio can be obtained from Fieller's theorem (FINNEY 1978). For nonlinear models, cell counts can be fitted using either non-linear regression (with the appropriate weightings) or maximum likelihood estimation.

16.3 Tumor Control Probability

Models of tumor control are generally based on the assumption that tumors regrow unless all clonogenic tumor cells are killed by the radiation. It is further assumed that cell kill obeys poisson statistics. Nearly all models also assume that given a sufficient time interval for repair, the percentage of cells surviving a given dose of radiation depends only on the dose at that fraction and not on the doses given previously. This time interval is uniformly assumed to be 24 h or less.

Under these assumptions, it is simple to derive the relationship between surviving fraction and tumor control probability (TCP). It is:

$$TCP = \exp(-K \cdot S^N),   (16.16)$$

where K is the number of clonogens in the tumor, S is the surviving fraction after a single dose, and N is the number of fractions. Further assumptions in this equation are that all cells respond identically. If this equation is to be applied to a population of patients, then it is also assumed that all individuals respond identically and that they have the same number of cells, K.

Using this model, it is simple to determine S for given values of N and K. Let us take a typical value of $N = 35$ for a curative course of radiation treatment. Then for TCP = 0.5, we have:

$$S = \left(-\frac{\ln(TCP)}{K}\right)^{1/N}   (16.17)$$

$$S = \left(\frac{\ln(2)}{K}\right)^{1/35}.   (16.18)$$

Table 16.1 lists values of S for various values of K to attain three different TCP levels. It is not unreasonable to assume that there is approximately 1 clonogen per 1000 cells in the tumor. However, this relationship is highly variable, difficult to determine, and the subject of some controversy. Note that for 10^7 clonogens, TCP goes from 1% to 99% over a cell survival range of only 10%. One may question whether in vivo cell survival data are sufficiently accurate to provide estimates that could distinguish between vastly different TCP values.

For fractionated radiation, the tumor control probability may be explored using the poisson distribution of cell kill. Under the assumption of equal effects from equal dose fractions, the overall cell survival curve for a series of equal dose fractions will be exponential in total dose, D_T. This is easily seen because $D_T = d \cdot N$, and therefore $S^N = S^{D_T/d}$. Using this expression in Eq. 16.16 yields:

$$TCP = \exp(-K \cdot \exp(-D_T/D_0)),   (16.19)$$

where D_0 is the effective slope of the fractionated cell survival curve (see Fig. 16.1). The slope of the TCP curve is:

$$\frac{\partial TCP}{\partial D_T} = -\frac{TCP \cdot \ln(TCP)}{D_0}.   (16.20)$$

D_0 can be eliminated from Eq. 16.20 by substituting from Eq. 16.19. For TCP = 0.5, we have:

Table 16.1. Single-dose cell survival to achieve various TCPs given 35 fractions and K clonogens

K	TCP		
	0.01	0.5	0.99
10^5	0.752	0.712	0.631
10^6	0.704	0.667	0.591
10^7	0.659	0.624	0.553
10^8	0.617	0.585	0.518

Table 16.2. Steepness in %/Gy of TCP curve at D_{50}

K	D_{50}(Gy)			
	55	60	65	70
10^4	6.0	5.5	5.1	4.7
10^5	7.5	6.9	6.3	5.9
10^6	8.9	8.2	7.6	7.0
10^7	10.4	9.5	8.2	8.8

$$\left.\frac{\partial TCP}{\partial D_T}\right|_{TCP=0.5} = -\frac{0.5 \cdot \ln(2) \cdot \ln\left(\frac{\ln(2)}{K}\right)}{D_{50}}. \quad (16.21)$$

Table 16.2 gives values of $\left.\dfrac{\partial TCP}{\partial D_T}\right|_{TCP=0.5}$ in %/Gy for

various values of K and D_{50}. An exceptionally steep (THAMES and HENDRY 1987) example of the steepness of the TCP curve was given by STEWART and JACKSON (1975), who reported TCP increasing from 35% to 65% when the dose increased from 52.5 to 57.5 Gy. This corresponds to a slope of 6%/Gy, and a D_{50} of about 55 Gy. From Table 16.2 we see that this would imply only 10^4 clonogens per tumor. Most other dose-response functions for tumor control are not nearly so steep, and investigators generally indicate this lack of steepness is a result of heterogeneity in sizes and in response among and within individual tumors (THAMES et al. 1992; ZAGARS et al. 1987) rather than being due to a failure of the assumption that tumor cell kill is a poisson process or the result of a very small clonogen number. It is because individual tumors may vary considerably from the population average of tumors that predictive assays are sought that indicate how a specific patient may respond to specific treatments.

Generally, it has been the practice in radiation oncology to attempt to deliver a uniform dose to the tumor. It has been assumed that the probability of tumor recurrence is controlled by the lowest dose to the tumor, and therefore delivering extra dose to part of the tumor will not increase local control but may increase complications. However, GOITEIN (1986) has argued that loss of TCP resulting from regions of lower dose within the tumor can be offset by increasing the dose in other areas of the tumor. This follows from the premise that tumor clonogen cell kill is a binary event, unrelated to the survival status of other clonogens. Thus, by analogy to Eq. 16.19, it can be seen that for an inhomogeneous dose distribution:

$$TCP = \prod_i \exp\left(-K_i \cdot e^{-D_i/D_0}\right), \quad (16.22)$$

where K_i is the number of clonogens receiving total dose D_i, and D_0 is the effective slope of the fractionated cell survival curve. Thus, increasing the dose to regions that can tolerate higher doses can partially compensate for lowering the dose in other regions.

A more general equation for clonogen cell survival (TCP) is:

$$TCP = \prod_i (1 - S(D_i))^{K_i}, \quad (16.23)$$

where $S(D_i)$ is the probability that a single clonogen survives a fractionated dose D_i. For large numbers of clonogens, Eq. 16.23 leads to the poisson model for TCP. However, because the number of clonogens per tumor is unknown, Eq. 16.23 is more generally justifiable.

An important factor in tumor response is the overall treatment time. Because tumors are growing, the radiation dose must be sufficient to kill all clonogens that are present at the start of treatment plus any additional clonogens produced by the tumor during treatment. Initially following a dose of radiation, cycling cells may be delayed during the S phase of the cell cycle. However, cells in G_0 may be induced back into the cell cycle. Furthermore, it has been hypothesized that after some period of treatment, tumor cells undergo accelerated proliferation or regrowth (TROTT 1990; WITHERS et al. 1988). For rapidly dividing cells, it is unwise to protract the radiation treatment because it may be difficult to deliver the dose in large enough daily fractions to counteract proliferation without exceeding normal tissue tolerance.

The above factors have resulted in the inclusion of a time factor in the dose metameter of radiation response. A metameter is a multivariate function of several parameters combined to form a single stimulus variable that is related to the response. Constant values of this metameter for different parameters values would result in the same response. Thus the dose metameter can be thought of as a biologic dose equivalent. During the early periods of the mathematical studies of radiation response, only the time and total dose were included in this dose metameter because dose per fraction was not appreciated to be of importance. Later, fraction number was explicitly included in the dose metameter, NSD, developed by ELLIS (1963). The NSD generalized to the variable exponent NSD

(**X** below) has the form:

$$\log(\mathbf{X}) = \beta_0 \log(D) + \beta_1 \log(N) + \beta_2 \log(T),$$
(16.24)

where **X** is the dose vector, N and T have their usual meaning, and $\boldsymbol{\beta}$ is a vector of fitted parameters. This form will be discussed below. For the NSD, $\beta_0 = 1$, $\beta_1 = 0.24$, and $\beta_2 = 0.11$ by definition. Because $\partial \mathbf{X}/\partial T$ is a decreasing function of time, the effect of extending overall treatment time is greater for short courses of treatment than for long courses. This feature of Eq. 16.24 has been criticized on the basis that little effect of overall treatment time is actually observed for treatment courses less than 4 weeks, but a decrease in control has been observed if treatment is extended beyond 8 weeks (WITHERS et al. 1988). Linear splines added to dose metameters have been proposed to account for the effect of overall treatment time (ROBERTS and HENDRY 1993; WITHERS et al. 1988). To date, no single model has been shown to fit experimental or clinical data by standard statistical criteria. However, analysis of several clinical data sets on different diseases has indicated that extending overall treatment time leads to the loss of local control (TROTT 1990).

16.4 Normal Tissue Dose Response

16.4.1 Target Cell Theory

The late response of normal tissues and organs to radiation is determined primarily by dose per fraction, number of fractions, time interval between fractions, and volume of tissue irradiated. Mechanistic models of normal tissue dose response are very difficult to generalize because the pathogenesis of each normal tissue's response is different. Nonetheless, a general theory has been espoused, the target-cell theory. In this theory, tissue failure is the result of the loss of either regenerative cells that ameliorate injury by replacing cells killed by radiation or functional cells that are responsible for maintaining physiological integrity of an organ. This theory has led to the search for target cells in various organs and corresponding radiation lesions. In tumors, the clonogenic cells are assumed to be the target cell for tumor control. However, distinction of clonogens from nonclonogenic tumor cells (if there is a distinction) has not been achieved. For some normal tissues, the identification of the target cell is relatively straightforward. These tissues include hematopoietic tissue, the immune system, testes, and skin. For other tissues, it is possible to identify cells

that are certainly involved in the pathogenesis of the lesion, but it is not possible to determine unequivocally whether these are the only target cells or whether there may be important indirect effects.

Using the target cell theory as a backbone for dose-response models, it is next assumed that a specific level of cell survival of target cells is responsible for a specific probability or level of normal tissue injury. Combinations of sets of values of d, N, overall treatment time, and any other relevant parameters that give a constant value of S are termed isoeffect relationships. The general form of these isoeffect relationships can usually be expressed as:

$$\ln(S) = f(d, N, T, D_T, \Delta t, \text{etc.}),$$
(16.25)

where T is the overall treatment time and Δt is the interfraction interval. For the LQ model this takes the simple form:

$$\ln(S) = -\alpha d N - \beta d^2 N$$
(16.26)

or

$$E = \alpha D_T + \beta D_T d,$$
(16.27)

where E is an unspecified level of effect related to target cell survival in vivo.

Many uses can be made of Eq. 16.27. One can relate total doses given with one dose per fraction to the biologically equivalent (under the requisite assumptions) total dose for a different dose per fraction. If D_1, d_1 and D_2, d_2 are the total doses, dose per fraction for fractionation regimens 1 and 2, respectively, then for equivalent endpoints:

$$D_1 = D_2 \frac{d_2 + \alpha/\beta}{d_1 + \alpha/\beta}.$$
(16.28)

Another feature of Eqs. 16.26 and 16.27 is a graphical method derived from them that is used to determine the ratio α/β. For Eq. 16.26 with $N = 1$, a graph of $-\ln(S)/d$ versus d will have a slope of β and an x-intercept of α. However, it is unwise to rely on this analysis when S is actually measured since a linear regression is much more informative, giving parameter and interval estimates as well as model adequacy statistics. For fractionated regimens, either clinical or experimental, a plot of $1/D_T$ versus d yields a line whose intercept/slope ratio is α/β. However, such an analysis is not subject to standard statistical validation and violates the fundamental requirement that the dependent variable is measurable and random. Indeed, there is a built-in and inescapable correlation between $1/D_T$ and d (HERBERT, 1993). More appropriate statistical analyses for determining α and β are available.

An important aspect of the LQ model and the use of Eq. 16.28 is that the ratio α/β is larger for early or acutely responding tissues (≥ 8 Gy) than for late responding tissues (approximately 2–3 Gy). One explanation for this has used Eq. 16.13 to show that if S has the form of Eq. 16.3, then $\langle S \rangle$ will have higher order terms in dose and fraction number (SCHULTHEISS et al. 1987). By ignoring these higher order terms, the estimates of α and β will be biased in such a way as to increase the ratio α/β. The heterogeneity that is responsible for this effect is likely to be greater in acutely responding tissues (with cells in the various stages of the cell cycle) than in late responding tissues, where the cells are not cycling.

Most mechanistic models of dose response rely on the clonogenic assay of cell survival being a true representation of the cellular response to radiation.

16.4.2 Empirical Dose-Response Models

Isoeffect relationships have been used in radiation oncology and biology for decades. In clinical radiation oncology, there is often a need to alter a fractionation schedule while still delivering a dose that is biologically equivalent to some standard schedule. However, for eliciting dose-response information from data, isoeffect relationships are not particularly helpful. Once a dose-response model has been fitted to data and the parameter estimates of the model have been made, an isoeffect relationship may be determined if necessary. However, in analyzing dose-response data, one should not start from a presumed isoeffect relationship.

An empirical model for dose response can be developed using the methods of generalized linear models (GLMs). GLMs are used to describe the responses that are not linear in the unknown parameters, but which can be linearized by some transform, g, called the link function. For bioassay, as applied to radiation responses, the dependent variable is the probability, P, that the endpoint (tumor control or normal tissue injury) is observed for a given set (vector) of dose parameters, $\mathbf{X}$. Applying the transform to P, we have $g(P) = Y$. Each different treatment regimen is a different vector of dose parameters. The set of these treatment regimens is represented by a matrix $\mathbf{X}$. It is common to express the dose variates in logarithmic form. Thus, the elements of $\mathbf{X}$ are:

$$x_{ij} = (\ln(d), \ln(N), \ln(T),...)_j, \tag{16.29}$$

where i refers to the ith element of the dose regimen $[x_1 = \ln(d), x_2 = \ln(N)$, etc.] and j refers to the jth dose regimen. The set of unknown parameters is represented by a vector $\boldsymbol{\beta}$. Then:

$$\mathbf{Y} = \mathbf{X} \cdot \boldsymbol{\beta}, \tag{16.30}$$

where $\mathbf{Y}$ is not a vector of subjects but of the linearized response at various dose regimens. $\mathbf{Y}$ *can* be a vector of individual subjects if each subject is treated with a different dose regimen. Note that $\mathbf{Y}$ is linear in $\boldsymbol{\beta}$. It can be seen that, with the appropriate choice of dose variates, the variable exponent NSD model for isoeffects is an example of a GLM (MCCULLAGH and NELDER 1989).

Common link functions in radiation response are the normal and logistic functions. For the latter:

$$Y_i = \ln\left(\frac{1}{P_i} - 1\right). \tag{16.31}$$

As an example of the application of this technique in experimental radiobiology, we use data from Wong et al. (1992) measuring paralysis in the rat after fractionated radiation to the cervical spinal cord. The dose was delivered in 20 fractions that followed an initial top-up dose of three fractions of 9 Gy. Because only the total dose (or dose per fraction) was varied we have:

$$Y = \beta_0 + \beta_1 \cdot \log(D), \tag{16.32}$$

from which

$$D_{50} = e^{-\beta_0/\beta_1}, \tag{16.33}$$

where D_{50} is the median tolerance dose. The dose-response function is:

$$P = \frac{1}{1 + e^{\beta_0 + \beta_1 \cdot \log(D)}}. \tag{16.34}$$

With $k = -\beta_1$, we obtain:

$$P = \frac{1}{1 + \left(\dfrac{D_{50}}{D}\right)^k}. \tag{16.35}$$

β_0 and β_1 were estimated using a weighted regression with the weights for each point being inversely proportional to the variance of the Y_i's. This variance is given by:

$$\sigma_Y^2 = \frac{S_P^2}{P^2(1-P)^2} \tag{16.36}$$

where $S_{P_i}^2 = r_i(n_i - r_i)/n_i^3$ is the variance of P_i using the binomial distribution with r_i animals responding out of n_i irradiated at the ith dose. Thus:

Table 16.3. Experimental results and analysis of paralysis in rats following 3 × 9 Gy plus the graded doses below given in 20 fractions

Dose	Response	Analytic method		
		GLM (last 4 points only)	MLE (last 4 points only)	MLE (all points)
20.0	0/9	$D_{50} = 29.95$	$D_{50} = 29.98$	$D_{50} = 29.96$
22.2	0/9	$k = 12.27$	$k = 12.40$	$k = 15.23$
23.7	0/9	$t = 6.28$	$\ln(L) = -19.8$	$\ln(L) = -20.3$
26.0	1/9		$LR = 0.24$	$LR = 1.23$
28.2	3/8		$\chi^2 = 0.24$	$\chi^2 = 0.90$
29.8	4/8			
32.0	6/9			

$$\sigma_{Y_i}^2 = \frac{n_i}{\left(r_i(n_i - r_i)\right)}. \qquad (16.37)$$

Fieller's theorem (FINNEY 1978) was used to estimate the 95% confidence interval of D_{50} using Eq. 16.33. The best estimate of D_{50} is not centrally located in its 95% confidence interval because β_0 and β_1 have a nonzero covariance and because the logarithm of dose was used as the independent variate.

The results of the experiment are shown in Table 16.3. Also given in this table are the estimated statistics so that the reader can duplicate the calculations. The resulting dose-response function is shown in Fig. 16.2. Note that a significant disadvantage of this method is that a minimum of three points with nonextreme values (between 0% and 100%) of the response are required.

Before discussing the analysis of clinical data that are subject to censoring, an alternative technique to

the use of GLMs will be presented. This alternative is the use of the maximum likelihood technique. If one has a model for estimating a certain response, then it is possible also to calculate the probability of achieving any given result, given values for the unknown parameters of the model. In the maximum likelihood method, the probability of obtaining the actual results is maximized as a function of the parameters in the model. The likelihood is defined as this probability or as differing only by a constant factor. For the experiment described above, the likelihood is given by:

$$L = \prod_i P_i^{r_i}\left[1 - P_i\right]^{n_i - r_i}, \qquad (16.38)$$

where all parameters have the same interpretations as above. Rather than using the likelihood, it is computationally easier and theoretically advantageous to use the logarithm of the likelihood. Thus the maximum likelihood estimates (MLE) of $\boldsymbol{\beta}$ are given by:

$$\frac{\partial \ln(L)}{\partial \beta_k} = \sum_i \left(\frac{r_i}{P_i} - \frac{n_i - r_i}{1 - P_i}\right) \frac{\partial P_i}{\partial \beta_k}$$

$$= \sum_i \left(\frac{r_i - n_i P_i}{P_i(1 - P_i)}\right) \frac{\partial P_i}{\partial \beta_k} = 0, \qquad (16.39)$$

where

$$\frac{\partial P_i}{\partial \beta_k} = \left.\frac{\partial P}{\partial \beta_k}\right|_{D = D_i} \qquad (16.40)$$

and $k = 0,1$. The matrix of second derivatives of $\ln(L)$ with respect to the elements of $\boldsymbol{\beta}$ is the information matrix, and the inverse of this matrix is the covariance matrix.

The $1 - \alpha$ confidence regions for the β_k's can be estimated in several ways (COX and SNELL 1989). The simplest uses the square root of the diagonals of the covariance matrix giving the confidence region as:

$$\hat{\beta}_k \pm z_{\alpha/2}\sqrt{v_{kk}(\hat{\beta})},$$

where $\hat{\beta}_k$ is the MLE of β_k, $z_{\alpha/2}$ is the $1/2\alpha$ point of the standard normal distribution, and $v_{kk}(\hat{\beta})$ is the kth diagonal of the covariance matrix evaluated with $\boldsymbol{\beta} = \hat{\beta}$. Using this method for estimating the confidence interval for $\hat{\beta}_k$ yields a symmetric interval. Other methods are available that depend on the exact shape of the log likelihood function. These methods are preferable when the log likelihood function is not well approximated by a quadratic

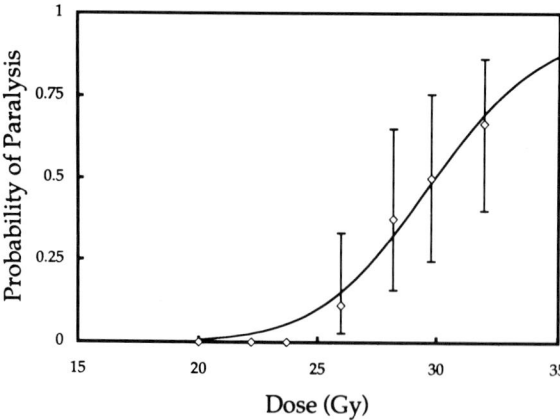

Fig. 16.2. Dose response of rats to spinal irradiation. The data are from WONG et al. (1992). The doses were given in 20 equal fractions following three fractions of 9 Gy. *Error bars* are 95% confidence limits based on binomial statistics. The *solid curve* represents Eq. 16.35 with $D_{50} = 29.95$ Gy and $k = 12.27$

function of β near its maximum. This is usually the case for the slope parameter of dose-response functions.

Finally, it is always advisable to examine the model adequacy (goodness-of-fit) parameters. In the case of binary dose response, the likelihood ratio and the generalized χ^2 statistic are both useful and are often close to each other in value (Cox and Snell 1989). It is also useful to test whether the inclusion of additional elements in the model (such as overall time or interfraction interval) improves the fit of the model to the data. Of course, any nested model's fit should improve or remain the same with the inclusion of additional parameters. However, it is possible and useful to determine whether the fit is *significantly* improved by the addition of new variates or effects. The fit of the models can be assessed by three methods: the generalized χ^2, the likelihood ratio (LR) (Cox and Snell 1989), and the Akaike Information Criterion (AIC) (Akaike 1977). These are given by the following equations:

$$\chi^2 = \sum_i \frac{(r_i - n_i P_i)^2}{n_i P_i (1 - P_i)} \tag{16.41}$$

$$\text{LR} = 2[\sum r_i \ln(r_i) + \sum (n_i - r_i)\ln(n_i - r_i) - \sum n_i \ln(n_i)$$
$$- \sum r_i \ln(P_i) - \sum (n_i - r_i)\ln(1 - P_i)] \tag{16.42}$$

$$\text{AIC} = -2\ln(L) + 2p \tag{16.43}$$
$$(p = \text{number of parameters}).$$

The decrease in the sum of squared residuals obtained by adding an additional parameter to a linear model is a measure of whether the additional parameter should be kept (Draper and Smith 1981). This technique is used for nested models. For non-nested models, the model of choice is the one which minimizes the AIC. The AIC statistic has components reflecting both accuracy [in the term $2\ln(L)$] and parsimony (in the $2p$ term).

16.4.3 Volume Effects

The model defined by Eq. 16.35 is a simple two-parameter model that could apply to a clinical situation where (a) the dose to the organ or tissue is homogeneous and (b) the irradiated volume of the organ is the same for all subjects in the study (or where there is no volume effect). However, these constraints rarely apply to a clinical situation. Methods have been developed to account for the volume effect and the effect of inhomogeneous doses on the probability of complication. Note that, if there is no volume effect, then the probability of complication is determined by the maximum dose delivered to the tissue.

The first model to account for both the volume of irradiated tissue and the inhomogeneity in the dose distribution was the probability model (Schultheiss et al. 1983). In the probability model, the probability of a complication in an organ is given by:

$$P(D, v) = 1 - [1 - P(D, 1)]^v$$
$$v = \frac{V_{\text{irr}}}{V_{\text{ref}}}, \tag{16.44}$$

where V_{irr} is the volume of the organ irradiated to dose D, V_{ref} is the reference volume (often the whole organ volume), and $P(D,1)$ is the probability of complication when V_{ref} is irradiated to dose D. The probability model is a probability transform, wherein changes in irradiated volume shift the dose-response function vertically (in the probability direction). For example, doubling the volume for a given probability of injury will result in a given change in the probability of injury that is independent of the mathematical form of the dose-response function, $P(D,1)$. That is, no specific dose-response model is needed in Eq. 16.44, only an estimate of the value of $P(D,1)$.

Most models of the volume effect include an additional parameter specifically to model the volume sensitivity. The parameter is usually a dose modifying factor that shifts a dose-response function horizontally (in the dose direction) when the volume is changed. Modeling the volume effect as a dose-modifying factor probably results from the common use of isoeffect equations in radiobiology, in which the biological consequences of variations in the radiation schema are modeled by describing the changes in dose required to maintain a constant biological effect. The probability model accounts for the change in the effect directly rather than by creating a biologic surrogate for dose. Because Eq. 16.44 models the effect on the probability, the steepness of the dose-response function determines the magnitude of the volume effect.

If we use the logistic dose-response function of Eq. 16.35, then:

$$D_v = D_{50} \left\{ \left[1 + \left(\frac{D_{v=1}}{D_{50}} \right)^k \right]^{1/v} - 1 \right\}^{1/k}, \tag{16.45}$$

where D_v is the dose to volume v that gives the same probability of complication as $D_{v=1}$ to the reference volume. For small probabilities of complication, we have:

$$D_v = v^{1/k} D_{v=1.} \qquad (16.46)$$

This equation has the same form as the power law model of the volume effect where the volume effect is modeled as a dose-modifying factor that shifts the dose-response function on the dose axis (KUTCHER et al. 1991; LYMAN and WOLBARST 1987). Some investigators build the power law model into the dose-response functions for normal tissues.

Equation 16.44 can be generalized to account for inhomogeneous dose distributions. If $\{D\}$ represents the dose distribution and Δv_i is the volume that receives dose D_i, then the probability of complication is given by:

$$P(\{D\},v) = 1 - \prod_i [1 - P(D_i,1)]^{\Delta v_i}, \qquad (16.47)$$

where the set of pairs $(\Delta v_i, D_i)$ is the differential dose-volume histogram.

In the probability model, the probability of injury is directly proportional to volume at small volume and small probability levels, less than about 15%. In Fig. 16.3, it is also clear that at clinically relevant probabilities, less than 1%, there is virtually no volume effect. This does not mean that the relative increase in the probability of injury is not proportional to the increase in volume. Rather, it means that the absolute increase in the probability of injury with increasing volume is imperceptible and clinically undetectable at these low probability levels. Figure 16.3 shows the low-dose dose response for a dose-volume response experiment carried out in

rhesus monkeys. Curves A, B, and C represent the dose response (in 2.2-Gy fractions) for radiation myelopathy for spinal cords lengths of 4, 8, and 16 cm, respectively (SCHULTHEISS et al. 1993). The curves were obtained by fitting the dose-volume response model defined by Eqs. 16.35 and 16.44 to the data using maximum likelihood estimation. The dose response depicted in this figure is an extrapolation beyond the actual data but using the parameter values obtained from the data analysis. It should be noted that the data obtained in this experiment represent a very steep dose-response function ($k=18$). Even so, below 60 Gy, it would not be possible to detect a dose or volume response under any reasonable clinical or experimental design. The curves begin to separate by more than about 1% only above 60 Gy. Even then, a large number of subjects would be required to exhibit a difference in the volume responses until one exceeded a 10% complication probability for the smallest volume used. Thus, under normal clinical circumstances (small probability of serious late effects) there is no observable volume effect. Furthermore, clinical protocols designed to detect differences in complication probabilities must recognize the flatness of the dose-response function at clinically acceptable complication rates.

The probability model for the volume effect is a direct consequence of binomial statistics and should be applicable to clinical normal tissue responses for which the observed endpoint is a result of the production of a discrete lesion. Responses to which this model should apply are radiation myelopathy and brain necrosis; productions of ulcers of the rectum, colon, bladder, or other mucosal surfaces; esophageal stricture; and a given level of skin reaction for fields sufficiently large that the response is not ameliorated cell migration (HOPEWELL et al. 1985). In these organs and tissues, the functional reserve is not relevant to the production of a lesion. For tissues and organs with excess capacity, the radiation injury depends more on the volume of tissue damaged and on the amount of viable tissue remaining than on the mere presence or absence of a specific level of damage. In such organs, the probability model may not hold and should be modified to account for the functional reserve of the organ. In these tissues, the volume effect is not so much a radiation effect as a volume-of-damage effect. Examples of these tissues are liver, lung, and kidney. In these tissues, the volume effect for radiation *damage* may be well described by the probability model, but damage is not equivalent to complication

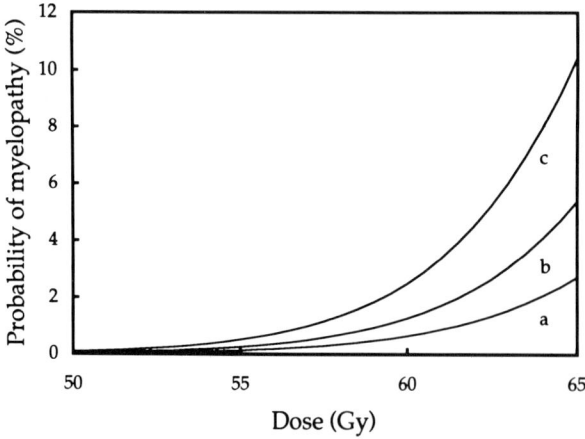

Fig. 16.3. The low dose response to 2.2-Gy fractions for rhesus monkey spinal cord. *Curve A* represents the dose response for 4 cm of cord, *B* for 8 cm, and *C* for 16 cm. At low doses, the separation of the dose-response curves would be clinically undetectable despite a fourfold change in spinal cord volume

because of the functional reserve. Volume effects in organs of this type should be modeled individually to reflect the pathogenesis of the injury. The dose-volume response functions of these organs will be more complex, and the response of experimental animals is less likely to be an accurate model of the human response.

The probability model has been recently extended by Jackson et al. (1993) by including a critical volume parameter. If the volume of tissue damaged by radiation is less than this critical volume, then no complication will occur. The probability of damage follows a sigmoid curve (either normal or logistic) that is described by two parameters, a location parameter and a slope parameter. In addition to the critical volume parameter, an additional parameter is included that accounts for the variation in this critical volume over the population of patients. Thus, this is a four-parameter model.

A model of the dose-volume response that has the power law model built into it has been developed by (Lyman and Wolbarst (1987). For this model, the normal distribution of tolerances is assumed, giving a dose-response function, P, of

$$P(D,v) = \frac{1}{\sqrt{2\pi}} \int_{-\infty}^{t} \exp\left(\frac{-u^2}{2}\right) du \qquad (16.48)$$

with

$$t = \frac{D - D_{50}v^{-n}}{mD_{50}v^{-n}}, \qquad (16.49)$$

where D, D_{50}, and v have the same meaning as above, and m and n define the steepness of the dose-response function and the magnitude of the volume effect respectively. As one would expect from the probability model, m and n are correlated. To account for dose inhomogeneities, Kutcher et al. (1991) have developed an algorithm that determines an effective volume that can be inserted into Eq. 16.48 to calculate P. The effective volume is the volume irradiated uniformly to D_{max} that would give the same probability of complication as that which results from the nonuniform dose distribution. The effective volume is given by:

$$v_{eff} = \sum_i \Delta v_i \left(\frac{D_i}{D_{max}}\right)^{1/n}, \qquad (16.50)$$

where the effective volume and D_{max} are used in Eqs. 16.48 and 16.49 to calculate P.

16.4.4 Effects of Censoring

To elicit dose-response information retrospectively from clinical studies (with either a prospective or a retrospective study design), the latent period must be considered. Patients may die of intercurrent disease or cancer or be lost of follow-up without surviving long enough to express a late injury. Several methods may be used to obtain dose-response information from data subject to this sort of censoring.

One method is a straightforward extension of the maximum likelihood method above (Schultheiss et al. 1986). The distribution of latent periods is denoted by $f(t)$. For a patient whose tolerance has been exceeded, the probability that the complication will be expressed between t and $t + dt$ following radiation treatment is given by $f(t)dt$. The likelihood function is then given by:

$$L = \prod_i P_i F(t_i) \prod_j [1 - P_j F(t_j)] \qquad (16.51)$$

or

$$L = \prod_i P_i f(t_i) \prod_j [1 - P_j F(t_j)], \qquad (16.52)$$

where $F(t)$ is the cumulative distribution of latencies corresponding to $f(t)$. In both equations, the product over i is for patients who express an injury, and the product over j is for patients who do not express an injury. For the patients who do not express the injury, the time t_j refers to the time of censoring or death. This is the interval for which they were followed and were known not to have expressed an injury at time t_j. For the patients who do express an injury, Eq. 16.52 is used if the time of the expression of injury is known to be t_i, and Eq. 16.51 is used if patients were known only to have expressed the injury some time prior to t_i. If the form of $f(t)$ is known, then the appropriate values of the function are used in the equations above. However, one can fit a model for $f(t)$ simultaneously with fitting a model for $P(D)$. When fitting models for $P(D)$ and $f(t)$ simultaneously, several caveats must be remembered. First, it is common in experimental radiobiology and has been reported in clinical data that the dose and the latency are correlated (Schultheiss et al. 1984). Therefore, one must consider whether there should be a dose term in $f(t)$. Second, for some radiation injuries, there are known to be more than one mechanism of pathogenesis giving rise to distributions of latent periods with more than one extremum, requiring more complex models of f with more parameters to be estimated. Finally, in analyzing data sets with few complications, increasing the number of unknown parameters by trying to model the latency distribution as well as the tolerance distribution will increase the confidence intervals of the estimates.

It may be possible to obtain estimates of $f(t)$ by combining reports from the literature to increase the number of cases from which f can be estimated or modeled (SCHULTHEISS et al. 1986). In such a case, it may even be possible to model correlations with dose and latency if the data are complete enough. However, it would be unwise to model these functions for use in the laboratory based on results from experimental animals. Latencies vary significantly within differing strains of the same species, and despite similarities among inbred animals, such strains are necessarily unstable and may not show identical responses from one experiment to the next.

An alternative method is available for which no estimate of the distribution of latencies is required and which addresses possible correlations between dose and latent period transparently (SCHULTHEISS et al. 1990). However, this method does not work well for widely varying fractionation schedules. In this method, patients are grouped by dose, D_i. To account for the effects of censoring during the latent period, P_i is determined by using the product limit method (the actuarial probability of remaining free of complication). It has been shown that for patients grouped according to dose regimen, the value of P obtained by the product limit method is identical to that obtained by maximizing the likelihood function with respect to the average values of $f(t)$ during the follow-up intervals that contain observed complications (SCHULTHEISS 1990). Having values of P_i for various D_i, it is possible to use regression techniques to estimate the parameters of the dose-response function if an appropriate link function is known and if one uses the appropriate weighting of each point. The weight of each point is again given by the inverse of the variance of the estimate of Y_i. In analogy with Eq. 16.36, the variance of Y_i is given by:

$$\sigma_Y^2 = \frac{S_G^2}{P_i^2 \cdot (1 - P_i^2)}. \qquad (16.53)$$

where S_G^2 is the variance of P_i for each dose regimen determined from Greenwood's formula (SCHULTHEISS et al. 1990). The advantages that this technique has are: (a) standard statistical packages may be used to analyze the data, (b) it is non-parametric with respect to the distribution of latencies, and (c) the problem that latency is correlated

with dose regimen is obviated. A disadvantage is that extreme values of P, i.e., 0 or 1, yield infinities in Y. Furthermore, one must ensure that all dose points have equivalent follow-up so that sufficient time has elapsed for valid estimates of P_i.

16.5 Other Measures of Association

16.5.1 Logistic Regression

Logistic regression can be defined as the fitting of a logistic function to data whose outcome is dichotomous. For continuous variables, methods of logistic regression were discussed in Sect. 16.4.2. In some cases, the independent variables are either naturally dichotomous, polychotomous, or they can be separated into two or more groups with meaningful ordinal values. Many radiation responses are analyzed using a multivariate logistic regression where dose is separated into two (or more) groups to determine whether the high-dose patients experienced a greater frequency of the endpoint under investigation. In this case, the independent variate, x, will be coded as 1 for patients who received a dose greater than some value and as 0 for patients whose dose was lower than this value. If this coding is used, odds ratio is given by e^β, where β is the coefficient of x in the regression equation. The odds ratio, Ψ, is defined as:

$$\Psi = \frac{\pi(1)/(1 - \pi(1))}{\pi(0)/(1 - \pi(0))}, \qquad (16.54)$$

where $\pi(1)$ is the probability of obtaining the outcome given that $x = 1$ and $\pi(0)$ is the probability for $x = 0$. The numerator and denominator of Eq. 16.54 are the odds for $x = 1$ and $x = 0$, respectively (HOSMER and LEMESHOW 1989).

Most investigators will report a significant association if the value of β is significantly different from 0. However, an assessment of model adequacy (goodness-of-fit) must precede statements regarding the significance of effects being tested in the model. It is meaningless to give significance levels for effects in models that do not fit the data. If one is using logistic regression to test the significance of a dose effect in a set of data, one must be certain that all patients were followed long enough to express the endpoint under study or that all patients are being evaluated at the same time after treatment. This method is not generally suitable for assessing late effects in a censored population.

16.5.2 Proportional Hazards

The proportional hazards model can be used to assess the importance of dose-related parameters in a population subject to censoring. In this model the hazard function is given by:

$$h(t \mid \mathbf{z}) = \psi(\mathbf{z} \mid \boldsymbol{\beta}) h_0(t), \qquad (16.55)$$

where t is the time after treatment, $\mathbf{z}$ is a vector of explanatory variables (including dose), and $\boldsymbol{\beta}$ is a vector of fitted parameters (Cox and Oakes 1984). The function, ψ, can take several forms, and if one is using a commercial software package for data analysis, it is important to know what functional form it takes. In analyzing the importance of dose-related factors, the survival may refer to absolute survival, disease-free survival, or complication-free survival. As with any survival analysis, it is important to be certain that the probability that an individual patient experiences a censoring event at time, t, is independent of the stratification group that the patient is in. This may be particularly important if treatment techniques change over time and more recently treated patients are more likely to be censored because of lack of follow-up. As always, model adequacy statistics should be evaluated. Finally, this method may not be suitable for constructing dose-response functions.

16.6 Conclusion

Models for various responses to radiation have been presented. In radiation oncology, we have used models such as these to predict the response to treatment. However, it has happened on several occasions during the history of radiation oncology that, because of invalid assumptions or omitted effects, extrapolation beyond the experience of clinical data (and therefore beyond the experience of the models) resulted in undesirable therapeutic results. Because the effects of increased dose per fraction were thought to be offset by lengthening overall treatment time (and vice versa), trials of split-course radiation therapy used high fractional doses and longer treatment times, and resulted in increased serious late effects. Similarly, overconfidence in the sparing effects of low dose per fraction led to hyperfractionation studies with insufficiently short interfraction intervals that also resulted in unexpected late effects. Because the natural tendency seems to be that clinical decisions are not based solely on clinical data, but also on experimental data, it is imperative that the analysis of both types of data be as informative as possible. This means that appropriate statistical techniques must be used, and that the model's ability to describe the data must be evaluated. If models are to be used to evaluate possible treatments prospectively, then it is imperative that these models have been tested prior to such an application and have been found to be useful and accurate in predicting outcomes of treatment. To do less cannot be ethically justified.

References

Akaike H (1977). On entropy maximization principle. In: Krishaiah PR (ed) Applications of statistics. North Holland, New York, pp 27–41

Chadwick KH, Leenhouts HP (1973) A molecular theory of cell survival. Phys Med Biol 18: 78–87

Cox DR, Oakes D (1984) Analysis of survival data. Chapman and Hall, London

Cox DR, Snell EJ (1989) Analysis of binary data. Chapman and Hall, New York

Curtis SB (1986) Lethal and potentially lethal lesions induced by radiation – a unified repair model. Radiat Res 106: 252–271

Douglas BG, Henkelman RM, Lau GKY, Fowler JG, Eaves CJ (1979) Practical and theoretical considerations in the use of the mouse foot system to derive epithelial stem cell survival parameters. Radiat Res 77: 453–471

Draper NR, Smith H (1981) Applied regression analysis, 2nd edn. Wiley, New York

Ellis F (1963) Fractionation and dose-rate. Br J Radiol 36: 153–162

Finney DJ (1978) Statistical method in biological assay. Charles Griffin, London

Goitein M (1986) Causes and consequences of inhomogeneous dose distributions in radiation therapy. Int J Radiat Oncol Biol Phys 12: 701–704

Herbert DE (1993) Quality assessment and improvement of dose response models: some effects of study weaknesses on study findings. "C'est magnifique?". Medical Physics Publishing Co

Hopewell JW, Hamlet R, Peel D (1985) The response of pig skin to single doses of irradiation from strontium-90 sources of differing surface area. Br J Radiol 58: 778–780

Hosmer DW, Lemeshow S (1989) Applied linear regression. Wiley, New York

Jackson A, Kutcher GJ, Yorke ED (1993) Probability of radiation induced complications for normal tissues with parallel architecture subject to non-uniform irradiation. Med Phys 20: 613–625

Kellerer AM, Rossi HH (1971) RBE and the primary mechanism of radiation action. Radiat Res 47: 15–34

Kutcher GJ, Burman C, Brewster L, Goitein M, Mohan R (1991) Histogram reduction method for calculating complication probabilities for three-dimensional treatment planning evaluations. Int J Radiat Oncol Biol Phys 21: 137–146

Lyman JT, Wolbarst AB (1987) Optimization of radiation therapy. III. A method of assessing complication probabilities from dose volume histograms. Int J Radiat Oncol Biol Phys 13: 103–109

McCullagh P, Nelder JA (1989) Generalized linear models. Chapman and Hall, New York

Neary GJ (1965) Chromosone aberrations and the theory of RBE. Int Radiat Oncol Biol Phys 9: 477–502

Nias AHW (1990) An introduction to radiobiology. Wiley, New York

Oliver R (1964) A comparison of the effects of acute and protracted gamma-radiation on the growth of seedlings of *Vicia faba*. II. Theoretical calculations. Int J Radiat Oncol Biol Phys 8: 475–488

Roberts SA, Hendry JH (1993) The delay before onset of accelerated tumor cell repopulation during radiotherapy: a direct maximum-likelihood analysis of a collection of worldwide tumor-control data. Radiother Oncol 29: 69–74

Schultheiss TE, Cohen L, Mansell J (1990) Normal tissue reactions and complications following high-energy neutron beam therapy. II. Complication rates adjusted for censoring. Int J Radiat Oncol Biol Phys 18: 169–171

Schultheiss TE, Orton CG, Peck RA (1983) Models in radiotherapy: volume effects. Med Phys 10: 410–415

Schultheiss TE, Higgins EH, El-Mahdi AM (1984) The latent period in radiation myelopathy. Int J Radiat Oncol Biol Phys 10: 1109–1115

Schultheiss TE, Thames HD, Peters LJ, Dixon DO (1986) Effect of latency on calculated complication rates. Int J Radiat Oncol Biol Phys 12: 1861–1865

Schultheiss TE, Zagars GK, Peters LJ (1987) An explanatory hypothesis for parameter values in the LQ model for early versus late effects. Radiother Oncol 9: 241–248

Schultheiss TE, Stephens LC, Ang KK, Price RE, Peters LJ (1994) Volume effects in rhesus monkey spinal cord. Int J Radiat Oncol Biol Phys 29: 67–72

Stewart JG, Jackson AW (1975) The steepness of the dose response curve both for tumour cure and normal tissue injury. Laryngoscope 85: 1107–1111

Thames HD (1985) An "incomplete-repair" model for survival after fractionated and continuous irradiations. Int J Radiat Oncol Biol Phys 47: 319–339

Thames HD, Hendry JH (1987) Fractionation in radiotherapy. Taylor & Francis, London

Thames HD, Schultheiss TE, Tucker SL, Dubray BM, Hendry JH, Brock WA (1992) Can modest escalations of dose be detected as increased tumor control? Int J Radiat Oncol Biol Phys 22: 241–246

Trott K-R (1990) Cell repopulation and overall treatment time. Int J Radiat Oncol Biol Phys 19: 1071–1075

Withers HR, Thames HD, Peters LJ (1983) A new isoeffect curve for change in dose per fraction. Radiother Oncol 1: 187–191

Withers HR, Taylor JMG, Maciejewski B (1988) The hazard of accelerated tumour clonogen repopulation during radiotherapy. Acta Oncol 27: 131–146

Wong CS, Mivkiv S, Hill RP (1992) Linear-quadratic model underestimates sparing effects of small doses per fraction in rat spinal cord. Radiother Oncol 23: 176–184

Zagars GK, Schultheiss TE, Peters LJ (1987) Inter-tumor heterogeneity and radiation dose control curves. Radiother Oncol 8: 353–362

17 Advances in Radiation Dosimetry

RAVINDER NATH and M. SAIFUL HUQ

CONTENTS

17.1 Introduction

Radiation dosimetry plays a central role in the physics of radiation oncology because potentially lethal amounts of radiation are necessary for the successful eradication of malignant tumors in humans. It has been estimated that an accuracy of ± 5% for dose delivered in the entire irradiated volume is necessary. Some of the steps in the radiation therapy procedure that influence the accuracy of dose delivered to tumor and healthy tissues are: (a) diagnosis, staging, localization of tumor volume,

RAVINDER NATH, PhD, Department of Therapeutic Radiology, Yale University School of Medicine, 333 Cedar Street, P.O. Box 3333, New Haven, CT 06510–8040, USA
M. SAIFUL HUQ, PhD, Department of Radiation Oncology and Nuclear Medicine, Thomas Jefferson University, 111 South 11th Street, Philadelphia, PA 19107-5097, USA

dose prescription, and fractionation schedule; (b) calculation of dose distributions resulting from an optimum geometric arrangement of radiation beams; (c) simulation and execution of the treatment plan; and (d) recording and verification of dose delivered. In this long chain of events, the dosimetry calibration of radiation fields used under reference conditions is the starting point for the accurate delivery of prescribed dose. Any errors in dose calibration of radiotherapy beams in a modern radiotherapy facility can result in underdosing or overdosing a large number of patients within a short time. Even a 1% error, if not detected for as little as a month in a busy clinic, can influence hundreds of patients. Small errors in dosimetry generally do not lead to major clinical sequelae in a patient population, but in some cases even small errors can have a significant impact on an individual patient.

In this paper, a review of some recent advances in radiation dosimetry pertaining to the calibration of radiotherapy beams is presented. We limit ourselves primarily to recent work reported after 1980, and primarily to photons and electrons used in radiation oncology. In what follows, we present an extensive review of recent dosimetry protocols followed by recent advances in instrumentation that is used in the calibration of radiation therapy equipment.

17.2 Dosimetry Protocols for Calibration of High-Energy Photon and Electron Beams

17.2.1 Pre-1980 Protocols

In the early part of 1960, high-energy photons from therapy machines were widely used to treat cancer patients throughout the world. However, at that time there were no accepted uniform methods for the measurement of absorbed dose at specified points in an absorbing medium for megavoltage x-rays and ^{137}Cs and ^{60}Co gamma-ray beams. In 1964, a sub-committee of the Hospital Physicists' Association (HPA) examined the various possible methods of dose and output measurements and recommended a code of practice for the determination of these quantities for 2- to 8-MV x-rays and ^{137}Cs and ^{60}Co gamma-ray beams (HPA 1964). Since the publication of this code of practice, a number of protocols and reports on the dosimetry of high-energy photon and electron beams have appeared in the literature. These include recommendations by the HPA (i.e., HPA 1969, 1971, 1975), by the subcommittee on radiation dosimetry (SCRAD) of the American Association of Physicists in Medicine (i.e., SCRAD 1966, 1971; AAPM 1975), by the German Standard Association (i.e., DIN 6800, German Standard Association 1975a, b), by the Nordic Association of Clinical Physicists (i.e., NACP 1972), and by the International Commission on Radiation Units and Measurements (ICRU) (i.e., ICRU 1969, 1972). In all of these reports and protocols, the method recommended for the determination of absorbed dose in a reference medium is based upon the use of an ionization chamber that has been exposure calibrated either at a National Standards Laboratory in a ^{60}Co gamma-ray beam or 2-MV x-ray beam or against a reference instrument which itself has been calibrated at the National Standards Laboratory. This procedure of linking a national standard to a local standard was given a position of prominence in all these protocols because it provides consistency, reproducibility, and uniformity in dose calibrations.

The possibility of using an exposure-calibrated ionization chamber for the determination of absorbed dose in a medium at a radiation quality which is different from that at which the chamber was exposure calibrated, was investigated by a number of authors (BARNARD 1964; BEWLEY 1963; DAVIES et al. 1963; GREENE 1962; GREENE and MASSEY 1966, 1967, 1968; HETTINGER et al. 1967a). A protocol based on this concept was first introduced by the HPA in 1964. The HPA-recommended approach is as follows: First obtain a 2-MV in-air exposure calibration factor for an ionization chamber from a National Standards Laboratory. For this calibration, measurement and for all subsequent measurements in any radiation quality in the user's beam, the chamber must be fitted with a standard buildup cap of adequate thickness (e.g., 4.6 mm Perspex) to provide electronic equilibrium. This chamber, fitted with the close-fitting watertight Perspex buildup cap, is then placed in a water phantom such that the central axis of the chamber is at 5-cm depth from the surface of the water phantom and lies on the central axis of the beam. The chamber is exposed to a megavoltage radiation of quality λ. It is assumed that the reading of the chamber will be proportional to the dose, in water, at the position of the chamber center. The constant relating the chamber reading to the dose will then be a function of λ only. Dose in rads to water at the position of the center of the chamber when the chamber is replaced by water and an identical monitor (or timer) exposure is made is given by

$$D = RNC_\lambda, \qquad (17.1)$$

where D = dose in rads to water at the position of the center of the chamber as described above; R = ionization chamber reading corrected to dry air at 22°C and 760 mm Hg; N = calibration factor given by the National Standards Laboratory to convert the ionization chamber reading to exposure in roentgens for 2-MV x-rays for dry air at 22°C and 760 mm Hg; and C_λ = overall conversion factor, which converts the ionization chamber reading to a statement of absorbed dose at the reference point in water and is a function of the radiation quality and the chamber plus Perspex cap used.

The protocol provided C_λ values for 2- to 8-MV x-rays and ^{137}Cs and ^{60}Co gamma rays for 0.6-cc Farmer-Baldwin and Victrometer type chambers only. It is assumed that these chambers behave in a uniform manner over this energy range. The values of C_λ were taken from the calculations of BARNARD (1964), who incorporated full allowance for the displacement effect for the above two chambers and used $W = 33.7$ eV per ion pair and WHYTE–LAURENCE (LAURENCE 1937; WHYTE 1954) stopping power data in the C_λ calculation. The displacement effect, mentioned above, results from the fact that when an ionization chamber of finite dimension is immersed in a medium, the radiation distribution at the center of the chamber will be slightly different from that in the medium which replaces the chamber after it is removed. To deduce the correct ionization in the medium from a measurement made with a chamber, it is necessary to correct the measured ionization for absorption in, and scatter from, the medium displaced by the chamber. The resulting correction factor is called a "displacement factor" and, as mentioned above, this effect was included in the C_λ calculation.

The protocol recommended that a depth of 5 cm in water be used as the reference depth for calibration measurements. There were three reasons for this: (a) The work of JOHNS et al. (1952) and BARNARD et al. (1959) suggested that a chamber will not be in electronic equilibrium when placed at the depth of maximum dose (d_{max}), (b) a small error in positioning the chamber at d_{max} can have a relatively large effect on the reading of the chamber, and (c) a depth of

5 cm is in the range of interest to the radiotherapists. Once the dose at 5 cm depth is known, then the dose at other depths can be determined by the application of appropriate percentage depth dose data (Br J Radiol, Suppl 10) or isodose charts (IAEA 1962).

In 1969 the HPA published another protocol for the dosimetry of high-energy photons. This protocol was essentially an extension of that published by the HPA previously (i.e., HPA 1964). The main change was that the upper limit of application of the protocol was extended from 8 MV to 35 MV. Furthermore, ionization chambers of suitable composition with an internal diameter less than 1 cm and calibrated against a primary standard of exposure in a 2-MV x-ray beam or ^{60}Co gamma ray beam were recommended for dose measurement. It should be noted that the HPA 1964 protocol was limited to the use of a Farmer-Baldwin and a Victrometer type chamber only.

The 1969 values of C_λ given for 2- to 8-MV x-rays are the same as those given in the 1964 protocol. The only difference is that the new values of C_λ were rounded off to two significant figures. Above 8 MV, GREENE and MASSEY (1968) calculated the values of C_λ by using BERGER and SELTZER's (1964) basic electron stopping power data and assuming that an ionization chamber of suitable composition with an internal diameter of less than 1 cm, when placed in a medium, could be regarded as an ideal Bragg-Gray cavity. In order to demonstrate that these assumptions are reasonable, it is necessary to show that when an ionization chamber of the above specification is placed in water and irradiated by megavoltage photons, nearly all of the observed ionization will be produced by fast electrons originating in the water, or conversely, only a very small proportion of ionization will be caused by fast electrons originating in the chamber wall or central electrode. Experimental evidence of this was given by GREENE (1962). He compared the response of a specially constructed chamber made entirely of Perspex (except for a thin film of carbon on the inner surface of the Perspex chamber wall and outer surface of the Perspex central electrode) and of the same dimensions as a Farmer-Baldwin chamber, with that of a Farmer-Baldwin chamber by placing both chambers in succession in a close-fitting hole in a Perspex phantom and irradiating them to radiation qualities ranging from 2 MV to 20 MV. The responses of the two chambers were found to be within 1% of each other over this energy range. GREENE and MASSEY (1966) compared the responses

of two Victrometer chambers with that of a Farmer-Baldwin chamber in a phantom at radiation qualities ranging from ^{131}Cs gamma rays to 15-MV x-rays. The ratio of the readings of these two types of chamber varied by less than ± 1.5% over the entire energy range. Similar experiments by HETTINGER et al. (1967b) showed that the ratios of the responses of four different types of commercial chamber at 34-MV x-rays to those at ^{60}Co gamma rays were nearly the same for each chamber. All these experiments suggest that ionization chambers of suitable composition with an internal diameter less than 1 cm can be regarded as Bragg-Gray cavities for radiation in the megavoltage range.

As the C_λ values given in the HPA 1969 protocol were calculated for an infinitesimal Bragg-Gray cavity, the displacement effect did not enter into the C_λ calculation. In contrast, the C_λ values given in the HPA 1964 protocol were calculated by making full allowance for the displacement effect. An error analysis for the product $N C_\lambda$ shows that the overall error to be expected on a dose determination can be as high as ± 3% at the highest energies. This is the relevant error that should be kept in mind when comparisons are made between the calculated values of C_λ and the experimental ones that are based on calorimetry or ferrous sulfate dosimetry. At this level of accuracy quoting C_λ to three significant figures is not meaningful and consequently the new values of C_λ were rounded off to two significant figures.

Values of C_λ depend on the secondary electron spectrum, which in turn depends on the beam quality, beam size, and depth of measurement. As the calculated values of C_λ were found to vary rather slowly with radiation quality, the nominal energy to which the electrons are accelerated in the x-ray generator was chosen to be the energy at which C_λ is specified.

New recommendations for calibration depths were made in the HPA 1969 protocol. One important reason for this is that the new C_λ values apply strictly to the situation where the ionization chamber is in electronic equilibrium. A calibration depth of 5 cm was thought to be within or even superficial to the region of peak dose for photons with energies greater than 10 MV. On the other hand, consideration had to be given to the depths which are of interest to radiotherapists. This resulted in the recommendation of a 5-cm depth in water for calibration measurements for radiation with energies up to and including 10 MV, a depth of 7 cm for energies between 11 and 25 MV, and a depth of 10 cm for higher energy radiations.

The use of a Fricke (ferrous sulfate dosimeter as the basis of a secondary standard of absorbed dose was also considered in the new protocol. However, because of many practical difficulties in achieving the required standard of precision and because many centers did not have the facilities for accurate chemical dosimetry, this form of dosimetry system was not recommended.

The ideas presented in the HPA 1964 and 1969 protocols were taken up by the ICRU, and in 1969 the commission made general recommendations on the dosimetric procedures for x-rays and gamma rays with maximum photon energies between 0.6 and 50 MeV. Four different dosimetric methods for determining the absorbed dose at a point in water were discussed in Report No. 14 (ICRU 1969). These were (a) an ionometric method based on the Bragg-Gray principle, (b) an ionometric method based on the exposure calibration of an ionization chamber at 2-MV x-rays or ^{60}Co gamma rays and the assumption that the chamber can be regarded as a Bragg-Gray cavity, (c) a chemical method based on Fricke ferrous sulfate dosimetry, and (d) a calorimetric method. The relative merits of the four methods were assessed by estimating the total uncertainties in absorbed dose determination for 2-MV and 30-MV x-rays. It was found that although the uncertainties had varied origins, the total uncertainty for each of the four methods was nearly the same. In view of the convenience, consistency, and reliability of ionization chamber dosimeters, method (b), which is based on the exposure calibration of an ionization chamber at 2-MV x-rays or ^{60}Co gamma rays, was recommended as an interim method for absorbed dose determination until calibration of dosimeters at energies higher than 2 MV is possible. It should be noted that this recommended method is essentially the same as that recommended by the HPA 1969 protocol.

In 1971, the Scientific Committee on Radiation Dosimetry (SCRAD) of the American Association of Physicists in Medicine (AAPM) published a protocol for the dosimetry of high-energy photons (SCRAD 1971). The energy range of photons covered in this protocol extended from 0.6 to 50 MeV. The intent of this protocol was to supplement reports written by the HPA (1969) and ICRU (1969). This was accomplished by recommending the C_λ method introduced by the HPA (1969) and adopted by the ICRU (1969) as the preferred method for determination of absorbed dose-to-war. SCRAD (1971) called this method the "calibrated cavity" method. New data were provided that enabled absorbed dose determination in water from measurements made in SCRAD polystyrene phantom (SCRAD 1966) and acrylic plastic phantoms. To check the performance of the calibrated cavity, alternative dosimetric methods based on (a) the Fricke ferrous sulfate dosimeter and (b) the absolute Bragg-Gray cazvity ionization chamber were recommended. These methods are applicable over the entire energy range covered in the protocol. An additional method based on the in-air measurements of exposure followed by calculation of dose was also suggested as an alternative dosimetric method for ^{137}Cs and ^{60}Co teletherapy units and 2-MV x-ray generators.

Neither the HPA (1969) nor the ICRU (1969) recommended the depths of dose maximum as the calibration depth. SCRAD (1971) points out that there are several advantages of placing a dosimeter at the depth of dose maximum (d_{max}). For example, the gradient of the depth-dose curve is zero at d_{max} so that small errors in positioning the chamber will have a negligible effect on the dosimeter response; the dose calibration at d_{max} is not dependent upon the central axis depth-dose data and d_{max} is the point at which central axis depth-dose data are usually normalized. If a depth other than d_{max} is employed for dose calibration, then dose at d_{max} is determined by the application of appropriate depth-dose data. Recognizing these advantages, SCRAD (1971) recommended d_{max} as the calibration depth for various photon energies.

If plastic phantom are used for dose calibrations, then the recommendation is to adjust the depth in plastic so that an equivalent recommended depth in water in g/cm^2 is obtained. Inverse-square corrections can be avoided by maintaining the same source-detector distance in a plastic phantom as would be used in a water phantom. Dose in water is obtained by first converting the ionization measured in a plastic phantom to dose-to-plastic and then by the application of the ratios of appropriate mass energy-absorption coefficients. The protocol provides these dose conversion factors for both polystyrene and acrylic phantoms.

Protocols for high-energy electron beam dosimetry were written by SCRAD in 1966, the ICRU in 1972, and the NACP in 1972. The energy range covered in the SCRAD (1966) protocol extended from 5 to 50 MeV whereas that in the ICRU (1972) and NACP (1972) protocols extended from 1 to 50 MeV. The goal of these protocols was to make recommendations which will bring uniformity to the dosimetry of high-energy electron beams.

The main features of the SCRAD (1966) protocol consist of recommendations for absorbed dose calibration, beam energy calibration, intercomparison of absorbed dose among various institutions, measurements of "output" in the SCRAD polystyrene phantom, and beam geometry specification. Three methods were considered for beam energy calibration: the magnetic deflection method, the electron disintegration thresholds method, and the range measurements method. Of these, the electron disintegration thresholds method was recommended as being the most generally available of the reliable methods. A simple method was recommended as a check for energy calibration. In this method, ionization measurements are made at a fixed depth and at the depth of maximum ionization by using a thimble chamber in the SCRAD polystyrene phantom. The ratio of these two readings can be plotted against energy to yield a straight line. Any serious deviation of this ratio from its normal value would indicate that there has been a change in the energy of the beam. Absorbed dose calibration is to be performed by using the Fricke ferrous sulfate dosimetric method with a "G" value of 15.4 molecules oxidized per 100 eV. A tissue-equivalent plastic phantom should be employed for these measurements. In an effort to standaridize the output of electron beam generators, a new measurable quantity called "output" was defined. This simply corresponds to a product of the reading of a thimble-type ionization chamber exposed in a polystyrene phantom at the position of the depth of d_{max} and the ^{60}Co exposure calibration factor of the chamber. This "output" convention has proved to be very useful because it is highly reproducible and in addition to comparing absorbed dose among various institutions engaged in electron beam therapy, one could intercompare "output" by using a recommended thimble chamber in a specified phantom.

In 1972 the NACP published their recommendations for the procedures in radiation therapy dosimetry for electrons with energies ranging from 5 to 50 MeV and photons with a maximum energy lying between 1 and 50 MeV (NACP 1972). These recommendations were based on experiences gained with the accelerators used in the Nordic Countries, that is betatrons below 50 MeV and linear accelerators below 10 MeV with photon beams only. The goal of this protocol was to give the hospital physicists of the Nordic countries a "code of practice" so that consistency and uniformity of dosimetry could be established among all the radiation therapy centers in the Nordic countries.

As the National Laboratories were not equipped to provide calibration factors based on absorbed dose measurements, it was recommended that the user's beam be calibrated by using an ionization chamber which has a ^{60}Co exposure calibration factor. For the determination of absorbed dose-to-water for photons, the C_λ concept developed by the HPA (1964, 1969) and adopted by the ICRU (1969) was also recommended by the NACP (1972). However, two changes were made in the applicability of the C_λ concept. These are an extension of the C_λ concept to other measurement depths and the use of a single standard depth (i.e., 5 cm) for the determintion of absorbed dose over the whole range of photon energies considered. Values of C_λ given in the protocol were adjusted above 10 MeV to allow for the changed depths by means of the experimental results of SVENSSON (1971), this adjustment being based on the concept of an effective point of measurement (DUTREIX and DUTREIX 1966; HETTINGER et al. 1967a; HARDER 1968). An effective point of measurement is one in which the effective location of the ionization chamber is displaced toward the radiation source by a fraction of the internal radius of the chamber. In other words, the ionization measured with the central axis of the chamber at a certain depth is assigned to another depth which is displaced toward the radiation source by a fraction "f" times the internal radius of the chamber. The recommended value of f is 0.75.

The reasons for choosing one depth as the standardized depth of absorbed dose measurements are (a) that it eliminates the risk of mistakes in measurement depths at centers that have several types of radiation equipment and (b) the overall simplicity in the construction of measurement phantoms. It was recognized that these advantages can be upset by the error in the conversion factor, C_λ that can result from the electron contamination of the beams at high photon energies. In the worst case (at high energies), the error in the absorbed dose determination was estimated to be less than 2%. In view of the simplicity and unambiguity in using a single depth of measurement, a 2% error in absorbed dose determination was considered acceptable.

For the determination of absorbed dose-to-water for electrons, a concept similar to C_λ was developed by the ICRU (1972) and adopted by the NACP (1972). Absorbed dose-to-water in this formalism is calculated from the expression

$$D_w = M N C_E, \tag{17.2}$$

where C_E is the absorbed dose conversion factor. The

protocol provided a table of C_E values as a function of initial electron energy and depth in water. These values of C_E were calculated from the expression

$$C_E = s_{w,air} \, W/e \, p_{air} \, A, \qquad (17.3)$$

where $s_{w,air}$ is the mean mass stopping power ratio of water to air calculated for different energies and depths; W/e is the quotient of the average energy expended to produce an ion pair by the electronic charge and has a value of 33.7 J/C; p_{air} accounts for lack of electron scattering in air inside the ionization chamber; and A is the attenuation factor in the buildup cap during ^{60}Co exposure calibration of the ionization chamber.

Values of C_E were calculated by taking $A = 0.985$, p_{air} for cylindrical chamber of internal diamter 5 mm from HARDER (1968), and $s_{w,air}$ times W/e from KESSARIS (1970). The calculated values of C_E are to be used when the absorbed dose is assigned to the effective point of measurement. Because of the slow variation of the depth-ionization curve near the depth of dose maximum, an error of about 0.5% would occur if the calculated values of C_E were to be used for the reference depth.

For a thimble ionization chamber of 5 mm diameter, the C_E values could also be calculated from the relationship (HARDER 1965)

$$C_E = C_3 - C_4 \log_{10}(\bar{E} C_5 + 1), \qquad (17.4)$$

where $C_3 = 0.987$ rad/"R," $C_4 = 0.120$ rad/"R," and $C_5 = 1$ MeV^{-1}.

This relationship could be employed in the energy range $3 \le \bar{E}_z \le 40$ MeV, where $\bar{E}_z$ is the average energy of the electrons at a depth z in a phantom.

Although ionization methods based on the ^{60}Co exposure calibration factor of an ionization chamber are recommended for absorbed dose determination, it is suggested that an independent check of the dose thus determined be performed by using methods based on calorimetry or ferrous sulfate dosimetry.

In addition to giving specific recommendations for the calibration measurements for photons and electrons, the NACP protocol also gave recommendations for beam energy calibration, beam alignment, and determination of absorbed dose-to-water at nonstandard points, and uniformity of the radiation beam in a given plane.

As the absorbed dose conversion factors C_λ and C_E are energy dependent and the possibility exists that standardized depth-dose tables may be used by therapy centers having accelerators similar in construction, it is important that the energy of the beams be determined in a uniform manner. For electron beams, the energy E_o at the surface of the phantom is determined from the following range-energy relationship (MARKUS 1964; SCRAD 1966; SVENSSON and HETTINGER 1971):

$$R_p = C_1 E_o - C_2, \qquad (17.5)$$

where $C_1 = 0.52$ cm (MeV)$^{-1}$ and $C_2 = 0.3$ cm. The electron energy determined from Eq. 17.5 is close to the most probable energy at the surface of the phantom. The average energy $\bar{E}_z$ of the electrons at a depth z in a phantom is determined from

$$\bar{E}_z = E_o(1 - z/R_p). \qquad (17.6)$$

For photons of maximum energy less than 10 MeV, a depth-ionization method is recommended for the evaluation of maximum photon energy in the spectrum. According to this method, ionization measurements are made by placing the effective point of measurement of the chamber at 5-cm and 15-cm depths in a water phantom. The ratio of these ionizations is then used to determine the maximum photon energy from Fig. 17.2 of the protocol, which is a plot of ionization ratios as a function of maximum photon energy. Photon energy thus determined can then be used for the determination of absorbed dose conversion factors and for the selection of depth-dose tables. If the maximum photon energy is higher than 10 MeV and the above ionization ratio method is used to determine the energy of the photon beam, then the energy thus determined can only be used for the calculation of the dose conversion factor. For the selection of depth-dose tables, the energy should be determined by electron disintegration methods discussed by SCRAD (1966).

17.2.2 NACP 1980 Protocol

Subsequent to the publication of the NACP 1972 protocol a number of developments occurred which led the NACP to revise it. A number of papers had been published in the literature giving new data suggesting that absorbed dose conversion factors used with ionization chamber dosimetry could be changed significantly. Hence, absorbed dose determined by following the methodology given in the NACP 1972 protocol could differ from that determined using the NACP 1980 protocol by as much as 5%. Another important reason for the revision of the 1972 protocol was that the recommendations given in the 1972 protocol were based on experiences gained with betatrons and linear accelerators that produced low-energy photons. As new kinds of accelerator with different beam qualities became available at various radiation therapy centers within the Nordic countries, it was felt that new recommendations based on experiences gained with the new accelerators should be given in order to facilitate uniformity of dosimetry.

So, in 1980 the NACP published a new protocol for the dosimetry calibration of radiotherapy beams. In this protocol a new formalism, based on the air-kerma calibration of an ionization chamber in a ^{60}Co gamma-ray beam, was recommended for the determination of absorbed dose. All correction and conversion factors that are necessary for the calculation of absorbed dose remain explicit in the new dose calculation algorithm. This is in sharp contrast to the recommendations of the NACP 1972 protocol in which all necessary corrections and conversions are combined into a single conversion factor, C_λ and C_E. According to the new formalism, absorbed dose at a point P in water is determined by following three steps:

1. Obtain an air kerma calibration factor N_k of an ionization chamber from the National Standards Laboratory. N_k is defined by

$$N_k = \frac{K_{air,c}}{M_c} \qquad (17.7)$$

where $K_{air,c}$ = kerma in air at the position of the center of the ionization chamber when the chamber is removed. Subscript c indicates that the measurements are carried out at the calibration quality beam. M_c = meter reading at calibration quality corrected for temperature, pressure, etc. $K_{air,c}$ is related to exposure X through the following relationship:

$$K_{air,c}(1-g) = X\,\overline{W}/e, \qquad (17.8)$$

where $\overline{W}/e = 33.85$ J/C and g = the fraction of energy of the secondary charged particles that is lost to bremsstrahlung in air for ^{60}Co gamma rays and has a value of 0.004 (BOUTILLON 1977).

2. Mean absorbed dose to air, $\overline{D}_{air,c}$ inside the air cavity of the ionization chamber is then obtained from

$$\overline{D}_{air,c} = K_{air,c}(1-g)k_{att}k_m, \qquad (17.9)$$

where k_{att} accounts for absorption and scatter of the primary photons in the chamber wall and buildup cap at the ^{60}Co gamma-ray beam quality and k_m accounts for lack of air equivalence of the chamber wall and buildup cap material at the ^{60}Co calibration beam.

Absorbed dose to air ionization chamber factor, N_D, is defined as

$$N_D = \frac{\overline{D}_{air,c}}{M_c} = N_k(1-g)k_{att}k_m. \qquad (17.10)$$

N_D could in principle be derived for any ionization chamber for which the factors in Eq. 17.10 are known. However, as the wall materials of the NACP 1980 recommended chambers should be homogeneous in graphite or tissue/water or air-equivalent material, N_D could only be calculated for these chambers. The recommended thickness for the graphite wall is about 0.5 mm. This is because perturbation factors needed for dose calculations given in the protocol (ALMOND and SVENSSON 1977; JOHANSSON et al. 1978) were determined for that wall thickness. For water-equivalent chambers, the wall thickness is not as important. The material of the buildup cap should be the same as that of the chamber wall. N_D is related to the ^{60}Co exposure calibration factor N_x through the following relationship:

$$N_D = N_x k_{att} k_m \overline{W}/ek_1, \qquad (17.11)$$

where $k_1 = 1.00$ with N_x in C kg^{-1}, and $k_1 = 2.58 \times 10^{-4}$ with N_x in RC^{-1}.

3. Absorbed dose $D_{w,u}$ at the reference point in water in the user beam quality is obtained from the Bragg-Gray relation:

$$D_{w,u} = \overline{D}_{air,u}(s_{w,air})_u\,p_u, \qquad (17.12)$$

where $\overline{D}_{air,u}$ = mean absorbed dose to air in the cavity of the chamber when measurements are made in water at the user's beam in Gy,

$(S_{w,air})_u$ = ratio of mean mass stopping power of water to air at the reference depth in the user's radiation quality [the stopping power ratios for photon beams were taken from the ICRU (1969), which are based on BERGER and SELTZER's (1964) electron stopping power data; for electrons $(S_{w,air})_u$ was taken from BERGER et al. 1975], and

P_u = perturbation correction to the Bragg-Gray equation and accounts for lack of water equivalence of the chamber wall at the user's beam qual-ity, perturbation of electron fluence that occurs because of replacement of water by air cavity, and location of the effective point of measurement of the cylindrical chamber due to the curved ionization chamber wall.

Assuming that

$$N_{\mathrm{D}} = \frac{\overline{D}_{\mathrm{air,c}}}{M_c} = \frac{\overline{D}_{\mathrm{air,u}}}{M_u}, \qquad (17.13)$$

absorbed dose-to-water at the reference point is obtained from

$$D_{\mathrm{w,u}} = M_u N_{\mathrm{D}} (s_{\mathrm{w,air}})_u p_u, \qquad (17.14)$$

where M_u is the meter reading at the user's radiation quality corrected for temperature, pressure, ionic recombination, polarity effect, etc.

When calibration measurements are made in an electron beam, the first step is to determine the mean energy $\overline{E}_o$ of the beam at the phantom surface. To determine $\overline{E}_o$, a depth-ionization curve is generated under reference conditions. In these measurements, the effective point of measurement of the chamber should be used. This point is located in front of the chamber center by a distance of $0.5\,r$, where r is the internal radius of the chamber. The next step is to generate a depth-absorbed dose curve by multiplying the depth-ionization curve by the appropriate $(S_{\mathrm{w,air}})_u$ values. $\overline{E}_o$ is then determined from Eq. 17.15, given below. The depth of the maximum absorbed dose thus determined is the reference depth at which the beam should be calibrated. During calibration measurements, the central axis of the cylindrical chamber should be placed at this depth.

The photon beam quality is estimated from the ratio J_{100}/J_{200}, where J_{100} and J_{200} are the relative ionizations measured at 100-mm and 200-mm depths in water for a field size of 100 mm × 100 mm and a source-to-surface distance (SSD) of 1 m. The protocol gives a plot of J_{100}/J_{200} versus the maximum photon energy for the energy range of 2–50 MV. This way of defining the beam quality is different from that given in the 1972 protocol, where measurements of ratios J_{50}/J_{150}, the relative ionizations at 50 mm and 150 mm, were recommended. After the energy of the beam is determined, the next step is to determine the reference depth from Table 3 of the protocol. The absorbed dose at the reference depth is then determined by placing the central axis of the chamber at that depth. The correction due to displacement of the effective point of measurement is included in the perturbation factor. Dose at d_{max} is determined from the ratio of the depth ionization at the reference depth and at d_{max}. This ratio is to be determined from the depth-ionization curve generated by using the effective point of measurement of the chamber. A value of $0.75\,r$ is recommended for the determination of the effective point of measurement.

Three parameters were defined for specifying the energy of electron beams: (a) the most probable energy in front of the accelerator window $E_{\mathrm{p,a}}$, (b) the most probable energy at the phantom surface $E_{\mathrm{p,o}}$, and (c) the mean energy at the phantom surface $\overline{E}_o$. Of these, $E_{\mathrm{p,o}}$ is used to indicate the energy of an absorbed dose distribution whereas $\overline{E}_o$ is used as an input parameter for stopping power ratios and perturbation correction factors. The protocol recommends that $\overline{E}_o$ and $E_{\mathrm{p,o}}$ be determined from the following relationships:

$$\overline{E}_o = 2.33 \times R_{50} \text{ MeV} \quad 5 \leq \overline{E}_o \leq 30 \text{ MeV} \quad (17.15)$$

and

$$\overline{E}_{\mathrm{p,o}} = C_1 + C_2 R_p + C_3 R_p^2 \text{ MeV} \qquad (17.16)$$
$$1 \leq E_{\mathrm{p,o}} \leq 50 \text{ MeV},$$

where $C_1 = 0.22$ MeV, $C_2 = 1.98$ MeV cm^{-1}, and $C_3 = 0.0025$ MeV cm^{-2} and R_p is the practical range. R_{50} in Eq. 17.15 is defined as the depth at which the absorbed dose in water is reduced to 50% of its maximum value. R_{50} should be determined from depth-absorbed dose measurements made with large field sizes. However, if depth-ionization measurements are made instead, then $\overline{E}_o$ can be obtained from Fig. 3 and Table 5 of the protocol, which gives $\overline{E}_o$ as a function of R_{50} determined from beam axis depth-absorbed dose curves with SSD = 1 m and ∞ and beam axis depth-ionization curves with SSD = 1 m. Figure 3 of the protocol should also be used for the determination of $\overline{E}_o$ for energies lying outside the range given by Eq. 17.15.

Water is the recommended phantom material for electron energies above 10 MeV. However, below 10 MeV, solid phantom materials such as PMMA, polystyrene, or A-150 are recommended. This is because plastic phantoms are sometimes more practical for measurements than water, especially if measurements have to be performed at a few millimeters' depth. If measurements are made in solid phantoms then the depths in the solid phantoms have to be scaled to the corresponding depths in water. The relation between R_p in water and that in plastic phantom is given by

$$\frac{R_{\mathrm{p,w}}}{R_{\mathrm{p,pl}}} = \frac{\rho_{\mathrm{pl}}}{\rho_w} \times \frac{(Z/A)_{\mathrm{eff,pl}}}{(Z/A)_{\mathrm{eff,w}}} \equiv b_m, \qquad (17.17)$$

where $(Z/A)_{\mathrm{eff}} = \Sigma f_i (Z_i/A_i)$, where f_i is the fraction by weight of the constituent element of atomic number Z_i and the relative atomic mass A_i. ρ is the density. Subscript "w" stands for water and "pl" stands for plastic. The above equation may be used for $(Z^2/A)_{\mathrm{eff}} < 4$.

If complete depth-ionization curves are measured in plastic phantoms, then conversion factors must be applied to obtain the corresponding curves in water. MATTSSON et al. (1981) have shown that for energies below 15 MeV and field sizes greater than 8×8 cm^2, Eq. 17.17 is a good approximation to scale the ionization curves measured in plastic to those in water. The largest deviation of up to 3 mm was obtained at the depth of R_{85} when measurements in polystyrene were converted to those in water.

The protocol recommends that both cylindrical and plane-parallel chambers be used for absorbed dose measurements for electron energies above 10 MeV. Below 10 MeV, only plane-parallel chambers are recommended. There are two reasons for this: (a) at low electron energies the cylindrical chambers produce an unacceptably large perturbation of the electron fluence at the point of interest and (b) the position of the effective point of measurement of a cylindrical chamber in relation to the shape of the central axis depth-dose curves is uncertain and is dependent on the energy of the beam. Experimental measurements show that the position of the effective point of measurement at the depth of dose maximum is displaced towards the radiation source by a fraction of the inner radius of the chamber. This fraction is a function of the energy of the beam and has a vlue between 0.3 and 0.7. In contrast to this, the plane-parallel chamber has a well-defined effective point of measurement which is located at the front surface of the air cavity.

Although plane-parallel ionization chambers were recommended for the dosimetry of electron beams for energies less than 10 MeV, necessary physical data to be used with these chambers were not available at the time the NACP 1980 protocol was written. So, it was suggested that a supplement, describing the dosimetric procedures for low-energy (< 10 MeV) electron beams, be published later. In order to acquire all the data to be used with plane-parallel ionization chambers, a comprehensive investigation was initiated by the Swedish National Institute of Radiation Protection. Their work resulted in the publication of a supplement in 1981 (NACP 1981) which describes procedures for absorbed dose determination for mean electron energies at the phantom surface, $\bar{E}_o$, below 15 MeV.

A special chamber, called NACP plane-parallel ionization chamber, was designed for use with this supplement. Absorbed dose-to-air chamber factor, $N_{D,pp}$, for this chamber was defined in a manner similar to that for a cylindrical chamber. Following Eq. 17.10, one can write

$$N_{D,pp} = \frac{\overline{D}_{air,c}}{M_{pp}} = N_{k,pp}(1-g)k_{pp},   (17.18)$$

where the factor k_{pp} has been determined for the NACP plane-parallel chamber by MATTSSON et al. (1981). k_{pp} in Eq. 17.18 is analogous to $k_{att} k_m$ in Eq. 17.10 for a cylindrical chamber plus buildup cap. The main difference is that k_{pp} includes a component to allow for backscatter from the phantom material behind the chamber.

The experimental determination of k_{pp} consists in determining $N_{D,pp}$ in a high-energy electron beam by using a cylindrical chamber for which $N_{D,cyl}$ is known. Both the cylindrical chamber and the plane-parallel chamber are placed in a PMMA phantom such that their effective points of measurement are located at the same reference depth in the PMMA. They are then exposed to high-energy ($\bar{E}_o \geq 18$ MeV) electrons such that they measure the same absorbed dose. $N_{D,pp}$ is then obtained from

$$N_{D,pp} = \frac{M_{cyl} N_{D,cyl} p_{u,cyl}}{M_{pp} p_{u,pp}},   (17.19)$$

where M_{cyl} and M_{pp} are the meter readings for the cylindrical and plane-parallel chambers, respectively, corrected for temperature, pressure, recombination losses, etc.; and $p_{u,cyl}$ is the perturbation correction factor for the cylindrical chamber, which is known with sufficient accuracy for high-energy electron beams. It was determined that $p_{u,pp}$ has a value equal to unity for the NACP-type chamber and for the chambers that meet the specifications given in the supplement. After determining $N_{D,pp}$ from Eq. 17.19, k_{pp} can be obtained from Eq. 17.18 from a knowledge of the air-kerma calibration factor. For the NACP chamber k_{pp} was found to have a value of 0.996.

Absorbed dose-to-water at the reference point is obtained from

$$D_{w,u} = N_{D,pp} M_{u,w} (s_{w,air})_u p_{u,pp},   (17.20)$$

where $M_{u,w}$ is the meter reading when the effective point of chamber measurement is placed at the reference depth in water at the user radiation quality. $M_{u,w}$ must be corrected for temperature, pressure, recombination loss, etc. If measurements are made in plastic phantoms then conversion factors must be applied in order to transfer dose-in-plastic to that in water. The protocol provides a table of such conversion factors, which are essentially a ratio of the signal measured at the ionization maximum on the central axis in water to that measured for the same accelerator monitor setting at the ionization maximum in other phantom materials. The correction factor is denoted by h_m and has been called a fluence correction factor by the AAPM (1983) protocol. The meter reading in plastic phantom is then related to the meter reading in water phantom through the equation

$$M_{u,w} = M_{u,m} \, h_m, \qquad (17.21)$$

where $M_{u,m}$ is the meter reading when the chamber is placed in the plastic phantom. For these measurements the depth in the plastic phantom $R_{100,m}$ is scaled to the corresponding depth of the reference point in water $R_{100,w}$ according to Eq. (17.17), i.e.,

$$R_{100,w} = R_{100,m} \, b_m. \qquad (17.22)$$

The stopping power ratios $(s_{w,air})_u$ used in the supplement are different from those recommended in the NACP 1980 protocol. These stopping power ratios are based on BERGER and SELTZER's (1982) electron stopping power data. For electron beams, differences in $(s_{w,air})_u$ between the values given in the supplement and those in the NACP 1980 protocol are never more than 1.5% for all E_o and phantom depths. Values of k_m are also affected by the new stopping power data. However, the effect is negligible for graphite and the value changes by only 0.6% for A-150. For photon beams, both $(s_{w,air})_u$ and $p_{u,wall}$ are affected. However, their product $(s_{w,air})_u \, p_{w,wall}$ (and hence the absorbed dose) changes by only 0.5% at most.

Two methods were recommended for the calibration of a plane-parallel chamber. The first method consists in comparison of the response of the plane-parallel chamber and that of a cylindrical chamber with known N_D in a high-energy ($\bar{E}_o > 18$ MeV) electron beam, and the second method consists in calibration in a ^{60}Co gamma-ray beam. The first method has been described earlier (i.e., Eq. 17.19). For the second method, one needs to know the air-kerma calibration factor $N_{k,pp}$ of the plane-parallel ionization chamber. This can be obtained from the National Standards Laboratory. Once $N_{k,pp}$ is known, then $N_{D,pp}$ can be obtained from Eq. 17.18.

17.2.3 HPA 1983 Protocol

Following the publication of the 1969 protocol, numerous studies indicated various inconsistencies among the absorbed dose conversion factors C_λ and C_E. NAHUM and GREENING (1976) reported that for an x-ray beam and an electron beam having the same average electron energy the ICRU (1969, 1972) recommended figures for absorbed dose conversion

factors C_λ and C_E differed from each other by 4%; this is in contrast to the expectation that the dose conversion factors will have the same value. WILLIAMS (1977) and HOLT and KESSARIS (1977) indicated that the discrepancy between C_λ and C_E could be due to the definitions. The work of ALMOND and SVENSSON (1977), ALMOND et al. (1978), McEWAN (1980), SHIRAGAI (1978), and the NCRP (1981) shows that values of C_λ depend on the size, shape, and construction of an ionization chamber. These findings are different from the assumptions made in the 1969 protocol, where values of C_λ were calculated by assuming that an ionization chamber could be regarded as an infinitesimal Bragg-Gray cavity and that ionization observed in the chamber cavity is caused by electrons originating in the surrounding medium. The effect of the chamber wall and central electrode on cavity ionization was neglected in this approximation. The papers referred to above have improved this approximation by estimating the fraction of ionization in the cavity due to electrons originating in the chamber wall and the surrounding medium. Measurements were made at 4, 6, 8, and 16 MV x-ray energies to determine the relative proportion of ionization in the cavity due to electrons originating in the chamber wall and in the surrounding medium. These fractions for other beam qualities have been calculated from these measured data assuming an exponential buildup for the secondary electrons. New sets of data were recommended in the literature for various physical parameters and interaction coefficients. These include the revised unrestricted electron stopping power data of BERGER and SELTZER (1982), mass-energy absorption coefficients for photons from HUBBELL (1977), a value of 33.85 J/C for W/e (ICRU 1979), which is different from the value of 33.7 J/C used previously in the 1969 protocol, and a value of 0.980 to correct for the displacement phantom material by the chamber assembly (a value of 0.985 was used in the 1969 protocol).

Because of all these developments in the dosimetry of photon beams, it was felt that new recommendations be given and changes be made to some of the earlier recommendations. So, in 1983 the HPA published a new protocol for the dosimetry of 2- to 35-MV x-ray and ^{137}Cs and ^{60}Co gamma-ray beams. The concept of a single conversion factor for the determination of absorbed dose in a medium, i.e., the C_λ approach developed in the HPA 1964 and 1969 protocols, was also followed in this new protocol. A new formulation of C_λ, proposed by SHIRAGAI (1978), was used in the 1983 protocol for the calculation of the absorbed dose conversion factors. This change of basis in the C_λ equation along with the use of improved values of the various interaction coefficients has resulted in the new values of C_λ being consistently higher than the old ones (HPA 1969) by as much as 3% at 35-MV x-radiation and an average of 1.5% over the entire energy range of 2–35 MV.

In contrast to the HPA 1969 protocol, which recommended the use of any commercial ionization chamber having a suitable composition and an internal diameter of less than 1 cm, the 1983 protocol recommended the use of only one type of reference chamber. This concept of the use of only one type of reference chamber in all radiotherapy departments is attractive because uncertainties in factors related to wall and buildup composition and conducting graphite films of unknown and variable thickness are avoided. The recommended chamber, NE 2561, was designed by the National Physical Laboratory (KEMP 1972) and the new C_λ values were calculated specifically for this chamber. For routine use, the protocol recommends that field chambers be calibrated against this reference chamber in the x-ray beams in which it is to be used.

It is recommended that the ionization chamber be calibrated in air in terms of air kerma or exposure by a National Standards Laboratory using either 2-MV x-rays or ^{60}Co gamma rays. Two formulas are given for the calculation of absorbed dose; one corresponds to an air kerma calibration of the ionization chamber in grays and the other corresponds to an exposure calibration in roentgens (R), both calibrations being performed by using 2-MV x-radiation. These formulas are

$$D = 1.139\, R N_k\, C_\lambda \qquad (17.23)$$

and

$$D = 0.01\, R N_x\, C_\lambda, \qquad (17.24)$$

where D is the absorbed dose-to-water in grays at the position of the chamber center when the chamber is replaced by water and an identical irradiation given. R is the meter reading corrected for temperature, pressure, etc. N_k is the air-kerma calibration factor of the chamber that converts the meter reading to air kerma in grays for 2-MV x-radiation (or ^{60}Co gamma radiation) for standard conditions. N_x is the exposure calibration factor that converts the meter reading in air to exposure in roentgens for 2-MV x-radiation for standard ambient conditions, and C_λ is the absorbed dose conversion factor. The values of C_λ given in the protocol are for an NE 2561 chamber

and are based on an in-air calibration of the chamber using 2-MV x-radiation.

17.2.4 AAPM 1983 Protocol

In 1983 the AAPM published a protocol (AAPM 1983) for the determination of absorbed dose-to-water from high-energy photon and electron beams. A Task Group (TG-21) of the AAPM Radiation Therapy Committee was formed to investigate the inconsistencies that existed among all existing protocols and to make revisions and recommendations as required. The AAPM 1983 protocol is the outcome of this effort.

In all protocols published before 1980 (HPA 1969; ICRU 1972; NACP 1972; SCRAD 1966, 1971) the formalism for the determination of absorbed dose-to-water is based on the 2-MV x-ray or ^{60}Co gamma-ray exposure calibration factor of an ionization chamber. Absorbed dose-to-water is obtained by the application of exposure calibration factor and various dose conversion factors C_λ and C_E to the chamber reading. In the C_λ approach to dose determination it is assumed that the chamber is water equivalent and that the same replacement correction can be applied to all chambers irrespective of the x-ray energy or the depth of measurement. Similarly, in the C_E approach, it is assumed that the chamber has an air-equivalent composition and that an electron perturbation correction is the only depth-dependent correction that is necessary for the determination of absorbed dose. In the course of evaluating these concepts the Task Group found the following inconsistencies:

1. The concept of applying an in-air exposure calibration factor to in-phantom measurements of exposure is questionable because this requires that the in-air calibration factor of an ionization chamber (plus buildup cap) remains constant regardless of the field size and depth of measurements in the phantom. When measurements are made in a phantom a significant portion of the signal is due to energy-degraded scattered photons that are not present in the calibration beam when in-air measurements are made.

2. As discussed earlier, the work of NAHUM and GREENING (1976, 1978) showed that for an x-ray beam and an electron beam having comparable electron spectra in a phantom, the dose conversion factors C_λ and C_E differed significantly from each other. This was contrary to the expectation that they would yield the same value.

3. The assumption that the same replacement correction can be applied to all chambers regardless of x-ray energy and depth of measurements is incorrect. Dependence of this factor on chamber dimension, energy, and depth of measurement needs to be considered.

4. Differences in the composition between the chamber wall and dosimetry phantoms need to be addressed if accuracy of dosimetry is to be improved.

All these inconsistencies were addressed in the AAPM 1983 protocol and new data and methods were recommended for the determination of absorbed dose-to-water. The approach followed in this protocol bears resemblance to that adopted in the NACP 1980 and 1981 protocols but is significantly different from the SCRAD 1971, ICRU 1972, or HPA 1969 and 1983 protocols. The main features of the protocol are as follows: (a) following the formalism developed by LOEVINGER (1981), it introduced the concept of a cavity gas calibration factor N_{gas} to characterize the response of an ionization chamber, (b) it introduced specific values for chamber specific factors, (c) it recommended the use of any water, polystyrene, and acrylic plastic phantom material as a primary dosimetry phantom for the calibration of high-energy photon and electron beams, and (d) it introduced a formalism in which all chamber-dependent and radiation-dependent parameters remain explicit in the dose calculation algorithm.

The cavity gas calibration factor N_{gas} is defined as dose to the gas (air) in the chamber per unit electrometer reading. Its value depends on various chamber-dependent parameters but it is unique to each chamber. It is related to the calibration beam through the ^{60}Co exposure calibration factor N_x. It is a constant for all radiation qualities for which the average energy required to produce an ion pair (W) is the same as that for ^{60}Co gamma rays. Once calculated for a chamber, its value remains the same unless the ^{60}Co exposure calibration factor changes.

According to this protocol absorbed dose measurements can be performed in water, PMMA, or polystyrene phantoms. However, the dose per monitor unit or per unit time should always be expressed in terms of dose-to-water. Absorbed dose measurements in a plastic phantom thus require an additional step of transferring the dose-to-plastic to dose-to-water. For x-ray dosimetry, if the photon fluence at a measurement point in a plastic phantom is the same as at a comparable point in a water phantom, then the transfer of dose-to-plastic to dose-to-water is accomplished by application of the ratio of average mass energy absorption coefficients of water to those of plastic. Equal photon fluences at the calibration depth in a water phantom and the corresponding depth in a plastic phantom can be obtained if the irradiation geometry in the plastic phantom is altered for x-ray dosimetry. The alteration of the irradiation geometry involves a modification of the SSD (while keeping the source–detector distance the same) so that equal attenuation of the incident x-ray beam would be obtained at the measurement points in the water and plastic phantoms. The protocol gives scaling factors, which when applied to the plastic phantoms yield the appropriate thickness of plastic that will result in equal x-ray attenuation. These scaling factors are simply ratios of linear plastic thickness to linear water thickness that result in equal x-ray attenuation.

For electron beam dosimetry, if measurements are made in a plastic phantom then the transfer of dose-to-plastic to dose-to-water is accomplished by the application of the ratio of the average unrestricted collision mass stopping power for water to that for plastic, and a correction factor that accounts for the differences in electron fluence at the measurement point (d_{max}) in plastic to that (d_{max}) in water.

Both cylindrical and plane-parallel chambers are recommended for the determination of absorbed dose-to-water in photon and electron beams. A detailed discussion of the specification and performance requirements of cylindrical and plane-parallel chambers has been given in the protocol. Plane-parallel chambers are recommended for measurements in electron beams in the energy range of 5–10 MeV. A judicious choice of chamber and phantom material for a given beam and energy is recommended because this can minimize the uncertainty in the determination of absorbed dose to the phantom material. For example, if a cylindrical chamber is used in a polystyrene phantom to determine the absorbed dose-to-water in a low-energy electron beam, then the uncertainty in the absorbed dose determination will be much higher than the situation where a plane-parallel chamber is used in a water phantom. This is because the replacement correction as well as the electron fluence correction that is required for the transfer of dose-to-plastic to dose-to-water for cylindrical chambers increases as the electron energy decreases. On the other hand, plane-parallel chambers in a water phantom do not require any of these corrections.

When an ionization chamber is used to measure absorbed dose in a phantom, the product of the ionization chamber reading corrected for temperature and pressure, and N_{gas} gives the dose to the gas in the chamber due to the radiation fluence at the measurement point. If the ionization chamber approximates a gas-filled Bragg-Gray cavity, then the dose to the phantom material can be obtained by the application of the Bragg-Gray formula. In the present protocol, an ionization chamber is assumed to behave like a Bragg-Gray cavity. Therefore, dose to the phantom material that replaces the chamber when the chamber is removed from the phantom is given by a product of the dose to the gas, the ratio of stopping powers of the phantom material to air, and a few correction factors that account for ionization recombination loss and replacement of the phantom material by the chamber wall and cavity.

If measurements are made in a plastic phantom then according to this protocol the absorbed dose-to-plastic $D^{plastic}$ in the user beam quality at a reference depth d_0 is given by

$$D^{plastic}(d_0) = MN_{gas}(\overline{L}/\rho)_{gas}^{plastic} P_{ion} P_{repl} P_{wall},$$
$$(17.25)$$

where M is the electrometer reading corrected for temperature and pressure from an ionization chamber with its center at the reference depth d_0; $(\overline{L}/\rho)_{gas}^{plastic}$ is the ratio of mean restricted collision mass stopping power of any plastic medium to that of the chamber gas averaged over the electron spectrum; P_{ion} accounts for the ionization recombination loss; P_{wall} is a factor that accounts for the differences in composition between the chamber wall and the medium; and P_{repl} is a replacement correction accounting for the change in photon and electron fluence that occurs because of the replacement of phantom material by the air cavity. P_{repl} has two parts: a gradient correction term, P_{gr}, and an electron fluence correction term, P_{fl}. The effect of P_{gr} is to move the point of measurement from the chamber center to an effective point in front of the chamber, thus correcting for the gradient of the electron fluence within the chamber cavity. The fluence correction term is required whenever the ionization chamber is at a location where charge particle equilibrium has not been established, i.e., in the dose buildup region in the high-energy x-ray beam or anywhere in an electron beam. For dose determinations made at or beyond d_{max} in a photon beam, fluence corrections P_{fl} are not required because the so-called transient equilibrium can be assumed to exist at these locations. Thus $P_{repl} = P_{gr}$ for photon beams. For electron beams, the recommendation is

to calibrate the beam at d_{max} where by definition $P_{gr} = 1$. Thus, for electron beams, $P_{repl} = P_{fl}$.

N_{gas} is obtained from the following expression:

$$N_{gas}$$
$$= N_x \frac{k(W/e)A_{ion}A_{wall}\beta_{wall}}{\alpha(\overline{L}/\rho)_{gas}^{wall}(\overline{\mu}_{en}/\rho)_{wall}^{air} + (1-\alpha)(\overline{L}/\rho)_{gas}^{cap}(\overline{\mu}_{en}/\rho)_{cap}^{air}},$$
$$(17.26)$$

where k is a constant equal to the charge produced in air per unit mass per unit exposure and has a value equal to 2.58×10^{-4} c kg^{-1} R^{-1}. (W/e) is the mean energy expended in air per ion pair formed and per electron charge; A_{ion} is the ion collection efficiency in the calibration quality beam; A_{wall} has the same meaning of K_{att} of the NACP 1980 protocol and accounts for the attenuation and multiple scatter of the primary photons in the chamber wall and build-up cap; β_{wall} is the ratio of absorbed dose to collision part of kerma; $(\overline{\mu}_{en}/\rho)_{wall}^{air}$ and $(\overline{\mu}_{en}/\rho)_{cap}^{air}$ are the average air/wall and air/buildup cap mass energy-absorption coefficient ratios averaged over the photon spectrum; α is the fraction of ionization due to electrons set in motion in the chamber wall; and $(1 - \alpha)$ is the fraction of ionization due to electrons coming from the buildup cap.

Absorbed dose-to-water is then calculated from

$$D^{water}(d_0) = D^{plastic}(d_0)(\overline{\mu}_{en}/\rho)_{plastic}^{water} ESC,$$
for photons $$(17.27)$$
and

$$D^{water}(d_0) = D^{plastic}(d_0)(\overline{S}/\rho)_{plastic}^{water} \phi_{plastic}^{water},$$
for electrons $$(17.28)$$

where ESC is a factor that corrects for the fractional increase in the scattered photons which occurs in polystyrene and PMMA phantoms compared to a water phantom. $(\overline{S}/\rho)_{plastic}^{water}$ is the ratio of mean unrestricted collision mass stopping power of water to that of plastic, and $\phi_{plastic}^{water}$ is the ratio of electron fluence at d_{max} in water to that in plastic.

The restricted stopping power ratios for electron beams have been calculated by using the BERGER and SELTZER 1982 electron stopping power data (data included in ICRU Report No. 37; ICRU 1984a), whereas those for photon beams appear to be based on the BERGER and SELTZER 1980 electron stopping power (data included in ICRU Report No. 35; ICRU 1984b) data. The mass energy absorption coefficient data have been taken from HUBBELL (1982). Tables of restricted stopping power ratios for electron beams are given in the protocol as a function of mean incident energy $\bar{E}_o$ on the phantom surface. The protocol recommends that $\bar{E}_o$ be determined from the relationship

$$\bar{E}_o = 2.33\, d_{50}, \qquad (17.29)$$

where d_{50} is the depth in water at which the ionization chamber reading is 50% of its maximum value. If the depth-ionization curve is generated by using a cylindrical chamber, then the point of measurement should be taken as displaced towards the radiation source by 0.75 times the internal radius of the chamber. For measurements made in plastic phantoms, the mean energy can be obtained from

$$\bar{E}_o = 2.33\, d_{50} f, \qquad (17.30)$$

where f is a scaling factor and has a value of 0.965 for polystyrene and 1.11 for PMMA phantom. The perturbation correction term P_{repl} is given as a function of mean energy E_z at a depth z. E_z can be obtained from Eq. 17.6.

For photon beams the beam quality is characterized by a new parameter called nominal accelerating potential (NAP). Ionization ratio measurements, made at a constant source–detector distance of 1 m and a field size of 10 cm × 10 cm at the detector for two different phantom thicknesses, are related to the NAP. This NAP in turn is related to the mean restricted collision mass stopping power ratios $(\bar{L}/\rho)_{air}^{med}$. $(\bar{L}/\rho)_{air}^{med}$ can then be looked up from a table that gives $(\bar{L}/\rho)_{air}^{med}$ as a function of the NAP of the beam.

Since the publication of the AAPM 1983 protocol, there has been a lot of criticism of various inconsistencies present in the protocol. ATTIX (1984a) and ROGERS et al. (1985) pointed out that the equation for N_{gas} given in the AAPM 1983 protocol is incorrect in that it incorrectly includes the factor β_{wall}. These authors show that the effect of β_{wall} has already been included in the determination of A_{wall}. ATTIX (1984b) has pointed out that inconsistencies exist in the relationship between ionization ratio and NAP. In a letter of clarification, SCHULZ et al. (1986) addressed these criticisms and modified the equation for N_{gas}. It was shown that use of the modified equation for N_{gas} and revised values for W/e results in cancellation of errors such that the revised values of N_{gas} are not different from the AAPM 1983 protocol values by more than 0.5%. Based on this letter of clarification and to maintain uniformity among the users, the Radiation Therapy Committee of the AAPM recommended that the AAPM 1983 protocol should be used in its original form as

presented in 1983. In 1988, ROGERS and ROSS published a paper in which they examined thoroughly the role of humidity and other correction factors in the AAPM 1983 protocol. They show that both the original (AAPM 1983) and the "clarified" (1986) equation for N_{gas} were incorrect and underestimate N_{gas} by 0.4% because of conceptual errors concerning humidity.

17.2.5 HPA 1985 Protocol

By the early part of 1980 it was clear that in order to improve the accuracy of the dosimetry of high-energy photon and electron beams the chamber dependence of the absorbed dose conversion factors C_λ and C_E and the most recently recommended values of various physical interaction coefficients needed to be incorporated in any formalism developed for the determination of absorbed dose-to-water. In 1983, the HPA revised its 1969 protocol for high-energy photon dosimetry and published a new protocol that incorporated all these changes. It was felt that the HPA 1971 and 1975 reports on high- and low-energy electron beam dosimetry should be updated and a revised protocol on electron beam dosimetry recommended. So, in 1985, the HPA published a new protocol for electron beams that retained some of the ideas developed in the 1971 and 1975 reports but also contained a number of modifications. Similar to the 1983 protocol for high-energy photon dosimetry, the 1985 protocol recommended that only two designated ionization chambers be used for electron beam dosimetry, i.e., a Farmer NE 2571 graphite-wall cylindrical chamber suitable for use with electron beams of 10 MeV energy or higher, and a Vinten Model 631 thin-window flat chamber (MORRIS and OWEN 1975) suitable for use with either high-energy or low-energy electrons. The choice of these chambers was dictated by the high frequency of their use in radiation dosimetry in the United Kingdom for a number of years. It was recommended that both of these chambers be calibrated in a ^{60}Co beam in a Perspex phantom against the secondary standard NE 2561 chamber. This calibration procedure is similar to that recommended in the 1975 report. However, there has been a significant change in the calculation of the calibration factor. In order to emphasize this change a new symbol, C_e, which replaced the old symbol, C_E, was introduced in the 1985 protocol. This protocol also provides a two-stage calibration procedure that enables determination of C_e for nondesignated chambers.

The goal of the protocol was to give its users a formalism and a procedure for the determination of absorbed dose-to-water in electron beams using the two desginated ionization chambers. In an effort to achieve this goal, the protocol gave a detailed description of how to select an ionization chamber for use with electron beams and, after its selection, how to use it, correct its readings, calibrate it against a secondary standard chamber in a ^{60}Co gamma-ray beam and use it to measure depth-ionization and depth-dose curves. According to the formalism developed in the protocol, determination of absorbed dose-to-water requires a knowledge of the energy of the electron beam at the point of measurement and the value of the absorbed dose conversion factor C_e. The protocol gives a detailed description of how to determine these parameters.

The protocol gave specific recommendations for the calibration of the electron chambers. The secondary standard against which the electron chambers are calibrated should be an NE 2561-type chamber. This chamber should have an air-kerma calibration factor N_s from a National Standards Laboratory. When measurements are made with this chamber in air in a beam of 2-MV x-ray or ^{60}Co gamma radiation under standard ambient conditions and a reading R_s is obtained, then the application of the calibration factor N_s converts this reading to air kerma in grays at a point in air that corresponds to the center of the chamber. If the National Standards Laboratories give the calibration factor N_s in roentgens, then an air-kerma calibration factor in grays for ^{60}Co gamma radiation can be obtained by multiplying N_s by 8.78×10^{-3} Gy R^{-1}. During the calibration of the electron chambers, both the NE 2561 chamber and the electron chambers should be placed in a Perspex phantom with their centers at the same depth of 5 cm for a field size of 10 cm $\times$ 10 cm at the phantom surface and an SSD of at least 50 cm. Under these conditions the chambers should be irradiated by a ^{60}Co gamma radiation. If R_f is the reading of the electron chamber, then $N_f = N_s k_s R_s / R_f$, where N_f is the air-kerma calibration factor for the electron chamber that converts its reading to air kerma at the point in Perspex corresponding to the center of the chamber under standard ambient conditions. k_s is a factor that converts the in-air calibration of the secondary standard chamber to an in-phantom calibration and has a value of 0.974.

If a ^{60}Co gamma-ray beam is not available for calibration, then the electron chambers can be calibrated in a high-energy x-ray beam. This is accomplished by multiplying the N_f values of the Vinten 631 chamber by some correction factors, which are energy dependent. The protocol provides a table of these correction factors as a function of energy. These factors are applicable only when the N_f values of the Vinten 631 chamber are obtained by comparison against an NE 2561 chamber in high-energy radiation in a Perspex phantom supplied by Vinten Instruments, Ltd. Such correction factors are not necessary for the NE 2571 chamber when calibrated against the NE 2561 chamber in the high-energy x-ray beam, because the two chambers are made of the same materials and are of same size and shape so that the ratio of the responses of the two chambers will not vary significantly with photon energy.

Absorbed dose-to-water, D_w, at the reference depth z in water is given by: $D_w = R_e N_f C_e$, where R_e is the chamber reading with its effective point of measurement at the depth z in water or an equivalent depth in some other medium corrected for any difference between the ambient air conditions affecting the chamber at the time of measurement and the standard ambient air conditions for which the calibration factor is applied. Corrections should also be made for ionic recombination and polarity effects. C_e is the air karma to absorbed dose-to-water conversion factor, which depends on the chamber type, energy of the electron beam at the point of measurement, and the medium in which measurements are made.

In order to derive absorbed dose-to-water, it is necessary to determine the reference depth in water and the value of the conversion factor C_e. The protocol gives C_e as a function of mean electron energy $\bar{E}_z$, which in turn is a function of $\bar{E}_o$. The mean incident energy $\bar{E}_o$ of an electron beam can be obtained from the empirical relationship $\bar{E}_o = 2.4$ HVD, where HVD is obtained from the depth-ionization curve generated with a large field size and source–chamber distance. HVD is the depth at which the ionization has a value that is 50% of its maximum value. The mean electron energy $\bar{E}_z$ at a depth z in the phantom can be obtained from Eq. 17.6. It is recommended that both $\bar{E}_o$ and $\bar{E}_z$ be determined from depth-ionization curves generated with the chamber at a fixed distance from the source. For these measurements, the effective point of chamber measurement should be used.

For a given electron beam energy, the recommended reference depth for dose calibration is at the depth of dose maximum. The protocol gives a step-by-step procedure for determining this depth and the relative depth-dose curve. The first step in this procedure is to generate a depth-ionization curve for a given energy by using a large field size and a fixed long source–chamber distance. From this curve $\bar{E}_o$ is calculated by using the formula given above, i.e., $E_o = 2.4$ HVD, and R_p is measured. The next step is to set the front surface of the phantom at the source-skin distance that is to be used subsequently for radiotherapy. For a given energy and field size, ionization chamber readings are then taken at a series of depths along the central axis in the phantom. If a cylindrical chamber is used for these measurements, then the effective point of measurement of the chamber should be used. If plastic phantoms are used for the measurements, then the water equivalent depth should be determined from a formula given below. From these measurements the value of the mean electron energy $\bar{E}_z$ at each depth is calculated by using Eq. 17.6 and the values of $\bar{E}_o$ and R_p are determined from the depth-ionization measurements. For each $\bar{E}_z$, the value of the conversion factor C_e is determined from the tables given in the protocol. The chamber readings are then converted to absorbed dose-to-water by using the equation $D_w = R_e\, N_f\, C_e$. The final step is to plot a graph of dose against depth. From this graph, the depth of the maximum dose is chosen and the whole graph is normalized to the value at this depth. It is this depth of maximum dose that should be chosen as the reference depth for dose calibration. The depth-dose curve thus generated can be converted into depth-ionization curves at a fixed source–chamber distance by applying an inverse square law correction, i.e., by multiplying each reading by $(f + d)^2/f^2$, where f is the distance of the front surface of the phantom from the source, and d is the depth of the effective point of measurement of the chamber beneath the phantom surface.

For lower energy electrons, it is recommended that the Vinten 631 flat chamber should be used in a Perspex or polystyrene phantom. If measurements are made in these plastic phantoms, then depths in plastic, d_{pl} in g cm^{-2}, should be converted to depths in water, d_w in g cm^{-2}, according to the relation $d_w = 0.97\, d_{pl}$. This is because differences in electron scattering and mass stopping power for electrons in these media will cause the depth-ionization curves to be different in the various media.

The protocol gave values of C_e for the Farmer NE 2571 graphite cylindrical chamber in water, Perspex, and polystyrene phantoms and for the Vinten 631 model chamber in Perspex and polystyrene phantoms. There are many factors that affect the values of C_e. These include the type of chamber, the electron

energy, and the phantom material. When measurements are made in different phantom materials, the readings of the ionization chamber at the depth of maximum ionization will be different in the different media. This is attributed to the effect the different media have on the electron fluence entering the cavity. Experimental determination of this factor was carried out for both chambers for the plastic phantoms recommended in the protocol and these values have been incorporated in the various C_e factors. The C_e factor for a particular chamber also depends on its response to electron beams in which measurements are made, as well as to ^{60}Co gamma radiation during its calibration. The C_e factors for the NE 2571 chamber in a water phantom were calculated by taking the ratios of restricted mass collision stopping powers for water-to-air and the electron fluence correction factors from the AAPM 1983 protocol. For plastic phantoms, the C_e factors for the same chamber were determined experimentally by making measurements in different media with the chamber at the depths of maximum ionization and then comparing these results to the C_e factors for water phantoms.

For the Vinten 631 model chamber the C_e factors are given for polystyrene and Perspex phantoms only. C_e factors for the polystyrene phantom were calculated by taking the restricted mass collision stopping power ratio for polystyrene-to-air and the unrestricted mass collision stopping power ratio for water-to-polystyrene from the AAPM 1983 protocol. Differences in electron fluence between polystyrene and water were determined experimentally and were found to range from 1.027 at $\bar{E}_z = 1$ MeV to 1.00 at 20 MeV and above. These correction factors have been included in the C_e values. In order to achieve good agreement between the calculated C_e values of the Vinten 631 model chamber and those of the NE 2571 chamber in water, all of the corrected C_e factors for the Vinten 631 chamber were normalized by a factor of 0.978 for the entire energy range covered in the new protocol. The C_e factors for the Vinten 631 chamber in the Perspex phantom were derived from the polystyrene factors by multiplying the C_e's for polystyrene by the ratio of chamber readings in polystyrene to those in Perspex.

There are two main reasons why the C_e factors given in the new protocol are significantly different from those given earlier in the two HPA guides for electron beam dosimetry (1971, 1975). The first is the change in calibration from exposure in roentgens in the earlier reports to air kerma in grays in the new protocol, while the second is the change in calibration procedures from in-air in the earlier reports to in-phantom in the new protocol. The old C_E factors converted a ^{60}Co exposure calibration factor in roentgens to a measurement of absorbed dose-to-water in rads for electrons, whereas the new C_e factor converts a ^{60}Co air-kerma calibration factor in grays to a measurement of absorbed dose-to-water in grays for electrons. This change alone increases the value of C_E by a factor of 1.139. This factor also appeared in the equation (i.e., Eq. 17.23) for the absorbed dose determination for photon beams in the HPA 1983 protocol, where an ionization chamber calibrated in air kerma was recommended for use in photon beams. The change from in-air to in-phantom calibration in the new protocol increases the value of C_E by 1.5%. In the original calculation of C_E in the guide for high-energy electron beam dosimetry (HPA 1971), the electron chamber was assumed to be calibrated in air. C_E thus included a correction factor for attenuation and scattering of photons in the chamber wall for ^{60}Co radiation during exposure calibration. This correction factor was assumed to have a value of 0.985 (SVENSSON and PETERSSON 1967). The same correction factor was also applied to the thin-window flat chamber in the HPA guide to low-energy electron beam dosimetry (HPA 1975) because its calibration factor was derived from the exposure calibration of the secondary standard chamber. Since in the new protocol the C_e factors relate the absorbed dose at a point in a phantom to air kerma at the same point in the phantom, the factor 0.985 no longer needs to be incorporated in the factor C_e. This, along with the change from exposure to air kerma, increases the overall value of C_E by a factor of 1.156.

17.2.6 NCS 1986 Protocol

In 1986 the Netherlands Commission on Radiation Dosimetry (NCS) published a new protocol for the dosimetry of high-energy photon beams. In this protocol the formalism for the determination of absorbed dose-to-water is based on the air-kerma calibration factor of an ionization chamber in a beam of ^{60}Co gamma rays or 2-MV x-rays. This is similar to that developed in the NACP 1980, HPA 1983, and HPA 1985 protocols. Although the formalisms in all these protocols are the same, the NCS 1986 protocol, in contrast to the other protocols, uses a consistent set of data for all the physical quantities that enter into the absorbed dose

equation. The numerical values of these physical quantities are based on the most recent data recommended by the ICRU and the Consultant Committee for the Measurements of ionization Radiation [CCEMRI(I)]. For example, the ICRU and CCEMRI(I) recommend the use of the BERGER and SELTZER (1982) basic electron stopping power data for dose calculation. The water-to-air stopping power ratios, the $W/e'(= 33.97$ J/C) value for dry air and the correction factors k_m and P_{wall} used in this protocol are all based on the BERGER and SELTZER (1982) electron stopping power data.

Following the approach introduced by the HPA (1964) and adopted by other protocols (ICRU 1969; SCRAD 1971; NACP 1972; HPA 1983, 1985), the NCS 1986 protocol also recommended the use of a single conversion factor $C_{w,u}$ that will convert the air kerma to absorbed dose-to-water. In this conversion factor all necessary corrections and conversions are combined. $C_{w,u}$ is a function of radiation quality and the protocol provided numerical values for $C_{w,u}$ for some commonly employed ionization chambers as a function of radiation quality.

Absorbed dose-to-water in the user's beam in this protocol is determined from the equation

$$D_{w,u} = MN_k C_{w,u}, \qquad (17.31)$$

where $D_{w,u}$ is the absorbed dose-to-water at the position of the chamber center when the chamber is replaced by water. M is the electrometer reading corrected for temperature p_t, pressure p_p, relative humidity p_{hum}, ion recombination loss p_{ion} and polarity effect p_{pol} ($M = M_{uncorr} p_t p_p p_{hum} p_{ion} p_{pol}$). N_k is the air-kerma calibration factor that converts the ionization chamber reading to air kerma for the calibration beam and geometry for standard air conditions and $C_{w,u}$ is the air kerma to absorbed dose-to-water conversion factor. $C_{w,u}$ is calculated from the equation

$$C_{w,u} = (1 - g)\, \pi k_i\, s_{w,air}\, \pi p_i, \qquad (17.32)$$

where $\pi k_i = k_{att} k_m k_{st} k_{ce}$, $\pi p_i = p_{wall} p_d p_{ce}$, $k_m = [\alpha\, s_{wall,air} (\bar{\mu}_{en}/\rho)_{air,wall} + (1 - \alpha)\, s_{cap,air} (\bar{\mu}_{en}/\rho)_{air,cap}]^{-1}$, and $p_{wall} = \alpha\, s_{wall,w} (\bar{\mu}_{en}/\rho)_{w,wall} + 1 - \alpha$.

k_{st} is a correction factor for the stem effect and k_{ce} is a correction factor for the lack of air equivalence of the central electrode material of the chamber, both in the calibration quality beam. The displacement correction factor p_d corrects for the displacement of the effective center of the ionization chamber in the user's beam. p_d is a function of both beam energy and chamber diameter. p_{ce} has the same meaning as k_{ce} but is applicable to the user's beam. The meanings of the other symbols have been described in Secs. 17.2.2 and 17.2.4.

The product $N_x(W/e)\pi k_i\beta_{wall}$, i.e., $N_x(W/e)k_{att}k_m$ $k_{ion}\beta_{wall}$ has been called the cavity gas calibration factor in the AAPM 1983 protocol and the product $N_k(1-g)k_{att}k_m$ has been called the absorbed dose-to-air chamber factor in the NACP 1980 protocol.

The NCS (1986) protocol recommended the use of NE 2505/3A, NE 2561, and NE 2571 graphite-walled ionization chambers as reference chambers. Chambers with A-150, nylon, or PMMA walls coated with graphite were not recommended for absorbed dose measurements at the reference depth because of the uncertainties involved in the determination of k_m and p_{wall}. However, it was proposed that such chambers, after suitable calibration, could be used as field instruments and the protocol provides numerical values of $C_{w,u}$ and other physical parameters for some commonly employed ionization chambers.

Water is the recommended phantom material for absorbed dose measurements. A depth of 5 cm is recommended as the reference depth of measurement for beams with a quality index up to and including 0.75, and a depth of 10 cm for beams with a quality index larger than 0.75. The quality index is obtained from ionization measurements made at 20-cm and 10-cm depths at a fixed source–detector distance for a 10 cm × 10 cm field at the geometrical center of the chamber and taking the ratio I_{20}/I_{10}. When measurements are made in a water phantom, the chamber should be enclosed in a waterproofing sheath made of PMMA. If the thickness of this sheath is more than 1 mm, then the ionization chamber reading has to be corrected. The protocol provides numerical values of this correction factor as a function of quality index of the beam and sheath thickness. These correction factors have been calculated using a formalism that is similar to the calculation of p_{wall}. In these calculations, the fraction of ionization due to electrons from the chamber wall as well as the PMMA sheath is taken into account. If solid phantoms are used for measurements, then the relationship between the measurements made in the solid phantom and the water phantom should be established experimentally. This can be done by comparing the readings of the ionization chamber in the two phantoms. Alternatively, the recommendation given in the AAPM 1983 protocol can be followed for the determination of absorbed dose-to-water.

A field chamber should be calibrated in terms of absorbed dose-to-water. This is accomplished by comparing the responses of the field chamber and the reference chamber in water. These measurements should be made with the geometric centers of the chambers at the reference depth. The absorbed dose calibration factor $N_{w,u}$ for the field chamber at the user's radiation quality is given by

$$N_{w,u} = \frac{D_{w,u}}{M_f} = \frac{(MN_kC_{w,u})_{\text{reference chamber}}}{M_f}, \qquad (17.33)$$

where M_f is the reading of the field instrument, corrected for temperature, pressure, relative humidity, ionization recombination loss, and polarity effects. It should be noted that for the reference chamber, the correction p_d due to the displacement of the effective point of measurement is already included in the $C_{w,u}$ factors. If the field chamber is a cylindrical chamber, then the effective measurement point of this chamber should be taken as displaced towards the radiation source by 0.75 times the internal radius of the chamber.

Although the response of an ionization chamber is dependent on the material of which the central electrode is made, the product of factors $p_{ce}\cdot k_{ce}$ is taken to be unity in this protocol because of the lack of data at beam energies other than ^{60}Co. Values of the water-to-air restricted stopping power ratios $s_{w,air}$ have been taken from the Monte Carlo calculations of ANDREO and BRAHME (1986a), who used the collision stopping power data for electrons at the reference depth from ICRU Report No. 37 (ICRU 1984). Values of $s_{wall,w}$ have been derived from the Monte Carlo calculated $s_{med,air}$ data of ANDREO and BRAHME (1986); data for $s_{wall,air}$ have been taken from ANDREO et al. (1986); values of $(\bar{\mu}_{en}/\rho)_{w,wall}$ have been taken from the AAPM protocol (AAPM 1983) and those of p_d from JOHANSSON et al. (1978). The p_{wall} values for ionization chambers having graphite-coated nylon/PMMA walls have been obtained by assuming that the wall is made of either pure graphite or pure nylon/PMMA, and taking an average of the p_{wall} values of both of these. The same procedure has been followed for the calculation of k_m for chambers that have graphite-coated nylon/PMMA walls.

MIJNHEER et al. (1987) have recently performed an experimental verification of the internal consistency of the air kerma to absorbed dose conversion factor $C_{w,u}$. Following the recommendations of the NCS (1986) protocol, they determined the absorbed dose-to-water with four different types of ionization chamber as a function of quality index of the beam and then compared the results with each other. The chambers used for measurements were the NE 2505/

3A, NE 2561, NE 2581, and NE 2503/3B. The two graphite-walled chambers NE 2505/3A and NE 2561 showed excellent consistency, while data for the other two chambers showed variations up to 0.8%. MIJNHEER et al. also determined the $C_{w,u}$ values for the NE 2561 chamber as a function of the quality index of the beam by following the calorimetric method and the recommendations given in the NCS 1986 protocol. The two methods gave results that are within the experimental uncertainty of each other.

17.2.7 IAEA 1987 Protocol

In 1987 the International Atomic Energy Agency (IAEA) published an international protocol for the determination of absorbed dose in photon and electron beams. Although a large number of codes of practice, documents, and protocols for the determination of absorbed dose from radiation beams exist in the literature, in general they serve the particular requirements and provisions of the country from which they originated. There was a need for a unified, simple, and easily executable protocol that could be used over the whole world, where a variety of dosimeters are used for measurement purposes. In response to this need, the IAEA formed an advisory committee and asked this committee to write a protocol that can be used by the radiation physicists of the member states of the IAEA and the World Health Organization (WHO). The new international protocol published by the IAEA is a result of this initiative.

During the later part of 1980 an important effort was made by the medical physics communities around the world to bring uniformity and consistency along the dosimetric chain. Consistency in characterization of the radiation beam can be achieved universally if both the National Standards Laboratories and the hospital physicists use the same protocol and data for various physical quantities. This was clearly not the case before 1986, when the National Standards Laboratories based their calibrations on the 1964 Berger and Seltzer stopping power data for electrons whereas various protocols recommended different sets of values for the basic electron stopping power data. The origin of the differences among these data sets lies in the different approximations applied in the calculation of density effect correction and mean excitation potential of elements and compounds. For example, the density effect correction in the Berger and Seltzer 1980

stopping power data (data included in Report No. 35, ICRU 1984b) is calculated according to the STERNHEIMER and PEIERLS (1971) approximation in which the density effect corrections are based on a general formula that gives a global fit to the density effect data. The density effect in the BERGER and SELTZER (1982) electron stopping power data has been calculated according to data published later by STERNHEIMER et al. (1984) in which individual density effect calculations for each material at each energy are carried out. Thus, restricted stopping power ratios for medium-to-air calculated by using the different databases for basic electron stopping powers will be different from each other. For example, the value of the ^{60}Co graphite-to-air stopping power ratio calculated by using the 1980 Berger and Seltzer electron stopping power data is about 1% higher than that obtained by using the BERGER and SELTZER (1982) electron stopping power data.

Another source of inconsistency among the various dosimetry protocols is that in a given protocol a database mixture of basic electron stopping power data and stopping power ratios has been used for the calculation of various dosimetric quantities pertaining to photon dosimetry and/or electron dosimetry. For example, in the AAPM 1983 protocol, the 1980 Berger and Seltzer electron stopping power data and Spencer-Attix stopping power ratios were used for photon dosimetry. In contrast, the BERGER and SELTZER (1982) electron stopping power data and Spencer-Attix stopping power ratios were used for electron dosimetry. And finally, the Bragg-Gray stopping power ratios and the 1980 Berger and Seltzer electron stopping power data were used for the calculation of chamber factors. Similarly, in the HPA protocol, BERGER and SELTZER (1982) stopping power data for electrons were used for both photon and electron dosimetry and the calculation of chamber factors; however, Spencer-Attix type stopping power ratios were used in electron dosimetry and Bragg-Gray stopping power ratios were used for the calculation of chamber factors and photon dosimetry. A similar mixture of basic electron stopping power data and Bragg-Gray and Spencer-Attix type stopping power ratios can be found in the NACP and the Spanish protocol SEFM (1984).

A third source of inconsistency is the inconsistent assignment of beam quality to stopping power ratios and mass energy absorption coefficients. Most protocols specify photon beam quality by dose or ionization ratios measured at two different depths in a phantom but the way these dose or ionization ratios are assigned to the stopping power ratios or mass energy absorption coefficient ratios is not consistent. For example, the NACP (1980) protocol specifies the beam quality by experimental ionization ratios obtained from different clinical accelerators but uses Bragg-Gray stopping power ratios that were calculated for thin target bremsstrahlung spectra (ICRU 1969). The AAPM protocol (1983) established a relationship between dose ratios and Spencer-Attix stopping power ratios but these parameters were derived independently. Similar inconsistency can also be found in the Spanish protocol (1984).

In 1985 the Consultant Committee for the Measurements of Ionizing Radiations [CCEMRI(I)] recommended that with effect from 1 January 1986, the National Standards Laboratories should adopt new values of W/e ($= 33.97$ J/C), stopping power data of electrons from ICRU Report No. 37 (ICRU 1984a), a g value of 3.2×10^{-3} for ^{60}Co in air, and mass energy absorption coefficient data from HUBBELL (1982). It was around this time that the advisory committee of the IAEA was formed to write a new protocol. So, historically, the IAEA was in a unique position to draw valuable information from all the national protocols, documents, and codes of practice and incorporate a uniform set of data for all the interaction coefficients in a new protocol. The values of various interaction parameters incorporated in the IAEA protocol are consistent with the recommendations of CCEMRI(I).

The protocol recommends water as the phantom material for both photon and electron beams. However, for electrons with mean incident energy $\bar{E}_o < 10$ MeV plastic phantoms are also recommended. The formalism developed in the protocol for the determination of absorbed dose-to-water bears a strong resemblance to that developed in the NACP 1980 and AAPM 1983 protocols. Similar to the NACP 1980 protocol, the IAEA recommends that a chamber be calibrated in terms of air kerma in a ^{60}Co beam and that the absorbed dose-to-air chamber factor N_D be determined from Eq. 17.10. However, in contrast to the NACP 1980 protocol, which recommends the use of a specific type of chamber having wall and buildup caps made homogeneously out of graphite, tissue-equivalent (A-150), or air-equivalent materials (for which the protocol provides values of the correction factors k_{att}, k_m, and p_u), the IAEA protocol recommends the use of a variety of chambers. This is similar to the recommendations given in the AAPM 1983 protocol. k_m for a chamber with different wall and buildup cap material can be obtained from

$$k_m = [\alpha\, s_{air,wall}\, (\mu_{en}/\rho)_{wall,\,air}$$
$$+ (1 - \alpha)\, s_{air,cap}\, (\mu_{en}/\rho)_{cap,air}]^{-1}. \qquad (17.34)$$

The meanings of these parameters have been described earlier. All these parameters have to be evaluated at the calibration quality (i.e., ^{60}Co) beam. The stopping power ratios given in the IAEA protocol were taken from the work of ANDREO and BRAHME (1986), who calculated these ratios by using the BERGER and SELTZER (1982) electron stopping power data and averaging over the total electron energy spectrum at the point of measurement in accordance with the Spencer-Attix theory.

Absorbed dose-to-water in the user's beam at the reference depth (d_o) is given by

$$D_w(d_o) = M_u N_D (s_{w,air})_u\, p_u p_{cel} h_m, \qquad (17.35)$$

where p_{cel} accounts for the non-air-equivalence of the central electrode material, which neither the NACP 1980 nor the AAPM 1983 protocol incorporated into their formalism; h_m is defined for electron beams only and has the same meaning as $\phi_{plastic}^{water}$ of the AAPM 1983 protocol. h_m is a measured quantity and its values have been taken from the NACP 1980 protocol. The perturbation correction term p_u accounts for the lack of water equivalence of the chamber wall (P_{wall} in the AAPM terminology) and the perturbation of electron fluence (P_{repl} in the AAPM terminology) that occurs because of water replacement by air cavity. p_u is defined in connection with the use of an effective point of measurement. This effective point of measurement is located at the reference depths (d_o) recommended in the protocol. Thus, if a cylindrical chamber is used for measurements, the central axis of the chamber should be placed at a depth d_o plus a fraction 'f' times the internal radius of the chamber where $f = 0.5$ for electron beams and 0.75 for photon beams. This is different from recommendations given in the NACP 1980 and AAPM 1983 protocols. According to these protocols the central axis of the chamber should be placed at the reference depth d_o. As reference depths for photon beams lie at or beyond d_{max}, where transient charged particle equilibrium is assumed to exist, fluence corrections are not required for dose determinations made at these depths. p_u in the IAEA protocol thus has the same meaning as P_{wall} in the AAPM 1983 protocol. For electron beams, dose determinations are made at d_{max} where by definition $P_{gr} = 1$. Since P_{wall} is assumed to have a value of unity in the protocol, p_u is equivalent to P_{repl} of the AAPM protocol and both these factors account for the fluence perturbation correction P_{fl} only.

For photon beams, the beam quality is specified by TPR_{10}^{20}, which corresponds to the ratio of absorbed doses at 20-cm and 10-cm depths in water for a constant source detector distance and a field size of 10 cm × 10 cm at the plane of the chamber. Values of TPR_{10}^{20} have been taken from the work of ANDREO and BRAHME (1986), who have calculated depth-dependent electron fluence spectra and central axis absorbed dose distributions for one and the same beam. This establishes a strict correlation between the stopping power ratios and the corresponding central axis depth dose distribution which in turn could be related to TPR_{10}^{20}. The TPR_{10}^{20} and the stopping power ratios in the IAEA protocol have been obtained in this manner. For electron beams, the stopping power ratios have been taken from the AAPM 1983 protocol.

According to the NACP 1980 and AAPM 1983 protocols the mean energy $\bar{E}_z$ at a depth z in a phantom should be determined from Eq. 17.6. The IAEA protocol points out that the approximations involved in the derivation of Eq. 17.6 are recommended only for use with electron energies $\bar{E}_0$ less than 10 MeV or for small depths at higher energies. Thus, instead of recommending Eq. 17.6 for the determination of $\bar{E}_z$, the IAEA provided a table of $\bar{E}_z/\bar{E}_0$ as a function of z/Rp. These values of $\bar{E}_z$ have been taken from the Monte Carlo calculations of ANDREO and BRAHME (1981) and ANDREO (1983). Once $\bar{E}_0$ and z/Rp are known, then $\bar{E}_z$ can be calculated from the tabulated values. For the determination of $\bar{E}_0$ one needs to generate a depth-ionization curve or depth-absorbed dose curve. The protocol gives values of $\bar{E}_0$ in tabular form as a function of half value depth (HVD) measured from absorbed dose and ionization curves at an SSD of 1 m and broad beams. This table has been taken from the NACP 1980 protocol.

If plastic phantoms are used for dose determination, then the recommendation is to scale the depth in plastic to an equivalent depth in water by a scale factor that corresponds to the ratios of linear continuous slowing down ranges in plastic-to-water. The protocol provides values of these scale factors as a function of mean energy $\bar{E}_0$.

SVENSSON et al. (1987) have performed a comparison of absorbed doses in photon and electron beams using ferrous sulfate dosimetry and the recommendations of the IAEA protocol. Excellent agreement, to within parts of 1%, was observed at all the beam qualities investigated.

17.2.8 HPA 1990 Protocol

In all the codes of practice and protocols discussed so far the recommended method for the determination of absorbed dose-to-water in high-energy photon and electron beams is based upon either exposure or air-kerma calibration factor of an ionization chamber in a ^{60}Co gamma-ray or 2-MV x-ray beam. Although this is the recommended approach, it was recognized in the late 1960s (HPA 1969) that direct absorbed dose calibrations of ionization chambers in terms of absorbed dose-to-water over the whole range of radiation qualities will eventually become available and replace the current recommended method. In keeping with the goal of recommending new techniques and data which provide an agreed method for the determination of absorbed dose with a high degree of consistency and with an accuracy and uncertainty appropriate to the present state of knowledge, the HPA in 1990 recommended a new protocol (HPA 1990) for high-energy photon therapy dosimetry based on the National Physical Laboratory (NPL) absorbed dose calibration service. In addition to providing 2-MV x-ray exposure and air-kerma calibration factors for an ionization chamber, the NPL initiated a new calibration service for high-energy photon beams in terms of absorbed dose-to-water. This new service is based on a graphite calorimeter as the primary standard (BURNS et al. 1988). The purpose of the HPA 1990 protocol is to recommend procedures for transferring calibration from an NPL absorbed dose calibrated ionization chamber to a field chamber which can then be used to measure the output from a ^{60}Co machine or a high-energy accelerator. Procedures for these output measurements are also given in the protocol. It is recommended that if an ionization chamber is calibrated in terms of exposure or air kerma then the HPA 1983 protocol should be used for absorbed dose determination in photon beams. On the other hand, if it is calibrated in terms of absorbed dose-to-water, then the procedures recommended in the HPA 1990 protocol should be followed for the determination of absorbed dose-to-water.

The NE 2561 chamber continued to be the recommended secondary standard chamber. The NPL will provide an absorbed dose-to-water calibration factor N_D for this chamber as a function of quality indices of the photon beams. The range of these indices extends from 0.57, corresponding to a ^{60}Co gamma-ray beam, to 0.79, corresponding to an x-ray beam of nominal accelerating potential of 19 MV.

These calibration factors are obtained from measurements made in a water phantom at the NPL. During these measurements the ionization chamber is enclosed in a close-fitting waterproof Perspex sheath specially designed for it. The calibration factor N_D is expressed in terms of absorbed dose-to-water through the equation.

$$D = RN_D, \tag{17.36}$$

where D is the absorbed dose-to-water at the position of the center of the chamber when the chamber and the sheath are replaced by water, and R is the reading of the instrument connected to the chamber, corrected to the standard conditions of a chamber air temperature of 20°C, an ambient air pressure of 1013.25 mbar, and a relative humidity of 50%. It is also corrected for ionization recombination loss. It is the responsibility of the user to determine the calibration factors for this chamber in the user's beam. For this value of the quality index for each radiation beam in the user's machine is needed. Recommended procedures for the measurement of quality index are given in this protocol. After the determination of the quality indices of the user's beams, the calibration factors for the secondary standard chamber as a function of user's beam quality is obtained by interpolating the NPL calibration factors versus the measured quality indices.

The next step in the transfer of calibration from the secondary chamber to the field chamber is to compare the responses of these chambers in a water or Perspex phantom under standard irradiation conditions given in the protocol. It is essential that this comparison be done using the same machine and radiation qualities as will be used later with the field chamber. The waterproof sheath that was used in conjunction with the secondary standard chamber during its NPL calibration should also be used during the comparison. Similarly, the waterproof sheath of the field chamber that has been used during the comparison should be used for any subsequent measurements even if the measurements are made in a Perspex phantom. The readings of both of these chambers should be corrected for ionization recombination loss and to the standard ambient air conditions that were present during the NPL calibrations. For a given radiation quality, a calibration factor for the field chamber can then be determined from the corrected chamber readings and the calibration factor of the secondary standard chamber at that quality.

It is suggested in the protocol that the calibrated field chamber may be used for standard output measurement, calibration, or routine checks of the user's beam. For calibration and output measurements a water phantom that provides full scatter condition is recommended, whereas for routine check measurements a Perspex phantom may be used. If the user decides to use Perspex phantom for routine check measurments then it will be necessary to establish a relationship between the measurements made in water and those made in the Perspex phantom. This can be accomplished by taking the routine phantom measurements at the same time when the standard output or calibration measurements are made. The advantages, disadvantages, and uncertainties that can result from the use of a Perspex phantom have been described in detail in the protocol.

Numerous examples of the determination of absorbed dose-to-water in photon beams over the energy range extending from ^{60}Co gamma radiation to nominal 19-MV x-radiation employing both the NPL 2-MV air-kerma calibration method used with the HPA 1983 protocol and the absorbed dose-to-water calibration method used with the HPA 1990 protocol show that differences of up to 1.6% or less exist between the two methods. The overall uncertainty in the absorbed dose-to-water calibration method is estimated to be ± 1.5% at the 95% confidence level. This uncertainty is much smaller than that in the HPA 1983 protocol, which gave an uncertainty of ± 2% at 2-MV x-radiation and ± 3.5% at 35-MV x-radiation.

17.2.9 AAPM 1991 Protocol

Our current understanding of clinical electron beam dosimetry is still far from complete. Over the last decade electrons have found increasing importance in radiation therapy. Because of this, extensive studies have been devoted to the scientific understanding of all aspects of electron beam dosimetry. This has resulted in the publication of a large number of documents on electron beam dosimetry. Of these, the one written by the International Commission on Radiation Units and Measurements (ICRU 1984b) is by far the most exhaustive and contains detailed information on all aspects of electron beam dosimetry. Although this report is an excellent reference, it was felt that it is not very practical for clinical use. So the Radiation Therapy Committee of the American Association of Physicists in Medicine formed a Task Group (No. 25) charged with producing a report that would address the needs of a clinical physicist involved in all aspects of clinical electron beam therapy. The recommendations of this task group are contained in the AAPM 1991 protocol. The scope of this protocol has been limited to the procedures and measurement techniques for the acquisition of basic dosimetric data necessary for treatment planning and acceptance testing of new electron accelerators and the utilization of these data for the calculation of monitor units. The energy range of the electron beams covered in this protocol extends from 5 to 25 MeV. The protocol deals with the measurements of dose distributions in both water and nonwater phantoms using ionization chambers, film, thermoluminescent dosimeters, and silicon diode dosimeters. It gives recommendations for the determination of energy of the electron beams, measurements of percent depth dose, isodose curves, output factors, virtual point source, and effective point source. Effects of air gap or extended SSD, oblique incidence, tissue heterogeneities, field shaping, and shielding on dose distribution are discussed. Finally, the protocol provides guidelines for dose specification for electron beams.

It is recommended that the AAPM 1983 protocol be followed for the calibration of electron beams. The AAPM 1991 protocol should be used for the measurements of dose relative to this calibration dose. For electron energies less than 10 MeV, plane-parallel chambers that meet the specifications given in the AAPM 1983 protocol should be used for dose calibrations. For energies greater than 10 MeV, either plane-parallel or cylindrical chambers can be used for dose calibrations. For measurements of depth dose, both plane-parallel and cylindrical chambers can be used. If plane-parallel chambers are used, the effective point of measurement should be taken at the front surface of the collecting volume. On the other hand, if cylindrical chambers are used, the effective point of measurement should be taken as displaced towards the radiation source from the geometric center of the chamer by 0.5 times the internal radius of the chamber. This recommendation is different from that given in the AAPM 1983 protocol, which recommended that the depth of measurement be shifted towards the radiation source by 0.75 times the internal radius of the chamber.

Water is the recommended phantom material for measurement. Solid phantoms that are water equivalent can also be used for dosimetric measurements. A solid phantom is considered to be water equivalent if it has the same collision stopping power and linear angular scattering power as water. This means that the solid phantom should have the same electron density and effective atomic number as water. If the solid phantom is not water equivalent and measurements are made in a solid phantom, then the measured central axis depth dose and off-axis dose distributions should be corrected for the effects of the phantom. The central axis depth dose can be corrected by properly scaling the depth in the plastic phantom to its water-equivalent depth and taking into account the differences in electron fluence that exist between the solid phantoms and the water phantom at equal water-equivalent depths. Although the protocol provides data for the fluence correction factor for high-impact polystyrene, polystyrene, and PMMA phantoms, it proposes that application of this correction factor for the determination of central axis depth dose is optional in a clinical environment. However, the scaling of the depth in solid phantom to its water-equivalent depth must be incorporated in the determination of depth dose.

The NACP 1980 protocol recommended that the ratio of electron density in the nonwater phantom to that in the water phantom can be used as the scaling factor to scale the practical range R_p in a nonwater phantom to that in a water phantom. The AAPM 1991 protocol points out that use of this scaling factor overestimates the penetration of the depth-dose curve near the therapeutic range. This is due to the neglect of the effect of the lower effective atomic number of the solid phantom on depth dose. The AAPM 1991 protocol recommends that depth in solid phantom be scaled to its water-equivalent depth by using the formula

$$d_{water} = d_{med} \times \rho_{eff} = d_{med} \times \left(\frac{R_{50}^{water}}{R_{50}^{med}} \right), \quad (17.37)$$

The protocol gives data for ρ_{eff} for various solid phantom materials.

If measurements are made in a solid phantom, then dose in the solid medium is related to dose-to-water through the equation

$$D_{water}(d_{water}) = D_{med}(d_{med}) \left[(\overline{S}/\rho)_{coll} \right]_{med}^{water} \phi_{med}^{water}, \quad (17.38)$$

where

$$D_{med}(d_{med}) = N_{gas} Q_{corr}(d_{med}) \left[(\overline{L}/\rho)_{coll} \right]_{air}^{med} P_{repl}. \quad (17.39)$$

$Q_{corr}(d_{med})$ is the reading of the ionization chamber corrected to a temperature of $22°C$ and a pressure of one standard atmosphere, and for polarity effect and ionization recombination loss. d_{med} is the corrected depth of measurement. The meaning of the other symbols has been described in detail in Sect. 17.2.4.

As discussed earlier, the fluence correction factor ϕ_{med}^{water} accounts for the difference in electron fluence between solid phantom and water phantom at water-equivalent depths. This difference is due to the differences in effective atomic number that exist between water and the commercially available solid phantoms. The commercially available solid phantoms have high carbon content, which causes their effective atomic number to be lower than that of water. This causes the linear scattering power in solid phantom to be different from that in water. When water and a solid medium have nearly equal electron densities but different effective atomic number, then the difference in linear scattering power between the two media will cause the root-mean-square angular distribution of the electrons to be different at equivalent depths. This is why a difference in electron fluence exists between water and solid phantoms at water-equivalent depths. As these water-equivalent depths are at different distances from the virtual source, ϕ_{med}^{water} also contains an inverse-square correction.

The mean energy $\bar{E}_o$ and the most probable energy $E_{p,o}$ are determined from Eqs. 17.15 and 17.16. A value of 2.33 MeV cm^{-1} has been recommended for the constant in Eq. 17.15 by both the AAPM 1983 protocol and the NACP 1981 protocol. The constant 2.33 MeV cm^{-1} has been determined by BERGER and SELTZER (1982) from an analysis of the depth-dose curves for broad beams of monoenergetic electrons incident upon a semi-infinite water phantom. This value is an average of values in the energy range 5–50 MeV. The depth of the 50% dose level for these monoenergetic electrons is related to their mean incident energy by the factor 2.33 MeV cm^{-1}. WU et al. (1984) have recommended a value of 2.381 for this constant for determination of mean energy from depth-ionization measurements. This value will be increased slightly if the effects of beam divergence are not taken into account. SCHULZ and MELI (1984) and the NACP 1981 protocol have shown that an error of less than 1% in dose calibration will occur if beam divergence is not taken into account in the determination of mean energy of the beam. The AAPM 1991 protocol recommends that for the calculation of relative dose, $\bar{E}_o$ can be determined from the product of either the AAPM 1983 recommended value of 2.33 or the HPA 1985 recommended value of 2.4 times R_{50}; R_{50} can be

determined from either depth-ionization or depth-dose curves, corrected or uncorrected for beam divergence.

R_p in Eq. 17.16 should be determined strictly from measured depth-dose data corrected for beam divergence. However, if R_p is determined from measured depth-ionization data uncorrected for beam divergence for an SSD $\geq$ 100 cm, then the difference in R_p is not clinically significant for the calculation of $E_{p,o}$. The AAPM 1991 protocol thus recommends that R_p be determined from depth-dose or depth-ionization data, corrected or uncorrected for beam divergence. If one is interested in correcting depth-dose or depth-ionization data for beam divergence, then it will be necessary to multiply the data by the factor

$$\left(\frac{SSD_{eff} + d}{SSD_{eff}} \right)^2 \qquad (17.40)$$

before the determination of R_p, where SSD_{eff} is the effective source-to-phantom distance.

17.3 Instrumentation for Dosimetry Calibration of Radiotherapy Beams

17.3.1 Ionization Chambers

A review of the ionization chambers used for radiation dosimetry has been provided by BOAG (1987).

17.3.1.1 Design of Ionization Chambers

JAYARAMAN et al. (1985) investigated how the physical principles outlined in the 1983 AAPM protocol could be utilized for the redesign of the therapy-level ion chambers in such a way that one can reduce the number of factors that need to be looked up in tables or graphs for the calibration of high-energy teletherapy photon beams. They concluded that one such design could be an ion chamber having a wall of acrylic or Bakelite of a thickness not exceeding 0.1 g/cm^2 and having an inner diameter of 6 mm, and used in conjunction with a ^{60}Co buildup cap of thickness 0.35 g/cm^2 and made of acrylic, Bakelite, or Tufnol. If a chamber of such design is used in a water phantom, the dosimetry is practically reduced to the simplicity of the former protocols (prior to 1983), with a single value of an energy-dependent multiplier (C_λ or C_E) needing to be obtained from a table.

KLEVENHAGEN (1991) developed a chamber which had an uncalibrated sensitive volume but which behaved as a Bragg-Gray cavity in high-energy radiation. The new type of chamber developed in the course of this study has a variable volume and was constructed from water-equivalent materials. It was used in a water phantom directly in a beam of a therapy megavoltage machine under clinical conditions. The chamber allowed absorbed dose to be determined from first principles. Good agreement was found between these dosimetric methods, and it was therefore concluded that the method developed in this work can be successfully employed for absolute dosimetry.

17.3.1.2 Chamber Response and Wall Correction Factors for ^{60}Co Gamma Rays

KRISTENSEN (1983) investigated the response to ^{60}Co radiation of a cylindrical chamber having wall and air cavity dimensions similar to those of the Farmer graphite chamber type 2505. Cylindrical aluminum and graphite electrodes of various diameters ranging from 1.35 to 5 mm were mounted in the cavity. The air volumes were accurately determined and the respective ionization yields were measured relative to the kerma. The following results were obtained with graphite central electrodes: The ionization per unit air volume, J/V, decreased by 0.4% ± 0.4% as the electrode diameter was increased from 1.35 to 5 mm; with aluminum electrodes J/V increased by 3.5% ± 0.4% for the same variation in diameter. The collision kerma response, defined as the absorbed dose to the air-filled cavity per unit collision air kerma, was found to be 0.983 ± 0.011 for the chamber with a 1.35-mm-diameter graphite electrode, a result close to the expected value. In 1981, NATH and SCHULZ reported and results obtained when response and wall correction factors for a variety of ionization chambers in a ^{60}Co gamma-ray field were calculated using a Monte Carlo photon-electron transport code. Among the chamber parameters studied were chamber wall material and its thickness, central electrode material and its dimensions, and the shape and size of the sensitive volume. The calculations showed that the response and wall correction factors are sensitive to the shape and volume of the ionization chamber, but relatively independent of the choice of material for the chamber wall and electrode when these are compared on the basis of electron density. Data were presented for cylindrical, plane-parallel, and spherical ionization chambers constructed from carbon, magnesium, aluminum, water, Lucite, polystyrene, and ICRU muscle, as well as for a number of commercially available

ionization chambers. NAHUM and KRISTENSEN (1982) pointed out that the calculated responses in the aforementioned work were not correct. In particular, the response of a 0.1-cm-radius spherical chamber was about 3.5% lower than that of a 10-cm-radius chamber of the same wall thickness, and the response of a cylindrical chamber was found to vary markedly with central electrode radius. Nahum and Kristensen postulated that these results may have been due to the perturbation effect of electron scattering, although such an effect had no theoretical basis.

McEWAN and SMYTH (1984) showed that when electron paths are stopped and rescattered at each interface a nonrandomness in the electron energy deposition pattern is introduced which can result in chamber responses being consistently about 2% lower than Bragg-Gray theory predicts.

These anomalies associated with interfaces were resolved by a systematic Monte Carlo simulation study by BIELAJEW et al. (1985). This effect was demonstrated for two computer codes, EGS and CYLTRAN/ETRAN, and the underlying physics was described. Details were also given of a Monte Carlo code suitable for the calculation of ion chamber response to ^{60}Co without calculational artifacts and of various variance reduction techniques that greatly reduce the computing time required. Using this modified version of EGS, ROGERS et al. (1985) calculated the responses and wall correction factors for ion chambers in broad parallel ^{60}Co beams. The calculated responses were in good agreement with Bragg-Gray cavity theory. In particular, the response divided by the wall correction factor A_{wall} was found to be independent of the detector's shape but dependent on the material used for the chamber wall in a manner predicted by Bragg-Gray cavity theory. A simple theory was proposed which predicts the increase in response of a Farmer ion chamber due to an electrode of an arbitrary material and radius. The change in chamber response as a function of buildup cap composition was in good agreement with analytic expressions. The effect of guard regions in plane-parallel chambers was found to be negligible. Embedding a plane-parallel chamber in a flat phantom during calibration was shown to increase the response by 1.0% ± 0.2%. Calculated values of A_{wall} were in good agreement with most experimental results and with those given in the 1983 AAPM protocol, but showed a considerably lower uncertainty of ± 0.2%.

In 1983 calculations of the perturbation correction to be applied to a plane-parallel cavity ionization chamber used as a primary standard of absorbed dose in graphite for ^{60}Co gamma rays were reported by BOUTILLON (1983). The good agreement between ionometric measurements and the calorimetric measurements of national laboratories of metrology gave strong support to the reliability of his calculated correction. Later, DE ALMEIDA et al. (1989) determined experimentally this perturbation correction to be applied to a graphite cylindrical (thimble-type) ionization chamber used for the measurement of absorbed dose in graphite. Their results showed that the magnitude of this correction increases with depth and depends on the size of the cavity. They concluded that knowledge of this correction makes possible the use of such a chamber as a standard for absorbed dose in graphite for ^{60}Co gamma rays.

In 1986 a consistent set of correction factors for ionization chambers of different wall and buildup cap composition was derived (ANDREO et al. 1986). Theoretically derived parameters k_m and k_{att}, which relate the exposure calibration of an ionization chamber to the absorbed dose to the air of the cavity, were compared with the experimentally derived product $k_m k_{att}$ showing generally good agreement but also significant discrepancies for plastic-walled chambers with inner graphite coatings. A table of k_m values was given for a large number of commercial ionization chambers. The new stopping power data were also used to evaluate the wall-dependent correction factor (p_{wall}) that enters into the determination of the absorbed dose-to-water in photon beams, results being given as a function of the quality of the beam. Their theoretical calculations of p_{wall} were consistent with existing experimental data.

O'CONNOR and MALONE (1987) presented a new method of measuring the wall contribution of an ionization chamber. The special equipment required consisted of a hollow shell of approximately water-equivalent material and small lead shields. The inner dimensions of the shell were large compared with the ionization chamber. The shell was thick enough to absorb all the contamination electrons. The principle of the method was described and the results of measurements on an ionization chamber were given.

In 1990 a theory for calculating the correction due to source nonuniformity which applies to thick-walled ionization chambers irradiated by point-source photon fields with arbitrary incident energy distributions was presented (BIELAJEW 1990a). The equations were derived within the framework of a fundamental theory of ionization chamber response and were suitable for Monte Carlo calculation.

Monte Carlo calculations for estimating the correction in plane-parallel, cylindrical, and spherical geometries were described and comparisons with the experimental results of Kondo and Randolph indicate agreement to better than 0.5%, demonstrating the viability of the theory under even the most extreme measurement conditions.

The theory developed by BIELAJEW (1990b) was compared with Monte Carlo calculations for chambers with pancake, cylindrical, and spherical geometries similar in size to the instruments employed by Standards laboratories (ROGERS and BIELAJEW 1990). The agreement between Monte Carlo calculations and the analytic theory was excellent and demonstrated the viability of the analytic theory at large and small source-to-chamber distances. The perturbations, which differ from those calculated or measured by some Standards laboratories, suggested that corrections of the order of 0.3% should be applied to typical plane-parallel geometries, smaller corrections of the order of 0.05% or less for typical Farmer-type chambers, and no correction for spherical chambers. The analytic theory predicted chamber geometries which can either minimize or maximize the effect of point source nonuniformity. An experiment was described that would measure the correction with good accuracy.

BIELAJEW (1990c) pointed out that wall correction factors can differ by as much as 1.0% for spherical chambers depending on whether they are obtained experimentally by extrapolation measurements or by Monte Carlo simulation. He demonstrated that linear extrapolation of experimental data for spherical chambers is inappropriate, owing to the curvature of the chamber walls. A simple nonlinear theory was constructed that resolves the difference. The Monte Carlo calculations and the nonlinear theory were compared with extrapolation measurements for the NIST (formerly NBS) spherical chambers. It was concluded that wall correction factors should be obtained by Monte Carlo calculation for spherical chambers and that linear-extrapolation techniques should be regarded with suspicion for all chambers.

In 1990 it was shown by using the EGS4 system that Monte Carlo-calculated wall correction factors predict relative variations in detector response with wall thickness which agree with all available experimental data within a statistical uncertainty of less than 0.1% from those obtained by extrapolation of these same measurements (ROGERS and BIELAJEW 1990). Use of calculated correction factors would imply increases of 0.7%–1.0% in the exposure and air-kerma standards based on spherical and large-diameter, large-length cylindrical chambers. Calculations were also shown to agree within 0.05% for standards based on large-diameter plane-parallel chambers. Calculations were also shown to agree within 0.05% with the measurements of Rocha and co-workers for clinical chambers. The final values of A_{wall} in the 1983 AAPM protocol agreed within 0.2% with the more accurate values calculated here.

In 1990 measurements of wall correction factors for different types of ionization chamber were done by WITTKÄMPER and MIJNHEER (1990). The procedure was to compare the reading of an unknown chamber with that of a reference chamber (graphite Farmer chamber with graphite buildup cap), both in air in a ^{60}Co gamma-ray beam and in a water phantom irradiated by a 20-MeV electron beam. A comparison of their values with other experimental values showed good agreement (within 0.3%) after normalization on the reference chamber values, for most chambers. However, a 0.9% lower value was observed for a nylon-walled Farmer chamber, suggesting that differences in thickness of the inner surface of the thimble might occur. The use of such an inhomogeneous ionization chamber as a reference instrument for absorbed dose determinations at the reference point in a phantom was therefore not recommended by the authors. Generally good agreement, within 0.5%, existed between their experimental and calculated wall correction factors. For two chambers constructed from A-150 plastic, larger deviations (0.6% and 1.4%) were observed. Due to these unexplained discrepancies the use of an average of calculated and measured values of wall correction factors in dosimetry protocols was recommended.

17.3.1.3 Perturbation Factors for Electrons

Perturbation factors for cylindrical chambers were measured by JOHANSSON et al. (1978) for various photon and electron beam energies. Their data have been adopted in most of the recent dosimetry protocols, including the 1983 AAPM protocol.

In 1980 a calculation of displacement factors in cylindrical chambers irradiated by photon beams was reported by CUNNINGHAM and SONTAG, who used a first scatter calculation.

In 1981 an experimental comparison of the perturbation correction in a parallel-plate and a cylindrical ion chamber was done by JONES (1981). The current from these ion chambers was measured in electron beams of energy in the range 7–12 MeV

and in a ^{60}Co gamma-ray beam. It was concluded that there is no difference between the perturbation correction factor for these two chambers. A significant polarity effect was also observed for the parallel-plate chamber, which was a function of energy and depth in the phantom. This work supports the ICRU recommendation for displacement correction for plane-parallel and cylindrical chambers. Jones concluded that "it would be inappropriate to suppose that the perturbation correction factor for this flat chamber is unity as recommended by ICRU 21." In 1982, SCHULZ presented arguments in defense of ICRU 21.

In 1987 an experimental measurement of the replacement correction factors P_{repl} to be applied to electron measurements with parallel-plate chambers was reported (CASSON and KILEY 1987). By comparison with a cylindrical chamber whose P_{repl} values at d_{max} were calculated from Task Group 21, the P_{repl} values for a PTW/Markus parallel-plate chamber were determined in the range of mean incident energies of 5–11 MeV. The P_{repl} values for this chamber were found to differ significantly from unity, assuming that the cylindrical chamber values are valid.

A theoretical treatment of perturbation correction factors for the plane-parallel chamber NE 2534 has been presented by GAJEWSKI and IZEWSKA (1987).

GOSWAMI and KASE (1989) reported a measurement of the replacement correction factors (P_{repl}) for a PTW/Markus parallel-plate chamber at mean incident electron energies of 3.1, 4.4, 8.9, 13.0, 16.3, and 18.8 MeV. The factors were significantly different from unity at low energies.

ANDREO et al. (1991) pointed out that for ^{60}Co gamma rays there is a significant discrepancy of about 0.5% between the calculated values of replacement factors for a Farmer chamber used in 1983 AAPM protocol and the experimental data of JOHANSSON et al. (1978). They concluded that further work is needed, using either theoretical or experimental techniques, to clarify the discrepancies in the existing data sets for replacement or displacement factors, especially if calibrations in terms of absorbed dose-to-water are adopted in the near future. Also, it is recommended by ANDREO et al. (1991) that uncertainties stated in dosimetry, based on cylindrical or plane-parallel ion chambers, should reflect the existence of differences in the available data.

BURNS (1992) disagreed with Andreo et al.'s statement that the underestimation of uncertainties in calibration of plane-parallel chambers may influence the NPL calibration factors. They pointed out that the NPL calibration factors do not depend critically on the values of replacement factors.

A systematic study of perturbation factors for ionization chambers in electron beam was presented by WITTKÄMPER et al. (1991). They determined values for the fluence perturbation correction factor in a number of electron beams for the PTW/Markus plane-parallel chamber and the cylindrical NE 2571 Farmer Chamber. These data were determined relative to the NACP plane-parallel chamber as a function of the mean energy at depth, $\bar{E}_z$, calculated according to the method recommended in modern dosimetry protocols. For the cylindrical chamber the results were in agreement with the data recommended in these protocols. For the PTW/Markus chamber a small but significant fluence perturbation correction was necessary: up to about 3% for a value of $\bar{E}_z = 2$ MeV. In order to check the value of these fluence perturbation correction values, dose measurements were performed ionometrically and compared with values determined with the Fricke dosimetry system. The ratio of dose values determined with the NACP and PTW/Markus chambers applying the NCS code of practice and the dose values determined with the Fricke dosimetry system was $1.007 \pm 0.5\%$ on average for a number of electron beams. This result was in agreement with the conclusions of other investigators if, for the Fricke dosimeter system, a value of $352 \times 10^{-6}\,\mathrm{m}^{-2}\,\mathrm{kg}^{-1}\,\mathrm{Gy}^{-1}$ for $\varepsilon_m G$ is applied.

17.3.1.4 Ion Recombination Correction

BOAG (1982) presented the development of a new theory for the determination of recombination correction.

WEINHOUS and MELI (1984) consolidated the available information, rectified certain omissions, and provided several convenient and readily implementable methods for determining P_{ion}. Computer programs, quadratic approximations, and data tables were presented to facilitate the determination of P_{ion} for continuous, pulsed, and pulsed-swept beams.

MAJENKA et al. (1982) reported a series of exposure measurements using an NPL dosimeter in the swept electron beam of a Sagittaire linear accelerator at energies of 6, 9, 13, 17, and 20 MeV, at two dose rates, and at four values of the collecting voltage on the ionization chamber. The true exposure value, corresponding to saturation current in the ionization chamber, was estimated using the theory developed by Boag. Good agreement was found among the various estimates made at several voltage ratios. The authors concluded that the measurements provide satisfactory confirmation of Boag's theory.

CONERE and BOAG (1984) pointed out the limitation of the two-voltage technique for determining the collection efficiency of an ionization chamber exposed to a pulsed and magnetically swept electron beam. The two-voltage method depends upon the form of the saturation curve agreeing closely with the formula based on ionic conduction and negligible space charge. The possible disturbing effects of charge transport by free electrons and of overlapping pulse were discussed and the conditions to be satisfied for the reliable use of the method were defined and demonstrated experimentally. In a separate paper BOAG (1984) presented a general saturation curve applicable for magnetically swept electron beams.

VAN DAM et al. (1985) observed that the maximum deviation between the TLD and "two-voltage" method was never more than 2% and mostly smaller than 1%.

In 1986 Boag's theory for the collection efficiency of a small ionization chamber in a pulsed swept beam was generalized by taking chamber size into account (MELI and WEINHOUS 1986). The collection efficiency was given in terms of the chamber radius, the Gaussian scale constant of the stationary beam, and the maximum distance between beam and chamber centers. It was shown that, for cases of practical interest, collection efficiency is independent of chamber size.

In 1986 the characteristics of swept electron beams produced by Sagittaire or Saturne type accelerators were investigated using film dosimetry (MARINELLO et al. 1986a,b). Beam sections perpendicular to the beam-limiting system (BLS) axis were fully described by Gaussian curves close to the center of the field defined by the BLS, but they became more and more distorted as the elementary beam approached the BLS. The swept area was also evaluated and was related to the magnitude of the current applied to the two orthogonal magnets of the sweeping system.

WEINHOUS and MELI (1988) investigated the consequences of the assumption in Boag's theory that the stationary beam has a Gaussian radial intensity distribution unperturbed by extra-phantom scatter. Consequently in Boag's theory a hyperbolic pulse-size distribution is expected on the central axis. Their measurements of pulse-size distributions at the isocenter of a Sagittaire accelerator were in accord with this model. However, measurements for large collimator settings yielded markedly nonhyperbolic pulse-size distributions for electron energies from 7 to 32 MeV. It was concluded that application of the model in such cases might result in significant errors depending on the inherent collection efficiency of the chamber.

In 1989 a simple method of determining recombination losses based on Boag's theory was developed (MÜLLER-SIEVERS and KOBER 1989). The deviation of these correction factors, compared with those of Boag's precise theory, did not exceed ± 0.5%.

BIELAJEW (1985) considered the effect of the collection of free electrons that had not formed negative ions in the gas of a parallel-plate ionization chamber for the case of continuous radiation. A theory and its solution, valid for the entire voltage range, was given. The solution was based upon the saturation theory of Townsend and free electrons were included in a perturbative fashion. The collection of free electrons was shown to become important at exposure rates above 430 R/min. It was concluded that since the electron contribution involves an extra dependence on exposure rate, some of the wide variation in the cavity gas parameter (m value) reported in the literature may be due to the presence of free electrons.

TAKATA and SAKIHARA (1989) measured values of m which are equal to $(\alpha/eK_+K_-)^{1/2}$, where α is the recombination coefficient, e the charge per ion, K_+ the mobility of positive ions and K_- that of negative ions using a double x-ray beam method. By this method, one value for m was obtained from a set of data on currents at each polarization voltage. Values obtained were for a recombination region 2.65 cm wide. It was shown that these values decrease from 1.85×10^6 to 1.65×10^6 V s$^{1/2}$ C$^{-1/2}$ with an increase in electric field strength from 4.8 to 48 V cm^{-1}. The authors concluded that the value of m increases with ion age. This was confirmed experimentally in a later study (TAKATA and MATIULLAH 1991) concerning the collection efficiency. Measurements were made for a parallel-plate ionization chamber which had a variable space between the polarizing electrode and collector. The chamber was exposed to ^{60}Co gamma rays at several exposure rates and the value of m was determined at various applied voltages. It was found that m depends upon the lifetime of ions in the chamber (i.e., it increases from 17.1 to 18.5 MVA$^{-1/2}$ m$^{-1/2}$ with a decrease in the ratio of the voltage to the square of the chamber space from 600 to 20 kV m^{-2}).

HOCHHÄUSER and BALK (1986) developed "fast" ionization chambers and, by means of these, directly measured the electron component of the chamber current. The actual number of electrons collected was significantly greater than that calculated when using attachment coefficient and drift velocity values found in the literature. For example, approximately four times the expected value was collected in room air. The deficit of negative ions resulting from the collection of electrons was incorporated within a modified version of Boag's recombination equation. It was concluded that the recombination loss up to an exposure per pulse of at least 2.58×10^{-2} C kg^{-1} (100 R) can be calculated in this manner.

In 1983 a semiempirical expression to describe the full saturation curve of parallel-plate ionization chambers was presented (FALLONE and PODGORSAK 1983). This empirically determined expression was shown to be in excellent agreement with measurements in the whole collection efficiency range from 0 to 1 for x-ray sources with effective energies from 20 to 150 keV and ^{60}Co gamma rays. In a separate paper, the same investigators presented results of measurements of saturation curves for air and nitrogen in the exposure rate range of 1 to 500 R/min (HEESE et al. 1986). Because of the differences between the ion–ion versus electron–ion recombination processes, the air and nitrogen saturation curves were found to differ considerably even when measured under identical chamber and exposure rate conditions.

HAYAKAWA et al. (1989) developed a method of compensation for fluctuations in beam intensity that may occur during measurement of P_{ion} by the two-voltage technique. They used a parallel-plate ionization chamber, whose P_{ion} was known, and a vacuum chamber to obtain signals that were proportional to the beam intensity. Experiments were conducted using pulsed proton beam providing doses that ranged from 0.16 to 0.01 cGy/pulse. The value of P_{ion} of a thimble ionization chamber was measured. With these measurements, the validity of their method for a pulsed proton beam was verified experimentally.

17.3.1.5 Polarity Effect

GERBI and KHAN (1987) investigated the polarity effect for three different commercially available plane-parallel ionization chambers: the Memorial Pipe chamber, the Victoreen/Nuclear Associates model 30–329 chamber manufactured by PTW, and the Capintec PS-033 thin-window ionization chamber. The polarity effect versus depth below the phantom surface for 6-, 10-, 18- and 24-MV x-ray beams and 9- and 22-MeV electron beams was investigated. Also the polarity effect in the region of nonelectronic equilibrium at the interface of two dissimilar materials, polystyrene and aluminum, was investigated as well as the effects of field size. For the group of plane-parallel ionization chambers studied, they found a polarity effect of only 1%–2% for electron beams at the depth of d_{max}. At depths greater than d_{max}, the polarity effect for electrons increased; it was as high as 4.5% for some chambers. When used in the buildup region of high-energy photon beams, these same chambers exhibited a difference of up to 30% in collected charge between one polarity and the other. This effect and its relationship to physical chamber characteristics were discussed further by Gerbi and Khan.

In 1991 the polarity effect in electron beams for several types of chamber and several irradiation conditions was investigated (AGET and ROSENWALD 1991). It was found that differences in readings for opposite polarities can be significant for cylindrical chambers (about 10%) as well as for plane-parallel chambers (20%). The effect was larger for large field sizes than for small ones. The observed effect generally included an appreciable stem and cable effect. Differences in reading with both polarities were related to the energy distribution of the electron beam and were greater for lower electron energies than for higher ones. Polarity effect and charge deposit within the chamber wall material appeared to be closely connected. This charge deposit, expressed as a proportion of the total collected charge, could be directly derived from double polarity measurements. It was concluded that careful investigation of the effect should be made to avoid significant error (over 5%) in the determination of the absorbed dose.

17.3.1.6 Effective Point of Measurement

ZOETELIEF et al. (1980) determined the effective point of measurement for spherical ionization chambers for ^{60}Co and ^{137}Cs gamma rays and for 300-kV x-rays inside a water phantom. For reasons of comparison, measurements were also performed with a thimble-type Baldwin-Farmer ion chamber. For spherical ion chambers the displacement correction factors were $1 - (0.37 \pm 0.04) r \cdot 10^{-2}$ for ^{60}Co gamma rays and $1 - (0.22 \pm 5) r \cdot 10^{-2}$ for ^{137}Cs gamma rays, where r is the radius of collection volume expressed in mm. No displacement correction was observed for x-rays.

AWSCHALOM et al. (1983) presented a new technique for determination of the effective point of measurement when cavity ionization chambers are used to measure the absorbed dose due to ionizing radiation in a dense medium. An algorithm was derived relating the effective point of measurement to the displacement correction factor. This algorithm related variations of the displacement factor to the radiation field gradient. The technique was applied to derive the magnitudes of the corrections for several chambers in a fast neutron therapy beam.

NIATEL (1983) has reported an investigation on the location of a plane-parallel ionization chamber for absorbed dose distribution for ^{60}Co gamma rays. Two choices were considered for the position of the ionization chamber with respect to the point of reference: (a) the middle of the cavity point and (b) the front plane of the cavity. The correction factor was determined for both positions of a graphite chamber at various depths in a graphite phantom. The experimental results confirmed the accuracy of Boutillon's theoretical calculations of the same effects (BOUTILLON 1983).

17.3.1.7 Environmental Effects

PEARSON et al. (1980) described some ionization chamber measurements which indicated a systematic anomalous behavior by graphite chambers due to environmental effects. They concluded that the results of these experiments indicate that the anomalous ionization yield for gases other than air in graphite chambers is due essentially to the diffusion of air through the graphite walls, thus contaminating and diluting the other gases. The air flows inward because its partial pressure is greater outside the wall than inside it – a situation that would pertain even if the total pressure inside exceeded that outside. The simple expedient of covering the chamber with a condom eliminates this effect. However, in the future, graphite chambers should always be constructed with a gas-impervious barrier such as a plastic or Al shell, properly vented to prevent backflow of air through the exit port. The authors suggest that all measurements made with bare-walled graphite ion chambers containing gases other than the ambient gas (usually air) are likely to be significantly in error. Also suspect are measurements with a covered graphite chamber, in water phantoms for example, which are interpreted on the basis of a gamma-ray calibration of the uncovered chamber with the same gas flow. The cause was shown to be the porosity of the graphite which makes the ionization chamber sensitive to environmental changes.

In 1984 the thermal characteristics of a polystyrene phantom of the SCRAD type was examined (BARISH 1984). In the case of a callibration phantom which is moved from one location to another, many hours may be required to achieve thermal equilibrium with the new environment. By using a correction term based on the temperature of the phantom, not of the room, they concluded that it is possible to calibrate a therapy unit without waiting for thermal equilibrium.

In 1985 it was shown that the volume of nylon-walled ionization chambers varies with relative air humidity due to the hygroscopic properties of nylon (MIJNHEER and WILLIAMS 1985). A 6% increase in exposure calibration factor of a commonly employed 0.6 cm^3 nylon-walled ionization chamber was observed when the relative air humidity decreased from 98% to 11%. The increase as well as the decrease in volume showed an initial fast change during the first day followed by an exponential change with half-lives of about 1.2 days and 2.4 days for water uptake and water loss, respectively.

VAN DER GIESSEN (1986) investigated further the rate of temperature change in a nylon thimble chamber provided with a temperature sensor when the chamber was inserted in different phantom media. It was found that the temperature time constant for the changes ranged from 40 s (chamber in water) to 515 s (chamber with buildup cap in air). Handling the detector for about 30 s caused a temperature rise of about 1°C. This temperature rise was reflected in dose-rate measurements with a graphite ionization chamber, but was not measurable with a nylon chamber.

In 1992 it was reported that the output of an open thimble chamber varied with pressure exactly as expected (MAYO and GOTTSCHALK 1992). The temperature dependence also followed the ideal gas law for a chamber with a graphite shell. However, the temperature dependence was some 20% less than the ideal gas law for a chamber constructed of A150 plastic. They concluded that the A-150 chamber expands with temperature, increasing the volume of gas and thus partially canceling the density decrease. This effect was not seen with the graphite chamber since the coefficient of expansion of graphite is much smaller than that of A150 plastic.

Data on the long-term use of an isotope check source for verification of ion chamber calibration were reported by BARISH and LERCH (1992). Eight years of data were analyzed for two ion chambers (and their associated electrometers) irradiated at fixed geometry in such a device. These dosimetry systems had also been calibrated every 2 years at a single Accredited Dosimetry Calibration Laboratory. The authors concluded that when a check source is used, and the results are consistent, the interval between formal calibrations can be lengthened.

17.3.1.8 Cable Effects in Ionization Chambers

SPOKAS and MEEKER (1980) investigated seven coaxial cables used for carrying currents generated in ionization chambers with reference to their suitability to this application. Included in this study were four low-noise triaxial cables and three low-noise two-conductor cables. For each cable the following characteristics were considered: inherent noise currents, currents produced by cable movements, polarization currents, the degree of electrostatic shielding of the central signal-carrying conductor, and radiation-induced cable currents. The study indicated that of the seven cables, two low-noise triaxial cables, both employing solid Teflon dielectric surrounding the central conductor, offer the best overall performance for use with ionization chambers.

CAMPOS and CALDAS (1991) reported a simple method, for possible use by hospital physicists, to evaluate the irradiation effects on cables and connectors during large-radiation-field dosimetry with ionization chambers and to determine correction factors for the used system or geometry. This method was based on the absorbed dose dependence of the correction factor for cable effects.

17.3.1.9 Protective Cap Effects

HANSON and TINOCO (1985) investigated the change in ionization within the chamber due to the presence of a protective acrylic cap for electron beams from 7 to 18 MeV. The change due to the cap was shown to be small, no more than 0.5% for x-rays and 0.7% for electrons. The change for polystyrene was seen to be as much as twice that for acrylic. Empirical correction factors to compensate for this effect were determined. A theoretical basis for photons was suggested by the authors by an extension of the theory in recent protocols. The effect for electrons was explained only qualitatively.

In 1985 the effects of thickness of the waterproofing sheath on the calibration of photon and electron beams were reported (GILLIN et al. 1985). This work investigated the effect of variations in the thickness of the waterproofing sheath from 0.5 to

5.5 mm on the calibration of photon beams ranging from ^{60}Co beams to 25-MV x-rays, and electron beams with nominal energies of 7–18 MeV. For photon beams, a maximum change of 1.2% was found for the 25-MV x-ray beam. For electron beams, a maximum change of 0.5% was found for 10-MeV electrons. It was concluded that the thickness of the waterproofing sheath is not a very sensitive variable, assuming the thickness is between 0.5 and 2.0 mm.

HANSON et al. (1988) reported effects of contamination by talcum powder in ionization chambers protected by thin rubber sheaths in water. Four Farmer-type ionization chambers contaminated with talcum powder were received for calibration by the Accredited Dosimetry Calibration Laboratory at the University of Texas M.D. Anderson Cancer Center. The chambers showed a marked energy dependence (5%–20%) for soft orthovoltage x-rays. The response of the contaminated chambers was compared with the chambers' response before contamination and after cleaning. Techniques for identifying contaminated chambers and suggestions for cleaning them were presented by the authors.

17.3.1.10 Plane-Parallel Ionization Chambers

MATTSSON et al. (1981) have described procedures for calibration and use of plane-parallel ionization chambers for determinations of dose in electron beams. In this paper, they discussed a number of possible procedures based on the main procedure described in the NACP (1980) protocol. They investigated the commercially available plane-parallel chambers and developed a new plane-parallel chamber for the dosimetry of electron beams. The design, construction, and performance of this chamber were described. Numerical values of the different factors to be used in the dosimetry with this plane-parallel chamber were determined.

In 1988 an investigation of the electron backscatter corrections for parallel-plate chambers was reported (HUNT et al. 1988). The authors measured electron backscatter from low atomic number materials with electrons from 6 to 20 MeV. The effect of the diameter and thickness of the backscattering material was studied. Based on these data, Lucite and polystyrene chambers in water phantoms were expected to underrespond by 1% and 2% at 6 MeV. The expected underresponse decreased to 0.8% and 0.4% for polystyrene and Lucite at 12 MeV and was insignificant above 16 MeV. Two commercially available parallel-plate chambers were compared

with a cylindrical chamber in electron beams from 6 to 20 MeV. Using the 20-MeV comparison, the expected chamber responses at the lower energies were calculated and compared with measurements. Both parallel-plate chambers underresponded by approximately 1% at 6 MeV and 0.5% at 9 MeV, which is qualitatively consistent with the electron backscatter data. The authors made several recommendations for minimizing electron backscatter effects through chamber design.

In 1988 the dosimetric characteristics of a Capintec parallel-plate ionization chamber were investigated in detail (KOOY et al. 1988). The Capintec chamber was used for dose measurements in a lead and polystyrene slab phantom irradiated with ^{60}Co gamma rays. The authors defined an enhancement ratio to quantify the dose measurements. The enhancement ratio equaled the ratio of dose measured with the lead slab present to dose measured under equilibrium conditions in polystyrene at equal primary beam attenuation. The measured enhancement ratio at the exit side of the lead/polystyrene interface was 25% lower than the Monte Carlo-predicted enhancement ratio. The authors proposed that geometric acceptance limitations of the Capintec chamber to large-angle, low-energy electrons are the cause of this difference. A Monte Carlo simulation of the Capintec chamber acceptance confirmed their hypothesis.

RUBACH et al. (1986) pointed out a systematic error in the determination of dose buildup curves for ^{60}Co gamma rays using a plane-parallel chamber.

GERBI and KHAN (1990) reemphasized the inaccuracies in the measurement of dose in the buildup region of normally incident photon beams when using fixed-separation plane-parallel ionization chambers. Data for ^{60}Co, 6-, 10-, 18-, and 24-MV photon beams were presented that show the magnitude of this overresponse in the buildup region for several commercially available plane-parallel ionization chambers versus results obtained using both an extrapolation chamber and LiF thermoluminescent detectors. Differences of >19% were found in the percent depth dose at the surface of a phantom for one of the chambers. All chambers overresponded in the buildup region to some degree based upon their internal dimensions. The appropriateness of published corrections for these chambers was evaluated and guidelines for the accurate measurement of dose in the buildup regions were presented.

MELLENBERG (1990) determined overresponse corrections for a widely used parallel-plate ionization chamber using contemporaneous measurement of buildup for 4-, 6-, 10-, and 18-MV photon beams utilizing a commercially available extrapolation chamber (PTW model 23392). The resultant over-response corrections were essentially independent of field size (5 × 5 to 30 × 30 cm) and were less for increased depth (from the surface) and higher x-ray energy. The over-response of the parallel-plate chamber (Markus-type, PTW model 329) was 13.8%, 10.7%, 6.2%, and 4.7% (absolute) at the surface for 4-, 6-, 10-, and 18-MV x-ray energies. At only 2-mm depth, the overresponse of the Markus chamber under investigation decreased to 50% of surface value overresponse. These corrections, so derived, were applied to Markus parallel-plate chamber buildup measurements by simple subtraction of the derived corrections. For example, a 6-MV percentage depth dose of 37.5% measured at the surface with the Markus chamber was reduced by 10.7% so as to agree with the surface dose indicated by the PTW extrapolation chamber for this x-ray field, i.e., 26.8%.

17.3.2 Radiation Calorimetry

DOMEN (1990) reviewed in literature on radiation calorimetry exhaustively. Some of the more recent developments are presented below.

17.3.2.1 Graphite-Core Calorimeters

By 1980 the calorimetry techniques for absolute determination of dose at a point in a water medium irradiated by a ^{60}Co beam had been well established. Because water calorimeters were not yet available, the conventional method was the two-step approach as recommended by ICRU Report No. 41. First, a graphite calorimeter was used to determine dose to graphite, which was subsequently transferred to dose to a point in water using a suitable transfer ionization chamber. An accurate transfer is only possible if the photon spectra in graphite are identical to those in water. The differences in spectra in depth, although quite similar, are sufficient to increase uncertainty in the transfer. To alleviate this problem, RAO and NAIK (1980) reported the development of a graphite calorimeter similar in size and shape to a Farmer-type 0.6-cm³ ionization chamber. It was developed for direct measurement of absorbed dose at a specified depth in a water phantom irradiated by a beam of ^{60}Co gamma radiation. The accuracy of the absorbed dose determined was esti-

mated to be ± 1.1% at a dose level of 4 cGy/s. The absorbed dose-to-water at the calibration depth of 5 cm was standardized for a ^{60}Co. The overall accuracy in calibrating the ionization chambers was ±1.2%.

To address the same problem, PRUITT et al. (1981) determined absorbed dose-to-water in a ^{60}Co gamma-ray beam using a thick-walled graphite ionization chamber. The chamber was calibrated in a graphite phantom against a graphite calorimeter, and the graphite calibration factor was converted to a water calibration factor using published energy absorption coefficient ratios and a measured replacement factor. Comparisons between the graphite and water measurements were made at pairs of points that were scaled in position according to the ratio of electron densities, so that the photon spectra were the same for the two points in a given pair. Measurements performed in graphite over a wide range of phantom depths, field sizes, and source distances showed that the calibration factor varies slowly with the phantom depth and field size, and probably has a negligible dependence on source distance. It was concluded that by comparison with the thick-walled chamber in a ^{60}Co gamma-ray beam, a secondary ionization chamber can be calibrated in terms of absorbed dose to water with an estimated uncertainty of about ± 1%.

Using a graphite calorimeter, SCHULZ and WEINHOUS (1985) determined N_{gas} for a Farmer-type ionization chamber using 4- and 25-MV x-rays. The procedure was to measure the dose to graphite using the calorimeter, and then to obtain the response of the chamber at the same depth in a graphite phantom. The AAPM 1983 protocol was used to calculate N_{gas}. The values of N_{gas} determined with the calorimeter are within 1% of N_{gas} calculated according to the AAPM protocol, using the ^{60}Co exposure calibration factor. This work thus confirmed the internal consistency between the results of the AAPM 1983 protocol using an exposure-calibrated ionization chamber and those obtained with a graphite calorimeter.

One of the assumptions in the use of graphite calorimeters for absolute dose in a ^{60}Co beam has been that perturbation effects of gaps between the various elements of a calorimeter are negligible. BOUTILLON (1989) used a Monte Carlo simulation for determination of the gap correction to be applied to the calorimetric measurement of absorbed dose in graphite in a ^{60}Co beam. His results show that this correction, which has been neglected so far by most national standards laboratories, is not negligible and depends on many parameters. A comparison with the available experimental data on gap corrections showed generally good agreement with Boutillon's calculations.

Owen and DuSautoy (1991) evaluated the corrections for gaps in the graphite calorimeter and concluded that the gaps around the core of a Domen-type graphite calorimeter (Domen 1990) have a small but significant effect on the measured dose. Adding graphite to the front of the calorimeter to compensate for the gaps in front of the core minimizes but does not eliminate the effect of the gaps. For the NPL calorimeter the total measured gap correction was $+ (0.53 \pm 0.14)\%$ for 4-MV x-radiation, the correction decreasing for higher x-ray energies.

17.3.2.2 A-150-Core Calorimeters

A-150 plastic is a mixture of 45.14% polyethylene, 35.22% nylon, 16.06% carbon black, and 3.58% calcium fluoride (Smathers et al. 1977). It is also described by Goodman (1978) and has been developed as a tissue-equivalent material for neutron dosimetry. In 1980, Domen reported an experimental determination of some thermal properties of A-150 tissue-equivalent plastic (Domen 1980b). His results were thermal diffusivity, 2.72×10^{-3} cm^2 s^{-1} $\pm 0.4\%$; specific heat, 1.72 J g^{-1} K^{-1} $\pm 1.3\%$; and thermal conductivity, 5.3×10^{-3} W K^{-1} cm^{-1} $\pm 1.4\%$.

Schulz et al. (1990b) determined the thermal defects of A-150 plastic and graphite referenced to aluminum for 800 keV protons scattered by a 2-μm nickel foil. Composite cores of Al, A-150, graphite/ A-150, and Al/graphite which could be irradiated from one side or the other were employed. The temperature increase of a core caused by 30–100 s of irradiation (3–6 nA of proton beam current) was detected by two thermistors mounted in opposite legs of a Wheatstone bridge. The thermal defect of A-150 plastic was determined to be 4.2% referenced to aluminum and 4.0% referenced to graphite. The thermal defect of graphite referenced to aluminum was negligible, with experimental error of about 0.3%. No change in the thermal defect of A-150 plastic was detected for accumulated doses up to 8×10^5 Gy.

17.3.2.3 Polystyrene-Core Calorimeter

Zeitz and Laughlin (1982) presented the development of a "nonisolated-sensor" solid polystyrene calorimeter to test the role of thermal diffusion in limiting the length of irradiation time during which temperature measurements with nonisolated sensors could be made sufficiently free of drift for determining dose with radiation fields such as gamma rays, x-rays, and high-energy electrons. From measured ratios of dose at 5.0 and 0.5 cm in polystyrene and comparisons to dose measurements with a polystyrene parallel-plate ion chamber, it was shown that thermal diffusion is sufficiently small in polystyrene to permit accurate measurements for irradiation periods of less than 20 min. Comparison of the absorbed dose measurements and depth-dose ratios with plane-parallel ion chambers and calorimeter showed that within the precision and accuracy of the two measuring systems, there was close agreement. It was concluded that the nonisolated-sensor solid polystyrene calorimeter has the interesting features of (a) simplicity of construction, (b) simplicity of operation without vacuum or feedback for temperature control, (c) capability of simultaneous measurements at several depths and off-axis positions, (d) very small thermal defect correction with polystyrene, and (e) operation with the calorimeter in any orientation.

Zeitz et al. (1986) presented an improved version of their earlier solid-polystyrene calorimeter and reported absorbed dose measurements with precision of less than 0.3%. The accuracy for obtaining absolute absorbed dose was estimated by comparisons with cavity ionization measurements. The calculation of absorbed dose with ionization chambers was carried out based upon the 1983 AAPM dosimetry protocol. Measurements in a ^{60}Co gamma-ray field with three different polystyrene parallel-plate ion chambers in a polystyrene phantom did not differ by more than 1.5% from measurements obtained with the polystyrene calorimeter. Measurements taken over a period of 247 days were compared with the expected values on the basis of the decay of ^{60}Co. The calorimeter system, with its capability of acquiring, printing, storing, plotting, and analyzing the data by computer, was described.

17.3.2.4 Water Calorimeter

In 1980a, Domen presented his original design of a water calorimeter which led to intense research activity in the 1980s. In the original Domen calorimeter an ultrasmall bead thermistor was sandwiched between two thin films stretched on polystyrene rings and immersed in an unregulated water bath. Advantage was taken of the low thermal diffusivity of water and the imperviousness of polyethylene film to water to construct a calorimeter for directly measuring absorbed dose in that medium. Following this feasibility study, details of a water calorimeter were presented by Domen in a later paper (DOMEN 1982). With his calorimeter design, Domen obtained reproducible measurements in distilled water supplies that had a wide range of impurities. Measurements, after saturating the water with nitrogen or oxygen, showed no difference. A difference of 0.6% would have been easily detectable. Tests with several chemicals added to water showed some unexpected results and changes in the measured absorbed dose rate versus accumulated dose. The measured absorbed dose rate in distilled water under the conditions described was 3.5% higher than that determined from measurements with a graphite calorimeter.

In 1984, an investigation of the effect of various dissolved gases on the heat defect of water was reported (ROSS et al. 1984). The authors constructed a small calorimeter (holding 100 ml of water) with which to measure the temperature rise in irradiated water saturated with various gases. The gases used were air, oxygen, argon, nitrogen, and hydrogen/oxygen mixtures. Irradiations were carried out with 20-MV x-rays at a dose rate of 0.41 Gy/s. Their results were consistent with model calculations, except for some differences for accumulated doses of less than 100 Gy. They concluded that the discrepancies they found at low doses and the discrepancies observed by others using water calorimeters may have arisen from impurities in the water.

SCHULZ and WEINHOUS (1985) demonstrated the presence of convection currents in a water calorimeter. Their calorimeter could be irradiated by vertical or horizontal beams, and operated at temperatures in the range 3°–40°C. When irradiated at 30°C with a vertically downward 19-MeV electron beam, the responses of the proximal and midline thermistors were in accordance with the depth-dose curve. When irradiated horizontally, the initial patterns of temperature rise were the same, but after about 30 s (4 Gy) the rate of temperature rise decreased at the proximal thermistors and increased at the midline thermistors. Shortly after irradiation, the temperature curve of the midline thermistors crossed that for the proximal thermistors, a pattern that suggests the presence of convection currents. To test this hypothesis, the calorimeter was operated at 4°C. The temperature patterns for horizontal irradiation became the same as those obtained with vertical beams, thus demonstrating the production of convection currents in water at a temperature of 30°C for temperature gradients as small as 10^{-3} °C cm^{-1}.

DOMEN (1986) speculated that the large power dissipation in the thermistors may have led to some artifacts in the SCHULZ and WEINHOUS (1985) experiment. To reduce this potential problem, KUBO et al. (1989) suggested the use of pulsed excitation in place of a conventional DC excitation to induce higher bridge output voltage while keeping the average self-heating of a thermistor to a reasonably low value. Performance evaluations of a prototype pulsed calorimeter were presented in their work.

SCHULZ et al. (1987) reported the construction of a flexible, temperature-regulated, water calorimeter which consisted of three nested cylinders. The innermost "core" was a 10×10 cm right cylinder made of glass, the contents of which were isolated from the environment. Surrounding the core was a "jacket" that provided approximately 2 cm of air insulation between the core and the "shield." The shield surrounded the jacket with a 2.5-cm layer of temperature-regulated water flowing at 5 l/min. The core was filled with highly purified water the gas content of which was established prior to filling. Convection currents, which may be induced by dose gradients or thermistor power dissipation, were eliminated by operating the calorimeter at 4°C. The response of the calorimeter to 4-MV x-rays was compared to that of an ionization chamber irradiated in an identical geometry. For nitrogen-saturated water, the grand mean of the calorimeter-to-ion chamber (Cal/Ion) dose ratio for five experiments conducted over a period of 6 months was 1.006 ± 0.001. Three experiments with oxygen-saturated water in the same core yielded a Cal/Ion ratio of 0.991 ± 0.001. These results were consistent with radiochemical models and refined experiments that used water saturated with various gases. It was concluded that nitrogen-saturated water calorimeters of the type described may be used to determine the dose-to-water without reference to radiation-dependent parameters or recourse to corrections for thermal defects.

ROSS et al. (1989) used water saturated with a 50/50 mixture of H$_2$ and O$_2$ gases, for which the heat defect was calculated to be –2.1%. As a test of this assignment, they compared the absorbed dose-to-water as measured using water calorimetry with that obtained for Fricke dosimetry. The water calorimeter consisted of a small sealed vessel containing 100 ml of stirred water saturated with a 50/50 mixture of H$_2$ and O$_2$ gases. It was irradiated with 20-MV x-rays at a dose rate of about 0.4 Gy s^{-1}. The same vessel was then filled with Fricke dosimeter solution and irradiated under identical conditions.

Their Fricke dosimetry was based on the SVENSSON and BRAHME (1979) value of εG (3.515×10^{-3} 1 cm^{-1} J^{-1}) and agrees to within 0.2% with the dose-to-water for ^{60}Co obtained via graphite calorimetry. They found that for 20-MV x-rays, the dose-to-water determined by water calorimetry was 1.006 ± 0.004 times the dose determined by Fricke dosimetry. Within 0.6 ($\pm$ 0.4)%, their result supported the calculated heat defect of -2.1% for water saturated with a 50/50 mixture of H$_2$ and O$_2$ gases.

Recently, SCHULZ et al. (1991) reported a comparison of ionization chamber and water calorimeter dosimetry for a wide range of beam energies, which include ^{60}Co and 4-, 6- and 25-MV x-rays. In this study a power level correction factor was employed to take into account the effect of change in the thermistor power level caused by the irradiation. They found the grand mean of the ratio of calorimeter to ion chamber measurements for the four beam energies to be 1.001 ± 0.001. As no significant trend with beam energy was detected, it is concluded that the calorimeter and ionization chamber yield equally accurate results. Because the calibration of the water calorimeter depends solely upon the accuracy with which water temperatures in the range 2°–10°C can be measured, and dose is given by the product of the specific heat of water and the temperature range produced by irradiation, the water calorimeter has the potential to place radiation dosimetry on a much firmer foundation than presently exists.

A water calorimeter was used by GALLOWAY et al. (1986) for the direct measurement of absorbed dose-to-water in a d(15) + Be neutron beam. The absorbed dose measured with the calorimeter was compared with that measured with an Exradin ionization chamber constructed of A-150 plastic. The doses measured by the ionization chamber were calculated according to the European protocol. Absorbed dose-to-tissue measured with the calorimeter was 4.3% lower than that measured with the ionization chamber. Relative to ionization chamber dosimetry, dose measurements with the calorimeter in the neutron beam were 9% lower than similar measurements in 4- and 9-MV photon beams. The significance of the results is discussed in terms of the heat defect in water. In a later publication, the authors revised and updated some of their conclusions.

17.3.2.5 Polystyrene-Water Calorimeters

DOMEN (1983) proposed a new type of calorimeter consisting of polystyrene discs in water. This calo-

rimeter was developed to provide an investigative tool for comparison with absorbed dose measurements in water. Measurements using ^{60}Co gamma radiation showed that converting the results to water agreed to within 1% with similarly determined results when using a graphite calorimeter. These results were ~3% lower than those determined with a water calorimeter, suggesting a negative heat defect in water of ~3%, when measuring an absorbed dose rate ~1 Gy/min that produced an accumulated dose up to 150 Gy.

In 1985, the thermal diffusivity of atactic amorphous polystyrene, from which polystyrene-water calorimeter detectors are made, was measured by KUBO (1985). Using the value of thermal diffusivity so determined, an estimate was made of the amount of thermal diffusion from the 2.5-cm-thick detector to the surrounding water during beam irradiation. The thermal diffusion was calculated as a function of different combinations of the following parameters: (a) the thermal diffusivity; (b) the size of the detector; (c) the absorbed dose rate; and (d) the duration of irradiation. His results indicated that all these parameters must be carefully examined, or else the polystyrene-water calorimeter may underestimate the absorbed dose by an appreciable amount because of thermal diffusion.

17.3.3 Chemical Dosimetry

The most commonly used chemical dosimeter for absolute determination is the ferrous sulfate Fricke dosimeter as described in an exhaustive review by FRICKE and HART (1966). More recently, a high-precision Fricke dosimetry system has been described by ROSS et al. (1989). This system is based upon a value of the product of extinction coefficient and the ferric ion yield (the G value) from SVENSSON and BRAHME's work (1979).

17.3.3.1 Fricke Dosimetry
Using UV Spectrometry

Before 1980, there had been considerable discussion about the dependence of ferric ion yield G on photon and electron energies. NAHUM et al. (1980) concluded from a semiempirical analysis of experimental data that the G value is nearly independent of energy in the range of photon and electron energies used in radiation therapy. This was challenged by LAW and NAYLOR (1984). They reanalyzed their earlier Fricke data relative to ionization chambers using new revised values of C_λ and C_E from the HPA and concluded that in place of the previous discrepancy of up to 8% for high-energy electrons, there was now close agreement between Fricke and ion chamber dosimetry systems. Their revised results showed a continuing upward trend of the G value with photon energy. The trend was closely in line with the variation from lower energy x-rays up to ^{60}Co radiation. Thus their results provide no support for the idea of a constant G value of 15.5 from the ^{60}Co energy upwards and suggest that the error in any such assumption could be about 2% at around 30 MV. For megavoltage electrons, their results were consistent with a constant value close to 15.75.

MOSSE et al. (1982) reported a G value of 1.63 μmol J^{-1} for the Fricke dosimeter for 35-MeV electrons using a graphite calorimeter.

For ultrasoft x-rays, FREYER et al. (1989) reported measurement of the G value of a modified ferrous sulfate solution for 1.5-keV x-rays. They modified the ferrous sulfate solution by the addition of benzoic acid and measured the relative G values for Al$_k$ characteristics x-rays (1.5 keV), ^{238}Pu α-particles (3.7 MeV), and ^{60}Co (1.17 MeV) and ^{137}Cs (0.66 MeV) gamma rays. This modified ferrous sulfate solution gave a fourfold increase in sensitivity relative to the conventional solution, making measurements with the Al$_k$ x-rays feasible. The relative ferrous-to-ferric conversions as a function of dose were similar for the two gamma-ray energies, yielding G values of 1.62 and 1.59 μmol J^{-1} for the ^{60}Co and ^{137}Cs radiations, respectively. The α-particle G value was 0.52 μmol J^{-1}, or 31% of that for the ^{60}Co gamma rays, in good agreement with previous measurements. The Al$_k$ x-rays had a G value of 0.92 μmol J^{-1} or 57% of that for the ^{60}Co radiation. This G value for the 1.5-keV x-rays is within 20% of the values predicted by current theories, and theoretical values are within the measurement error.

HOSHI et al. (1992) measured the G values for 8.86- and 13.55-keV x-rays produced by synchrotron radiation. Monoenergetic x-rays were produced using a silicon crystal monochromator. Doses in the solution were determined using a thin-window, parallel-plate chamber calibrated against a primary standard free-air chamber at the Electrotechnical Laboratory (Osaka, Japan). Yields (G) of 1.50 ± 0.06 and 1.43 ± 0.06 μmol J^{-1} were obtained for 8.86- and 13.55-keV x-rays respectively.

SHORTT (1989) reported a new determination of the temperature dependence of $G(Fe^{3+})$ using precision spectrophotometry and a state-of-the-art Fricke dosimetry system (ROSS et al. 1989).

A recent comparison of the Fricke dosimeter with ionization chambers using the NACP protocol was presented by MATTSSON et al. (1982), and one using the 1983 AAPM protocol and the National Research Council of Cancer protocol by KWA and KORNELSON (1990). Both comparisons show agreement within 1%.

COTTENS et al. (1982) presented a study of the effect of chloride ion on the ferric ion yield of the Fricke dosimeter in the absence of impurities. Significant chloride ion effects on $G(Fe^{3+})$, not related to the suppression of impurity effects, were measured. This observation necessitates a modification of an earlier recommendation to use the absence of a Cl$^-$ effect as a purity criterion. Data in this work revealed a cubic root relationship with the sodium chloride concentration. This relationship was shown to hold up to 1 mol l^{-1} Cl$^-$, after suitable correction for the increased density of the dosimeter solution and for the effect of chloro-Fe^{3+} complexes at high Cl$^-$ concentration. It was shown that the measured G value decrease corresponds exactly to the decrease in the initial H_2O_2 yield in the presence of Cl$^-$, i.e., it equals $2\Delta G_{H_2O_2}$.

17.3.3.2 Fricke Dosimetry Using NMR Spectrometry

GORE et al. (1984) described a method for determining the spatial distribution of radiation dose in a tissue-equivalent phantom using NMR imaging. Their method is based upon the conversion of ferrous ions to ferric ions by ionizing radiation that alters the magnetic moment and electron spin relaxation times of the metal ion. The spin relaxation times (T1 and T2) of the hydrogen nuclei in an aqueous solution of a ferrous salt are consequently reduced substantially. These changes in T1 and T2 were measured using standard NMR techniques. The same conversion was used as in conventional Fricke dosimetry, which was employed to calibrate the technique.

APPLEBY et al. (1987) irradiated agarose gels containing ferrous sulfate, sulfuric acid, and benzoic acid with ^{137}Cs gamma rays and 6- to 14-Mev electrons, to doses of up to 20 Gy. The dose distributions were imaged by NMR, making use of the effect on the T1 proton relaxation times of the radiolytic Fe^{3+}. The image intensity was proportional to doses of up to 10 Gy, and images were stable for at least 24 h postirradiation. The G value for Fe^{3+} production was about 100 (molecules per 100 eV absorbed).

OLSSON et al. (1989) have reported the development of two gels that were suitable for loading with ferrous sulfate solution. In these soft tissue equivalent phantoms, the absorbed dose distribution was measured after irradiation in clinically used MRI equipment. A ferrous sulfate solution, 0.05 M with respect to sulfuric acid, was gelled with 4% gelatin to give a dosimeter which had a response which was linearly correlated ($r = 0.998$) with the absorbed dose in the interval 0–40 Gy. Ferrous sulfate solution was also gelled with 1% agarose, but this gel has to be purged with oxygen to obtain a linear relationship ($r = 0997$) in the same absorbed dose interval. The ferrous sulfate-loaded gels had a sensitivity which was a factor of 2.2 or 4.0 times higher for gelatin and agarose, respectively, than the ordinary dosimeter solution. Because the standard deviation of background measurements was higher for the gels than for the dosimeter solution, the minimum detectable absorbed dose was about the same, or 1.0 Gy, for the two gels and the dosimeter solution. The sensitivity of the ferrous sulfate-loaded gels showed no dependence on dose rate if the mean dose rate and the absorbed dose per pulse were within the limits normally used by accelerators for radiotherapy.

In a later study OLSSON et al. (1990) reported the measurement of absorbed dose distributions using this dosimeter gel and MRI. Absorbed depth-dose curves and profiles measured with this new technique showed good agreement with corresponding measurements using diodes. This was proven in a ^{60}Co beam as well as an electron beam. The dosimeter gel was made of agarose and ferrous sulfate solution. The sensitivity was higher than that of ordinary ferrous sulfate solution by a factor of about 6.

A dose-response curve for Fricke-infused agarose gels as obtained by nuclear magnetic resonance has been reported (SCHULZ et al. 1990a). It was concluded by SCHULZ et al. that:

1. Oxygen saturation assures consistent and maximum sensitivity.
2. Agarose concentrations in the range 1.0%–2.0% have no effect upon sensitivity.
3. The initial G value is 150 Fe^{3+}/100 eV for gels containing 0.5 mM Fe^{2+} ions.
4. Increasing NMR frequencies only causes a moderate increase in sensitivity.
5. The gel dosimeters are dose rate independent in the range 4.7–24.2 Gy min^{-1}.
6. Sensitivity is pH dependent, being zero at pH 7.
7. Freshly prepared gels are slightly more sensitive than those more than 24 h old.

8. The diffusion coefficient for ferric ions in a 1.0% agarose gel containing 0.0125 M H_2SO_4 is 1.83×10^{-2} cm^2 h^{-1}, and this will require consideration for the NMR imaging of dose distributions.

The spin-spin relaxation rate has been shown by PRASAD et al. (1991) to be a more sensitive parameter than the spin-lattice relaxation rate. In this work it was also demonstrated that the addition of chemical sensitizers could improve the dose sensitivity of the measured NMR parameters. As measured by this technique the two features characterizing a photon beam, depth-dose relationship and beam profile, were shown to be in good agreement with the measurements using conventional methods, ionization chambers, and film dosimetry.

A new method of producing gels in which the distributions of radiation dose can also be visualized as a color change was developed by APPLEBY and LEGHROUZ (1991). The color developed depended qualitatively and quantitatively on the concentrations of solutes in the gel. To detect ferric ions in irradiated ferrous solutions they used the metal ion indicator xylenol orange, which forms a colored complex with ferric ions. This is the basis of a sensitive radiation dosimetric system known as the FBX dosimeter, which is a water solution of ferrous ions, sulfuric acid, xylenol orange, and benzoic acid. APPLEBY and LEGHROUZ report that using their method radiation dose distributions can be visualized directly by eye, as well as by MRI.

HAZLE et al. (1991) investigated the NMR longitudinal relaxation rate R1 dose-response characteristics of a ferrous-sulfate-doped chemical dosimeter system (FeMRI) immobilized in a gelatin matrix. Samples containing various concentrations of the ferrous sulfate dosimeter were irradiated to absorbed doses of 0–150 Gy. R1 relaxation rates were determined by imaging the samples at a field strength of 1.5 T (^{1}H Lamor frequency of 63.8 MHz). The response of the system was found to be approximately linear up to doses of 50 Gy for all ferrous sulfate concentrations studied (0.1–2.0 mM). Changing concentrations in the range of 0.1–0.5 mM affected both the slope and the intercept of the dose-response curve. For concentrations of 0.5–2.0 mM, the slope of the dose-response curves remained constant at approximately 0.0423 s^{-1} Gy^{-1} in the dose range 0–50 Gy. However, the intercept of the curve continued to increase in that region, as expected, because of the additional paramagnetic ions. The reproducibility of the absorbed dose estimates for measurements made over a 22-cm field of view was found to be 5% in the range 20–50 Gy (an uncertainty of 0.81 Gy on average), degrading to approximately 10% in the dose range 5–10 Gy.

17.4 Phantoms for Dosimetry Calibration of Radiotherapy Beams

17.4.1 Charge Deposition Effects in Solid Phantoms

GALBRAITH et al. (1984) showed that plastic dosimetry phantoms used in electron beams can store charge and produce internal electric fields large enough to measurably alter the electron dose distribution in the plastic. The reading per monitor unit from a cylindrical ion chamber embedded in a polymethylmethacrylate (PMMA) or polystyrene phantom was observed to increase with accumulated electron dose, the increase being detectable after about 20 Gy of 6-MeV electrons. The magnitude of the effect also depends on the type of the plastic, the thickness of the plastic, the wall thickness of the detector, the diameter and depth of the hole in the plastic, the energy of the electron beam, and the dose rate used. The authors concluded that the charging effects can occur in plastics at the dose levels encountered in therapy dosimetry, where ion chamber or other dosimeter readings could easily increase by 5%–10% and where a phantom, once charged, would also affect subsequent readings taken in ^{60}Co beams and high-energy electron and x-ray beams for periods of several days to many months. It was recommended that conducting plastic phantoms should replace PMMA and polystyrene phantoms in radiation dosimetry.

These findings were confirmed by MATTSSON and SVENSSON (1984), who also showed that the electron fluence inside a cylindric ionization chamber used in these materials increases when the number of electrons trapped in the phantom increases. The influence of various parameters, such as electron energy, accumulated absorbed dose, type and purity of the phantom material, radiation-induced conductivity, and ionization chamber construction were discussed. MATTSSON and SVENSSON also concluded that the effect can cause errors of several percent in the dosimetry. In agreement with the conclusion of GALBRAITH et al. (1984), they also recommended that insulating phantom materials should be avoided in electron beam dosimetry.

Later, RAWLINSON et al. (1984) attributed this effect to the generation of large electric fields in the phantom by charge storage causing alteration of electron trajectories and an increase in the measured dose. They calculated the change in response to radiation of an ion chamber in a cylindrical cavity in an electron-irradiated PMMA phantom. The electric field distribution was determined using a model that allows for charge leakage by radiation-induced conductivity, and the dose in the cavity was determined by a Monte Carlo simulation using the EGS (electron gamma shower) code modified to account for electron trajectories in the electric field. The theoretical results were shown to agree well with new and previously published experimental dose enhancement data. The agreement was taken as confirmation of the reported explanation of the effect. The use of conducting phantoms in radiation dosimetry was advocated.

THWAITES (1984) carried out experiments at 10 MeV on the effect of charge storage on chamber readings in polystyrene, following previous irradiation with electron beams. In contrast to previous studies, he concluded that such effects have a negligible influence on ionization ratios. This was corroborated by BRUINVIS et al. (1985), who did not observe differences larger than 0.5% in the chamber readings in polystyrene irradiated by 6- and 19-MeV electrons.

17.4.2 Comparisons of Photon Dose Calibrations Using Different Phantoms

CONSTANTINOU et al. (1982) presented the results of a test on the equivalence of Solid Water to water. The formation, manufacture, and testing of an epoxy resin-based solid substitute for water was presented. This "solid water" had radiation characteristics very close volumetrically to those of water. Relative transmission measurements showed that for x-rays and gamma rays the transmission through 10 cm of Solid Water is within 0.2% of that through an equal thickness of water. It was concluded that the use of this material for calibration phantoms can help to ensure that radiotherapy beam calibrations are within ± 1.0% of the true dose rate.

ALMOND (1985) presented a comparison of x-ray dose calibrations using different phantom materials. Measurements were made in water, polystyrene, and acrylic phantoms using six commercially available cylinder ionization chambers (PTW N2333 A008, PTW N2333 A275P, Nuclear Enterprises NE 2505/ 3A, NE 2505/3B, Exradin TE and AE) in ^{60}Co gamma ray and 2-, 4-, and 8-MV x-ray beams.

KUBO and CHENG (1988) compared photon dose calibrations in polystyrene and acrylic phantoms using the AAPM 1983 protocol. The ionization chambers used were a Farmer graphite chamber, a PTW acrylic chamber, a homemade polystyrene chamber, and an Exradin air-equivalent chamber, all of cylindrical type. A Memorial parallel-plate polystyrene chamber was also included. ^{60}Co gamma rays and 4-, 6-, and 18-MV x-rays were investigated.

REFT (1989) tested the accuracy of calculated values for the restricted mass stopping powers and mass energy absorption coefficients that have been published for Solid Water (HO and PALIWAL 1986). He made output measurements, following the 1983 AAPM protocol, with a Farmer-type chamber in four materials for ^{60}Co gamma-ray and 4-, 6-, 10-, 18-, and 24-MV photon beams. The results showed that the scaled dose-to-water for the different media agrees to better than 1%, and the analysis supports the methodology of the protocol for obtaining the dose-to-water from the different media.

17.4.3 Comparisons of Electron Dose Calibrations Using Different Phantoms

BRUINVIS et al. (1985) determined correction factors for dose calibrations in plastic phantoms using a Farmer 0.6 cm^3 graphite-walled chamber for electron beams with mean energies at the phantom surface between 6 and 19 MeV. Experiments with white polystyrene yielded corrections for the measured ionization ranging from 0.3% to 2.4%. For clear polystyrene, 0.6%–1% higher corrections were found. For beams with the same mean energy at the phantom surface, but with different beam-flattening and collimation systems, variations of up to 1.2% were observed in this correction.

THWAITES (1985) carried out measurements of central axis depth-ionization curves and ionization at ionization maximum in water, clear polystyrene, and a commercially available Solid Water phantom material. Flat and cylindrical chambers were used for electron beams of 5–10 MeV. Displacements for the cylindrical chambers were determined, indicating a recommended value of 0.55 times cavity radius (without perturbation corrections). The use of a single scaling parameter was considered for converting depth-ionization curves obtained in plastic to those in water. This was shown to be valid within 2 mm at these energies and for these materials. Ionization ratios between water and polystyrene were presented, i.e., correction factors for converting

ionization readings in the plastic to readings in water required for electron dosimetry determinations. These showed 3% differences on average at these energies, ionization in water being higher. Variations were observed with chamber wall material and chamber type. Measurements in the Solid Water material showed it to be a better water substitute than polystyrene for electron dosimetry in this energy range, although still significantly different from a true water phantom.

KUBO and CHENG (1988) reported an absorbed dose comparison among various commercial ionization chambers in polystyrene and acrylic phantoms irradiated by 9- and 20-MeV electrons. The ionization chambers used were a Farmer graphite chamber, a PTW acrylic chamber, a homemade polystyrene chamber, and an Exradin air-equivalent chamber, all of cylindrical type. A Memorial parallel-plate polystyrene chamber was also included.

17.4.4 Scaling Theories

PRUITT and LOEVINGER (1982) presented a method of scaling photon fluence from one scattering material to another when the photon energies are such that the dominant mode of interaction is Compton scattering. The theorem established a one-to-one correspondence between points in the two scattering media where the spectra of primary and scattered photons have the same distribution in energy and angle, and where the fluence ratio equals the square of the electron density ratio. Experimental tests were made with ^{60}Co gamma radiation using ionization chamber measurements in graphite, acrylic plastic, polystyrene, and water phantoms. The experimental results were consistent with the equality of photon spectral shapes and angular distributions at corresponding points. They concluded that the fluence ratios may differ by a few percent from the predicted values, depending on distance from the source.

BJÄRNGARD (1987) presented a generalization of Fano's and O'Connor's theorems. Fano's theorem states that the fluence of particles, emitted uniformly per unit mass, is constant throughout an infinite medium of uniform composition but varying density. O'Connor's scaling theorem says that the ratio of the fluence of secondary particles to that of primary particles, caused by an external source irradiating a medium in a collimated beam, is the same in two uniform media of the same composition but different density, provided geometric distances are scaled inversely to density. Bjärngard showed that these two theorems follow one line of reasoning and presented a more general formulation.

17.5 Concluding Remarks

In this review we have presented recent advances in radiation dosimetry as they relate to calibration of radiotherapy beams of high-energy photons and electrons. We have not included the special considerations necessary for medium- and low-energy photons or electrons. Also, the calibration of proton, heavy ion, neutron, and pion beams and the dosimetry of brachytherapy sources are not covered here. Other major areas of radiation dosimetry that we did not include are the dosimetry procedures necessary for diagnostic imaging equipment and for radiation protection. Other chapters in the present book address some of these areas.

Acknowledgments. The authors would like to thank Anjali Nath for preparing an initial draft of this manuscript and Deanna Jacobs for editing and preparing the final version. We would also like to thank Dr. Kazi Motakabbir for his careful comments and review of this manuscript.

References

AAPM, American Association of Physicists in Medicine (1966) Protocol for the dosimetry of high energy electrons. Phys Med Biol 11: 505–520

AAPM, American Association of Physicists in Medicine (1971) Protocol for the dosimetry of X- and gamma-ray beams with maximum energies between 0.6 and 50 MeV. Phys Med Biol 16: 379–396

AAPM, American Association of Physicists in Medicine, Task Group No. 10, Radiation Therapy Committee (1975) Code of practice for x-ray therapy linear accelerators. Med Phys 2: 110–121

AAPM, American Association of Physicists in Medicine, Task Group No. 21, Radiation Therapy Committee (1983) A protocol for the determination of absorbed dose from high-energy photon and electron beams. Med Phys 10: 741–771

AAPM, American Association of Physicists in Medicine, Task Group 21, Radiation Therapy Committee (1984) Erratum: a protocol for the determination of absorbed dose from high-energy photon and electron beams [Med. Phys. 10: 741 (1983)]. Med Phys 11: 213

AAPM, American Association of Physicists in Medicine, Task Group 25, Radiation Therapy Committee (1991) Clinical electron-beam dosimetry: report of AAPM Radiation Therapy Committee Task Group no. 25. Med Phys 18: 73–109

Aget H, Rosenwald JC (1991) Polarity effect for various ionization chambers with multiple irradiation conditions in electron beams. Med Phys 18: 67–72

Almond PR (1967) The physical measurement of electron beams from 6 to 18 MeV: absorbed dose and energy calibrations. Phys Med Biol 12: 13–24

Almond PR (1985) A comparison of x-ray dose calibration using different phantom materials. Radiother Oncol 4: 319–323

Almond PR, Svensson H (1977) Ionization chamber dosimetry for photon and electron beams. Acta Radiol Ther Phys Biol 16: 177–186

Almond PR, Mendez A, Behmardf M (1978) In: National and international standardization of radiation dosimetry, vol II. STI/PUB/471, IAEA, Vienna, pp 271–289

Andreo P (1983) DOSIS: a computer program for the calculation of absorbed dose in photon and electron beams from ionization measurements in a phantom. Nucl Instrum Methods 211: 481

Andreo P, Brahme A (1981) Mean energy in electron beams. Med Phys 8: 682–687

Andreo P, Brahme A (1984) Restricted energy-loss straggling and multiple scattering of electrons in mixed Monte Carlo procedures. Radiat Res 100: 16–29

Andreo P, Brahme A (1986) Stopping power data for high-energy photon beams. Phys Med Biol 31: 839–858

Andreo P, Nahum AE (1985) Stopping-power ratio for a photon spectrum as a weighted sum of the values for monoenergetic photon beams. Phys Med Biol 30: 1055–1065

Andreo P, Nahum AE, Brahme A (1986) Chamber-dependent wall correction factors in dosimetry. Phys Med Biol 31: 1189–1199

Andreo P, Nahum AE, Svensson H (1987) Recent developments in basic dosimetry. Radiother Oncol 10: 117–126

Andreo P, Brahme A, Nahum A, Mattsson O (1989) Influence of energy and angular spread on stopping-power ratios for electron beams. Phys Med Biol 34: 751–768

Andreo P, Lindborg L, Medin J (1991) On the calibration of plane-parallel ion chambers using ^{60}Co beams. Med Phys 18: 326–327

Andreo P, Lindborg L, Medin J (1992) Comments to "Chamber replacement correction in absorbed dose calibrations," by J.E. Burns. Med Phys 19: 211

Appleby A, Leghrouz A (1991) Imaging of radiation dose by visible color development in ferrous-agarose-xylenol orange gels. Med Phys 18: 309–321

Appleby A, Christman EA, Leghrouz A (1987) Imaging of spatial radiation dose distribution in agarose gels using magnetic resonance. Med Phys 14: 382–384

Ashley JC (1982a) Density effect in liquid water. Radiat Res 89: 32–37

Ashley JC (1982b) Stopping power of liquid water for low-energy electrons. Radiat Res 89: 25–31

Attix FH (1984a) Presentation of nominal accelerating potential as a function of the ionization ratio in the new AAPM dosimetry protocol. Med Phys 11: 565–566

Attix FH (1984b) Determination of A_{ion} and P_{ion} in the new AAPM radiotherapy dosimetry protocol. Med Phys 11: 714–716

Attix FH (1984c) A simple derivation of N_{gas}, a correction in A_{wall}, and other comments on the AAPM Task Group 21 protocol. Med Phys 11: 725–728

Attix FH (1989) Equations for N_{gas} and N_{air} in terms of N_x and N_k. Med Phys 16: 803–806

Attix FH (1990) A proposal for the calibration of plane-parallel ion chambers by accredited dosimetry calibration laboratories. Med Phys 17: 931–933

Austerlitz C, Sibata CH, de Almeida CE (1987) A graphite transmission ionization chamber. Med Phys 14: 1056–1059

Awschalom M, Rosenberg I, Ten Haken RK (1983) A new look at displacement factor and point of measurement corrections in ionization chamber dosimetry. Med Phys 10: 307–313

Barish RJ (1984) Thermal characteristics of a common polystyrene phantom. Med Phys 11: 214–215

Barish RJ, Lerch IA (1992) Long-term use of an isotope check source or verification of ion chamber calibration. Med Phys 19: 203–205

Barnard GP (1964) Dose-exposure conversion factors for megavoltage x-ray dosimetry. Phys Med Biol 9: 321–332

Barnard GP, Axton EJ, Marsh ARS (1959) A study of cavity ion chambers for use with 2 MV x-rays: equilibrium wall thickness: wall absorption correction. Phys Med Biol 3: 366

Berger MJ, Seltzer SM (1964) Studies in penetration of charged particles in matter. National Research Council Publication 1133, Washington National Academy of Sciences

Berger MJ, Seltzer SM (1982) Stopping powers and ranges of electrons and positrons. NBS Report IR 82–2550. US Dept. of Commerce, Gaithersburg, MD

Berger MJ, Domen SR, Lamperti PJ (1975) Stopping power ratios for electron dosimetry with ionization chambers. In: Biomedical dosimetry. IAEA Vienna. pp 589–609

Berkley LW, Gagnon WF, Hanson WF, Weaver KA, Shalek RJ (1980) A review of the discrepancy between the in-air and in-water calibration of cobalt-60 machines. Med Phys 7: 520–524

Bewley DK (1963) The measurement of locally absorbed dose of megavoltage x-rays by means of a carbon calorimeter. Br J Radiol 36: 865–878

Bielajew AF (1985) The effect of free electrons on ionization chamber saturation curves. Med Phys 12: 197–200

Bielajew AF (1990a) Correction factors for thick-walled ionisation chambers in point-source photon beams. Phys Med Biol 35: 501–516

Bielajew AF (1990b) An analytic theory of the point-source non-uniformity correction factor for thick-walled ionisation chambers in photon beams. Phys Med Biol 35: 517–538

Bielajew AF (1990c) On the technique of extrapolation to obtain wall correction factors for ion chambers irradiated by photon beams. Med Phys 17: 583–587

Bielajew AF, Rogers DWO, Nahum AE (1985) The Monte Carlo simualtion of ion chamber response to ^{60}Co resolution of anomalies associated with interfaces. Phys Med Biol 30: 419–427

Bistrovic M, Viculin T (1987) Comments on the comparison of the new and old C_E factors listed in the 1985 HPA code. Phys Med Biol 32: 905–906

Bjärngard BE (1987) On Fano's and O'Connor's theorems. Radiat Res 109: 184–189

Bjärngard BE, Kase KR (1985) Replacement correction factors for photon and electron dose measurements. Med Phys 12: 785–787

Bjärngard BE, Tsai J-S, Rice RK (1989) Attenuation in very narrow photon beams. Radiat Res 118: 195–200

Bloch P (1988) A unified electron/photon dosimetry approach. Phys Med Biol 33: 373–377

Boag JW (1982) The recombination correction for an ionization chamber exposed to pulsed radiation in a 'swept beam' technique. Phys Med Biol 27: 201–211

Boag JW (1984) Dosimetry in a magnetically swept electron beam. Radiother Oncol 2: 37–40

Boag JW (1987) Ionization chambers. In: Kase KR, Bjärngard BE, Attix FH (eds) The dosimetry of ionizing radiation, vol II. Academic Press, 169–244

Boese HR, Cormack DV (1985) Detection of a leak in a "sealed" monitor chamber. Med Phys 12: 377–378

Böhm J (1980) The perturbation correction factor of ionisation chambers in β-radiation fields. Phys Med Biol 25: 65–75

Boutillon M (November 1977) Some remarks concerning the measurement of kerma with a cavity ionization chamber. Bureau International des Poids et Mesurer (CCEMRI (I))/ 77–114

Boutillon M (1983) Perturbation correction for the ionometric determination of absorbed dose in a graphite phantom for ^{60}Co gamma rays. Phys Med Biol 28: 375–388

Boutillon M (1989) Gap correction for the calorimetric measurement of absorbed dose in graphite with a ^{60}Co beam. Phys Med Biol 34: 1809–1821

Boutillon M, Perroche-Roux AM (1987) Re-evaluation of the W value for electrons in dry air. Phys Med Biol 32: 213–219

Brahme A, Andreo P (1986) Dosimetry and quality specification of high energy photon beams. Acta Radiol Oncol 25: 213–223

Bruinvis IAD, Heukelom S, Mijnheer BJ (1985) Comparison of ionisation measurements in water and polystyrene for electron beam dosimetry. Phys Med Biol 30: 1043–1053

Burns JE (1992) Chamber replacement correction in absorbed dose calibration. Med Phys 19: 209–211

Burns JE, Rosser KE (1990) Saturation correction for the NE 2560/1 dosemeter in photon dosimetry. Phys Med Biol 35: 687–693

Burns JE, Dale JWG, DuSautoy AR, Owen B, Pritchard DH (1988) New calibration service for high-energy x-radiation at NPL. In: Proceedings of sympium on dosimetry in radiotherapy, vol 2. IAEA, Vienna, pp 125–132

Campos LL, Caldas LVE (1991) Absorbed dose dependence of the correction factors for ionization chamber cable irradiation effects. Phys Med Biol 36: 339–344

Carlsson GA (1985) Theoretical basis for dosimetry. In: Kase KR, Bjärngard BE, Attix TM (eds) The dosimetry of ionizing radiation, vol I. Academic Press, London, pp 2–77

Casson H, Kiley JP (1987) Replacement correction factors for electron measurements with a parallel-plate chamber. Med Phys 14: 216–217

Comité de Dosimetría en Radioterapia, Sociedad Española de Física Médica, Brosed A, Andreo P, Gómez D, Gultresa J, Mincholé JL, Serrano C, Vivanco J (1985) The Spanish dosimetry protocol. Radiother Oncol 4: 305–308

Conere TJ (1986a) Some dosimetric discrepancies obtained using a guarded parallel-plate ion chamber with a high input impedance electrometer in measurements involving a pulsed and magnetically swept electron beam. Phys Med Biol 31: 1157–1160

Conere TJ (1986b) Variation in collection efficiency of two ion chambers of the same model type. Radiother Oncol 6: 77–78

Conere TJ, Boag JW (1984) The collection efficiency of an ionization chamber in a pulsed and magnetically swept electron beam: limits of validity of the two-voltage technique. Med Phys 11: 465–468

Constantinou C (1982) Phantom materials for radiation dosimetry. I. Liquids and gels. Br J Radiol 55: 217–224

Constantinou C, Attix FH, Paliwal BR (1982) A solid water phantom material for radiotherapy x-ray and γ-ray beam calibrations. Med Phys 9: 436–441

Cottens E, Janssens A, Eggermont G, Buysse J (1982) Study of the effect of chloride ion on the ferric ion yield of the Fricke dosemeter in the absense of impurities. Phys Med Biol 27: 597–602

Cunningham JR, Johns HE (1980) Calculations of the average energy absorbed in photon interactions. Med Phys 7: 51–54

Cunningham JR, Schulz RJ (1984) On the selection of stopping-power and mass energy-absorption coefficient ratios for high-energy x-ray dosimetry. Med Phys 11: 618–623

Cunningham JR, Sontag MR (1980) Displacement corrections used in absorbed dose determinations. Med Phys 7: 672–676

Cunningham JR, Woo M, Rogers DWO, Bielajew AF (1986) The dependence of mass energy absorption coefficient ratios on beam size and depth in a phantom. Med Phys 13: 496–502

Davies JV, Greene D, Keene JP, Law J, Massey JB (1963) A comparison of ionization, calorimetric and ferrous sulphate dosimetry. Phys Med Biol 8: 97–102

Day MJ (1990) Radiation dosimetry using nuclear magnetic resonance: an introductory review. Phys Med Biol 35: 1605–1609

de Almeida CE, Perroche-Roux A-M, Boutillon M (1989) Perturbation correction of a cylindrical thimble-type chamber in a graphite phantom for ^{60}Co gamma rays. Phys Med Biol 34: 1443–1449

Domen SR (1980a) Absorbed dose water calorimeter. Med Phys 7: 157–159

Domen SR (1980b) Thermal diffusivity, specific heat, and thermal conductivity of A-150 plastic. Phys Med Biol 25: 93–102

Domen SR (1982) An absorbed dose water calorimeter: therapy, design, and performance. J Res Natl Bur Stand 87: 211–235

Domen SR (1983) A polystyrene-water calorimeter. Int J Appl Radiat Isot 34: 643–644

Domen SR (1986) Comment on 'convection current in a water calorimeter.' Phys Med Biol 31: 1166–69

Domen SR (1988) Further comments on convection currents in a water calorimeter. Phys Med Biol 33: 1083–1086

Domen SR (1990) Advances in calorimetry for radiation dosimetry. In: Kase KR, Bjärngard BE, Attix FH (eds.) The dosimetry of ionizing radiation, vol II. Academic Press, London, pp 245–320

Dutreix A (1985) The French dosimetry protocol. Radiother Oncol 4: 301–304

Dutreix A, Bridier A (1985) Dosimetry for external beams of photon and electron radiation. In: Kase KR, Bjärngard BE, Attix FH (eds) The dosimetry of ionizing radiation, vol I. Academic Press, London, pp 164–229

Dutreix J, Dutreix A (1966) Etude comparée d'une série de chambres d'ionisation dans des faisceaux d'électrons de 20 et 10 MeV. Biophysik 3: 249–258

Dutreix A, Mijnheer B, Svensson H (1985) New protocols for the dosimetry of high-energy photon and electron beams (introduction). Radiother Oncol 4: 289–290

Engler MJ, Jones GL (1984) Small-beam calibration by 0.6– and 0.2-cm^3 ionization chambers. Med Phys 11: 822–826

Fallone BG, Podgorsak EB (1983) Saturation curves of parallel-plate ionization chambers. Med Phys 10: 191–196

Freyer JP, Schillaci ME, Raju MR (1989) Measurement of the G-value for 1.5 keV X-rays. Int J Radiat Biol 56: 885–892

Fricke H, Hart EJ (1966) Chemical dosimetry, in Radiation Dosimetry vol II edited by FH Attix and WC Roesch, Academic Press 1966, p. 167–240

Gajewski R, Izewska J (1987) Perturbation correction factors for the plane-parallel chamber NE 2534. In: Proc. dosimetry in radiotherapy IAEA-SM-298/82, vol 1. IAEA, Vienna, pp 187–193

Galbraith DM, Rawlinson JA, Munro P (1984) Dose errors due to charge storage in electron irradiated plastic phantoms. Med Phys 11: 197–203

Galloway G, Greening JR, Williams JR (1986) A water calorimeter for neutron dosimetry. Phys Med Biol 31: 397–406

Gastorf RJ, Hanson WF, Shalel RJ, Berkley LW (1984) The implementation of the AAPM Task Group 21 protocol by the Radiological Physics Center and its implications. Med Phys 11: 547–551

Gastorf RJ, Humphries L, Rozenfeld M (1986) Cylindrical chamber dimensions and the corresponding values of A_{wall} and $N_{gas}/(N_x A_{ion})$. Med Phys 13: 751–754

Gerbi BJ, Khan FM (1987) The polarity effect for commercially available plane-parallel ionization chambers. Med Phys 14: 210–215

Gerbi BJ, Khan FM (1990) Measurement of dose in the buildup region using fixed-separation plane-parallel ionization chambers. Med Phys 17: 17–26

German Standard Association (1975a, draft) Procedures in dosimetry; principles of photon and electron dosimetry with probe-type detectors. In: DIN 6800/1

German Standard Association (1975b, draft) Procedures in dosimetry; ionization dosimetry. In: DIN 6800/2

German Standard Association (1976) Clinical dosimetry; therapeutical application of x-ray, gamma-ray and electron beams. In: DIN 6809/1

Gillin MT, Kline RW, Niroomand-Rad A, Grimm DF (1985) The effect of thickness of the waterproofing sheath on the calibration of photon and electron beams. Med Phys 12: 234–236

Goodman LJ (1978) Density and composition uniformity of A-150 tissue-equivalent plastic. Phys Med Biol 23: 753–758

Gore JC, Kang YS, Schulz RJ (1984) Measurement of radiation dose distributions by nuclear magnetic resonance (NMR) imaging. Phys Med Biol 29: 1189–1197

Goswami GC, Kase KR (1989) Measurement of replacement factors for a parallel-plate chamber. Med Phys 16: 791–793

Greene D (1962) The use of an ethylene-filled polythene chamber for dosimetry of megavoltage x-rays. Phys Med Biol 7: 213–224

Greene D, Massey JB (1966) The use of Farmer-Baldwin and Victrometer ionization chambers for dosimetry of high energy x-radiation. Phys Med Biol 11: 569–575

Greene D, Massey JB (1967) The use of Farmer-Baldwin and Victrometer ionization chambers for dosimetry of high energy x-radiation. Phys Med Biol 12: 257–258

Greene D, Massey JB (1968) The use of Farmer-Baldwin and Victrometer ionization chambers for dosimetry of high energy x-radiation. Phys Med Biol 13: 287–288

Hanson WF, Tinoco JAD (1985) Effects of plastic protective caps on the calibration of therapy beams in water. Med Phys 12: 243–248

Hanson WF, Arnold DJ, Shalek RJ, Humphrines LJ (1988) Contamination of ionization chambers by talcum powder. Med Phys 15: 776–777

Harder D (1965) Berechnung der Energiedosis aus Ionisationsmessungen bei Sekundärelektronen-gleichgewicht. In: Zuppinger A, Poretti G (eds) Symposium on high-energy electrons. Springer, Berlin Heidelberg New York, pp 260

Harder D (1968) Einfluss der Vielfachstreuung von Elektronen auf die Ionisation in gasgefüllten Hohlräumen. Biophysik 5: 157–164

Hayakama Y, Schechtman H (1988) Comments on the value of the average energy expended per ion pair formed in air for a proton beam recommended by the American Association of Physicists in Medicine. Med Phys 15: 778

Hayakama Y, Loch CP, Tada J, Inada T (1989) Compensation for beam intensity fluctuation in determination of P_{ion}, the ion-recombination correction factor for ionization chambers, by the two-voltage technique. Med Phys 16: 346–351

Hazle JD, Hefner L, Nyerick CE, Wilson L, Boyer AL (1991) Dose-response characteristics of a ferrous-sulphate-doped gelatin system for determining radiation absorbed dose distributions by magnetic resonance imaging (FeMRI)*. Phys Med Biol 36: 1117–1125

Hermann K-P, Geworski L, Hatzky T, Lietz R, Harder D (1986) Muscle- and fat-equivalent polyethylene-based phantom materials for x-ray dosimetry at tube voltages below 100 kV. Phys Med Biol 31: 1041–1046

Heese RN, Podgorsak EB, Fallone BG (1986) Approximations to saturation curves in gas-filled parallel-plate ionization chambers. Med Phys 13: 93–98

Hettinger G, Pettersson C, Svensson H (1967a) Displacement effect of thimble chambers exposed to a photon or electron beam from a betatron. Acta Radiol Ther 6: 61–64

Hettinger G, Pettersson C, Svensson H (1967b) Calibration of thimble chambers in a 34 MV roentgen beam. Acta Radiol Ther 6: 214–218

Heukelom S, Lanson JH, Mijnheer BJ (1991) Comparison of entrance and exit dose measurements using ionization chambers and silicon diodes. Phys Med Biol 36: 47–59

Ho AK, Paliwal BR (1986) Stopping-power and mass energy-absorption coefficient ratios for solid water. Med Phys 13: 403–404

Hochhäuser E, Balk OA (1986) The influence of unattached electrons on the collection efficiency of ionisation chambers for the measurement of radiation pulses of high dose rate. Phys Med Biol 31: 223–233

Hogstrom KR, Almond PR (1982) The effect of electron multiple scattering on dose measured in non-water phantoms (abstract). Med Phys 9: 607

Holt JG, Kessaris ND (1977) Discrepancy between C and C_E. Phys Med Biol 22: 538–540

Hoshi M, Uehara S, Yamamoto O, et al. (1992) Iron (II) sulphate (Ficke solution) oxidation yields for 8.9 and 13.6 keV X-rays from synchrotron radiation. Int J Radiat Biol 61: 21–27

Houdek PV (1983) Dosimetry of small radiation fields for 10-MV x-rays. Med Phys 10: 333–336

HPA, Hospital Physicists' Association (1960) A code of practice for x-ray measurements. Br J Radiol 33: 55–59

HPA, Hospital Physicists' Association (1964) A code of practice for the dosimetry of 2 to 8 MV X-ray and caesium-137 and cobalt-60 γ-ray beams. Phys Med Biol 9: 457–463

HPA, Hospital Physicists' Association (1969) A code of practice for the dosimetry of 2 to 35 MV X-ray and caesium-137 and cobalt-60 gamma-ray beams. Phys Med Biol 14: 1–8

HPA, Hospital Physicists' Association (1971) A practical guide to electron dosimetry 5–35 MeV. In: HPA Report Series No. 4

HPA, Hospital Physicists' Association (1975) A practical guide to electron dosimetry below 5 MeV for radiotherapy purposes. In: HPA Report Series No. 13

HPA, Hospital Physicists' Association (1983) Revised code of practice for the dosimetry of 2 to 25 MV x-ray, and of caesium-137 and cobalt-60 gamma-ray beams. Phys Med Biol 28: 1097–1104

HPA, Hospital Physicists' Association (1985) Code of practice for electron beam dosimetry in radiotherapy. Phys Med Biol 30: 1169–1194

HPA, Hospital Physicists' Association (1990) Code of practice for high-energy photon therapy dosimetry based on the NPL absorbed dose calibration service. Phys Med Biol 35: 1355–1360

Hubbell JH (1977) Photon mass attenuation and mass energy-absorption coefficients for H, C, N, O, Ar and seven mixtures from 0.1 keV to 20 MeV. Radiat Res 70: 58–81

Hubbell JH (1982) Photon mass attenuation and energy-absorption coefficients for 1 keV to 20 MeV. Int J Appl Radiat Isot 33: 1269–1290

Hunt MA, Malik S, Thomason C, Masterson ME (1984) A comparison of the AAPM "Protocol for the determination of absorbed dose from high-energy photon and electron beams" with currently used protocols. Med Phys 11: 806–813

Hunt MA, Kutcher GJ, Buffa A (1988) Electron backscatter corrections for parallel-plate chambers. Med Phys 15: 96–103

IAEA, International Atomic Energy Agency (1962) Single field isodose charts: an international guide. IAEA, Vienna

IAEA, International Atomic Energy Agency (1987) Absorbed dose determination in photon and electron beams: an international code of practice. Technical Reports Series No. 277, IAEA, Vienna, pp 1–98

ICRU, International Commission on Radiation Units and Measurements (1969) Radiation dosimetry: x-rays and gamma rays with maximum photon energies between 0.6 and 50 MeV. ICRU Report No. 14, Washington, DC

ICRU, International Commission on Radiation Units and Measurements (1972) Radiation dosimetry: electrons with initial energies between 1 and 50 MeV. Report No. 21, ICRU, Washington, DC

ICRU, International Commission on Radiation Units and Measurements (1973) Measurement of absorbed dose in a phantom irradiated by a single beam of X or gamma rays. Report No. 23, ICRU, Washington, DC

ICRU, International Commission on Radiation Units and Measurements (1979) Average energy required to produce an ion pair. Report No. 31, ICRU, Washington, DC

ICRU, International Commission on Radiation Units and Measurements (1984a) Stopping powers for electrons and positrons. Report No. 37, ICRU, Bethesda, MD

ICRU, International Commission on Radiation Units and Measurements (1984b) Radiation dosimetry: electron beams with energies between 1 and 50 MeV. Report No. 35, ICRU, Bethesda, MD

IPSM, Institute of Physical Sciences in Medicine (1990) Code of practice for high-energy photon therapy dosimetry based on the NPL absorbed dose calibration service. Phys Med Biol 35: 1355–1360

IPSM, Institute of Physical Sciences in Medicine (1991) Report of the IPSM working party on low- and medium-energy x-ray dosimetry. Phys Med Biol 36: 1027–1038

Janssens A (1984) The fundamental constraint of cavity theory. Phys Med Biol 1157–1158

Jayaraman S, Rozenfeld M, Lanzl LH, Chung-Bin A (1985) Can the AAPM Task Group 21 protocol lead to optimum ion chamber designs? Med Phys 12: 373–376

Johansson K-A, Svensson H (1982) Liquid ionization chamber for absorbed dose determinations in photon and electron beams. Acta Radiol Oncol 21: 359–367

Johansson K-A, Mattsson LO, Lindborg L, Svesson H (1978) Absorbed dose determination with ionization chambers in electron and photon beams having energies between 1 and 50 MeV. In: Proceedings of international symposium on national and international standardization of radiation dosimetry 2. IAEA, Vienna, pp 243–270

Johansson K-A, Horiot JC, Van Dam J, Lepinoy D, Sentenac I, Sernbo G (1986) Quality assurance control in the EORTC cooperative group of radiotherapy. 2. Dosimetric intercomparison. Radiother Oncol 7: 269–279

Johns HE, Epp ER, Cormack DV, Fedoruk SO (1952) II. Depth dose data and diaphragm design for the Saskatchewan 1000 Curie cobalt unit. Br J Radiol 25: 302

Jones D (1981) Comparison of the perturbation correction in a parallel plate and a cylindrical ion chamber. Med Phys 8: 239–241

Kase KR, Adler GJ, Björngard BE (1982) Comparisons of electron beam dose measurements in water and polystyrene using various dosimeters. Med Phys 9: 13–19

Kearsley E (1984) A new general cavity theory. Phys Med Biol 29: 1179–1187

Kemp LAW (1972) The NPL secondary standard therapy-level x-ray exposure meter. Br J Radiol 45: 775–778

Kessaris ND (1970) Absorbed dose and cavity ionization for high-energy electron beams. Radiat Res 43: 288–301

Khan FM, Doppke KP, Hogstrom DR, et al. (1991) Clinical electron-beam dosimetry: report of AAPM Radiation Therapy Committee Task Group No. 25. Med Phys 18: 73–109

Klevenhagen SC (1991) Determination of absorbed dose in high-energy electron and photon radiation by means of an uncalibrated ionization chamber. Phys Med Biol 36: 239–253

Kooy HM, Simpson LD, McFaul JA (1988) Parallel-plate ionization chamber response in cobalt-60 irradiated transition zones. Med Phys 15: 199–203

Kristensen M (1983) Measured influence of the central electrode diameter and material on the response of a graphite ionisation chamber to cobalt-60 gamma rays. Phys Med Biol 28: 1269–1278

Krithivas G (1984) A study of the efficacy of a single voltage electrometer-chamber system in determining the ion collection efficiency. Phys Med Biol 29: 1265–1269

Krithivas G, Rao SN (1986) N_{gas} determination for a parallel-plate ion chamber. Med Phys 13: 674–677

Kubo H (1985) Estimate of the amount of thermal diffusion from a polystyrene-water calorimeter detector to surrounding water during irradiation. Phys Med Biol 30: 785–798

Kubo H (1990) Reply to 'Comments on construction of a calorimeter prototype with a high sensitivity pulsed signal detection circuit.' Phys Med Biol 35: 1029–1030

Kubo H, Cheng P (1988) Absorbed dose comparison among commercial ionization chambers in polystyrene and acrylic phantoms. Med Phys 15: 269–272

Kubo H, Brown DE, Russell MD (1985) A thermoregulated enclosure for controlling thermal drift in a radiation calorimeter. Med Phys 12: 344–346

Kubo H, Kent LJ, Krithivas G (1986) Determinations of N_{gas} and P_{repl} factors from commercially available parallel-plate chambers: AAPM Task Group 21 protocol. Med Phys 13: 908–912

Kubo H, Kageyama Y, Lo KK (1989) Construction of a calorimeter prototype with a high sensitivity pulsed signal detection circuit. Phys Med Biol 34: 1119–1123

Kwa W, Kornelson RO (1990) Comparison of ferrous sulfate (Fricke) and ionization dosimetry for high-energy photon and electron beams. Med Phys 17: 602–606

Laurence GC (1937) Can J Res A 15: 67, as cited in: Principles of radiation dosimetry by GN Whyte, John Wiley, New York (1959) pp 71

Law J, Foster CJ (1987) Calibration of radiotherapy dosemeters against secondary standard dosemeters: an anomalous result. Phys Med Biol 32: 1039–1043

Law J, Naylor GP (1984) Ferrous sulphate G-values for megavoltage photons and electrons derived from ionisation dosimetry using C_λ and C_E values. Phys Med Biol 29: 749–750

Liu P, Kruger RA (1984) Comments on "quantum noise in detectors." Med Phys 11: 561

Loevinger R (1981) A formalism for calculation of absorbed dose to a medium from photon and electron beams. Med Phys 8: 1–12

Loevinger R (1985) The new AAPM protocol. Radiother Oncol 4: 295–296

Ma CM, Nahum AE (1991) Bragg-Gray theory and ion chamber dosimetry for photon beams. Phys Med Biol 36: 413–428

Mach H, Rogers DWO (1983) An absolutely calibrated source of 6.13 MeV gamma-rays. IEEE Trans Nucl Sci NS-30: 1514

Mach H, Rogers DWO (1984) A measurement of absorbed dose to water per unit incident 7 MeV photon fluence. Phys Med Biol 29: 1555–1570

Majenka I, Rostkowska J, Derezinski M, Paz N (1982) The recombination correction for an ionization chamber exposed to pulsed radiation in a 'swept beam' technique. II. Experimental. Phys Med Biol 27: 213–221

Markus B (1964) Beiträge zur Entwicklung der Dosimetrie Schneller Elecktronen, Teil III. Strahlentherapie 124: 33

Marinello G, Valero M, Delplanque JM (1986a) The study of a swept electron beam in order to apply Boag's theory for calculation of the collection efficiency. I. Beam and swept area characteristics. Phys Med Biol 31: 859–868

Marinello G, Valero M, Bellec-Pollack J (1986b) The study of a swept electron beam in order to apply Boag's theory for calculation of the collection efficiency. II. Application to different ionisation chambers and comparison with other methods. Phys Med Biol 31: 869–878

Mattsson O (1985) Comparison of different protocols for the dosimetry of high-energy photon and electron beams. Radiother Oncol 4: 313–318

Mattsson O (1990) Comparison of absorbed dose determinations using the IAEA dosimetry protocol and the ferrous sulphate dosimeter. Med Phys World 6

Mattsson O, Svensson H (1984) Charge build-up effects in insulating phantom materials. Acta Radiol Oncol 23: 393–399

Mattsson O, Johansson K-A, Svensson H (1981) Calibration and use of plane-parallel ionization chambers for the determination of absorbed dose in electron beams. Acta Radiol Oncol 20: 385–399

Mattsson O, Johansson K-A, Svensson H (1982) Ferrous sulphate dosimeter for control of ionization chamber dosimetry of electron and ^{60}Co gamma beams. Acta Radiol Oncol 21: 139–144

Mattsson O, Svensson H, Wickman G, Domen SR, Pruitt JS, Loevinger R (1990) Absorbed dose in water. Acta Oncol 29: 235–240

Mayo CS, Gottschalk (1992) Temperature coefficient of open thimble chambers. Phys Med Biol 37: 289–291

McEwan AC (1980) A theoretical study of cavity chamber correction factors for photon beam absorbed dose determination. Phys Med Biol 25: 39–50

McEwan AC, Smyth VG (1984) Comments on "calculated response and wall correction factors for ionization chambers exposed to ^{60}Co gamma-rays." Med Phys 11: 216–218

Meli JA, Weinhous MS (1986) Collection efficiency of an ionisation chamber in a pulsed swept beam: chamber size effects. Phys Med Biol 31: 1139–1146

Mellenberg DE Jr (1990) Determination of build-up region over-response corrections for a Markus-type chamber. Med Phys 17: 1041–1044

Mijnheer BJ (1985a) Variations in response to radiation of a nylon-walled ionization chamber induced by humidity changes. Med Phys 12: 625–626

Mijnheer BJ (1985b) Summary of the discussion on the practical use and comparison of new protocols for the dosimetry of high-energy photon and electron beams. Radiother Oncol 4: 325–328

Mijnheer BJ, Chin LM (1989) The effect of differences in data base on the determination of absorbed dose in high-energy photon beams using the American Association of Physicists in Medicine protocol. Med Phys 16: 119–122

Mijnheer BJ, Williams JR (1985) Comments on dry air or humid air values for physical parameters using in AAPM protocol for photon and electron dosimetry. Med Phys 12: 656–658

Mijnheer BJ, Wittämper FW (1986) Comparison of recent codes of practice for high-energy photon dosimetry. Phys Med Biol 31: 407–416

Mijnheer BJ, Aalbers AHL, Visser AG, Wittämper FW (1986) Consistency and simplicity in the determination of absorbed dose to water in high-energy photon beams: a new code of practice. Radiother Oncol 7: 371–384

Mijnheer BJ, Wittämper FW, Aalbers AHL, van Dijk E (1987) Experimental verification of the air kerma to absorbed dose conversion factor $C_{w,u}$. Radiother Oncol 8: 49–56

Morris WT, Owen B (1975) An ionisation chamber for therapy-level dosimetry of electron beams. Phys Med Biol 20: 718–727

Mosse D, Cance M, Steinschaden K, Chartier M, Ostrowsky A, Simoen JP (1982) Détermination du rendement du dosimètre au sulfate ferreux dans un faisceau d'électrons de 35 MeV. Phys Med Biol 27: 583–596

Müller-Sievers K, Kober B (1989) Considerations on recombination losses in ionization chambers using pulsed electron beams with beam scanning. Int J Radiat Oncol Biol Phys 17: 1323–1325

NACP, Nordic Association of Clinical Physics (1972) Procedures in radiation therapy dosimetry with 5 to 50 MeV electrons and roentgen and gamma rays with maximum photon energies between 1 and 50 MeV. Acta Radiat Ther 11: 603–624

NACP, Nordic Association of Clincial Physics (1980) Procedures in external radiation therapy dosimetry with electron and photon beams with maximum energies between 1 and 50 MeV. Acta Radiol Oncol 19: 55–79

NACP, Nordic Association of Clinical Physics (1981) Electron beams with mean energies at the phantom surface below 15 MeV. Acta Radiol Oncol 20: 401–415

Nahum AE (1975) Ph.D. Thesis. University of Edinburgh, Univ. Micofilm Int. Order No. 77–70,006

Nahum AE (1978) Water/air mass stopping power ratios for megavoltage photon and electron beams. Phys Med Biol 23: 24–38

Nahum AE, Greening JR (1976) Inconsistencies in derivation of C_λ and C_E. Phys Med Biol 21: 862–864

Nahum AE, Greening JR (1978) A detailed re-evaluation of C_λ and C_E with application to ferrous sulphate G-values. Phys Med Biol 23: 894–908

Nahum AE, Kristensen M (1982) Calculated response and wall correction factors for ionization chambers exposed to ^{60}Co gamma rays. Med Phys 9: 925–927

Nahum AE, Svensson H, Brahme A (1980) The ferrous sulfate G-value for electron and photon beams: a semi-empirical analysis and its experimental support. In: Proceedings of the seventh symposium on microdosimetry. Harwood, New York, pp 841–851

Nahum AE, Thwaites DI, Andreo P (1988) An analysis of the revised HPA dosimetry protocols. Phys Med Biol 33: 923–938

Nath R, Schulz RJ (1981) Calculated response and wall correction factors for ionization chambers exposed to ^{60}Co gamma-rays. Med Phys 8: 85–93

NCRP, National Council on Radiation Protection and Measurements (1981) Dosimetry of x-ray and gamma-ray beams for radiation therapy in the energy range 10 keV to 50 MeV. Report No. 69, Washington, DC

Niatel M-T (1983) On the location of a flat ionisation chamber for absorbed dose determination. Phys Med Biol 28: 407–410

Niatel MT, Perroche-Roux AM, Boutillon M (1985) Two determinations of W for electrons in dry air. Phys Med Biol 30: 67–75

Nilsson B, Brahme A (1983) Relation between kerma and absorbed dose in photon beams. Acta Radiol Oncol 22: 77–85

Nilsson B, Montenlius A (1986) Fluence perturbation in photon beams under nonequilibrium conditions. Med Phys 13: 191–195

O'Connor JE, Malone DE (1987) A method of measuring the wall contribution of an ionisation chamber. Phys Med Biol 32: 1603–1607

Olsson LE, Petersson S, Ahlgren L, Mattsson S (1989) Ferrous sulphate gels for determination of absorbed dose distributions using MRI technique: basic studies. Phys Med Biol 34: 43–52

Olsson LE, Fransson A, Ericsson A, Mattsson S (1990) MR imaging of absorbed dose distributions for radiotherapy using ferrous sulpate gels. Phys Med Biol 35: 1623–1631

Owen B, DuSautoy AR (1991) Correction for the effect of the gaps around the core of an absorbed dose in graphite calorimeter in high energy photon radiation. Phys Med Biol 36: 1699–1704

Paul JM, Koch RF, Philip PC (1985) AAPM Task Group 21 protocol: dosimetric evaluation. Med Phys 12: 424–430

Pearson DW, Attix FH, DeLuca PM Jr, Goetsch SJ, Torti RP (1980) Ionisation error due to porosity in graphite ionisation chambers. Phys Med Biol 25: 333–338

Perris A, Zarris G (1989) Specific primary ionisation for electrons, protons and alpha particles incident on water. Phys Med Biol 34: 1113–1118

Pitchford WG (1985) The HPA photon protocol and proposed electron protocol. Radiother Oncol 4: 297–300

Prasad PV, Nalcioglu O, Rabbani B (1991) Measurement of three-dimensional radiation dose distributions using MRI[1]. Radiat Res 128: 1–13

Pruitt JS, Loevinger R (1982) The photon-fluence scaling theorem for Compton-scattered radiation. Med Phys 9: 176–179

Pruitt JS, Domen SR, Loevinger R (1981) The graphite calorimeter as a standard of absorbed dose for cobalt-60 gamma radiation. J Res Nat Bur Stand (U.S.) 86: 495–502

Rao ISS, Naik SB (1980) Graphite calorimeter in water phantom and calibration of ionization chamber in dose to water for ^{60}Co gamma radiation. Med Phys 7: 196–201

Rawlinson JA, Bielajew AF, Munro P, Galbraith DM (1984) Theoretical and experimental investigation of dose enhancement due to charge storage in electron-irradiated phantoms. Med Phys 11: 814–821

Reft CS (1989) Output calibration in solid water for high energy photon beams. Med Phys 16: 299–301

Reich H (1979) Choice of the measuring quantity for therapy-level dosemeters. Phys Med Biol 24: 895–900

Rogers DWO (1984) Fluence to dose equivalent conversion factors calculated with EGS3 for electrons from 100 keV to 20 GeV and photons from 11 keV to 20 GeV. Health Phys 46: 891–914

Rogers DWO (1989) Fundamentals of the AAPM's TG-21 dosimetry protocol. Refresher course RC-9, 26 July 1989, AAPM Annual Meeting, Memphis, Tenn., PIRSO 198

Rogers DWO (1991) Fundamentals of high energy x-ray and electron dosimetry protocols and new dosimetry standards. In: Purdy J (ed) Advances in Radiation Oncology Physics. AAPM, New York, pp 181–223

Rogers DWO, Bielajew AF (1990) Wall attenuation and scatter corrections for ion chambers: measurements versus calculations. Phys Med Biol 35: 1065–1078

Rogers DWO, Ross CK (1988) the role of humidity and other correction factors in the AAPM TG-21 dosimetry protocol. Med Phys 15: 40–48

Rogers DWO, Bielajew AF, Nahum AE (1985) Ion chamber response and A_{wall} correction factors in a ^{60}Co beam by Monte Carlo simulation. Phys Med Biol 30: 429–443

Ross CK, Klassen NY, Smith GD (1984) The effect of various dissolved gases on the heat defect of water. Med Phys 11: 635–658

Ross CK, Klassen NV, Shortt KR, Smith GD (1989) A direct comparison of water calorimetry and Fricke dosimetry. Phys Med Biol 34: 23–42

Roy SC, Apfel RE (1984) Semi-empirical formula for the stopping power of ions. Nucl Instrum Meth Phys Res B4: 20–22

Rubach A, Conrad F, Bichsel H (1986) Dose build-up curves for cobalt-60 irradiation: a systematic error occurring with pan-cake chamber measurements. Phys Med Biol 31: 441–448

Scharf K (1971) Spectrophotometric measurement of ferric ion concentration in the ferrous sulphate (Fricke) dosemeter. Phys Med Biol 16: 77–86

Schulz RJ (1982) Concerning the perturbation correction in electron-beam dosimetry. Med Phys 9: 131

Schulz RJ (1986) Reply to comments of Rogers et al. Med Phys 13: 965–966

Schulz RJ (1990) Comments on construction of a calorimeter prototype with a high sensitivity pulsed signal detection circuit. Phys Med Biol 35: 467–469

Schulz RJ, Meli JA (1984) Reply to comments of Wu et al. Med Phys 11: 872–874

Schulz RJ, Weinhous MS (1985) Calorimeteric determination of the cavity-gas calibration factor N_{gas}. Med Phys 12: 166–168

Schulz RJ, Almond PR, Kutcher G, et al. (1986) Clarification of the AAPM Task Group 21 protocol. Med Phys 13: 755–759

Schulz RJ, Wuu CS, Weinhous MS (1987) The direct determination of dose-to-water using a water calorimeter. Med Phys 14: 790–796

Schulz RJ, deGuzman AF, Nguyen DB, Gore JC (1990a) Dose-response curves for Fricke-infused agarose gels as obtained by nuclear magnetic resonance. Phys Med Biol 35: 1611–1622

Schulz RJ, Venkataramanan N, Huq MS (1990b) The thermal defect of A-150 plastic and graphite for low-energy protons. Phys Med Biol 35: 1563–1574

Schulz RJ, Huq MS, Venkataramanan N, Motakabbir KA (1991) A comparison of ionization chamber and water calorimeter dosimetry for high energy x rays. Med Phys 18: 1229–1233

SCRAD, Sub-Committee of Radiation Dosimetry of the American Association of Physicists in Medicine (1966) Protocol for the dosimetry of high energy electrons. Phys Med Biol 11: 505–520

SCRAD, Sub-Committee on Radiation Dosimetry of the American Association of Physicists in Medicine (1971) Protocol for the dosimetry of x- and gamma-ray beams with maximum energies between 0.6 and 50 MeV. Phys Med Biol 16: 379–396

SEFM, Sociedad Española de Fisica Medica (1984) Procedimientos recomendados para la dosimettría de fotones y electrones de energías comprendidas entre 1 MeV y 50 MeV en radioterapia de haces externos. Publication No. 1–1984, Madrid, Spain

Seuntjens J, Thierens H, Van der Plaetsen A, Segaert O (1987) Conversion factor f for x-ray beam qualities, specified by peak tube potential and HVL value. Phys Med Biol 32: 595–603

Seuntjens J, Thierens H, Van der Plaetsen A, Segaert O (1988) Determination of absorbed dose to water with ionisation chambers calibrated in free air for medium-energy x-rays. Phys Med Biol 33: 1171–1185

Shiragai A (1978) A proposal concerning the absorbed dose conversion factor. Phys Med Biol 23: 245–252

Shiragai A (1984) A comment on a modification of Burlin's general cavity theory. Phys Med Biol 29: 427–432

Shiragai A (1991) A formulation for high-energy photon and electron beam dosimetry. Phys Med Biol 36: 633–642

Shortt KR (1989) The temperature dependence of G (Fe^{3+}) for the Fricke dosemeter. Phys Med Biol 34: 1923–1926

Smathers JB, Otte VA, Smith AR, et al. (1977) Composition of A-150 tissue-equivalent plastic. Med Phys 4: 74–77

Smyth VG, McEwan AC (1984) Verification of a result of Kristensen by Monte Carlo modelling. Phys Med Biol 29: 1279–1282

Spokas JJ, Meeker RD (1980) Investigation of cables for ionization chambers. Med Phys 7: 135–140

Sternheimer RM, Peierls RF (1971) General expression for the density effect for the ionization loss of charged particles. Phys Rev B3: 3681

Sternheimer RM, Berger MJ, Seltzer SM (1984) Density effect for the ionization loss of charged particles in various substances. Atomic Data Nucl Data Tables 30: 261–271

Svensson H (1971) Dosimetric measurements at the Nordic Medical Accelerators. II. Absorbed dose measurements. Acta Radiol Ther Phys Biol 10: 631–654

Svensson H (1985) The new NACP- and ICRU-dosimetry protocols for dosimetry of high-energy photon and electron radiation. Radiother Oncol 85: 291–294

Svensson H (1990) Presentation of TRS No. 277 "Absorbed dose determination in photon and electron beams. An international code of practice." Med Phys World 6

Svensson H, Brahme A (1979) Ferrous sulfate dosimetry for electrons. A re-evaluation. Acta Radiol Oncol 18: 326–36

Svensson H, Hettinger G (1971) Dosimetric measurements at the Nordic medical accelerators. I. Characteristics of the radiation beam. Acta Radiol Ther Phys Biol 10: 369–384

Svensson H, Petersson S (1967) Absorbed dose calibration of thimble chambers with high-energy electrons at different phantom depths. Ark Fys 34: 377–384

Svensson H, Andreo P, Cunningham J, Hohlfeld K (1987) Code of practice for absorbed dose determination in photon and electron beams. In: Radiotherapy in developing countries. IAEA, Vienna, p 333

Takata N, Matiullah (1991) Dependence of the value of m on the lifetime of ions in parallel-plate ionization chambers. Phys Med Biol 36: 449–459

Takata N, Sakihara K (1989) The dependence of the m value on applied voltage in the collection efficiency of ionisation chambers. Phys Med Biol 34: 589–597

Thomas SJ, Palmer N (1989) The use of carbon-loaded thermoluminescent dosimeters for the measurement of surface doses in megavoltage x-ray beams. Med Phys 16: 902–904

Thwaites DI (1984) Charge storage effect on dose in insulating phantoms irradiated with electrons. Phys Med Biol 29: 1153–1156

Thwaites DI (1985) Measurements of ionisation in water, polystyrene and a 'solid water' phantom material for electron beams. Phys Med Biol 30: 41–53

Van Dam J, Rijnders A, Ang KK, Mellaerts M, Grobet P (1985) Determination of ionisation chamber collection efficiency in a swept electron beam by means of thermoluminescent detectors and the "two-voltage" method. Radiother Oncol 3: 363–370

van der Giessen PH (1986) About the rate of temperature changes in a thimble chamber. Radiother Oncol 7: 287–291

Vandyk J, Macdonald JCF (1972) Charge desposition from high energy electron beams. Radiat Res 50: 20–32

Waiter GD, Lerski RA (1991) The variation of proton density in agarose gels used as NMR test substances through the use of glass beads. Phys Med Biol 36: 541–546

Weinhous MS, Meli JA (1984) Determining P$_{ion}$, the correction factor for recombination losses in an ionization chamber. Med Phys 11: 846–849

Weinhous MS, Meli JA (1988) Collection efficiency of an ionisation chamber in a pulsed swept beam: collimator scattered effects. Phys Med Biol 31: 1147–1155

White GA, Gibbs GL (1985) Comments on "A protocol for the determination of absorbed dose from high-energy photon and electron beams." Med Phys 12: 114

Whyte GN (1954) Nucleonics 12: 18, as cited in: Principles of radiation dosimetry by G.N. Whyte, John Wiley, New York (1959), pp 71

Wielopolski L, Pai S, Mlyn M (1991) Semianalytical expressions for L/P and P$_{repl}$ for electron beams. Med Phys 18: 559–564

Williams JR (1987) Dosimetry with a water calorimeter in a p(62) + Be neutron beam. Phys Med Biol 32: 403–406

Williams PC (1977) Discrepancy between C$_{\lambda}$ and C$_E$. Phys Med Biol 22: 535–538

Williams PC (1985) The selection of stopping power and mass energy absorption coefficient data for the HPA Code of Practice for dosimetry. Phys Med Biol 30: 707–708

Williams PC, Jordan TJ (1984) Extra-cameral volume effects in ionisation chambers for electron beam dosimetry. Phys Med Biol 29: 277–286

Wittkämper FW, Mijnheer BJ (1990) Experimental determination of wall correction factors. Part I. Cylindrical ionisation chambers. Phys Med Biol 35: 835–846

Wittkämper FW, Mijnheer BJ, van Kleffens HJ (1987) Dose intercomparison at the radiotherapy centers in the Netherlands. 1. Photon beams under reference conditions and for prostatic cancer treatment. Radiother Oncol 9: 33–44

Wittkämper FW, Mijnheer BJ, van Kleffens HJ (1988) Dose intercomparison at the radiotherapy centers in the Netherlands. 2. Accuracy of locally applied computer planning systems for external photon beams. Radiother Oncol 11: 405–414

Wittkämper FW, Thierens H, Van der Plaetsen A, de Wagter C, Mijnheer BJ (1991) Perturbation correction factors for some ionization chambers commonly applied electron beams. Phys Med Biol 36: 1639–1652

Woo MK, Cunningham JR (1988) Comments on a unified electron/photon dosimetry approach. Phys Med Biol 33: 981–982

Woo MK, Cunningham JR, Jezioranski JJ (1990) Extending the concept of primary and scatter separation to the condition of electronic disequilibrium. Med Phys 17: 588–595

Wu A, Kalend AM, Zwicker RD, Sternick ES (1984) Comments on the method of energy determination for electron beams in TG-21 protocol. Med Phys 11: 871–872

Zeitz L and · Laughlin JS (1982) "Nonisolated-sensor" solid polystyrene absorbed dose measurements. Med Phys 9: 763–768

Zeitz L (1989) Design of apparatus for precise x-ray dose chamber calibrations. Med Phys 16: 644–647

Zeitz L, Ulin K, Caley R (1986) Improved "nonisolated-sensor" solid polystyrene calorimeter. Med Phys 13: 399–402

Zoetelief J, Engels AC, Broerse JJ (1980) Effective measuring point of ion chambers for photon dosimetry in phantoms. Br J Radiol 53: 580–583

Zoetelief J, Eisenhauer CM, Coyne JJ (1990) Calculations of displacement corrections for in-phantom measurements with ionisation chambers for mammography. Phys Med Biol 35: 1287–1299

18 Quality Assurance in Conformal Radiation Therapy

Isaac I. Rosen

18.1 Introduction

The delivery of conformal treatments, in which the high-dose region of the dose distribution closely conforms to the shape of the target volume in three dimensions, has been a long-standing goal of radiation therapy. In its broadest sense, conformal radiation therapy includes many conventional therapy techniques: external beam photon therapy using shaped beams, beam modifiers, and rotations to focus the radiation onto well-defined target volumes; stereotaxic radiosurgery; electron beam therapy using bolus and compensation; and many brachytherapy procedures. Usually, however, conformal radiation therapy refers to external beam photon treatments using complex, technologically sophisticated arrangements of customized static beams or dynamic irradiations.

Advances in conformal radiation therapy have resulted from tremendous improvements in computer technology. Faster, affordable computers with more memory and disk storage have improved the quality of computed tomography (CT), magnetic resonance imaging (mRI), and other imaging studies. Better images and faster computers have improved computer models of patient anatomy and dose deposition. These improved computer models have, in turn, led to better targeting of the tumor volume

and to innovative designs for radiation delivery techniques. New therapy machines with computer control systems which have the capability to execute these new types of treatments are now commercially available.

As new conformal therapy techniques find their way into clinical use, quality assurance (QA) procedures must expand to include the greater capabilities of treatment machines and treatment planning systems. Just as new conformal therapy techniques will generally require more time and effort in treatment planning and treatment delivery, they will also require more time and effort in QA. Conformal radiation therapy QA will require measurements and tests in addition to those currently demanded for conventional treatments. Some of those additional tests and measurements will be common to all conformal techniques. Others will need to be tailored to the specific details of the conformal treatment technique in use.

The overall goals of QA in radiation therapy are to ensure that patients accurately receive the doses prescribed by their physicians, that errors are discovered quickly so that they can be corrected in a timely manner, and that treatments are delivered safely for both patients and staff. The need for QA is universally acknowledged and there now exist general guidelines and recommendations for proper QA programs (AAPM 1984; ACMP 1986; ACR 1990; HORTON 1987; KUTCHER and PURDY 1992; NATH et al., to be published; PURDY et al. 1987; STARKSCHALL and HORTON 1991; SVENSSON 1983; VAN DYKE et al. 1993; WOOTTON et al. 1975). QA tests are generally based on the philosophy that (a) there are no known potentially hazardous flaws in the design of the treatment equipment or in the implementation of that design; (b) there may be undetected errors of unknown consequence in the equipment design and/or its manufacture; (c) errors can occur in the manufacture and installation of the equipment; (d) performance of the equipment hardware will change with use and age; and (e) changes in hardware can affect performance of

Isaac I. Rosen, PhD, University of Texas, M.D. Anderson Cancer Center, Department of Radiation Physics, 1515 Holcombe Blvd., Houston, TX 77030 USA

software in computerized equipment. Comprehensive QA programs include measurements of the accuracy and precision of the equipment used for the QA measurements; measurements of the accuracy, precision, and safety of the equipment used to plan, deliver, and verify treatments; and ongoing evaluation of the procedures for checking the work of personnel involved in the planning and delivery of treatments. QA test results should be reviewed frequently. Displays of the QA data in tabular and graphic from are useful for detecting subtle changes in performance over time. While the details of QA programs vary among institutions (the frequency of tests, the personnel performing the tests, and details of how the tests are performed), there is a generally accepted body of tests to be performed and the variations in frequency are relatively minor.

The areas in which conformal therapy QA may differ from the QA for conventional treatments are: acceptance testing of equipment and installations, ongoing monitoring of equipment performance, detailed measurements and checks of individual patient treatments, and regular monitoring of a patient's overall course of treatment. For conformal therapy, the equipment subject to QA procedures includes treatment machines, simulators, CT scanners, treatment planning systems, and treatment verification systems. Each of these devices contributes to the design and delivery of the patient treatments. Because undetected errors and malfunctions from the therapy machine and treatment planning system have the greatest potential for harm, the greatest amount of QA effort is devoted to these systems. Improper operation of simulation and imaging equipment tends to be more readily apparent and generally has less serious consequences.

18.2 Computer-Controlled Accelerators

Verifying the safe and proper operation of a computer-controlled accelerator is more difficult than it is for a machine with purely electromechanical control systems (PURDY et al. 1993; ROSEN and PURDY 1992; WEINHOUS et al. 1990). The complexity of interactions between hardware and software in a real-time environment can mask design flaws. Confidence in proper operation gained through experience delivering conventional treatments may not apply when new treatment strategies are initiated and new modes of operation or use may uncover problems previously undetected. A series of severe radiation accidents in the late 1980s with a commercial computer-controlled accelerator (LEVENSON and TURNER 1993) demonstrated the difficulty of producing safety-critical software and underscored the need for special QA procedures for computer-controlled machines.

Quality assurance for a computer-controlled accelerator starts with acceptance testing to verify the proper and safe operation of the particular machine purchased and installed at the user's site. All of the tests traditionally performed on noncomputerized accelerators are performed plus tests to verify operation of software, communications, and hardware/software/user interfaces. As a machine ages, its performance degrades through mechanical wear, radiation damage, and other environmental/operational stresses. Periodic routine QA procedures and measurements monitor the operation of the machine and guide the user in keeping it running according to its mechanical and radiation specifications. While acceptance and periodic QA tests are often easier to perform in a "service" mode, on computer-controlled machines such tests may not always accurately predict the behavior of the equipment in clinical modes of operation. Therefore, it is recommended that whenever possible, tests be performed in the operation modes used to treat patients. Safety interlocks especially should be tested in clinical modes.

Acceptance testing begins with tests of the performance of the machine against its specifications. Published recommendations, specifications from the purchase order, and documentation from the manufacturer should guide the selection and evaluation of tests. At a minimum, acceptance testing should include the following traditional performance measurements:

1. Mechanical accuracy and precision of gantry rotation, collimator rotation, photon beam jaw positions, and table positions
2. Accuracy and precision of isocenter position
3. Coincidence of light field and radiation field
4. Output, energy, flatness, and symmetry of each beam
5. Profiles of beams with wedges
6. Stability of the dose monitoring systems
7. Proper operation of safety interlocks (emergency off switches, anti-collision devices, excess dose rate, excess dose per pulse)

During acceptance testing, the values of machine operating parameters, range limits for parameters, and safety interlock settings should be documented. These values are useful for later comparison with the results of routine QA tests and tests following machine repairs and software modifications.

On computer-controlled accelerators some additional tests should be performed. Some of the safety interlocks may be activated by input to the computer control system rather than directly hard-wired. Therefore, sensitivity to activation and speed of response (beam-off and motion termination) should be measured. Proper reactivation of interlocks should be verified after any "by-pass" in service mode. The user interface should be carefully examined. All screen display control devices (function keys, trackball/mouse, cursor arrows, etc.) should be tested for proper operation. Lock-out functions, key switches, and passwords should be verified for proper functioning and integrity. Special functions and control keys accessible by the operator should be carefully tested. User interface tests should be performed with the machine in all of its operational states (clinical modes, service modes, "beam on" condition, etc.) and from all available input devices (operator keyboard, service keyboard, special keypad, etc.). If the accelerator has computer-assisted setup capability, the safety and accuracy of operation should be confirmed. If there is no power conditioning for the computer systems, then computer and machine operation should be carefully monitored for adverse effects due to power transients. The effects of hardware failures in computer peripherals (e.g., printers) on accelerator operation should be assessed if possible. Finally, isolation of the different operating modes of the machine should be verified to prevent treatment of a patient in a service mode.

Once the accelerator is put into clinical use, a routine QA program is initiated to uncover and correct performance changes due to normal wear and aging. In a computer-controlled machine, changes to the computer hardware may affect correct software operation by corrupting essential data or the programs themselves. Furthermore, because hardware malfunctions put the machine into abnormal operational states, latent software problems may appear as the hardware ages, even though acceptance testing revealed no errors.

The schedule for periodic testing of most operating parameters is based primarily on the stability of those parameters over time. General recommendations are based on experiences with "normally" operating machines. Accelerators exhibiting unusual or unstable performance will require special QA and maintenance schedules. Radiation therapy technologists, machine engineers, dosimetrists, and physicists all participate in the QA process. Daily tests, usually performed by the radiation therapy technologist or maintenance engineer, include:

1. Radiation output constancy for each energy
2. Patient audiovisual communications
3. Radiation monitor lights
4. Treatment door interlock
5. Accuracy of the laser localization lights
6. Accuracy of optical and mechanical distance indicators

On a computer-controlled accelerator, internal self-diagnostic tests for the computer systems should execute each time the machine is turned on. It is very useful to record daily values of operating parameters which affect machine output. This information can help to uncover slow systematic changes in performance, can help to diagnose problems when they occur, and can help in anticipating hardware failures. On a weekly basis, a dosimetrist or physicist should repeat the daily tests and add light-radiation field coincidence, field flatness and symmetry, and an independent measurement of radiation output using a method different from the constancy check. Monthly measurements by a dosimetrist or physicist should include at least the following:

1. Proper operation of all electrical and mechanical interlocks and "emergency off" switches
2. Beam energy
3. Physical integrity of all accessories
4. Accuracy of all mechanical and electronic read-outs (gantry, collimator, table)

Accuracy of table positions is especially important for conformal techniques in which the table position is automatically changed during treatment. Finally, annual QA testing by a physicist should include:

1. Full calibration of each beam
2. Position of mechanical isocenter
3. All mechanical alignments and axes of rotation (gantry, collimator, couch)
4. The stability, flatness, symmetry, and output of the beams at different gantry angles
5. Wedge, tray, and off-axis factors
6. Total dose, dose rate, and geometric accuracy of gantry angles for arc therapy modes
7. Dose linearity of monitor chambers
8. Radiation safety measurements
9. Effective source position (inverse square law)

The complexity of computer-controlled machines suggests the need for more stringent testing following routine and preventive maintenance and following nontrivial repairs. Proper operation of all safety interlocks should be confirmed. The integrity of software and data should be verified if possible using appropriate tools supplied by the manufacturer. When repairs require beam tuning procedures which change the values of nondynamic parameters controlling machine operation, then all treatment beam characteristics should be verified. If repairs are extensive or involve critical components, full acceptance testing may again be indicated.

One of the theoretical advantages of computer-controlled accelerators is that new functionality can be added through software changes without accompanying hardware modifications. For the physicist, however, software improvements must be treated in the same manner as hardware upgrades. In some respects, dealing with software changes is more difficult because the physicist cannot visually inspect the changes and independently assess their consequences. Therefore, the integrity of all safety interlocks, of the software itself, and of the patient and machine parameter databases should be verified following installation of a software upgrade. All tests required and suggested by the manufacturer to confirm correct operation of the new software should be performed. Any treatment beam parameters potentially affected by the software changes should be verified. Depending on the nature and extent of the software changes, full acceptance testing may be advisable.

18.3 Simulators and CT Scanners

Simulators are used to localize target volumes and critical normal structures, to select beam directions, to produce radiographs for designing beam apertures, and to confirm computer-generated treatment plans. The mechanical accuracy and precision of the simulator movements and images are of prime importance for radiation therapy. As for therapy machines, acceptance testing for simulators begins with testing the performance of the machine against its specifications. These tests include measurement of the accuracy and precision of all mechanical motions which control the position of the radiation beam relative to the patient. Generally mechanical tolerances for the gantry and collimator motions of a simulator should be tighter than for a therapy machine in order not to introduce additional errors into the treatment setup. All electronic and mechanical read-outs should be checked for accuracy

and recalibrated if necessary. Door interlocks, anticollision devices, and emergency off interlocks should all be tested. Two new International Electrotechnical Commission documents are currently being finalized which define functional performance tests for simulators and recommend tolerance values (IEC, to be published, a, b).

Ongoing QA for simulators consists of periodic measurements of the accuracy and precision of the mechanical motions of the couch, gantry, collimator, and field delineation wires. Localization lasers, field definition wires, distance indicators, light field/radiation field congruence, position and size read-outs, and all interlocks should be checked weekly. Isocenter location should be measured on a monthly basis. The performance of the imaging system should also be checked monthly. A more thorough mechanical assessment should be done semiannually and a complete acceptance procedure is appropriate on an annual basis.

For conformal therapy, much of the function of the simulator has been taken over by treatment planning software using anatomic data from a CT scanner. Inaccuracies in determination of electron density and geometric errors will result in dose errors in regions which contain significant tissue heterogeneities. Geometric distortions can also cause targeting errors when beams are designed from three-dimensional anatomic reconstructions. Quality assurance of CT scanners for therapy purposes is done primarily using phantoms of well-defined geometry and internal densities. They are used to measure the geometric accuracy of the reconstructed images and to calibrate the conversion of Hounsfield units (HU) to electron density. TEN HAKEN et al. (1991) suggested the relationship between Hounsfield number and electron density be verified and that the sensitivity of those numbers to phantom size, shape, and position be examined. However, they also pointed out that the computation of dose is relatively insensitive to small errors in geometry or electron density. They estimated that a 20 HU calibration error would result in a dose change of approximately 0.4%–1.6% while the dose change from a 5-mm error in pathlength would be about 1.6%–2.1%. Incorrectly applying a low-density linear HU to electron density conversion function to high-density bone (rather than using a piecewise linear function over the range of densities) would produce dose errors larger than these. The impact of errors on targeting of beams could be more significant. However, typical uncertainties in scanner output are expected to be much less than the error estimates used in the calculations above. An initial study should be performed to establish a baseline for accuracy of the CT data and for comparison with later periodic checks.

18.4 Treatment Planning Systems

Conformal therapy treatments rely heavily on treatment planning systems to design individual patient treatments. Treatment planning software is used to create beam outlines based on the three-dimensional shape of the target volume and normal tissues, to select beam directions which miss critical structures, to design beam modifiers or beam modulation functions for shaping the high-dose region and/or to compute the optimal dose to be delivered from each beam. For these reasons, geometric and dosimetric errors in the treatment planning system can adversely affect the quality of the treatment and defeat the gains from using a conformal technique.

There are four distinct areas of QA for treatment planning systems: dose model evaluation, algorithm verification, accuracy of hardware peripherals, and database accuracy. Errors in any of these areas can result in treatment plans which do not accurately predict doses in patients.

Dose model evaluation measures the inherent limits of accuracy in computing doses while algorithm verification confirms that the model has been properly transformed into software, that correct parameter values have been input, and that the algorithm is being applied appropriately. For users of treatment planning systems, it is virtually impossible to examine the accuracy of the dose model alone because the final computed dose values also depend on the algorithm implementation of the dose model. Similarly, with regard to computing doses, the algorithm implementation cannot be entirely verified because differences between computed and measured doses are expected due to limitations of the dose model and beam data set. Implementation errors which produce small dose errors may, therefore, be overlooked. Nevertheless, a careful comparison of computed and measured doses for a wide variety of cases should provide confidence in the overall system and in the parameter values describing the beams.

Because beam outlines, directions, and doses may be determined from the anatomic model in the computer, the geometric accuracy of the internal model and of the input/output devices is important. Geometric accuracy is generally limited by the accuracy and precision of digitizing devices, graphics displays, and hardcopy plotters. The software algorithms selected and their implementation may also affect geometric accuracy. Of cource, hardware malfunctions and software errors may produce serious errors in beam shapes drawn from beam's-eye view projections of the anatomy and in beam directions chosen on the basis of locations of critical structures relative to the target. Testing with phantoms of known geometry should be used to determine the overall geometric accuracy of the system.

VAN DYKE et al. (1993) describe a comprehensive program for commissioning and QA of treatment planning systems. They recommend weekly tests of the accuracy of input and output peripheral devices and semi-annual constancy tests using many of the tests performed during the initial commissioning. Testing of the transfer of data from CT scanners to the treatment planning system should be done at least quarterly. CURRAN and STARKSCHALL (1991), on the other hand, recommend monthly tests of the accuracy of peripheral devices (plotters, digitizers, screen recorders, and graphic screens), of the constancy of standard plans, and of the integrity of data files. They suggest that algorithm tests and data set tests, which compare the output of the treatment planning software to measured dose values, should be done on an annual basis. Because the precision of conformal treatments accentuates the impact of errors, it is prudent to err on the side of testing too frequently rather than not often enough. Finally, new versions of software should be treated as new products and, therefore, undergo complete acceptance testing.

18.5 Quality Assurance of Patient Treatments

Quality assurance for individual patient conformal therapy treatments should include independent checks of beam-on time settings and computer plans, verification of initial treatment setups, frequent portal images and verification films, chart checks, and dry-run testing. Record-and-verify systems and adequate staffing can also reduce the probability of serious errors. LEUNENS et al. (1991) reported a notable decrease in the rate of large ($\geq 5\%$) dose deviations for conventional tangential breast treatments on a linear accelerator with a verification system and staffed by three radiotherapists (2.3%

large deviations) versus a cobalt machine staffed by a single radiotherapist (15% large deviations). For complex conformal treatments the benefits of an automated verification system may be even greater.

Careful verification of the first treatment setup is especially important since errors in the first setup will probably be repeated throughout the course of therapy. MITINE et al. (1993) analyzed the setup variation in a series of head and neck patients immobilized with plastic masks by measuring the differences in field placement on port films versus simulator films. They found that reproducibility of setup on a daily basis was good, as indicated by a standard deviation of 2 mm, and that most of deviations from the simulation setup were related to transfer errors from the simulator to the treatment table. Similarly, BIJHOLD et al. (1992) measured setup variations for conformal prostate treatments (without immobilization) and found that the random variances for individual patients were about 2 mm but that there were also systematic displacements of 2–4 mm in the inferior-superior direction. Verification of the first treatment setup should eliminate transfer errors for individual patents and could reveal systematic errors which affect all patient treatments.

At a minimum, patient and beam positions should be checked weekly using portal or verification films. In a pilot study covering a variety of treatment sites, DENHAM et al. (1993) measured systematic setup errors ranging from 3 mm to 8.5 mm. They suggested more frequent port films during the first week of treatment because their analysis indicated that modest systematic errors could go undetected for 5–6 weeks if only one port film per week were obtained. Optical densities measured from verification films could also theoretically be used as a coarse check on the delivered doses from the individual beams (VAN DAM et al. 1992). Commercial electronic portal imaging devices (EPIDs) are now available that rapidly produce images of a quality comparable to that of film (BOYER et al. 1992). Images from these devices can be modified and enhanced in seconds and can be mathematically compared to other reference images. Methods for automatic analysis of EPID images are actively being studied by many investigators. EPIDs with automatic evaluation of images offer the potential to adjust patient alignment before each treatment.

The entire course of treatment should be reviewed in weekly chart checks. Because chart review can be tedious and repetitive, a check list is a valuable aid in avoiding oversights. While this is standard QA

procedure for all radiation therapy, it is especially important for conformal treatments where complex and/or unusual beam arrangements or machine motions may easily hide errors in treatment plan interpretation, prescription entry, or setup.

Conformal techniques should include a "dry run" for each patient prior to the start of the course of therapy. In a dry run, the entire treatment is executed without the patient in position. Ion chambers, film, and/or other dosimeters should be used to verify beam output and profile characteristics. Visual inspection and position templates should be used to verify proper machine motions. The dry run not only ensures that the machine is operating correctly but also that the prescription has been entered correctly. An error in prescription entry is especially dangerous since an incorrect prescription will be faithfully repeated each day. Subsequently, for each patient setup, the full range of machine motions expected during treatment should be executed with the patient in position to verify that no collisions will occur. MAGERAS et al. (1992) use this "virtual treatment" to check for collisions for every multisegment conformal treatment. MORGAN (1992) describes how these QA techniques have been applied to two conformal techniques using variable couch position during treatment delivered by a computer-controlled cobalt machine.

The greater precision in targeting and the use of innovative beam directions lead to more stringent requirements for patient immobilization and repositioning. A wide variety of immobilization and repositioning devices are available for various parts of the body: bite blocks and temperature-sensitive plastic for the head and neck, vacuum-form and polyurethane immobilizers for the body, arm boards, tilt boards, breast bridges, and squeeze devices for the breast. Immobilization devices permit the patient to be planned in the treatment position using CT and MRI, improve the reproducibility of patient position, and can be indexed to the treatment table. Identifying the location of the immobilization device on the treatment couch can help reduce setup time and will reduce the probability of collision after clearance has been checked for the first treatment.

Some conformal techniques utilize nontraditional beam directions. Verification of these beams by simulation and portal films is often difficult because the view of the anatomy is unfamiliar and may be impossible for some beam directions. Digitally reconstructed radiographs should be generated by the treatment planning system as a standard for comparison to the simulator and treatment films.

18.6 Summary

Conformal therapy treatments offer the potential for better local control of disease through the delivery of higher tumor doses and/or the reduction of treatment morbidity by reduced normal tissue doses. However, the complexity of conformal treatments increases the possibility of errors in the placement and/or amount of radiation delivered. The danger of collision accidents also exists for dynamic methods. Each type of conformal therapy technique will require some special QA procedures but all should use automated machine setting verification patient immobilization, dry run testing, and frequent port images using film or an electronic imaging device. Stringent QA procedures for all equipment involved in conformal therapy are required for accurate clinical evaluation of conformal techniques and to ensure the best possible results for individual patients.

References

AAPM (1984) Report no. 13: Physical aspects of quality assurance in radiation therapy. American Institute of Physics, New York

ACMP (1986) Report no. 2: Radiation control and quality assurance in radiation oncology, a suggested protocol. American College of Medical Physics

ACR (1990) Physical aspects of quality control. American College of Radiology, New York

Bijhold J, Lebesque JV, Hart AAM, Vihlbrief RE (1992) Maximizing setup accuracy using portal images as applied to a conformal boost technique for prostatic cancer. Radiother Oncol 24: 261–271

Boyer AL, Antonuk L, Fenster A et al. (1992) A review of electronic portal imaging devices (EPIDs). Med Phys 19: 1–17

Curran B, Starkschall G (1991) A program for quality assurance of dose planning computers. In: Starkschall G, Horton J (eds) Quality assurance in radiotherapy physics, proceedings of an American College of Medical Physics symposium. Medical Physics Publishing, Madison, p 207

Denham JW, Dally MJ, Hunter K, Wheat J, Fahey PP, Hamilton CS (1993)

Objective decision-making following a portal film: The results of a pilot study. Int J Radiat Oncol Biol Phys 26: 869–876

Horton JL (1987) Handbook of radiation therapy physics. Prentice-Hall, Englewood Cliffs

IEC (to be published, a) Technical Report 1168: Medical electrical equipment. Functional performance characteristics of radiotherapy simulators. International Electrotechnical Commission, Vienna

IEC (to be published, b) Technical Report 1170: Medical electrical equipment. Radiotherapy simulators. Guidelines for functional performance characteristics. International Electrotechnical Commission, Vienna

Kutcher GJ, Purdy JA (1992) Comprehensive quality assurance. In: Purdy JA (ed) Advances in radiation oncology physics – dosimetry, treatment planning, and brachytherapy. American Institute of Physics, Woodbury, p 224

Leunens G, Verstraete J, van Dam J, Dutreix A, van der Schueren E (1991) In vivo dosimetry for tangential breast irradiation: role of the equipment in the accuracy of dose delivery. Radiother Oncol 22: 285–289

Levenson NG, Turner CS (1993) An investigation of the Therac-25 accidents. Computer, July, pp 18–41

Mageras GS, Podmaniczky KC, Mohan R (1992) A model for computer-controlled delivery of 3-D conformal treatments. Med Phys 19: 945–953

Mitine C, Dutreix A., van der Schueren E (1993) Black and white in accuracy assessment of megavoltage images: the medical decision is often grey. Radiother Oncology 28: 31–36

Morgan HM (1992) Quality assurance of computer controlled radiotherapy treatments. Br J Radiol 65: 409–416

Nath R, Biggs PJ, Bova FJ, Ling CC, Purdy JA, van de Geijn J, Weinhous MS (to be published) AAPM code of practice for radiotherapy accelerators: report of AAPM radiation therapy task group no. 45

Purdy JA, Harms WB, Gerber RL (1987) Report on a long-term quality assurance program. In: Kereiakes JG, Elson HR, Born CG (eds) Radiation oncology physics – 1986. American Institute of Physics, New York, p 91

Purdy JA, Biggs PJ, Bowers C et al. (1993) Medical accelerator safety considerations: Report of AAPM radiation therapy committee task group no. 35. Med Phys 20: 1261–1275

Rosen II, Purdy JA (1992) Computer controlled medical accelerators. In: Purdy JA (ed) Advances in radiation oncology physics – dosimetry, treatment planning, and brachytherapy. American Institute of Physics, Woodbury, p 1

Starkschall G, Horton J (eds) (1991) Quality assurance in radiotherapy physics, proceedings of an American College of Medical Physics Symposium. Medical Physics Publishing, Madison

Svensson GK (1983) Quality assurance in radiation therapy treatment planning. In: Wright AE, Boyer AL (eds) advances in radiation therapy. American Institute of Physics, New York, p 244

Ten Haken RK, Kessler ML, Stern RL, Ellis JH, Niklason LT (1991) Quality assurance of CT and MRI for radiation therapy treatment planning. In: Starkschall G, Horton J (eds) Quality assurance in radiotherapy physics, proceedings of an American College of Medical Physics symposium. Medical Physics Publishing, Madison, p 74

Van Dam J, Vaerman C, Blanckaert N, Leunens G, Dutreix A, van der Schueren E (1992) Are port films reliable for in vivo exit dose measurements? Radiother Oncol 25: 67–72

Van Dyk J, Barnett RB, Cygler JE, Shragge PC (1993) Commissioning and quality assurance of treatment planning computers. Int. J. Radiat. Oncol. Biol. Phys. 26: 261–273

Weinhous MS, Purdy JA, Granda CO (1990) Testing of medical linear accelerator's computer-controlled system. Med Phys 17:95–102

Wootton P, Almond PR, Holt JG, Hughes DB, Jones D, Karzmark CJ, Schulz RJ (1975) Code of practice for x-ray therapy linear accelerators. Med Phys 2: 110–121

Subject Index

List of Contributors

MARTIN D. ALTSCHULER, PhD
Hospital of the University of Pennsylvania
Department of Radiation Oncology
3400 Spruce Street, 2 Donner
Philadelphia, PA 19104
USA

PETER BLOCH, PhD
Hospital of the University of Pennsylvania
Department of Radiation Oncology
3400 Spruce Street, 2 Donner
Philadelphia, PA 19104
USA

FRANCIS J. BOVA, PhD
Associate Professor
University of Florida
College of Medicine
Department of Radiation Oncology
UF Shands Cancer Center
P.O. Box 100385
Gainesville, FL 32610-0385
USA

ANDERS BRAHME, PhD
Department of Medical Radiation Physics
Karolinska Institute and University of Stockholm
P.O. Box 260
S-17176 Stockholm
Sweden

DONALD J. BUCHSBAUM, PhD
University of Alabama – Birmingham
Department of Radiation Oncology
619 S. 19th Street
Birmingham, AL 35233
USA

GEORGE T.Y. CHEN, PhD
Department of Radiation and Cellular Oncology
The University of Chicago Medical Center
5841 S. Maryland Avenue, MC 0085
Chicago, IL 60637
USA

PETER FESSENDEN, PhD
Stanford University School of Medicine
Department of Radiation Oncology, Room S-044
300 Pasteur Drive
Stanford, CA 94305
USA

BENEDICK A. FRAASS, PhD
University of Michigan Medical Center
Radiation Oncology, Rm. B2C490
1500 E. Medical Center Drive
Ann Arbor, MI 48109-0010
USA

ZVI FUKS, MD
Department of Radiation Oncology
Memorial Sloan-Kettering Cancer Center
1275 York Avenue
New York, NY 10021
USA

JAMES M. GALVIN, PhD
New York University Medical Center
Department of Radiation Oncology
566 First Avenue
New York, NY 10016
USA

JEFFREY W. HAND, PhD
Department of Medical Physics
Royal Postgraduate Medical School
Hammersmith Hospital
Du Cane Road
London W12 ONN
United Kingdom

M. SAIFUL HUQ, PhD
Department of Radiation Oncology
and Nuclear Medicine
Thomas Jefferson University
111 South 11th Street
Philadelphia, PA 19107-5097
USA

DAVID JETTE, PhD
Executive Director
The Lawrence H. Lanzl Institute of Medical Physics
3876 Bridge Way N, Suite 300
Seattle, WA 98103-7951
USA
and
Professor of Medical Physics
Rush-Presbyterian-St. Luke's Medical Center
Chicago, IL 60612
USA

GERALD J. KUTCHER, PhD
Department of Medical Physics
Memorial Sloan-Kettering Cancer Center
1275 York Avenue
New York, NY 10021
USA

DENNIS D. LEAVITT, PhD
Division of Radiation Oncology
University of Utah Medical Center
50 N. Medical Drive
Salt Lake City, UT 84132
USA

STEVEN A. LEIBEL, MD
Department of Radiation Oncology
Memorial Sloan-Kettering Cancer Center
1275 York Avenue
New York, NY 10021
USA

C. CLIFTON LING, PhD
Department of Medical Physics
Memorial Sloan-Kettering Cancer Center
1275 York Avenue
New York, NY 10021
USA

DANIEL L. MCSHAN, PhD
University of Michigan Medical Center
Radiation Oncology, Rm. B2C490
1500 E. Medical Center Drive
Ann Arbor, MI 48109-0010
USA

RADHE MOHAN, PhD
Department of Medical Physics
Memorial Sloan-Kettering Cancer Center
1275 York Avenue
New York, NY 10021
USA

RAVINDER NATH, PhD
Department of Therapeutic Radiology
Yale University School of Medicine
333 Cedar Street, P.O. Box 3333
New Haven, CT 06510-8040
USA

CHARLES A. PELIZZARI, PhD
Department of Radiation and Cellular Oncology
The University of Chicago Medical Center
5841 S. Maryland Avenue, MC 0085
Chicago, IL 60637
USA

ISAAC I. ROSEN, PhD
University of Texas M.D. Anderson Cancer Center
Department of Radiation Physics
1515 Holcombe Blvd.
Houston, TX 77030
USA

MICHAEL C. SCHELL, PhD
University of Rochester Cancer Center
Department of Radiation Oncology
601 Elmwood Avenue, Box 647
Rochester, NY 14642-8647
USA

TIMOTHY E. SCHULTHEISS, PhD
Fox Chase Cancer Center
Department of Radiation Oncology
7701 Burholme Avenue
Philadelphia, PA 19111
USA

SHLOMO SHALEV, PhD
Department of Medical Physics
Manitoba Cancer Treatment and
Research Foundation
100 Olivia Street
Winnipeg, Manitoba R3E OV9
Canada

KEITH A. WEAVER, PhD
University of California
Department of Radiation Oncology
Long Hospital, Room L-75
Parnassus Avenue
San Francisco, CA 94143-0226
USA

BARRY W. WESSELS, PhD
George Washington University
Medical Center, Department of Radiation Oncology
901 Twenty Third St. NW
Washington, DC 20037
USA

JEFFREY F. WILLIAMSON, PhD
Mallinckrodt Institute of Radiology
Radiation Oncology Center, Physics Section
510 S. Kingshighway
St. Louis, MO 63110
USA

ANDREW WU, PhD
Division of Radiation Oncology
Allegheny General Hospital
Medical Center of Pennsylvania
Pittsburgh, PA 15212
USA

Titles in the series already published

Springer-Verlag
and the Environment

We at Springer-Verlag firmly believe that an international science publisher has a special obligation to the environment, and our corporate policies consistently reflect this conviction.

We also expect our business partners – paper mills, printers, packaging manufacturers, etc. – to commit themselves to using environmentally friendly materials and production processes.

The paper in this book is made from low- or no-chlorine pulp and is acid free, in conformance with international standards for paper permanency.